AF341984

Rotating Hydraulics

Nonlinear Topographic Effects in the Ocean and Atmosphere

Larry J. Pratt

Woods Hole Oceanographic Institution
Woods Hole, MA, U.S.A.

and

John A. Whitehead

Woods Hole Oceanographic Institution
Woods Hole, MA, U.S.A.

 Springer

Larry J. Pratt
Woods Hole Oceanographic Institution
Physical Oceanography Department,
  MS#21
Woods Hole, MA 02543
USA
lpratt@whoi.edu

John A. Whitehead
Woods Hole Oceanographic Institution
Physical Oceanography Department,
  MS#21
Woods Hole, MA 02543
USA
jwhitehead@whoi.edu

Library of Congress Control Number: 2007922184

ISBN-13: 978-0-387-36639-5   e-ISBN-13: 978-0-387-49572-9

Printed on acid-free paper.

9 8 7 6 5 4 3 2 1

springer.com

For our wives,
Mindy and Lin.

Contents

Preface

What is "rotating hydraulics" and why would anyone wish to read a book on the subject? Over the past three decades, the term has come to describe the physics of overflows and other choked flows of the ocean and atmosphere that are broad enough to be influenced by Earth's rotation. The currents and winds in question typically have high speeds, subcritical-to-supercritical transitions, shocks, and other objects familiar to open-channel or aeronautical engineers. Bores, intrusions, steepening waveforms and separation phenomena are considered part of the subject because they tend to arise within these flows. Mixing with neighboring fluid often occurs as the result of wave breaking or of instabilities associated with the high velocities. Interest in the field is often excited by the dramatic and strongly nonlinear character of the features in question and by the mixing and its downstream consequences. The subject is also important for the study of the Earth's climate because of the special opportunities for observation and long term monitoring made possible as a result of the choking effect.

This book is concerned primarily with the theory of rotating hydraulics. However, the Introduction contains an overview of the observations that have motivated much of the theoretical development, and more detailed case studies appear later in the book. Though both the atmosphere and the ocean are covered, the latter is the source of the most numerous examples. Laboratory experiments have also played a key role in the development of the field and many of these are described. Our intent is to provide the reader with the material necessary to develop a solid grasp of the fundamental ideas and physical processes as well as a general familiarity with geophysical applications. We will also introduce the reader to a range of mathematical techniques that have proved useful in dealing with the types of nonlinear problems that arise the field.

An introduction and review of classical hydraulics appears in Chapter 1. The prospective reader should have a good understanding of basic fluid dynamics and be familiar with the shallow water equations and the approximations behind them. A grasp of the basics of linear wave propagation in fluids, including the concepts of phase speed and group speed, is also desirable. Beginning with Chapter 2, where the effects of rotation are first discussed, the reader will need to know about Coriolis acceleration and geostrophic flow. Thorough discussions of all of these topics appear in the texts of Gill (1982), Pedlosky (1987), Cushman-Roisin (1994), Salmon (1998) and Vallis (2006).

The notation and conventions used in this book are largely standard for geophysical fluid dynamics. However there are two departures worth mentioning.

The first is the use of y, in place of the more common x, to denote the predominant direction of flow. This convention stems from early models of currents in deep ocean straits and along coasts, which are often aligned in the north-south (y-) direction. The second matter concerns the representation of dimensional *vs* dimensionless variables. Most of the book (Chapters 2–6) makes use of a common convention in which a star ($*$) superscript signifies a dimensional quantity, at least where an ambiguity might arise. Stars are used to indicate the dimensional form of common variables such as y^* that also have nondimensional counterparts (y). Stars are omitted, however, for well-known dimensional parameters such as the gravitational acceleration g and the Coriolis parameter f. Stars are also omitted for dimensional scales, indicated by capital letters, that do not have a nondimensional counterpart. Examples include the generic depth scale D and length scale L. There is one exception to this scheme: nearly all the variable used in Chapter 1 are dimensional and it would have been cumbersome to place stars on every one. The star notation is therefore not used at all there. We have tried to avoid any confusion by placing reminders where ambiguities might arise.

Finally, to avoid exotic notation, we sometimes use the same symbol to denote different quantities in different places. One example is the symbol α, which is given a thorough workout. The context usually makes the meaning clear, but a list of variables (Appendix A) can be consulted should a questions arise concerning the meaning of a certain symbol in a certain section.

A number of texts explore the hydraulics of nonrotating fluids in much more depth than is present here. At the time of this printing, the most scientific and up-to-date book is Baines' *Topographic Effects in Stratified Flows*.Treatments of linear and nonlinear waves in shallow water systems can also be found in Stoker's *Water Waves* and Whitham's *Linear and Nonlinear Waves*. Engineering texts such as Chow's *Open Channel Hydraulics* and Henderson's *Open Channel Flow* present the traditional engineering perspective.

Acknowledgments

The authors gratefully acknowledge financial support from the Office of Naval Research, the National Science Foundation, and the Woods Hole Oceanographic Institution Mellon Awards. In addition, we wish to thank the following who made valuable suggestions or who have otherwise contributed to the preparation of the manuscript: Larry Armi, Karin Borenäs, Ruth Curry, Claudia Cenedese, Elin Darelius, Heather Deese, Clive Dorman, Ted Durland, Rafaelle Ferrari, Chris Garrett, Frank Gerdes, Melinda Hall, Karl Helfrich, Rob Hetland, Greg Ivey, Ted Johnson, Allan Kuo, Janek Laanearu, Greg Lawrence, Sonya Legg, Susan Lozier, Peter Lundberg, Mike McCartney, Tom McClimans, Terry McKee, Peter Monkmeyer, Andrew Mosedale, Anne Nikolopoulos, Joe Pedlosky, Jim Price, Leif Thomas, Mary-Louise Timmermans, John Toole, and Anna Wåhlin.

Introduction

Hydraulic Effects in the Ocean and Atmosphere

"Hydraulics" is a nebulous term that evokes images of pumps, dams, brake fluid and lifting machines. In geophysics the term has been applied to wind or current systems that exhibit behavior found in spillways, aqueducts, dams and other open channel engineering applications. Many oceanic and atmospheric flows are topographically constrained in the same way that rivers and reservoirs are, and so it is not surprising that similar physical features arise. For instance, the spillage of dense air over a mountain range and the resultant strong down-slope winds are visually and dynamically similar to the flow of water over a dam or weir. Overflows of dense water flowing along the ocean bottom or in sea straits exhibit similar behavior. The term *rotating hydraulics* has been used to describe the peculiar physical features that arise when hydraulic behavior occurs in flows sufficiently broad to be influenced by the earth's rotation. Examples may include large-scale oceanic flows such as jets and coastal currents that bear little resemblance to open-channel flows. Precisely what do geophysicists mean when they talk about hydraulics and why should one listen?

A good starting point in the understanding of hydraulic phenomena is the principle of signal or information propagation along a conduit. There are a variety of ways that fluids can transmit information, including advection by a velocity field and wave propagation. Most of what we think of as hydraulic behavior is manifested by signal propagation due to waves. As an example, consider the steady flow of water over a dam (Figure I.1). The flow originates from a deep reservoir and spills across the crest or sill of the dam and down the spillway. At the base of the spillway is a hydraulic jump, an abrupt increase in the fluid depth accompanied by intense turbulence. Although the flow is steady, we can imagine the effect of temporarily disturbing the fluid at some location in order to observe where the resulting waves travel. The travel paths tell us something about how information propagates through the system. If the flow is disturbed upstream of the dam crest, the waves that are generated can propagate in either direction, as suggested by the wavy arrows in the figure. We call this type of flow *subcritical*. On the other hand, the flow immediately downstream of the crest is so rapid that waves are prevented from propagating in the upstream direction. We call this flow *supercritical*. As the fluid passes through the hydraulic jump, it is returned to a subcritical state. At the crest of the dam the flow is *critical*, meaning that the wave that would otherwise propagate towards the reservoir is frozen. Clearly,

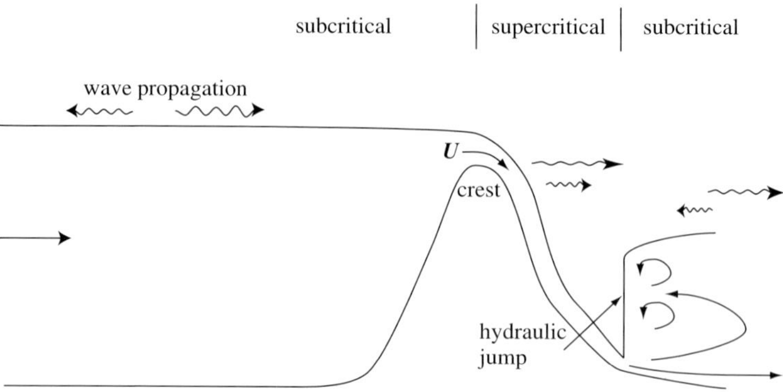

FIGURE I.1. Hydraulically controlled, free-surface flow over a dam.

the reservoir flow cannot be affected by a change in conditions downstream of the dam crest; the information generated would never reach the reservoir. On the other hand, the reservoir *is* influenced by the characteristics of the dam itself. In fact, we would say that the dam *is* the downstream source of information. The outflow from the reservoir is said to be *choked* or *hydraulically controlled* by the dam. The exact meaning of these terms will be discussed at length in Chapter 1, but for now we simply think of 'control' as the ability of the dam to regulate the volume flow rate and the reservoir level.

As an example of flow regulation, suppose that the reservoir is fed by river runoff and drained at the same rate by the discharge over the dam. The whole system is in a steady state. If the elevation of the dam crest is then raised, the height of the reservoir surface above the dam is diminished and the volume outflow decreases. The river runoff now exceeds the outflow and the excess is stored in the reservoir. The reservoir level rises and eventually the original outflow rate is restored. The actual sequence of events may be a little more complicated than what we have described, but the result is essentially correct: the dam influences the time history of the discharge.

There are many fluid systems that experience choking or regulation in an analogous manner. One is the transonic flow of a compressible gas in a wind tunnel. The flow is subsonic in the wider, upstream section of the tunnel, meaning that sound waves can propagate upstream and downstream. The tunnel narrows to its most constricted area at midsection and there a transition to supersonic flow occurs. The tunnel widens farther downstream but the flow remains supersonic. There, sound wave propagation can occur only in the downstream direction.

a. *The Defining Characteristics of Hydraulic Behavior in Geophysics*

We can now identify two conditions that typify hydraulic behavior and serve as criteria for the examples included in this book. The first is that large variations in the flow properties occur along the predominant flow direction. In the example sketched in Figure I.1, large variations in fluid depth and velocity occur as the fluid passes over the dam and moves from a subcritical to a supercritical state. Large variations also occur where the flow passes through the hydraulic jump. 'Large' generally means that the alongstream variations in the depth and velocity are as large as the mean values of these quantities. There are some exceptions in which traditional features such as subcritical-to-supercritical transitions occur without dramatic changes in structure.

The second (and most important) characteristic feature of hydraulic behavior is that the flow in question develops velocities large enough to arrest the propagation of information by waves. This in itself is not a particularly strong condition since the propagation speed of most types of waves depends on wavelength and a wide range of speeds may occur as the length is varied. For a given current speed, it is often possible to find some wavelengths that permit upstream propagation and others that do not. However, the condition becomes more stringent if attention is restricted to nondispersive waves, i.e. ones whose speed is independent of wavelength. Nondispersive behavior usually occurs at the limit of long wavelengths, although there are some exceptions. For reasons to be explained in later chapters, nondispersive waves are most efficient at altering the streams in which they propagate and are of primary interest in hydraulics. The condition that the propagation of information be arrested therefore applies only to such waves. For the case of a surface gravity wave propagating on a one-dimensional stream of velocity U and depth D, wavelengths that are $\gg D$ have speed $U \pm c'$, where $c' = (gD)^{1/2}$. Hydraulic effects of interest do not occur if U remains $\ll (gD)^{1/2}$ over the whole fluid domain since the long waves are everywhere free to propagate in both directions. The situation becomes much more interesting if U becomes as large as $(gD)^{1/2}$ somewhere; in fact, this is what happens at the crest of the dam in Figure I.1.

If several classes of waves exist, then hydraulic behavior with respect to a certain class is possible if $U/c' \cong 1$ for the nondispersive waves in that class. If the typical c values are much less, or much greater than U, for other classes, then hydraulic effects are not expected with respect to these classes. In the example of the overflow shown in Figure I.1, the fluid can support both sound waves and free surface gravity waves. However, the former travel much faster than the typical fluid speeds and are inconsequential to hydraulic behavior. In ocean straits, where the depth D can range from several hundred to several thousand meters, the value of $(gD)^{1/2}$ exceeds $50\,m/s$, which is far greater than the typical velocities ($< 2\,\mathrm{m/s}$). So free-surface, long gravity waves in the ocean are largely irrelevant for hydraulic behavior in major straits. However, the ocean and atmosphere are density stratified as the result of variations in temperature, salinity and humidity. Stratification gives rise to the presence of *internal* gravity waves, in which gravity remains the restoring force but where the effective value of g is reduced in proportion to the vertical density (or potential density) gradient. The corresponding long wave speeds lie in the range of the fluid velocities observed in the ocean and atmosphere and hydraulic effects are therefore possible. Such behavior is sometimes referred to as *internal hydraulics*.

In summary, hydraulic effects arise in flows that are rapid enough to arrest nondispersive waves and that undergo large transitions of velocity and layer depth along the predominant direction of the current. An important consequence of these properties is nonlinearity. Processes such as hydraulic control and features like jumps and bores are fundamentally nonlinear. Nonlinearity gives rise to multiple solutions and part of the challenge is to identify the solutions that are physically robust. Nonlinearity also forces one to stretch his or her intuition and, in some cases, take a leap beyond traditional linear thinking. This is what makes the study of hydraulic phenomena fun.

b.   *The Coastal Atmospheric Marine Layer*

The atmosphere exhibits hydraulic behavior, the most familiar examples being the severe downslope winds generated when dense air masses spill over mountain ranges. The Chinook winds of the American Rocky Mountains, the Santa Anas of southern California, the Mistrals of Provence, and the Boras of the Adriatic Sea are all well documented. Rotational effects in such cases tend to be limited by the relatively small spatial extent of the spilling flows. A case in which rotation is more important involves the winds in the California coastal marine layer, a relatively dense and well-mixed slab of moist air that occupies the lower 300–600 m of the atmosphere above the sea surface in that region. The interface that separates the moist air from the overlying lighter and drier air can be quite sharp. The marine layer itself can be seen in images of low-level cloud cover (Figure I.2a). Mountain ranges along the California coast steer the marine layer winds along to the coast, here towards the southeast direction. Point Arena and other promontories constrict or choke the winds, causing them to accelerate and become supercritical with respect to an internal Kelvin wave. The latter is an internal gravity wave that propagates on the upper boundary of the marine layer and is trapped to the coast. The supercritical flow is marked by an area of clear air to the southeast of Point Arena. The clear area terminates abruptly near Bodega Bay in a cloudy region that contains streaks or undulations that sweep far offshore. The leading edge of this feature is thought to be a type of hydraulic jump. The velocity arrows and isopleths for a similar event are shown in Figure I.2b. In this case the hydraulic jump occurs at Stewarts Point. In contrast to the picture of nonrotating hydraulic control (Figure I.1), the choking effect in these examples is due to the protrusion of a single sidewall into the path of the current.

When the prevailing northwesterly winds relax, the marine layer may reverse direction and flow towards the northwest (Figure I.2c). The intrusion of cloudy moist air that moves along the coast resembles a *gravity current*, a flow created in a rotating environment when dense fluid is allowed to spill into a less dense environment. If a wall is present, the gravitationally driven current will remain trapped to it. For the case shown in Figure I.2c, the moist air of the marine layer plays the role of the dense flow. Gravity currents and other coastal flows will be discussed further in Chapter 4.

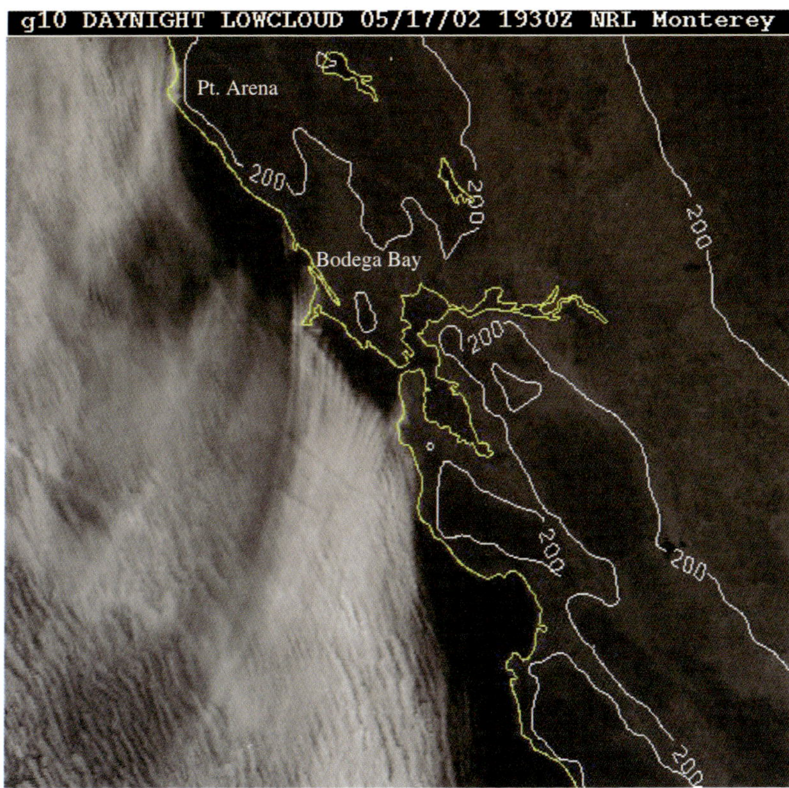

FIGURE I.2a. Cloud cover near the N. California Coast on 05/17/02. (Image courtesy of Clive Dorman).

c.   Examples from the Oceans

Many examples of rotationally influenced hydraulic phenomena arise within the abyssal, basin-to-basin flow in the oceans. This circulation is sometimes thought of as the lower limbs of a great 'conveyor belt', a schematic representation of broad transports at different levels between the major oceans (Figure I.3). Changes in the strength or direction of the overturning cells within this scheme have been linked to rapid climate change. A resident of Northern Europe might begin her description of the conveyor in the North Atlantic area with a (white) northward surface current consisting of the Gulf Stream and its extension, the North Atlantic Current. These currents transport relatively warm, saline water into the Nordic Seas north of Iceland and Scotland. There, concentrated distillation, cooling, freezing and mixing at the sea surface cause some of the surface water to become unstably stratified and to overturn, sometimes resulting in sinking to great depth. The geographic distribution of mixing and deep sinking is not completely understood, but it is known that the resulting deepwater masses move away from their convective origin. They spill out of the Nordic Seas and flow equatorward in a deep western boundary current

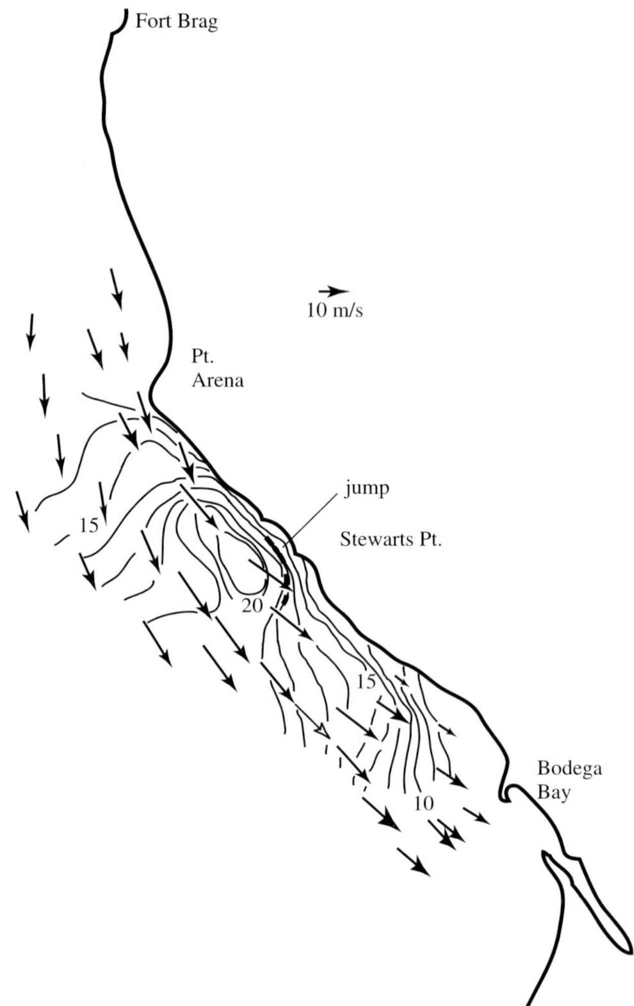

FIGURE I.2b. Velocity arrows and speed contours in the coastal marine layer from Pt. Arena to Bodega Bay, California. This is an earlier event than the one shown in I.2a but there is a rough correspondence between the features. (Based on a figure from Winant *et al.*, 1988).

(colored grey in Figure I.3). Before it reaches the equator, this *North Atlantic Deep Water* (hereafter NADW) detaches from the bottom, riding up over a water mass of Antarctic origin (black path). Some portion of NADW continues southward, enters the Antarctic Circumpolar Current, and eventually makes its way into the Indian and Pacific Oceans. It gradually wells up and becomes part of the (white) warmer and fresher surface currents that transport water from the Pacific to the Indian Ocean and eventually back into the Atlantic, completing the meridional overturning cell.

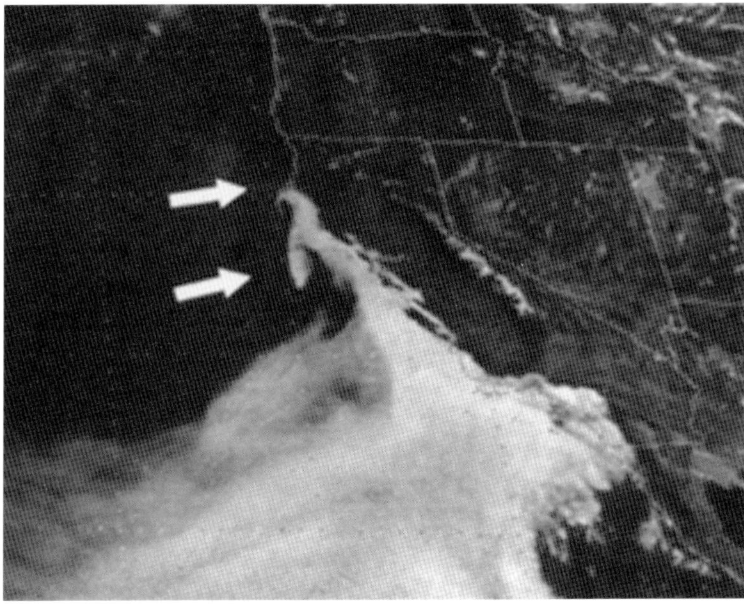

FIGURE I.2c. A marine-layer gravity current (light, cloudy area) flowing northward along the California coastline. The arrows mark the leading edge and an eddy formed behind it. (Image courtesy of Clive Dorman).

The second main contributor to the abyssal circulation is the aforementioned *Antarctic Bottom Water* (AABW). This water mass is formed in the Weddell and Ross Seas and over the greater continental shelves of Antarctica. This water flows northward along the bottom into the Atlantic, Indian, and Pacific Oceans

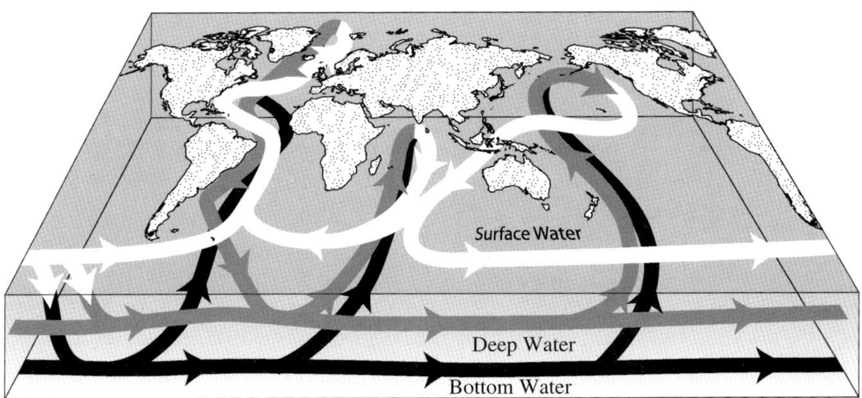

FIGURE I.3. A three-layer version of the ocean thermohaline circulation. White bands indicate mean circulation within and above the main thermocline, gray represents deep circulation, and black represents bottom circulation.

where it mixes and wells up. It is believed that the upward movement and mixing of AABW and NADW is enhanced over regions of rough topography.

A viewer of Figure I.3 should be aware of several caveats. One is that the ocean circulation is strongly time dependent and the transport of water and other properties can be strongly affected by fluctuating eddy processes. Not all the pathways suggested should be interpreted as persistent currents, nor do they always suggest the direct paths of typical water parcels. For example, a parcel of NADW entering the Circumpolar Current from the Atlantic may spin around Antarctica multiple times before exiting into the Indian or Pacific Oceans, or it may simply re-enter the Atlantic. In addition, the figure does not acknowledge some sinking regions that lie at the margins of the Atlantic and Southern Oceans and in marginal seas. For example, deep convection in the Labrador Sea is known to contribute to NADW. Dense water masses are also produced in the Red and Mediterranean Seas. The salty outflows from these seas spread and circulate in the Indian and Atlantic Oceans, respectively, in ways that are not fully understood.

Hydraulic effects occurring within the lower limbs of the conveyor are due to interactions with bottom topography. The bottom is a bumpy collection of old tectonic plates surrounded by hotspot tracks, swells, flood basalts, pieces of

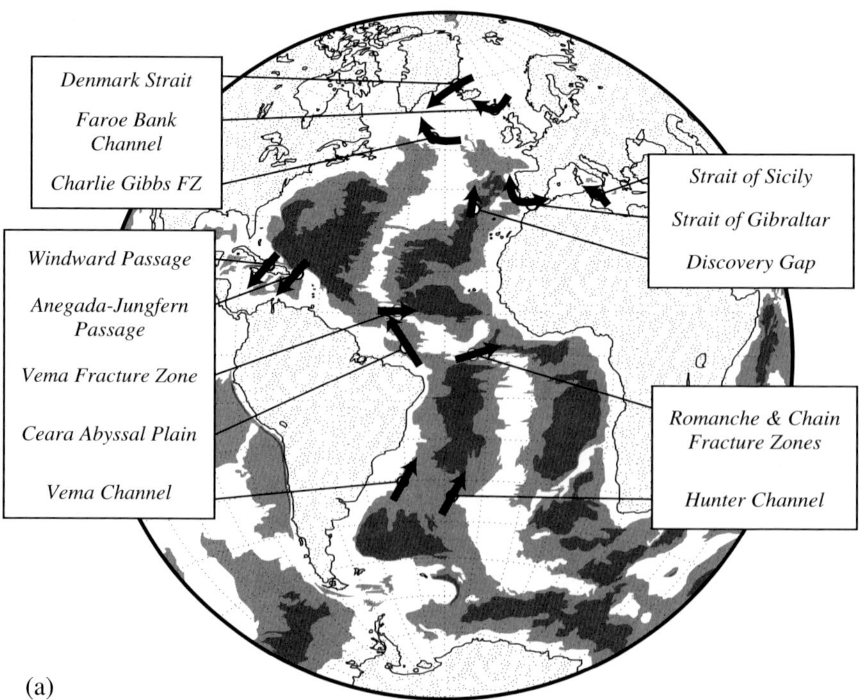

FIGURE I.4a. Map of the Atlantic Ocean with smoothed versions of the 4000 m and 5000 m isobaths. Some well-known passages and the direction of flow through these passages are indicated. Bidirectional arrows indicate an exchange flow.

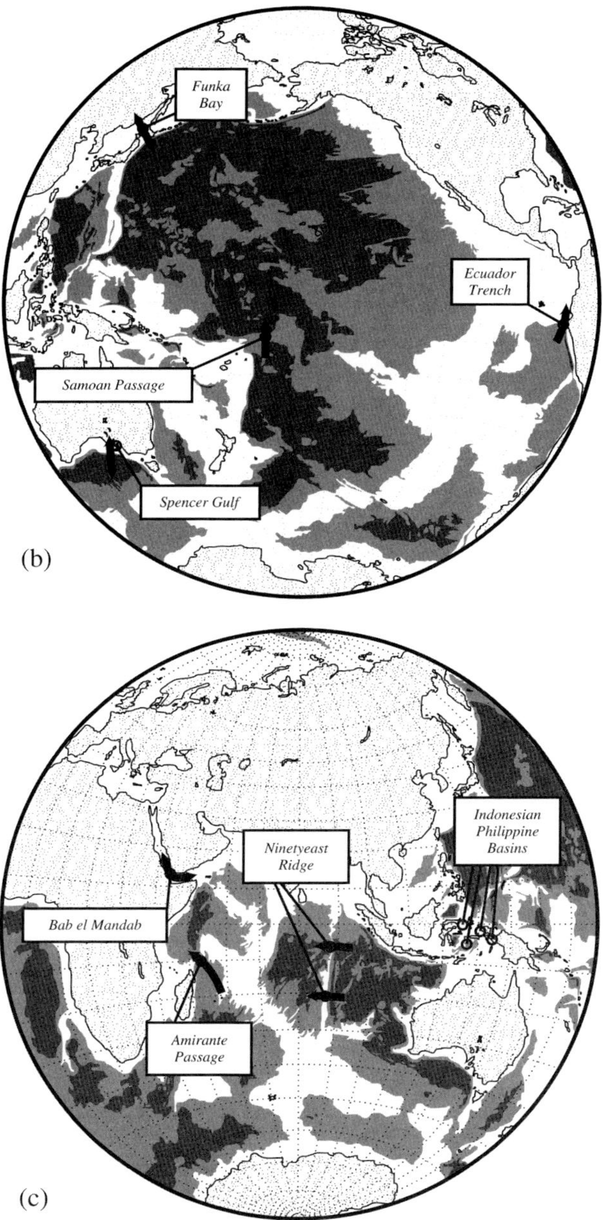

FIGURE I.4b,c. The deep passages and topography of the Pacific and Indian Oceans.

thickened crust, and ridges. These all subdivide the major oceans into numerous basins (Figures I.4a, b, c). The abyssal currents that form the lowest limbs of the conveyor make their way from basin to basin, seeking out the deepest connecting strait or sill. Some of these deep flows are observed to spill into the downstream basin in roughly the same manner as the flow over a dam. The volume flux of the spilling flows can double or triple as the result of mixing with and entrainment of overlying water. A deep current will sometimes flow into a terminal basin; there, it mixes with overlying water and gradually rises to shallower levels.

Some of the major passages in which overflows are observed have been indicated in Figures I.4a,b,c. Each example has been documented by the direct measurement of deep velocities. Also shown are the locations of several prominent straits connecting marginal seas to the ocean proper. These shallower passages generally contain exchange flows (a double arrow), often with an inflowing surface current overlying a deep outflow. Many of the passages, shallow or deep, are strategically advantageous locations for the measurement of property fluxes relevant to the ocean circulation and to global climate. They funnel massive flows through a relatively small area and they may be candidates for the same type of hydraulic monitoring that engineers use to keep track of the outflow from a reservoir. The overflows can have distinct chemical distributions and a time-history of this chemistry can be extracted from deep sediments and used to infer properties of the thermohaline circulation through geological time.

The most thoroughly documented deep overflows are those of the North Atlantic and these are now described in more detail. The two main water masses involved, NAWD and AABW, can be identified in a north-south section of potential density (σ_4)[1] in the western Atlantic (Figure I.5). The section track is shown in I.6a, which also gives a plan view of the ocean bottom and the 1.8°C potential temperature[2] surface. The geographical distribution of this surface gives some indication of the spreading of the deep water masses away from their source. The reader should interpret both figures with caution: the thinning or dilution of a water mass in a particular direction need not indicate a flow in that direction.

AABW enters the western Atlantic across the south boundary of the Figure I.5. The overall northward movement is known from the temperature, salinity, silicate and other properties suggesting an Antarctic origin, and from observations of northward flow in deep straits such as the Vema Channel, where spilling is also observed. After passing through the Vema and Hunter Channels, AABW enters the Brazil Basin. Along this path, the deep isopycnal surfaces deepen in a way that suggests spilling. This deepening could also be the result of mixing with overlying water or the presence of transverse (east-west) geostrophic flow. However, direct measurements show that AABW exits the basin in two directions. To the west is the Ceara Abyssal Plain that serves as a complex and very broad sill leading into the Western North Atlantic Basin. The Figure I.5 transect, which crosses this plain, shows that the isopycnals surfaces with $\sigma_4 \geq 45.9$ progressively deepen, and

---

[1] σ_4 is the density that a water parcel would have if that parcel were moved to a pressure of 4000 decibars (db), roughly equivalent to 4000 m depth. The 4000 db reference level is used to avoid difficulties that arise in connection with nonlinearities in the equation of state. For example, use of the sea surface as a reference pressure would lead in some cases to the conclusion that the deep stratification is hydrostatically unstable.

[2] In all cases, potential temperature refers to the temperature a fluid parcel would have if that parcel were moved adiabatically to ocean surface pressure (zero db).

evcntually ground, to the north and across the Western North Atlantic Basin. The second exit lies to the east and through the mid-Atlantic Ridge, which is transected by the Romanche and Chain Fracture Zones (Figure I.4a). Some AABW passes across the sills of these passages and into the Eastern Atlantic, while some continue northward along the western flank of the ridge. Some of this northward flow turns east and passes through the ridge within the Vema Fracture Zone at 11 °N. All of the eastward flows that cross the ridge spill and mix into the basins of the eastern Atlantic. In the case of the Vema Fracture Zone, some of the overflow water has been mixed to the point where $\sigma_4 < 46.0$. The diluted water makes its way through Discovery Gap at about 36 °N and eventually terminates.

The deep waters entering the Atlantic from the north spill across the Greenland-Iceland-Scotland ridges (Figure I.5) through two main passages: the Denmark Strait (650 m sill depth) and Faroe Bank Channel (880 m), as indicated in Figure I.4a. The two outflows merge to form NADW, which moves southward in the aforementioned deep western boundary current. Its transport is augmented by deep convection in the Labrador Sea. The overall southward flow is complicated by branching, eddying, and mixing of the boundary current. As a result, the NADW becomes spread over the bottom of the western North Atlantic basin

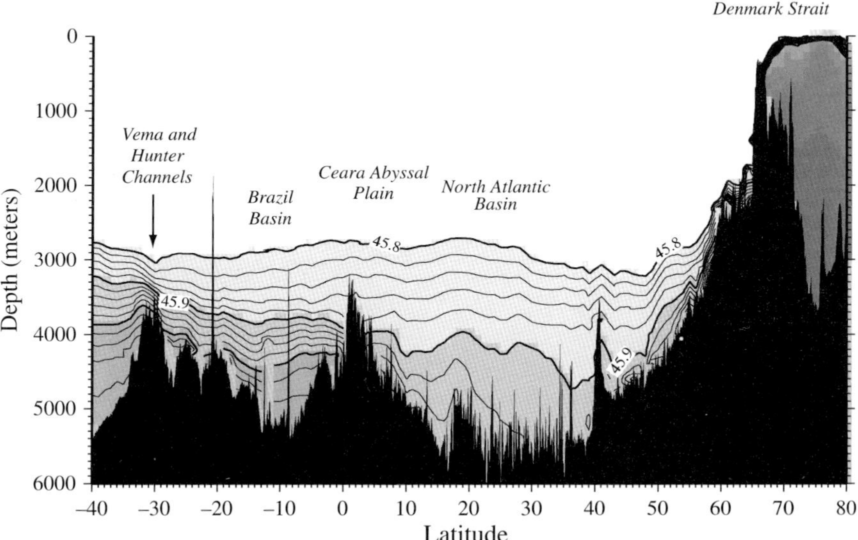

FIGURE I.5. A roughly north-south section through the western Atlantic showing selected deep potential density (σ_4) surfaces, given in (kg/m^3) minus 1000. The section track is shown in figure I.6a. Water denser than 46.0 is considered Antarctic Bottom Water, and North Atlantic Deep Water has the density range $45.5 < \sigma_4 < 45.9$. The topographic spikes are isolated seamounts that do not impede the overall deep currents. Individual sill depths in passages such as the Vema Channel are deeper than what is indicated along the track of the section. This and Figure I.6 were made by T. McKee and R. Curry using the Hydrobase data archive (http://www.whoi.edu/science/PO/hydrobase).

southward until it encounters AABW and leaves the bottom. The presence and spreading of both NAWD and AABW are also suggested by the Figure I.6a map of water with potential temperatures greater than 1.8°C. Deep spreading also occurs in the Pacific and Indian Oceans (Figures I.6b and I.6c).

The Denmark Strait provided some of the first observations of deep-ocean hydraulic behavior. The draw-down of isopycnals associated with the southward

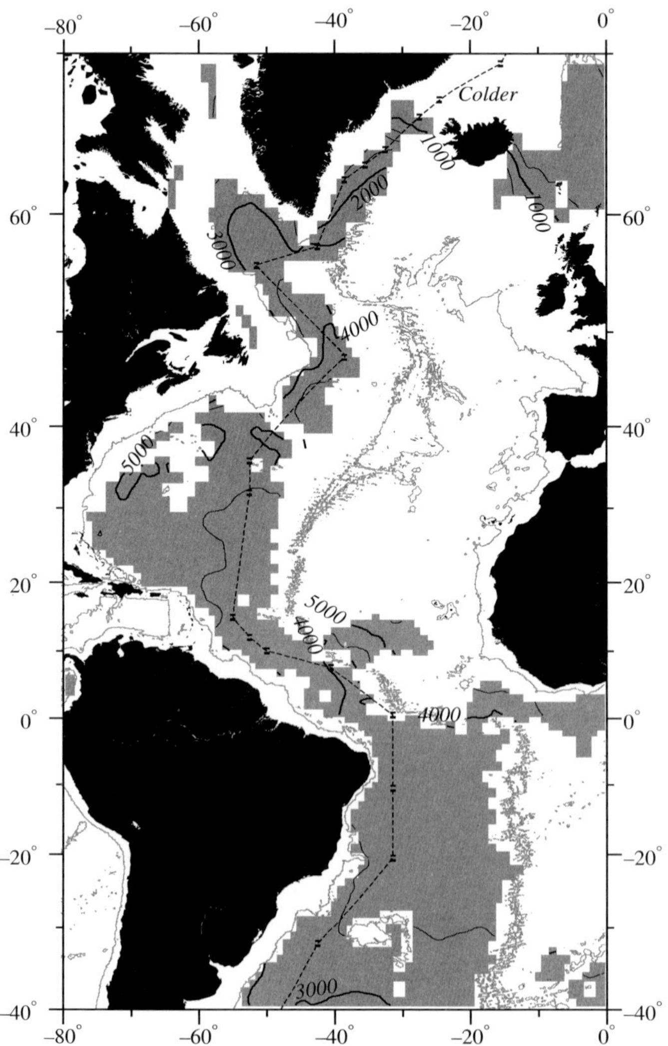

FIGURE I.6a. The shaded area shows those portions of the Atlantic Ocean containing bottom water with potential temperature greater than 1.8°C. The edge of the shaded area is the intersection of the 1.8°C isothermal surface with the bottom. The contours overlain on the 1.8°C surface show its depth. The thin contours elsewhere show the 3000 m isobath.

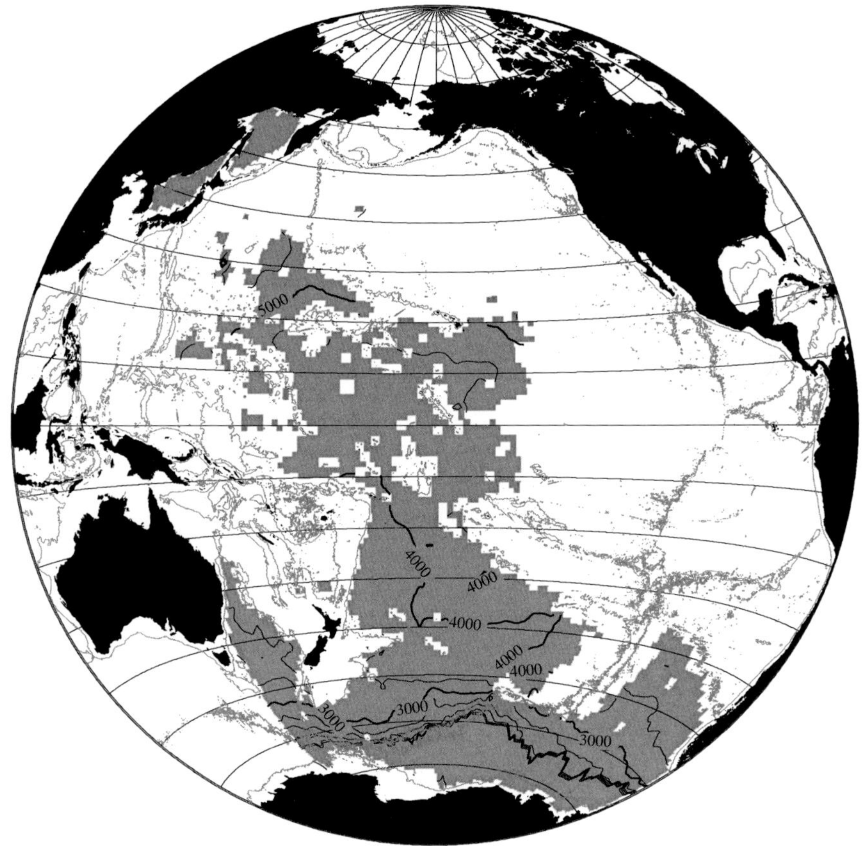

FIGURE I.6b. Similar to Figure I.6a, but for the 1.0°C potential temperature surface in the Pacific Ocean.

spilling, apparent in Figure I.5, is shown in more detail (Figure I.7). Most of the mixing and entrainment that occurs in this and other major overflows takes place in a 'plume' or 'outflow' region, roughly corresponding to the descending region downstream of the sill. The transverse structure of the flow (Figure I.8) is strongly influenced by the earth's rotation, which causes the isopycnals to tilt and the overflow water ($\sigma_\theta > 27.6$) to bank against the Greenland continental slope. The details of this structure can be quite variable in time.

The overflow in the Faroe Bank Channel exhibits similar features. The deepest water in this passage is denser than that of the Denmark Strait. The overflow descends into the Iceland Basin to depths of about 3500 m. The water then flows southward along the flanks of the mid-Atlantic Ridge until it encounters the Charlie-Gibbs fracture zone. Most of the flow continues westward through this gap, where it joins the Denmark Strait overflow. A more detailed examination of this overflow appears in Section 2.11.

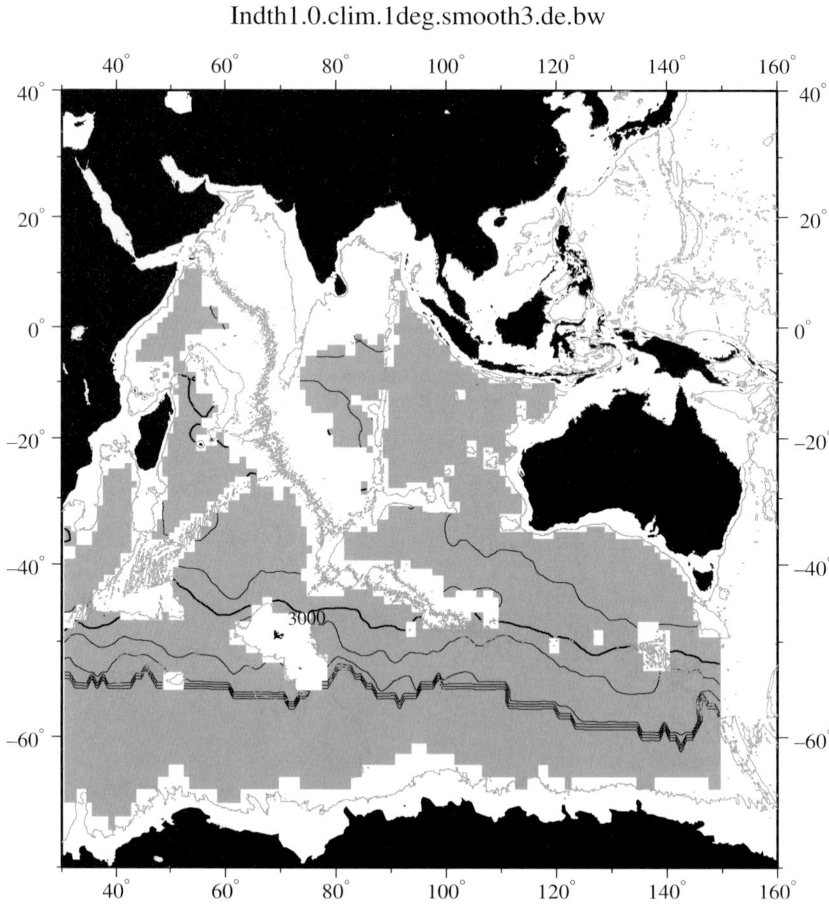

FIGURE I.6c. Similar to Figure I.6a, but for the 1.0°C potential temperature surface in the Indian Ocean.

Other oceanographically important, deep passages with overflows exist in the Pacific and Indian Oceans. In the Pacific, the Samoan passage (Figure I.4b) has clearly defined currents that carry a mixture of North Atlantic Deep Water and Antarctic Bottom Water north. The inlet to the Panama basin near South America contains a bottom current feeding the deep basin. To the west, there are many sub-basins in the Indonesian-Philippine Basin, each with a deep inlet and an inflow of dense water. In the Indian Ocean (Figure I.4c) there are overflows across the Ninety-East Ridge and in the Amirante Passage. The spreading of dense water masses in these locations is suggested in Figures I.6b and I.6c, which map the 1.0°C potential temperature surface.

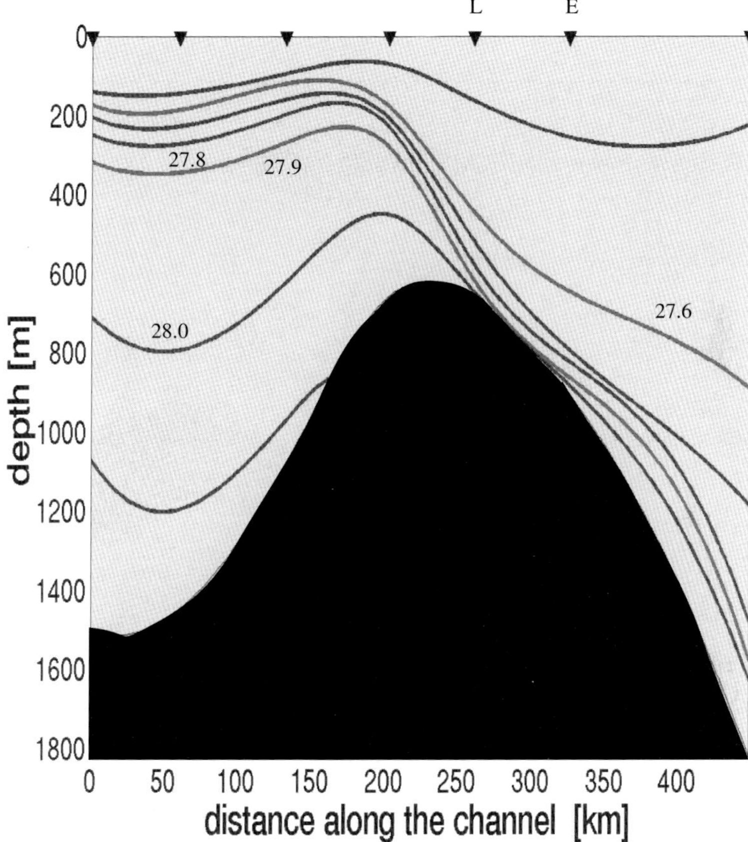

FIGURE I.7. Longitudinal section of surface-referenced potential density (σ_θ), taken along the middle of the Denmark Strait. (From Figure 4 of Nikolopoulos et al., 2003).

d. Exchange Flows

To this point, we have treated overflows as unidirectional and isolated, paying little attention to interactions with the overlying fluid. Indeed, the neglect of interactions with the overlying layers is an assumption held by many models. However relevant this view may be to the deep ocean, it rarely applies in the shallow straits separating marginal seas from the major oceans. Bodies such as the Red and Mediterranean Seas act as *inverse estuaries* by losing more fresh water to evaporation than is gained by river runoff and precipitation. Salt is concentrated in the surface layers and this makes the water more susceptible to deep convection due to localized atmospheric cooling events. For example, deep convection has been observed in the Gulf of Lyon and in the far northern reaches of the Red Sea. The dense waters accumulate and spill out through relatively

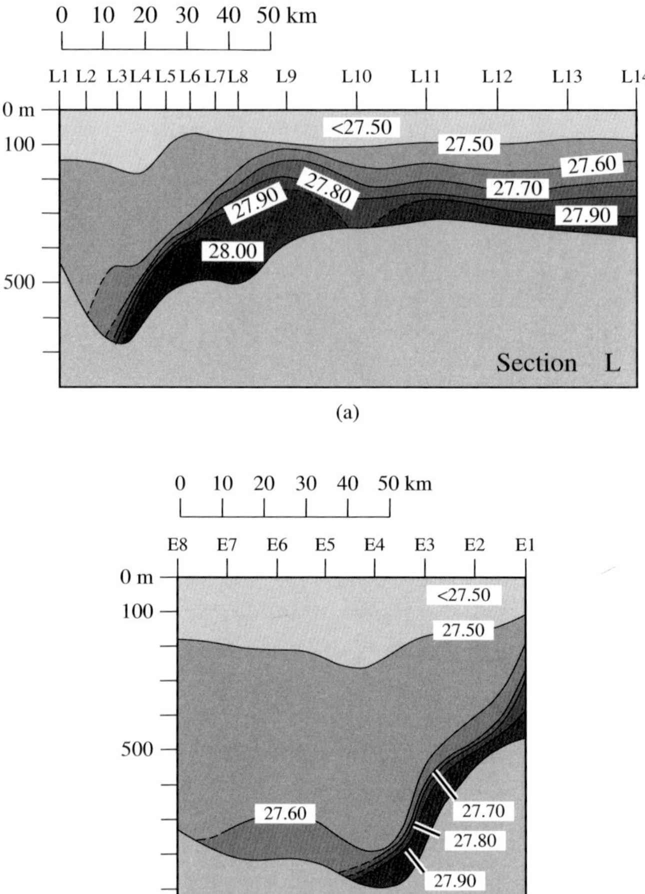

FIGURE I.8. Potential density (σ_θ) across the Denmark Strait at (a) the sill and (b) approximately 50 km downstream of the sill. The position of the section (E and L) are also indicated in Figure I.7. (From Nikolopoulos et al., 2003).

shallow straits such as the Bab al Mandab and Strait of Gibraltar. Mass and salt conservation require that the dense waters be replaced and this occurs in the form of surface inflow. *Exchange flows* with opposing upper and lower layers of comparable depth and velocity are thereby set up in the connecting straits. An exchange flow also occurs in the Bosporus, a narrow strait separating the Mediterranean and Black Seas.

One of the earliest references to exchange flows in sea straits was a set of remarks by Smith (1684) on the Strait of Gibraltar. Smith knew about the surface inflow:

"There is a vast draught of water poured continually out of the Atlantic into the Mediterranean; the mouth or entrance of which between Cape Spartel or Sprat, as the seamen call it, and Cape Trafalgar, may be near 7 leagues wide, the current setting strong into it, and not losing its force till it runs as far as Malaga, which is about 20 leagues within the Straits."

There was, at the time, much speculation about the destination of this inflow, including:

"subterraneous vents, cavities, and indraughts, exhalations by the sun beams, the running out of the water on the African side, as if there were a kind of circular motion of the water, and that it only flowed in upon the Christian shore."

Smith considered these conjectures fanciful and contrary to observation. His hypothesis was that the surface inflow must be balanced by a deep outflow in the Strait:

"an under-current, by which as great a quantity of water is carried out as comes flowing in."

Smith's theory was not based on mass conservation alone. He had heard a report from a sailor aboard a vessel that had attempted to enter the Baltic Sea through the Öresund, the strait that separates Denmark and Sweden:

"He told me that, being there in one of the king's frigates, they went with their pinnance into the mid stream, and were carried violently by the current; that soon after they sunk a bucket with a large cannon ball, to a certain depth of water, which gave check to the boat's motion, and sinking it still lower and lower, the boat was driven ahead to windward against the upper current: the current aloft, as he added, not being 4 or 5 fathom deep, and that the lower the bucket was let fall, they found the under current the stronger."

The Gibraltar outflow descends the continental slope west of Portugal down to about 1000 m to form a saline plume in the Atlantic known as the Mediterranean salt tongue. This water mass can be detected throughout the North Atlantic at intermediate depths. As documented by Armi and Farmer (1988), the Strait of Gibraltar acts as a choke point or hydraulic control on the inflow and outflow. Figure I.9 shows a longitudinal acoustic image taken near the 250 m deep Camerinal Sill at the western end of the Strait. The Atlantic Ocean is to the right and the Mediterranean Sea is to the left. The ship-mounted acoustic transponder generates sound waves that are scattered from fish, small particles, bubbles, and temperature microstructure. The wavy band that lies near the surface in the left-hand portion of the image approximates the interface between the outflowing (left-to-right) lower layer and the incoming upper layer. Superimposed temperature profiles confirm that the inferred interface level lies in the depth range of most rapid temperature change. Note the descent of the interface into deeper water as the lower layer spills over the sill. Further downstream are some dark undulations that have been interpreted as an internal hydraulic jump. Although rotation is probably not important within the strait proper, the outflows in either direction become strongly influenced by rotation past the exits.

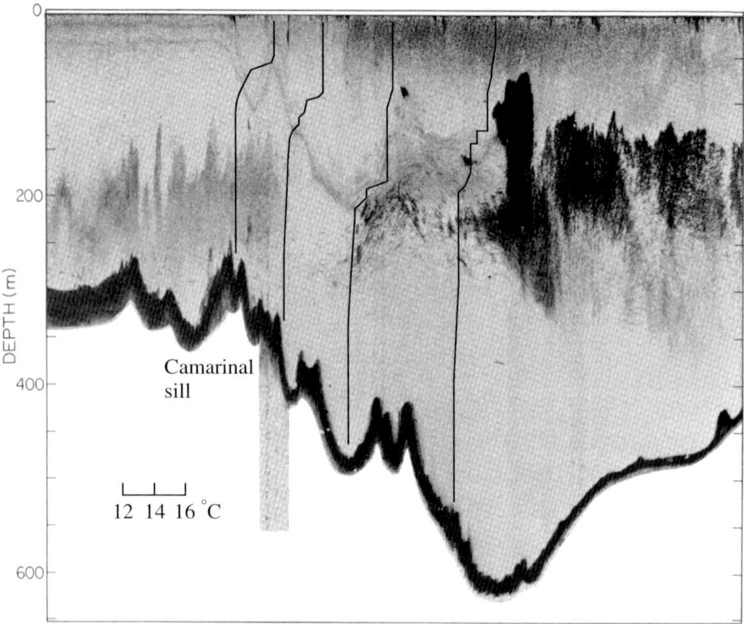

FIGURE I.9. Acoustic image with temperature profiles from Camerinal Sill in the Strait of Gibraltar. (Constructed using two figures from Armi and Farmer, 1988).

e. *Other Examples*

The world's marginal seas and coastal areas also have an abundance of other straits and gaps that may be important regionally or globally and that are wide enough to feel the Earth's rotation. One example is the Strait of Sicily, which contains a current of relatively saline bottom water that flows from the eastern to the western Mediterranean (Astraldi *et al.* 2001). At a sill depth of 1600 m, the Anegada-Jungfern Passage is the deepest inlet into the Caribbean basin (Figures I.3a and I.10). Dense inflows through this and other inlets supply all deep waters for the Caribbean, filling it from the bottom to about this sill depth. The passage actually contains two parallel channels, the Jungfern Passage and the Grappler Channel. The earth's rotation is evident in the intensification of the flow along the right-hand walls of both channels (Figure I.11). The Baltic Sea, which acts as a giant estuary, is separated from the North Sea by a series of straits that carry fresh water seaward and allow an intermittent inflow.

Smaller scale estuaries and inverse estuaries may experience hydraulic effects as well. Another important body that experience estuary dynamics but is large enough to be influenced by the earth's rotation is Chesapeake Bay, North America (Figure I.12a, b). Although there is no evidence for hydraulic control of the exchange flow that occurs at the mouth, the bay is the source of an important rotating gravity current. The dynamics of such currents are often considered as belonging to hydraulic phenomena and are treated later in this book. In the case of Chesapeake Bay, the fresh surface outflow can be halted by upwelling-

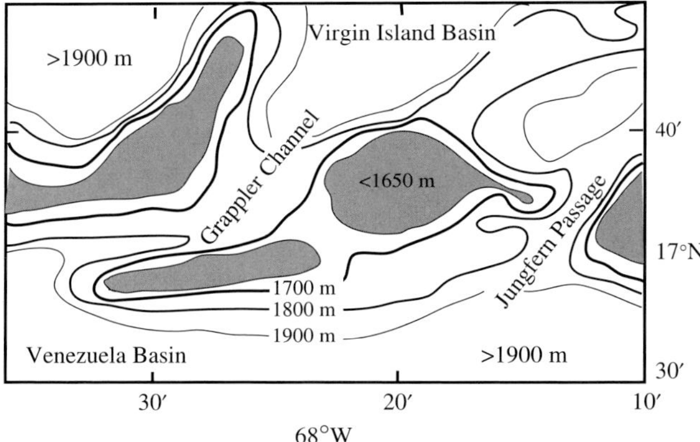

FIGURE I.10. Map showing Jungfern and Grappler Passages. The deep throughflow is to the north. (from Fratantoni et al., 1997).

favorable winds blowing across the mouth. When these winds relax, the mass of released fresh water forms a buoyant current that exits the mouth, turns to the right, and flows southward along the coastline. The current remains trapped to the coastline by the effects of rotation and can have a striking blunt leading edge or nose that can be seen with the naked eye (see Figure 4.5.2).

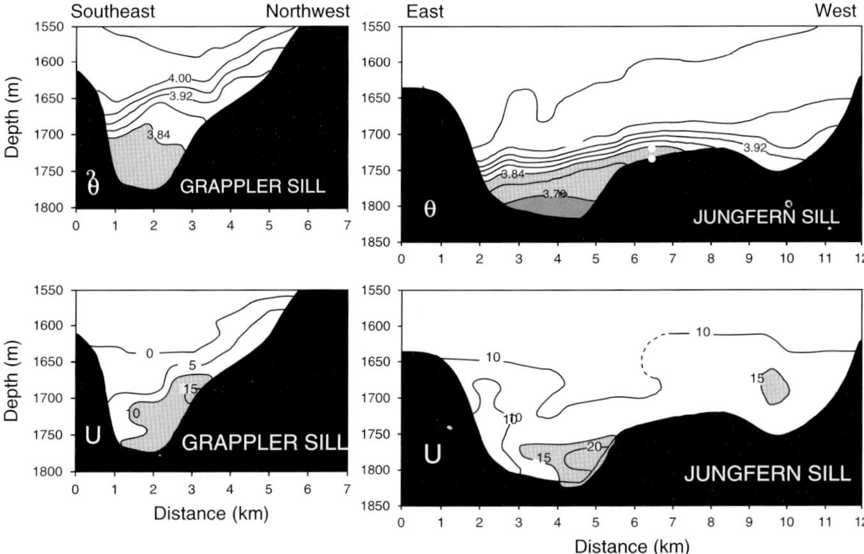

FIGURE I.11. Cross-sections of potential temperature and along-strait velocity across the Grappler and Jungfern Passages (from Fratantoni *et al.*, 1997).

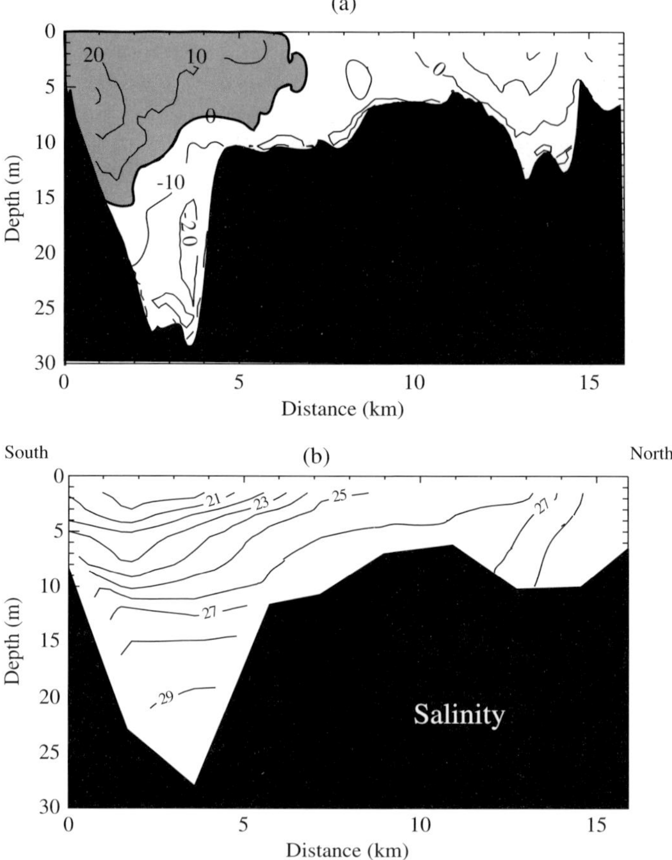

FIGURE I.12. a) Normal velocity across the mouth of Chesapeake Bay. Positive values indicate outflow. b) Salinity distribution across the mouth of Chesapeake Bay. (From Levinson et al., 1998).

This book explores the theories, idealized models, and laboratory experiments that have been developed in an attempt to understand the dynamics of these applications and to answer fundamental questions about their global significance. For example, does the horizontal circulation of deep water in a particular basin depend on whether the outflow is hydraulically controlled? Suppose that the world oceans were devoid of basins and deep passages. Would the production rates of various deep waters be unchanged? How would the rate of mixing between deep and shallow water masses be altered? The answers to these and other broad questions about the upstream and downstream influence of ocean and atmospheric choke points are not known, but the idealized models presented herein give some clues. We hope that readers will be able to build on the basic material in order to develop new thinking and new approaches to long-standing problems.

f. Further Reading

There is an abundance of literature on the applications mentioned above. Specific citations will follow in the discussions of deep ocean overflows (Sections 2.11 and 2.14), the California coastal marine layer (Sections 4.3 and 4.4), the Strait of Gibraltar (Sections 5.6 and 5.7) and others. For the reader seeking a more general discussion of the ocean deep circulation, Warren (1981) and the 2001 WOCE book: *Ocean Circulation and Climate: Observing and Modeling the Global Ocean* are good starting points. A brief history of the discovery of deep overflows can be found in Pratt and Lundberg (1991). An important landmark in this history is the Worthington and Wright (1970) atlas. Its maps of deep potential temperature surfaces suggesting penetration of dense intrusions into the deep North Atlantic inspired our Figure I.6. The website "Sills of the Global Ocean" at www.noc.soton.ac.uk/JRD/OCCAM/sills.html contains an extensive list of ocean sills that are potentially important in the distribution of deep water in the oceans. The oft-cited global conveyer belt is described by Broecker (1991) and on numerous websites.

1
Review of the Hydraulics of Nonrotating, Homogeneous Flow

Hydraulic behavior is closely related to wave propagation tendencies. This does not imply that waves need to be present; indeed, most 'textbook' examples are based on steady flow. Rather, hydraulic behavior in a steady flow is related to the routing of information, potentially carried by waves through the fluid domain. Analyses of steady flows that appear throughout this book are often preceded or accompanied by discussions of linear and nonlinear waves. Linear wave dynamics are important in understanding the structure of steady flows and of regions of influence. Nonlinear wave dynamics are essential in the formation of hydraulic jumps and bores, and can be important in the establishment of hydraulically controlled steady states as the result of evolution from simple initial conditions. Some of our students skip through the material on waves and proceed directly to the discussions of steady flows; we encourage you to resist this temptation.

1.1.　The Long-wave Approximation

One of the elusive aspects of hydraulic phenomena is that nonlinearity is inherent to much of the important behavior. In order to make models that retain the nonlinear essence, but remain tractable, it is necessary to introduce a number of simplifying assumptions. Most are standard and will be familiar to readers having experience with geophysical fluid dynamics. For example, the flow is generally considered inviscid, nondiffusive, incompressible, and homogeneous (or at least stacked in homogeneous layers). However, the assumption that is most emblematic of hydraulics is that of gradual variations along the predominant direction of flow. Sometimes referred to as the *long-wave* approximation, the assumption can be illustrated by considering a current confined to a channel aligned in the y-direction (Figure 1.1.1a). The channel guides the flow more or less along the y-axis. The fluid depth $d(x, y, t)$ and bottom elevation $h(x, y)$ may vary across the channel (Figure 1.1.1b). The y- and z- velocity components are given by $v(y, z, t)$ and $w(y, z, t)$. If the scale L of variations of d, v, and w in the y-direction is large compared to the scale of cross-sectional variations of these variables, then we say that the flow is gradually varying. The scale L might

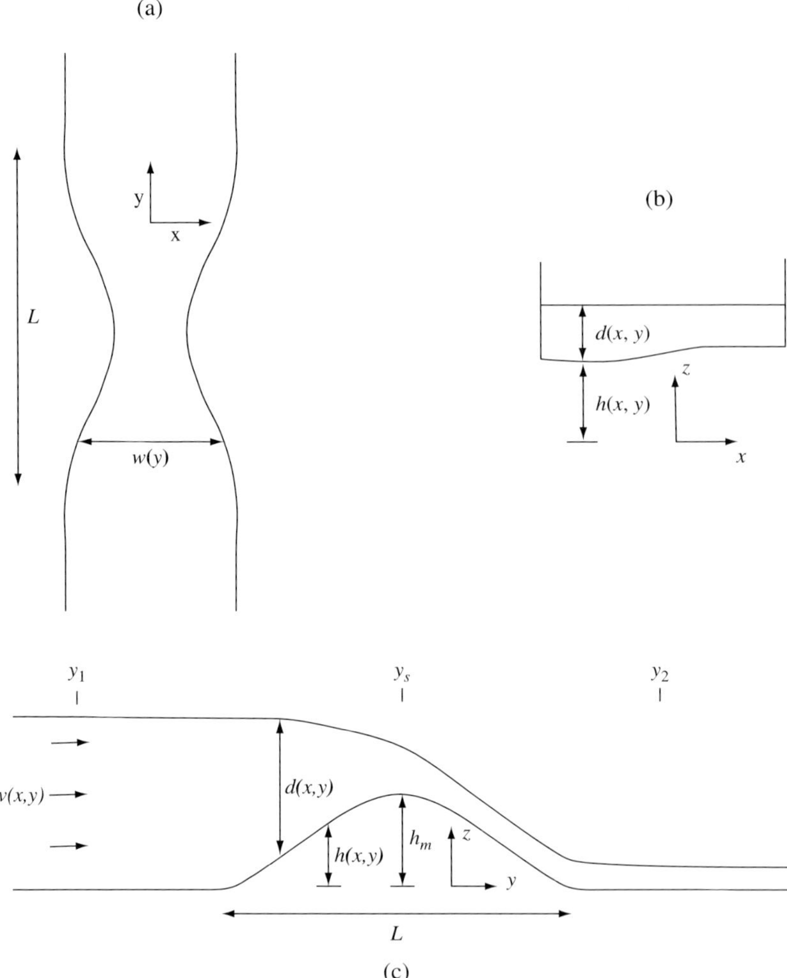

FIGURE 1.1.1. Definition sketch showing the channel geometry and the flow variables.

be associated with width variations (Figure 1.1.1a) or topographic variations (Figure 1.1.1c), or L might simply be the wavelength of a disturbance propagating along a uniform channel. Scales characterizing cross-sectional variations of the flow might include the mean depth and the channel width. In coastal geometries, rotating flows can be trapped to a coast within a distance set by internal dynamics. For example, currents may be trapped within the Rossby radius of deformation, to be defined later. Other oceanic and atmospheric flows that move freely of lateral boundaries have their own width scales and a definition of gradual variations may be made accordingly. In all cases, the long-wave approximation is satisfied when the *largest transverse scale is much less than the smallest along-channel scale*. The chief mathematical simplification gained is separation: the transverse

(x- and z-) structure of the flow may be resolved, at least in a general form, before the y- or t-dependence need be considered.

Shallow-water theory for a homogeneous layer will serve as the basis for many of the examples that follow. The shallow-water equations are themselves a product of the long-wave approximation. Consider their derivation for the case of a two-dimensional, inviscid, incompressible flow with a free surface. Begin with the Euler equations in two-dimensions:

$$\frac{\partial v}{\partial t} + v\frac{\partial v}{\partial y} + w\frac{\partial v}{\partial z} = -\frac{1}{\rho}\frac{\partial p}{\partial y}, \tag{1.1.1}$$

$$\frac{\partial w}{\partial t} + v\frac{\partial w}{\partial y} + w\frac{\partial w}{\partial z} = -\frac{1}{\rho}\frac{\partial p}{\partial z} - g, \tag{1.1.2}$$

and

$$\frac{\partial v}{\partial y} + \frac{\partial w}{\partial z} = 0. \tag{1.1.3}$$

Here ρ denotes the (uniform) density, p the pressure, and g the gravitational acceleration.

Subtraction of $\partial(1.1.1)/\partial z$ from $\partial(1.1.2)/\partial y$ and use of (1.1.3) lead to

$$\left(\frac{\partial}{\partial t} + v\frac{\partial}{\partial y} + w\frac{\partial}{\partial z}\right)\left(\frac{\partial w}{\partial y} - \frac{\partial v}{\partial z}\right) = 0, \tag{1.1.4}$$

showing that the vorticity $\partial w/\partial y - \partial v/\partial z$ is conserved following fluid elements.

The long-wave approximation for this flow is valid if variations in y, which occur over scale L, are gradual in relation to the depth scale D. The continuity equation (1.1.3) then suggests that $w/v = O(D/L) \ll 1$ and this leads to two simplifications involving the cross-sectional (z-) structure of the flow. First, the fluid vorticity is dominated by the factor $-\partial v/\partial z$. If the vorticity is zero, as would occur if motion had been conservatively generated from a state of rest, then $\partial v/\partial z = 0$ for all space and time. The horizontal velocity therefore remains independent of z :[1]

$$v = v(y, t). \tag{1.1.5}$$

The second simplification is that the pressure p becomes hydrostatically balanced:

$$\partial p/\partial z = -\rho g.$$

This result can be deduced from (1.1.2) by showing that each term on the left-hand side becomes small compared to g when $w/v = O(D/L) \ll 1$. If the bottom

---

[1] More general structures are also possible. If the vorticity is initially finite but uniform then $v(y, z, t) = az + b(y, t)$.

lies at $z = h(y)$, the free surface at $z = h(y) + d(y, t)$, and the surface pressure is zero, then

$$p(y, z, t) = \rho g(d(y, t) + h(y) - z). \tag{1.1.6}$$

The transverse (z-) structure of the flow field has now been determined. The y-velocity is constant, the pressure varies linearly, and it is left as an exercise for the reader to show w also varies linearly. One may now proceed to the task of determining the y- and t- dependence of the variables. To do so, one must first develop the shallow water equations from the above expressions, a task left as an exercise (see #2 below) to the reader who has not seen this done. The general point is that that the cross-sectional structure of the flow can be determined at the outset, allowing one to concentrate on other aspects of the problem. We will return to this theme repeatedly as part of the consideration of flows with stratification or with structure in the cross-channel (x-) direction.

Exercises

(1) Consider the streamfunction $\psi(x, y, t)$ for the flow described by (1.1.4), noting that

$$\frac{\partial \psi}{\partial y} = w \quad \text{and} \quad \frac{\partial \psi}{\partial z} = -v.$$

(a) When the flow is steady, show that

$$J_{y, z}(\psi, \partial w/\partial y - \partial v/\partial z) = 0;$$

that is, the fluid vorticity is conserved along streamlines.
(b) Deduce from these relations the result

$$\nabla^2 \psi = F(\psi)$$

where $F(\psi)$ specifies the value of the vorticity along each streamline $\psi = \text{constant}$.
(c) Show that, in the long-wave limit, this last relation reduces to

$$\frac{\partial^2 \psi}{\partial z^2} = F(\psi).$$

(d) Suppose that at some upstream section $\psi = \alpha z^2/2$. Show that at any other section, the velocity v is described by $\alpha z + \beta(y)$. Note that our ability to write down a general form for the z-structure of the flow at an arbitrary location is made possible by the long-wave approximation.

(2) *Derivation of the shallow water equations.* This exercise makes use of the kinematic boundary conditions at the bottom and at the free surface of the fluid layer:

$$w(y, h, t) = v(y, t)\frac{\partial h}{\partial y} \tag{1.1.7a}$$

$$w(y, h+d, t) = \frac{\partial d}{\partial t} + v(y, t)\frac{\partial(h+d)}{\partial y}. \tag{1.1.7b}$$

(a) Integrate the continuity equation (1.1.3) over the depth of the fluid and use (1.1.7) to obtain

$$\frac{\partial d}{\partial t} + \frac{\partial(vd)}{\partial y} = 0.$$

(b) Substitute the expression for the long-wave pressure and velocity into (1.1.1) to obtain

$$\frac{\partial v}{\partial t} + v\frac{\partial v}{\partial y} = -g\left(\frac{\partial d}{\partial y} + \frac{\partial h}{\partial y}\right).$$

The results obtained in (a) and (b) are the shallow water equations.

1.2. The Shallow Water Equations and One-dimensional Wave Propagation

Traditional discussions of hydraulic effects such as those found in engineering textbooks are often based on analyses of steady flows. At the same time, interpretation of these effects almost always involves waves and wave propagation. We therefore preface our discussion of steady hydraulics with a conversation about wave propagation in shallow water. Attention is restricted to flows governed by the shallow water equations in one spatial dimension:

$$\frac{\partial v}{\partial t} + v\frac{\partial v}{\partial y} + g\frac{\partial d}{\partial y} = -g\frac{\partial h}{\partial y} \tag{1.2.1}$$

and

$$\frac{\partial d}{\partial t} + d\frac{\partial v}{\partial y} + v\frac{\partial d}{\partial y} = 0. \tag{1.2.2}$$

For a homogeneous fluid, g is the ordinary gravitational acceleration. In oceanic and atmospheric models, we often consider flow in a layer that has uniform density and that is overlain by a much thicker and inactive layer of slightly lower (but still uniform) density. In such cases, the above equations remain valid provided that g is interpreted as the *reduced gravity*: the ordinary g multiplied by the fractional difference in density between the two layers. Such a model is alternately referred to as having a '$1\frac{1}{2}$-layer', 'reduced gravity' or 'equivalent barotropic' stratification. The reduced gravitational constant is usually denoted g', but we will allow the original g to be interpreted as either.

Now consider a steady flow with uniform velocity V and depth D over a horizontal bottom $(dh/dy = 0)$. Infinitesimal disturbances to this flow, denoted v' and η, can be introduced by setting $v = V + v'$ and $d = D + \eta$, where $v' \ll V$ and $\eta \ll D$. Substitution into (1.2.1) and (1.2.2) and neglect of quadratic terms in the primed variables, leads to the linear shallow water equations

$$\frac{\partial v'}{\partial t} + V\frac{\partial v'}{\partial y} + g\frac{\partial \eta}{\partial y} = 0, \tag{1.2.3}$$

and

$$\frac{\partial \eta}{\partial t} + V\frac{\partial \eta}{\partial y} + D\frac{\partial v'}{\partial y} = 0. \tag{1.2.4}$$

The most general solution is given by

$$\eta = f_+(y - c_+ t) + f_-(y - c_- t) \tag{1.2.5}$$

and

$$v' = \left(\frac{g}{D}\right)^{1/2}[f_+(y - c_+ t) - f_-(y - c_- t)] \tag{1.2.6}$$

where

$$c_+ = V + (gD)^{1/2} \tag{1.2.7}$$

and

$$c_- = V - (gD)^{1/2}. \tag{1.2.8}$$

Equations (1.2.5) and (1.2.6) describe waveforms traveling at speeds c_+ and c_-. The propagation speeds are made up of two components: the unidirectional background flow velocity V and a bidirectional propagation velocity $\pm(gD)^{1/2}$, equivalent to the propagation speed in a fluid at rest. For $V > 0$ the two types of disturbance travel in opposite directions when $(gD)^{1/2} > V$. We call the background flow *subcritical* in this case. The quiet and smooth currents common in rivers away from dams or rapids are generally subcritical. Propagation in the positive y-direction occurs for both waves when $(gD)^{1/2} < V$, in which case the background flow is called *supercritical*. Flows in spillways, waterfalls, and in parts of rapids are supercritical. If $(gD)^{1/2} = V$, one wave propagates in the direction of the background flow and the other is stationary: $c_- = 0$. In this case the background flow is *critical* and can support stationary disturbances.

Critical flow is normally a local phenomenon that occurs near the crests of dams and spillways. In the long wave limit, there is no distinction between the speeds of phase propagation and energy propagation.

The Froude number, F_d, defined by

$$F_d{}^2 = \frac{V^2}{gD},$$

is often used to characterize the relative importance of inertia and gravity in the dynamics of a particular flow. It is seen that F_d is the ratio of the advective component to the intrinsic 'propagation' component of the phase speed and is $< 1, = 1, > 1$ for subcritical, critical, and supercritical flow, respectively.

With more complicated flows, it may be difficult to unambiguously define *upstream* and *downstream*. Such is the case when the fluid is stratified and has positive and negative horizontal velocities at different depths. In such cases, we reserve the term *subcritical* to mean that signal speeds c_+ and c_- belonging to a particular pair of waves are of opposite sign: $c_+ c_- < 0$. Information carried by the waves can therefore travel in both directions. Supercritical flow is defined by $c_+ c_- > 0$ and corresponds to information flow in one direction only. Critical flow is defined by $c_+ c_- = 0$ and corresponds to the arrest of one or both of the waves. Note that this definition applies to the homogeneous flow under consideration and is independent of the sign of V.

A simple example of wave generation that will be built upon throughout this book is the linear dam-break problem. Consider two bodies of resting fluid with slightly different depths $D \pm a$, separated by a barrier located at $y = 0$ (Figure 1.2.1a). At $t = 0$ the barrier is removed, allowing the deeper fluid to move towards positive y. Assuming $a \ll D$, the subsequent motion can be approximated by solving (1.2.3) and (1.2.4) with $V = 0$ and subject to the initial conditions

$$\eta(y, 0) = -a \ \mathrm{sgn}(y) = a \begin{cases} -1 \, (y > 0) \\ +1 \, (y < 0) \end{cases}$$

and

$$v'(y, 0) = 0.$$

As shown in Figure 1.2.1b, the solution

$$\eta(y, t) = -\frac{1}{2} a [\mathrm{sgn}(y - c_+ t) + \mathrm{sgn}(y - c_- t)]$$

and

$$v'(y, t) = \frac{1}{2} a \left(\frac{g}{D}\right)^{1/2} [-\mathrm{sgn}(y - c_+ t) + \mathrm{sgn}(y - c_- t)]$$

consists of two step-like wave fronts propagating away from $y=0$ at the speeds $c_\pm = \pm(gD)^{1/2}$. Left behind is a uniform stream with velocity $a(g/D)^{1/2}$ and

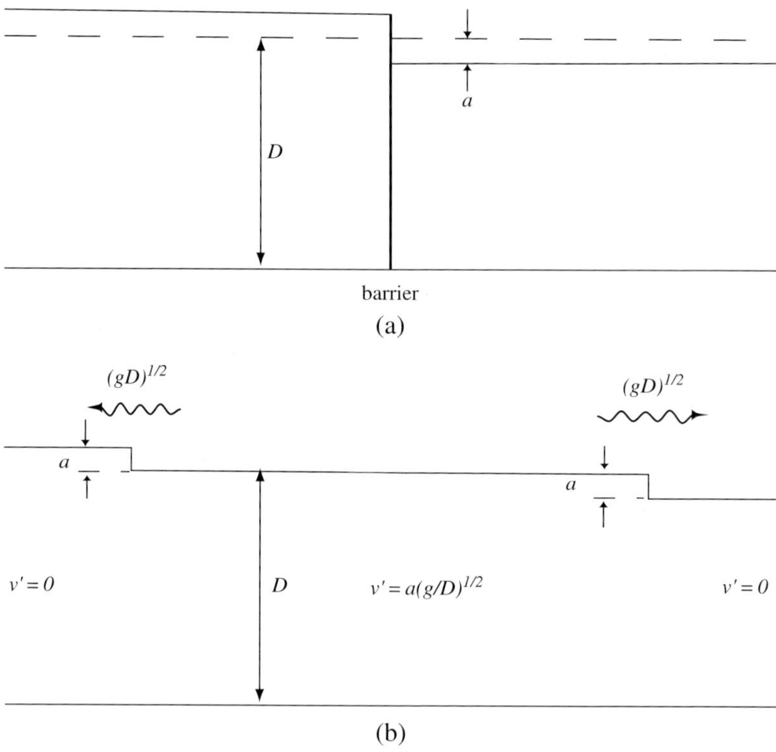

FIGURE 1.2.1. The linear dam-break problem.

with depth equal to the mean initial depth. It is apparent that, between the two wave fronts, the available potential energy associated with the initial mismatch in fluid depths has been entirely converted to kinetic energy (see Exercise 1). The complete removal of available potential energy is a feature that does not persist in the presence of rotation.

Another view of linear, long-wave propagation, one that will be helpful in understanding nonlinear waves, comes from the method of characteristics. A formal discussion of this method appears in Appendix B, but many readers will be satisfied with the less formal derivations that appear in the main text. If (1.2.4) is multiplied by $(g/D)^{1/2}$ and the product is added to (1.2.3), the resulting equation can be arranged in the form:

$$\left(\frac{\partial}{\partial t} + \left(V + (gD)^{1/2} \right) \frac{\partial}{\partial y} \right) \left(v' + \left(\frac{g}{D} \right)^{1/2} \eta \right) = 0. \tag{1.2.9}$$

Subtraction of the two results in

$$\left(\frac{\partial}{\partial t} + \left(V - (gD)^{1/2} \right) \frac{\partial}{\partial y} \right) \left(v' - \left(\frac{g}{D} \right)^{1/2} \eta \right) = 0. \tag{1.2.10}$$

The operator in (1.2.9) can be interpreted as the time derivative seen by an observer moving at the wave speed $c_+ = V + (gD)^{1/2}$. To that observer, the value of the *linearized Riemann invariant* $(v' + (g/D)^{1/2}\eta)$ remains fixed. A similar interpretation holds for (1.2.10), with an observer moving at speed $c_- = V - (gD)^{1/2}$ seeing a fixed value of $(v' - (g/D)^{1/2}\eta)$. In this context, c_- and c_+ are called *characteristic speeds*. The general solutions (1.2.5) and (1.2.6) can be deduced directly from the characteristic forms (1.2.9) and (1.2.10) of the linear shallow water equations.

The Riemann invariants can be used to measure the distribution of 'forward' and 'backward' propagating waves in a time-dependent flow field. Consider a single, forward wave (with speed c_+) with the form $\eta = \sin[y - c_+ t]$. In view of (1.2.6) the corresponding perturbation velocity is given by $v' = (g/D)^{1/2}\sin[y - c_+ t]$. The value of the 'forward' Riemann invariant $(v' + (g/D)^{1/2}\eta)$ over this wave form varies from $2(g/D)^{1/2}$ at a wave crest to $-2(g/D)^{1/2}$ at a trough, whereas the value of $(v' - (g/D)^{1/2}\eta)$ is uniformly zero over the same interval. The reverse is true for a 'backward' wave (with speed c_-). One could use this property to decompose a more complicated wave field into backward and forward components (see Exercise 3); forward waves project entirely onto the forward Riemann invariant and vice versa. It will be important to keep this interpretation in mind when reading the next section, where nonlinear generalizations of the Riemann functions will be introduced.

Now consider an initial value problem for which v' and η are specified for all y at $t = 0$. In determining a solution for $t > 0$, it is useful to think about the propagation of this information forward in time. Consider the space $-\infty < y < \infty$ and $t \geq 0$, also known as the *characteristic plane*. An observer moving at the speed $c_+ = V + (gD)^{1/2}$ travels through this space along one of the *characteristic curves* (or *characteristics*) indicated by a '+' in Figure 1.2.2a. The value of $(v' + (g/D)^{1/2}\eta)$ is conserved along such curves. A similar result holds for the characteristic curves labeled '$-$', along which $(v' - (g/D)^{1/2}\eta)$ is conserved. The characteristic curves therefore represent paths along which specific information travels.

As an example of the use of the method of characteristics, reconsider the dam-break problem. The initial conditions are sketched below the characteristic plane in Figure 1.2.2a. Begin by considering a '+' characteristic curve originating at a point e on the y-axis at $t = 0$. Here the initial conditions are $v' = 0$ and $\eta = -a$. The value of the Riemann invariant that is carried forward in time along the curve ef is given by

$$v' + (g/D)^{1/2}\eta = -(g/D)^{1/2}\, a \;\text{(along } ef\text{)}. \tag{1.2.11}$$

The same is true for all the solid curves originating from the positive portion of the y-axis. A similar argument establishes the values of $(v' - (g/D)^{1/2}\eta)$, which are carried along the dashed characteristics. For example, the value along the curve $e'f$ is determined from the initial conditions as

$$v' - (g/D)^{1/2}\eta = (g/D)^{1/2}\, a \;\text{(along } e'f\text{)}. \tag{1.2.12}$$

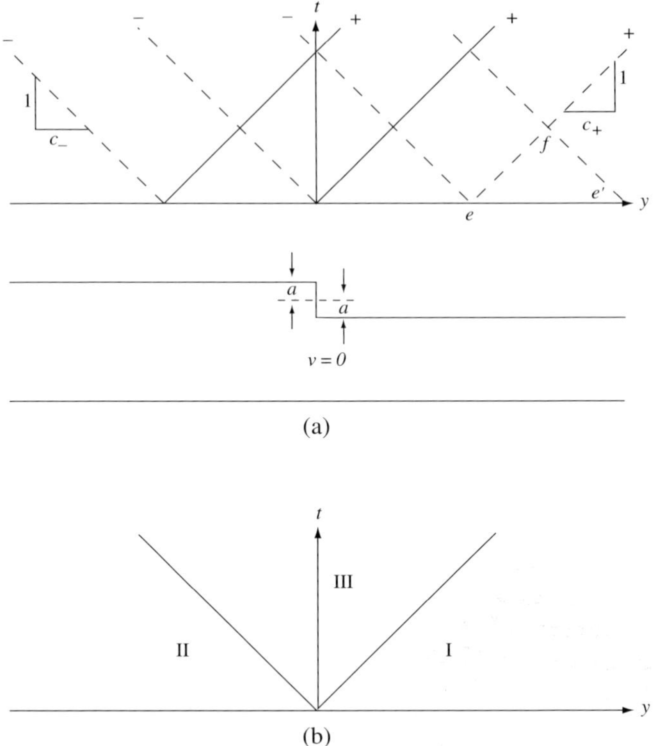

FIGURE 1.2.2. Characteristic curves (a) and regions of influence (b) for the linear dam break.

To determine the individual values of v' and η at a point in the characteristic plane, we use the values of the Riemann invariants given along the characteristic curves that intersect that point. At the intersection point f, for example, Equations (1.2.11) and (1.2.12) lead to $v' = 0$ and $\eta = -a$. This result will hold at all points within Region I of the characteristic plane, as indicated in Figure 1.2.2b. This region of the flow has not yet been reached by the forward propagating wave front that is generated by the step in surface elevation. The reader may wish to verify that a similar result holds in Region II, which lies to the left of the wave front advancing to the left and where the values $v' = 0$ and $\eta = a$ remain equal to the initial values.

Each point in Region III of the characteristic plane is intersected by dashed characteristic curves emanating from the positive y-axis and by solid curves emanating from the negative y-axis. The corresponding Riemann invariants are given by

$$v' + (g/D)^{1/2}\eta = (g/D)^{1/2}\, a \quad \text{(in region III)}$$

and

$$v' - (g/D)^{1/2}\eta = (g/D)^{1/2} a \quad \text{(in region III)}$$

and therefore $\eta = 0$ and $v' = (g/D)^{1/2}a$. Thus the passage of the wave fronts leaves behind a steady flow with velocity $(g/D)^{1/2}a$. The paths of the fronts themselves are the characteristic curves that form the boundaries between the three regions.

Exercises

(1) *Energy conversion in the linear dam-break problem.* Multiply (1.2.3) by Dv' and (1.2.4) by $g\eta$ and add the results to obtain the energy equation

$$\left(\frac{\partial}{\partial t} + V\frac{\partial}{\partial y}\right)\left(\frac{g\eta^2}{2} + \frac{Dv'^2}{2}\right) = -gD\frac{\partial(v'\eta)}{\partial y}.$$

For the solution to the linear dam-break problem (Figure 1.2.1b), integrate the above equation (with $V = 0$) with respect to y over any fixed interval $I = (-y_o < y < y_o)$. Then integrate the resulting relation with respect to t from 0 to ∞. Show from the final result that the available potential energy in I is converted entirely into kinetic energy. This finding is consistent with the fact that the energy radiated away from I by the gravity waves (as measured by $\int_0^\infty [(v'\eta)_{y_o} - (v'\eta)_{-y_o}]dt$) is zero.

(2) Consider the initial condition $v = 0$ and

$$d(y, 0) = \begin{array}{ll} d_o & (|y| > L) \\ d_o + a(1 - \dfrac{|y|}{L}) & (|y| \leq L) \end{array}.$$

Discuss the evolution of this disturbance according to linear theory.

(3) Using Riemann invariants, decompose the following flow field into 'forward' and 'backward' waves:

$$\eta(y, t) = -\sin[y]\cos[t] \quad \text{and} \quad v' = (g/D)^{1/2}\cos[y]\sin[t].$$

(4) *Linear wave speeds in the presence of vertical shear.* Consider the wave problem for a free surface flow with uniform depth D and velocity $V(z)$. Define $v = V(z) + v'(y, z, t)$ and show that the linearized y-momentum and continuity equations in the long wave limit are

$$\frac{\partial v'}{\partial t} + V\frac{\partial v'}{\partial y} + w\frac{\partial V}{\partial z} + g\frac{\partial \eta}{\partial y} = 0 \quad \text{and} \quad \frac{\partial v'}{\partial y} + \frac{\partial w}{\partial z} = 0.$$

(a) Assuming the waveforms $(v', w, \eta) = \text{Re}(\tilde{v}(z), \tilde{w}(z), \tilde{\eta})e^{il(y-ct)}$, obtain the relation

$$(c - V)\frac{\partial \tilde{w}}{\partial z} + \tilde{w}\frac{dV}{dz} + igl\tilde{\eta} = 0.$$

Divide this equation by $(c-V)^2$ and integrate the result from the bottom (here $z=0$) to the free surface ($z=D$ by the linear approximation) to obtain

$$\left[\frac{\tilde{w}}{c-V}\right]_{z=D} - \left[\frac{\tilde{w}}{c-V}\right]_{z=0} = -igl\tilde{\eta}\int_0^D \frac{dz}{(c-V)^2}.$$

(b) Apply the kinematic boundary conditions at the bottom and free surface to obtain the result.

$$g\int_0^d \frac{dz}{(V-c)^2} = 1$$

Show that the case $V=$ constant results in $c = V \pm (gD)^{1/2}$. For nonconstant V observe that real values of c must lie outside the range of variation of V.

(c) Finally, if the variations of V are weak: $V = V_0 + \varepsilon \hat{V}(z)$ with $\varepsilon \ll 1$ and $\int_0^D \hat{V} dz = 0$, show that

$$(V_0 - c)^2 = gD + \frac{3\varepsilon^2}{D}\int_0^D \hat{V}^2 dz,$$

and therefore a section at which $V_0 = (gD)^{1/2}$ allows upstream propagation.

Further discussion of the implications of these results can be found in Garrett and Gerdes (2003). The derivation of the wave speeds appears in Freeman and Johnson (1970).

1.3. Nonlinear Steepening and Rarefaction

A basic knowledge of the hydraulic properties of a steady flow requires that one understand the characteristics of linear disturbances that propagate on that flow. However, some grasp of the elements of nonlinear propagation are crucial in understanding how hydraulic jumps and other types of shock waves are formed. This subject will also be of assistance when we explore the formation of steady solutions in laboratory or numerical experiments. A feature common to most nonlinear disturbances that arise in hydraulic models is that they are governed by hyperbolic partial differential equations. The defining characteristics of quasi-linear hyperbolic systems in two dimensions are described in detail in Appendix B, as are the methods for transforming the governing equations into standard forms. However, a heuristic definition would center on the properties that two independent types of disturbances (waves) exist and that these waves propagate through the physical domain at finite speeds. In the example of the previous section, the disturbances consisted of two linear gravity waves with speeds $c_\pm = V \pm (gD)^{1/2}$, propagating on a uniform background flow. As we now show, standard methodology also allows one to deal with wave amplitudes sufficiently large to destroy the distinction between the wave and the background flow.

To begin, it is helpful to rewrite the one-dimensional shallow water equations (1.2.1) and (1.2.2) in the form

$$\frac{d_\pm R_\pm}{dt} = -g \frac{dh}{dy} \tag{1.3.1}$$

where

$$\frac{d_\pm}{dt} = \frac{\partial}{\partial t} + [v \pm (gd)^{1/2}] \frac{\partial}{\partial y} \tag{1.3.2}$$

and

$$R_\pm = v \pm 2(gd)^{1/2}. \tag{1.3.3}$$

The procedure for obtaining this new form is discussed in Exercise 1 and the reader who seeks a more general discussion can consult Appendix B or look at standard texts such as Courant and Friedricks (1976) or Whitham (1974).

To interpret (1.3.1–1.3.3) first note that the operators $\frac{d_\pm}{dt}$ are time derivatives seen by observers traveling with *characteristic speeds*

$$\frac{dy_\pm}{dt} = v \pm (gd)^{1/2}. \tag{1.3.4}$$

These speeds are nothing more than the linear wave speeds with V and D replaced by the local velocity and depth, v and d. As before, it is helpful to think of the characteristic speeds as defining individual signals that move through the fluid and that compose general wave forms. Since the characteristic speeds vary throughout the flow field, different parts of a wave form move at different rates, leading to steepening (convergence) or rarefacation (spreading) of this form. If the bottom slope is zero $(dh/dy = 0)$, an observer moving at one of the characteristic speeds sees a fixed value of the corresponding Riemann invariant R_+ or R_-. The latter are nonlinear generalizations of the functions introduced in the previous section. Among other things, they serve as indicators of the presence of 'forward' and 'backward' wave forms. If, for example, R_- is uniform in y, then the flow field contains no 'backward' wave forms (i.e. those propagating at speed $v - (gd)^{1/2}$). The forward propagating waves in such a field are sometime called *simple waves*. A simple physical interpretation of Riemann invariants in terms of energy or momentum has proved to be elusive, but perhaps the reader has a suggestion.

The characteristic speeds have real and unequal values for all flows in which the depth is nonzero, implying that (1.2.1) and (1.2.2) are hyperbolic. The importance of this property is that solutions to the initial-value problems can be constructed using the method of characteristics. Suppose that one is given the initial conditions $v(y,0)$ and $d(y,0)$ for all y and asked to compute the evolution of the flow for $t > 0$. The initial values of the Riemann invariants are given by $R_+ = v(y,0) + 2[gd(y,0)]^{1/2}$ and $R_- = v(y,0) - 2[gd(y,0)]^{1/2}$ and, provided that $dh/dy = 0$, these values are conserved along *characteristic curves* (or 'characteristics'), paths traced out in the $(y,\ t)$-plane by moving at the appropriate characteristic speed. Unlike the case for the linear waves considered in the previous section, the slopes of the characteristic curves are generally not constant. They depend on the local values of the velocity and depth within an evolving flow field.

We can now lay out a procedure for solving any initial-value problem involving smooth initial conditions $v(y,0)$ and $d(y,0)$. As in the previous section, the solution is described in the characteristic $(y,\ t)$ plane (Figure 1.3.1a). Let $y_+(y_o,\ t)$ and $y_-(y_1,\ t)$ represent the characteristic curves originating from $y = y_o$ and $y = y_1$ on the y-axis. The slopes of these curves are determined by (1.3.4) and can be calculated at $t = 0$ from the initial conditions. For $t > 0$ the slopes depend on the solution itself and remain to be determined. Suppose for the moment that these slopes, and thus the curves $y_+(y_o,\ t)$ and $y_-(y_1,\ t)$ themselves are known, and that the curves intersect at point p. Then the velocity and depth at p can be computed from the values of the Riemann invariants that are carried along the curves. The value of R_+ is the value carried along the curve $y_+(y_o,\ t)$ and is equal to its value at $y = y_o$ and $t = 0$. The value of R_- at p is that specified at $y = y_1$ and $t = 0$. Once the two are known, the velocity and depth are computed from

$$v = \frac{1}{2}(R_+ + R_-), \qquad (1.3.5)$$

$$d = [(R_+ - R_-)/4]^2/g. \qquad (1.3.6)$$

More generally, the shapes of the characteristics are not known in advance and there is no immediate way of knowing the origin of the characteristic curves passing through p. In practice, this problem is dealt with by calculation of the initial slopes of the characteristics from the values of c_- and c_+ all along the y-axis. Straight-line approximations of the characteristic curves having these initial slopes are then projected forward a time increment Δt. A provisional solution is then computed at $t = \Delta t$ by carrying the initial values of R_+ and R_- forward along these curves. The characteristic speeds that follow from the provisional solution will generally be different than the initial estimates, implying that the characteristic curves are not straight. However, a correction can be made and the whole process repeated. Once satisfactory values of v and d have been found at $t = \Delta t$, the solution may be advanced further in time through reiteration.

The method will continue to work as long as the '+' curves (or the '−' curves) do not begin to intersect each other. Should the latter occur, as at q in Figure 1.3.1a, multiple values of R_+ (or R_-) would apply at the same point and the solution would be overdetermined. This situation is associated with the formation of shocks, meaning discontinuities in v and/or d, a circumstance to be explored later. Note that when the channel bottom contains topography, $R_\pm$

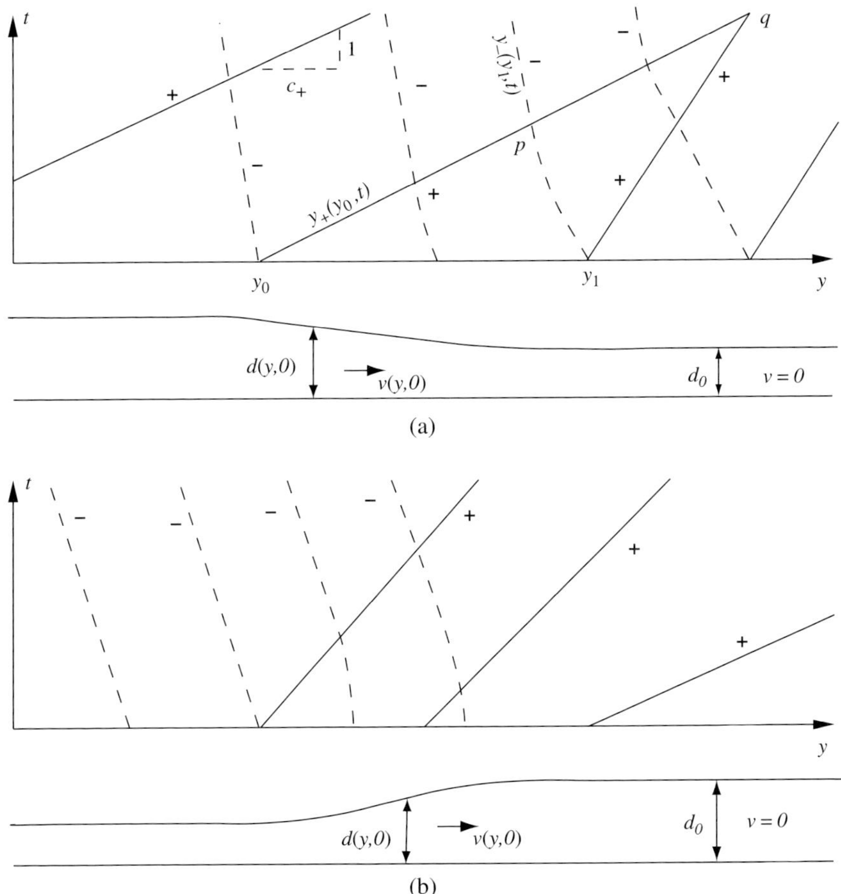

FIGURE 1.3.1. Characteristic curves for two initial value problems, one with deeper water to the left (a) and the second with deeper water to the right (b). The solid and dashed curves represent 'plus' and 'minus' characteristic curves corresponding to $dy_{\pm}/dt = c_{\pm}$.

are no longer conserved and must be computed by integration of (1.3.1) along characteristic curves. In either case the curves may be interpreted as paths along which information travels.

Elementary examples of nonlinear evolution can be constructed through the consideration of a simple wave, as generated from an initial condition with uniform R_- or R_+. Consider the initial condition shown at the base of Figure 1.3.1a, with shallow water to the right and deeper water to the left. Suppose further that the shallower region has uniform depth d_o and is motionless ($v = 0$). Then choose the initial velocity to the left of the shallow region such that R_- is uniform. The value of R_- can be found by evaluating $v - 2(gd)^{1/2}$ in the shallow, quiescent region, leading to $R_- = -2(gd_o)^{1/2}$. R_- must have this value for all y and therefore for all y and t reached by '$-$' characteristics, provided they do not intersect. An immediate consequence is that v and d become linked by the relation

$$v = 2(gd)^{1/2} - 2(gd_o)^{1/2},$$

which follows from the definition of R_-. The definition of R_+ then leads to

$$R_+ = 4(gd(y, t))^{1/2} - 2(gd_o)^{1/2},$$

and thus $d(y,\ t)$ itself is conserved along each '+' characteristic curve. Since both R_+ and d are conserved, v must also be conserved along each such curve and the characteristic speed must be constant and equal to its initial speed:

$$c_+ = v(y, t) + (gd(y, t))^{1/2}$$
$$= 3(gd(y, 0))^{1/2} - 2(gd_o)^{1/2}. \tag{1.3.7}$$

The slope $1/c_+$ of each '+' curve is therefore constant, though different curves have different slopes. For the disturbance shown in Figure 1.3.1a, characteristics emanating from the deeper part of the disturbance are tilted more steeply than those emanating from the shallower portion. The tilt of each curve is an indication of how rapidly the signal corresponding to a particular part of the disturbance travels. The signal itself can be identified as a particular value of the depth d. Here the larger depths on the left propagate to the right more rapidly than shallower depths on the right. The slope $|\partial d/\partial y|$ of the free surface will therefore increase in what is called *nonlinear steepening*. We leave it as an exercise for the reader to show that a disturbance of the type shown in Figure 1.3.1b would spread or *rarefy*[2] (provided R_- remains uniform). In other words, the left-hand (shallower) d-values would propagate more slowly than those to the right.

The steepening wave form in the above example formally leads to a singularity corresponding to the intersection point q in Figure 1.3.1a. As more rapid signals overtake slower ones, the free surface slope increases without bound and eventually multiple values of d occupy the same y. The formation of a singularity is not proof of a real world catastrophe but rather an indication of breakdown in the shallow water approximation. This breakdown occurs when the horizontal length of the steepening wave becomes as small as the fluid depth. Beyond this point the steepening may or may not be arrested due to the intervention of nonhydrostatic effects or possibly other processes not captured in inviscid shallow water theory. If the length of the ultimate wave form or shock remains comparable to the depth, it is possible to represent it as a discontinuity in depth within a shallow water model and to approximate its amplitude and propagation speed. The ideas involved, collectively known as *shock joining theory* will be discussed in later sections.

---

[2] The term originates from gas dynamics and refers to the decreasing gas density that occurs when the wave form spreads.

A further illustration of the power of the method of characteristics is provided by a nonlinear version of the dam-break problem explored in the previous section. We now consider the motion resulting from the destruction of a barrier separating a resting fluid of depth D from a region with no fluid (Figure 1.3.2). The initial conditions are

$$d(y, 0) = \begin{cases} D(y > 0) \\ 0(y < 0) \end{cases} \tag{1.3.8}$$

and

$$v(y, 0) = 0. \tag{1.3.9}$$

The solution to this problem, as posed, is nonunique. Different results are obtained depending upon how one deals with the discontinuity in initial depth

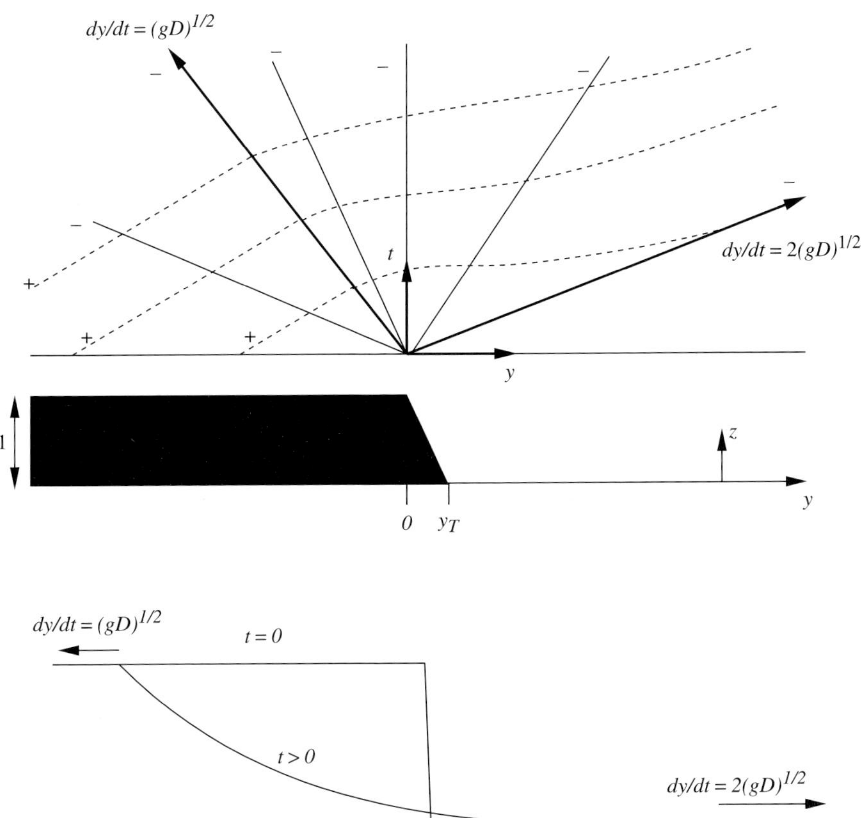

FIGURE 1.3.2. The full-dam break problem as visualized with a gradual initial change in depth, rather than a discontinuity, near $x = 0$. The characteristic curves are shown in the upper frame and the rarefying surface disturbance and intrusion are shown in the lower frame.

at $y = 0$. A reasonable way to resolve this difficulty is to replace this discontinuity with a smooth, but abrupt, transition over $0 < y < y_T$, as shown in the figure. One must specify the initial values of d and v within this short region and the corresponding characteristic speeds and Riemann invariants can be used to compute the evolution. Different specifications lead to different outcomes and this is the source of the nonuniqueness. The calculation of the evolution becomes quite simple if d and v are chosen such that either R_+ or R_- is uniform in the abrupt region and has the same value (either $2(gD)^{1/2}$ or $-2(gD)^{1/2}$) as in the region $y < 0$. Then one of the Riemann invariants will be initially uniform throughout the fluid, allowing application of the simplifications described above for 'simple' waves. The limit $y_T \to 0$ may be taken later in order to approach the original step geometry.

Following this idea further, suppose that R_- is initially uniform in the transitional interval $0 < y < y_T$. Its value must therefore be $-2(gD)^{1/2}$ in order to match that in $y < 0$. It follows from the definition of R_- in the transition region that

$$v(y, 0) - 2(gd(y, 0))^{1/2} = -2(gD)^{1/2}. \qquad (0 < y < y_T)$$

However, $d < D$ in $0 < y < y_T$, implying that $v(y, 0) < 0$. In other words, the fluid in the vicinity of the barrier will initially move *to the left* after the barrier is removed. Obviously, the assumption of uniform R_- is not one that leads to a physically realistic evolution. On the other hand, the choice $R_+ = $ uniform leads to

$$v(y, 0) + 2(gd(y, 0))^{1/2} = 2(gD)^{1/2} \qquad (0 < y < y_T) \qquad (1.3.10)$$

so that $v(y, 0) > 0$ in the vicinity of $y = 0$, as expected.

It is now easy to make a sketch of the characteristic curves for all y and t, as is done at the top of Figure 1.3.2. Since R_+ is uniform, all of the '$-$' characteristics are straight. Their slope is determined by the initial value of the characteristic speed:

$$c_-(y, 0) = \begin{cases} -(gD)^{1/2} & (y < 0) \\ v(y, 0) - (gd(y, 0))^{1/2} = 2(gD)^{1/2} - 3(gd(y, 0))^{1/2}. & (0 < y < y_T) \end{cases}$$

Since $d(y, 0)$ decreases monotonically from D to 0 as y increases from 0 to y_T, c_- increases from $-(gD)^{1/2}$ to $2(gD)^{1/2}$ over the transitional interval. The '$-$' characteristic curves originating from this interval therefore fan out as shown in Figure 1.3.2. Since d and v are constant along these curves, the developing flow consists of a rarefaction wave. The leading edge ($d = 0$) of this wave moves to the right at speed $2(gD)^{1/2}$ whereas the rear edge moves to the left at speed $(gD)^{1/2}$. The leading edge speed is also the fluid velocity at the leading edge. One of the fanning characteristic curves has $c_- = 0$ and therefore points directly upwards. In the limit $y_T \to 0$ this curve lies at the position ($y = 0$) of the barrier. Thus, the flow at $y = 0$ immediately becomes steady and critical after the barrier is removed. The flow at all other y approaches this same critical state as $t \to \infty$. The depth d_∞ and velocity v_∞ of this final state are determined by the condition of criticality $[v_\infty = (gd_\infty)^{1/2}]$ and by the uniformity of $R_+ = v_\infty + 2(gd_\infty)^{1/2} = 2(gD)^{1/2}$, leading to $v_\infty = \frac{2}{3}(gD)^{1/2}$ and $d_\infty = (\frac{2}{3})^2 D$. The volume transport per unit width of channel is therefore given by

$$v_\infty d_\infty = \left(\frac{2}{3}\right)^3 g^{1/2} D^{3/2}. \tag{1.3.11}$$

If the initial depth in $y > 0$ is finite, then the advancing edge of the wave forms a shock. Calculation of the solution for this case requires knowledge of shock joining theory. A reader interested in the solution can consult Stoker (1957) for the full solution.

Exercises

(1) *Derivation of Riemann Invariants.* Obtain the homogeneous form of (1.3.1) from the shallow water equations (1.2.1) and (1.2.2) by the following procedure:

(a) Try to write the homogeneous versions of (1.2.1) and (1.2.2) in the *characteristic form*

$$\left(\frac{\partial}{\partial t} + c_\pm \frac{\partial}{\partial y}\right) v + \alpha_\pm(v, d)\left(\frac{\partial}{\partial t} + c_\pm \frac{\partial}{\partial y}\right) d = 0$$

by multiplying (1.2.2) by a factor $\alpha_\pm(v, d)$, adding the result to (1.2.1), and calculating $\alpha_\pm(v, d)$ and $c_\pm(v, d)$ such that the above form is achieved.

(b) Use this result to find the functions $R_\pm(v, d)$ satisfying $\left(\frac{\partial}{\partial t} + c_\pm \frac{\partial}{\partial y}\right) R_\pm = 0.$

(2) Linearize the Riemann invariants $R_\pm$ about a uniform background flow $v = V$ and $d = D$. How do the resulting expressions relate to the traveling wave functions $f_+(y - c_+ t)$ and $f_-(y - c_- t)$ defined in Section 1.2?

(3) Consider the initial condition $v = 0$ and

$$d(y, 0) = \begin{cases} d_o & (|y| > L) \\ d_o + a(1 - \dfrac{|y|}{L}) & (|y| \le L) \end{cases}.$$

Although this initial condition does not formally give a 'simple wave' solution, a simple-wave character emerges in parts of the domain after a finite time has elapsed. Use this behavior to discuss the qualitative features of the nonlinear evolution of this disturbance and compare it with the linear result (Exercise 2 of Section 1.2).

(4) For the example shown in Figure 1.3.1a, at what time does wave breaking (shock formation) *first* occur? [Hint: do not necessarily be satisfied with the obvious answer.]

(5) Consider the following twist on the classical dam-break problem with initial conditions (1.3.8) and (1.3.9). Suppose that at $t = 0$ the barrier is not destroyed but instead is made to recede from the reservoir at a constant speed $c_o < 2(gD)^{1/2}$. Use the method of characteristics to sketch the solution.

1.4. The Hydraulics of Steady, Homogeneous Flow over an Obstacle

We are now in a position to review one of the simplest examples of hydraulic behavior: that of a steady, homogeneous, free-surface flow passing over an obstacle or through a sidewall contraction. The channel will continue to have rectangular cross section with gradually varying width w and bottom elevation h. The governing steady shallow water equations are

$$v\frac{\mathrm{d}v}{\mathrm{d}y}+g\frac{\mathrm{d}d}{\mathrm{d}y}=-g\frac{\mathrm{d}h}{\mathrm{d}y} \tag{1.4.1}$$

and

$$d\frac{\mathrm{d}v}{\mathrm{d}y}+v\frac{\mathrm{d}d}{\mathrm{d}y}=-vdw^{-1}\frac{\mathrm{d}w}{\mathrm{d}y}. \tag{1.4.2}$$

Some of the general properties of the flow can be deduced by elimination of $\frac{\mathrm{d}v}{\mathrm{d}y}$ between the first two equations in favor of $\frac{\mathrm{d}d}{\mathrm{d}y}$. There follows

$$\frac{\mathrm{d}d}{\mathrm{d}y}=\frac{\dfrac{\mathrm{d}h}{\mathrm{d}y}-F_d^2\dfrac{d}{w}\dfrac{\mathrm{d}w}{\mathrm{d}y}}{F_d^2-1}, \tag{1.4.3}$$

where $F_d^2=v^2/gd$. This expression gives the rate of change of the fluid depth along the channel in terms of v and d and in terms of the rate of change of the geometrical parameters w and h. Positive values of the numerator on the right-hand side are associated with constrictions of the geometry due to increasing bottom elevation or to decreasing width. If the flow is subcritical ($F_d^2 < 1$), the denominator is negative and the fluid depth decreases in response to contractions. This is the situation when flow in a reservoir approaches a dam. Supercritical flow ($F_d^2 > 1$) experiences increases in depth in response to constrictions, a situation that can be observed in river rapids; where the water passes over a boulder, the depth increases and the free surface bulges out. Finally, critical flow ($F_d^2 = 1$) with a finite free-surface slope requires that the rate of contraction be zero: $\frac{\mathrm{d}h}{\mathrm{d}y}=\frac{d}{w}\frac{\mathrm{d}w}{\mathrm{d}y}$. This *regularity condition* holds where $\frac{\mathrm{d}h}{\mathrm{d}y}$ and $\frac{\mathrm{d}w}{\mathrm{d}y}$ are both zero, as at the crest or sill of an obstacle in a constant-width channel, at a narrows of a constant-elevation channel, or at a section where the minimum width coincides with a sill. Critical flow can also occur where increases in bottom elevation coincide with increases in width, or vice versa, such that the rate of geometrical contraction is zero according to the above criterion. Locations of critical flow are called *critical* or *control sections*.

Now consider the class of steady flows that arises when w is constant and the channel contains a single obstacle of height h_m, as shown in Figure 1.1.1c. Normally, computation of the flow is carried out using statements of conservation of energy:

$$\frac{v^2}{2}+gd+gh=B, \tag{1.4.4}$$

and conservation of volume transport:

$$vdw = Q, \tag{1.4.5}$$

obtained through the integration of (1.4.1) and (1.4.2) with respect to y. The constants B and Q represent the Bernoulli 'head' and volume flow rate. The former is the energy per unit mass of a fluid parcel and is always independent of depth in our slab-like, shallow water system. For steady flow B is independent of y as well.

Solutions for the fluid depth can be found by eliminating v between (1.4.4) and (1.4.5), with the result:

$$\frac{Q^2}{2w^2 d^2} + gd = B - gh. \tag{1.4.6}$$

The quantity $B\text{-}gh$, sometimes called the *specific energy*, is the total energy minus the potential energy provided by the bottom elevation. It represents the intrinsic energy of the flow. Changing the bottom elevation alters the specific energy, forcing the depth to adjust to new values. One approach to the steady flow problem is to imagine that Q and B are predetermined, say, by conditions set far upstream of the obstacle, so that one can march along the channel, using (1.4.6) to calculate the depth at each point along the way. Of course (1.4.6) is cubic and there may be more than one value of d for each h, a situation which can be clarified by plotting h (or, more conveniently, $B\text{-}gh$) as a function of d. To make such a plot as general as possible, we first render (1.4.6) dimensionless by dividing by gD, where D is a scale chosen for convenience as $(Q/w)^{2/3} g^{-1/3}$. Then let $\tilde{d} = d/D$, $\tilde{h} = h/D$, and $\tilde{B} = B/gD$. The resulting relation

$$\frac{1}{2(\tilde{d})^2} + \tilde{d} = \tilde{B} - \tilde{h} \tag{1.4.7}$$

is plotted in Figure 1.4.1a. To construct a solution at a particular y, first set the normalized energy $\tilde{B}$ and note the bottom elevation $\tilde{h}$ at that y. This fixes a point on the ordinate $\tilde{B} - \tilde{h}$. If the latter is $> 3/2$, two possible solutions for $\tilde{d}$ can be found. One corresponds to the left-hand branch and the other to the right-hand branch of the curve. There is one solution for $\tilde{B} - \tilde{h} = 3/2$ corresponding to the minimum of the curve. Here

$$\frac{\partial}{\partial \tilde{d}} \left(\frac{1}{2\tilde{d}^2} + \tilde{d} \right) = -\frac{1}{\tilde{d}^3} + 1 = 0. \tag{1.4.8}$$

and therefore $\tilde{d} = 1$ or $d = (Q/w)^{2/3} g^{-1/3} = (vd)^{2/3} g^{-1/3}$ or, finally, $F_d^2 = v^2/gd = 1$. The solution at the minimum of the curve therefore corresponds to critical flow. The left hand branch of the curve is associated with smaller depths and, since the flow rate is the same, larger velocities. Therefore the left-hand branch corresponds to supercritical ($F_d^2 > 1$) flow, while the right-hand branch corresponds to subcritical ($F_d^2 < 1$) flow. Constructing a solution requires choosing between the right- and left-hand branches, and there is nothing yet to suggest how this choice is to be made.

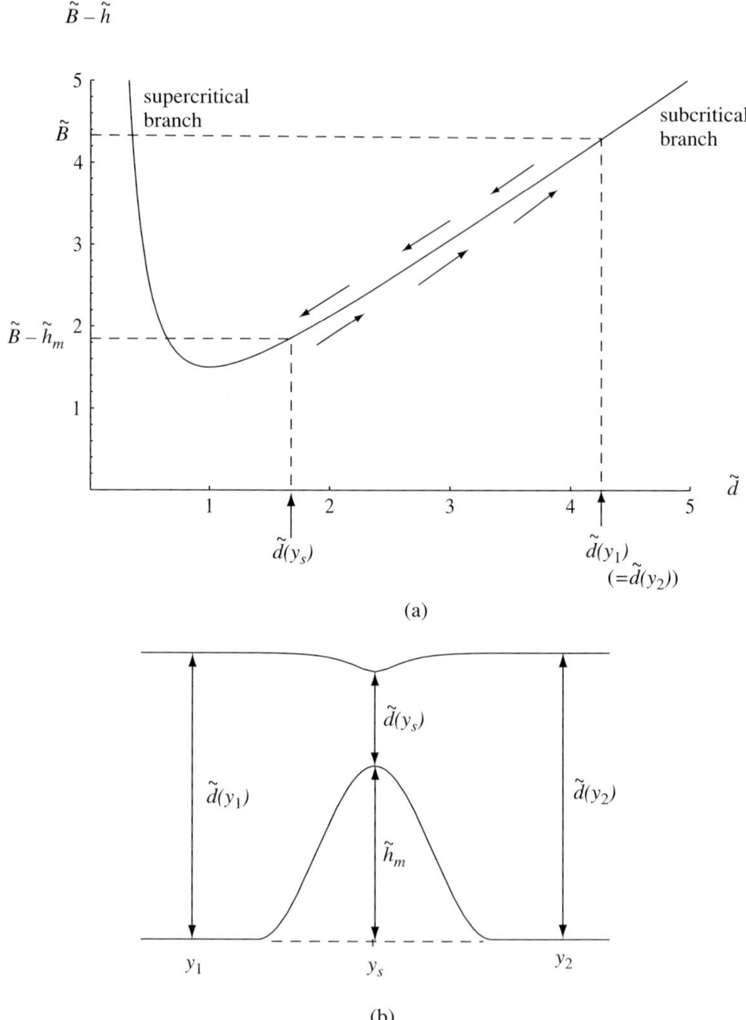

FIGURE 1.4.1. (a) Plot of equation (1.4.1), with arrows indicating the route traced out by a subcritical solution. (b) Profile of a subcritical solution corresponding to the trace shown in (a).

Ignoring, for the moment, the dilemma of being forced to choose between two possible solutions, we arbitrarily begin on the subcritical branch of the solution curve. To construct a solution over a particular obstacle, begin at the section $(y = y_1)$ upstream of the obstacle, where $\tilde{h} = 0$. To find the depth $\tilde{d}$ at this section, go to Figure 1.4.1a and read off the value $\tilde{d}(y_1)$ corresponding to $\tilde{B} - \tilde{h}(y_1) = \tilde{B}$. Next, move forward along the channel to where the bottom elevation $\tilde{h}$ begins to increase, causing $\tilde{B} - \tilde{h}$ to decrease. Remaining on the subcritical branch of the solution curve leads to lower values of $\tilde{d}$ as indicated by the arrows drawn above the curve. We can continue in this way until we reach the obstacle's sill at $y = y_s$ and $\tilde{h} = \tilde{h}_m$. If the sill elevation $\tilde{h}_m$ is sufficiently small that $\tilde{B} - \tilde{h}_m > 3/2$ the minimum of the solution curve is not reached and the depth $\tilde{d}(y_s)$ will exceed the critical depth $\tilde{d} = 1$. Continuing further downstream causes one to retrace the solution curve in Figure 1.4.1a as indicated by the arrows drawn underneath. After the obstacle is passed $(y = y_2)$, the depth returns to its upstream value. It is left as an exercise to show that where the depth decreases, the elevation $h + d$ of the free surface also decreases, so that the free surface will appear as shown in Figure 1.4.1b. Note that the solution is symmetrical in the sense that equal bottom elevations upstream and downstream of the sill see the same fluid depth. If the left-hand branch of the solution curve had been traced for the same topographic variations, a symmetrical supercritical solution with $\tilde{d}$ *increasing* over the obstacle would have resulted. We will refer to these solutions as pure subcritical or pure supercritical flow.

Next suppose that $\tilde{B} - \tilde{h}_m = 3/2$ so that the minimum of the solution curve is just reached at the sill. If the approach to the sill had been along the subcritical solution branch, there are two choices in constructing the downstream solution. First, one retraces the subcritical solution branch as in the above example. Second, one precedes onto the supercritical branch and thereby traces out an asymmetrical solution with the fluid depth *decreasing* in the downstream direction. This situation is depicted in Figure 1.4.2. As it turns out, the first of these scenarios results in a solution with a discontinuity in the free-surface slope at the sill and can be ruled out. The proof of this result is the subject of Exercise 1 below. The preferred solution is thus the one with subcritical flow upstream, supercritical flow downstream, and critical flow at the sill. This type of flow, which resembles flow over a dam or spillway, is often described as being *hydraulically controlled*. The meaning of the term 'control' will soon become apparent. For now, we simply note that small disturbances generated downstream of the sill are unable to propagate upstream. The subcritical flow upstream of the sill is therefore immune to weak forcing imposed downstream of the sill.

It is also natural to ask what happens when $\tilde{B} - \tilde{h}_m < 3/2$, in which case no solution exists at the obstacle's crest. This situation occurs when the energy $\tilde{B}$ is insufficient to allow the fluid to surmount the obstacle. For example, one might start with the hydraulically controlled flow described above and raise the elevation of the sill a small amount, creating a small region about the sill for which no steady solution exists. Under these conditions a time-dependent adjustment must take place leading to a new upstream flow with a larger $\tilde{B}\,(= B(gQ/w)^{-2/3})$. This process is known as *upstream influence* and will be illustrated further in Sections 1.6 and 1.7. As we shall show, the value of $\tilde{B}$ has altered the minimal amount required to allow the flow to continue, implying that the new steady state is hydraulically controlled. Note that the change can be effected by increasing the Bernoulli function B or by decreasing

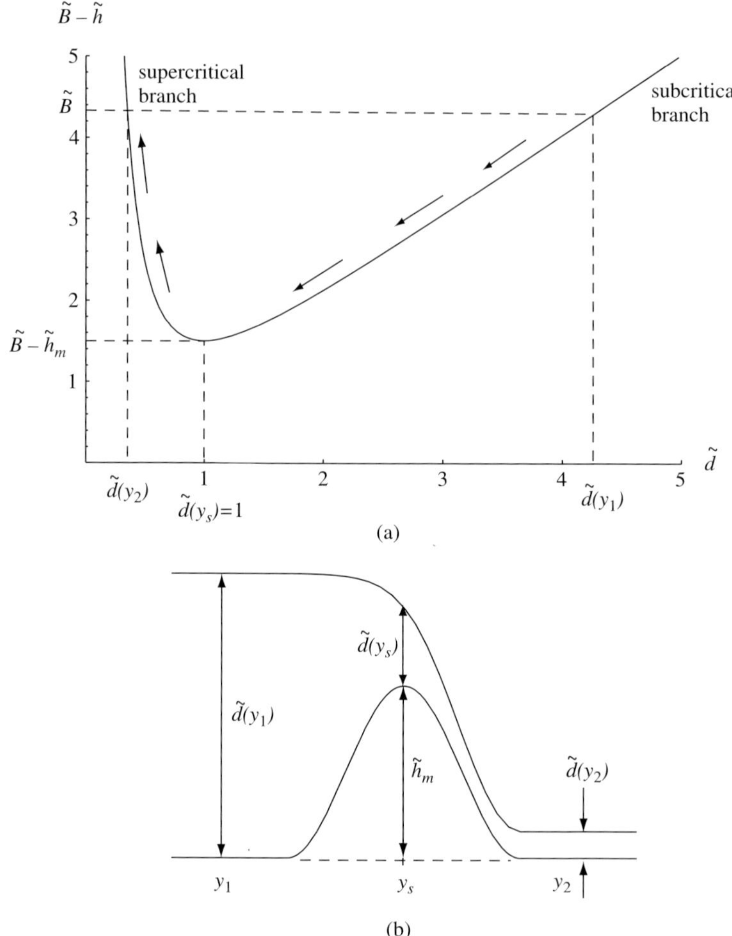

FIGURE 1.4.2. Similar to Figure 1.4.1, but now showing a trace of a hydraulically controlled solution.

the transport Q. Upstream influence over these quantities is an essential aspect of hydraulic control.

So far, we have constructed various solutions by fixing the energy parameter $\tilde{B}$ and varying the sill height of the topography. For a different perspective, consider the family of solutions obtained for a fixed topography by varying $\tilde{B}$. Figure 1.4.3 shows the free-surface profiles of the solutions over an obstacle of unit dimensionless height. Each value of $\tilde{B}$ shown is associated with two solutions, one having supercritical and one subcritical flow upstream of the obstacle. For $\tilde{B} > 2.5$ the curves are the symmetrical, purely sub- or supercritical solutions discussed before. For $\tilde{B} = 2.5$ the two solutions intersect each other at the sill; one of these is the hydraulically controlled solution discussed above and

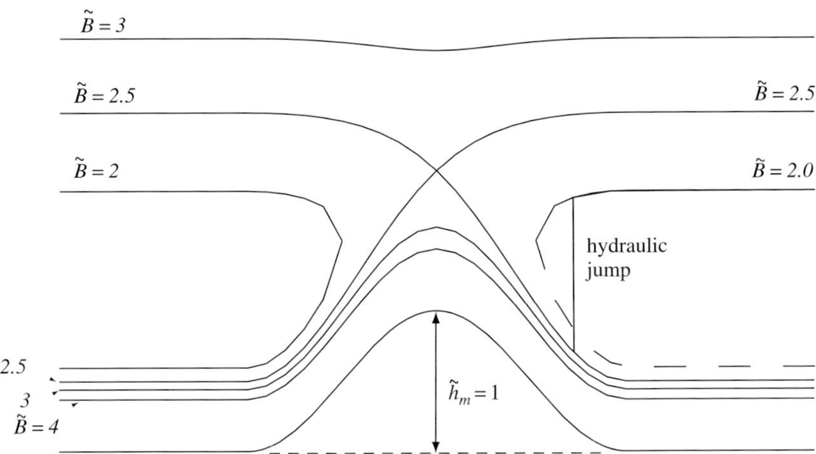

FIGURE 1.4.3. Free-surface profiles for flow with different values of $\tilde{B}$ over the same obstacle.

the other, its mirror image, is supercritical upstream and subcritical downstream of the obstacle. For $\tilde{B} < 2.5$ the solutions are not continuous across the sill.

The asymmetrical solution that is supercritical upstream and subcritical downstream of the sill is unstable and probably unrealizable in most laboratory or field settings. A heuristic demonstration of the instability can be made through consideration of a small-amplitude disturbance imposed on the flow at the sill (Figure 1.4.4). This disturbance may be synthesized using the two linear wave modes of the system, which propagate at speeds $v - (gd)^{1/2}$ and $v + (gd)^{1/2}$. Since the slower wave propagates in the downstream direction *upstream* of the sill and in the upstream direction *downstream* of the sill, any energy carried in the slower mode will become focused and amplified at the sill. This situation will henceforth be referred to as the *shock forming instability*.

Some insight into the special requirements for the permissible location of critical flow can be gained through a consideration of the physics of resonance in a linear system. In general, an external force that translates along the channel with speed c_F tends to excite waves that have phase speed c_F. The efficiency of the excitation depends on how well the spatial structure of the disturbance projects on the free wave in question. In the steady, shallow flow under consideration, the forcing is due to the bottom topography (or width variations) and is clearly stationary. Therefore, resonant excitation can only occur when the wave in question is itself stationary; that is, the flow must be critical. Since the topography and flow have been constrained to vary gradually in the y-direction, the spatial

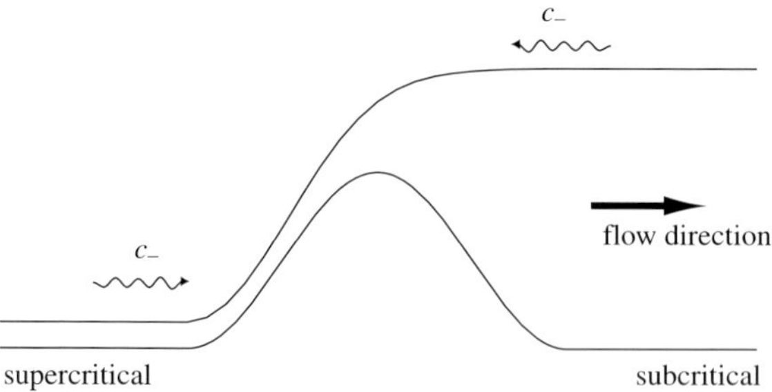

FIGURE 1.4.4. Conditions leading to the shock forming instability. The flow supports two linear waves with speeds $v + (gd)^{1/2}$ and $v - (gd)^{1/2}$. In the supercritical-to-subcritical transition shown, the latter propagate towards the critical sill section from both upstream and downstream (wavy arrows).

structure of the forcing projects perfectly onto the wave. All of this implies that a steady critical flow cannot exist in the presence of forcing, a statement consistent with (1.4.3). Unless the forcing is zero (i.e. $\partial h/\partial y - (d/w)\partial w/\partial y = 0$), the steady flow becomes singular. The reader might wonder why such behavior does not occur in connection with well-known stationary disturbances such as mountain lee waves. The answer is that such disturbances have finite wavelengths and are therefore nonhydrostatic. The significance of a lack of long-wave character is that the waves are dispersive, meaning that energy propagates at a different speed than phase. Thus the terrain may excite stationary waves, but radiation of energy away from the terrain limits local growth. Long gravity wave disturbances are characterized by equal phase and energy propagation (group) speeds.

If it is known in advance that the flow is hydraulically controlled, one can derive a transport or 'weir' relation that facilitates measurement of the volume transport. The goal is to monitor the discharge through measurements of the free-surface elevation at some convenient location upstream of the control section. The procedure circumvents the need to directly measure the fluid velocity. Oceanographers would like to apply the same methodology to deep overflows, allowing them to avoid expensive and technically difficult velocity measurements. In such cases the deep flow is bounded above by an isopycnal (constant density) surface and the goal is to relate the deep transport to the elevation of this surface at some upstream location.

Derivations of weir formulae begin at the control section, where the flow is critical, and use conservation of volume flux and energy in order to link conditions there to those at the monitoring location. As an example, consider a reservoir of width w_1 that drains across a sill of width w_s. The condition of criticality $v_c = (gd_c)^{1/2}$ at the sill can be used to write the volume transport $Q = vdw$ as $g^{1/2}d_c^{3/2}w_s$. Equating energy and volume transport at the reservoir $y = y_1$ and sill sections leads to

$$\frac{v_1^2}{2} + gd_1 = \frac{v_c^2}{2} + gd_c + gh_m, \tag{1.4.9}$$

and

$$v_1 d_1 w_1 = v_c d_c w_s. \tag{1.4.10}$$

Eliminating the velocities by combining these relations and using the critical condition leads to

$$\frac{3}{2}\left(\frac{gQ}{w_s}\right)^{2/3} - \frac{Q^2}{2d_1^2 w_1^2} = g(d_1 - h_m) = g\Delta z, \tag{1.4.11}$$

where Δz is the difference in elevation between the free surface at $y = y_1$ and the sill. Measuring Δz and d_1 allows Q to be determined from the above formula. In many cases, y_1 can be chosen in a location where the depth d_1 or width w_1 is sufficiently large that the second term on the left-hand side can be neglected, resulting in the approximation

$$Q = \left(\frac{2}{3}\right)^{3/2} w_s g^{1/2} \Delta z^{3/2}. \tag{1.4.12}$$

For the reduced gravity analog of the current model, weir formulas would permit the calculation of volume transport based on the interface elevation upstream of the critical section.

Although equation (1.4.11) was motivated by the practical necessity of measuring volume flux, it has a deeper meaning that bears on the concept of hydraulic control. In a controlled state, there is a fixed relationship between the parameters governing the flow, in this case Q and Δz, and the geometrical parameters describing the control section, in this case the sill height h_m. For noncontrolled solutions no such relationship exists, implying that one has more freedom to manipulate these flows. We will elaborate on this point further. In addition, it is easy to show that critical flow is associated with a number of variational properties of steady flows. For fixed Q and h, the energy B of the flow is minimized over all possible values of d, which can be seen from Figure 1.4.3 or from taking $\partial B/\partial d = 0$ in (1.4.6). Similarly, it can be shown that for fixed B and h, Q is maximized over all d. Hydraulically controlled solutions thus minimize the energy available at a given volume transport, which is consistent with the idea that the fluid is barely able to surmount the obstacle. In addition, these solutions tend to maximize the transport available at a given energy level.

Exercises

(1) Using l'Hôpital's rule in connection with (1.4.3), derive an expression for the slope of the free surface at a sill under critical flow conditions. You may assume that the channel width is constant. From the form of the result, show that critical flow can occur over a sill ($d^2 h/dy^2 < 0$) but not a trough ($d^2 h/dy^2 > 0$). Also show that a solution passing through a critical state at a sill generally cannot be subcritical (or supercritical) both upstream and downstream of the sill without incurring a discontinuity in the slope of the free surface.

(2) Construct a nondimensional solution curve akin to that of Figure 1.4.1 or 1.4.2 for the case of a channel of constant h but variable w.
(3) Consider a 100 m-deep reservoir that is drained by spillage over a dam of height 99 m. Both the dam and reservoir have width $w = 100$ m. Approximate

   (a) The volume flow rate from the reservoir
   (b) The depth and velocity at the sill of the dam
   (c) If you used an approximation to answer (a), estimate the error.

(4) Suppose that the channel has a triangular cross section. The width w at any z is given by

$$w(z) = 2\alpha(z - h(y))$$

where $h(y)$ is the elevation of the bottom apex. The along-channel velocity v and surface elevation are independent of x.

   (a) Taking $h = $ constant, find the speed of long surface gravity waves in the channel.
   (b) For steady flow, formulate a solution curve like that of Figure 1.4.1a or 1.4.2a showing how the fluid depth varies with h.
   (c) Show that the condition obtained at the extrema of the curve is the same critical condition that can be deduced from part (a).
   (d) Write down the weir formula for the case in which the fluid originates from an infinitely deep reservoir and spills over a sill.

1.5.   Hydraulics in Abstract

In the example of the previous section, solutions in terms of the fluid depth d were obtained using conservation laws for the volume transport Q and the energy per unit mass B. For set values of these parameters, d depends only on the *local* bottom elevation h and channel width w, though several choices of d might be possible. There is no dependence on the values of h or w, or on the flow itself, at neighboring sections. As pointed out by Gill (1977), these elements are shared by a wide class of "hydraulic-type" systems. By taking advantage of the common elements, it is possible to develop machinery that allows a wide class of flows to be analyzed systematically.

The first hydraulic-type flow to be formally analyzed was probably the movement of a compressible gas through an orifice. The crucial result that the fluid velocity v in the orifice equals the speed of sound was derived independently by Reynolds (1886) and Hugoniot (1886). The statements of conservation of mass and energy are given by

$$\rho v A = M \quad \text{and} \quad \frac{v^2}{2} + \int^{p} \frac{dp}{\rho} = B.$$

These are supplemented by an equation of state $\rho = F(p)$. Here A is the cross-sectional area of the conduit and M is the (constant) mass flux. The equations are analogous to our shallow water model, with d playing the role of density ρ. Hugoniot was aware of experiments in which the velocity of the gas was observed to monotonically increase through an orifice, where A first decreases and then increases. This upstream/downstream asymmetry with respect to A is analogous to the asymmetry of d with respect to h and is characteristic of hydraulic transitions. Reynolds knew of an experiment in which the upstream propagation of information appeared to be blocked within the orifice:

"Amongst the results of Mr. Wilde's experiments on the flow of gas, one, to which attention is particularly called, is that when gas is flowing from a discharging vessel through an orifice into a receiving vessel, the rate at which the pressure falls in the discharging vessel is independent of the pressure in the receiving vessel until this becomes greater than about five tenths the pressure in the discharging vessel."

The critical condition in the orifice was derived by both authors, essentially by considering the pressure decrease in a continuously narrowing conduit. They showed that a minimal possible pressure (Reynolds) or maximum possible ρv (Hugoniot) is reached when A is sufficiently small and they both observed that the implied v is equal to the speed of sound in the gas. (The details of the derivation are explored in Exercise 1.) The minimum in pressure found by Reynolds is analogous to the minimum in specific energy $\tilde{B} - \tilde{h}$ exhibited by the curves in Figure 1.4.2. The existence of a minimum or maximum implies that more than one v is possible for a given A, at least within a certain range of A. The minimization or maximization of properties as a way of obtaining a control criterion is sometimes referred to as a Hugoniot condition. The existence of more than one possible solution at a given cross section is characteristic of hydraulics problems in general.

a.   Gill's Original Approach

In the gas dynamics model and the shallow water analogy, the state of the flow at any section of the channel can be specified in terms of a *single* dependent variable. This variable, which we denote γ, is related to the local geometry h, w, etc. along with the parameters Q, B, etc. by a conservation law of the form[3]

$$\mathcal{G}(\gamma(y); h(y), w(y), \ldots; B, Q, \ldots) = const. \tag{1.5.1}$$

B and Q could be the energy and flow rate, or they could represent other conserved properties of the system. The value of γ at a particular y determines all other attributes of the flow at that section. The position y does not appear explicitly. The constant on the right-hand side, which appears in Gill's original

---

[3] Perhaps out of modesty, Gill used the symbol J to represent the function in (1.5.1). To honor him, and to avoid confusion with the Jacobian operator, we use the symbol $\mathcal{G}$.

formulation, may be disposed by redefining $\mathcal{G}$. We may therefore take the constant to be zero with no loss of generality.

In the shallow water example of the Section 1.4, $\gamma(y)$ is the fluid depth d and the Bernoulli equation (1.4.6) may be written as

$$\mathcal{G} = \frac{Q^2}{2\gamma^2 w^2} + g\gamma + gh - gB.$$

Other forms of $\mathcal{G}$ could be written down by using variables like v instead of d.

A useful identity

$$\frac{d\mathcal{G}}{dy} = \frac{\partial\mathcal{G}}{\partial\gamma}\frac{d\gamma}{dy} + \frac{\partial\mathcal{G}}{\partial h}\frac{dh}{dy} + \frac{\partial\mathcal{G}}{\partial w}\frac{dw}{dy} + \ldots = 0. \tag{1.5.2}$$

is obtained by differentiation of (1.5.1). This result is often just the differential form of a momentum or continuity equation. The reader may wish to verify that application of (1.5.2) to the previous shallow-water example leads back to the y-momentum equation.

Now consider the conditions under which free, stationary long waves of small amplitude exist. By 'long' we mean disturbances that vary gradually in the y-direction, just as the steady flow does. By 'free' we mean disturbances that occur spontaneously and are independent of any forcing mechanism such as bottom slope. When a steady flow becomes hydraulically critical at a particular section $y = y_c$, it can support a free, stationary disturbance at that section. In other words, the steady state can be locally altered without changing either the conduit geometry or the upstream conditions. The altered flow must therefore have the same volume flux, energy, etc. as the undisturbed flow. Let γ_c represent the undisturbed state at the critical section and let γ' represent the disturbance. Then the altered flow $\gamma_c + \gamma'$ must also satisfy (1.5.1):

$$\mathcal{G}(\gamma_c + \gamma'; h(y_c), w(y_c), \cdots ; B, Q, \cdots) = 0.$$

Taylor's expansion of this relation leads to

$$\mathcal{G}(\gamma_c + \gamma'; h(y_c), w(y_c), \cdots) = \mathcal{G}(\gamma_c; h(y_c), w(y_c), \cdots) + \gamma'\left(\frac{\partial\mathcal{G}}{\partial\gamma}\right)_{\gamma=\gamma_c} + \cdots$$

$$= 0.$$

Since the undisturbed flow must satisfy (1.5.10) the first term on the right-hand side is zero. It follows that

$$\frac{\partial\mathcal{G}}{\partial\gamma} = 0 \tag{1.5.3}$$

at the critical section. In plain words, criticality implies that the steady flow at a fixed location (fixed h, w, etc.) can be altered by an infinitesimal amount $\delta\gamma$ such at that (1.5.1) remains satisfied ($\mathcal{G}$ remains zero).

One of the important aspects of (1.5.3) is that it formally links the minimization (or maximization) used by Reynolds and Hugoniot. In the example of the previous section, (1.5.3) implies that

$$\frac{\partial \mathcal{G}}{\partial d} = \frac{\partial}{\partial d}\left[\frac{Q^2}{2d^2w^2} + gd + gh - gB\right] = \frac{\partial}{\partial d}\left[\frac{Q^2}{2d^2w^2} + gd\right] = 0$$

and thus 'specific energy' $Q^2/2d^2w^2 + gd$ is minimized when the flow is critical. Engineering texts often used this minimization as a basis for defining critical flow, even though the physical motivation is not always transparent.

The flow state at a particular section can be computed by solving (1.5.1) for the values of γ. In hydraulic applications, more than one root is possible; the cubic equation (1.4.7) admits as many as two real, positive roots for the depth of the shallow flow at each h. The two roots correspond to the two depths (Figure 1.4.3) for a given value of the Bernoulli constant $\tilde{B}$. The condition (1.5.3) implies that the roots coalesce, as occurs at the sill. All of the behavior described above is thus linked to the tendency of the hydraulic function $\mathcal{G}$ to admit multiple roots. It is important to note that this property will be lost when the shallow water (or other) governing equations are linearized.

A further constraint implied by flow criticality follows from setting $\partial \mathcal{G}/\partial \gamma = 0$ in (1.5.2) leading to

$$\left(\frac{\partial \mathcal{G}}{\partial h}\frac{dh}{dy} + \frac{\partial \mathcal{G}}{\partial w}\frac{dw}{dy} + \cdots\right)_{y=y_c} = 0. \tag{1.5.4}$$

This condition restricts the locations $y = y_c$ at which critical flow can occur. The locations at which critical flow actually occurs are sometimes called *control* sections. To obtain (1.5.4), it has been assumed that the flow remains smooth at $y = y_c$, so that $d\gamma/dy$ is finite there. Thus (1.5.4) is often referred to as a *regularity* condition. In fact, the satisfaction of (1.5.4) is equivalent in shallow water theory to the requirement that the numerator in (1.4.3) vanishes. It can readily be seen from that equation that the requirement is a necessary condition that the slope of the free surface remain bounded.

As in Figure 1.4.3, critical flow generally occurs at a section (or sections) $y = y_c$ marking the transition between states supporting wave propagation in different directions. Strictly speaking, the flow is able to support stationary disturbances only at y_c and not at points immediately upstream and downstream. The stationary disturbances are therefore possible in theory but are difficult to visualize in most applications. They should not be confused with stationary lee waves, which involve waves of finite length.

The purely local dependence of the functional $\mathcal{G}$ on y is a product of the conservative nature of the flow and of the gradually varying geometry. When dissipation or rapid variations are present, the y-dependence generally becomes nonlocal. Such systems can still exhibit forms of hydraulic behavior. Examples are discussed in Exercise 4 of this section and in Section 3.8.

We have seen that critical flow can form at a maximum in h (a sill) and it is natural to ask whether the same is true of a minimum in h. Guidance comes from differentiating (1.5.1) twice with respect to y and applying the critical condition (1.5.3), leading to

$$\frac{\partial^2 \mathcal{G}}{\partial \gamma^2}\left(\frac{d\gamma}{dy}\right)^2 = -\frac{\partial}{\partial y}\left(\frac{\partial \mathcal{G}}{\partial h}\frac{dh}{dy} + \frac{\partial \mathcal{G}}{\partial w}\frac{dw}{dy} + \cdots\right).$$

In order for real values of $\frac{d\gamma}{dy}$ to exist at the critical section, $\frac{\partial}{\partial y}\left(\frac{\partial \mathcal{G}}{\partial h}\frac{dh}{dy} + \frac{\partial \mathcal{G}}{\partial w}\frac{dw}{dy} + \cdots\right)$ and $\frac{\partial^2 \mathcal{G}}{\partial \gamma^2}$ must have opposite signs. This condition generalizes the concepts of expansions and contractions in the channel geometry. In the example of the previous section, $\frac{\partial^2 \mathcal{G}}{\partial \gamma^2} = \frac{3Q^2}{d^4} > 0$, whereas $\frac{\partial}{\partial y}\left(\frac{\partial \mathcal{G}}{\partial h}\frac{dh}{dy} + \frac{\partial \mathcal{G}}{\partial w}\frac{dw}{dy} + \cdots\right) = g\frac{d^2 h}{dy^2}$, so the bottom curvature $d^2 h/dy^2$ must be < 0. Negative curvature is characteristic of a sill but not a depression in the bottom and the implication is that physically meaningful critical solutions require a sill geometry. At the sill, $\left(\frac{d\gamma}{dy}\right)^2 = \frac{gd^4}{3Q^2}\frac{d^2 h}{dy^2}$, indicating two possible free-surface slopes. The two slopes are simply those of the intersecting solutions (both with $\tilde{B} = 2.5$) shown in Figure 1.4.3. Computation of a continuous solution through a critical section therefore requires a *hydraulic transition* in which subcritical upstream flow joins with supercritical downstream flow (or vice versa). One may not move through the critical point and remain on the subcritical branches.

b. *Multiple Variables*

Reduction of the problem to the single-variable format envisioned by Gill (1977) is not always easy. It is often more convenient, and sometimes necessary, to work with two independent relations in two variables γ_1 and γ_2:

$$\mathcal{G}_1(\gamma_1, \gamma_2; h, w, \cdots; B, Q, \cdots) = C_1 \tag{1.5.5}$$

and

$$\mathcal{G}_2(\gamma_1, \gamma_2; h, w, \cdots; B, Q, \cdots) = C_2. \tag{1.5.6}$$

The approach to dealing with this system is laid out by Pratt and Armi (1987) and Dalziel (1991) and the generalization to an arbitrary number of variables is discussed by Lane-Serff *et al.* (2000) and Pratt and Helfrich (2005).

For the system (1.5.5 and 1.5.6), the existence of a stationary wave requires that small perturbations γ_1' and γ_2' of the flow exist at a fixed location such that the new altered flow remains a solution. Taylor expansion of $\mathcal{G}_1$ and $\mathcal{G}_2$ for fixed h, w, etc. about the unperturbed state leads to

$$\frac{\partial \mathcal{G}_1}{\partial \gamma_1}\gamma_1' + \frac{\partial \mathcal{G}_1}{\partial \gamma_2}\gamma_2' = 0 \tag{1.5.7}$$

$$\frac{\partial \mathcal{G}_2}{\partial \gamma_1}\gamma_1' + \frac{\partial \mathcal{G}_2}{\partial \gamma_2}\gamma_2' = 0. \tag{1.5.8}$$

The critical condition is just the solvability condition for this pair:

$$\frac{\partial \mathcal{G}_1}{\partial \gamma_1} \frac{\partial \mathcal{G}_2}{\partial \gamma_2} - \frac{\partial \mathcal{G}_1}{\partial \gamma_2} \frac{\partial \mathcal{G}_2}{\partial \gamma_1} = 0. \tag{1.5.9}$$

Stationary waves then involve the displacement $(d\gamma_1,\ d\gamma_2)$ as given by (1.5.7) or (1.5.8):

$$\vec{\gamma}\,' = \gamma_1' \left(1, -\left[\frac{\partial \mathcal{G}_1/\partial \gamma_1}{\partial \mathcal{G}_2/\partial \gamma_2} \right]_{y=y_c} \right), \tag{1.5.10}$$

where $d\gamma_1$ is small but arbitrary. The displacement vector contains information about the structure of the stationary wave (see Exercise 3).

The generalization of the regularity condition (2.5) can be found by writing out the identities $d\mathcal{G}_1/dy = 0$ and $d\mathcal{G}_2/dy = 0$:

$$\frac{d\mathcal{G}_1}{dy} = \frac{\partial \mathcal{G}_1}{\partial \gamma_1} \frac{d\gamma_1}{dy} + \frac{\partial \mathcal{G}_1}{\partial \gamma_2} \frac{d\gamma_2}{dy} + \frac{\partial \mathcal{G}_1}{\partial h} \frac{dh}{dy} + \frac{\partial \mathcal{G}_1}{\partial w} \frac{dw}{dy} + \cdots = 0$$

$$\frac{d\mathcal{G}_2}{dy} = \frac{\partial \mathcal{G}_2}{\partial \gamma_1} \frac{d\gamma_1}{dy} + \frac{\partial \mathcal{G}_2}{\partial \gamma_2} \frac{d\gamma_2}{dy} + \frac{\partial \mathcal{G}_2}{\partial h} \frac{dh}{dy} + \frac{\partial \mathcal{G}_2}{\partial w} \frac{dw}{dy} + \cdots = 0$$

Solving for $d\gamma_1/dy$ leads to

$$\frac{d\gamma_1}{dy} = \frac{\dfrac{\partial \mathcal{G}_1}{\partial \gamma_2} \left(\dfrac{\partial \mathcal{G}_2}{\partial y} \right)_{\gamma_1, \gamma_2} - \dfrac{\partial \mathcal{G}_2}{\partial \gamma_2} \left(\dfrac{\partial \mathcal{G}_1}{\partial y} \right)_{\gamma_1, \gamma_2}}{\dfrac{\partial \mathcal{G}_1}{\partial \gamma_1} \dfrac{\partial \mathcal{G}_2}{\partial \gamma_2} - \dfrac{\partial \mathcal{G}_1}{\partial \gamma_2} \dfrac{\partial \mathcal{G}_2}{\partial \gamma_1}}$$

where $\left(\dfrac{\partial ()}{\partial y} \right)_{\gamma_1, \gamma_2} = \dfrac{\partial ()}{\partial w} \dfrac{\partial w}{\partial y} + \dfrac{\partial ()}{\partial h} \dfrac{\partial h}{\partial y} + \cdots$ is a derivative taken with γ_1 and γ_2 held constant. Critical flow requires that the denominator vanish and the numerator must then vanish if the flow is to remain well-behaved. The regularity condition is thus

$$\frac{\partial \mathcal{G}_1}{\partial \gamma_i} \left(\frac{\partial \mathcal{G}_2}{\partial y} \right)_{\gamma_1, \gamma_2} - \frac{\partial \mathcal{G}_2}{\partial \gamma_i} \left(\frac{\partial \mathcal{G}_1}{\partial y} \right)_{\gamma_1, \gamma_2} = 0 \quad (i = 1 \text{ or } i = 2), \tag{1.5.11}$$

(The $i = 2$ version, which follows from developing an expression for $d\gamma_2/dy$, is not independent of the $i = 1$ version.)

The machinery is easily extended to problems governed by N relations for N independent variables:

$$\mathcal{G}_1(\gamma_1(y), \cdots) = C_1$$

$$\mathcal{G}_2(\gamma_1(y), \cdots) = C_2 \tag{1.5.12}$$

$$\vdots$$

$$\mathcal{G}_N(\gamma_1(y), \cdots) = C_N.$$

The condition for stationary waves is now

$$\sum_{j=1}^{N} \frac{\partial \mathcal{G}_i}{\partial \gamma_j} \gamma_j' = 0 \ (i = 1, 2, \cdots N), \tag{1.5.13}$$

and the corresponding solvability condition is the vanishing of the generalized Jacobian:

$$\det \left(\frac{\partial \mathcal{G}_i}{\partial \gamma_j} \right)^T = 0, \tag{1.5.14}$$

where

$$\left(\frac{\partial \mathcal{G}_i}{\partial \gamma_j} \right)^T = \begin{pmatrix} \partial \mathcal{G}_1 / \partial \gamma_1 & \cdots & \partial \mathcal{G}_1 / \partial \gamma_N \\ \cdot & \cdot\cdot & \cdot \\ \cdot & \cdot\cdot & \cdot \\ \partial \mathcal{G}_N / \partial \gamma_1 & \cdots & \partial \mathcal{G}_N / \partial \gamma_N \end{pmatrix}. \tag{1.5.15}$$

The tangent displacement vector $(d\gamma_1, d\gamma_2, \ldots)_c$, which is computed from any member of (1.5.13), again determines the transverse structure of the stationary wave.

It can also be shown (see Exercise 6) that the generalized regularity condition is

$$\det \left(\left(\frac{\partial \mathcal{G}_i}{\partial \gamma_j} \right)^T \middle| \left(\frac{\partial \mathcal{G}_i}{\partial y} \right)^T_\gamma \right) = 0 \tag{1.5.16}$$

where $\left(\frac{\partial \mathcal{G}_i}{\partial \gamma_j} \right)^T \middle| \left(\frac{\partial \mathcal{G}_i}{\partial y} \right)^T_\gamma$ is the matrix obtained by replacing column i of $\left(\frac{\partial \mathcal{G}_i}{\partial \gamma_j} \right)^T$ by

$$\left(\frac{\partial \mathcal{G}_i}{\partial y} \right)^T_\gamma = \begin{pmatrix} (\partial \mathcal{G}_1 / \partial y)_\gamma \\ (\partial \mathcal{G}_2 / \partial y)_\gamma \\ \vdots \\ (\partial \mathcal{G}_N / \partial y)_\gamma \end{pmatrix}.$$

When formulating hydraulic functionals $\mathcal{G}_1$, $\mathcal{G}_2$, etc. for a particular system, there is a disadvantage in reducing the system to a single functional in a single unknown. Namely, certain kinds of critical states may be missed in the evaluation of the critical condition (1.5.3) for the single-variable formulation. This difficulty arises when the stationary wave in question has no displacement in terms of the chosen single variable. That is, the tangent displacement vector $(\gamma_1', \gamma_2', \cdots)$ for a particular stationary disturbance may have a zero constituent, say γ_2'. If the system is reduced such that γ_2 is the only variable, then the critical condition for this disturbance will not be identified by (1.5.3). The missing critical condition will, however, be identified by the multivariate formula (1.5.14). An example will be given in Section 2.4.

Exercises

(1) *Transonic flow in an isotropic gas.* Consider an inviscid and diffusion-free, compressible gas whose motion is governed by the following equations:

$$\rho\frac{d\mathbf{u}}{dt} = -\nabla p + \rho\chi$$

$$\frac{d\rho}{dt} + \rho\nabla\cdot\mathbf{u} = 0$$

$$p = \rho RT$$

$$\rho c_v\frac{dT}{dt} + p\nabla\cdot\mathbf{u} = 0$$

where T is the absolute temperature, c_v is the specific heat at constant volume, and χ is the body force per unit mass.

(a) Show that $\frac{d}{dt}\left(\frac{p}{\rho^{\hat{\gamma}}}\right) = 0$, where $\frac{c_v}{R} = \frac{1}{\hat{\gamma}-1}$. [*Hint: one starting point is elimination of $\nabla\cdot\mathbf{u}$ from the second and fourth equations.*]

(b) Next consider the generalized form of the Bernoulli function for steady compressible flow:

$$\int\frac{dp}{\rho} + \Omega + \frac{|\mathbf{u}|^2}{2} = \text{constant along streamlines}$$

where Ω is the body force potential. Applying this and the steady form of the result in (a) to a one-dimensional flow in a wind tunnel of slowly varying cross-sectional area $A(y)$, derive a hydraulic functional of the form $\mathcal{G}(\rho; A) = C$. (The body force potential may be neglected.)

(c) From the result of (b), obtain a critical condition and deduce that the intrinsic signal speed (here the speed of sound) is $[\hat{\gamma}p/\rho]^{1/2}$.

(2) *Homogeneous, free-surface flow with shear.* Following Garrett and Gerdes (2003) consider a steady, shallow, homogeneous flow with vertical shear ($\partial v/\partial z \neq 0$). The flow is described by a stream function $\psi(y, z)$ such that $\partial\psi/\partial z = v$, $\psi(y, h(y)) = 0$, and $\psi(y, h(y) + d(y)) = Q$. The Bernoulli function is given by

$$B(\psi) = \frac{v^2}{2} + gd + gh.$$

Construct a hydraulic functional for the flow by following these steps:

(a) Show that $d = \int_0^Q\frac{d\psi}{v}$ and therefore

$$d = \frac{1}{\sqrt{2}}\int_0^Q\frac{d\psi}{[B(\psi) - g(d+h)]^{1/2}}.$$

(b) Define a hydraulic functional $\mathcal{G}_2(d(y), v(y); h(y), w(y), \cdots; B, Q, \cdots) = 0$ based on the above relation. Show that setting $\partial\mathcal{G}/\partial d = 0$ leads to the critical condition:

$$g \int_h^{h+d} \frac{dz}{v^2} = 1$$

and compare this with the result of Exercise 4 of Section 1.2.

(3) Cast the hydraulic problem for homogeneous, free-surface flow in terms of two functionals $\mathcal{G}_1(d(y), v(y); h(y), w(y)) = Q$ and $\mathcal{G}_2(d(y), v(y); h(y), w(y)) = B$ representing the continuity and Bernoulli equations. Show that the critical and regularity conditions obtained using the two-variable (d, v) machinery is the same if the single-variable representation were used. Using (1.5.10), show that the displacement vector specifies a relationship between the depth and velocity perturbations, and that this relationship is the same as that implied by (1.2.5) and (1.2.6) for the '$-$'wave.

(4) *Non local dependence on y.* Consider the functional

$$G_P\left(d(y), \int_{y_0}^{y} f(d)dy'; h(y); q, y_0 \cdots \right) = \frac{Q^2}{2d^2 w^2}$$

$$+ d + h + \alpha \frac{Q^2}{w^2} \int_{y_0}^{y} d^{-3} dy' = B(y_0) \qquad (1.5.17)$$

governing a shallow flow under the influence of bottom drag (Pratt, 1986). Fixed parameters include the drag parameter α. The presence of drag introduces an integration from an upstream location $y = y_0$, where the depth and velocity v are known, to the section under consideration. Consider the possibility that a free, small amplitude, stationary disturbance exists at a section at $y = y_c$ but at no other upstream point. Show that a necessary condition for its existence is

$$\lim_{\delta d \to 0} \left[\frac{G_P(d + \delta d, \int_{y_0}^{y_c} f(d + \delta d)dy'; h(y_c), q, y_0 \cdots) - G_P(d, \int_{y_0}^{y_c} f(d)dy'; h(y_c), q, y_0 \cdots)}{\delta d} \right] = 0,$$

Show that evaluation of this limit leads to the critical condition $v = d^{1/2}$.

(5) Suppose that the dimensionless obstacle height in (1.4.7) $\tilde{h}(y) = a\hat{h}(y)$, where $a \ll 1$. Let

$$\tilde{d} = \tilde{d}^{(0)} + a\tilde{d}^{(1)} + O(a^2)$$

where $\tilde{d}^{(0)}$ is the solution for $\tilde{h} = 0$. Formulate a Gill function for the variable $\tilde{d}^{(1)}$, but show that there can be no hydraulic transitions. Why does linearization of the problem prevent this phenomenon?

(6) By taking the y-derivatives of (1.5.12) and applying the critical condition, show, using Cramer's rule, that

$$\frac{d\gamma_i}{dy} = -\frac{\det\left(\left(\frac{\partial \mathcal{G}_i}{\partial \gamma_j}\right)^T \middle| \left(\frac{\partial \mathcal{G}_i}{\partial y}\right)^T_\gamma\right)}{\det\left(\frac{\partial \mathcal{G}_i}{\partial \gamma_j}\right)^T}.$$

Deduce the regularity condition (1.5.16).

1.6. Hydraulic Jumps, Bores, Rarefaction Waves, and Long's Experiment

One of the traditional challenges in learning about hydraulics is to reach an understanding of why controlled (subcritical-to-supercritical) solutions arise and how they are established. Calculations of steady flows merely show the existence of hydraulically controlled solutions for special values of the governing parameters (e.g. $\tilde{B} = 2.5$ in Figure 1.4.3) and this gives the impression that such a state might be difficult to realize in nature. On the other hand, observations and laboratory experiments show that controlled solutions prevail when topography is sufficiently high. It is valuable to observe how steady flows are established as the result of time-dependent adjustment from a simple initial state, or as the result of varying the upstream conditions. A classical example is the experiments of Long (1954, 1955 and 1970), who towed an obstacle through a laboratory tank containing a fluid initially at rest. The initial fluid depth d_o is constant and the obstacle is towed at a fixed speed v_0 until a translating steady state is achieved in the vicinity. For a frictionless system, the experiment is equivalent to the sudden introduction of an obstacle into a moving stream of depth and velocity d_0 and v_0 (Figure 1.6.1). This is the viewpoint we will use. The outcome depends crucially on the height $h_{\rm m}$ of the obstacle relative to a threshold value $h_{\rm c}$. A variety of experiments have confirmed that $h_{\rm c}$ is simply the obstacle height associated with a hydraulically controlled *steady* state whose upstream depth and velocity are d_o and v_o. This is exactly the height that appears in (1.4.11) if Q/w is interpreted as $v_0 d_0$. A nondimensional form of this relation is

$$\frac{h_c}{d_0} = 1 - \frac{3}{2}F_0^{2/3} + \frac{F_0^2}{2} \tag{1.6.1}$$

where $F_0 = \frac{v_0}{(gd_0)^{1/2}}$.

For $h_{\rm m} < h_{\rm c}$ the sudden appearance of the obstacle generates disturbances that propagate away from the obstacle and leave behind an uncontrolled steady solution, either completely supercritical or completely subcritical. When the initial state is subcritical $v_0 < (gd_o)^{1/2}$, a subcritical steady state with a dip in the upper surface is established (Figure 1.6.1b). Note that the disturbances propagating away from the obstacle are *isolated* in the sense that they do not permanently alter the flow into which they propagate. For supercritical initial flow and $h_{\rm m} < h_{\rm c}$, a supercritical steady state is established, this time with the two isolated disturbances propagating downstream.

When $h_{\rm m} > h_{\rm c}$, the situation is quite different. The obstacle now generates an upstream *bore*: a propagating wave consisting of an abrupt increase in depth. As shown in Figure 1.6.1c, the upstream bore increases the depth from d_0 to d_1. In practice, the bore can vary from a nearly discontinuous, turbulent transition to a gradual, and perhaps oscillatory, change (Peregrin, 1968). The latter is called an *undular bore*. Here, the bore is represented as a simple depth discontinuity. Downstream of the obstacle, the adjustment is caused by a bore and a rarefaction

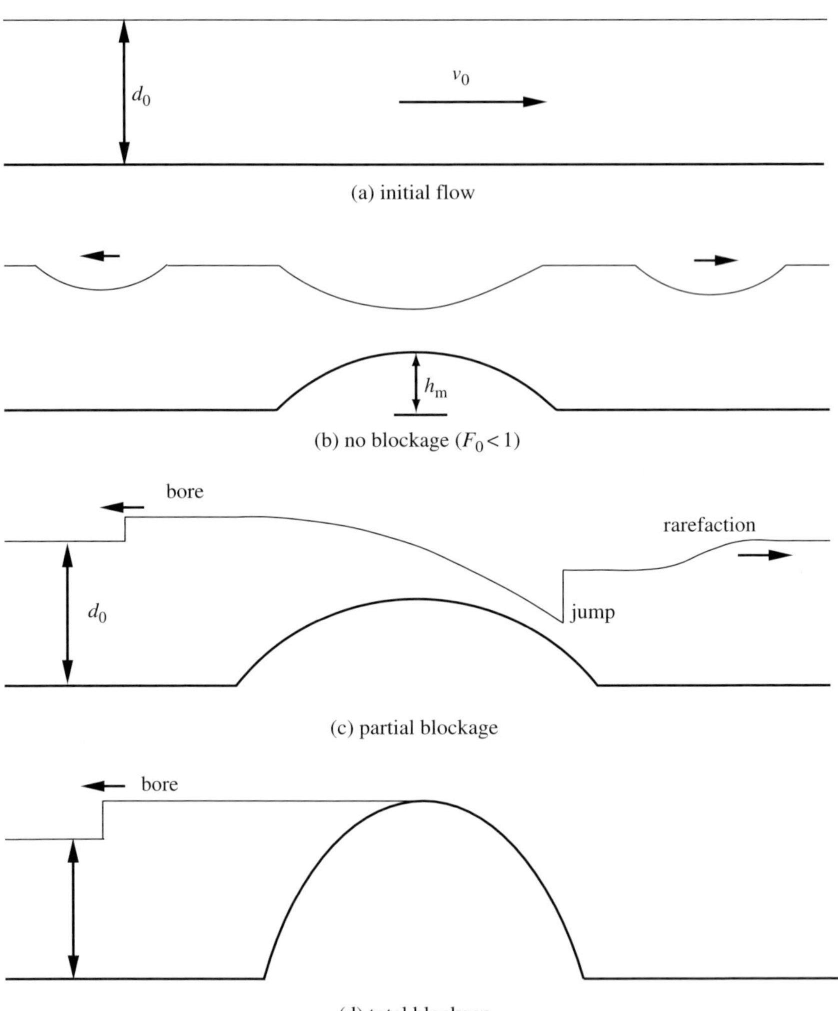

FIGURE 1.6.1. Schematic depiction of the various types of shallow water adjustment caused when an obstacle is introduced into a uniform, subcritical stream (a). In (b) the obstacle height is less than the critical value and the flow remains subcritical. In (c) the obstacle exceeds its critical height and a hydraulically controlled flow with a jump emerges. In such cases the jump may also propagate downstream as a bore. In (d) the obstacle has exceeded the height required for complete blocking. (The downstream disturbances are not shown for this case.)

wave. In some cases the downstream bore may become stationary on the downslope of the obstacle forming a hydraulic jump (Figure 1.6.1c). Over the obstacle a hydraulically controlled steady state develops. The final steady state thus has subcritical flow upstream, supercritical flow downstream (perhaps terminating in a hydraulic jump) and critical flow at the sill. There also exists a second threshold height $h_b(> h_c)$ that, when exceeded, results in complete blockage of the flow (Figure 1.6.1d).

Long's experiments give a particular view of the concept of hydraulic control, one in which the obstacle gains the ability to permanently alter the far field flow. When $h_m < h_c$ the long-term influence of the obstacle is local; when $h_m > h_c$ this influence is global. In the latter case, it is often said that the obstacle exerts *upstream influence* (even though the downstream flow is also altered). Another virtue of Long's experiment is that the final steady state can be predicted from the initial conditions. To do so, one must analyze the time-dependent flow that has developed long after the obstacle is introduced. In particular, sufficient time must have elapsed to allow the transients to move away from the obstacle and developed into bores and/or rarefaction waves. The analysis makes use of *shock-joining* conditions linking the uniform flows on either side of the transients. The full solution to the adjustment problem will be presented in the next section; first we must develop a theory for shock joining.

a. *Shock Joining*

Bores and hydraulic jumps are nonhydrostatic and often highly turbulent. Both produce changes in the thickness and velocity that take place over a distance of the order of the fluid depth. This distance is very short in the context of our long-wave model and can formally be treated as a discontinuity in d and v. Away from the discontinuity the pressure is hydrostatic and the velocity independent of depth. As an example, consider a hydraulic jump consisting of a stationary discontinuity between two steady flows (Figure 1.6.2). Let (d_u, v_u) and (d_d, v_d) denote the depth and velocity immediately upstream and downstream of the jump. In practice, one must measure these end-state values far enough away from the jump that the fluid is hydrostatic. Then it is immediately clear from mass conservation that

$$v_u d_u = v_d d_d \tag{1.6.2}$$

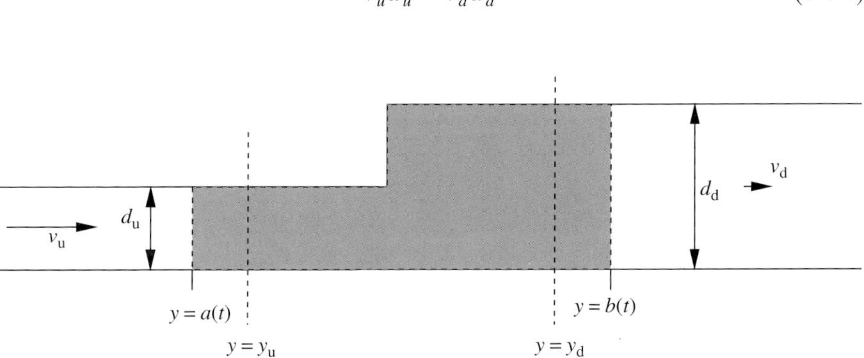

FIGURE 1.6.2. An abstraction of a hydraulic jump.

Although the channel width w may vary with y, the assumed abrupt nature of the jump means that the gradually varying w is essentially the same on each side of the jump. Hence w does not enter the above mass balance.

A second matching condition can be obtained from the observation that no external forces in the y-direction act on the fluid at the discontinuity. In practice, there might be a frictional stress acting along the bottom or a pressure component in the y-direction resulting from a nonzero bottom slope; however, the force arising from this stress will be negligible if the length of the shock is sufficiently short. Hence the difference in the pressure forces on either side of the jump must equal the change in the momentum flux of fluid entering and leaving the jump. The total pressure force acting over a section of the flow is the integral of the hydrostatic pressure p over that section, $\rho w g d^2/2$. The total momentum flux across the section is $\rho w v^2 d$. Our momentum budget therefore requires

$$d_u v_u^2 + g d_u^2/2 = d_d v_d^2 + g d_d^2/2, \tag{1.6.3}$$

The value of w has again been considered equal on either side of the jump. The quantity $\rho w(dv^2 + gd^2/2)$ is sometimes called the *flow force* and (1.6.3) shows that it is conserved across a jump.

If the discontinuity translates steadily at speed c_1, the above analysis can be repeated in a frame of reference moving with the discontinuity. Since the flow appears steady in this frame, and since the governing equations are invariant with respect to steady translation, (1.6.2) and (1.6.3) are again obtained, but with v_d and v_u interpreted as moving frame velocities. To return to the rest frame, replace these velocities by $v_d - c_1$ and $v_u - c_1$, where v_d and v_u now denote the rest-frame velocities. The general shock-joining relations are therefore given by:

$$(v_u - c_1)d_u = (v_d - c_1)d_d \tag{1.6.4}$$

and

$$d_u(v_u - c_1)^2 + g d_u^2/2 = d_d(v_d - c_1)^2 + g d_d^2/2 \tag{1.6.5}$$

If the end states are unsteady, the shock speed will vary with time. In this case it is possible to show that (1.6.4) and (1.6.5) continue to hold, but we leave the proof as an exercise for the reader.

Equations (1.6.2) and (1.6.3) allow the downstream state of a hydraulic jump to be calculated given a known upstream depth and velocity. These relations also show that the energy of the fluid crossing the jump is not conserved. Since $B_u = \frac{v_u^2}{2} + g d_u$ is the energy per unit mass of any fluid element entering the jump, the total energy influx is QB_u and the total outflux is QB_d. The difference between these two is proportional to the rate of energy dissipation $-\dot{E}$ (per unit mass) within the jump. Using (1.6.2) and (1.6.3) it can be shown that

$$-\dot{E} = \frac{gQ}{4}\frac{(d_d - d_u)^3}{d_d d_u}. \tag{1.6.6}$$

For a bore, the above expression is valid if Q is interpreted as $(v_u - c_1)d_u w$, the transport seen in the moving frame of the bore. In either case, energy dissipation $(-\dot{E} > 0)$ for positive Q requires that $(d_d - d_u)$ also be positive. More generally, energy dissipation requires that the depth of fluid increase as the fluid passes through the jump or bore. It is remarkable that the rate of dissipation can be calculated independently of viscosity or even the form of internal dissipation.

Since a bore or jump contains no internal sources of energy, the fluid depth must increase in the direction of the flow passing through. This is an important constraint as (1.6.4) and (1.6.5) admit solutions with positive *and* negative dissipation. An example can be found through elimination of $c_1 - v_d$ from (1.6.4) and (1.6.5), yielding

$$(v_u - c_1)^2 = gd_d(\frac{d_d + d_u}{2d_u}) \tag{1.6.7}$$

The left-hand side of this relation is the square of the velocity of the fluid to the left, as seen in the moving frame of the bore. For given d_d and d_u, two values of this velocity can be found corresponding to the positive and negative square roots of the right-hand side. The negative root corresponds to fluid entering the bore from the right while the positive root corresponds to fluid entering from the left. If $d_d > d_u$ the positive root must be selected.

Returning temporarily to the case of a stationary jump, a bit of manipulation of (1.6.2) and (1.6.3) leads to

$$\frac{d_d}{d_u} = \frac{-1 + \sqrt{1 + 8F_u^2}}{2} \tag{1.6.8}$$

where $F_u = v_u/\sqrt{gd_u}$, the Froude number of the approach flow. Since the fluid depth must increase in the direction of the flow, $d_d/d_u > 1$ and thus F_u must exceed unity. The approach flow must be supercritical. Since the subscripts u and d can be interchanged without affecting (1.6.2) and (1.6.3), an expression involving the downstream Froude number $F_d = v_d/\sqrt{gd_d}$ can be obtained simply by interchanging the subscripts in (1.6.8). Thus

$$\frac{d_u}{d_d} = \frac{-1 + \sqrt{1 + 8F_d^2}}{2}, \tag{1.6.9}$$

showing that the downstream flow must be subcritical. For a flow with positive v, waves with speeds $v - (gd)^{1/2}$ must move towards the jump from both upstream and downstream. A similar interpretation is possible for a bore, which overtakes linear waves propagating against the upstream flow but is overtaken from the rear by the same type of linear waves. The convergence of waves at the discontinuity is closely related to the nonlinear steepening process discussed in Section 1.3 and is instrumental in maintaining the bore. The same mechanism is related to the shock-forming instability depicted in Figure 1.4.4.

The hydraulic jump provides a mechanism by which a supercritical flow can join to a downstream subcritical flow with the same Q but lower B. For the steady solutions sketched in Figure 1.4.3, this means that the hydraulically controlled flow ($\tilde{B} = 5/2$) could connect to one of the solutions for which $\tilde{B} < 5/2$. The connection would occur in the form of a hydraulic jump on the down-slope of the obstacle, and one possibility is indicated in the figure.

The above analysis takes for granted that the jump or bore occurs over a horizontal distance short enough that bottom friction and other external sources or sinks of momentum are insignificant. For hydraulic jumps this assumption is valid as long as the Froude number of the approach flow is greater than about 1.7 (Chow, 1959). Then the depth change occurs over a horizontal distance of the order of the fluid depth. Such a change is tantamount to a discontinuity in the gradually varying framework of shallow water dynamics. For Froude numbers < 1.7 however, the jump becomes undular (wavelike) and the depth changes occur over a much longer distance. Nonhydrostatic effects are essential to the wavy structure of the jump and the increased horizontal length may necessitate consideration of additional sources of momentum. The reader is referred to Baines (1995) for a deeper discussion.

Some of the best places to observe bores are over gently sloping beaches such as those of southern California (Figure 1.6.3). On the left-hand side of the photo is a turbulent bore caused by the shallow surge of a wave running towards the beach. The middle of the photo shows a fairly quiescent, V-shaped region in which the water depth is just a few inches. To the right is the smooth, wavy front of a surge that is running away from the beach (right-to-left). The latter was generated by a previous wave that ran up on the beach and is now spilling back. This reverse surge is a good example of an undular bore.

b. *Discontinuities and Matching Conditions*

Discontinuities, real or contrived, are encountered quite often in fluid dynamics. In many situations, matching conditions are found through integration across the discontinuity of the equations governing the flow away from it. Of course, this procedure is only valid when the governing equations hold at the discontinuity as well. One must take great care in applying this method to free-surface jumps and bores, for which the shallow water equations do not hold. For example, (1.6.5) cannot be derived by integrating the shallow water momentum equation (1.2.1) across the discontinuity. Doing so would lead to the incorrect conclusion that the Bernoulli function B is conserved across the shock.

The safest approach in such cases is to formulate property budgets for a fixed control volume surrounding the shock. This is essentially the approach used to derive (1.6.3). An alternative form of the momentum budget

$$\frac{d}{dt} \iiint_V \rho v \, dxdydz = \sum_{\partial V} F^{(y)} \tag{1.6.10}$$

is valid if a material control volume V is used. The right-hand side is the sum of forces $F^{(y)}$ in the y-direction around the bounding surface ∂V. [The derivation of (1.6.3) using (1.6.10) is described in Exercise 2.]

FIGURE 1.6.3. The foamy wave front is a bore, formed by the leading edge of a wave propagating onto a gently sloping beach in southern California. The wavy feature to the right is an undular bore that is propagating in the opposite direction (right-to-left). The latter is formed at the leading edge of a long wave that has been reflected from the beach. (L. Pratt photo.)

A useful form of the shallow water momentum equation can be derived from (1.6.10) if it is temporarily assumed that the flow fields are smooth. If V is made infinitesimal and the shallow water approximations are applied, the result is the so called *flux* form of the y-momentum equation:

$$\frac{\partial(vd)}{\partial t} + \frac{\partial}{\partial y}[v^2 d + g d^2/2] = 0 \tag{1.6.11}$$

This result can also be obtained by multiplying (1.2.1) by d and using the continuity equation (1.2.2). Although it is formally invalid within the jump, (1.6.11) yields the correct matching condition when integrated across a discontinuity in depth. Numerical solutions of the shallow water equations based on the finite-difference method (e.g. Helfrich *et al.*, 1999) frequently use (1.6.11) in place of (1.2.1) since the resulting solutions better approximate the correct matching conditions when jumps and bores are present.

c. *Entrainment from an Overlying Layer*

The above discussion has assumed a single layer with a free upper surface, but most ocean and atmospheric applications will involve an overlying or underlying fluid with slightly different density. Experiments by Wilkinson and Wood

(1971) reveal the anatomy of a jump when the second fluid is relatively deep and inactive (Figure 1.6.4). The jump consists of two stages, an upstream region in which overlying fluid is entrained down into the moving layer, and a 'roller' region with a large counterclockwise eddy. The Froude number based on reduced gravity remains >1 in the entrainment region and jumps to below unity downstream of the roller. Entrainment is produced by shear instabilities at the interface between the two fluids. At the top of the roller, where the horizontal velocity is negative and the vertical shear is reduced relative to upstream values, entrainment is not observed. By traditional definition the entrainment region and the roller comprise the hydraulic jump, even though the entraining region may be much longer than the roller.

The presence of entrainment gives rise to a significant departure from the single-layer case considered earlier. One of the consequences is that for a given upstream state there is no unique downstream state. As demonstrated by Wilkinson and Wood, a range of downstream states may be found by varying the height h_m of an obstacle placed downstream of the jump (Figure 1.6.4). Lowering h_m causes the roller region to migrate downstream, lengthening the entraining regions and increasing the total amount of entrainment. For sufficiently small h_m the roller disappears and the jump consists entirely of a gradually deepening region of entrainment. This is the state of maximum entrainment. If h_m is increased, the roller moves upstream and eats up the entrainment region. For sufficiently large h_m the entrainment region disappears and the jump consists only of the roller. A further increase in h_m causes the roller to come into contact with the vertical wall beneath which lower layer fluid is injected. The jump at this point is said to be flooded. Photographs of the three cases (no roller, combination of entrainment region and roller, and flooded jump) are shown (Figure 1.6.5) for the Wilkinson and Wood experiment, an upside-down version of the scheme we have been discussing.

Entrainment gives rise to a lack of conservation of mass and volume flux in the lower layer. If E is the volume flux per unit width introduced into the lower layer

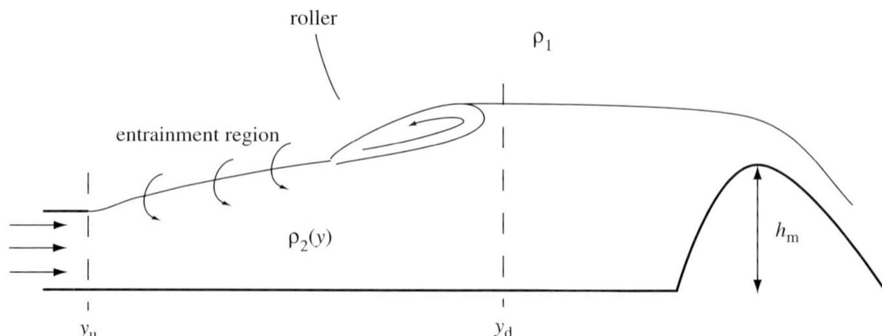

FIGURE 1.6.4. A schematic view of the two-fluid jump observed by Wilkinson and Wood (1971).

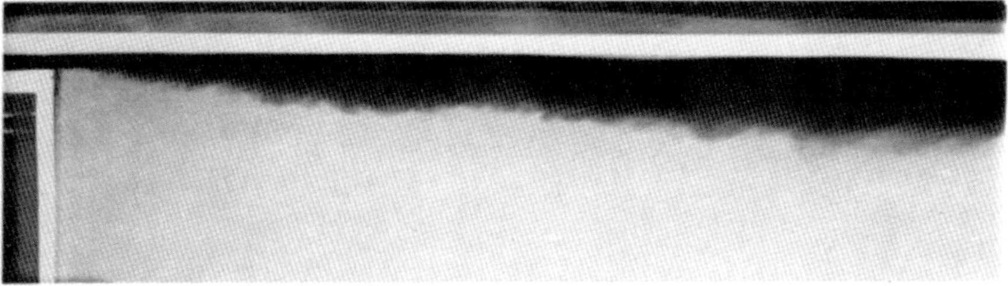

A density jump of the maximum entraining type with a free overfall downstream. Entrainment is occurring along the length of the jump.

. A density jump controlled by a broad crested weir downstream. Entrainment is occurring only at the far upstream end of the jump, the remainder consists of a roller zone.

Photograph of a flooded density jump.

FIGURE 1.6.5. Photographs of the laboratory experiment of Wilkinson and Woods (1971).

by entrainment, then the mass and volume budgets for the lower layer between sections immediately upstream and downstream of the jump (Figure 1.6.5) are

$$v_u d_u + E_n = v_d d_d \qquad (1.6.12)$$

and

$$\rho_u v_u d_u + E_n \rho_1 = \rho_d v_d d_d,$$

where ρ_1 is the upper layer density and ρ_u and ρ_d are upstream and downstream values of the lower layer density. Subtraction of the product of ρ_1 and the first equation from the second leads to

$$(\rho_u - \rho_1)v_u d_u = (\rho_d - \rho_1)v_d d_d,$$

which is often written in the form

$$g'_u v_u d_u = g'_d v_d d_d, \tag{1.6.13}$$

where $g'_u = g(\rho_u - \rho_1)/\rho_1$ and similarly for g'_d. The quantity $g'vd$ is called *buoyancy flux* and its conservation is a consequence of the conservation of mass for the two layers as a whole.

Further complicating the problem of shock joining is the fact that a horizontal pressure force, exerted by the overlying fluid, now exists on the upstream face of the roller and the top of the entraining region. However, the flow force for the two layers as a whole remains conserved provided that contributions from the bottom slope and frictional bottom drag are negligible. To find the total flow force, we assume that the upper layer is motionless, implying that the free surface ($z = D$) is level. Integrating the hydrostatic pressure over the whole depth of the layer then leads, after some rearrangement, to

$$\frac{\rho_u}{\rho_1} d_u v_u^2 + g'_u \frac{d_u^2}{2} = \frac{\rho_d}{\rho_1} d_d v_d^2 + g'_d \frac{d_d^2}{2}. \tag{1.6.14}$$

If the entrainment E_n is known, then (1.6.12–1.6.14) provide three relations allowing the downstream velocity, layer depth, and density to be calculated from their upstream values. Of course E_n is not known in advance nor, as shown by the experiment, can it be predicted solely on the basis of the upstream state. Some sort of downstream information, or an assumption about the downstream flow, must be made. An approach taken by Wilkinson and Wood (1971) is to assume that the downstream flow is hydraulically controlled by an obstacle of height h_m, as in the experiment. It is further assumed that no entrainment or dissipation occurs between the downstream section and the sill. Although two additional unknowns (the velocity and layer thickness at the sill) are introduced, there are three constraints. These include conservation of energy and volume flux between y_d and the sill, as well as the critical condition at the sill. For given h_m the entrainment can be calculated and the problem closes.

Although this last procedure is elegant, it is difficult to apply in geophysical settings due to the general lack of a clearly defined downstream obstacle or h_m value. Supercritical flows often spill out onto vast terrestrial or abyssal plains and the factors controlling the downstream layer thickness are complex. Alternatives to the Wilkinson and Wood procedure use turbulence closure assumptions to predict the energy dissipation or entrainment in the jump. The reader is referred to the work by Jiang and Smith (2001a, b) and Holland *et al.* (2002) and references contained therein.

d.   Form Drag

We close this section with a brief description of *form drag*, a property that is strongly influenced by the presence and location of a hydraulic jump. Consider a steady, free-surface flow over an obstacle in a channel of uniform width (Figure 1.6.6a). Integration of the steady flux form of the momentum equation (1.6.11) between sections a and b leads to

$$\left[v^2 d + g d^2/2\right]_a^b = -\int_a^b \left(g d \frac{\mathrm{d}h}{\mathrm{d}y}\right) \mathrm{d}y \qquad (1.6.15)$$

Thus the difference between the flow force at either end of the obstacle is equal to the integral of the horizontal component of bottom pressure gd over the obstacle. For the hydraulically controlled flow shown in the figure, the pressure on the upstream face of the obstacle is generally greater than on the downstream face and thus the obstacle exerts a net force on the flow in the upstream direction. This type of 'drag' requires no bottom friction or viscosity. If the hydraulic jump lies closer to the sill in the figures, the depth over the downstream part of the obstacle increases and the form drag is reduced. The maximum drag occurs when the jump is absent. When the flow has upstream or downstream symmetry with respect to the topography, such as in a purely subcritical or supercritical state, the form drag is zero.

The concept of form drag is most meaningful when the object in question is isolated. If the topography in the example begins at one elevation $h = 0$ but ends at another $h = h_o$ (Figure 1.6.6b), then even a resting fluid experiences a form drag as computed by the integral in (1.6.15). This difficulty can be removed by performing the integration between two sections of equal bottom elevation, say a' and b in the figure. However, the resulting form drag is more arbitrary and may or may not be of interest. Geophysical applications typically involve two-dimensional topographic variations, making it even more difficult to define isolated objects. If the channel axis bends substantially, the use of a linear (y-) momentum balance itself becomes less meaningful. Edwards *et al.* (2004) describe an example and suggest ways of dealing with some of these complications.

Exercises

(1) Derive the shock-joining conditions for a hydraulic jump in a channel with the same triangular cross section as that given in problem 4 of Section 1.4. The fluid is homogeneous and has a free surface.

(2) Consider a bore propagating in a homogeneous, free-surface flow with spatially and temporally varying velocity and depth. The speed of the bore is unsteady: $c_1 = c_1(t)$. Define a material volume V bounded by the free surface, the sidewalls of the channel, and by material fluid columns located at position $a(t) < y < b(t)$ as shown in Figure 1.6.2. Also, let y_u and y_d be fixed positions lying within the volume as shown in the figure.

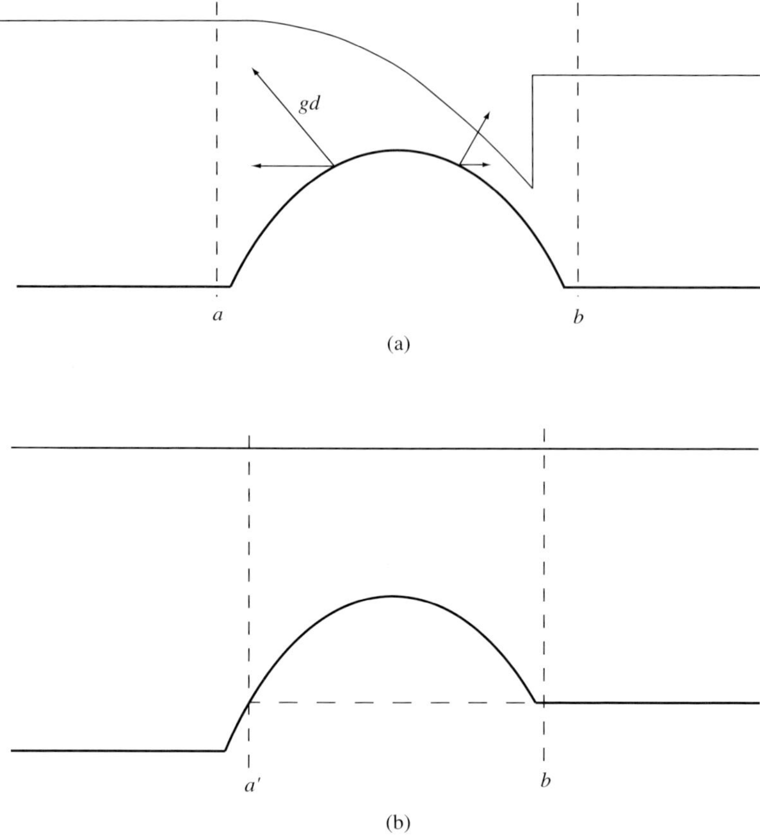

FIGURE 1.6.6. (a) The horizontal distribution of bottom pressure leading to form drag. The arrows show the normal and horizontal components of bottom pressure gd at two points of equal elevation. The resisting pressure force on the upstream face of the obstacle exceeds the enhancing force on the downstream face. (b) If the obstacle is not isolated, it will exert a net horizontal pressure force even if the fluid is at rest. This effect can be removed by relocating the upstream and downstream sections to lie at equal elevation, as indicated by a' and b.

(a) Show that

$$\frac{d}{dt}\iiint_V v\,dxdydz = w\left(\frac{d}{dt}\int_{a(t)}^{y_u}(dv)dy + \frac{d}{dt}\int_{y_u}^{y_d}(dv)dy + \frac{d}{dt}\int_{y_d}^{b(t)}(dv)dy\right)$$

(b) Note that the above equation also applies in a steadily translating frame of reference. Let the speed of translation be $c_1(0)$, so that the frame speed matches the bore speed at $t = 0$. By shrinking the distances between $a(t)$, y_u, y_d, and $b(t)$ to zero, show that at $t = 0$

$$\frac{d}{dt}\int_{a(t)}^{y_u}(dv)dy \to -v_u^2 d_u$$

$$\frac{d}{dt}\int_{y_u}^{y_d}(dv)dy \to 0$$

$$\frac{d}{dt}\int_{y_d}^{b(t)}(dv)dy \to v_d^2 d_d$$

where $v_u = (da/dt)_{t=0}$ and $v_d = (db/dt)_{t=0}$.

(c) By applying (1.6.10) and evaluating the forcing terms on the right-hand side using the hydrostatic pressure at $y = a(0)$ and $y = b(0)$, show that (1.6.3) is recovered. Note that (1.6.5) follows by transformation back to a rest frame.

(d) Perform the same series of operations starting with a primitive statement of mass conservation in order to recover (1.6.4) for an unsteady shock.

(3) Consider two sections a and b of a channel in which the width $w(y)$ and topographic elevation $h(x, y)$ are identical. Between the two sections the topography and width vary. Define a generalized form drag between a and b.

(4) *Form drag and energy dissipation.* Consider the situation shown in Figure 1.6.6a: a 2D flow over an obstacle with a hydraulic jump in the lee.

(a) Show that the form drag and energy dissipation between the sections a and b can be written in terms of the upstream Froude number and the ratio of downstream to upstream depths as

$$\tilde{D}_f = F_a^2 + \frac{1}{2} - F_a^2\left(\frac{d_b}{d_a}\right)^{-1} - \frac{1}{2}\left(\frac{d_b}{d_a}\right)^2 \qquad (1.6.16)$$

$$\tilde{E} = \frac{1}{2}F_a^2 + 1 - \frac{1}{2}F_a^2\left(\frac{d_b}{d_a}\right)^{-2} - \left(\frac{d_b}{d_a}\right) \qquad (1.6.17)$$

where

$$\tilde{D}_f = \frac{D_f}{gd_a^2}, \quad \tilde{E} = \frac{-\dot{E}}{gv_a d_a^2},$$

and where $F_a^2 = \frac{v_a^2}{gd_a}$ is the Froude number based on the upstream state.

(b) Show that the Jacobian $J_{F_u,(d_b/d_a)}(\tilde{E}, \tilde{D}_f)$ is nonzero and thus there is no direct functional relation between $\tilde{D}_f$ and $\tilde{E}$.

(c) However, for a fixed F_u, we know that the maximum values of $\tilde{D}_f$ and $\tilde{E}$ occur when the hydraulic jump lies right at the foot of the obstacle. As the jump is moved towards the sill and its amplitude decreases, so do $\tilde{D}_f$ and $\tilde{E}$. Show, in fact, that the maximum value of the Jacobian in (b) is numerically small (< 0.05) over the permissible range of F_a and d_b/d_a. (Note that $0 \le F_a \le 1$ and for each F_a there is a range $(d_b/d_a)_{\min} \le d_b/d_a \le 1$ corresponding to various positions of the hydraulic jump. This range varies from the sill, for which $d_b/d_a = 1$ to the foot of the obstacle, for which d_b/d_a has a minimum value.) Thus $\tilde{E}$ may be considered a function of $\tilde{D}_f$ to a first approximation. In fact, Ms. Christie Wood has shown that $\tilde{E} = 0.9744\tilde{D}_f + 0.8608\tilde{D}_f^2$ with in an error less than 0.016.

1.7.   Solution to the Initial-Value Problem

The shock-joining relations developed in the previous section make it possible to solve the initial-value problem posed by Long's experiment. The term 'solve' is used advisedly here for we do not actually calculate the evolving flow during its early development. Instead, we wait until the various transients have separated from one another, at which point the flow field consists of steady segments separated by isolated bores and rarefaction waves. The formal solution is thereby guided by the experiment. Piecing together the different steady segments of flow permits a solution to be constructed and, more importantly, allows the calculation of the obstacle heights required to initiate partial or total blockage or establishment of a hydraulic jump. The calculations herein are due to Long (1954, 1970), Houghton and Kasahara (1968) and Baines and Davies (1980).

Let us continue to view the problem as the adjustment to the sudden introduction of a stationary obstacle into a uniform stream. As noted in the previous section, permanent upstream effects (partial blockage) occur when the obstacle is sufficiently high that the initial flow has insufficient energy to ascend the crest or sill, at least according to a steady-state calculation. The critical obstacle height h_c is given by (1.6.1). Figure 1.7.1a shows the developing upstream flow for $h_m > h_c$. The flow state (v_0, d_0) far upstream, also the initial flow, is approached by a bore that moves at speed c_1 and establishes a new upstream state (v_a, d_a). Equations (1.6.4) and (1.6.5) can be used to link the two steady flows across the bore, leading to

$$(v_0 - c_1)d_0 = (v_a - c_1)d_a \tag{1.7.1}$$

and

$$(v_0 - c_1)^2 d_0 + g d_0^2/2 = (v_a - c_1)^2 d_a + g d_a^2/2. \tag{1.7.2}$$

In addition, conservation of energy and mass connect the sill flow with the steady flow immediately upstream of the obstacle according to

$$v_a d_a = v_c d_c \tag{1.7.3}$$

and

$$\frac{v_a^2}{2} + g d_a = \frac{v_c^2}{2} + g(d_c + h_m). \tag{1.7.4}$$

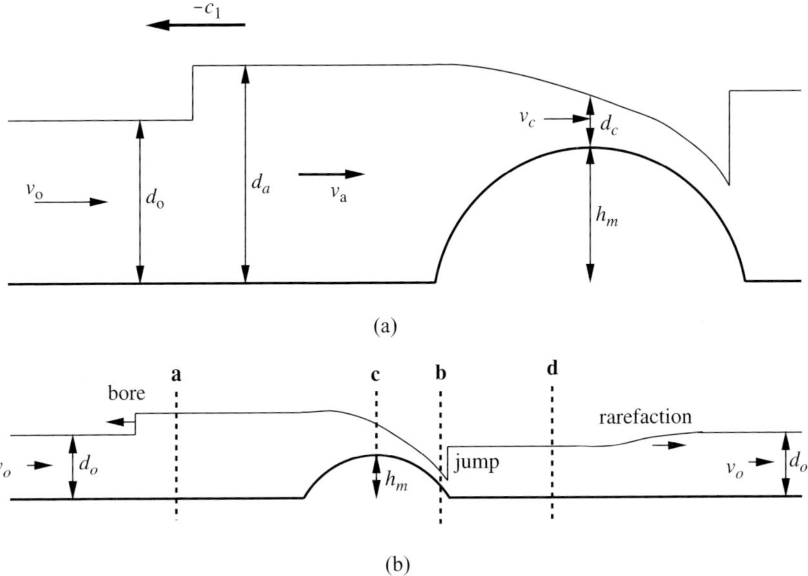

FIGURE 1.7.1. The various transients generated by the introduction of an obstacle into a uniform stream when h_m exceeds the critical value h_c for upstream influence.

Adding the condition that the sill flow is critical,

$$v_c = (gd_c)^{1/2}, \qquad (1.7.5)$$

results in five equations for the unknowns c_1, d_a, v_a, v_c, and d_c.

The locations of the different solution regimes can be plotted (Figure 1.7.2) in terms of the dimensionless obstacle height h_m/d_0 and initial Froude number F_0. The curve BAE gives the critical obstacle height h_c/d_0 in terms of F_0 and is determined by (1.6.1). To the left of this curve the obstacle is lower than the critical height and the steady flow established is completely supercritical or subcritical, depending on the initial Froude number. No upstream influence exists. To the right of this curve, upstream (and downstream) influence occurs and the flow adjusts to a hydraulically controlled steady state. As we have shown, the upstream influence takes the form of a bore that partially blocks the flow. Note that *any* bore that propagates upstream must decrease the volume transport, a property that can be deduced from conservation of mass (1.7.1) in the form:

$$v_a d_a = v_0 d_0 + c_1(d_a - d_0) \qquad (1.7.6)$$

Since $c_1 < 0$ and $d_a > d_0$ the final transport is less than the initial transport ($v_a d_a < v_0 d_0$) and we say that the flow is partially blocked. Various properties of the solution including the bore speed and the final transport can be obtained

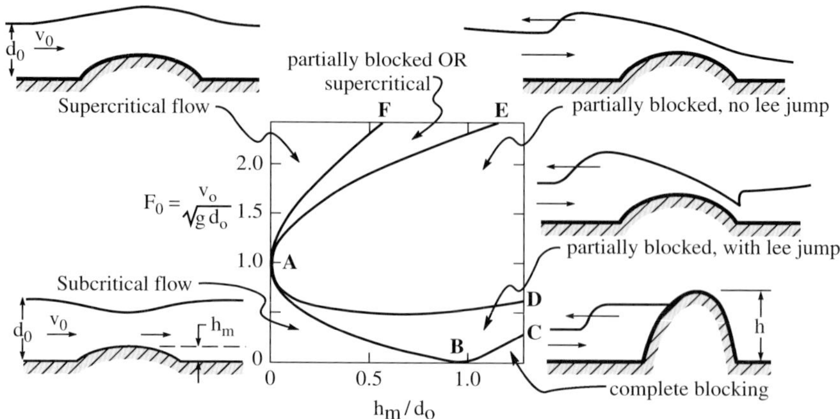

FIGURE 1.7.2. The various asymptotic regimes of the Long-type initial-value experiment in terms of the initial conditions. (From Baines, 1995).

by solving (1.7.1–1.7.5) and some of these properties are presented in Baines (1995, Figures 2.10 and 2.12).

Further to the right in the diagram, curve BC gives the value of h_m/d_0 needed to completely block the flow. The governing relation (see Exercise 1) is given by

$$F_0 = \left(\frac{h_m}{d_0} - 1\right)\left(\frac{1 + h_m/d_0}{2h_m/d_0}\right)^{1/2} \tag{1.7.7}$$

The wedge-shaped region EAF in Figure 1.7.2 represents special initial conditions for which two final steady states are possible, depending on how the experiment is carried out. Consider the curve AF, which indicates upstream values of F_d and h_m/d_0 where a stationary bore is possible in the flow approaching the obstacle. For these upstream conditions the steady flow near the obstacle can either be entirely supercritical, or have the stationary bore upstream of the obstacle leading to hydraulically controlled flow over the obstacle. The curve is obtained by setting $c_1 = 0$ in (1.7.1)–(1.7.5), resulting in

$$\frac{h_m}{d_0} = \frac{(8F_0^2 + 1)^{3/2} + 1}{16F_0^2} - \frac{1}{4} - \frac{3}{2}F_0^{2/3}.$$

If one performs the original version of Long's experiment in EAF, no upstream bore is found and the final steady state is the entirely supercritical flow, as in the upper left inset of Figure 1.7.2. The other alternative can be realized by starting with an obstacle of height $h_m > h_c$ (to the right of curve AE) and waiting until a hydraulically controlled flow is established. If the obstacle height is then gradually reduced to a value in the region EAF, the hydraulically controlled solution will persist.

A numerical demonstration of the implied hysteresis is shown in Figure 1.7.3. In frame (a) the obstacle of height $h_m > h_c$ is introduced, exciting an upstream bore. In (b) the obstacle has been lowered to a height $h_m < h_c$ such that h_m/d_o lies in region EAF. Here the bore continues to propagate upstream and the flow over the sill remains critical. Next the obstacle is lowered to point to the *left* of curve AF, causing the bore to reverse directions and move downstream towards the obstacle (c). Eventually the bore moves past the obstacle (d) and a supercritical state is achieved.

Finally, the curve AD (Figure 1.7.2) separates flows with and without hydraulic jumps attached to the downstream slope of the obstacle. For initial conditions lying below AD the jump would be positioned on the downstream slope of the obstacle. Above AD the jump would move downstream leaving supercritical

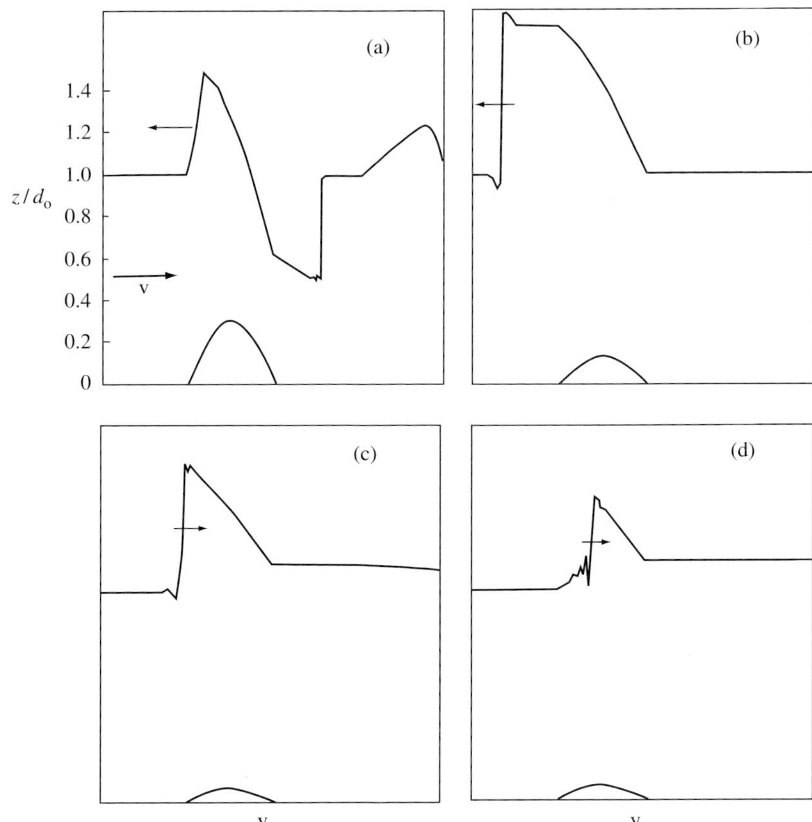

FIGURE 1.7.3. Frame *a* shows the evolution of a shallow stream when an obstacle of height h_m is introduced into a moving stream of depth d_o, such that the initial conditions lie to the right of curve AE in Figure 1.7.2. The obstacle height is then lowered (Frame *b*) so that h_m/d_o lies in region EAF. Later h_m/d_o is decreased so as to lie to the left of curve AF (*c* and *d*). (From Pratt, 1983a).

flow behind. On AD the hydraulic jump will become stationary at the foot of the obstacle, as shown in Figure 1.7.1b. In order to find the obstacle height at which this last situation occurs one must piece together the segments of steady flow shown at sections 'a', 'b', 'c' and 'd' in the figure. There are 10 unknowns, including the depths and velocities at these four sections, the upstream bore speed, and the obstacle height. Four constraints are provided by the shock-joining conditions across the bore and hydraulic jump. Also volume transport and energy (Bernoulli function) are conserved between sections 'a' and 'c' and between 'c' and 'b', providing 4 additional constraints. The final two constraints are provided by the condition of critical flow at the sill and the conservation of $R_- = v_0 - 2(gd_0)^{1/2}$ across the rarefaction wave that moves downstream of the

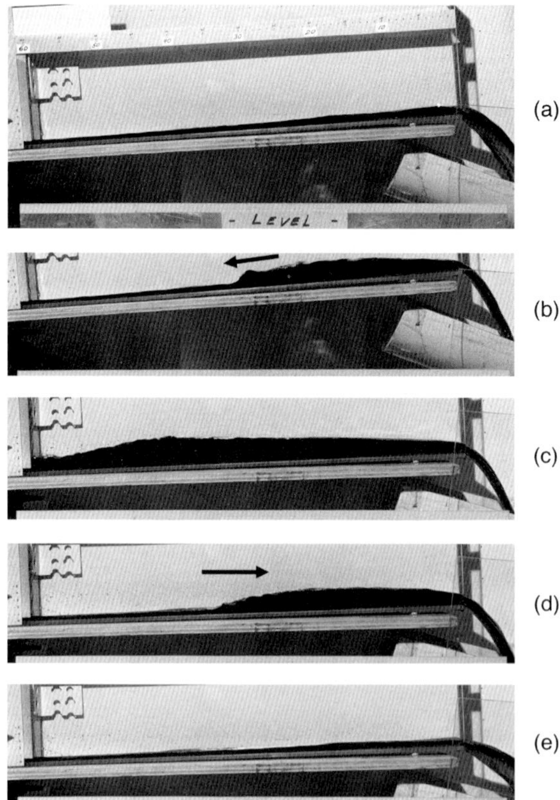

FIGURE 1.7.4. Photograph of dyed water flowing up a sloping channel and spilling out at the right-hand end. The water is fed by a sluice gate with $F_0 = 5.6$ from a reservoir on the left. (a) Supercritical flow with $h_m/d_0 = 7.9$. (b) A bore moving upstream trailed by subcritical flow $h_m/d_0 = 8.0$. (c) Subcritical flow in the entire channel $h_m/d_0 = 6.0$. (d) A bore moving downstream trailed by supercritical flow $h_m/d_0 = 4.4$. (e) Supercritical flow in the entire channel a few seconds later. (From Baines and Whitehead, 2003).

obstacle. The algebra involved in the determination of the obstacle height from these ten relations is formidable.

The same sequence of events is seen in a laboratory demonstration that directs a supercritical current up a sloping channel with an open end (Figure 1.7.4a). For a small channel slope, the current of dyed water remains in the supercritical state. Slowly tilting the channel to progressively greater slopes is equivalent to gradually increasing h_m/d_0 with F_0 constant. If the slope is increased to the point where $h_m > h_c$, a bore forms at the edge of the open end (Panel b). The bore propagates to the left, down the slope, and establishes subcritical flow in the channel with critically controlled flow at the exit (Panel c). If the slope is then gradually decreased, this subcritical state persists. Eventually the slope is reduced to the point where a bore forms at the source (Panel d). The bore moves to the right, up the slope, and reestablishes supercritical flow in the channel (Panel e). This experiment is easy to set up in the classroom. All that is required is a small, hand-held channel, and a system for circulating the water. The apparatus can be used to show that the stationary upstream jump predicted along the curve AD of Figure 1.7.2 is unstable but can be manually balanced in the sloping channel with a little practice.

Exercises

(1) Obtain equation 1.7.7 for the blocking height of the obstacle.

1.8.   Wave Reflections and Upstream Influence in Time-Dependent Flows

Although Long's experiment provides intuition into blocking and upstream influence, it does not tell us how these processes occur in the real ocean or atmosphere. There the heights of the sills are fixed and adjustments occur in response to temporally varying water mass formation and other time-variable forcing. To gain some perspective, consider a channel flow that is established and that is subject to slow variations in the upstream or downstream state. If the flow is hydraulically controlled, it will be immune to disturbances generated downstream of the controlling sill or narrows as long as those disturbances remain small, and we will therefore concentrate on disturbances generated upstream. How is hydraulic control manifested in such a situation? The guiding principle here is that *control* establishes a relationship between the parameters determining the upstream flow and those describing the channel geometry at the critical section. If we choose the flow rate Q and depth d_o to represent the upstream flow, and w is constant, then this relationship is given by (1.4.11), which links Q and d_o to the sill height h_c. In a laboratory experiment with fixed h_m, one would be free to vary Q alone or d_o alone, but not both.

It is natural to ask what would happen if the upstream flow were altered so as to violate this relationship. A numerical simulation along these lines begins with a steady, hydraulically controlled solution (Figure 1.8.1a). The upstream depth d_o is then increased to a new value d_1 (Figure 1.8.1b) creating a wave of elevation that approaches the obstacle from upstream. The new values of d_1 and Q (to the left of the wave) do not satisfy the relationship (1.4.11) required by critical control. The subsequent evolution is shown in Figures 1.8.1c and 1.8.1d. The incident wave strikes the obstacle generating a reflected wave that moves upstream and establishes a new steady flow of depth $d_2 > d_1$ and a new Q. These new values satisfy (1.4.11) and thus the reflection process re-establishes the essential relationship between the upstream variables.

If the initial flow in the above experiment is *not* hydraulically controlled, the outcome is quite different, as shown in Figure 1.8.2. Here the reflected wave is isolated and does not alter the new steady state established by the disturbance. Thus the final upstream depth is d_1 rather than d_2. In this case one is clearly free to vary the upstream parameters independently.

The above experiments shows how hydraulic control is exercised and suggests a means of distinguishing controlled from uncontrolled flows using data time series at a fixed instrument. Figure 1.8.3 shows the difference in the time-histories of d measured at a fixed location upstream of the obstacle. In the uncontrolled flow the reflected disturbance results in only a temporary change in d, while the controlled case gives a permanent change in d.

Some of these ideas can be exploited in order to parameterize the upstream effects of a sill or width contraction in a numerical model (Pratt and Chechelnitsky, 1997). The grid scale of such models is often too coarse to resolve the controlling topographic feature. Since the upstream effect of the sill or narrows is communicated by a reflected wave, it may be sufficient to know the reflection coefficient. For the incident wave shown in Figure 1.8.1b, which has

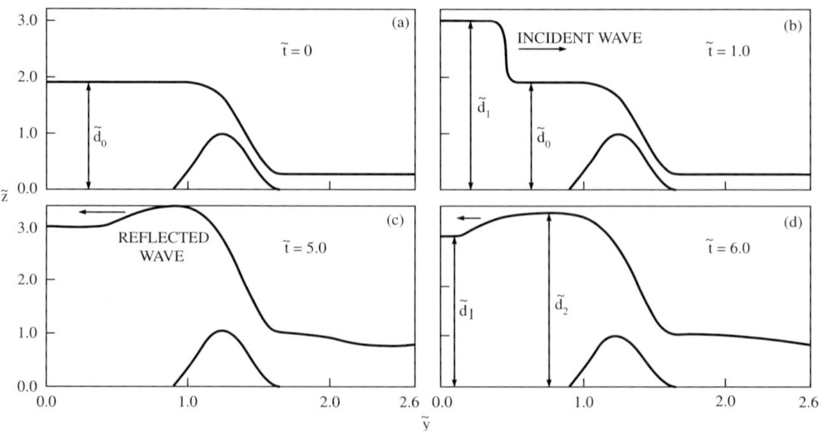

FIGURE 1.8.1. The wave reflection process for a hydraulically controlled flow. The nondimensional quantities shown are $(\tilde{d}, \tilde{z}) = (d, z)/h_o$, $\tilde{v} = v/\sqrt{gd_o}$ and $\tilde{t} = t\sqrt{gd_o}/L$ where d_o is the initial upstream depth, h_o is the height of the obstacle, and L is the obstacle length. (from Pratt 1984b).

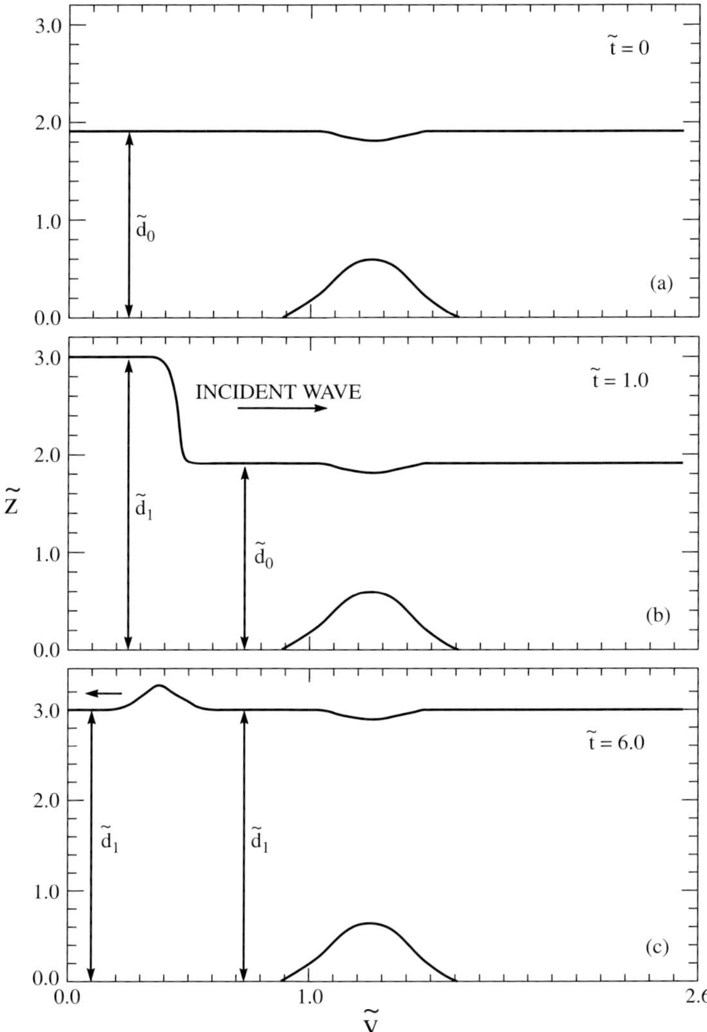

FIGURE 1.8.2. The wave reflection process for a purely subcritical flow. The notation is as in Figure 1.8.1. (from Pratt 1984b).

a height amplitude $d_1 - d_o$, the ultimate upstream depth d_2 established after wave reflection would be given by the reflection coefficient $(d_2 - d_o)(d_1 - d_o)$. Consider a linearized version of this problem in which the incident wave has the form

$$\eta_I = f_I(y - c_+ t), \quad \text{and}$$

$$v'_I = \left(\frac{g}{d_o}\right)^{1/2} f_I(y - c_+ t)$$

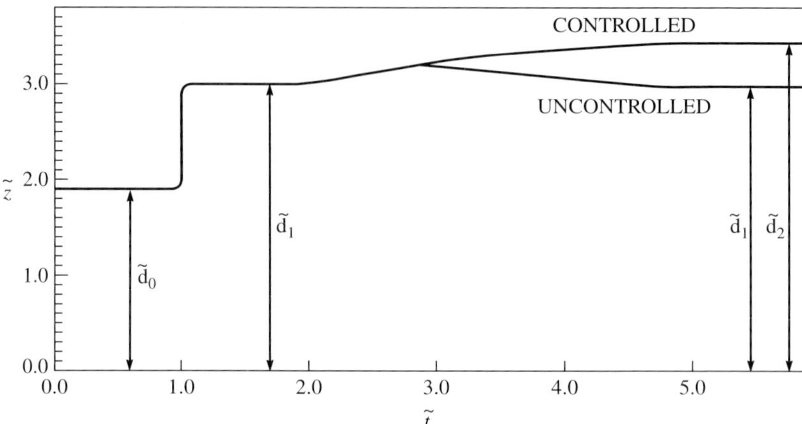

FIGURE 1.8.3. The time-history of the upstream surface level during the wave reflection process for the controlled flow of Figure 1.8.1 and the subcritical (uncontrolled) flow of Figure 1.8.2. (from Pratt 1984b).

where $c_+ = v_o + (gd_o)^{1/2}$ and v_o and d_o are the undisturbed upstream flow. The reflected wave is of the form

$$\eta_R = f_R(y - c_- t), \quad \text{and}$$

$$v'_R = -\left(\frac{g}{d_o}\right)^{1/2} f_R(y - c_- t),$$

with, $c_- = v_o - (gd_o)^{1/2}$.

If the wave length is much longer than the length of the obstacle, then the flow over the obstacle can be approximated as steady at any given instant. Thus the relationship (1.4.11) holds at any instant even though the flow itself is evolving in time. In the present context this relationship can be written

$$\frac{v^2(0, t)}{2} - \frac{3}{2}\left(\frac{gv(0, t)d(0, t)w(0)}{w_s}\right)^{2/3} + gd(0, t) = gh_m \tag{1.8.1}$$

where $v(0, t)$, $d(0, t)$ and $w(0)$ are the velocity, depth and width at the upstream edge (here $y = 0$) of the obstacle, h_m is the obstacle height and w_s is the minimum width (here assumed to coincide with the sill).

If the values $d_o + f_I(-c_+ t) + f_R(-c_- t)$ and $v_o + (g/d_o)^{1/2}(f_I(-c_+ t) - f_R(-c_- t))$ are substituted for $d(0, t)$ and $v(0, t)$ in (1.8.1) and the resulting equation linearized, it follows that

$$R_c = \frac{\eta_R(0, t)}{\eta_I(0, t)} = \frac{(1 + F_d)[1 - F_d^{1/3}(w(0)/w_s)^{2/3}]}{(1 - F_d)[1 + F_d^{1/3}(w(0)/w_s)^{2/3}]} \tag{1.8.2}$$

where $F_d = v_o/(gd_o)^{1/2}$.

In order to apply this relation, suppose that an upstream disturbance is created in which the free surface level is raised from the value d_o to $d_o + a$. The disturbance propagates downstream and eventually reaches the obstacle where it is reflected with amplitude aR_c, with R_c given by (1.8.2). The reflected disturbance travels upstream and established a new state with depth $d_o + aR_c$ and velocity $v_o - (g/d_0)^{1/2} aR_c$. This new state is guaranteed to satisfy the upstream conditions consistent with a hydraulically controlled flow, at least to $O(a/d_o)^2$.

Hydraulic control is often equated exclusively with regulation of the flow rate Q, but this this is an oversimplification. Suppose that the drain in a kitchen sink is closed and the faucet is left running, causing the sink to fill up and water to spill out onto the floor. At the lip of the sink the flow will be critical and the flow will therefore be hydraulically controlled. However Q in this case is set by the faucet and is independent of the height of the lip or sill. In this case it is the depth of water in the sink (d_0) that is controlled: Q and h_m are set and d_o is then determined by something like (1.4.11).

1.9.   Friction and Bottom Drag

Fluid viscosity and frictional drag have been tacitly ignored to this point, an omission that speaks more to the difficulty of including such effects than to their lack of importance. For example, the no-slip condition ($v = 0$) at the bottom of the channel ruins the possibility that the velocity v can be z-independent, or even x-independent if channel sidewalls are considered. The computation of bottom and sidewall viscous boundary layers generally requires numerical methods even when the flow is laminar. Most geophysical and engineering applications involve Reynolds numbers that are much larger than the $[O(10^3)]$ threshold required for turbulence. These difficulties have led civil engineers to parameterize the effects of friction through the use of drag laws that date back to the nineteenth century and were obtained through observations of the Mississippi River and various rivers in Europe (Chow, 1959). We will concentrate less on the empirical forms of drag used by engineers and more on the physics of the flow in the presence of friction. The main ideas discussed below are presented in detail by Pratt (1986), Garrett and Gerdes (2003), Garrett (2004) and Hogg and Hughes (2006).

Drag laws introduce a depth-averaged frictional stress into the y-momentum equations. The horizontal velocity v remains z- and x-independent as before. The most common drag law employed in oceanography and meteorology involves a body force in a direction opposite to the fluid motion and proportional to the square of the fluid velocity. The momentum equation (1.3.1) is replaced by

$$\frac{\partial v}{\partial t} + v \frac{\partial v}{\partial y} + g \frac{\partial d}{\partial y} = -g \frac{\partial h}{\partial y} - C_d \frac{v|v|}{d}, \tag{1.9.1}$$

where C_d is a dimensionless drag coefficient, nominally of order 10^{-3} in sea straits.

If the flow is steady, a solution can be found by integrating (1.9.1) from an upstream point y_o of known velocity and depth to the point y where the solution is desired. The result of this integration can be written

$$\frac{Q^2}{2d(y)^2 w(y)^2} + g[d(y) + h(y)] = \frac{v(y_o)^2}{2} + g[d(y_o) + h(y_o)]$$

$$-C_d \int_{y_o}^{y} \frac{v(y')|v(y')|}{d(y')} dy' \qquad (1.9.2)$$

The continuity relation $Q = v(y)d(y)w(y) = v(y_o)d(y_o)w(y_o)$ has been used to replace $v(y)$ on the left-hand side. The presence of the integral means that the flow state at y depends on the entire history of the flow between y_o and y, and not just the values of the geometric variables h and w at y. The nonlocal nature of the relationship between the flow and the topography means that (1.9.2) is not of the form sought by Gill (1977) in his generalization of governing relations.

In view of the failure of Gill's formalism, we might ask whether any of the concepts we have developed, including subcritical and supercritical flow, hydraulic control and the like, have any meaning or importance when friction is present. Some insight into this question can be gained by writing (1.9.1) and the continuity equation (1.3.1) in characteristic form. Following the method established in Appendix B, the characteristic equations are

$$\frac{d_\pm R_\pm}{dt} = -g\frac{dh}{dy} - C_d\frac{v|v|}{d} \mp \frac{(gd)^{1/2}v}{w}\frac{dw}{dy} \qquad (1.9.3)$$

where

$$\frac{d_\pm}{dt} = \frac{\partial}{\partial t} + c_\pm \frac{\partial}{\partial y},$$

$R_\pm = v \pm 2(gd)^{1/2}$, and $c_\pm = v \pm (gd)^{1/2}$ as usual. Solutions to initial value problems can be constructed by integrating (1.9.3) along characteristic curves given by $dy_\pm/dt = c_\pm$, just as described in Section 1.3. Although the Riemann functions $R_\pm$ are not conserved, the characteristic curves still represent paths along which information travels. The characteristic speeds $c_\pm$ continue to represent speeds at which information travels and it therefore remains meaningful to classify the flow as being critical, supercritical, or subcritical according as $v - (gd)^{1/2} > 0, = 0$, or < 0. This reasoning falls apart if the frictional term involves derivatives of the flow variables in the y-direction.

A geometrical constraint on the location of a critical section in a steady flow can be found by dividing the steady form of (1.9.3) for R_- by c_-, leading to

$$\frac{\partial R_-}{\partial y} = \frac{-g\dfrac{dh}{dy} - C_d\dfrac{v|v|}{d} + \dfrac{(gd)^{1/2}v}{w}\dfrac{dw}{dy}}{c_-}.$$

The existence of a well-behaved solution at a critical section requires that the denominator vanish, and therefore

$$-(dh/dy)_c - C_d + v_c|v_c|(gw_c)^{-1}(dw/dy)_c = 0, \qquad (1.9.4)$$

where the subscript 'c' indicates evaluation at the critical section. If w is constant, (1.9.4) reduces to the simple condition that the critical section must lie where the bottom slope equals the negative of the drag coefficient. Friction therefore tends to shift the control section from the sill to a point downstream. If the bottom is horizontal and only the width varies, then critical flow must occur where the channel widens ($dw/dy > 0$).

Some indication of the importance of friction can be gained by comparing the drag and advective terms in (1.9.1). For flow with characteristic depth D passing over an obstacle or through a contraction with y-length L,

$$\frac{C_d \dfrac{v|v|}{d}}{v \dfrac{\partial v}{\partial y}} \approx O(C_d L/D)$$

and thus friction is significant when $C_d L/D = O(1)$. Friction is typically ignored in simple models of deep ocean overflows and it is an embarrassing fact that estimates of $C_d L/D$ for these flows often exceed unity, even when conservative values of C_d are used. The accompanying table contains some examples.

Table of values of $C_d L/H$ for 9 oceanographically important straits. L is the strait length, D is the average thickness of the overflowing layer, and C_d is assigned the conservative value 10^{-3}.

Sea Strait	$D(m)$	$L(m)$	$C_d L/H$
Strait of Gibraltar Outflow*	2×10^2	$(2-5) \times 10^4$	$0.1 - 0.3$
Vema Channel	3×10^2	2×10^5	0.7
Bornholm Strait	30	2.5×10^5	0.8
Bab al Mandab Outflow	10^2	1.5×10^5	4.5
Denmark Strait	5×10^2	5×10^5	1.0
Ecuador Trench	3×10^2	3×10^5	1.0
Faroe Bank Channel	3×10^2	6×10^5	2.0
Bosporus	20	2×10^4	1.0

* Depends on how the strait proper is defined.

Bottom drag can lead to some interesting departures from the steady behavior we have previously discussed. Some of these changes are evident in Figures 1.9.1a, b, which give a comparison between two sets of steady solutions, the first with $C_d = 0$ and the second with $C_d > 0$. Each solution has the same volume flux and the channel width is constant. Solutions are obtained by choosing y_0 as the upstream edge of the obstacle, specifying the value B of the Bernoulli

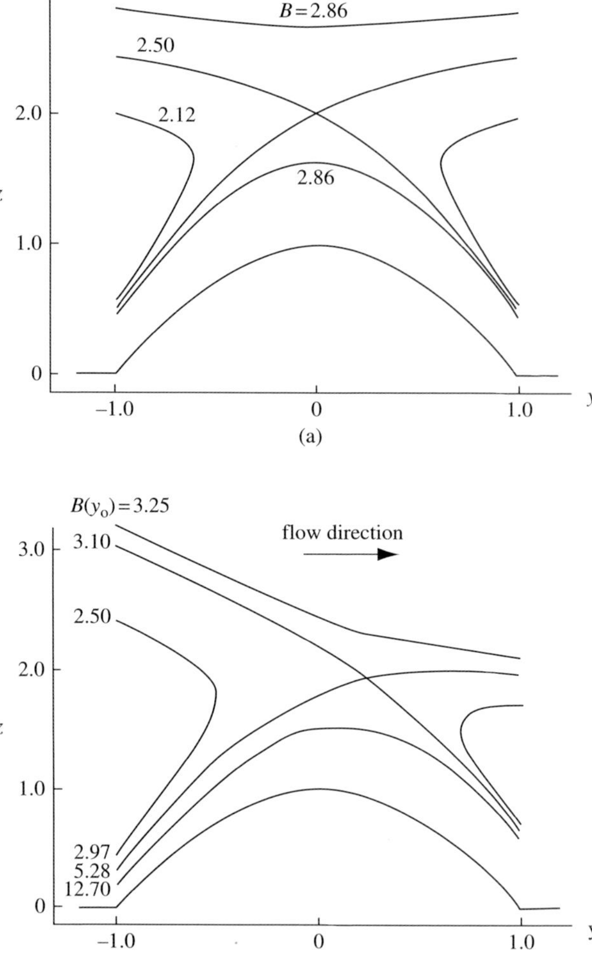

FIGURE 1.9.1. Steady solutions for flow over an obstacle with height h_m with constant volume flux $(Q/gh_m^{3/2}w = 1)$ and various values of the Bernoulli function B/gh_m. The solutions in (a) have no bottom drag whereas those in (b) have a drag equivalent to $C_dL/h_m = 0.5$. (From Pratt, 1986).

function there, and solving (1.9.1) for the fluid depth at successively larger values of y. Each curve is labeled with the nondimensional upstream value of B. The family of solutions with finite drag has a subcritical-to-supercritical and a supercritical-to-subcritical flow. The flow is critical where the two curves cross each other and, as suggested above, this point lies downstream of the sill. Purely subcritical and supercritical solutions also exist, but these no longer have the upstream/downstream symmetry of their inviscid counterparts. Note that the subcritical solution suffers a reduction in depth as it passes the obstacle, creating the impression of fluid spilling over the sill. The reduction in depth is a consequence of the loss of energy that the fluid experiences as it crosses the topography. Under subcritical conditions the Bernoulli function is dominated by the potential energy $g(d+h)$ and thus a significant depletion of energy must come at the cost of potential energy. The spilling character that a subcritical flow can take on when bottom drag is significant can lead one to mistake the solution for a hydraulically controlled flow.

Some channels contain flow that remains subcritical throughout and evolves mainly due to frictional processes. In fact, a large drag coefficient or sufficiently weak variation in channel geometry may preclude (1.9.4) from ever being satisfied. A simple example would be a constant-width channel in which the maximum negative value of the bottom slope is less than C_d. Such cases are sometimes referred to as being *frictionally controlled*, though the term 'control' in this context is ambiguous. Simple models of such flows assume that the channel cross-section and elevation are uniform, in which case analytical solutions may be found. An example is presented in Exercise 1.

Another case that can be analyzed simply is that of flow down a uniform slope $dh/dy = -S$ in a channel of constant width. A useful relation governing the Froude number of such a flow is

$$\frac{\partial F_d^{\,2}}{\partial y} = \frac{3F_d^{\,2}(S - C_d F_d^{\,2})}{(F_d^{\,2} - 1)d}, \tag{1.9.5}$$

which can be derived from (1.9.1) and the continuity equation. It can be seen that any positive S will support a uniform $(\partial/\partial y = 0)$ flow, and that the Froude number of this flow is given by $F_d^{\,2} = S/C_d$. The uniform flow is critical when $S = C_d$, in agreement with (1.9.4).

Suppose that the $S < C_d$, so that the uniform flow is subcritical (Figure 1.9.2a). Then suppose that the flow at some y is perturbed by causing $F_d^{\,2}$ to decrease slightly below the value $F_d^{\,2} = S/C_d$. The right-hand side of (1.9.5) now becomes negative, requiring that $F_d^{\,2}$ further diminish in the downstream direction. It is easily shown, in fact, that $F_d^{\,2}$ decreases to zero as $y \to \infty$, so that the fluid becomes infinitely deep and stagnant. If the perturbation instead consists of an increase in $F_d^{\,2}$, then the right-hand side of (1.9.5) becomes positive and the $F_d^{\,2}$ increases in the downstream direction. At the point where $F_d^{\,2}$ reaches unity, $\partial F_d^{\,2}/\partial y \to \infty$ and the solution cannot be continued further. The key feature in either case is that uniform subcritical solution is unstable. It is left as an exercise for the reader to argue that a supercritical uniform flow $(S < C_d)$ is stable in the sense that a steady perturbation will diminish in amplitude in the downstream direction (Figure 1.9.2b). Note, however, that the supercritical solution can be unstable to time-dependent perturbations, resulting in a phenomenon known as *roll waves*. Baines (1995) reviews this topic.

It is possible to move beyond the 'slab', in which the bottom drag is distributed equally over the otherwise inviscid water column, to a more realistic setting with vertical shear. The assumption of gradual variations in y is maintained and

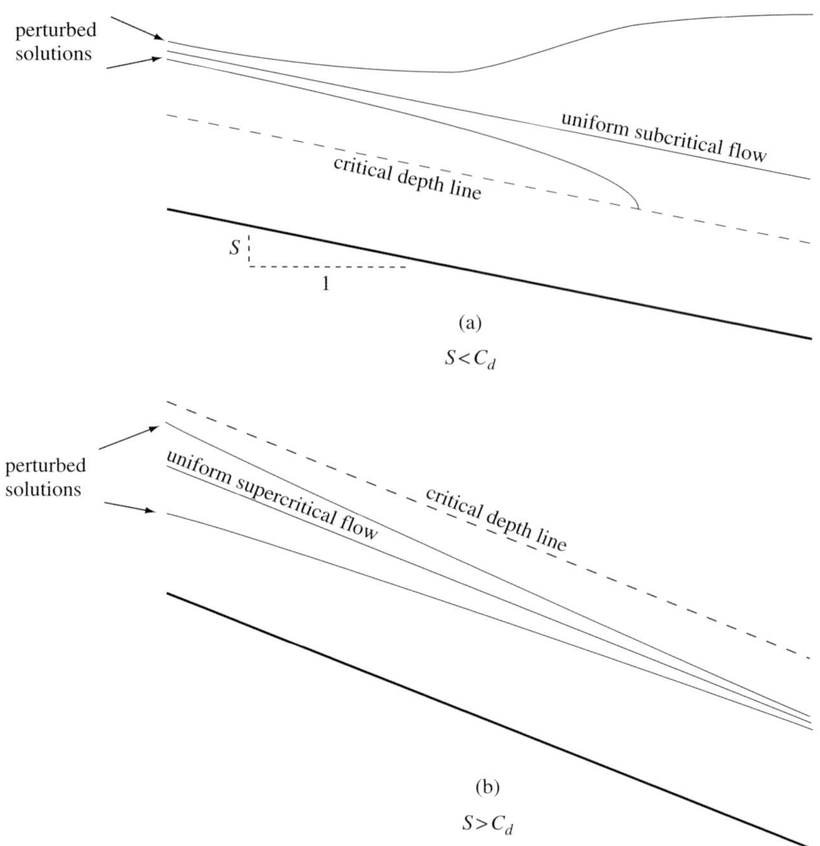

FIGURE 1.9.2. The stability of uniform flow down a constant slope. In (a), $S < C_d$ and so the uniform flow is subcritical. The critical depth for a flow with the same volume flux is indicated by the dashed line. If the solution is perturbed at some upstream point, the flow will depart from the uniform state and tend towards a deep quiescent state or towards the critical depth (thinner curves). The free surface slope becomes infinite when the critical depth is reached. In (b), $S > C_d$ and the uniform flow is therefore supercritical. Steady perturbations decay in the downstream direction, thought the flow may still be unstable to roll waves.

thus the pressure remains hydrostatic, but now vertical shear is allowed. The horizontal momentum equation becomes

$$v\frac{\partial v}{\partial y} + w\frac{\partial v}{\partial z} = -g\frac{\partial d}{\partial y} - g\frac{\partial h}{\partial y} + \frac{\partial \tau}{\partial z},\qquad(1.9.6)$$

where τ is the horizontal shear stress per unit mass. The local condition of incompressibility

$$\frac{\partial v}{\partial y} + \frac{\partial w}{\partial z} = 0$$

implies the existence of a streamfunction ψ such that $\partial\psi/\partial y = -w$ [not to be confused with width] and $\partial\psi/\partial z = v$.

It is possible to express (1.9.6) in the form

$$\frac{\partial B(\psi, y)}{\partial y} = \frac{\partial \tau}{\partial z},\tag{1.9.7}$$

as described in Exercise 3. The Bernoulli function $B(\psi, y) = \frac{1}{2}v^2 + gd + gh$ now varies throughout the fluid, though it is conserved along streamlines if the frictional term on the right-hand side is absent. Following Garrett (2004) we may attempt to formulate a Gill type functional for the flow beginning with the trivial relation

$$d = \int_h^{h+d} dz = \int_0^Q \frac{d\psi}{v},$$

where we have assumed the boundary conditions $\psi = (0, Q)$ at $z = (h, d+h)$. Use of the definition of the Bernoulli function to substitute for v allows this relation to be expressed as

$$d - \int_0^Q \frac{d\psi}{2^{1/2}[B(\psi, y) - gd - gh]^{1/2}} = 0.\tag{1.9.8}$$

If the fluid is inviscid, B is a function of ψ alone and may be prescribed by the upstream conditions. Under this condition the only remaining dependent variable is the depth d and the right-hand side is of the desired form. Setting its derivative with respect to d to zero leads to

$$1 = \int_0^Q g\frac{d\psi}{2^{3/2}[B(\psi, y) - gd - gh]^{3/2}} = \int_0^Q g\frac{d\psi}{v^3} = \int_h^{h+d} g\frac{dz}{v^2}$$

and thus the average over the water column of the square of the inverse Froude number must be unity for the flow to be hydraulically critical

$$\frac{1}{d}\int_h^{h+d}\left(\frac{gd}{v^2}\right) dz = 1.\tag{1.9.9}$$

A remarkable aspect of this condition is that it apparently applies to any stationary wave, including waves that propagate on the vertical vorticity gradients in the flow. However, the coordinate transformation that makes the derivation possible assumes a one-to-one relationship between ψ and z, and this holds only when v does not change sign. It is possible that critical conditions with respect to certain wave modes require reversals in the background flow.

The introduction of frictional dissipation means that B varies along streamlines and can no longer be prescribed by conditions far upstream. Because of the unknown y-dependence in $B(\psi, y)$, the left side of (1.9.8) no longer fits Gill's definition of a hydraulic function. However we may still use this relation to formulate a critical condition, provided that the dissipation takes a particular form. Consider a hypothetical flow over varying topography that becomes critical at a particular section $y = y_c$. Criticality specifically means that the flow at $y = y_c$ can support a stationary, infinitesimal disturbance and that this disturbance can exist only at $y = y_c$. This definition is consistent with the inviscid examples considered elsewhere in this chapter, but it has yet to be shown that the postulated state is dynamically consistent in the presence of dissipation. In order for it to be so, the disturbance at $y = y_c$ must clearly be isolated and cannot contaminate the flow upstream. This assumption can be supported if the dissipation depends on the local properties of the flow at y_c and not, say, on the derivatives of the flow fields with respect to y. Thus if $\partial \tau / \partial z$ in (1.9.6) takes the form $v \partial^2 v / \partial z^2$, where v is a molecular viscosity, the assumption is justified. In this case the disturbed flow at $y = y_c$ has the same $B(\psi, y_c)$ as the undisturbed flow, the latter being set by conditions occurring in $y < y_c$ where the disturbance is not present. The stationary wave at y_c then involves a perturbation in d that satisfies (1.9.6) for a fixed $B(\psi, y_c)$. The critical condition in this case is therefore identical to the inviscid condition (1.9.9). On the other hand, a dissipation form that contains derivatives in y or otherwise gives rise to nonlocal influences may invalidate the assumptions. We will proceed on the assumption that this is not the case.

A compatibility condition for critical flow may be derived by differentiating (1.9.8) with respect to y and applying the result at a critical section. The result can be written

$$\int_h^{h+d} \frac{\partial \tau}{\partial z} v^{-2} dz - \frac{dh}{dy} = 0.$$

after application of (1.9.7) and (1.9.9). Integration by parts of the first term leads to

$$\frac{dh}{dy} = \left[\frac{\tau}{v^2} \right]_{z=h+d} - \left[\frac{\tau}{v^2} \right]_{z=h} + 2 \int_h^{h+d} \frac{\tau}{v^3} \frac{\partial v}{\partial z} dz.$$

If the stress at the free surface is zero, the first term on the right-hand side vanishes. The bottom stress term is simply what is parameterized by the drag coefficient C_d in slab models. The expression $\tau \partial v / \partial z$ may be regarded as the internal rate of energy dissipation and is denoted by ε. With these substitutions

$$\frac{dh}{dy} = -C_d + 2 \int_h^{h+d} \frac{\varepsilon}{v^3} dz. \tag{1.9.10}$$

It follows that the action of bottom drag alone causes the control section to lie where the bottom slope is the negative of the drag coefficient, as in a slab model. However, internal dissipation gives rise to the opposite tendency.

Hogg and Hughes (2006) have calculated numerical solutions for free surface flows with constant molecular viscosity and an example is shown in Figure 1.9.3. The usual no-slip boundary condition at the bottom is replaced by specification of the bottom stress in the form

$$\tau_{z=h} = C_d v_{z=h}^2. \tag{1.9.11}$$

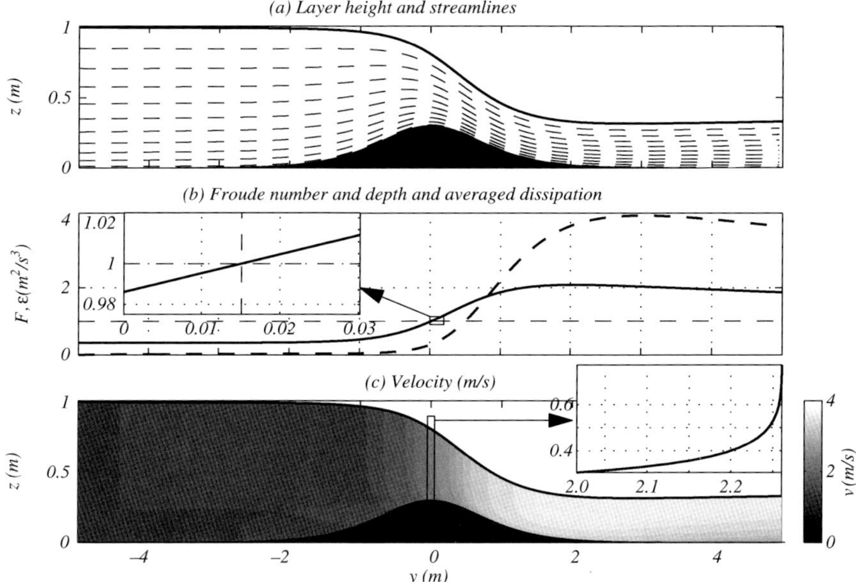

FIGURE 1.9.3. Numerical solution for a viscous free-surface flow over an isolated obstacle with $\nu = 10^{-2} m^2/s$, $C_d = 10^{-2}$ and uniform upstream velocity. Streamlines are shown in (a) while the Froude number (right-hand term in 1.9.9, solid line) and depth average internal dissipation ε (dashed line) are shown in (b). The inset shows the Froude number in the vicinity of the critical section. Panel (c) shows the velocity v and, in the inset, the velocity profile at the critical section. (from Hogg and Hughes, 2006).

The fluid is therefore free to slip over the bottom with horizontal velocity $v_{z=h}$ and the drag coefficient and molecular viscosity are specified independently. This artificial setting is concession to more realistic applications in which the viscosity is a parameterization of turbulence and where the exact form of the bottom boundary condition is unknown. The numerical solution shown has uniform velocity upstream of the obstacle and has the appearance of an inviscid, hydraulically controlled flow (panel a of Figure 1.9.3). The flow passes through a critical section at a point slightly downstream of the sill where the left-hand side of (1.9.9), which can be interpreted as a generalized Froude number, passes though unity (solid curve in b). The velocity field and the velocity profile at the control section (c) shows the development of vertical shear as the fluid spills over the sill. The development of shear leads to higher rates of depth-averaged internal dissipation (dashed line in b).

An illuminating exercise in assessing the validity of slab models is to fix the drag coefficient, vary the viscosity, and note the behavior of the resulting velocity profiles. If the upstream conditions are fixed as in the previous experiment, C_d is held fixed at value 10^{-2}, and ν is varied over six decades, a set of differing critical-section velocity profiles is obtained (Figure 1.9.4). For small viscosity, the shear is concentrated in a thin bottom boundary layer (a). As ν is increased the boundary layer grows (b) and the shear becomes distributed over the whole depth (c and d).

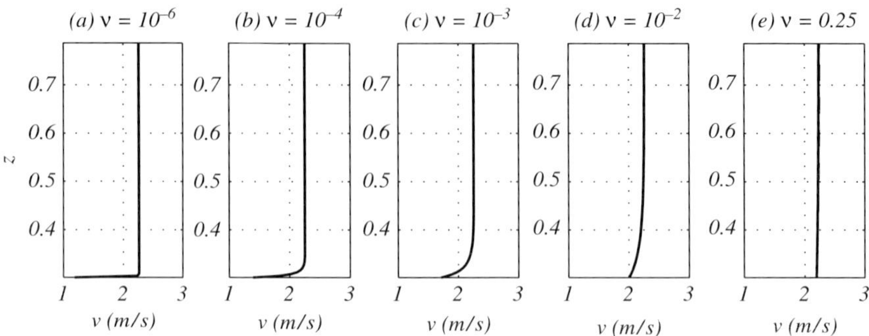

FIGURE 1.9.4. A sequence of velocity profiles at the critical section and obtained from numerical experiments of the type shown in Figure 1.9.3. The upstream conditions and the drag coefficient $C_d = 10^{-2}$ are fixed. The viscosity is varied as indicated in each frame. (from Hogg and Hughes, 2006).

Even larger values of v smooth the velocity over the whole water column leading to a depth-independent profile (e). The flow is therefore slab-like in the limits of low and high viscosity. Hogg and Hughes also find that the position of the control is generally dominated by the bottom drag term in (1.9.10).

Exercises

(1) For steady flow in a channel with constant h and w, show that bottom friction causes the flow to evolve in the downstream direction towards criticality.

(2) Consider a strait with constant w and h connecting two infinitely wide reservoirs. The flow in the strait is subcritical and subject to quadratic bottom drag but no entrainment.

    (a) Assuming that the strait extends from $y = 0$ to $y = L$, find a general algebraic expression relating the depth d to the position y. Calculate the drop in the level of the surface (or interface) between the ends of the strait as a function of $d(0)$ and the transport $Q > 0$.

    (b) Show that the only possible location for critical flow must be at the right end ($y = L$) of the strait, where w changes from a finite value to infinity.

    (c) Find the solution that is critical at $y = L$ and sketch the profile of the interface through the strait. (Note that the surface slope becomes infinite as y approaches L.)

    (Further discussion and an application of this procedure to two-layer flow can found in Assaf and Hecht, 1974.)

(3) For the vertically sheared flow described by equation (1.9.6) suppose that the variables v and w are expressed in terms of the coordinates ψ and y (rather than z and y). By transforming the right-hand side to the new variables, show that (1.9.7) holds.

1.10.   Entrainment

When a dense layer of fluid spills over a sill and accelerates, the shear between the moving layer and the overlying fluid increases. These conditions favor the formation of shear instabilities and turbulent mixing about the interface. Turbulence in the bottom boundary layer can also intensify and sometimes penetrate up to the interface. In cases where the turbulence is localized near the level of the interface, the mixing gives rise to an intermediate region whose thickness increases with downstream distance. A numerical simulation of this process is shown in Figure 1.10.1. The flow upstream of the obstacle consists of homogeneous upper and lower layers separated by a thin region of rapidly varying stratification (upper panel). The upper layer is relatively quiescent but the lower layer flows towards the obstacle and spills over the sill in a familiar way. Downstream of the sill the interfacial region thickens and eventually spreads to the bottom. The streamlines in the upper fluid (bottom panel) suggest a slow subduction of fluid into the dense overflow.

Following the ideas discussed by Gerdes *et al.* (2002), it is possible to formulate a model for the lower layer that incorporates turbulent mixing but retains the layer formalism. Consider an idealization of the intermediate region as

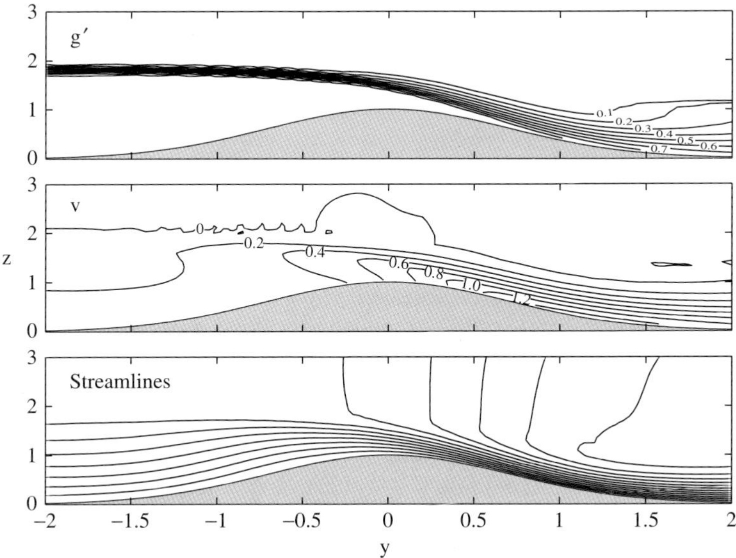

FIGURE 1.10.1. Continuously stratified exchange flow as computed by a nonhydrostatic, two-dimensional model. The upper panel shows the stratification in terms of g'/g'_0, where $g' = g(\rho(y, z) - \rho_1)/\rho_1$, ρ_1 is the density of the overlying fluid, and g'_0 is the upstream value of g' based on the two homogeneous layers there. The middle panel shows contours of horizontal velocity while the lower panel shows the streamlines. Other scales include the half-length L of the obstacle and the height h_m of the obstacle. (From Nielsen, 2004).

typically produced in laboratory experiments (Figure 1.10.2). Two homogeneous layers of different density and velocity are brought into contact at the left-hand boundary. Mixing between the two results in the formation of a wedge-like intermediate region. Suppose that all of the fluid lying below the upper boundary of the wedge is treated as a single layer. Then the effect of mixing is to cause upper layer fluid to be *entrained* into this lower layer. Should the interface be defined to coincide with the lower boundary of the wedge, the lower layer fluid would be *detrained*. In the first scenario, the mass flux in the lower layer increases with downstream distance; in the second scenario it decreases. The loss or gain of fluid by a particular layer can be accounted for by introducing an *entrainment velocity* w_e that is normal to the interface and that carries fluid parcels, and the properties of those parcels, across the interface. Since the shallow water model assumes the interface to be nearly horizontal, the entrainment velocity is nearly vertical and will be approximated as such. We will concentrate on the process of entrainment as depicted in Figure 1.10.1. By convention, w_e is positive in the direction of entrainment, here downwards.

A second assumption required to retain the layer-model formalism is that the entrained mass and momentum are instantly mixed all the way to bottom, so that the lower layer density and velocity depend only on y (Figure 1.10.2b). The resulting 'slab' model is most convincing when the interfacial or bottom-generated turbulent eddies are comparable in size to the layer thickness. With these idealizations, the equations of volume and mass conservation for a one-dimensional lower layer are

$$\frac{\partial(v_2 d_2)}{\partial y} = w_e \qquad (1.10.1)$$

and

$$\frac{\partial(\rho_2 v_2 d_2)}{\partial y} = w_e \rho_1. \qquad (1.10.2)$$

The subscripts 1 and 2 denote the upper and lower layer, respectively, and the width of the channel is assumed constant. If the first equation is multiplied by ρ_1 and subtracted from the second equation, it follows that $(\rho_2(y) - \rho_1)v_2 d_2$ is independent of y. A more common form for this quantity is the *buoyancy flux* $g' v_2 d_2$, where $g'(y) = (\rho_2(y) - \rho_1)/\rho_1$.

The entrainment process carries momentum from the overlying fluid into the lower layer and consequences for the lower layer momentum equation must be dealt with carefully. Consider a control volume drawn about a column of lower layer fluid extending from the bottom to the interface, as shown in Figure 1.10.3. The sum of the horizontal forces acting on the four faces of the volume must equal the sum of the fluxes of horizontal momentum into the volume. Thus

$$\int_h^{h+d_2} (\rho_2 v_2^2 + p_2)\big|_{y+dy} dz - \int_h^{h+d_2} (\rho_2 v_2^2 + p_2)\big|_y dz$$

$$\cong p\big|_{h+d} \frac{\partial(h+d_2)}{\partial y} dy - p\big|_h \frac{\partial h}{\partial y} dy + w_e \rho_1 v_1 dy.$$

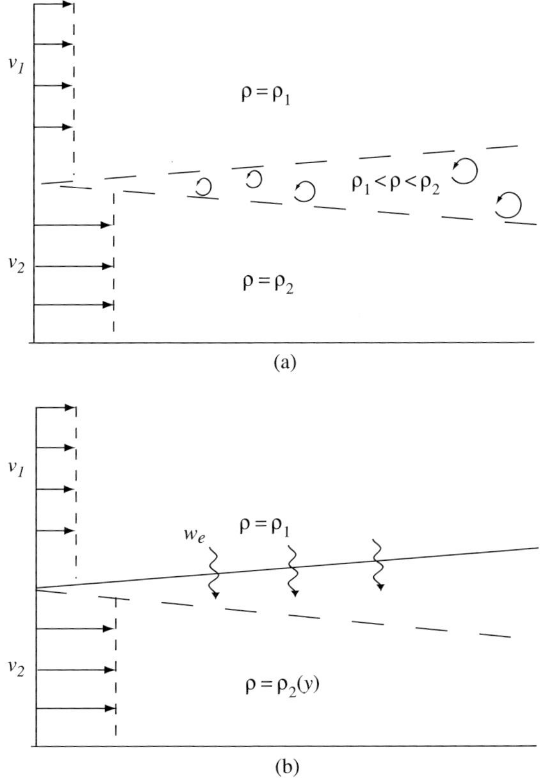

FIGURE 1.10.2. (a): The intermediate layer formed due to interfacial instability and mixing between two layers of different density moving at different speeds. (b): An idealization of the flow in which all fluid below the top of the mixing wedge in (a) is considered to be a single layer and where the transfer of mass into that layer is represented by an entrainment velocity w_e.

The terms on the left hand side are the integrals over the left and right control volume faces of the momentum flux and pressure force. The first two terms on the right-hand side represent first-order approximations to the horizontal component of the pressure force exerted at the top and bottom surfaces of the control volume. The final term represents the flux of horizontal momentum across the interface by the entrainment velocity. Dividing the above relation by dy and taking $dy \to 0$ results in the differential relation

$$\frac{\partial}{\partial y} \int_h^{h+d_2} (\rho_2 v_2^2 + p_2)\,dz \cong p\big|_{h+d}\frac{\partial(h+d_2)}{\partial y} - p\big|_h\frac{\partial h}{\partial y} + w_e\rho_1 v_1. \qquad (1.10.3)$$

The integral of $\rho_2 v_2^2$ is just $d_2\rho_2 v_2^2$ and the derivative of the pressure integral can be written as $\int_h^{h+d_2} \frac{\partial p_2}{\partial y}\,dz + p_{h+d_2}\frac{\partial(h+d_2)}{\partial y} - p_h\frac{\partial h}{\partial y}$, the final two terms of which

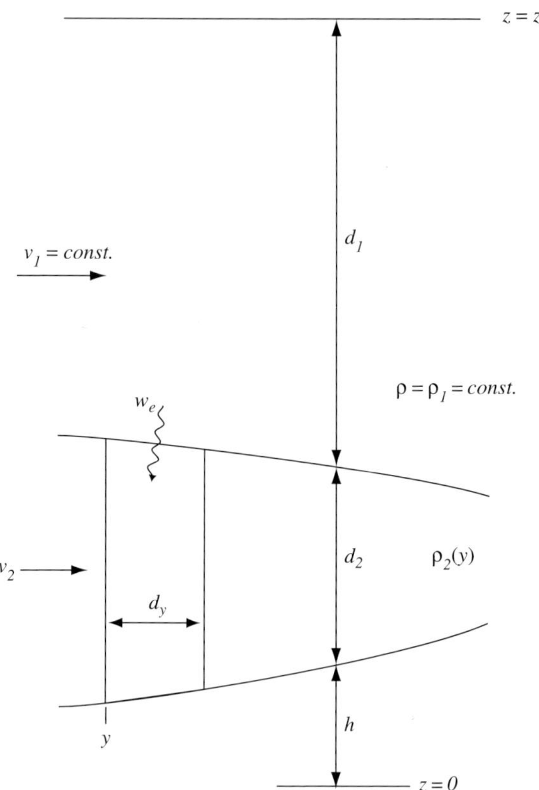

FIGURE 1.10.3. The control volume used as a basis for mass and momentum budgets for the lower layer.

negate identical terms on the right hand side of (1.10.3). With these modifications, (1.10.3) reduces to

$$\frac{\partial}{\partial y}(\rho_2 d_2 v_2^2) + \int_h^{h+d_2} \frac{\partial p}{\partial y} dz = w_e \rho_1 v_1 \qquad (1.10.4)$$

The integral can be evaluated by calculating the hydrostatic pressure in the lower layer, an exercise left to the reader. It is here that the inactive character of the upper-layer is enforced. The upper-layer depth is assumed to be so much greater than d_2 that the pressure at the upper lid $(z = z_T)$ and the upper layer velocity v_1 remain constant. Equation (1.10.4) now becomes

$$\frac{\partial}{\partial y}\left[\rho_2 d_2 v_2^2 + g(\rho_2 - \rho_1)\frac{d_2^2}{2}\right] = -(\rho_2 - \rho_1)g d_2 \frac{dh}{dy} + w_e \rho_1 v_1 \qquad (1.10.5)$$

In this 'flux' form of the momentum equation [a generalization of (1.6.11)] the flow 'force' $\rho_2 d_2 v_2^2 + g(\rho_2 - \rho_1)\frac{d_2^2}{2}$ can only be altered by bottom pressure drag or by fluxes of horizontal momentum across the interface.

In order to investigate the effects of entrainment on the hydraulic properties of the flow, it is convenient to work with the equation for horizontal momentum per unit mass. We assume that $(\rho_2 - \rho_1)/\rho_2 \ll 1$ and apply the Boussinesq approximation, meaning that density variations of $0[(\rho_2 - \rho_1)/\rho_2]$ are ignored unless multiplied by g (also see Section 5.1). Expansion of the y-derivative of the terms on the left, division of the result by d_2, and use of (1.10.1) and (1.10.2) lead to the modified momentum equation:

$$v_2\frac{\partial v_2}{\partial y} + g'\frac{\partial d_2}{\partial y} = -g'\frac{dh}{dy} + w_e\frac{(v_1 - v_2)}{d_2} + g'\frac{w_e}{2v_2}. \qquad (1.10.6)$$

The interfacial flux of horizontal momentum per unit mass is proportional to the difference in layer velocities. If the upper layer is at rest ($v_1 = 0$), the corresponding term reduces to $-w_e v_2/d_2$. The second entrainment term ($g'w_e/2v_2$) has a more subtle interpretation. It originates from the y-derivative of $g\rho_2(y)$ in (1.10.5), leading to $\frac{1}{2}gd_2^2\partial\rho_2/\partial y$, the contribution to the pressure gradient due to variations of the lower layer density. [Use of (1.10.1) and (1.10.2) allows this term to be rewritten in the form that appears in (1.10.6).] The entrainment of upper layer fluid causes the lower layer density to decrease in the direction of flow. In terms of pressure, the effect is the same as if the interface elevation decreased in the direction of flow.

As discussed by Pedlosky (1996, Sec. 4.2) there is an alternate model for w_e that holds in cases of thermal forcing. Direct cooling is imagined to trigger a convection process in which the density of an upper layer parcel increases from ρ_1 to ρ_2, causing it to sink across the interface. In this setting the lower layer density is preserved and the final term in (1.10.6) is absent.

Some of the effects of entrainment on hydraulic properties are revealed by consideration of the evolution of the Froude number of the flow:

$$\frac{\partial F_d^2}{\partial y} = \frac{\partial}{\partial y}\left(\frac{v_2^2}{g'd_2}\right) = F_d^2\left[\frac{2}{v_2}\frac{\partial v_2}{\partial y} - \frac{1}{d_2}\frac{\partial d_2}{\partial y} - \frac{1}{\rho_2 - \rho_1}\frac{\partial \rho_2}{\partial y}\right] \qquad (1.10.7)$$

$$= \frac{3F_d^2}{(F_d^2 - 1)d_2}\left\{-\frac{dh}{dy} + \frac{w_e}{v_2}(F_d^2\frac{v_1}{v_2} - F_d^2 - \frac{1}{2})\right\}$$

This relation was first derived by Gerdes *et al.* (2002), who also show that inclusion of width variations and quadratic bottom drag adds the term

$$-\frac{F_d^2}{F_d^2 - 1}[\frac{3F_d^2 C_d}{d_2} - \frac{(2 + F_d^2)}{w}\frac{dw}{dy}]$$

to the right-hand side.

Generalizations are straightforward for the case in which the velocity of the upper layer is less than or equal to that of the lower layer ($v_1/v_2 \leq 1$). In this case, the entrained momentum flux tends to retard the flow. Then the terms proportional to w_e in (1.19.12) make a positive contribution to $\partial F_d^2/\partial y$ when the flow is subcritical and a negative contribution when the flow is supercritical. In this case, entrainment drives the flow towards a critical state. It also tends to shift the point of hydraulic control downstream of the sill, to a location given by

$$\frac{dh}{dy} = \frac{w_e}{v_2}\left(\frac{v_1}{v_2} - \frac{3}{2}\right). \tag{1.10.8}$$

If entrainment adds momentum to the flow ($v_1 > v_2$) it is harder to make generalizations. However, (1.19.13) does show that entrainment will move the control section to a point upstream of a sill provided the value of v_1/v_2 exceeds 3/2. This shift would only occur if mixing (and corresponding finite values of w_e), takes place upstream of the sill.

A standard parameterization for the entrainment velocity is that due to Ellison and Turner (1959):

$$w_e = \begin{cases} |v_1 - v_2|\left(\dfrac{0.08 - 0.1R_i}{1 + 5R_i}\right) & (R_i < 0.8) \\ 0 & (R_i \geq 0.8) \end{cases}, \tag{1.10.9}$$

where

$$R_i = F_d^{-2}\left(\frac{v_1}{v_2} - 1\right)^{-2}$$

is called the *bulk Richardson number*. If the upper layer is motionless ($v_1 = 0$) then $F_d = R_i^{-1/2}$ and the requirement $R_i < 0.8$ means that entrainment only occurs for supercritical flows.

In our formulation of the shallow water equations with entrainment, the normal velocity w_e may be considered a vertical velocity. In ocean overflows, where outflow bottom slopes are of the order 10^{-2} or smaller, this approximation is justified. In the Ellison-Turner formulation, and in other situations involving nonnegligible interface tilts, w_e must be considered as directed normal to the interface. The Froude number in such cases is based on the velocity component parallel to the bottom and on the layer thickness measured normal to the bottom. In some of these cases, the interface more or less parallels the bottom and w_e is then taken to be normal to the bottom.

One consequence of using a large bottom slope in an experiment is that the Froude numbers obtained tend to be larger than those observed in the ocean (Figure 1.10.4). The entrainment rates also tend to be unrealistically large. Recent experiments (e.g. Cenedese *et al.*, 2004) designed to achieve lower Froude numbers have reproduced more realistic entrainment rates. Estimates of w_e for the outflows of the Mediterranean, the Denmark Strait, and the Faroe-Bank Channel, as well as a dense gravity plume in Lake Ogawara, are shown in the figure along with data from three laboratory experiments. The entrainment velocity in the low Froude number oceanographically relevant range increases roughly in proportion to the eighth power of the Froude number (Price, *private communication*, solid line).

There are also examples of measurements in the atmosphere (e.g. Princevac *et al.*, 2005) involving flows with oceanic Froude numbers but comparatively large Reynolds numbers. The turbulence in such cases is more fully developed and the entrainment rates are larger (upper left data points in Figure 1.10.4). A parameterization based solely on the Froude number is clearly inadequate to explain all cases. Another questionable practice in the formulation of parameterizations of turbulence, here and at large, is the reliance on local properties of the background flow. Turbulent eddies generated at a particular location may

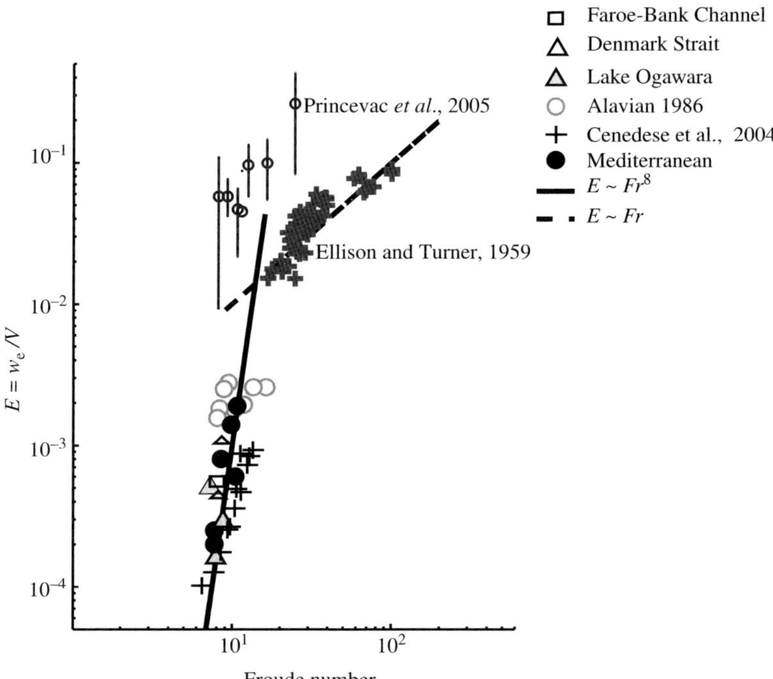

FIGURE 1.10.4. Entrainment coefficient w_e/V as a function of Froude number. The entrainment velocity is directed normal to the bottom, V represents the velocity parallel to the bottom and δV is the jump in V across the interface. The Froude number is based on this δV and on the layer thickness measured normal to the bottom. Data from laboratory experiments of Ellison and Turner (1959), Alavian (1986) and Cenedese *et al.* (2004) are indicated, as are observations in the Mediterranean, Denmark Strait and Faroe-Bank Channel (all from Baringer and Price, 1999), and from Lake Ogawarra (Dillimore *et al.*, 2001). The Princevac *et al.* (2005) data are from an atmospheric gravity current with higher Reynolds numbers than the ocean and laboratory examples. (Based on a figure from Wells and Wettlaufer, 2005 and on M. Wells and J. Price, *private communications*).

intensify or grow as they are advected by the mean velocity field. The value of w_e at a certain y may therefore depend on the background flow at and upstream of that y.

If the Ellison-Turner parameterization is used to specify w_e, the resulting solutions (Figures 1.10.5 and 1.10.6) show some of the features anticipated earlier in this discussion. The solutions are obtained by fixing the upstream values of $v_2 d$ and g' and varying the upstream value of d. Equations (1.10.1), (1.10.2) and (1.10.7) are then integrated forward in y to obtain the solutions at points downstream. The solutions should be compared to the conservative family of solutions shown in Figure 1.4.3. When the upper layer is motionless ($v_1 = 0$, Figure 1.10.5), w_e is finite only when the Froude number exceeds unity. In this case the subcritical solution (upper solid curve in Frame a) is unaffected. On the other hand, the supercritical (dashed) solutions are greatly altered. For example, the solution with upstream depth $d(-3) = 0.05$ immediately experiences entrainment causing its volume flow rate and depth to rapidly increase over much of the domain. The depth ($\cong 4.0$) that this solution reaches at the downstream end of the domain is greater than all other solutions shown, despite the fact that its upstream depth is less than all the other solutions.

Critical flow at the sill is obtained when the upstream flow is subcritical and has value $d(-3) \cong 2.41$ or when the upstream flow is supercritical and has value $d(-3) \cong 0.26$. In each case, the subcritical and supercritical branches of the solution that occur downstream of the critical section are shown. The appropriate choice of downstream solution is the one that allows the fluid to pass smoothly through the critical section. For example, one would follow the subcritical (solid) curve beginning at $d(-3) \cong 2.41$ and continue on to the supercritical (dashed) branch downstream of the sill. (There is an upstream continuation of the downstream subcritical branch; however, this solution is associated with different upstream values of $v_1 d$ and g' than those used to generate the family of curves in Figure 1.10.5).

Intersections between different solution curves do not carry the same significance as in a conservative system. In the latter, intersections imply the existence of two solutions with the same depth and volume fluxes, but different interface slopes. Such behavior is indicative of critical flow since it implies that stationary disturbances can exist at the section in question. An example is the intersection point corresponding to critical sill flow in Figure 1.9.1a. For the (nonconservative) solutions shown in Figure 1.10.5 or 1.10.6, an intersection implies only that the depths of the two solutions, and not necessarily the fluxes or values of g', are equal. For example, the intersection between the dashed curves near $x = -1.9$ in Figure 1.10.5a involves two solutions with identical depths but different Froude numbers (as shown in the Figure 1.10.5b).

The previous case involved $v_1 = 0$, so entrainment occurred only when the flow was supercritical. One consequence is that critical flow can only occur at the sill. We next consider a case with finite upper velocity, $v_1 = -1$ (Figure 1.10.6). A reverse upper layer velocity is characteristic of outflows from marginal seas, a subject treated in Chapter 5. Inspection of Figure 1.10.6a shows that critical

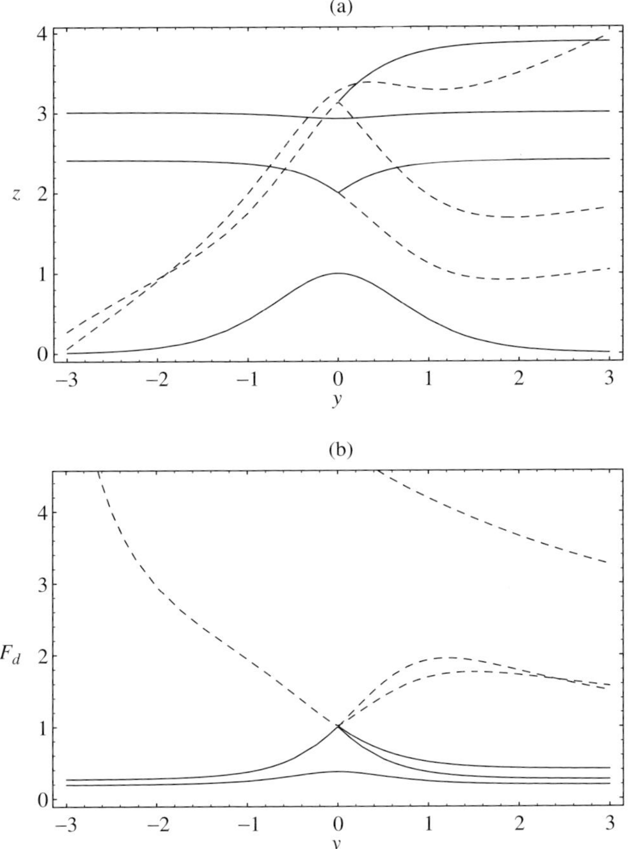

FIGURE 1.10.5. A family of steady solutions, all having the same upstream values of volume flux and g' but different lower-layer thicknesses. Entrainment is parameterized using the Ellison-Turner formulation (1.10.9) and the velocity v_1 in the overlying fluid is zero. The z-coordinate in the upper panel has been normalized using the obstacle height and the obstacle height-to-length ratio is 0.2. The lower panel shows the Froude numbers. L and h_m are the obstacle half width and height. (From Nielsen, *et al.* 2004).

transitions occur downstream of the sill as predicted by (1.10.8). As in the previous case, entrainment tends to push the solutions towards a critical state (Figure 1.10.6b) and, in the case of some of the supercritical curves, results in the formation of an infinite interface slope corresponding to a hydraulic jump. Jumps are represented in the figure by vertical terminations of the dashed curves.

One of the strongest assumptions made in connection with entraining layer models is that density and momentum carried across an interface are instantly mixed over the thickness of the target layer. In reality this mixing is rarely complete and the resulting distribution of v and ρ within the layer are vertically nonuniform. One of the most striking examples is the exchange flow in the Strait

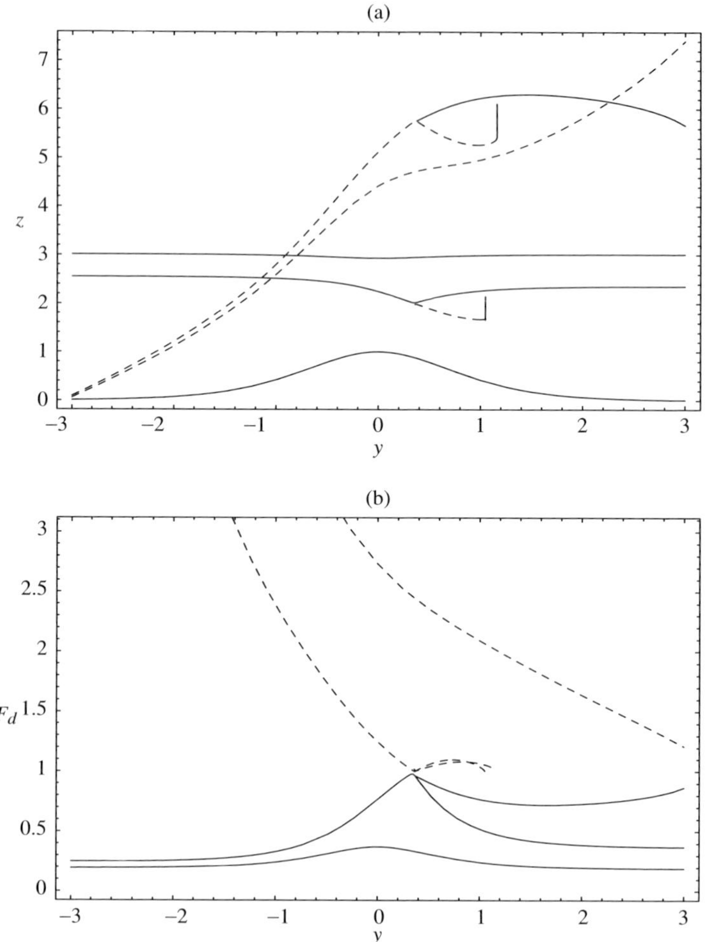

FIGURE 1.10.6. Same as Figure 1.10.5 except that the upper layer velocity is negative, in this case $v_1/(g'h_m) = -1$. (Frim Nielsen et al. 2004).

of Gibraltar (Figure I.9) where mixing between the inflowing and outflowing layers is often confined to a relatively thin interfacial layer. In the Bab al Mandab (Figure 1.10.7) the interfacial region is much thicker, but v and ρ remain strongly nonuniform. Further discussion of this subject can be found in Nielsen *et al.* (2004).

Exercises

(1) For the case of entrainment with no bottom friction or variations in w and h, derive an equation for the rate of change of d_2 with y (comparable to 1.10.7).

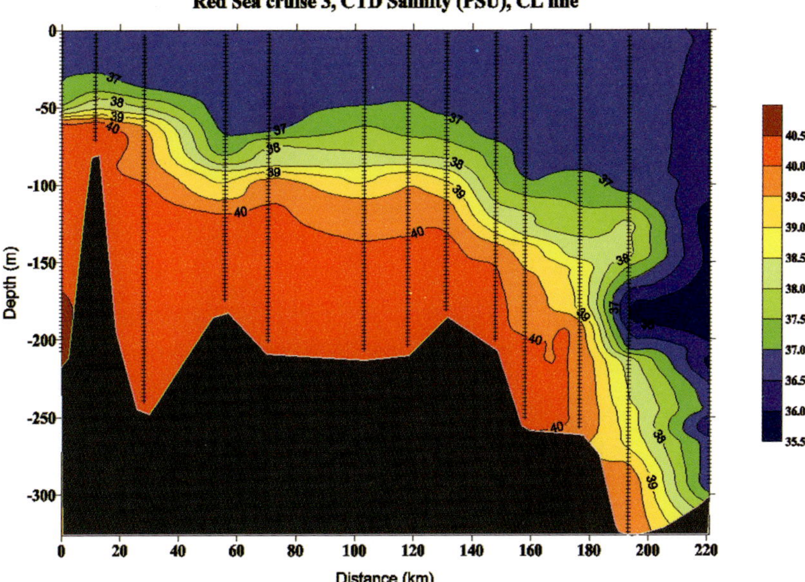

FIGURE 1.10.7. A temperature section along the central axis of the Bab al Mandab, with the Red Sea to the left. (Courtesy of Dr. S. Murray).

For subcritical flow, comment on possible circumstances in which d_2 can increase while F increases.

1.11. Nonlinear Dispersion

The hydrostatic approximation is central to everything discussed to this point. Nonhydrostatic effects associated with vertical accelerations of the fluid remain small as long as the ratio of the depth scale to horizontal length scale is small. However, special circumstances may exist that allow nonhydrostatic effects to become important even when this scale separation exists. For example, when Long's experiment is performed for $h_{\mathrm{m}}/d_{\mathrm{o}} \ll 1$ and $F_{\mathrm{o}} \cong 1$, wave-like free surface effects arise in certain parts of the flow field. In fact, certain values of $h_{\mathrm{m}}/d_{\mathrm{o}}$ and F_{o} produce a situation in which the flow refuses to settle into a steady state (Baines, 1995, Sec. 2.4). Another example is a hydraulic jump with an upstream Froude number less than about 1.7. Instead of an abrupt transition between supercritical and subcritical flow extending over a few depth scales, the jump is undular and extends over a much longer distance.

In both of these examples, changes in v and d along the channel are relatively small and corresponding advective terms like $v\partial v/\partial y$ are weak. Small terms like $w\partial v/\partial z$, which have been neglected as a consequence of the nonhydrostatic approximation, may now be as large as the retained terms. The conservation laws for momentum and mass now consist of delicate balances between weak hydro-static and nonhydrostatic terms. Some insight into the form of the governing equation can be gained through consideration of the special case of a wave propagating into an undisturbed fluid with uniform depth D and positive velocity V. Let us assume that the wave attempts to propagate against the current so that, within the context of shallow water theory, its evolution is governed by (1.3.1):

$$\left[\frac{\partial}{\partial t} + [v - (gd)^{1/2}]\frac{\partial}{\partial y}\right][v - 2(gd)^{1/2}] = 0 \tag{1.11.1}$$

The value of the Riemann invariant $R_+ = v + 2(gd)^{1/2}$ is equal to its value $V + 2(gD)^{1/2}$ in the undisturbed fluid and thus v in the above equation can be replaced by $V - 2(gd)^{1/2} + 2(gD)^{1/2}$. If the depth in the wave is only slightly different than D, we can write $d = D + \eta$, where $\eta/D \ll 1$. Then $(gd)^{1/2} = (gD)^{1/2}[1 + (\eta/2D) + \cdots]$ and substitution into (1.11.1) yields

$$\left[\frac{\partial}{\partial t} + \left(c_o - \frac{3g^{1/2}\eta}{2D^{1/2}}\right)\frac{\partial}{\partial y}\right]\eta + O\left(\frac{\eta}{D}\right)^3 = 0, \tag{1.11.2}$$

where $c_o = V - (gD)^{1/2}$.

The correction introduced into this equation by nonhydrostatic effects can be anticipated through consideration of the dispersion relation $\omega = [gl \tanh(lD)]^{1/2}$ for a surface gravity wave propagating in a resting fluid of uniform depth D. The wave has the form $\eta = ae^{i(ly - \omega t)}$, where ω denotes the frequency and $2\pi/l$ the (arbitrarily short) wavelength. If the latter is long compared to D, this relation may be expanded:

$$\omega = Vl - [gl \tanh(lD)]^{1/2} = Vl - (gD)^{1/2}l[1 - \frac{(lD)^2}{6} + O(lD)^4]$$

The linear equation that would produce the two leading terms in this expansion is

$$\left[\frac{\partial}{\partial t} + c_o\frac{\partial}{\partial y}\right]\eta - \frac{(gD)^{1/2}D^2}{6}\frac{\partial^3\eta}{\partial y^3} = 0$$

and thus the nonhydrostatic correction should be $-\frac{1}{6}(gD)^{1/2}D^2\partial^3\eta/\partial y^3$.

If (1.11.2) is modified to include this factor, the result is the celebrated *Korteweg-de Vries* (KdV) *equation*

$$\left[\frac{\partial}{\partial t} + \left[c_o - \frac{3g^{1/2}\eta}{2D^{1/2}}\right]\frac{\partial}{\partial y}\right]\eta - \frac{(gD)^{1/2}D^2}{6}\frac{\partial^3\eta}{\partial y^3} = 0, \tag{1.11.3}$$

as can be verified by a more systematic analysis (Whitham, 1974).

According to (1.11.3), the wave propagates at the base speed c_o and evolves slowly in response to weak nonlinearity and dispersion. The competition between the two processes can be isolated by expressing the equation in a frame of reference moving at the base speed. With $y' = y - c_o t$, we have

$$\left\{ \frac{\partial}{\partial t} - \frac{3g^{1/2}\eta}{2D^{1/2}} \frac{\partial}{\partial y'} \right\} \eta - \frac{(gD)^{1/2}D^2}{6} \frac{\partial^3 \eta}{\partial y'^3} = 0. \tag{1.11.4}$$

It is possible to find steadily-propagating solutions, one of which is the *soliton*:

$$\eta = \eta_o \text{sech}^2 \left[\left(\frac{3\eta_o}{4D^3} \right)^{1/2} (y + \hat{c}t) \right]$$

where $\hat{c} = c_o \left(1 + \frac{\eta_o}{2D} \right)$. In this case a balance between steepening and dispersion has achieved an isolated disturbance of permanent form that propagates with an amplitude-dependent speed. A class of periodic disturbances ('Cnoidal waves') is also admitted, as explored in Exercise 1.

The KdV equation and its extensions have been successfully used in the analysis of undular bores (Peregrine, 1966 and Fornberg and Whitham, 1978). For hydraulic applications, a topographically forced version of (1.11.4) may be used. To remain consistent with the assumption of unidirectional propagation, any forcing that is added must move at the base speed c_0 of the disturbance. Stationary forcing therefore requires that c_0 is zero: that is, the flow is critical to leading order. The obstacle height must also be small in order to preserve consistency with the assumption of weak nonlinearity. The evolution equation is then obtained through introduction of the term $\frac{1}{2}(gD)^{1/2}dh/dy$ on the right-hand side of (1.11.4). The result can be used to solve Long-type adjustment problems with $h_\text{m}/d_\text{o} \ll 1$ and $F_\text{o} \cong 1$ (Cole, 1985 and Grimshaw and Smyth, 1986), and in other weakly nonlinear hydraulic applications. The reader is referred to Baines 1995 for a more thorough summary.

One interesting and simple application is to the problem of steady, shallow flow over consecutive obstacles of identical height (Figure 1.11.1). According to shallow water theory, there is no solution that is hydraulically controlled and everwhere stable. If the approach flow is subcritical (solution *ab* in the figure), a subcritical-to-supercritical transition occurs over the first sill. The approach to the second obstacle is now supercritical and an (unstable) transition back to a subcritical state is required. This transition is shown as a dashed section of the *ab* curve. It is also possible for the flow approaching the first obstacle to

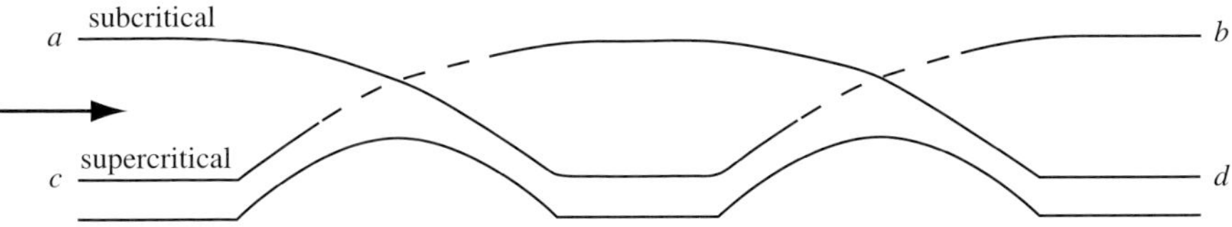

FIGURE 1.11.1. Long-wave solutions for hydraulically controlled flow over two consecutive obstacles of identical heights. Dashed curves show segments where the flow is vulnerable to a shock-forming instability.

FIGURE 1.11.2. Laboratory simulation of shallow flow over two obstacles of nearly the same height. The flow is from right to left. (From Pratt, 1984).

be supercritical (solution *cd*) but then an unstable transition is forced over the first sill. Nor is it possible to avoid the unstable transitions by introducing a hydraulic jump between the obstacles: the resulting energy loss would prevent the flow from surmounting the next obstacle. It would seem, then, that shallow water theory fails to provide a satisfactory steady solution.

Laboratory experiments (Figure 1.11.2) have shown, in fact, that the spilling flow occurs over the second obstacle and that the flow between the two obstacles is wavelike. The heights of the obstacles do not need to be identical for this behavior to occur, and the phenomena appears to be more than just a curiosity. A solution with the observed properties can be found to the forced KdV equation for nearly critical flow. The reader is referred to Exercise 1 for more details.

Exercises

1) As described in the text, the equation governing steady, weakly nonlinear, weakly dispersive flow over a small obstacle is obtained by setting $c_o = 0$ and adding the term $\frac{1}{2}(gD)^{1/2} dh/dy$ to the right-hand side of the steady form of (1.11.3). The result is

$$\frac{3g^{1/2}\eta}{2D^{1/2}}\frac{d\eta}{dy} + \frac{(gD)^{1/2}D^2}{6}\frac{d^3\eta}{dy^3} + \frac{(gD)^{1/2}}{2}\frac{dh}{dy} = 0. \tag{1.11.5}$$

(a) Integrate (1.11.5) once and show that the result can be written as the following dynamical system:

$$\frac{d\hat{\zeta}}{dy} = C - \frac{9}{2}\hat{\eta}^2 - 2\hat{h} \tag{1.11.6}$$

and

$$\frac{d\hat{\eta}}{dy} = \hat{\zeta}, \tag{1.11.7}$$

where C is a constant.

(b) For the case of a flat bottom, note that for each $C > 0$ there are two uniform $(d/dy = 0)$ flows corresponding to $\hat{\zeta} = 0$ and $\hat{\eta}_{\pm} = \pm\sqrt{2C/9}$. Draw a picture of the phase plane $(\hat{\zeta}, \hat{\eta})$ and locate the points corresponding to the two uniform flows. Show that the solution trajectories near $(0, \hat{\eta}_+)$ are closed, corresponding to a set of stationary periodic waveforms. These are the 'cnoidal' waves referred to in the text. Also show that the solution $(0, \hat{\eta}_-)$ is unstable, corresponding to trajectories that diverge away.

(c) Also show that one of the trajectories that diverges from $(0, \hat{\eta}_-)$ forms a closed obit that circumnavigates $(0, \hat{\eta}_+)$. This solution corresponds to a stationary solitary wave. The addition of the topographic term in (1.11.6) allows the actual solution to cross the trajectories of the unforced flow and can lead to a satisfactory solution for the two-obstacle problem. The reader is referred to Pratt (1984a) for more details.

2
The Hydraulics of Homogeneous Flow in a Rotating Channel

The original models of rotating, hydraulically driven currents were motivated by observations of deep overflows. The spillage of dense fluid over the sills of the Denmark Strait, the Faroe Bank Channel and other deep passages is suggestive of hydraulic control and one hope was that formulae used to estimate the volume outflow from a reservoir might be extended to these settings. To this end the whole volume of dense, overflowing fluid is treated as a single homogeneous layer with reduced gravity. In the Denmark Strait overflow example (Figures I.7 and I.8) this layer typically includes all fluid denser than $\sigma_\theta = 27.9$. The layer experiences strong, cross-channel variations in thickness and velocity, complications that can arise in engineering applications but are unavoidable where the earth's rotation is important. Much of the development of the theory of rotating hydraulics consists of attempts to come to grips with this extra degree of freedom. We shall trace this development beginning with early models of rotating-channel flow and show that hydraulic control and many of the other features reviewed in the first chapter remain present in one form or another. A number of novel features will also arise, including boundary layers, flow reversals, and sidewall separation. In this presentation, we will use northern hemisphere flows as paradigms.

Another distinctive aspect of rotating hydraulics concerns the permissible waves. Under the usual assumption of gradual variations of the flow along its predominant direction, three types of waves arise. The first is the Kelvin wave, an edge wave closely related to the long gravity waves of the last chapter. The second is the frontal wave, which replaces the Kelvin wave when the edge of the flow is free to meander independently of sidewall boundaries. Frontal waves are sometimes referred to as Kelvin waves in the literature. The third is the potential vorticity wave, a disturbance that exists when gradients of potential vorticity, defined in this chapter, exist within the fluid. Nearly all analytical models of deep overflows assume that the potential vorticity is uniform within the flow, thereby eliminating this wave. We will consider only one model that does not. Free jets in the ocean and atmosphere are more dependent on potential vorticity dynamics and will be covered in Chapter 6.

In contrast to Chapter 1, where nearly all variables were dimensional, the present Chapter (and the remainder of the book) will primarily make use of dimensionless variables, and will frequently cite the dimensional representation of particular results. It becomes necessary to distinguish between the two formats and we do so by assigning stars to dimensional quantities. There are some exceptions to this convention. Stars are not used, for example, for certain universally recognized dimensional quantities such as the Coriolis parameter, f, or for generic scales such D (for depth) and L (for length).

2.1.   The Semigeostrophic Approximation in a Rotating Channel Flow

We consider homogeneous flows confined to a channel rotating with constant angular speed $f/2$ in the horizontal plane. The coordinates (x^*, y^*) denote cross-channel and along-channel directions, (u^*, v^*) the corresponding velocity components, and (d^*, h^*) the fluid depth and bottom elevation. Provided the scale of x^*- and y^*-variations of d^* are large compared to the typical depth, the shallow water equations continue to apply. The dimensional version of these equations is

$$\frac{\partial u^*}{\partial t^*} + u^* \frac{\partial u^*}{\partial x^*} + v^* \frac{\partial u^*}{\partial y^*} - f v^* = -g \frac{\partial d^*}{\partial x^*} - g \frac{\partial h^*}{\partial x^*} + F^{(x)*} \qquad (2.1.1)$$

$$\frac{\partial v^*}{\partial t^*} + u^* \frac{\partial v^*}{\partial x^*} + v^* \frac{\partial v^*}{\partial y^*} + f u^* = -g \frac{\partial d^*}{\partial y^*} - g \frac{\partial h^*}{\partial y^*} + F^{(y)*} \qquad (2.1.2)$$

$$\frac{\partial d^*}{\partial t^*} + \frac{\partial (u^* d^*)}{\partial x^*} + \frac{\partial (v^* d^*)}{\partial y^*} = 0. \qquad (2.1.3)$$

Unspecified forcing and dissipation is contained in $\mathbf{F} = (F^{(x)*}, F^{(y)*})$. For positive f, the channel rotation is counterclockwise looking down from above, as in the northern hemisphere. These equations apply to a homogeneous layer with a free surface or to the active lower layer of a '1½-layer' or 'equivalent barotropic' model. In the latter, g is reduced in proportion to the fractional density difference between the two layers. In such cases the upper boundary of the active layer will be referred to as 'the interface'.

For large-scale oceanic and atmospheric flows away from the equator and away from fronts and boundary layers, the forcing and dissipation terms and the terms expressing acceleration relative to the rotating earth are generally small in comparison to the Coriolis acceleration. The horizontal velocity for these types of flows is approximately geostrophic, or

$$f v^* \cong g \frac{\partial (d^* + h^*)}{\partial x^*} \quad \text{and} \quad f u^* \cong -g \frac{\partial (d^* + h^*)}{\partial y^*}$$

in the context of our shallow water model. These relations suggest that geostrophic flow moves parallel to lines of constant pressure, with high pressure to the right in the northern hemisphere. This situation was quite different for the flows treated in Chapter 1, in which the velocity is aligned with the pressure gradient and flow is accelerated from high to low pressure. For the deep overflows and strong atmospheric down-slope winds the acceleration of the flow down the pressure gradient is also a characteristic feature, suggesting a departure from the geostrophic balance.

To explore this issue further it is helpful to nondimensionalize variables. Define D as a scale characterizing the typical depth and L as a measure of the horizontal distance over which along-channel variations take place. Also take $(gD)^{1/2}$ as a scale for v^*, anticipating that the gravity wave speed will continue to be a factor in the dynamics of hydraulically controlled states and that such states will require velocities as large as this speed. A natural scale for t^* is therefore $L/(gD)^{1/2}$. As a width scale, we pick $(gD)^{1/2}/f$, which is the Rossby radius of deformation based on the depth scale D. For readers not familiar with the theory of rotating fluids, $2\pi(gD)^{1/2}/f$ is the distance a long gravity wave [with speed $(gD)^{1/2}$] will travel in an inertial period $2\pi/f$. It is the distance the wave must travel before it is influenced by the earth's rotation. Motions with much smaller length or timescales are generally not influenced by rotation. The Rossby radius appears as a natural width scale for boundary currents and boundary-trapped waves. With these choices, the cross-channel velocity scale $(gD)/fL$ is suggested by balancing the second and third terms in (2.1.3). The dimensionless variables are therefore

$$x = \frac{x^* f}{(gD)^{1/2}}, \quad y = \frac{y^*}{L}, \quad t = \frac{t^*(gD)^{1/2}}{L} \tag{2.1.4}$$

$$v = \frac{v^*}{(gD)^{1/2}}, \quad u = \frac{fLu^*}{gD}, \quad d = \frac{d^*}{D}, \quad h = \frac{h^*}{D}, \quad \mathbf{F} = \frac{L\mathbf{F}^*}{gD}.$$

Substitution into (2.1.1–2.1.3) leads to

$$\delta^2\left(\frac{\partial u}{\partial t} + u\frac{\partial u}{\partial x} + v\frac{\partial u}{\partial y}\right) - v = -\frac{\partial d}{\partial x} - \frac{\partial h}{\partial x} + \delta F^{(x)} \tag{2.1.5}$$

$$\frac{\partial v}{\partial t} + u\frac{\partial v}{\partial x} + v\frac{\partial v}{\partial y} + u = -\frac{\partial d}{\partial y} - \frac{\partial h}{\partial y} + F^{(y)} \tag{2.1.6}$$

$$\frac{\partial d}{\partial t} + \frac{\partial(ud)}{\partial x} + \frac{\partial(vd)}{\partial y} = 0 \tag{2.1.7}$$

where $\delta = (gD)^{1/2}/fL$ is the ratio of the width scale of the flow to L: a horizontal aspect ratio.

The limit $\delta \to 0$ leads to a geostrophic balance in the cross-channel (x-) direction but not the along-channel direction. The along-channel velocity v is geostrophically balanced but the cross-channel velocity u is not. The flow in this limit is therefore referred to as *semigeostrophic*. The semigeostrophic approximation requires that variations of the flow along the channel are gradual in comparison with variations across the channel. In particular, the interface may slope steeply across the channel but can do so only mildly along the channel. The along-channel velocity component v is therefore directed nearly perpendicular to the pressure gradient. As (2.1.6) suggests, the (weaker) along-channel pressure gradient *does* lead to acceleration in the same direction, but this occurs over a distance L large compared to the cross-stream scale δL.

Semigeostrophic and *quasigeostrophic* models should not be confused. In the latter, both of the horizontal velocity components are geostrophically balanced, at least to a first approximation, and variations in the depth or layer thickness are required to be slight. Time variations occur on a scale much longer than $1/f$. Quasigeostophic models form the basis for much of the theory of broad scale waves and circulation in the ocean and atmosphere (e.g. Pedlosky, 1987). Hydraulic effects with respect to gravity waves cannot occur because these waves are filtered by the quasigeostrophic approximation.

Vorticity and potential vorticity are conceptually and computationally central to rotating flows. For shallow homogeneous flow, the discussion is simplified by the fact that the horizontal velocity is z-independent, so that the fluid moves in vertical columns. Vorticity and potential vorticity are therefore assigned to fluid columns as a whole. If the curl of the shallow water momentum equations (i.e. $\partial(2.1.2)/\partial x^* - \partial(2.1.1)/\partial y^*$) is taken and (2.1.3) is used to eliminate the divergence of the horizontal velocity from the resulting expression, the following conservation law for potential vorticity can be obtained:

$$\frac{d^* q^*}{dt^*} = \frac{\mathbf{k} \cdot \nabla^* \times \mathbf{F}^*}{d^*}. \tag{2.1.8}$$

Here $\dfrac{d^*}{dt^*} = \dfrac{\partial}{\partial t^*} + u^* \dfrac{\partial}{\partial x^*} + v^* \dfrac{\partial}{\partial y^*}$, $\mathbf{k}$ is the vertical unit vector, and

$$q^* = \frac{f + \zeta^*}{d^*}. \tag{2.1.9}$$

The *relative vorticity* $\zeta^* = \frac{\partial v^*}{\partial x^*} - \frac{\partial u^*}{\partial y^*}$ is the vorticity of a fluid column as seen in the rotating frame of reference. The *absolute vorticity* is the total vorticity $\zeta^* + f$ of the column. The *potential vorticity* q^* is simply the absolute vorticity divided by the column thickness d^*. If the forcing and dissipation have no curl ($\nabla^* \times \mathbf{F}^* = 0$) the potential vorticity of the material column remains constant. Conservation of potential vorticity is a consequence of angular momentum conservation; if the column thickness d^* increases, conservation of mass requires the cross-sectional area of the column to decrease, and the column must spin more rapidly to compensate for a decreased moment of inertia.

It is sometimes convenient to represent the potential vorticity as

$$q^* = \frac{f + \zeta^*}{d^*} = \frac{f}{D_\infty}$$

where D_∞ is known as the *potential depth*. In the absence of forcing or dissipation, each fluid column owns its own time-independent potential depth. To interpret this quantity, consider a column with relative vorticity ζ^* (also $= q^* d^* - f$ by the definition of q^*). Next alter the column thickness d^* to the value f/q^*, so that ζ^* vanishes. This new thickness *is* the potential thickness D_∞. This interpretation is limited by the fact that D_∞ may be negative, making it physically impossible to remove ζ^* by stretching. Most of the applications we will deal with have positive potential depth.

The nondimensional versions of (2.1.8) and (2.1.9) are

$$\frac{dq}{dt} = \frac{\dfrac{\partial F^{(y)}}{\partial x} - \delta \dfrac{\partial F^{(x)}}{\partial y}}{d} \qquad (2.1.10)$$

and

$$q = \frac{1 + \dfrac{\partial v}{\partial x} - \delta^2 \dfrac{\partial u}{\partial y}}{d}. \qquad (2.1.11)$$

In the semigeostrophic limit $\delta \to 0$:

$$v = \frac{\partial d}{\partial x} + \frac{\partial h}{\partial x}. \qquad (2.1.12)$$

and

$$q = \frac{1 + \dfrac{\partial v}{\partial x}}{d}. \qquad (2.1.13)$$

The last two relations can be combined, yielding an equation for the x-variation in depth

$$\frac{\partial^2 d}{\partial x^2} - qd = -1 - \frac{\partial^2 h}{\partial x^2}. \qquad (2.1.14)$$

If $q = $ constant, the above equation can easily be solved and one can then focus their attention on the x- and t- structure of the flow. This situation arises if q is initially uniform throughout the fluid and no forcing or dissipation is present.

Two other forms of the shallow water momentum equations will prove very helpful. One is

$$\frac{\partial \mathbf{u}^*}{\partial t^*} + (f + \zeta^*)\mathbf{k} \times \mathbf{u}^* = -\nabla B^* + \mathbf{F}^*, \qquad (2.1.15)$$

where

$$B^* = \frac{u^{*2} + v^{*2}}{2} + g(d^* + h^*) \qquad (2.1.16)$$

is the two-dimensional Bernoulli function. The dimensionless form of the latter is $B = B^*/gD = \frac{1}{2}(\delta^2 u^2 + v^2) + d + h$. In the semigeostrophic limit B formally reduces to its one-dimensional equivalent $v^2/2 + g(d + h)$. The second version of interest is the depth-integrated or 'flux' form, obtained by multiplication of (2.1.1) and (2.1.2) by d^*, rearrangement of some derivatives, and use (2.1.3). The results:

$$\frac{\partial(d^*u^*)}{\partial t^*} + \frac{\partial}{\partial x^*}(d^*u^{*2} + \tfrac{1}{2}gd^{*2}) + \frac{\partial}{\partial y^*}(u^*v^*d^*) - fv^*d^* = -gd^*\frac{\partial h^*}{\partial x^*} + d^*F^{(x)*}$$

$$(2.1.17a)$$

and

$$\frac{\partial(d^*v^*)}{\partial t^*} + \frac{\partial}{\partial y^*}(d^*v^{*2} + \tfrac{1}{2}gd^{*2}) + \frac{\partial}{\partial x^*}(u^*v^*d^*) + fd^*u^* = -gd^*\frac{\partial h^*}{\partial y^*} + d^*F^{(y)*}$$

$$(2.1.17b)$$

are used in the analysis of hydraulic jumps, form drag and other applications where the total momentum over the water column is at issue.

If the flow is steady ($\partial/\partial t^* = 0$), the continuity equation (2.1.3) implies the existence of a *transport stream function* $\psi^*(x, y)$ such that

$$v^*d^* = \frac{\partial \psi^*}{\partial x^*} \quad \text{and} \quad -u^*d^* = \frac{\partial \psi^*}{\partial y^*}$$

The total volume transport Q^* is the value of ψ^* on the right-hand edge of the flow (facing positive y^*) minus ψ^* on the left wall. If, in addition, there is no forcing or dissipation ($\mathbf{F}^* = 0$) then (2.1.15) can be written

$$\frac{(f + \zeta^*)}{d^*}\mathbf{k} \times \mathbf{u}^*d^* = -\nabla B^* \tag{2.1.18}$$

or $q^*\nabla\psi^* = \nabla B^*$. Thus the Bernoulli function is conserved along streamlines:

$$B^* = B^*(\psi^*)$$

and

$$q^* = \frac{dB^*}{d\psi^*}. \tag{2.1.19}$$

This remarkable link between energy and potential vorticity, which can be traced back to Crocco (1937), is one of the central constraints used in hydraulic theories for two-dimensional flow.

In the steady sill flows discussed in Chapter 1, the reservoir state is specified by the values of Q^* and B^*, the fundamental conserved quantities of the one-dimensional flow. Discussion of the present generalization often centers on three conserved quantities: the *functions* $B^*(\psi^*)$, $q^*(\psi^*)$ and the constant Q^*. Crocco's theorem shows that these three are not independent; specification of $B^*(\psi^*)$ and of the range of ψ^* allows $q^*(\psi^*)$ to be completely determined.

We have already touched on the different types of long (semigeostrophic) waves that arise in rotating channel flows. Kelvin waves and their frontal relatives depend on the combined effects of rotation and gravity and are important to the hydraulics of gravity-driven flows. Potential vorticity waves can exist in flows with neither gravity nor background rotation. Their dynamics involve vortex induction mechanics that can arise when the potential vorticity of the fluid flow varies spatially. If the long-wave assumption is relaxed, inertia-gravity (Poincaré) waves come into play. They are not important in traditional models of rotating hydraulics, but they are important for a range of transient phenomena generally considered to be part of hydraulics. We now discuss some of the linear properties of these waves where they arise as small perturbations from a resting state. Nonlinear steepening and other finite amplitude effects will be treated in later sections.

Consider the shallow water equations, linearized about a state of rest with $d = 1$ and $\boldsymbol{F} = h = 0$. Take $d = 1 + \eta$, with $|\eta| \ll 1$; assume $u \ll 1$ and $v \ll 1$; and neglect terms quadratic in η, v etc. in (2.1.5–2.1.7) to obtain

$$\delta^2 \frac{\partial u}{\partial t} - v = -\frac{\partial \eta}{\partial x} \tag{2.1.20}$$

$$\frac{\partial v}{\partial t} + u = -\frac{\partial \eta}{\partial y} \tag{2.1.21}$$

and

$$\frac{\partial \eta}{\partial t} + \frac{\partial u}{\partial x} + \frac{\partial v}{\partial y} = 0. \tag{2.1.22}$$

The corresponding potential vorticity equation, which can be obtained directly from the above or simply by linearizing the nondissipative version of (2.1.8), is

$$\frac{\partial}{\partial t} \left[\frac{\partial v}{\partial x} - \delta^2 \frac{\partial u}{\partial y} - \eta \right] = 0$$

or

$$\frac{\partial v}{\partial x} - \delta^2 \frac{\partial u}{\partial y} - \eta = \frac{\partial v_o}{\partial x} - \delta^2 \frac{\partial u_o}{\partial y} - \eta_o, \tag{2.1.23}$$

where $()_o$ indicates an initial value. The last equation indicates that the linearized potential vorticity, equal to the relative vorticity $\partial v/\partial x - \delta^2 \partial u/\partial y$ plus the stretching contribution $-\eta$, is conserved at each (x, y).

The left-hand side of (2.1.23) can be expressed in any of the three variables u, v, or η by using (2.1.20–2.1.22) to eliminate the remaining two. For example the equation for η is

$$\frac{\partial^2 \eta}{\partial x^2} + \delta^2 \frac{\partial^2 \eta}{\partial y^2} - \delta^2 \frac{\partial^2 \eta}{\partial t^2} - \eta = \frac{\partial v_o}{\partial x} - \delta^2 \frac{\partial u_o}{\partial y} - \eta_o. \tag{2.1.24}$$

For an arbitrary initial disturbance the resulting flow will consist of two parts. The first is a steady flow whose potential vorticity is given by the potential vorticity of the initial disturbance. This flow is obtained by finding a steady solution to (2.1.24). The second component consists of waves that are generated as a result of the unbalanced part of the initial flow. Individually, these waves are solutions to the homogeneous version of (2.1.24) subject to the boundary condition

$$\frac{\partial^2 \eta}{\partial x \partial t} = -\frac{\partial \eta}{\partial y}, \qquad (x = \pm w/2) \tag{2.1.25}$$

obtained by evaluating (2.1.20) and (2.1.21) at the sidewalls, where $u = 0$, and eliminating v from the result.

Assuming traveling waves of the form $\eta = \mathrm{Re}[aN(x)e^{i(ly-\omega t)}]$, where ω is the frequency and l is the longitudinal wave number, one finds two distinct solutions (Gill, 1982; Pedlosky, 2003), both of which were discovered by Kelvin (1879). The first, named after Poincaré (1910), has an oscillatory structure in x:

$$N_n(x) = \cos(k_n x) + b_n \sin(k_n x) \tag{2.1.26}$$

where $k_n = n\pi/w$, and $b_n = -\omega_n k_n/l$ ($n =$ odd) or $b_n = l/\omega_n k_n$ ($n =$ even). The frequency satisfies the dispersion relation

$$\delta^2 \omega^2 = \frac{n^2 \pi^2}{w^2} + \delta^2 l^2 + 1 \ (n = 1, 2, 3, \ldots), \tag{2.1.27a}$$

the dimensional form of which is

$$\omega^{*2} = gD\left(\frac{n^2 \pi^2}{w^{*2}} + l^{*2}\right) + f^2 \tag{2.1.27b}$$

where D is the background depth.

Poincaré waves can be better understood by first considering a long gravity wave propagating in an arbitrary direction on an infinite, nonrotating plane. The form of the wave is given by $\eta^* = \mathrm{Re}[a^* e^{i(k^* x^* + l^* y - \omega^* t^*)}]$, where k^* and l^* represent the wave numbers. The dispersion relation for this wave is given in dimensional terms by (2.1.27b) with $f = 0$ and with $\dfrac{n^2 \pi^2}{w^{*2}}$ replaced by k^{*2}. Next consider a second wave with wave numbers $(-k^*, l^*)$ and therefore having the same frequency as the first wave. If the second wave has the same amplitude a as the first, a superposition of the two leads to a u^* field proportional to $\mathrm{Re}[a(e^{i(k^* x^* + l^* y - \omega^* t^*)} - e^{i(-k^* x^* + l^* y - \omega^* t^*)})] = \mathrm{Re}[2aie^{i(l^* y - \omega^* t^*)} \sin k^* x^*]$. Since u^* is zero whenever $k^* x^*$ is an integer multiple of π, the waves satisfy the sidewall boundary conditions in a channel with sidewalls at $x^* = \pm w^*/2$ provided that k^* is chosen to be $2n\pi/w^*$. These waves are sometimes called oblique gravity waves and their cross-channel structure is said to be *standing*. Poincaré waves are rotationally modified versions of these waves.

The second class consists of edge waves named after Kelvin himself. The cross-channel structure and dispersion relation are given by

$$N_{\pm}(x) = \frac{\sinh(x) \pm \cosh(x)}{\sinh(\frac{1}{2}w)} \tag{2.1.28}$$

and

$$\omega_{\pm} = \pm l \text{ or}$$

$$\omega^{*}{}_{\pm} = \pm(gD)^{1/2}l^{*} \tag{2.1.29}$$

Kelvin waves have a boundary layer structure that becomes apparent when the channel width is much wider than the deformation radius. Taking the limit $w \gg 1$ (equivalently $w^{*} \gg (gD)^{1/2}/f$) in (2.1.28) leads to

$$N_{+}{}^{*}(x^{*}) \propto N^{*}(\tfrac{1}{2}w^{*})e^{(x^{*}-\frac{1}{2}w^{*})f/(gD)^{1/2}}$$

and

$$N_{-}{}^{*}(x^{*}) \propto N^{*}(-\tfrac{1}{2}w^{*})e^{-(x^{*}+\frac{1}{2}w^{*})f/(gD)^{1/2}}.$$

The first solution corresponds to a wave propagating in the positive y-direction at speed $c^{*}=\omega^{*}_{+}/l^{*}=(gD)^{1/2}$ and trapped to the wall at $x^{*} = w^{*}/2$. The trapping distance is the Rossby radius of deformation based on the background depth D. The other wave moves in the opposite direction and is trapped to the wall at $x^{*} = -w^{*}/2$. In the limit of weak rotation, $N_{\pm}{}^{*}$ becomes constant and the Kelvin waves reduce to x-independent, long gravity waves propagating along the channel. A further distinguishing property of linear Kelvin waves is that the cross-channel velocity u is identically zero.

Kelvin waves are nondispersive, meaning that the phase speed c^{*} does not depend on the wave number l^{*}. The wave frequency $\omega^{*} = c^{*}l^{*}$ is proportional to l^{*} and therefore the group velocity $\partial\omega^{*}/\partial l^{*}$ is equal to c^{*}. In Chapter 1, we described the topographic resonance that can occur when a background flow is critical $c^{*} = 0$ with respect to a nondispersive wave. A bottom slope or other stationary forcing introduces disturbance energy that cannot propagate away. The disturbance amplitude grows and becomes large and sufficiently nonlinear to break away, leading to fundamental changes in the upstream flow. We expect that Kelvin waves will play an important role in the upstream influence of rotating channel flows.

Poincaré waves are not admitted under semigeostrophic dynamics, a result that can be shown by taking $(\delta \to 0)$ in (2.1.27a). The resulting condition cannot be satisfied for real n. Since most simple models of the hydraulics of rotating flow in a channel or along a coast use the semigeostrophic approximation, Poincaré waves do not arise. Section 3.8 will explore an exception to this traditional picture.

The restoring mechanism for Poincaré and Kelvin relies on gravity and a free surface or interface. Potential vorticity waves, on the other hand, rely on gradients of potential vorticity within the fluid. One can describe this effect by modifying the above example to include a lateral bottom slope $\partial h^*/\partial x^* = -S = constant$. For simplicity, we will eliminate the gravitational restoring mechanism by placing a rigid lid on the top of the fluid. The resting basic state now contains a potential vorticity gradient associated with the variable depth alone. If D is the layer thickness at midchannel ($x^* = 0$) and if the bottom and surface tilt lead to only slight variations of h^* about D, then the potential vorticity of the ambient fluid is

$$q^* = \frac{f + \partial v^*/\partial x^*}{d^*} = \frac{f}{D + Sx^*} \cong \frac{f}{D} - \left(\frac{Sf}{D^2}\right) x^*.$$

Under these conditions the flow will support potential vorticity waves with phase speeds given by

$$c^* = -\left(\frac{Sf}{D}\right) \frac{1}{(n^2 \pi^2/w^{*2}) + l^{*2}}, \quad n = 1, 2, 3, \cdots$$

In the long-wave limit ($w^* l^* \to 0$) the waves are nondispersive:

$$c^* = -\frac{Sf w^{*2}}{n^2 \pi^2 D} = \left(\frac{dq^*}{dx^*}\right) \frac{w^{*2}}{n^2 \pi^2} \frac{D}{}, \quad n = 1, 2, 3, \cdots \tag{2.1.30}$$

where w^* is the channel width and $\frac{dq^*}{dx^*} = -\frac{Sf}{D^2}$. This example is discussed fully by Pedlosky (2003). For positive S, $\partial q^*/\partial x^* < 0$ and higher potential vorticity is found on the left-hand side (facing positive y^*) of the channel. In this case the propagation tendency of the waves is towards negative y^*.

The waves produced in the last example are called topographic Rossby waves since the background potential vorticity gradient was created by a sloping bottom. More generally, steady flows with nontrivial depth and vorticity distributions have potential vorticity gradients and will support potential vorticity waves, although some of these may be unstable. The nondispersive character of the long waves is indicative of their ability to transmit upstream influence, an effect that will be demonstrated in later sections.

Exercises

(1) *Dissipation and vorticity flux.*

    (a) By taking the curl of the shallow water momentum equation (2.1.15) obtain the vorticity equation

$$\frac{\partial \zeta_a^*}{\partial t^*} + \nabla \cdot (\mathbf{u}^* \zeta_a^*) = \mathbf{k} \cdot (\nabla \times \mathbf{F}^*), \tag{2.1.31}$$

    where $\zeta_a^* = f + \zeta^*$ is the total (or *absolute*) vorticity of a fluid column.

(b) Define $\mathbf{J}_n^* = \mathbf{k} \times \mathbf{F}^*$ and write $\mathbf{k} \cdot (\nabla \times \mathbf{F}^*) = -\nabla \cdot \mathbf{J}_n^*$, so that (2.1.31) becomes

$$\frac{\partial \zeta_a^*}{\partial t^*} + \nabla \cdot (\mathbf{u}^* \zeta_a^* + \mathbf{J}_n^*) = 0. \tag{2.1.32}$$

The quantity $\mathbf{u}^* \zeta_a^* + \mathbf{J}_n^*$ may be interpreted as the total flux of absolute vorticity, the term $\mathbf{u}^* \zeta_a^*$ accounting for the advective part of the flux and the term $\mathbf{J}_n^*$ accounting for the dissipative flux.

(c) By taking the cross product of $\mathbf{k}$ with the steady version of (2.1.15) obtain the relation

$$\mathbf{k} \times \nabla B^* = \mathbf{u}^* \zeta_a^* + \mathbf{J}_n^*. \tag{2.1.33}$$

By comparing this with the relation $\mathbf{k} \times \nabla \psi^* = \mathbf{u}^*$ interpret B^* as a streamfunction for the total vorticity flux. Further show that the derivative of B^* along streamlines gives a vorticity flux that is entirely due to dissipation, whereas the derivative of B^* in the direction normal to streamlines gives a flux that is partly due to dissipation and partly due to advection.

[The main ideas developed in this exercise are due to Schär and Smith (1993).]

(2) Equation (2.1.24) paves the way for solution to the linear shallow water equations in terms of η. Show that the equivalent equations for u and v are given by

$$\frac{\partial^2 u}{\partial x^2} + \delta^2 \frac{\partial^2 u}{\partial y^2} - \delta^2 \frac{\partial^2 u}{\partial t^2} - u = -\frac{\partial}{\partial y} \left[\frac{\partial v_o}{\partial x} - \delta^2 \frac{\partial u_o}{\partial y} - \eta_o \right] \tag{2.1.34}$$

and

$$\frac{\partial^2 v}{\partial x^2} + \delta^2 \frac{\partial^2 v}{\partial y^2} - \delta^2 \frac{\partial^2 v}{\partial t^2} - v = \frac{\partial}{\partial x} \left[\frac{\partial v_o}{\partial x} - \delta^2 \frac{\partial u_o}{\partial y} - \eta_o \right]. \tag{2.1.35}$$

2.2.   Uniform Potential Vorticity: Boundary Layers and Kelvin Waves

The models of hydraulic behavior in rotating channels that first appeared in the 1970s consider flows with uniform potential vorticity:

$$q^* = \frac{f + \partial v^* / \partial x^*}{d^*} = \frac{f}{D_\infty} = const. \tag{2.2.1}$$

If q^* is materially conserved, such flows arise as a result of evolution from an initial state of uniform q^*. Or, a steady flow with uniform q^* is justified if all streamlines can be traced back into a reservoir or other source region with uniform q^*. Either case must be free of potential vorticity-altering dissipation or forcing, a situation that is probably not representative of deep-ocean overflows. There is no compelling reason to believe that the potential vorticity of such flows should be uniform and, indeed, the observation evidence is that it is not. Despite these objections, uniform potential vorticity models are quite tractable and give valuable insight into a dynamically restricted but intriguing arena. The main deficiency of such models is that they lack the ability to support potential vorticity waves. However, it is not clear that the latter are important for flows that are driven primarily by gravity.

Before entering into a detailed discussion of benchmark theories for steady flow (particularly Sections 2.4 and 2.5), we make some observations about uniform potential vorticity flow in general. First note that the potential depth D_∞ is now constant throughout the fluid and that the potential vorticity, in nondimensional terms, is $q = (D/f)q^* = D/D_\infty$. The depth scale D is simply a measure of the typical or average depth in a region of interest: often a section of the channel near a sill or narrows. The dimensionless potential vorticity is therefore a measure of the departure of this scale depth from the potential depth. In many hydraulics problems the average depth varies significantly between the sill or narrows and the upstream reservoir. If the average depth in the sill region is chosen as D then the nondimensional depth d may be regarded as O(1) there. In this case a *small* value of q would indicate that the typical depth at the sill is $<< D_\infty$, implying the sill region to be one of significant relative vorticity. In addition the small value of q can be exploited to further simplify the governing equations in what is called the 'zero potential vorticity' approximation. Caution is advised, however, as these properties only hold where d remains O(1). The values of d in the reservoir might be $\gg 1$, implying a completely different flow regime.

In most of what follows, the rotating channel is assumed to have a rectangular cross-section with vertical sidewalls at $x = \pm w(y)/2$, as shown in Figure 2.2.1. For this geometry equation (2.1.14) becomes

$$\frac{\partial^2 d(x, y, t)}{\partial x^2} - qd(x, y, t) = -1. \tag{2.2.2}$$

The boundary conditions at the edges of the stream depend upon whether the fluid depth remains nonzero over the entire cross section, as in Figure 2.2.1a, or whether the depth vanishes at one or more values of x, as in Figure 2.2.1b. As it turns out, there are a limited number of possible configurations. These may be found by first noting that if $d \to 0$ at some point (x_o, y) then one of the following conditions must hold:

(1) (x_o, y) lies at a side wall, as in Figure (2.2.2a),
(2) d is also zero in a neighborhood to the left ($x < x_o$) or to the right ($x > x_o$) of (x_o, y), as in Figure (2.2.2b), or
(3) $\partial d/\partial x$ is discontinuous at (x_o, y), as in Figure (2.2.2c).

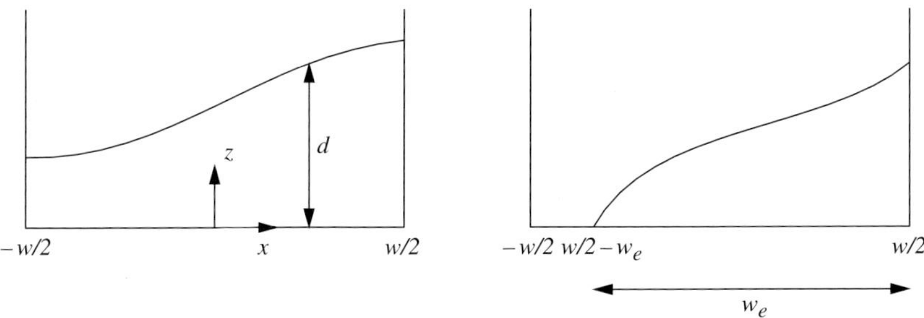

FIGURE 2.2.1. Cross sections for attached and separated flows, facing downstream. The notation corresponds to dimensionless quantities.

Most importantly, d cannot vanish smoothly in the interior of the current as shown in Figure (2.2.2d). The curvature $\partial^2 d/\partial x^2$ is clearly positive at such a point, whereas (2.2.2) indicates that it must be negative. In other words the fluid depth cannot smoothly vanish in the interior of the current; it must first vanish at a sidewall. Gill (1977) first showed this result for steady channel flow with uniform q but the same clearly applies to time-dependent flows with nonuniform q since (2.2.2) continues to apply. For northern hemisphere rotation, there are only two configurations we will generally have to consider and they are the ones shown in Figures 2.2.1a and 2.2.1b.

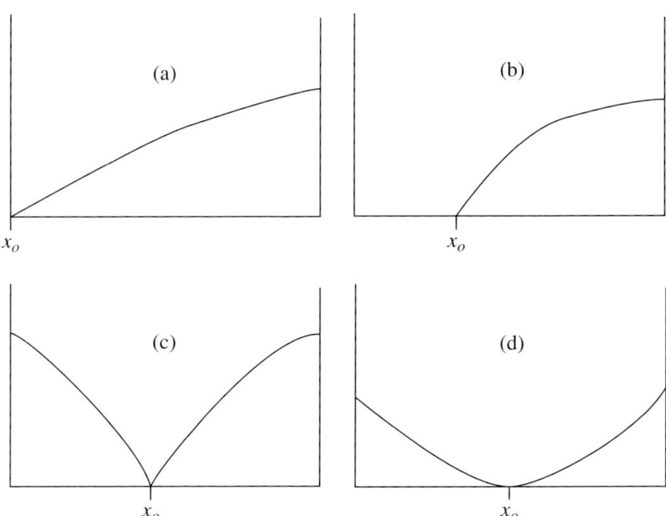

FIGURE 2.2.2. Hypothetical states of separation.

For $q =$ constant (2.2.2) can easily be solved and a convenient representation of the solution is given by

$$d(x, y, t) = q^{-1} + \hat{d}(y, t)\frac{\sinh(q^{1/2}x)}{\sinh(\frac{1}{2}q^{1/2}w)} + (\bar{d}(y, t) - q^{-1})\frac{\cosh(q^{1/2}x)}{\cosh(\frac{1}{2}q^{1/2}w)}, \quad (2.2.3)$$

following a form first used by Gill (1977). The geostrophic velocity associated with this depth is

$$v = q^{1/2}\hat{d}(y, t)\frac{\cosh(q^{1/2}x)}{\sinh(\frac{1}{2}q^{1/2}w)} + q^{1/2}(\bar{d}(y, t) - q^{-1})\frac{\sinh(q^{1/2}x)}{\cosh(\frac{1}{2}q^{1/2}w)}, \quad (2.2.4)$$

in view of (2.1.12).

The quantities $\bar{d}$ and $\hat{d}$ represent half the sum and difference of the depth along the sidewalls:

$$\bar{d} = \frac{1}{2}\left[d\left(\tfrac{1}{2}w, y, t\right) + d\left(-\tfrac{1}{2}w, y, t\right)\right] \quad (2.2.5)$$

and

$$\hat{d} = \frac{1}{2}\left[d\left(\tfrac{1}{2}w, y, t\right) - d\left(-\tfrac{1}{2}w, y, t\right)\right]. \quad (2.2.6)$$

Use of (2.2.4) allows these variables to be related to the average and difference of the wall velocities:

$$\bar{v} = \frac{1}{2}\left[v\left(\tfrac{1}{2}w, y, t\right) + v\left(-\tfrac{1}{2}w, y, t\right)\right] = q^{1/2}T^{-1}\hat{d} \quad (2.2.7)$$

and

$$\hat{v} = \frac{1}{2}\left[v\left(\tfrac{1}{2}w, y, t\right) - v\left(-\tfrac{1}{2}w, y, t\right)\right] = q^{1/2}T\left(\bar{d} - q^{-1}\right), \quad (2.2.8)$$

where

$$T = \tanh\left(\tfrac{1}{2}q^{1/2}w\right) \quad (2.2.9)$$

The dimensional form of the depth profile (2.2.3) is

$$d^* = dD = D_\infty + \hat{d}^*\frac{\sinh(x^*/L_d)}{\sinh(w^*/2L_d)} + (\bar{d}^* - D_\infty)\frac{\cosh(x^*/L_d)}{\cosh(w^*/2L_d)}, \quad (2.2.10)$$

where

$$L_d = \frac{(gD_\infty)^{1/2}}{f} \qquad (2.2.11)$$

is the Rossby radius of deformation based on the potential depth D_∞. In the limit $w^*/L_d \to \infty$ (2.2.10) becomes

$$d^* = D_\infty + [d(w^*/2, y^*, t^*) - D_\infty]e^{(x^* - \frac{1}{2}w^*)/L_d}$$

$$+ [d^*(-w^*/2, y^*, t^*) - D_\infty]e^{-(x^* + \frac{1}{2}w^*)/L_d}, \qquad (2.2.12)$$

and thus the solution takes on a boundary layer structure in which the interior depth has uniform value D_∞ and depth variation occurs within a distance L_d of the walls. The situation is depicted in Figure 2.2.3.

In the other extreme, a sufficiently narrow channel should force a return to the one-dimensional equations describing nonrotating flow. This limit is more subtle than what one might guess. Consider a channel whose width w^* is $\leq L_d$ so that w^* becomes the cross-channel length scale. Then a good indication of the strength of rotation is the change in the interface elevation across the channel

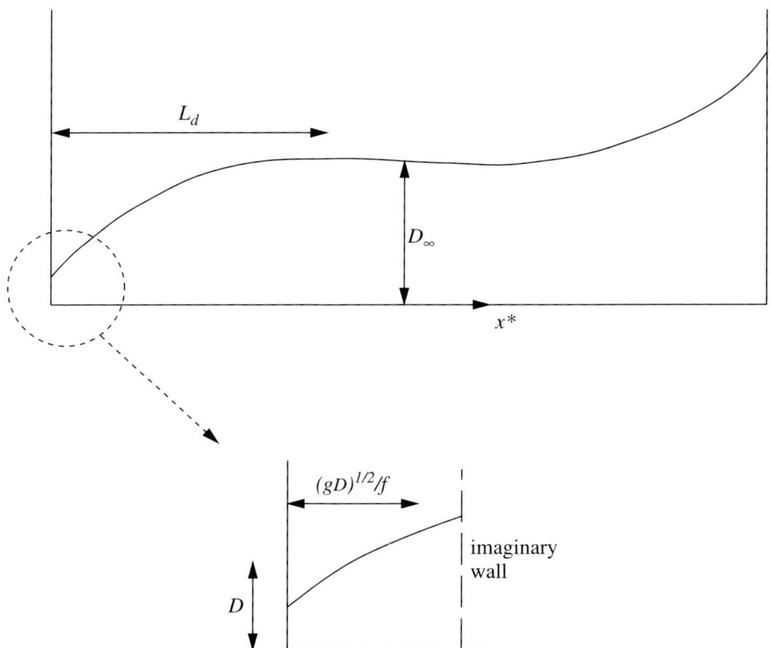

FIGURE 2.2.3. The cross section for the case in which the deformation radius L_d based on D_∞ is small compared to the channel width. The inset shows a segment of the flow that has been cut out and placed in an imaginary channel. The new flow is clearly influenced by rotation even though its width is $\ll L_d$.

divided by the average depth, equal to $\hat{d}^*/\overline{d}^*$ in the present model. An estimate for this quantity can be obtained by approximating $\partial^2 d/\partial x^2$ by $\hat{d}/w^2$ in (2.2.2) and writing the result in dimensional form. After a little rearrangement, one finds

$$\frac{\hat{d}^*}{\overline{d}^*} = O\left(\frac{w^{*2}}{L_d^2}\right) + O\left(\frac{w^{*2}f^2}{gD}\right),$$

implying that the channel width must be much less that two different scales in order for rotation to be negligible. The first is the Rossby radius of deformation based on the potential depth D_∞, also with width of the boundary layers discussed above. The second is the Rossby radius based on the local depth scale D. In most applications D will be $O(D_\infty)$ or less and therefore a good rule of thumb is to ignore rotation if w^* is $\ll (gD)^{1/2}/f$. In nondimensional terms, the latter is equivalent to $w \ll 1$.

The importance of two deformation radii may seem confusing to the reader who has observed that L_d appears to be the only intrinsic lateral length scale in the depth and velocity profiles. The situation can be clarified by reference to Figure 2.2.3 which shows a cross section for a case where w^* is somewhat larger than L_d. As suggested by (2.2.12) the depth near the center of the channel is $\cong D_\infty$. The overall depth scale D is also $\cong D_\infty$. Rotation is clearly important here as both L_d and $(gD)^{1/2}/f$ are $< w^*$. However, we could consider a second flow consisting of a short section of the depth profile near the left wall, as shown in the inset. An imaginary right wall is inserted a short distance from the left wall so that a new channel is formed. Although the width of the hypothetical channel is $\ll L_d$ rotation continues to be important because the *new* depth scale D is $\ll D_\infty$ and the deformation radius based on the new D approximates this width.

Recalling that the argument $q^{1/2}w$ in the velocity and depth profiles (2.2.3, 2.2.4) is equivalent to w^*/L_d, and that $w = w^*f/(gD)^{1/2}$, we can now distinguish between two 'narrow' channel limits. In the first, $w^* \ll L_d$ but rotation is still important $[w = O(1)]$, so that

$$q \ll 1 \text{ and } w = O(1).$$

This case is sometimes called the *zero potential vorticity limit* and will be explored shortly. The second, more severe limit is that of negligible rotation:

$$q \le O(1) \text{ and } w \ll 1.$$

The distinguishing features of these limits are hidden by the fact that $q^{1/2}w$ is vanishingly small in each case.

The y- and t- dependence of the solutions can be obtained by first evaluating the y-momentum equation (2.1.6) at both channel sidewalls. After some manipulation, which is left as Exercise (2.2.3), the wall version of the momentum equations can be written as

$$\frac{\partial v(\pm w/2, y, t)}{\partial t} + \frac{\partial B(\pm w(y)/2, y, t)}{\partial y} = 0 \qquad (2.2.13)$$

where

$$B = \left[\tfrac{1}{2}v^2 + d + h\right] \tag{2.2.14}$$

is the semigeostrophic Bernoulli function.

The difference and sum of the two sidewall equations in (2.2.13) are next taken and the relations $\bar{v} = q^{\frac{1}{2}}T^{-1}\hat{d}$ and $\hat{v} = q^{1/2}T(\bar{d} - q^{-1})$ used to obtain

$$2q^{-1/2}\frac{\partial(T\bar{d})}{\partial t} + \frac{\partial Q}{\partial y} = 0 \tag{2.2.15}$$

and

$$q^{1/2}\frac{\partial(T^{-1}\hat{d})}{\partial t} + \frac{\partial \bar{B}}{\partial y} = 0, \tag{2.2.16}$$

where

$$\begin{aligned}
\bar{B} &= [B(w/2, y, t) + B(-w/2, y, t)]/2 \\
&= \tfrac{1}{2}q[T^{-2}\hat{d}^2 + T^2(\bar{d} - q^{-1})^2] + \bar{d} + h
\end{aligned} \tag{2.2.17}$$

is the average of the Bernoulli function on the two side walls and

$$Q = 2\bar{d}\hat{d} \tag{2.2.18}$$

is the volume transport over the channel cross section. The latter follows from multiplication of the geostrophic relation by v and integration of the result across the channel:

$$Q = \int_{\frac{-w}{2}}^{\frac{w}{2}} (vd)dx = \int_{\frac{-w}{2}}^{\frac{w}{2}} d\frac{\partial d}{\partial y}dx = \tfrac{1}{2}[d^2(\frac{w}{2}, y, t) - d^2(\frac{-w}{2}, y, t)] = 2\hat{d}\bar{d}. \tag{2.2.19}$$

Thus (2.2.15) is a statement of mass conservation and (2.2.16) is an expression of momentum conservation. (The latter is made more evident if (2.2.7) is used to show that the first term in (2.2.16) is just $\partial\bar{v}/\partial t$.)

The notation and algebra should not distract the reader from the close relation that (2.2.15) and (2.2.16) bear to the one-dimensional momentum and continuity equations (1.2.1 and 1.2.2). As with the latter, the essential aspects of wave propagation and other transient behavior can be discussed by casting the equations in characteristic form. When the channel width and bottom elevation are constant, conserved Riemann invariants can be found by following a general procedure laid out in Appendix B (or Exercise 1 of Section 1.3). If the channel width is assumed constant, (2.2.15) and (2.2.16) can be can written as

$$\frac{d_{\pm}R_{\pm}}{dt} = -\frac{dh}{dy}, \tag{2.2.20}$$

where

$$\frac{d_{\pm}}{dt} = \frac{\partial}{\partial t} + c_{\pm}\frac{\partial}{\partial y}, \tag{2.2.21}$$

$$\begin{aligned} c_{\pm} &= q^{1/2}T^{-1}\hat{d} \pm \overline{d}^{1/2}[1 - T^2(1 - q\overline{d})]^{1/2} \\ &= \overline{v} \pm \overline{d}^{1/2}[1 - T^2(1 - q\overline{d})]^{1/2}, \end{aligned} \tag{2.2.22}$$

$$R_{\pm} = q^{1/2}T^{-1}\hat{d} \pm \int^{\overline{d}} r(\alpha)\,d\alpha, \tag{2.2.23}$$

and

$$r(\overline{d}) = \overline{d}^{-1/2}[1 - T^2(1 - q\overline{d})]^{1/2}. \tag{2.2.24}$$

The procedure used to obtain these expressions is discussed in Appendix B and the derivation itself appears in Pratt (1983b).

As before, there are two wave modes characterized by the Riemann invariants $R_{\pm}$. If the channel bottom is horizontal $(dh/dy = 0)$, $R_{\pm}$ are conserved following the corresponding characteristic speeds $c_{\pm}$. The presence of a particular mode is indicated by variations with y in its Riemann invariant. If R_{+} is uniform, for example, there is no 'forward propagating' wave; that is, no wave with speed c_{+}. The term 'wave' should be interpreted in a very general way to mean signals that propagate within a flow field that may be rapidly changing and strongly nonlinear. If one follows a particular signal (that is a particular value of R_{+} or R_{-}) the speed of that signal depends only on the local properties of the flow field and may itself vary rapidly in space and time.

The expression (2.2.22) for $c_{\pm}$ tempts one to interpret $\overline{v}$ as an advective speed due to the current, and $\pm\overline{d}^{1/2}[1 - T^2(1 - q\overline{d})]^{1/2}$ as a propagation speed relative to the current. This interpretation is not entirely correct, as can be seen by taking the wide-channel limit $q^{1/2}w \to \infty$ (or $T \to 1$), resulting in $c_{\pm} = \overline{v} \pm \overline{d}q^{1/2}$. Substituting $\hat{v} + q^{-1/2}$ for $\overline{d}q^{1/2}$, which follows from (2.2.8), leads to $c_{\pm} = \overline{v} \pm (\hat{v} + q^{-1/2})$, or $c_{+} = v(w/2, y, t) + q^{-1/2}$ and $c_{-} = v(-w/2, y, t) - q^{-1/2}$. The expression $\pm\overline{d}^{1/2}[1 - T^2(1 - q\overline{d})]^{1/2}$ therefore contains a hidden advection component.

If (2.2.4–2.2.6) are used to express the wall velocities in terms of the wall depths, the characteristic speeds for wide channels can be further simplified and written, in dimensional form, as

$$c_{+}^{*} = v^{*}(w^{*}/2, y^{*}, t^{*}) + (gD_{\infty})^{1/2} = \left(\frac{g}{D_{\infty}}\right)^{1/2} d^{*}(w^{*}/2, y, t) \tag{2.2.25}$$

and

$$c_{-}^{*} = v^{*}(-w^{*}/2, y^{*}, t^{*}) - (gD_{\infty})^{1/2} = -\left(\frac{g}{D_{\infty}}\right)^{1/2} d^{*}(-w^{*}/2, y, t) \tag{2.2.26}$$

The two modes can now be identified as nonlinear Kelvin waves trapped along the two sidewalls. As shown by the cross-sectional depth profile (2.2.12) the trapping scale is L_d. For the wave trapped on the right-hand wall (facing in the positive y-direction) the characteristic speed is positive provided the depth at that wall is nonzero (the flow is attached). The left wall counterpart propagates towards negative y provided the flow there is attached. All wide-channel flows of finite depth are therefore subcritical in the sense that the two long waves propagate in opposite directions. In order to reverse the propagation speed of one of the waves, it is therefore necessary for the flow to separate or for the channel width to be less than several L_d, forcing the Kelvin waves to overlap. Finally, note that the linear case can be obtained by allowing $d^*(-w^*/2, y^*, t^*)$ or $d^*(w^*/2, y^*, t^*)$ to approach D_∞. This limit causes the velocity field to vanish, so that the waves propagate on a uniform, quiescent state with depth D_∞ at speeds $\pm(gD_\infty)$.

The right-wall Kelvin wave will steepen if the wall depth, and therefore c_+, decreases with positive y. It can be shown that $R_+ = c_+$ (and $R_- = c_-$) and therefore the right-wall depth is conserved following the characteristic speeds. From (2.1.6) it can also be shown that the cross-channel velocity in an evolving right-wall wave is given by

$$u^* = \frac{g}{2fd^*}(1 - e^{(x^*-w^*/2)/L_d})e^{(x^*-w^*/2)/L_d}\frac{\partial[d^*(w^*/2, y, t) - D_\infty]^2}{\partial y}. \qquad (2.2.27)$$

The quantity $d^*(w^*/2, y, t) - D_\infty$ is the height of the free surface at the right wall above the resting interior depth, and can be thought of as a wave amplitude. The fact that the square of this amplitude appears in (2.2.27) is an indication that the cross-channel velocities are generated by nonlinear advection. In fact, it is well known that linear Kelvin waves have $u \equiv 0$. As a wave steepens, the y-derivative of the square of the amplitude increases and strong cross-channel velocities are generated. For a wave of elevation propagating into a quiescent region, the square of the amplitude decreases with y and the steepening generates strong u^* away from the wall. The combination of diminishing along-channel scale and increasing u^* may lead to violations of the semigeostrophic (and perhaps even the hydrostatic) approximation.

For channel widths of the order of L_d or less, each Kelvin wave is felt all across the channel. A limiting case of this behavior is that of zero potential vorticity: $q \to 0$ and $w = O(1)$. With $q = D/D_\infty \ll 1$, the limit is approached when the channel flow is fed by fluid originating from a relatively deep, quiescent reservoir. The reservoir depth is therefore the potential depth D_∞ and the process of squashing fluid columns to the scale depth D implies that

$$\frac{f + \partial v^*/\partial x^*}{f} = O\left(\frac{D}{D_\infty}\right) \ll 1, \qquad (2.2.28)$$

in view of (2.2.1). Thus the relative vorticity $\partial v^*/\partial x^*$ of the fluid in the channel is nearly equal and opposite to the planetary vorticity f, and the absolute vorticity is much less than f.

The cross-channel structure for this case can be found by taking the limit $q \to 0$ while assuming that w, $\hat{d}$, and $\overline{d}$ remain O(1) in (2.2.3) and (2.2.4), or simply by solving the cross-channel structure equation (2.1.14) with $q = 0$. In either case one obtains

$$d = \overline{d}(y, t) + \frac{2\hat{d}(y, t)x}{w} - \frac{x^2 - (w/2)^2}{2} \tag{2.2.29}$$

and

$$v = \overline{v}(y, t) - x, \tag{2.2.30}$$

where $\overline{v} = 2\hat{d}/w$. The velocity therefore varies linearly across the channel while the depth variation is quadratic. It is important to realize that (2.2.29) and (2.2.30) are not uniformly valid over the entire length of channel. If one moves upstream into the reservoir, $\overline{d} \to \infty$ the assumption that d remains O(1) as $q \to 0$ no longer holds. (The term qd in the cross-channel structure equation remains finite and the full depth and velocity profiles (2.2.3) and (2.2.4) would hold.)

The zero potential vorticity characteristic wave speeds and Riemann invariants are

$$c_{\pm} = \overline{v} \pm \overline{d}^{1/2} \tag{2.2.31}$$

and

$$R_{\pm} = \overline{v} \pm 2\overline{d}^{1/2}, \tag{2.2.32}$$

the same as the expressions for nonrotating, one-dimensional flow if the v and d are replaced by the average of their wall values. Steepening and rarefacation of disturbances can therefore be treated the same as in the nonrotating limit. However, when a shock (intersection of characteristic curves) occurs, the subsequent development is potentially much different.

Exercises

(1) If q is uniform and nonnegative, and the channel flow is semigeostrophic, show that there can be at the most one point of flow reversal ($v = 0$ where $\partial v/\partial x \neq 0$) across any channel section.

(2) Derive equation (2.2.15) by integrating the continuity equation across the channel. (Be sure to allow for the possibility that w varies with y.)

(3) Derive (2.2.13) by writing the semigeostrophic momentum equations along the sidewalls of the channel. (Hint: use the kinematic boundary conditions $u(\pm w/2, y, t) = v(\pm w/2, y, t)\partial(\pm w/2)/\partial y$. Also make the replacement

$$\left(\frac{\partial v(x, y, t)}{\partial y}\right)_{x=\pm w/2} = \frac{\partial v(\pm w(y)/2, y, t)}{\partial y} \mp \frac{\partial v}{\partial x}\frac{\partial(w/2)}{\partial y}, \tag{2.2.33}$$

which expresses the y-derivative of v at constant x in terms of the y-derivative of the wall value of v.)

(4) Derive (2.2.20–2.2.24) by following the same procedure for obtaining characteristics and Riemann invariants laid out in Exercise 1 of Section 1.3.

2.3. Flow Separation and Frontal Waves

If the fluid depth goes to zero over a finite portion of the channel cross section, the interface or free surface forms a free edge or 'front'. Arguments presented in the previous section prove that this grounding must first occur at the left sidewall (Figure 2.1b), provided that the potential vorticity is nonnegative. Separation of the layer from the left wall means that the corresponding Kelvin wave is replaced by a new 'frontal' wave whose properties are quite different from its predecessor. This twofold behavior is an artifact of the rectangular channel geometry; real ocean straits have continuously varying h and layer thickness that always vanishes at the edges. However, the inconvenience in treating attached and detached flows separately is minor compared to the technical difficulties in dealing with smooth cross sections (e.g. Section 2.8).

For detached flow it is not necessary to re-derive the depth and velocity profile; one can simply modify the old ones. In doing so, the reader should keep in mind that (2.2.3) and (2.2.4) remain valid when the depth just vanishes at the left wall. For a flow that is separated, one can place an imaginary wall at the free left edge and adjust the coordinate to conform to the imaginary channel. Since (2.2.3) and (2.2.4) assume symmetry in the position of the channel walls about $x = 0$, we replace x by $x - x_c$ where $x_c = (w - w_e(y, t))/2$ is the midpoint of the separated current. The edges of the current now lie at $x - x_c = \pm w_e/2$ and the condition of vanishing depth at the left edge implies $\hat{d} = \overline{d} = \frac{1}{2}d(w/2, y, t)$. The new depth and velocity are then given by

$$d(x, y, t) = q^{-1} + \overline{d}(y, t)\frac{\sinh[q^{1/2}(x - x_c(y, t))]}{\sinh\left(\frac{1}{2}q^{1/2}w_e(y, t)\right)}$$

$$+ (\overline{d}(y, t) - q^{-1})\frac{\cosh[q^{1/2}(x - x_c(y, t))]}{\cosh\left(\frac{1}{2}q^{1/2}w_e(y, t)\right)}. \qquad (2.3.1)$$

and

$$v(x, y, t) = q^{1/2}\overline{d}(y, t)\frac{\cosh[q^{1/2}(x - x_c(y, t))]}{\sinh\left(\frac{1}{2}q^{1/2}w_e(y, t)\right)}$$

$$+ q^{1/2}(\overline{d}(y, t) - q^{-1})\frac{\sinh[q^{1/2}(x - x_c(y, t))]}{\cosh\left(\frac{1}{2}q^{1/2}w_e(y, t)\right)}. \qquad (2.3.2)$$

From these expressions follow modified versions of (2.2.7) and (2.2.8):

$$\bar{v} = q^{1/2} T_e^{-1} \bar{d} \tag{2.3.3}$$

and

$$\hat{v} = q^{1/2} T_e (\bar{d} - q^{-1}), \tag{2.3.4}$$

where $T_e = \tanh(\frac{1}{2} q^{1/2} w_e(y, t))$.

One may now repeat the steps outlined in the previous section to obtain equations governing the evolution of the free edge $x = \frac{1}{2} w(y) - w_e(y, t)$. Begin by writing the y-momentum equation at the free edge and using the kinematic condition

$$\frac{\partial v(\frac{1}{2} w(y) - w_e(y, t), y, t)}{\partial t} = \left[\frac{\partial v}{\partial t} \right]_{x=const.} - \left[\frac{\partial v}{\partial x} \right]_{x=\frac{1}{2} w - w_e} \frac{\partial w_e}{\partial t}. \tag{2.3.5}$$

The result is

$$\frac{\partial v_e}{\partial t} - \frac{\partial w_e}{\partial t} + \frac{\partial}{\partial y} \left(\frac{v_e^2}{2} + h \right) = 0, \tag{2.3.6}$$

where v_e is the value of v at the free edge.

To the momentum equation written along the right wall (2.2.13 with the '+' sign) one now subtracts or adds (2.3.6), resulting in

$$\frac{\partial [q^{1/2} T_e (\bar{d} - q^{-1})]}{\partial t} + \frac{1}{2} \frac{\partial w_e}{\partial t} + \frac{1}{2} q \frac{\partial Q}{\partial y} = 0 \tag{2.3.7}$$

and

$$\frac{\partial (q^{1/2} T_e^{-1} \bar{d})}{\partial t} - \frac{1}{2} \frac{\partial w_e}{\partial t} + \frac{\partial \bar{B}}{\partial y} = 0, \tag{2.3.8}$$

where

$$\bar{B} = \frac{1}{2} q [T_2^{-e} \bar{d}^2 + T_e^2 (\bar{d} - q^{-1})^2] + \bar{d} + h \tag{2.3.9}$$

and

$$Q = 2\bar{d}^2 \tag{2.3.10}$$

Note that the two dependent variables are now $\bar{d}$ (equivalent to half the depth at the right wall) and the stream width w_e (as contained in T_e). Equations (2.3.7) and (2.3.8) can be interpreted as momentum and continuity relations for the cross-sectional flow as a whole.

It is possible to write (2.3.7) and (2.3.8) in characteristic form (Stern *et al.*, 1982) and to show that the characteristic speeds are given by (2.2.22) with $\hat{d} = \overline{d}$ and $w = w_{\mathrm{e}}$ (Kubokawa and Hanawa, 1984a). The Riemann invariants cannot be obtained in closed form and must be determined numerically (Stern *et al.*, 1982). In the interest of simplicity, we will explore two limiting cases, those of narrow and wide stream widths compared with L_{d}. The narrow-channel limit again corresponds to $q \to 0$, now with w_{e} fixed, and was first described by Stern (1980). The depth and velocity profiles can be obtained as limiting cases of (2.3.1) and (2.3.2), or simply by solving (2.1.14) with $q = 0$. The resulting depth and velocity profiles are given by

$$d = \frac{\overline{d}(x - \tfrac{1}{2}w + w_e)}{\tfrac{1}{2}w_e} - \tfrac{1}{2}(x - \tfrac{1}{2}w)(x - \tfrac{1}{2}w + w_e), \qquad (2.3.11)$$

and

$$v = \overline{v}(y, t) - x + \tfrac{1}{2}w - \tfrac{1}{2}w_e, \qquad (2.3.12)$$

with

$$\overline{v} = 2\overline{d}/w_e \qquad (2.3.13)$$

A more natural representation of the depth can be found in terms of the distance $x' = x - \tfrac{1}{2}w + w_e$ measured from the free edge of the current (see Figure 2.3.1a). This replacement leads to

$$d = x'(v_e - \tfrac{1}{2}x'), \qquad (2.3.14)$$

where

$$v_e = \overline{v} + \tfrac{1}{2}w_e. \qquad (2.3.15)$$

The velocity is given by $v(x, y) = v_e(y) - x'$. When extended past the wall, as shown in Figure 2.3.1a, the depth profile is a parabola with a maximum value $d_{\max} = \tfrac{1}{2}v_e^2$. The shape of the profile is independent of the position of the wall and variations of the profile with y or t can be thought of as a combination of lateral displacement with respect to the wall (due to changes in w_e) and uniform expansion or contraction of the profile (due to changes in v_e). When $w_e > v_e$ (Figure 2.3.1b) the depth has a maximum to the left of the wall implying negative v along the wall and positive v further offshore. When $w_e < v_e$, as in Figure 2.3.1c, there is no depth maximum and $v > 0$ across the entire stream.

The equations governing the evolution of the above profiles in y and t can be obtained from (2.3.7) and (2.3.8) in the limit of small q. Use of the expansion

$$T_e \approx q^{1/2}\frac{w_e}{2} - \tfrac{1}{3}\left(q^{1/2}\frac{w_e}{2}\right)^3 + \cdots$$

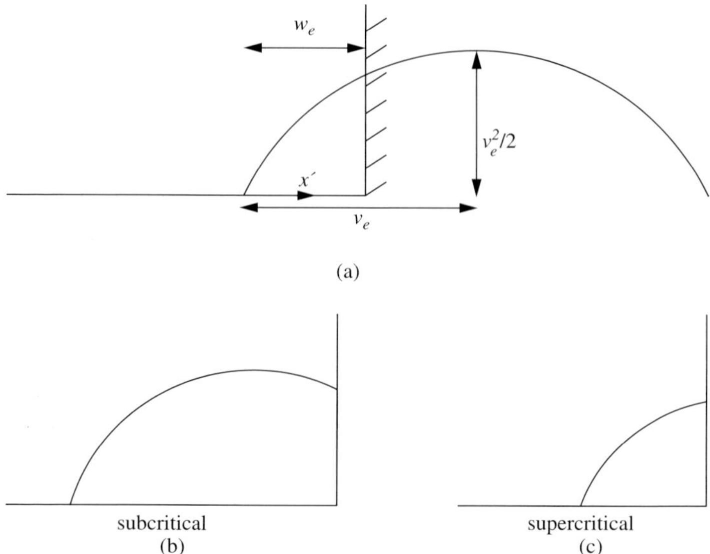

FIGURE 2.3.1. Possible cross sections for separated flow with zero potential vorticity.

in (2.3.7) along with the relation $\overline{d} = \frac{1}{2}w_e(v_e - \frac{1}{2}w_e)$ leads to

$$\frac{\partial}{\partial t}[w_e^2(v_e - \frac{1}{3}w_e)] + \frac{\partial}{\partial y}2Q = 0 \tag{2.3.16}$$

where $Q = \frac{1}{2}\left[w_e(v_e - \frac{1}{2}w_e)\right]^2$. In addition, the value of B is uniform in x for $q = 0$ and this allows (2.3.6) to be written as

$$\frac{\partial}{\partial t}(v_e - w_e) + \frac{\partial}{\partial y}\overline{B} = 0 \tag{2.3.17}$$

where $\overline{B} = \frac{v_e^{\,2}}{2} + h$.

Equations (2.3.16) and (2.3.17) may be written in the standard characteristic form[1] with

$$c_{\pm} = \frac{v_e - w_e}{1 \mp [w_e/(2v_e - w_e)]^{1/2}} \tag{2.3.18}$$

and

$$R_{\pm} = \int^{v_e/w_e} \frac{1}{1 - \xi - \dfrac{2\xi}{1 \pm (2\xi - 1)^{1/2}}}\,d\xi - \ln(w_e). \tag{2.3.19}$$

---

[1] The characteristic speeds can also be written as $c_{\pm} = \overline{v} \pm \overline{d}^{1/2}$, which was the expression obtained for the characteristic speeds in a nonseparated flow. The $\pm$ sign has been reversed from how it appears in Stern (1980) so that the '−' waves are the only ones with the ability to propagate upstream.

$R_\pm$ is conserved following the characteristic speed $c_\pm$, provided that $h = $ constant. Note that both characteristic speeds are zero for $w_e = 2v_e$, corresponding to the case in which the right wall depth is zero (the flow is separated from both walls). The right-wall depth is finite for $0 < w_e < 2v_e$ and careful evaluation of (2.3.18) shows that c_+ is always > 0 over this range. The behavior of c_- is more complicated, with $c_- < 0$ for $v_e < w_e < 2v_e$ and $c_- > 0$ for $w_e < v_e$. Thus, the flow is subcritical ($c_+ > 0$ and $c_- < 0$) when $v_e < w_e < 2v_e$, in which case the depth profile resembles that of Figure 2.3.1b (with reverse velocity along the right wall). The flow is supercritical ($c_\pm > 0$) when $w_e < v_e$, in which case the depth profile resembles that of Figure 2.3.1c and the velocity is unidirectional. The different possibilities are shown as insets in Figure 2.3.2.

Insight into various wave forms and transient motions can be gained by looking at a plot of the Riemann functions in (w_e, v_e) space (Figure 2.3.2). The solid

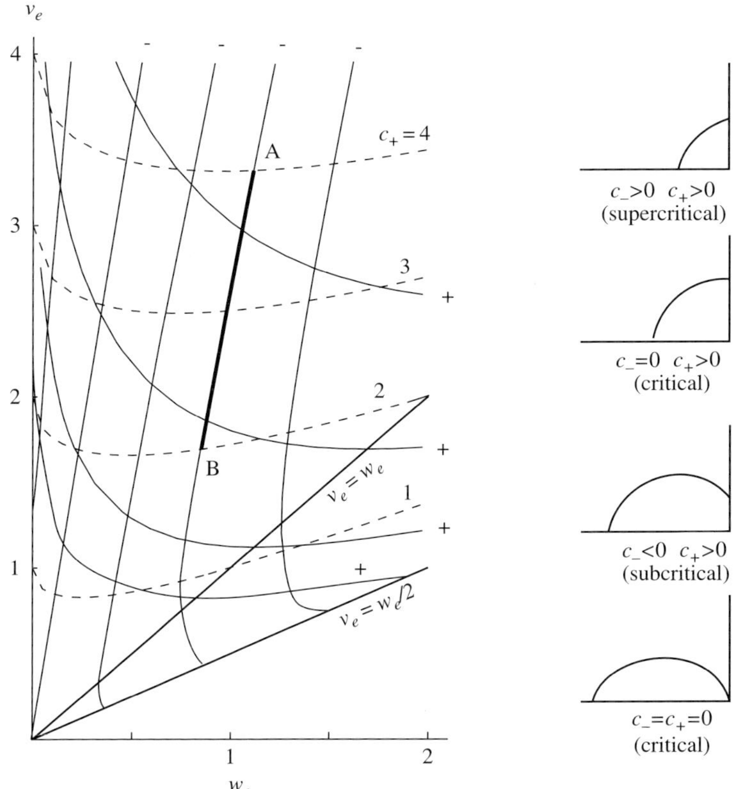

FIGURE 2.3.2. Riemann invariants and characteristic speeds for separated, zero potential vorticity flow. Solid contours show constant values of R_+ and R_-, as indicated by a '+' or '−'. The dashed curves show values of the characteristic wave speed c_+. The insets show different states of criticality corresponding to particular cross sections. (Based on a similar figure in Stern, 1980).

contours correspond to constant R_+ or R_-, as labeled. The curves terminate at the diagonal $v_e = w_e/2$, along which $c_\pm = 0$ and below which the flow is separated from both walls. Paldor (1983) has shown that this doubly separated state is unstable.[2] Slightly above is a second diagonal $v_e = w_e$ along which the flow is critical, $c_- = 0$ (and $c_+ > 0$). In the wedge-shaped region between these two lines the flow is subcritical and above it the flow is supercritical. Contours of constant c_+ correspond to dashed contours. Figure 2.3.3 shows part of the same parameter space with contours of constant c_-.

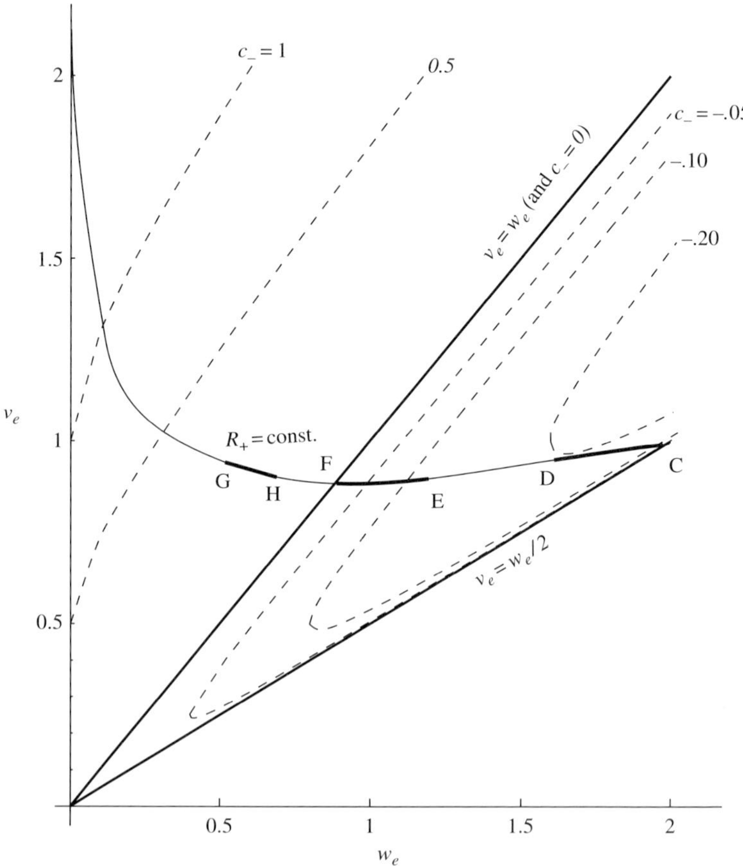

FIGURE 2.3.3. Same as Figure 2.3.2 but showing contours of the characteristic speed c_-.

---

[2] Paldor has shown that the flow is provably stable for $v_e > 3w_e/2$. Also, by his direct numerical calculation of eigenfunctions, he has suggested that the flow is stable for $3w_e/2 > v_e > w_e/2$. At $v_e = w_e/2$ the entire flow separates and, as we have shown, both long waves take on the same speed (zero). Resonance between these two waves produces an instability. This subject is taken up in a more general context in Section 3.9.

The orientation of the curves of constant R_+ and R_- gives clues concerning the differences between the '+' and '−' waves. Over most of the (w_e, v_e) -plane, the $R_+ = $ constant curves tend to be horizontal whereas the $R_- = $ constant curves tend to be vertical. Variations in R_- therefore tend to be associated with variations in the stream width; that is, lateral shifts of the fixed depth profile relative to the right wall. We refer to the corresponding disturbances as *frontal waves*. Variations in R_+ are more closely associated with variations in v_e that, in turn, are linked to uniform expansions of the depth profile. Plots of Riemann invariants for finite values of q (e.g. Stern *et al.*, 1982) display the same tendencies. Frontal waves are sometime referred to as Kelvin waves in the literature, but their signature lateral motion is more characteristic of potential vorticity waves.

Now consider a transient disturbance that has $R_- = $ constant and therefore consists only of forward propagating signals-those assigned the '+' sign. An initial state for this 'simple wave' can be constructed by choosing $v_e(y, 0)$ and calculating $w_e(y, 0)$ by tracing along a particular $R_- = $ constant curve in Figure 2.3.2. Suppose that we use the initial distribution shown in Figure 2.3.4a, where v_e decreases with increasing y. This distribution might span the segment **AB** of Figure 2.3.2. Since both $R_-(w_e, v_e)$ and $R_+(w_e, v_e)$ are constant following

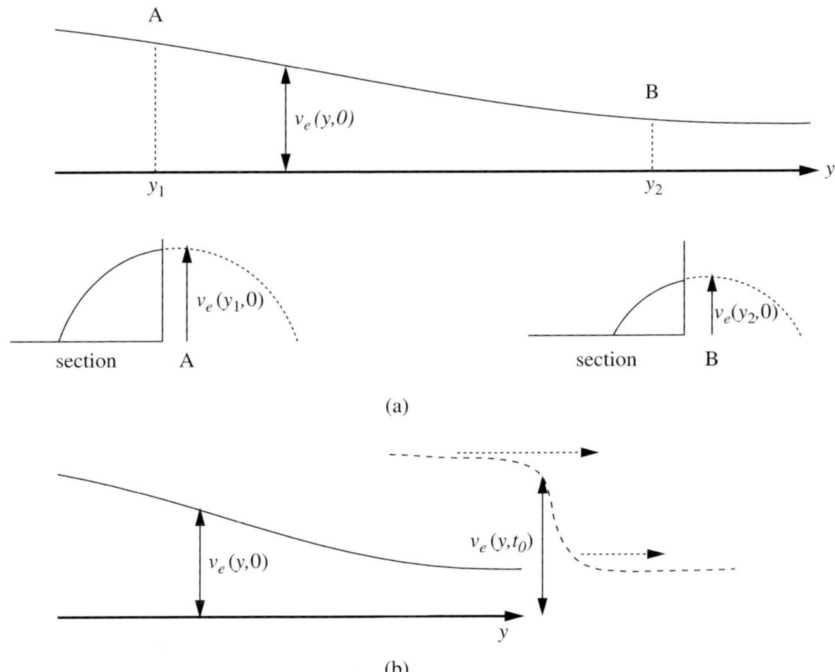

FIGURE 2.3.4. The evolution of a modified gravity wave (with uniform R_-) for which the initial distribution of the free-edge velocity is specified.

the characteristic speed c_+ for this initial condition, w_e and v_e must be individually conserved along the corresponding characteristic curve. An observe moving at speed c_+ sees fixed values of v_e and w_e and therefore a fixed cross-sectional profile. The value of w_e is nearly constant along **AB** and therefore the variation in cross section from **A** to **B** is primarily one in wall depth (Figure 2.3.4a). From the contours of c_+ shown in Figure 2.3.2, all of which have positive values, the characteristic speed corresponding to **A** is larger than that associated with **B**. Therefore the *entire profile* at **A** will move more rapidly in the positive y-direction than the **B** profile, resulting in wave steepening (Figure 2.3.4b).

Next consider a case in which the '+' waves are filtered out of the flow by a choice of initial condition with $R_+ = \text{constant}$. The remaining frontal waves are associated more with variations in w_e than in v_e and it therefore makes sense to choose $w_e(y, 0)$ and calculate the corresponding $v_e(y, 0)$. The latter can be accomplished by tracing along the $R_+ = \text{constant}$ curve shown in Figure 2.3.3. Consider the initial condition shown in Figure 2.3.5a, with $dw_e(y, 0)/dy < 0$. As shown by the dashed contours of Figure 2.3.3, the behavior of c_- is somewhat

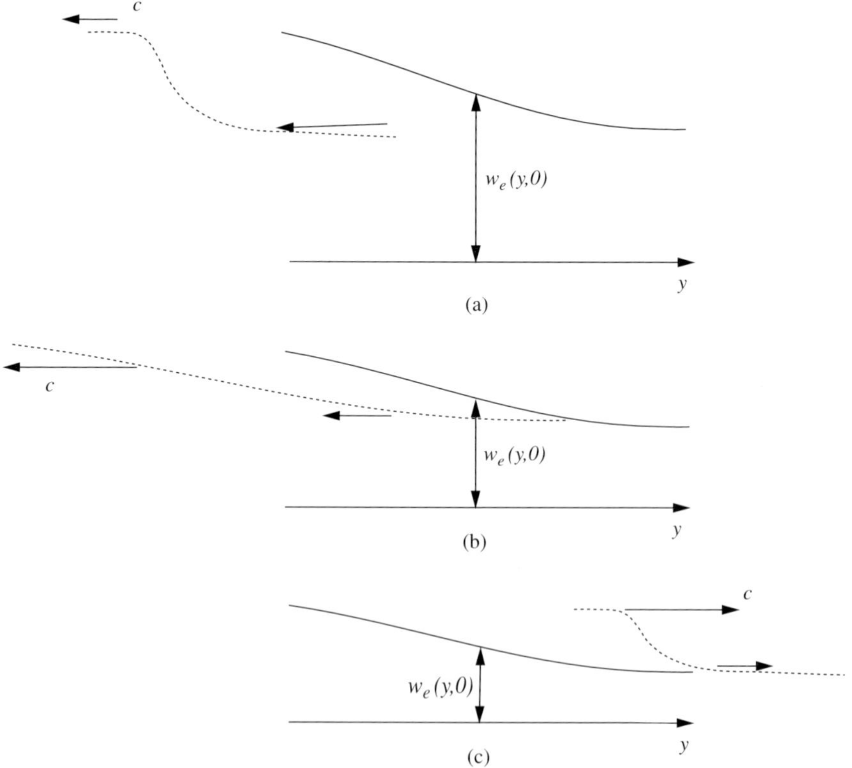

FIGURE 2.3.5. The evolution of a frontal disturbance (with uniform R_+) for which the initial distribution of the free-edge position is specified.

more complicated than that for c_+. If the initial condition spans the segment **CD**, then c_- is negative with larger absolute values associated with smaller widths. In this case the frontal wave will propagate to the left and steepen, as in Figure 2.3.5a. On the other hand, an initial condition of the same general shape and spanning the segment **EF** will rarefy, as suggested in Figure 2.3.5b.

An example of steepening of the 'frontal' wave is shown in Figure 3.4.12. The wave is generated in the region $4 < y < 7$ of the $t = 10$ frame, where the current widens. The current is supercritical and the narrow and wide end states correspond to something like points **G** and **H** in Figure 2.3.3. The narrower, upstream end state propagates forward and the greater speed in this case overtakes the wider portion ($t = 20$ frame near $y = 10$) eventually leading to the near detachment of a blob of fluid ($t = 40$).

The other limiting case is that of a relatively wide stream, $w_e{}^* \gg L_d$ (or $q^{1/2}w \gg 1$). Here the Kelvin wave trapped to the right wall of the channel is isolated from the free edge of the stream and therefore the propagation speed is given by the formula (2.2.25) for attached flow. The frontal wave is trapped to the free edge and has properties quite different from those of the left-wall Kelvin wave that it replaces. These new features are revealed by examining (2.3.6), the momentum equation for the flow at the free edge. The velocity v_e can be found by taking the limit of (2.3.2) as $q^{1/2}w \to \infty$ and evaluating the result at the free edge, leading to $v_e = q^{-1/2}$. The free-edge velocity is constant and thus (2.3.6) gives

$$\frac{\partial w_e}{\partial t} = 0 \ (h = \text{constant}).$$

Any initial distribution $w_e(y, \ 0)$ therefore remains frozen in the flow, implying that the characteristic speed for such disturbances, c_-, is zero. A wide, separated flow over a horizontal bottom is therefore always critical with respect to a frontal wave.[3]

2.4.   Steady Flow from a Deep Basin: The WLK Model

Following discussions with H. Stommel in the early 1970s, Whitehead *et al.*, (1974), hereafter WLK, developed the first analytical model of hydraulic behavior in a steady, rotating-channel flow with topography. Their model utilizes rectangular cross-sectional geometry and is based on the zero potential vorticity limit [$q \to 0$ with $w = 0(1)$]. Since $q = D/D_\infty$ and $w = w^* f/(gD)^{1/2}$, the channel width is comparable to the Rossby deformation radius based on the local depth

---

[3] Cushman-Roisin, Pratt and Ralph (1993) have explored the slow evolution of the frontal waves in a wide flow when weak dispersive effects are introduced. Expansion in powers of the aspect ratio δ shows that the free edge of the stream is governed by the modified Korteweg-de-Vries equation. As it turns out, only propagation in the positive y-direction is permitted.

scale D, while D itself is much less than the potential depth. The situation envisioned by WLK is that the flow is fed from a reservoir that is deep and quiescent, and in which the layer thickness is therefore D_∞. Fluid is drawn up from the reservoir and over a relatively shallow sill, where the depth scale $D \ll D_\infty$.[4] With $q = 0$ the absolute vorticity of the fluid is zero ($\partial v/\partial x + 1 = 0$) and the depth profiles are given by (2.2.29) for attached flow and (2.3.14) for separated flow. These profiles are valid only as long as the local depth remains much smaller than the reservoir depth. The calculation therefore cannot be extended indefinitely far upstream from the shallow section of channel.

For the case of attached flow, y-variations of the current are governed by the steady versions of (2.2.15) and (2.2.16), which express conservation of the volume flow rate Q and average $\overline{B}$ of the sidewall Bernoulli functions. The flow rate is defined in terms of average $(\overline{d})$ and difference $(\hat{d})$ of the wall depths by

$$Q = 2\hat{d}\overline{d}. \tag{2.4.1}$$

In the limit $q \to 0$, the average Bernoulli function (2.2.17) is

$$\overline{B} = \frac{2\hat{d}^2}{w^2} + \frac{w^2}{8} + \overline{d} + h. \tag{2.4.2}$$

Elimination of $\hat{d}$ between the last two relations yields

$$\frac{Q^2}{2\overline{d}^2 w^2} + \frac{w^2}{8} + \overline{d} + h - \overline{B} = 0, \tag{2.4.3}$$

which is of the form of the standard hydraulic relation $\mathcal{G}(\overline{d}; w, h; Q, \overline{B}) = 0$ sought by Gill (1977). Here $\overline{d}$ represents the single variable characterizing the flow cross section; if $\overline{d}$ is known, $\hat{d}$ can be computed from (2.4.1) and the remaining cross-sectional properties from (2.2.29) and (2.2.30). Critical states are found by taking $\partial \mathcal{G}/\partial \overline{d} = 0$, resulting in

$$Q = \overline{d}_c^{3/2} w, \tag{2.4.4}$$

where the subscript c denotes a critical value. From (2.4.1) it follows that $2\hat{d}_c/w = \overline{d}_c^{1/2}$, or

$$\overline{v}_c = \overline{d}_c^{1/2}, \tag{2.4.5}$$

in view of the relation $2\hat{d}/w = \overline{v}$ derived in Section 2.2. As expected, Gill's criterion for critical flow matches the relation based on the direct propagation speed calculation (2.2.31).

---

[4] It is not necessary that D_∞ be uniform from one fluid column to the next, only that D_∞ for each column be much larger than D. Thus the potential vorticity in the reservoir need not be uniform.

Possible locations where critical flow can occur are found by taking $[\partial \mathcal{G}/\partial y]_{\bar{d}=const} = 0$, which leads to

$$\left(\frac{w}{4} - \frac{Q^2}{\bar{d}_c^{\,2} w^3} \right) \frac{\partial w}{\partial y} + \frac{\partial h}{\partial y} = 0. \tag{2.4.6}$$

In the WLK model w is constant and critical flow therefore requires that $\partial h/\partial y = 0$, as at a sill. In a channel of constant elevation h and variable w, critical flow requires that either $\partial w/\partial y = 0$, as at a narrows, or that the expression in parentheses vanish. In the latter case (2.4.1) and (2.4.4) imply separation of the flow from the left wall $(\bar{d}_c = \hat{d}_c)$. However this possibility can be eliminated, as explored in Exercise 1 of this section.

It is possible to obtain a 'weir' formula relating Q to the reservoir conditions. In the nonrotating example of Section 1.4 the formula was obtained by equating the Bernoulli functions at the sill and reservoir. Following the same procedure, we use (2.4.4) to evaluate (2.4.3) at the sill, leading to

$$\frac{3}{2} \left(\frac{Q}{w} \right)^{2/3} + \frac{w^2}{8} = \bar{B} - h_m, \tag{2.4.7}$$

where h_m is the sill elevation. Next, we evaluate $\bar{B}$ in the quiescent reservoir, being careful to avoid using the definition (2.4.2), which is not valid there. Instead we simply note, from primitive definition, that the Bernoulli function must be

$$B = D_\infty + h_\infty \tag{2.4.8}$$

where h_∞ is the reservoir bottom elevation. Since B is uniform throughout the reservoir, $\bar{B} = B$ and therefore

$$\frac{3}{2} \left(\frac{Q}{w} \right)^{2/3} + \frac{w^2}{8} = \Delta z \tag{2.4.9}$$

where $\Delta z = D_\infty + h_\infty - h_m$ is the elevation of the reservoir surface (or interface) above the channel bottom at the critical section. Rearranging (2.4.9) and writing the result in dimensional form gives

$$Q^* = \left(\frac{2}{3} \right)^{3/2} w^* g^{1/2} \left[\Delta z^* - \frac{w^{*2} f^2}{8g} \right]^{3/2}. \tag{2.4.10}$$

As $f \to 0$ the limit (1.4.12) for nonrotating flow from a deep reservoir is realized. The reader is reminded that g represents either the full or reduced gravity, depending on whether the upper surface is interpreted as a free surface or deep interface.

If the flow in the channel becomes separated, we switch to the natural variables v_e and w_e (see Figure 2.3.1). The y-structure of the flow is then described by the steady forms of (2.3.16) and (2.3.17):

$$\tfrac{1}{2}w_e^{\,2}\left(v_e - \tfrac{1}{2}w_e\right)^2 = Q \qquad (2.4.11)$$

and

$$\frac{v_e^{\,2}}{2} + h = \overline{B}. \qquad (2.4.12)$$

Note that the channel width w does not enter these relations. Changes in the position of the right-hand wall cause lateral displacements of the entire flow with no change in the shape of the interface.

Equation (2.4.12) expresses energy conservation along the free edge of the separated current. Since the depth is zero there, changes in kinetic energy must be directly balanced by changes in bottom elevation. It is tempting to treat the left-hand side of this equation as a Gill-type hydraulic function $\mathcal{G}(v_e, h)$ since it contains the single flow variable v_e. However, taking $\partial \mathcal{G}/\partial v_e = 0$ results in $v_e = 0$, whereas the true critical condition (see 2.3.18) based on wave speed calculation is $v_e = w_e$. On the other hand, one can use the multivariate critical condition:

$$\frac{\partial \mathcal{G}_1}{\partial w_e}\frac{\partial \mathcal{G}_2}{\partial v_e} - \frac{\partial \mathcal{G}_1}{\partial v_e}\frac{\partial \mathcal{G}_2}{\partial w_e} = 0, \qquad (2.4.13)$$

based on (1.5.9). With $\mathcal{G}_1$ and $\mathcal{G}_2$ given by the left-hand sides of (2.4.11) and (2.4.12), this condition leads to the desired result

$$v_{ec} = w_{ec}. \qquad (2.4.14)$$

It is left as an exercise to show that the multivariate regularity condition (1.5.11) leads to the condition that $dh/dy = 0$ at a critical section.

The failure of the criterion $\partial \mathcal{G}/\partial v_e = 0$ to yield the correct critical condition in its application to (2.4.12) is tied into the peculiar dynamics of the frontal wave and the choice of v_e as the dependent variable. Consider the depth profile under critical conditions, as shown by the solid curve in Figure 2.4.1. The slope of the free surface is zero at the wall and $v_e = w_e$. Suppose that the profile is slid an infinitesimal distance to the right or left without changing its shape, as suggested by the dashed line. Since $\partial d/\partial x = 0$ at the wall the altered flow has the same wall depth, and therefore the same volume flux, as before. In addition, the sideways displacement does not alter the value of B at the free edge, since v_e is unchanged. Since B is uniform when $q = 0$, the value of $\overline{B}$ remains unchanged as well. In summary, neither $\overline{B}$ nor Q is altered by the sideways displacement and the disturbance, which involves only changes in w_e. The disturbance therefore

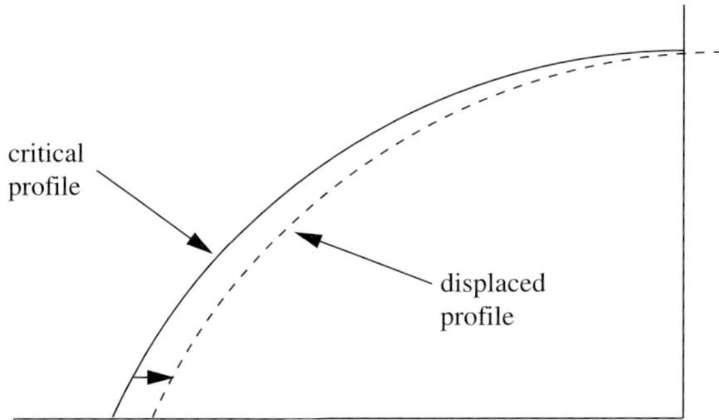

FIGURE 2.4.1. Cross section (surface profile) of a critical, separated current (solid curve) and a new steady flow with the same Bernoulli function and volume flux, created by a sideways displacement of the current (dashed line).

qualifies as a stationary wave.[5] The condition $\partial \mathcal{G}/\partial v_e = 0$, which checks for disturbances in v_e that leave $\mathcal{G}$ unchanged, misses the critical condition. In essence, satisfaction of Gill's criterion for a $\mathcal{G}$ written in terms of a particular dependent variable is a sufficient, but not necessary, condition for criticality. To avoid such cases one must be sure to use *all* the constraints available in the formulation. The multivariate version (1.5.14) of the critical condition therefore provides the safest route.

The weir formula for this case may be obtained using a similar procedure as above, resulting in

$$Q^* = \frac{g(\Delta z^*)^2}{2 f} \tag{2.4.15}$$

for the separated case.

Equation (2.4.14) suggests that the separation first occurs at the critical section when $w_c = v_{ec}$ or, in view of (2.4.5), when $w_c = 2\overline{d}_c^{-1/2}$. Furthermore, since $v = 0$ at the right wall in this case, energy conservation implies that the level of the interface at the right wall is the same as that of the reservoir, so that $2\overline{d}_c = \Delta z$. Elimination of $\overline{d}_c$ between these last conditions leads to $\Delta z = w_c^2/2$, and therefore

$$w_c^* \begin{cases} < \sqrt{2}(g\Delta z^*)^{1/2}/f & \text{(nonseparated)} \\ > \sqrt{2}(g\Delta z^*)^{1/2}/f & \text{(separated)} \end{cases} \tag{2.4.16}$$

---

[5] It should also be noted that the same argument is applicable to a separated flow with an arbitrary potential vorticity distribution. Such a flow is hydraulically critical if the velocity at the right wall vanishes.

That is, the flow at the critical section is separated if the channel width is greater than $\sqrt{2}$ times the Rossby Radius of deformation based on Δz^*. Thus, a decrease in reservoir surface elevation relative to the sill encourages critical separation of the flow.

Whitehead *et al.* (1974) (WLK) carried out an experiment designed to test the transport relations (2.4.10 and 2.4.15) and the separation criterion (2.4.16). Shen (1981) conducted further experiments using the same type of apparatus, which consists of a cylindrical tank divided into two basins by a vertical wall (Figure 2.4.2). Well above the bottom, a short channel with rectangular cross section is fit through an opening in the wall. An alcohol-water mixture is filled up to the level of the bottom of the channel in both basins, and above this lies a layer of kerosene with slightly lower density. A pump transfers the lower fluid from the left-hand basin to the right, where it wells up through a packed bed of rocks. This fluid rises, passes through the channel, and spills into the left-hand basin. Photos of the overflow as seen looking upstream into the channel as well as from above appear in Figure 2.4.3. For the cases shown, the flow in the channel is nonseparated, even though separation from the right wall (looking upstream) is predicted. The height Δz^* of the upstream interface above the channel bottom is measured by an optical device. The value of Q^* was measured directly by Shen, but not by WLK.

The experiment is initiated by establishing a hydraulically controlled flow with $f = 0$ and measuring the corresponding $\Delta z^* = \Delta z_0{}^*$. In principle, $\Delta z_0{}^*$ should equal $\frac{3}{2} Q^{*2/3} g^{-1/3} w^{*-2/3}$. The turntable is then spun up to a particular f and, once a new steady state had been established, a new Δz^* is measured. The transport Q^* is determined only by the pumping rate and remains constant throughout the spin-up. The reservoir interface elevation is forced to adjust to drive the same amount of fluid across the sill.

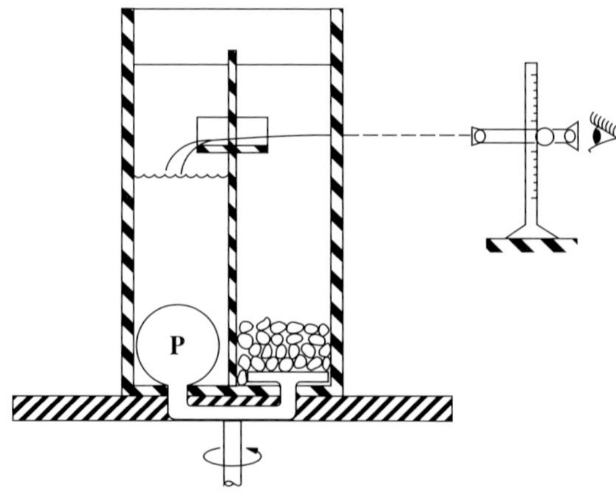

FIGURE 2.4.2. Cross section of the cylindrical tank used in the WLK experiments. (From Whitehead *et al.*, 1974).

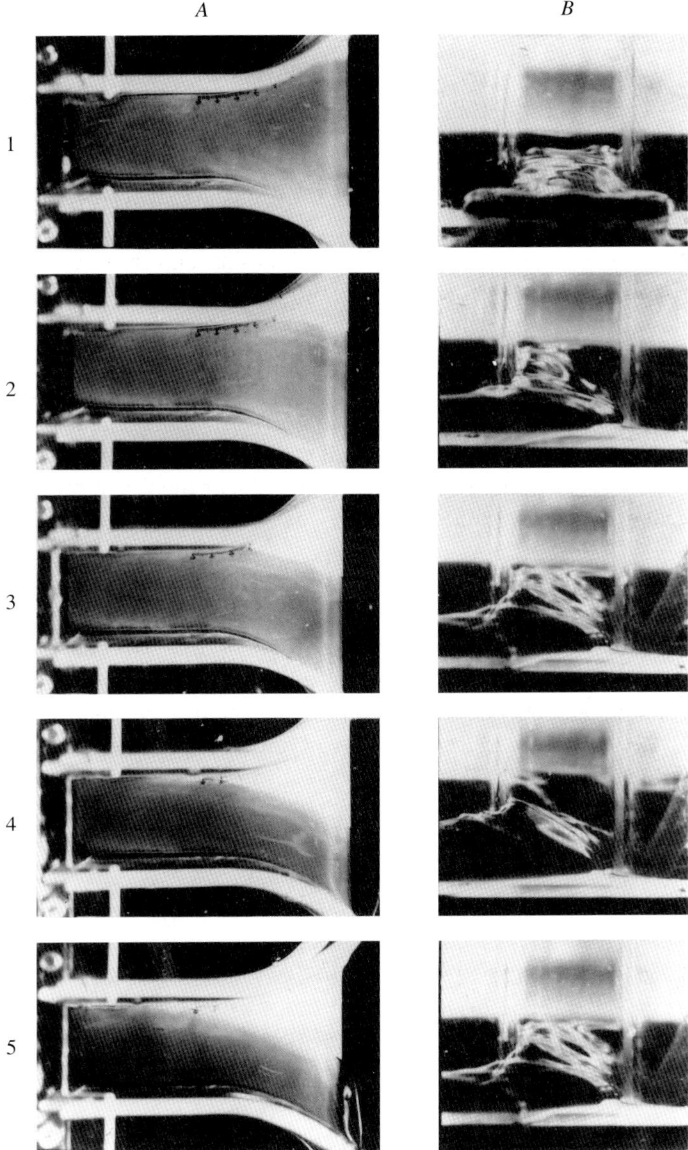

FIGURE 2.4.3. Top view (column A) and front view (column B) of overflow through a rectangular channel. The ends are flared to permit smoother flow. Rotation rate is progressively greater starting from zero in rows 1–5, so $\sqrt{2g\Delta z^*}/wf$ takes the following values ∞, 3.0, 1.57, 0.75, and 0.68. (From Shen, 1981).

For attached flow, the ratio $\Delta z^*/\Delta z_o^*$ can be determined using (2.4.9). The dimensional version of the result is

$$\frac{\Delta z^*}{\Delta z_o^*} = \frac{f^2 w^{*2}}{8g\Delta z_o^*} + 1. \tag{2.4.17}$$

For the separated sill flow, the value of Δz^* is given by (2.4.15) and thus

$$\frac{\Delta z^*}{\Delta z_o^*} = \left(\frac{2}{3}\right)^{3/4} \left(\frac{4w^{*2} f^2}{g\Delta z_o^*}\right)^{1/4}. \tag{2.4.18}$$

The transition between the two cases occurs when $w^* = w_e^* = (2g\Delta z^*/f^2)^{1/2}$, which corresponds to $\frac{3w^{*2}f^2}{8g\Delta z_o^2} = 1$ or $\frac{\Delta z^*}{\Delta z_o^*} = \frac{4}{3}$. A plot of the relation between $\frac{\Delta z^*}{\Delta z_o^*}$ and $\left[3w^{*2}f^2/8g\Delta z_o^*\right]^{1/2}$ along with the experimental data from WLK and Shen (1981) shows good agreement to the left of the transition (Figure 2.4.4). The data and theory in this region agree to within 5%. To the right of the transition the agreement is within 20%. However, there was no clear evidence of separation in either experiment.[6] Although the transport formulas (2.4.10) and (2.4.15) suggest that increasing f leads to smaller Q^*, this conclusion is only valid if the upstream interface level remains fixed. In reality, the effect of rotation on transport depends on how the flow is driven; in the WLK experiment Q^* is maintained at a fixed rate while f is varied.

Shen (1981) also investigated flows that remain subcritical over the entire domain and are therefore not hydraulically controlled. Transport relations like (2.4.10) and (2.4.15) are clearly no longer valid for such flows and one may ask whether there is another strategy for measuring the volume flux. The chief obstacle is that the critical condition over the controlling weir or sill is lost and the flow is therefore less constrained. Civil engineers encounter this problem when a weir that normally produces a subcritical-to-supercritical transition becomes 'submerged', meaning that the transition region has become flooded and subcritical.

According to our inviscid models, there is no difference between the surface heights of a subcritical flow upstream and downstream of an isolated obstacle (Figure 2.4.5). In reality, the surface level always drops slightly due to a frictional energy loss over the obstacle. Real world flows also tend to separate from the channel sidewalls and/or the bottom, if broadening of the channel downstream of the contraction or sill is abrupt. In the case of sidewall separation, pools containing weak recirculations exist along the sidewalls of the flow. These features also enhance the asymmetry between the upstream and downstream. One strategy for measuring the volume flux of such a flow is to measure the upstream and downstream surface elevations and relate Q to their difference.

---

[6] If fact, more recent laboratory and numerical experiments have largely failed to produce flow separation. We will mention many of these experiments throughout the remainder of the book.

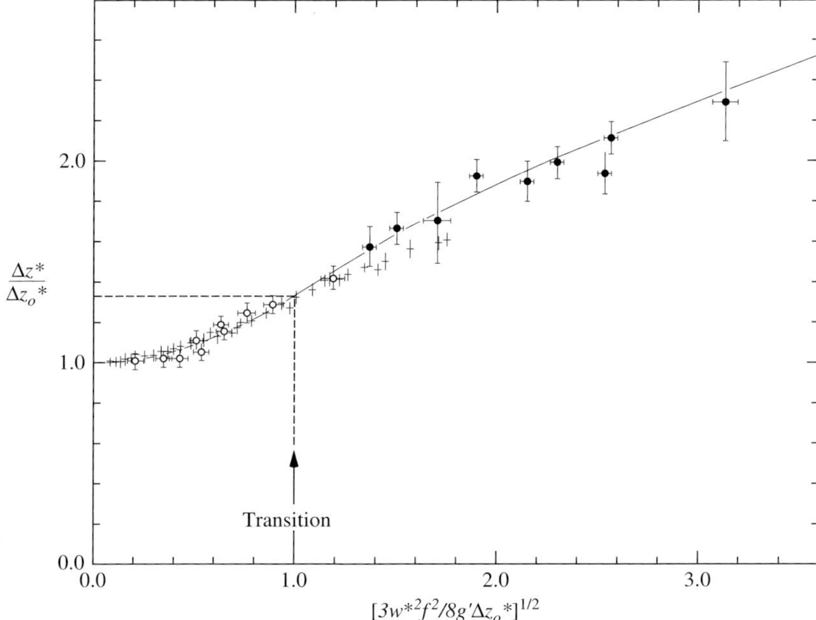

FIGURE 2.4.4. Comparison of the WLK theory (curves given by equations 2.4.17 and 2.4.18), with experimental data from WLK (circles with error bars) and Shen (crosses denoting error bars). (Adapted from Whitehead *et al.*, 1974 and Shen, 1981).

Shen describes a procedure along these lines for a rotating, subcritical flow with uniform potential vorticity. He assumes that separation occurs from the left wall of the channel as the subcritical flow passes the most constricted section. (The constriction in his experiment takes the form of a uniform length of channel separating two deep reservoirs, as in Figure 2.4.2.) The geostrophic flux in the channel is given by $Q^* = \frac{g}{2f}\left(d^{*\,2}_R - d^{*\,2}_L\right)$, where d^*_R and d^*_L are right and left wall depths. A key assumption is that the downstream separation causes the surface elevation along the left wall to be nearly equal to that downstream of the channel and in the stagnant pool on the left wall. Thus $d^*_L = D_{-\infty} - h_m^*$, where $D_{-\infty}$ is the downstream depth in the pool and h_m^* is the elevation of the channel bottom relative to the downstream bottom. Also d^*_R can be related to the upstream interior depth D_∞ using constraints provided by the Bernoulli function, which is assumed to be conserved in the upstream region, and the assumption of uniform potential vorticity. In this way Q^* is related to D_∞ and $D_{-\infty}$. Computation of the volume flux therefore requires two depth measurements in quiescent areas, one upstream and the other downstream of the sill or channel. The reader should consult Shen's (1981) paper for the formulae, which are quite involved.

Shen carried out laboratory experiments to test the submerged weir theory in the limit of zero potential vorticity (Figure 2.4.6). One complicating factor is that some of the solutions exhibit flow reversal ($v < 0$) along the right wall

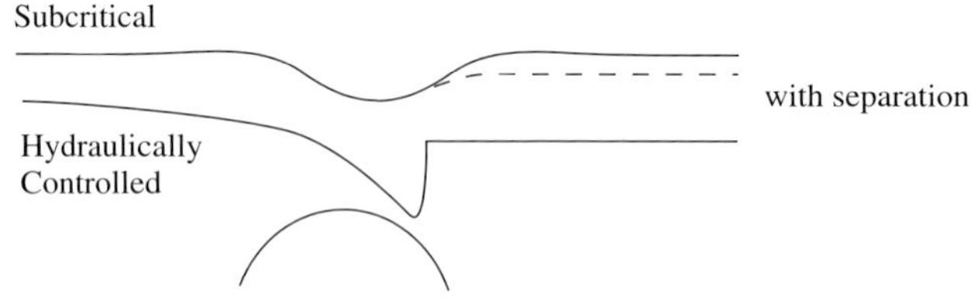

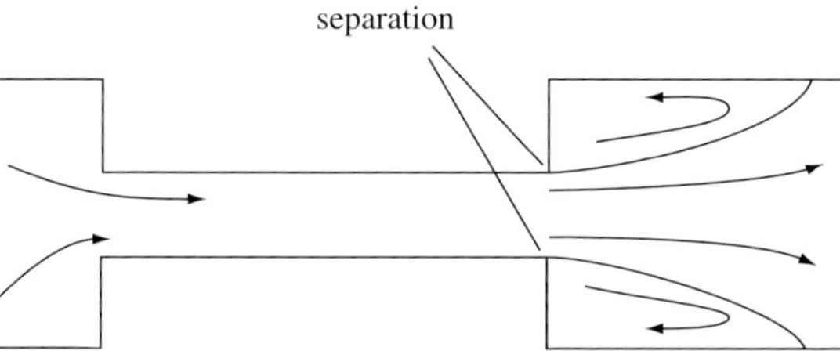

FIGURE 2.4.5. The top frame shows the surface profile for an idealized subcritical flow with no friction and the modification (dashed extension) caused by frictional effects, possibly associated with flow separation downstream of the sill. The lower curve shows a hydraulically controlled flow with a jump. The lower frame shows sidewall separation and the closed gyres with weak circulation that can be produced as a consequence.

of the channel. In such cases, the predicted flux is based only on that part of the flow with $v > 0$. The possible flow regimes include critical flow without reversal (I); submerged flow without reversal (II); submerged flow with reversal (III); and critical flow with reversal (IV). As indicated by the crosses in the figure, there are few observations of flow in regime III. The theory, which is represented by four curves for four channel widths, relates the upstream level Δz of the interface (relative to the channel bottom) to a nondimensional flux Q'. The dependence on the downstream flow is hidden in the scaling factor $D = \frac{1}{2}(D_\infty + D_{-\infty} - 2h_m{}^*)$, which is used in the nondimensionalization. It is evident that the agreement between theory and experiment is quite good for the cases shown. It is not known how accurate the formulae are for other geometries or separation scenarios.

The experiments of WLK and of Shen (1981) were designed to approximate the zero potential vorticity limit by causing fluid to be drawn from a deep, quiescent reservoir. Clearly, the long wave approximation is violated at the entrance of the channel, where an abrupt change in geometry occurs. Also, a gyre generally forms in the deep upstream fluid, making the assumption of quiescence doubtful. Despite the violations of underlying assumptions, agreement between predicted and observed transports is generally good. In fact, the models of WLK and Shen count among the very few that have been subjected to careful laboratory

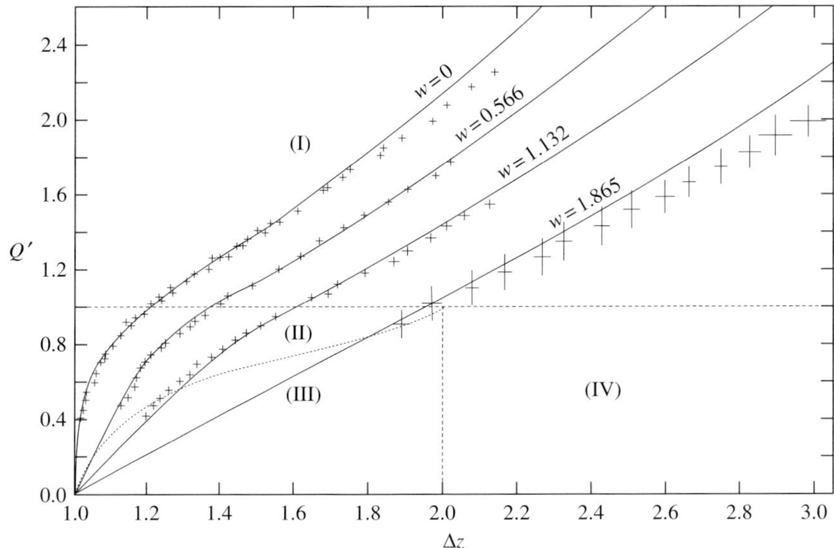

FIGURE 2.4.6. Transport $Q' = Q/Q_c$ versus the dimensionless upstream elevation $\Delta z = \Delta z^*/D$ above the sill for various channel widths $w = w^* f/(gD)^{1/2}$. Downstream information is contained in the scale factor $D = (D_\infty + D_{-\infty} - 2h_m^*)$. Q_c refers to the flux that occurs for the same value of w at the point where the flow is marginally submerged. The solid curves correspond to the equations appearing in Table 1 of Shen (1981). The dashed lines are the boundaries of the four regimes: (I) critical flow without reversal, (II) Submerged flow without reversal, (III) submerged flow with reversal, and (IV) critical flow with reversal. The bars denote the measurements and their uncertainty. (From Shen, 1981).

verification. As we will show in Section 2.6, the relation (2.4.15) provides the transport under more general conditions, provided Δz^* is suitably interpreted. It will also be shown (Section 2.10) that the same relation provides a bound on the transport in even more general circumstances.

Exercises

(1) Consider a channel with variable w and constant h. Equation (2.4.6) suggests that a critical section in such a channel can occur where $\partial w/\partial y = 0$ or where $w^2 = 2Q/\overline{d}_c$.

(a) Show that the latter implies $\overline{d}_c = \hat{d}_c$ (the flow is separating from the left wall).

(b) Suppose that separation occurs at a section $y = y_s$ where the width is changing and suppose, with no loss in generality, that $dw/dy > 0$. Now consider a section slightly upstream of $y = y_s$, where the flow is attached. Show by conservation of Q that the right wall depth at the upstream section must be greater than that at y_s. Using the fact that the right-wall velocity at y_s must be zero, show that the value of the right-wall Bernoulli function cannot be conserved between the two sections and therefore a continuous solution is not possible.

(2) Suppose that the channel draining the reservoir in the WLK model has constant w. Further suppose that the flow separates from the left wall upstream of the sill. Given the values of w, h_{m}, and Δz, at what value of h does separation occur?

(3) Equations (2.4.11) and (2.4.12) can be cast as two hydraulic functions $\mathcal{G}_1$ and $\mathcal{G}_2$ in the two variable space $(w_e,\ v_e)$. Show that the stationary wave displacement vector defined by (1.5.10) is aligned in the direction $(1,\ 0)$. Can you use this result to explain the failure of (2.4.12), when treated as a single function in the single variable v_e, to provide the correct critical condition?

2.5.   Uniform Potential Vorticity Flow from a Wide Basin: Gill's Model

The Whitehead, Leetma and Knox (WLK) model was followed three years hence by a more elaborate treatment due to Gill (1977). In addition to his model, detailed below, Gill introduced a unifying framework for treating hydraulics problems. We have made repeated use of his formalism, particularly in the derivation of conditions for hydraulic criticality. This material was reviewed and generalized in Section 1.5. The model developed by Gill was based on his particular view of the upstream basin and is rather more involved than that of WLK. Some investigators have found Gill's scaling and choice of upstream parameters unintuitive and have developed their own versions of his basic model. In consideration of the historical importance of Gill's paper, our preference in presenting the work is to first describe the model as originally formulated. The next section will discuss some insights that are gained from alternative formulations.

a.   Basics

The depth and velocity profiles predicted by zero potential vorticity models such as WLK are valid near the sill, where the local depth (scaled by D) is small compared to the reservoir depth D_∞. However, these expressions do not apply in the reservoir, where by hypothesis the depth *equals* D_∞. It is therefore difficult to verify the self-consistency of the model, in particular the hypothesis that a quiescent, infinitely deep upstream state can be linked to the sill flow in a dynamically consistent way. In thinking about the character of the upstream flow, one might also wish to consider other possible states. Observations from deep straits such as the Faroe Bank Channel suggest the bulk of the overflow comes from intermediate water masses, which span the relatively wide upstream basin but may not be significantly thicker than the layer depth at the sill. Some or all of these realities led Gill (1977) to consider nonzero (but still uniform) values of $q(=D/D_\infty)$. The depth and velocity profiles across the channel are given by the more general forms (2.2.3) and (2.2.4), which show that the flow is confined to sidewall boundary layers of width $L_d = (gD_\infty)^{1/2}/f$. In the WLK model this width is much larger than the channel width $[fw^*/(gD_\infty)^{1/2} = q^{1/2}w \ll 1]$ due to the fact that $q \ll 1$. Most of the novel features of Gill's model can be linked to the boundary layers.

The model employs rectangular cross-sectional geometry and we analyze the case of nonseparated flow first. The steady forms of (2.2.15) and (2.2.16) then require conservation of the volume transport Q and the average $\overline{B}$ of the Bernoulli function on the two walls. Since we no longer care about special values or limits of D (such as $D \ll D_\infty$) we are free to set it to any convenient value. The choice $D = (fQ^*/2g)^{1/2}$ is convenient as it is equivalent to setting $Q = 2$ (see Exercise 1) in the statement of conservation of volume flux (2.2.18). Therefore

$$\overline{d}\,\hat{d} = 1. \tag{2.5.1}$$

In addition, conservation of the energy (2.2.16 and 2.2.17) for the steady flow implies

$$\tfrac{1}{2}q[T^{-2}\hat{d}^2 + T^2(\overline{d} - q^{-1})^2] + \overline{d} + h = \overline{B}, \tag{2.5.2}$$

where again $T = \tanh(\tfrac{1}{2}q^{1/2}w)$. Eliminating $\hat{d}$ from these two equations gives

$$T^2(\overline{d} - q^{-1})^2 + \frac{1}{T^2\overline{d}^2} + 2q^{-1}(\overline{d} + h) = 2q^{-1}\overline{B}. \tag{2.5.3}$$

The parameter $\overline{B}$ is generally neither a convenient nor intuitive measure of the reservoir state. If the reservoir is much wider than L_d, the flow there will be confined to sidewall layers (Figure 2.5.1). The physical separation of the boundary layers makes it difficult to see how $\overline{B}$ would be specified in a laboratory experiment or oceanic setting. Furthermore, the velocity along each wall is generally nonzero (in the inviscid model) and the Bernoulli functions there may no longer be dominated by the potential energy terms $h + d$, as assumed in the WLK model. Only in the interior of the reservoir, at a distance $\gg L_d$ from either wall, will the velocity be small. There, the dimensional depth is D_∞ (or $d = q^{-1}$) according to (2.2.12). With these ideas in mind, Gill (1977) suggested that a new parameter measuring the partitioning of the volume transports of each boundary layers would be more descriptive than $\overline{B}$. Some other choices are discussed in the next section.

Let the transport streamfunction ψ have the value ± 1 on the sidewalls $x = \pm w/2$, so that the total transport is 2, as assumed. Further, let ψ_i denote the value of ψ in the quiescent interior separating the two upstream boundary

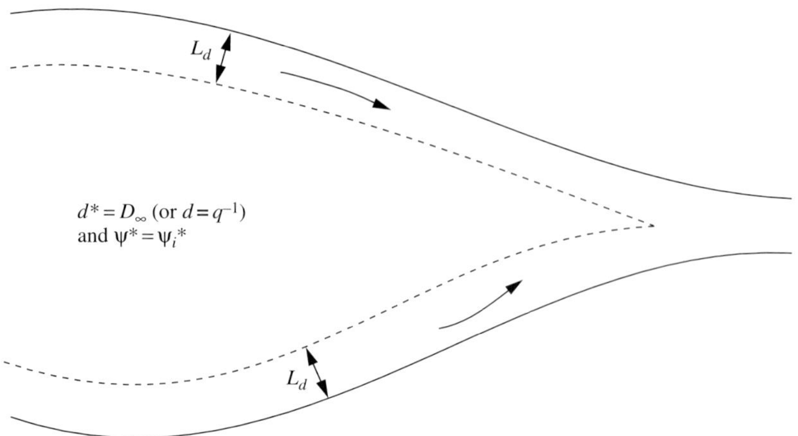

FIGURE 2.5.1. Gill's (1977) ideal of the upstream basin or reservoir.

currents. The transports in the right- and left-hand boundary currents (facing downstream) are therefore $1 - \psi_i$ and $1 + \psi_i$. Included is the possibility that $|\psi_i| > 1$ in which case one of the boundary layer transports will be greater than 2 and the other will be negative. We can write ψ_i in terms of $\overline{B}$ by first integrating $dB/d\psi = q$, yielding $B = \overline{B} + q\psi$. Then note that $B(= \frac{1}{2}v^2 + d + h)$ has the value q^{-1} along $\psi = \psi_i$, as follows from the evaluation of B in the quiescent region, where $v = 0$, $d = q^{-1}$, and where we will take h to be zero. Thus

$$\overline{B} = q^{-1} - q\psi_i, \tag{2.5.4}$$

and substitution into (2.5.3) results in

$$\mathcal{G}(\overline{d}; T, h) = T^2(\overline{d} - q^{-1})^2 + \frac{1}{T^2\overline{d}^2} + 2q^{-1}(\overline{d} + h) - 2(q^{-2} - \psi_i) = 0. \tag{2.5.5}$$

The function $\mathcal{G}(\overline{d}; T, h)$ relates the single flow variable $\overline{d}$ to the local geometric parameters $T(w(y))$ and $h(y)$, and is therefore of the desired form. The parameters describing the upstream flow are ψ_i and the interior reservoir depth q^{-1}. In light of the particular choice of D this last parameter can also be written as $(2gD_\infty^2/fQ^*)^{1/2}$ leading to an alternative interpretation. For a fixed interior depth D_∞ the maximum possible geostrophic transport in the left-wall boundary layer occurs when the depth along the left wall is zero. This transport is given by $Q_{max} = gD_\infty^2/2f$ and therefore $q = 2(Q^*/Q_{max}^*)^{1/2}$. In summary, it is possible to think about the reservoir parameters entirely in terms of volume transports: ψ_i measures the partitioning between boundary layers and q measures the total transport relative to the maximum possible value in the left-hand boundary layer.

b. Critical States

Critical states are found by taking $\partial \mathcal{G}/\partial \bar{d} = 0$, resulting in

$$(1 - T_c^{\,2})q^{-1} + T_c^{\,2}\bar{d}_c = \bar{d}_c^{-3}T_c^{\,-2}, \qquad (2.5.6)$$

where the subscript 'c' denotes quantities evaluated at a critical section. According to (2.2.22), the characteristic speed c_- of a Kelvin wave propagating along the left-hand $(y = w/2)$ wall is

$$c_- = q^{1/2}T^{-1}\hat{d} - \bar{d}^{\,1/2}[1 - T^2(1 - q\bar{d})]^{1/2} = 0,$$

and it is simple to show that $c_- = 0$ is equivalent to (2.5.6).

Gill (1977) also defined a Froude number

$$F_d = \frac{q^{1/2}T^{-1}\hat{d}}{\bar{d}^{1/2}[1 - T^2(1 - q\bar{d})]^{1/2}} = \frac{\bar{v}}{\bar{d}^{1/2}[1 - T^2(1 - q\bar{d})]^{1/2}} \qquad (2.5.7)$$

such that $F_d(< 1, = 1, > 1)$ indicates (subcritical, critical, and supercritical) flow corresponding to c_- $(< 0, = 0, > 0)$. As pointed out in Section 2.2, this Froude number should not be interpreted as the ratio of an advection to relative propagation speed. However it does measure the ability of a Kelvin wave, trapped to the left wall of the channel, to propagate upstream. If $F = 1$ this wave is stationary; if $F > 1$ it propagates downstream.

The geometric requirements for critical flow are obtained by setting $d\mathcal{G}/dy = 0$ in (2.5.5). If the channel width is constant, critical flow can only occur where $dh/dy = 0$ as before. When h is constant the requirement becomes

$$[T_c^{\,4}(\bar{d}_c - q^{-1})^2 - \bar{d}_c^{-2}]dw/dy = 0, \qquad (2.5.8)$$

implying that $dw/dy = 0$, as at a contraction, or that the coefficient in brackets is zero. As in the WLK model, the latter implies that $v_c(w/2, y) = d_c(-w/2, y) = 0$ meaning that the flow is in the process of separating from the left wall. However, this possibility can be rejected on the same grounds as discussed in Exercise 1 or Section 2.4.

We now turn to the case of separated flow. Here $\hat{d} = \bar{d} = 1$ in view of (2.5.1) and the only dependent variable is the width parameter $T_e = \tanh(q^{1/2}w_e/2)$, where w_e is the separated stream width. As shown by (2.3.7) and (2.3.8), the steady equations relating the flow to the geometry are identical to those describing nonseparated flow, but with T replaced by T_e. With this replacement and with $\bar{d} = 1$, (2.5.5) leads to an altered hydraulic function:

$$\mathcal{G}(T_e; h) = T_e^{\,2}(1 - q^{-1})^2 + \frac{1}{T_e^{\,2}} + 2q^{-1}(1 + h) - 2(q^{-2} - \psi_i) = 0. \qquad (2.5.9)$$

The channel width $w(y)$ does not enter this relation and thus the separated current width responds only to changes in bottom elevation h. If h remains constant, changes in the position of the right wall lead to identical changes in the position of the left edge of the separated flow. This property clarifies the condition implied by the vanishing of the bracketed term in (2.5.8). Along a horizontal bottom, critical separation of the flow can occur where dw/dy is nonzero since the actual width w_e of the flow becomes stationary $dw_e/dy = 0$ at that point.

The conditions for critical flow are obtained by setting $\partial \mathcal{G}/\partial T_e = 0$ with $\mathcal{G}$ defined by (2.5.9) and this leads to

$$q^{-1} = 1 + T_{ec}^{-2} \qquad (2.5.10)$$

Since T_{ec} must be $< T_c$ for the critical flow to be separated, (2.5.10) requires

$$q^{-1} \geq 1 + T_c^{-2}. \qquad (2.5.11)$$

It can also be shown that separated critical flow has $v = 0$ on the right wall (see Exercise 2), a property that could have been anticipated on the basis of remarks surrounding Figure 2.4.1.

It can also be shown that the long wave speeds in this case are given by

$$c_{\pm} = q^{1/2} T_e^{-1} \pm [1 - T_e^{2}(1-q)]^{1/2} = 0,$$

which is just the expression for attached flow (*cf* 2.2.22) but with $\hat{d} = \overline{d} = 1$ and T replaced by T_e. The corresponding Froude number is

$$F = T_{ec}^{-1}[(1 - T_{ec}^{2})q^{-1} + T_{ec}^{2}]^{-1/2}. \qquad (2.5.12)$$

c.   *Examples of Solutions*

Before discussing actual solutions it is worth noting several results regarding flow separations and reversals. Following the remarks made in connection with Figure 2.2.2, we know that a continuous wetted band of current at some upstream section cannot, at some downstream section, split into multiple bands separated by dry bottom. If fluid depth in the reservoir is nonzero across the reservoir width, then the current downstream will remain in one continuous band across each section of channel. The depth may go to zero at the left wall and the current may separate there, but it may not ground at some point interior to the fluid. In addition, the along-channel velocity may reverse signs only once in the interior of the flow (see Exercise 1 of Section 2.2). Finally, it can be shown (Exercise 3 of this section) that v must remain nonnegative at a critical section.

Trying to develop a detailed understanding of Gill's model over all parametric variations and channel geometries is nearly impossible. Instead we will attempt to illustrate the features of the solutions that are interesting and exhibit behavior different from that of the WLK model. To begin with, consider the case when the channel bottom is horizontal and the flow is forced only by width contractions. Equation (2.5.5) can then be solved to obtain plots of $\overline{d}$ as a function of T for various values of the interior reservoir depth q^{-1} (Figure 2.5.2). All curves have

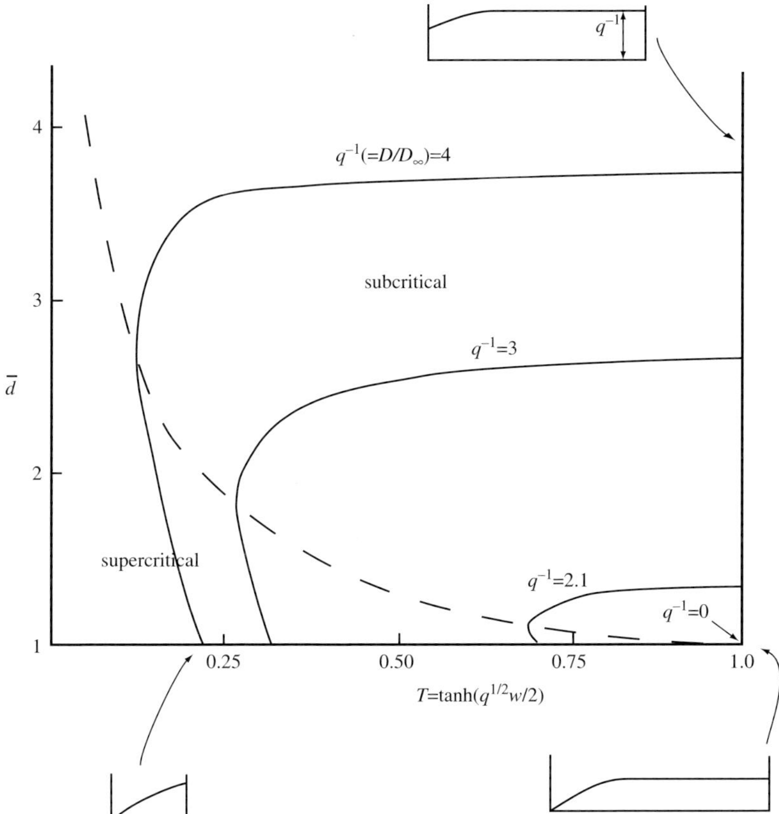

FIGURE 2.5.2. Solution curves for flow through a pure contraction. Note that $T = \tanh(q^{1/2}w/2)$. [From Gill (1997)].

$\psi_i = 1$ so that the reservoir is drained entirely by the left-wall boundary layer. This upstream state is sometimes motivated by consideration of a dam-break problem. Imagine a barrier that is located in the channel and that separates two resting bodies of fluid, the deeper fluid extending back into our upstream reservoir. Removal of the barrier will excite a Kelvin wave that propagates into the reservoir along the left wall and sets up the draining flow. (There are a number of complicating factors that arise in such experiments. For example, a finite reservoir would allow the Kelvin wave to propagate around the perimeter and reenter the channel. However, the draining flow along the left wall would at least persist for some finite time.)

The solution space of Figure 2.5.2 has been restricted to $\bar{d} \geq 1$ (nonseparated flows) since changes in the properties of separated flows can only be forced by bottom topography. The curves $q^{-1} = constant.$ can be used to construct particular solutions for different upstream states. To determine the appropriate value of

$$q^{-1} = 2(Q^*_{max}/Q^*)^{1/2} = 2(gD^{*\,2}_\infty/2fQ^*)^{1/2}$$

one would need to know the flow rate Q^* and the interior reservoir depth D_∞. The values of $\overline{d}$ for a range of channel widths are then traced out by following the corresponding curve. Note that all the curves extend between the right edge ($T = 1$) of the diagram, corresponding to the reservoir ($w \to \infty$), and the lower boundary $\overline{d} = 1$, corresponding to a point of separation. Since the slopes of the curves near the lower boundary are negative, w increases as the separation point is approached. If further increases in width occur downstream of that point the stream will separate and continue at the same width with no further changes in properties. Each solution has a supercritical branch and a subcritical branch that merge at a point determined by the critical condition (2.5.6), indicated by the dashed line. Note that this line lies above $\overline{d} = 1$, indicating that all separated flows are supercritical for $\psi_i = 1$.

Once a particular q^{-1} is selected, it is natural to follow the solution by beginning in the reservoir ($T = 1$) and tracing along the appropriate curve in Figure 2.5.2 until the narrowest section of the channel is reached. (Two of the reservoir states are drawn in the figure insets at the right.) If the narrowest section is reached before the dashed line is encountered, the solution is subcritical with no hydraulic transition. Downstream of the narrows, the solution is obtained by retracing the solution curve back towards $T = 1$ as the channel widens. All such solutions are nonseparated. If T at the narrows happens to be the critical T_c, then the dashed curve is crossed there and the downstream flow is supercritical. All supercritical branches of the solution curve terminate on the line $\overline{d} = 1$ indicating flow separation for sufficiently large w. If the narrows is sufficiently constricted that $T < T_c$ for that curve, a complete steady solution cannot be constructed. In this case a time-dependent adjustment must occur, perhaps resulting in a change in q, ψ_i, or both. Figure 2.5.2 suggests that, in the absence of changes in ψ_i, the upstream depth must *increase* to accommodate the narrower width.

The limiting case $q = 2$ corresponds to separation of the reservoir flow from the left wall. Here the outflow transport is the maximum that can be carried by the left boundary layer ($q^* = Q^*_{max}$). Higher transports are possible in general, but these require flow in the right boundary layer. When the flow in the reservoir is separated it is also critical, as suggest by the figure or by (2.2.26). Downstream of the reservoir, the channel would have to remain infinitely wide to sustain a solution.

Next consider the opposite case of variable topography with constant width. Since we have already assumed the reservoir to be infinitely wide, it is convenient to imagine the reservoir narrowing to a finite value, during which h remains zero, followed by a constant-width section containing a sill. Figure 2.5.2 is used to track the solution over the variable-width section of channel and Figure 2.5.3, which shows solution curves for variable h and fixed width ($w^*/L_d = .75$ or $T = .63$), is then used to continue further. The solution space of Figure 2.5.3 is divided into two regions: an upper portion ($\overline{d} > 1$), for which the flow is nonseparated and the dependent variable is $\overline{d}$, and a lower portion ($\overline{d} < 1$), for which the flow is separated and T_e is the dependent variable. As before, $\psi_i = 1$ and critical flow is marked by a dashed line.

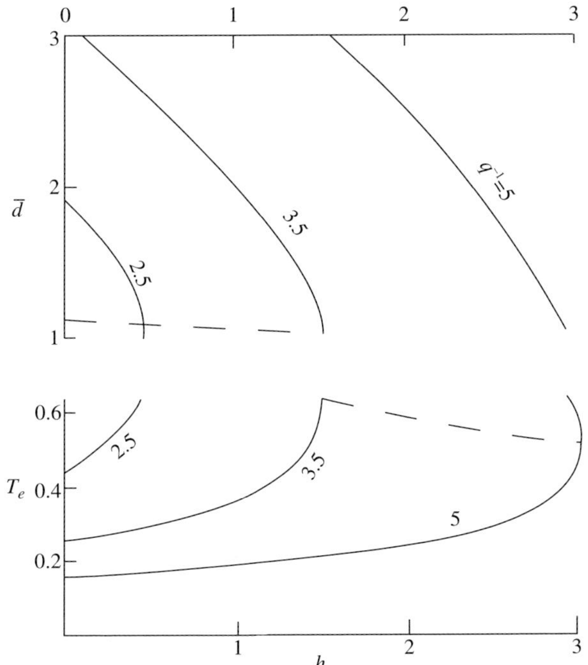

FIGURE 2.5.3. Solution curves for flow over a sill in a constant width channel. The lower half of the diagram applies to separated flow, with $T_e = \tanh(q^{1/2}w_e/2)$. [From Gill (1977)].

If one begins at the upstream end of the uniform width section, where $h = 0$ and where the flow is subcritical, the solution lies along the upper left hand border of Figure 2.5.3. Increases in h cause $\overline{d}$ to decrease as one follows the appropriate q^{-1} = constant curve. There are now two scenarios depending on the value of the interior reservoir depth. If $q^{-1} < 3.5$ the flow will become critical *before* the separation point $\overline{d} = 1$ is reached, so that separation will occur downstream of the sill. This behavior occurs for relatively low sills ($h_m < 1.5$). If $D_\infty/D > 3.5$ the flow separates upstream of the sill (while it is still subcritical) and remains subcritical until it reaches the sill, where it becomes critical. This situation, which is predicted when the interior reservoir surface elevation and the sill elevation are relatively high, has proven difficult to produce in laboratory or numerical experiments (e.g. Shen 1981; Pratt *et al.,* 2000).

d. *Transport Relations*

The essential nature of upstream influence in a hydraulic model is expressed as a relationship between the parameters that characterized the basin flow and the control section geometry. In the nonrotating models discussed earlier, and in the

WLK model, this relationship takes the form of a 'weir formula' in which the volume transport Q^* is written in terms of the basin surface elevation Δz^* above the sill. The situation in the Gill model is more complicated; for one thing the surface elevation varies across the upstream basin. The weir relationship is most easily expressed for the case of separated flow at the critical section. If (2.5.9) is applied there and (2.5.10) is used to eliminate T_{ec} from the resulting equation, one obtains

$$h_c = q^{-1} - 2 + (1 - \psi_i)q. \tag{2.5.13}$$

Because of Gill's choice of the scaling factor $D = (fQ^*/2g)^{1/2}$, the volume flux is hidden in the nondimensionalization. The scaling relations

$$h_c = \frac{h_c^*}{(fQ^*/2g)^{1/2}}, \quad q^{-1} = \left(\frac{2gD_\infty^2}{fQ^*}\right)^{1/2}, \quad \text{and} \quad \psi_i = \frac{2\psi_i^*}{Q^*}$$

allow (2.5.13) to be recast as a formula for the transport:

$$\left(\frac{fQ^*}{2g}\right)^{1/2} = D_\infty - \left(D_\infty h_c^* + \frac{f\psi_i^*}{g}\right)^{1/2} \tag{2.5.14}$$

(see Exercise 4.) In contrast to the nonrotating case and the zero potential vorticity case, two measurements in the reservoir are now needed to compute the volume flux Q^*. A depth measurement in the reservoir interior gives D_∞ while a depth measurement along either wall and use of the geostrophic relation gives ψ_i^*. Of course, depth measurements on both sidewalls would give the geostrophic transport directly and thus the utility of (2.5.14) is called into question. An alternative is discussed in the next chapter.

For nonseparated flow the situation is more difficult. Applying (2.5.5) at the critical section, adding $\overline{d}_c$ times (2.5.6), and multiplying the result by $q/2$ gives

$$h_c = (1 - \tfrac{1}{2}T_c^2)q^{-1} - 1 - \tfrac{3}{2}(1 - T_c^2)\overline{d}_c - (\psi_i + T_c^2\overline{d}_c^{\,2})q, \tag{2.5.15}$$

or, in dimensional terms:

$$h_c^* = (1 - \tfrac{1}{2}T_c^2)D_\infty + \tfrac{3}{2}(T_c^2 - 1)\overline{d}_c^{\,*} - D_\infty^{-1}(T_c^2\overline{d}_c^{\,*2} + \psi_i^* f/g) \tag{2.5.16}$$

In addition, the dimensional version of (2.5.6) is

$$(1 - T_c^2)D_\infty\overline{d}_c^{\,*} + T_c^2\overline{d}_c^{\,*2} = \frac{f^2Q^{*2}}{4g^2\overline{d}_c^{\,*2}T_c^2} \tag{2.5.17}$$

If the algebraic complexity were not prohibitive, a 'weir' relation could be obtained by eliminating $\overline{d}_c^{\,*}$ between the last two equations. In general, the relationship between Q^*, D_∞ and ψ_i^* for a given h_c^* must be determined numerically. This subject is pursued further in Section 2.6, where different choices of scales and of the upstream parameters lead to more elegant formulations.

e. *Wide Channel Limit*

Another way to gain intuition about the behavior of controlled solutions over the space of the parameters q and ψ_{I} is to consider the limiting case of a wide channel ($w \to \infty$, or $T \to 1$), containing variable h. Many of the novel features of the full problem are captured in this setting. Critical flow must occur at the sill and we first examine the case in which the sill flow is separated. The nondimensional relationship between the sill height and the upstream variables is given by (2.5.13). In addition (2.5.11) requires that $q^{-1} \geq 2$ with marginal separation corresponding to $q^{-1} = 2$. In this regime it is also possible for the upstream flow to be separated and the value of h_c at marginal separation can be calculated by evaluating (2.5.9) in the reservoir ($h = 0$) and setting $T_e = 1$. If (2.5.13) is then used to eliminate ψ_i from the resulting relation, one finds

$$h_c = \frac{3 + (1 - q^{-1})^2}{2q^{-1}} - 1 \qquad (2.5.18)$$

The case of attached flow is more subtle. In Section 2.2 we showed that the characteristic speed of a left-wall Kelvin wave for $T = 1$ is proportional to the negative of the depth at the left wall (see 2.2.26). The flow must therefore be subcritical if it is attached at the left wall, a finding that rules out critical control of attached flow in the problem under consideration. If however, the channel width is considered to be large but finite, a class of attached, critically controlled flows arises. These solutions are described by expanding (2.5.6) in powers of $1 - T_c^2$:

$$\overline{d}_c = 1 + \tfrac{1}{4}(1 - T_c^2)(2 - q^{-1}) + O(1 - T_c^2)^2, \qquad (2.5.19)$$

showing that marginal separation ($\overline{d}_c \to 1$) occurs as $1 - T_c^2 \to 0$ as anticipated. However, the first correction to this limit allows the possibility of attached flows $\overline{d}_c > 1$ provided that $q^{-1} < 2$. These flows are close to separation at the critical section and the relationship between the upstream variables and h_c is obtained by substituting (2.5.19) into (2.5.15) and expanding the results. The end product is

$$h_c = \tfrac{1}{2}q^{-1} - q(1 + \psi_i) + O(1 - T_c^2). \qquad (2.5.20)$$

Equations (2.5.13) and (2.5.20) relate the sill height to the upstream conditions for a hydraulically controlled flow for the cases of separated and nonseparated flow at the control section. The dimensional versions of these relations could be recast as transport (or 'weir') formulas. Figure 2.5.4 shows the solutions to the two relations, with ψ_i plotted as a function of h_c and q^{-1}. Each point in the diagram represents a specific, hydraulically controlled flow. In the upper part of the figure ($q^{-1} > 2$) the flow is separated at the critical section, here the sill. The effective width w_{ec} of the separated flow is determined completely by q^{-1} and these widths have been indicated along the right-hand border of the figure. The dashed curve is determined by (2.5.18) and the region lying to the left

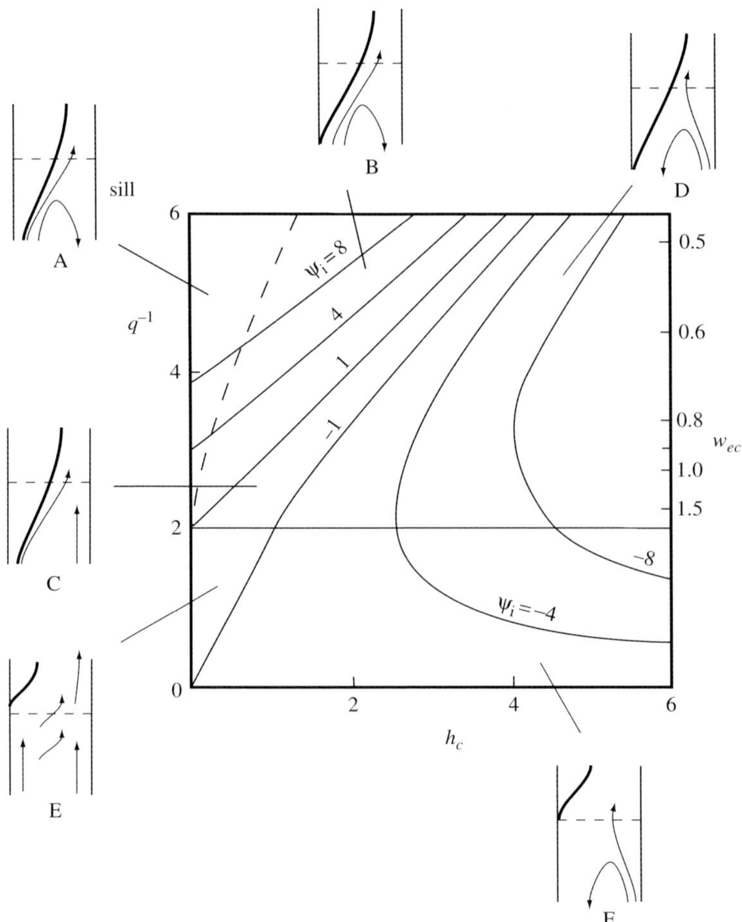

FIGURE 2.5.4. Regime diagram showing various states of separation and recirculation for flow in an infinitely wide channel with a sill. Solutions to the left of the dashed line are entirely separated from the left wall. Those lying below the line $q^{-1}=2$ are attached at and upstream of the sill. [Based on a figure from Gill (1977)].

corresponds to flows that are also separated in the upstream basin. These flows experience no contact with the left wall along the entire length of channel. All such solutions have $\psi_i > 1$, implying that the approach flow in the reservoir is along the left-hand free edge and that some of this flow returns upstream along the right wall before reaching the sill, as shown in Inset A. Such a solution could be considered a coastal flow forced by along-shore changes in topography. To the immediate right of the dashed region the upstream flow is nonseparated but the flow at the sill is separated. In addition the approach flow is concentrated in the left-hand boundary layer, as sketched in Inset B. Continuing to move to the right into regions of higher sill elevation, one enters a region where $-1 \leq \psi_i \leq 1$, so that the approach flow is unidirectional, as shown in Inset C. One of the interesting aspects of cases A, B, and C is that the critical width w_{ec} is often $O(1)$ or less. Thus, approach flow along the left-hand edge can cross the channel and be carried close to the right-hand boundary at the sill. The remaining region in the upper part of the figure ($\psi_i < -1$) corresponds to approach flow along the right-hand wall with some return flow along the left-hand wall, as sketched in Inset D.

In the lower part ($q^{-1} < 2$) of the figure, the flow is marginally attached at the sill. Since $\psi_i \leq 1$ in this lower region, the upstream flows are either unidirectional or approach along the right-hand wall and partially return along the left-hand wall, as sketched in Insets E and F. One of the interesting characteristics of the type F flows is that the surface or interface elevation in the interior of the reservoir can be lower than the sill elevation ($D_\infty < h_c$). [This can be shown by holding $q^{-1} (= D_\infty/D)$ constant in (2.5.19) and taking ψ_i sufficiently negative and large.] Only the interior interface elevation is below the sill; the elevation along the right-hand wall remains above it.

It is also natural to inquire after the dynamics that allow the upstream flow to cross from the right to the left side of the channel before the sill is reached. What happens, in fact, is that a weak along-channel pressure gradient exists in the interior, supporting a cross-channel geostrophic flow. Since $d = q^{-1}$ and $v \cong 0$ in the channel interior, the y-momentum equation reduces to

$$u \cong -\frac{\partial h}{\partial y}.$$

On the upstream face of the obstacle $\partial h/\partial y > 0$, a negative (right-to-left) geostrophic flow exists, whereas the opposite situation occurs on the downstream face.

The Gill model is rather difficult to digest and it is worth recapping some of the highlights. These include the introduction of the concept of *potential depth* D_∞ and the appearance of a global deformation radius $L_d = (gD_\infty)^{1/2}/f$, which is uniform throughout the fluid regardless of the local depth. Another novel feature is the containment of the flow in boundary layers of width L_d. Exploitation of this structure in the wide upstream reservoir allows one to use ψ_i as a parameter in place of the less intuitive $\overline{B}$. Critical control of the flow is exercised by Kelvin waves or their frontal counterparts, both of which are trapped to sidewalls or free edges. Another new feature of Gill's model is that three-dimensional parameters (Q^*, D_∞, and ψ_i^*) are needed to specify the upstream state. If the flow is hydraulically controlled, so that Q^* is a function of D_∞ and ψ_i^*, then just the latter two are needed. Thus, a 'weir' formula relating Q^* to a single upstream depth is not possible without further approximation. Finally, some of Gill's solutions exhibit interesting new behavior including counterflows, crossing of the fluid from one side of the channel to the other over great distances, and instances in which the interior reservoir interface level lies below the sill level.

f. Experiments

As described in the Section 2.4, Shen (1981) set up a series of experiments designed to produce steady, rotating channel flow with uniform potential vorticity. The upstream basin has a horizontal bottom of adjustable depth, allowing the nominal value of the potential vorticity to be varied. Cases of hydraulically controlled flow and submerged weir flow are reproduced and compared to theories. The fundamentals of the submerged weir theory were described in the previous section and the resulting formulas for volume flux are listed by Shen only for the case of zero potential vorticity. For controlled flow, the flux prediction is based on a version of the Gill (1977) that employs a different scaling and that assumes that all flow enters the channel along the left wall ($\psi_i = 1$). The volume flux can then be scaled in such a way as to depend only on the single parameter $w^* f/(gD_\infty)^{1/2}$. We have already shown that this parameter is a measure of the potential vorticity, the 'zero potential vorticity' limit corresponding to vanishingly small values. As shown in Figure 2.5.5, agreement between experiment and theory is good and is best for small or moderate values of $w^* f/(gD_\infty)^{1/2}$. Shen suggests that the disagreement of about 10% for moderate values is due to the effects of nonuniform upwelling in the upstream reservoir, producing nonuniform potential vorticity.

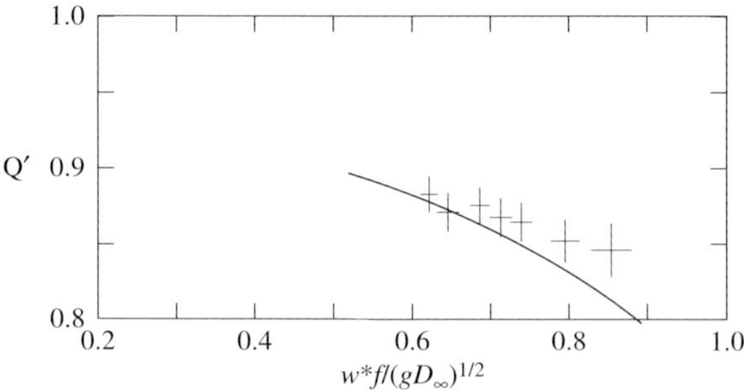

FIGURE 2.5.5. The curve gives the predicted volume flux Q' as a function of $w^* f/(gD_\infty)^{1/2}$ for critically controlled flows observed in the Shen (1981) experiment. The crosses give the experimental values. The transport has been scaled by its value at $w^* f/(gD_\infty)^{1/2} = 0.3$, corresponding to the deepest upstream depth D_∞ used in the range of experiments.

Exercises

(1) Show that setting $D = (fQ^*/2g)^{1/2}$ is equivalent to setting $Q = 2$.

(2) In connection with (2.5.8) show that $[T_c^4(\overline{d}_c - q^{-1})^2 - \overline{d}_c^{-2}] = 0$ implies that $v_c(w/2, y) = d_c(-w/2, y) = 0$.

(3) By following the steps outlined below, show that nonseparated flow at a critical section must be unidirectional in $-w/2 < y < w/2$ provided that the (uniform) potential vorticity is nonnegative. Further show that separated critical flow must have $v(w/2, y) = 0$.

(a) Use the result of Problem 1 of Section 2.2 to argue that the flow is unidirectional at any y provided that $v(y, w/2)$ and $v(y, -w/2)$ do not differ in sign.

(b) Introduce the quantity $r = \hat{v}/\bar{v}$ and argue that the flow is unidirectional for $|r| < 1$ and has $v(y, w/2) = 0$ for $r = -1$. Further show that $r = T^2\bar{d}(\bar{d} - q^{-1})$

(c) Using the critical condition (2.5.6) along with (2.5.1), show that

$$r = \frac{T_c^2 \bar{d}_c^2 - \bar{d}_c^{-2}}{1 - T_c^2}$$

and deduce that $r = -1$ when the flow is separated from the left wall $(\bar{d}_c = 1)$.

(d) For attached flow $(\bar{d}_c > 1)$ show that $r > -1$. Then show that the requirement of nonnegative potential vorticity and the result of (c) lead to $r \leq 1$.

(4) Obtain the transport relation

$$\left(\frac{fQ^*}{2g}\right)^{1/2} = D_\infty \pm \left(D_\infty h_c^* + \frac{f\psi_i^*}{g}\right)^{1/2}$$

by writing (2.5.13) in dimensional units. Next, show that only the '$-$'sign is appropriate. (Hint: one way to do this is to consider the case of an infinitely wide channel and with no flux in the right-wall boundary layer.)

(5) *The limit of small potential vorticity*: $(q \ll 1)$.
Since $q = (fQ^*/2gD_\infty^2)^{1/2}$ this limit can be achieved by fixing Q^* and increasing D_∞.

(a) Show that the critical condition for attached flow (2.5.6) requires that

$$\bar{d}_c = \left(\tfrac{1}{2}w_c\right)^{-2/3} + O(q).$$

(b) Using (2.5.15) and the above result, show that

$$h_c = q^{-1} - \tfrac{1}{2}\left(\tfrac{1}{2}w_c\right)^2 - \tfrac{3}{2}\left(\tfrac{1}{2}w_c\right)^{-2/3} + O(q).$$

provided ψ_i remains fixed. Thus the sill height must (to lowest order) increase in proportion to the interior reservoir depth q^{-1} (dimensionally D_∞).

(c) Show that dimensionalization of the result in (b) leads to the WLK transport formula (2.4.10) for attached flow.

2.6.   Uniform Potential Vorticity Flow Revisited

Some aspects of the Gill's formulation for uniform potential vorticity flow from a wide basin can be unintuitive. One hobgoblin is the scaling choice $D = (fQ^*/2g)^{1/2}$, which causes the volume flux to be hidden in the dimensionless forms of the governing equations. Another difficulty involves the choice of ψ_i as a parameter for the upstream flow. Although intuitively satisfying, this quantity (or its dimensional version) is difficult to measure in typical deep-sea settings. More recent investigators have explored other choices of upstream parameters and have shown that certain choices can lead to a simplified, and in some cases more data-friendly, formulation.

If the depth scale D is left unspecified for the moment, the continuity equation (2.5.1) reverts to its earlier form

$$2\hat{d}\overline{d} = Q, \tag{2.6.1}$$

while the Bernoulli relation (2.5.3) becomes

$$T^2(\overline{d} - q^{-1})^2 + \frac{Q^2}{4T^2\overline{d}^2} + 2q^{-1}(\overline{d} + h) = 2q^{-1}\overline{B}. \tag{2.6.2}$$

The left-hand side of (2.6.2) may be treated as a Gill-type functional and the critical condition may be found by setting its derivative with respect to $\overline{d}$ to zero. The result is a slightly modified form of (2.5.6):

$$(1 - T_c^2)q^{-1} + T_c^2\overline{d}_c = Q^2\overline{d}_c^{-3}T_c^{-2}/4. \tag{2.6.3}$$

The parameter $\overline{B}$, the average of the sidewall Bernoulli functions, was vanquished by Gill in preference to ψ_i. One alternative (Whitehead and Salzig, 2001) is to use the value $B = B_R$ along the right-side wall. We will generalize their discussion and also consider the value B_L on the left wall. Integration of the relation $dB/d\psi = q$ across the channel then yields

$$\overline{B} = \tfrac{1}{2}[B_R + B_L] = B_R - \tfrac{1}{2}qQ = B_L + \tfrac{1}{2}qQ. \tag{2.6.4}$$

If (2.6.2) is now evaluated at the sill ($h = h_c$), (2.6.4) can be used to write the result in the various forms

$$T_c^2(\overline{d}_c - q^{-1})^2 + \frac{Q^2}{4T_c^2\overline{d}_c^2} + 2q^{-1}\overline{d}_c = \begin{cases} 2q^{-1}(B_R - h_c) - Q \\ \text{or } 2q^{-1}(\overline{B} - h_c) \\ \text{or } 2q^{-1}(B_L - h_c) + Q \end{cases} \tag{2.6.5a,b,c}$$

So far the depth scale D has remained arbitrary, but it is now possible to select it in a way that reduces the dependence of Q on the upstream state to a single parameter. For example, suppose that the first form of the above relation is chosen and that $B_R - h_c$ is set to unity, which is equivalent to

$$D = g^{-1}B_R^* - h_c^*. \tag{2.6.6}$$

The only remaining flow parameter is q. Elimination of $\overline{d}_c$ between (2.6.3) and the newly scaled form of (2.6.5a) determines Q in terms of only w_c and q. In dimensional terms the volume flow rate is $Q^* = gD^2 f^{-1} Q(q, w_c) = (B_R^* - gh_c^*)^2 (gf)^{-1} Q(q, w_c)$. A further conceptual simplification can be made by imagining that the flow stagnates along the right-hand (northern hemisphere) wall at some point upstream of the sill. The surface elevation above the sill at this location is $\Delta z_R^* = g^{-1} B_R^* - h_c^*$, just the depth scale D. The transport relation may therefore be written in the form

$$Q^* = \frac{g(\Delta z_R^*)^2}{f} Q(q, w_c). \tag{2.6.7a}$$

The function $Q(q, w_c)$ is contoured in Figure 2.6.1. One facet that stands out is the insensitivity of Q to the potential vorticity q when the sill width is moderate or small ($w_c < 0.8$).

The discussion thus far has been restricted to attached flow. However, the form of Q for separated flow at the control section can be deduced from a simple argument. We first note that the dimensional geostrophic transport at a separated section, regardless of the potential vorticity distribution, is $Q^* = g(d^*(w^*/2))^2/2f$, where $d_c^*(w^*/2)$ is the depth at the right wall. It has also been shown (Exercise 2 of Section 2.5) that the velocity at the right wall is zero

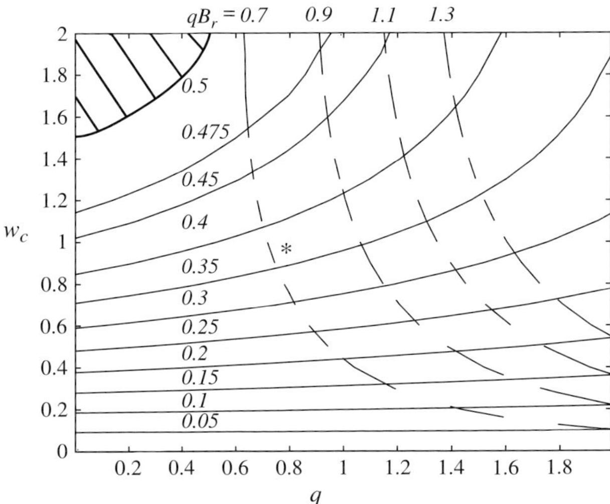

FIGURE 2.6.1. Contours of dimensionless volume flux Q as a function of sill width $w_c = w_c^* f/(gD)^{1/2}$ and upstream potential vorticity $q = D/D_\infty$. In contrast to the Gill (1977) model, D has been chosen as $g^{-1} B_R^* - h_c^*$. The hatched region has separation of the current from the left-hand sill wall, where $Q = 1/2$. The dashed contours indicate the values of qB_R below which separation occurs from the left wall of the upstream basin. The star indicates the values of q and w_c used in Figure 2.6.2. (From Whitehead and Salzig, 2001).

when the flow is both critical and separated and the potential vorticity is uniform. The value of the right-wall Bernoulli function, generally $g(h_c^* + \Delta z_R^*)$, must then be $g(h_c^* + d_c^*(w^*/2))$. The right wall depth is therefore Δz_R^* and the transport is given by (2.6.7a) with $Q = 1/2$, or

$$Q^* = \frac{g(\Delta z_R^*)^2}{2f} \qquad (2.6.7b)$$

In fact, this formula is valid for a larger class of solutions than those with uniform q. It was argued in Section 2.3 that any separated flow with $v = 0$ along the right wall is hydraulically critical. The arguments leading to (2.6.7b) remain valid for any such flow. We also leave it to the reader to prove (see Exercise 1) that the deceptively complex relation (2.5.14) obtained by Gill is equivalent to (2.6.7b). The reader may also wish to note the similarity of (2.6.7b) to the zero potential vorticity transport relation (2.4.15).

One could have made similar simplifications, using the second or third forms of (2.6.5) in connection with different choices in D. As shown by Iacono (2006), an advantage in choosing the second form, which utilizes the average Bernoulli function, is that it leads to a closed form analytical expression for the transport (see Exercise 4). However, there are reasons to prefer the right-wall Bernoulli function B_R^*, or equivalently Δz_R^*, as the upstream parameter. One is that stagnant or sluggish upstreams regions are found in models and laboratory experiments along the right wall. This topic is discussed in detail in Sections 2.7 and 2.14. Also, since Kelvin wave propagation in the basin is positive (downstream) along the right wall, changes in the flow far upstream are communicated to the strait along this wall. Information propagation along the left wall must be upstream since the flow there is subcritical. Information is therefore routed counterclockwise around the edge of the basin, making it reasonable to believe that right-wall information can be specified independently. Of course, these ideas require modification when the upstream basin is closed.

In the Gill model, specification of the upstream state requires two dimensionless parameters, q and Gill's ψ_i. With the present scaling and parameter choices, upstream information is formally specified by q alone. This treatment is more elegant, but it hides the fact that a particular Q corresponds to a whole range of upstream states, each with its own distribution of boundary layer fluxes. That is, each point in Figure 2.6.1 corresponds to a range of upstream flows with the same Q. To specify the full upstream state at such a point, one must know the second upstream parameter B_R, which is hidden in the scaling. It is advantageous to use the closely related parameter qB_R which, in dimensional terms, is the ratio of B_R^* to the Bernoulli function gD_∞ in the basin interior. Large values of qB_R therefore have relatively energetic right-wall boundary currents, whereas $qB_R = 1$ has no right-wall current at all. Consider the range of upstream states possible for the setting $q = 0.8$ and $w_c = 1$, indicated by a star in Figure 2.6.1. The case $qB_R = 7.0$ (Figure 2.6.2a) confirms the expected energetic nature of the right-wall flow. In fact the boundary layer transport is

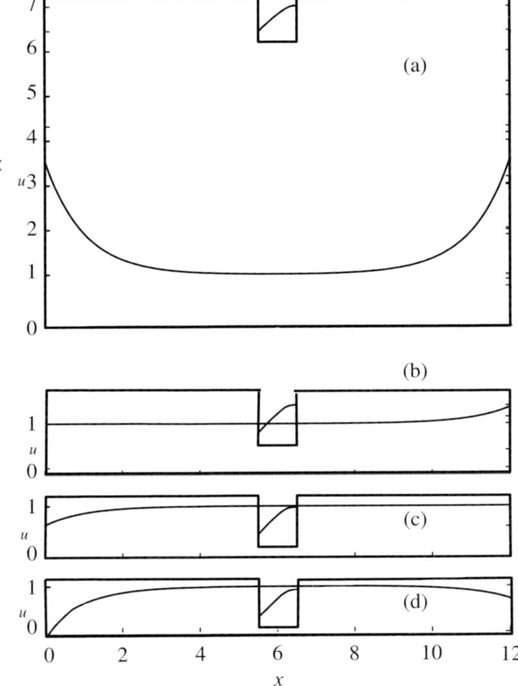

FIGURE 2.6.2. Profiles of the dimensionless surface elevation $z^*/D_\infty (= qz)$ in the upstream basin and at the sill, all for $q = 0.8$ and $w_c = 1$. The dimensionless transport Q equals 0.375 in each case. (a) $qB_R = 7$, (b) $qB_R = 1.375$ (c) $qB_R = 1.0$ (d) $qB_R = 0.74$. (From Whitehead and Salzig, 2001).

more than what can squeeze through the critical section; most of it returns along the left wall. The high kinetic energy along the right wall allows the flow to rise up and pass over a sill whose elevation is much greater than the interior surface elevation. As qB_R decreases, the right-wall boundary layer weakens (frame b) and disappears (frame c for $qB_R = 1$). At the lowest value of qB_R (frame d), the right-wall flow reverses and the left layer carries all of the positive flux. When qB_R reaches its minimum possible value of $q^2Q + \frac{1}{2}$ (see Exercise 2 and the dashed contours in Figure 2.6.1), the surface at the left wall grounds and the flow separates.

To apply Figure (2.6.1) in laboratory or field situations, it is helpful to write out the dimensional form of (2.6.7)

$$Q^* = \frac{g(\Delta z_R^*)^2}{f} Q\left(\frac{\Delta z_R^*}{D_\infty}, \frac{f w_c^*}{(g\Delta z_R^*)^{1/2}}\right) \quad (\Delta z_R^* = g^{-1}B_R^* - h_c^*) \qquad (2.6.8)$$

and thereby acknowledge that two dimensional upstream parameters (B_R^* and D_∞) are required, along with the sill height and width, to determine the dimensional transport. (Gill requires ψ_i^* and D_∞ as upstream parameters.)

Determination of the value of $B_R{}^*$ generally means that a velocity measurement must be made. However this problem is alleviated if a region of quiescent flow along the right wall can be found: there the surface elevation above the sill is exactly $\Delta z_R{}^*$. Alternatively, $B_R{}^*$ can be related to other properties that may be more easily measured. Some of these can be derived from the expressions for the boundary layer flow in the vicinity of the right wall:

$$d^*(x, y) = D_\infty + (d^*(w/2, y) - D_\infty)e^{(x^*-w^*/2)f/(gD_\infty)^{1/2}} \qquad (2.6.9)$$

and

$$v^*(x^*, y^*) = \left(\frac{g}{D_\infty}\right)^{1/2} (d^*(w/2, y) - D_\infty)e^{(x^*-w^*/2)f/(gD_\infty)^{1/2}}, \qquad (2.6.10)$$

which follow from 2.2.12. For example, it can be shown that the transport in the right-hand boundary layer is $Q_R{}^* = (g/f)(d^{*2}(w^*/2, -\infty) - D_\infty^2)$, where $d^{*2}(w^*/2, -\infty)$ denotes the right wall depth in the basin. The Bernoulli function on the right wall is $B_R{}^* = \frac{1}{2}(g/D_\infty)[D_\infty{}^2 + d^{*2}(w^*/2, -\infty)]$. Complete knowledge of the right-wall basin flow therefore requires any two of $Q_R{}^*$, $B_R{}^*$, $d^*(w^*/2, -\infty)$, or D_∞.

There does not appear to be a simple analytical expression for $Q(w_c, q)$ (though see Exercise 4). Figure 2.6.3 shows some plots of Q as a function of w_c for various values of q. For fixed w_c it is apparent that the transport decreases as q increases

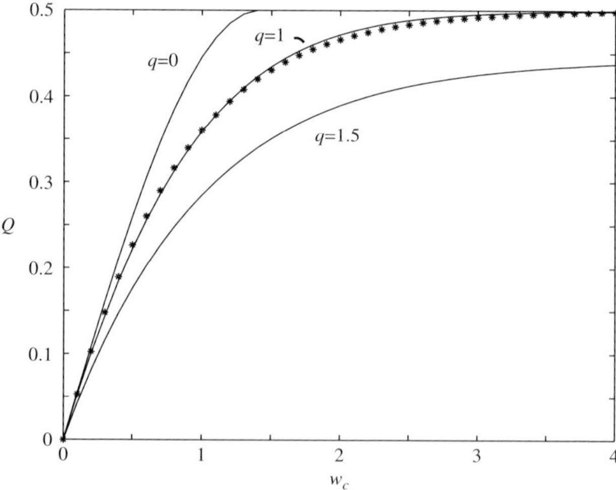

FIGURE 2.6.3. The dimensionless flow rate Q as a function of the dimensionless sill width w_c for various values of q. The $q = 0$ curve merges with the constant value $Q = 1/2$ (see 2.4.15) for separated flow $w_c \geq \sqrt{2}$. The solid lines are exact values, calculated by Whitehead (2005) and equivalent to the information shown in Figure 2.6.1. The starred curve is an approximation to $Q(1, w_c)$ that will be discussed in section 2.14. (From Whitehead and Salzig, 2001).

and therefore the largest transport occurs for zero potential vorticity. One should exercise caution in interpreting this result, however. If the dimensional critical width w_c^* is held fixed, then fixed w_c means that D and therefore B_R^* is fixed. That is, if the sill geometry is regarded fixed, the maximum in Q for zero q occurs when B_R^* is held fixed. As shown by Iacono (2006), the same is not true when the scaling for D is based on B^*. The volume transport then has its maxima at finite values of q.

Exercises

(1) By writing (2.5.14) in dimensional form and introducing Δz_R^* as a depth scale, show that the much simpler transport relation (2.6.7b) is obtained.

(2) Among the upstream states that are possible at a given location in Figure 2.6.1, show that the limiting case of separated upstream basin flow occurs when qB_R falls below $q^2Q+\frac{1}{2}$.

(3) *Asymptotic properties of the function* $Q(w_c, q)$.

(a) Using (2.6.3) and the form of (2.6.5a) with $(B_R - h_c) = 1$, show that

$$\lim_{w_c \to 0} Q = \left(\frac{2}{3}\right)^{3/2} w_c,$$

and thus the slope of all the Figure 2.6.3 curves at the origin is $(2/3)^{3/2}$ regardless of the value of q.

(b) Next show that for a given $q =$ constant curve in Figure 2.6.3, that separation of the flow at the sill section first occurs where $Q = 1/2$, corresponding to $q = 2T_c^2/(1+T_c^2)$ or

$$w_c = 2q^{1/2}\tanh\left(\frac{q}{2-q}\right)^{1/2}.$$

Note that separation can occur only for $0 < q < 1$. Hint: Use the velocity profile (2.2.4) along with the fact that the right-wall value of v is zero when the flow is critical.

(c) Note that the results in (a) and (b) provide endpoints for the curves with $0 \leq q \leq 1.0$. For $q > 1$ show that

$$\lim_{w_c \to \infty} Q = q^{-1}(1 - \frac{1}{2}q^{-1}).$$

Hint: Use the same equations as in part (a).

(4) *Iacono's (2006) solution.* A closed formula for the constant-potential vorticity transport from a wide reservoir can be obtained if the upstream conditions are chosen in a particular way. Proceed as follows:

(a) Evaluate the Bernoulli relation (2.6.2) at the sill, where the flow is assumed to be critical, leading to

$$T_c^2(\overline{d}_c - q^{-1})^2 + \frac{Q^2}{4T_c^2\overline{d}_c^2} + 2q^{-1}\overline{d}_c = 2q^{-1}\Delta z_I,$$

where $\Delta z_I = \overline{B} - h_c$.

(b) Use the critical condition (2.6.3) to substitute for the second term and rearrange the result, eventually obtaining a quadratic equation for $\overline{d}_c$. Show that the physically meaningful solution to this equation can be written as

$$\overline{d}_c = \frac{3}{4qS_c^2}\left[-1 + \left(1 - \frac{8}{9}S_c^4 + \frac{16}{9}S_c^2C_c^2q\Delta z_I\right)^{1/2}\right]$$

where $S_c = \sinh[\tfrac{1}{2}q^{1/2}w]$ and $C_c = \cosh[\tfrac{1}{2}q^{1/2}w]$.

(c) Note that with (2.6.3) rearranged to give

$$Q^2 = 4\overline{d}_c^3 T_c^2[(1 - T_c^2)q^{-1} + T_c^2\overline{d}_c]$$

one now has a direct formula for the transport in terms of the new upstream parameter Δz_I. Make some plots of Q vs q for fixed Δz_I and fixed sill geometry and thereby show that the maximum flux is obtained at a finite q. Contrast this to the result obtained in Figure (2.6.1).

2.7.   Flow Reversals and Recirculation

Counterflows and closed circulations are commonly observed in experiments with hydraulically active rotating flows and have also been seen in deep passages such as the Hunter Channel (Zenk *et al.*, 1999). Closed gyres tend to occur upstream of a controlling sill, typically along the northern hemisphere right wall, or downstream along the left wall. The latter case occurs in connection with a hydraulic jump and will be discussed further in Chapter 3. Borenäs and Whitehead (1998) present examples of right-wall gyres, one of which is shown in Figure 2.7.1. The flow is confined to the lower layer of a two-fluid (water and kerosene) system in a rotating, rectangular channel vertical sidewalls. The channel is fitted with an obstacle that smoothly reaches a maximum height midway through the channel. The water is pumped into an upstream reservoir (to the left) where it collects and passes through a porous filter into the channel. The flow is critical at or very near the sill, the position of which is indicated by a dashed line. The gyre can be seen as a semicircular region of fluid that remains clear and free of dark dye introduced upstream.

   The typical gyre geometry as seen in cross-sectional and plan views is sketched in Figure 2.7.2. The region of closed streamlines lies between two right-wall

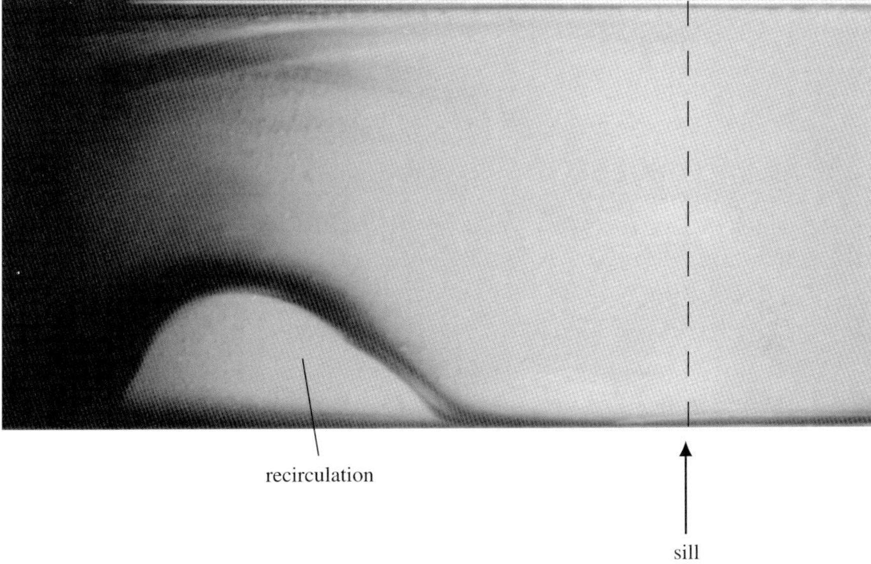

FIGURE 2.7.1. A gyre in a channel flow with a parabolic sill (from Borenäs and Whitehead, 1998). The left-to-right flow is marked by dark dye, which is introduced at the upstream end of the channel and is deflected around the gyre. The value of $wf/2\sqrt{g'D_\infty}$ is varied between 0.25 and 0.39, and the case shown has value 0.35. The velocity of the flow entering the left end of the channel has been rendered approximately uniform by a filter. The potential vorticity is therefore not uniform.

stagnation points at $y = y_1$ and $y = y_2$. The streamline joining the two points and located a distance w_g from the right wall will be called the stagnation streamline. The interface elevation along this contour must remain the same as that along the right wall, else the gyre would have a net geostrophic volume flux. The presence of a closed gyre gives rise to a number of questions concerning the physics and analysis of the flow. How is potential vorticity specified along the closed streamlines? Does the gyre choke the flow the same way that a contraction in channel width would? Before addressing these matters we first attempt to determine the conditions under which the right-wall gyre can form.

Counterflow is integral to the gyre and a mechanism that can work in favor of right-wall flow reversals is vortex squashing. As fluid columns leave a relatively deep reservoir and move towards a relatively shallow sill, their thickness d decreases and their vorticity ($\cong \partial v/\partial x$) must decrease, perhaps becoming negative. This process can contribute to small or negative velocities along the right wall. Of course, the fact that counterflows are not observed in the downstream supercritical flow suggests that vortex squashing is not the whole story. Such flows tend to be separated from the left wall and therefore to be much narrower than the upstream flow. Even though $\partial v/\partial x$ may be negative, the narrowness of the current and the strength of the (positive) mean velocity prevents v from becoming negative on the right wall.

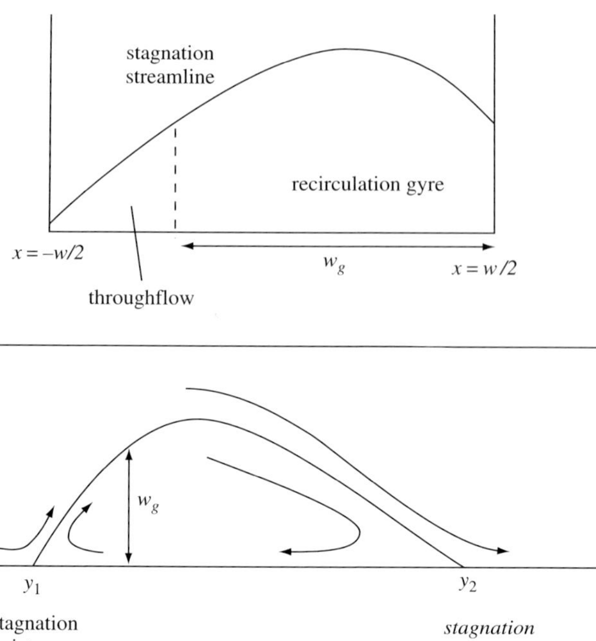

FIGURE 2.7.2. Diagram showing the recirculation and throughflow in cross section (top) and in plan view (bottom).

A more careful examination of stagnation point formation and flow reversal may be made by appealing to earlier results on the formation of a counterflow. This subject was discussed in the context of a separated, zero potential vorticity flow, as summarized by the insets in Figure 2.3.2. Should this flow be critical at a sill, the interface will be level at the right wall (second inset from top) and v will vanish there. Immediately upstream from this section the flow will be subcritical (second inset from bottom), with counterflow ($v < 0$) along the right wall. In the laboratory example in Figure 2.7.1, which is not separated, the counterflow begins a finite distance upstream of the sill.

The presence of a second, upstream stagnation point will terminate the counterflow and give rise to a closed gyre, as in the experiment. The existence of the second stagnation point may be anticipated using the expression for the right-wall velocity (2.2.4) based on uniform potential vorticity. If (2.2.18) is used this velocity may be expressed as

$$v(w/2, y) = q^{1/2}T\left[\frac{Q}{2\overline{d}T^2} + \overline{d} - q^{-1}\right], \tag{2.7.1}$$

A right-wall stagnation point occurs where

$$\frac{Q}{2\overline{d}T^2} + \overline{d} - q^{-1} = 0,$$

or

$$\bar{d} = \frac{q^{-1}}{2} \pm \sqrt{\frac{q^{-2}}{4} - \frac{Q}{2T^2}} \qquad (2.7.2)$$

Two stagnation points are possible for $Q < T^2 q^{-2}/2$ and both will be encountered if $\bar{d}$ passes through the two indicated roots. This situation is favored by weak transports ($Q \ll 1$), small values of the potential vorticity ($q \ll 1$), or wide channels ($T = tanh[q^{1/2}w/2] \cong 1$). In addition, strong shoaling of the bottom encourages a greater range of $\bar{d}$ along the channel, and this increases the probability that two stagnation points will occur.

Can a counterflow exist at a control section? Under conditions of uniform potential vorticity and rectangular cross section this possibility is ruled out by the theorem constraining critical flow to be unidirectional (Section 2.5, Exercise 3). However, this restriction does not generally hold. We will identify examples showing that counterflows can occur at a critical section when the geometry is nonrectangular (Section 2.8) or when the potential vorticity is nonuniform (Section 2.9).

In dynamical terms, one of the distinguishing characteristics of a closed gyre is that the potential vorticity distribution is no longer imposed by upstream conditions. Fluid parcels are free to circulate indefinitely and dissipative effects, while arguably negligible for the throughflow, become paramount. An instructive constraint may be written down by integrating the tangential component of (2.1.15) around the circuit Γ_ψ formed by any closed streamline $\psi =$ constant within the gyre. For steady flow, the result is

$$\oint_{\Gamma_\psi} \mathbf{F}^* \cdot \mathbf{l} \, ds = 0, \qquad (2.7.3)$$

where $\mathbf{l}$ and ds are the tangential unit vector and differential arc length along Γ_ψ. The vector $\mathbf{F}^*$, which generally contains all momentum forcing and dissipation terms, might in the present context consist of a bottom drag term of the form $C_d \mathbf{u}/d$ (consistent with an Ekman layer on the bottom) and a lateral stress term S. In the laminar laboratory flow (Figure 2.6.1) S is presumably dominated by the lateral viscous stress generated by the throughflow moving along the left side of the gyre. In geophysical applications the lateral stress would be dominated by turbulent momentum fluxes. One of the difficulties in using (2.7.3) to solve for the circulation is that the shape of the latter is generally not known in advance. The immediate importance of (2.7.3) is that forcing and dissipation cannot be ignored once the streamlines are closed.

In order to incorporate a gyre into a hydraulic model for the flow as a whole, it is necessary to know something about the potential vorticity distribution along the closed streamlines. Borenäs and Whitehead (1998) explored two approaches, the first based on the assumption that the gyre potential vorticity has the same (constant) value as the throughflow and the second that the gyre is stagnant. The first approach has the virtue of simplicity; solutions may be calculated from the relations laid out in the Gill model (Section 2.5). The second approach is more consistent with observations of the laboratory flow, which show the gyre circulation to be relatively weak. The novel features of the calculation are explored in Exercise 1. Neither approach is easily motivated by dynamical principles. Determination of the true distribution of potential vorticity within the gyre remains an unresolved issue.

A comparison between gyres with uniform potential vorticity and stagnant fluid, both with the same upstream conditions, appears in Figure 2.7.3. The former is distinguished mainly by its strong anticyclonic circulation. Note that the positions of the stagnation points for the two cases are identical, as they must be (Exercise 2). The overall shape and size are similar. The gyre length tends to be moderately shorter than what is observed in the laboratory (Figure 2.7.4). The observed gyre widths (not shown in the figure) are also somewhat smaller than those predicted by either theory. A possible reason for the discrepancy is that the observed gyres often contained small cyclonic features. The experiment agrees with the theoretical prediction of a minimum width below which no gyre forms, though the threshold values are somewhat different.

Since a closed gyre carries no net volume transport, there is a temptation to think of the gyre edge as being equivalent to a solid wall. If the flow is steady and lateral viscous effects are ignored, the throughflow is apparently unaffected

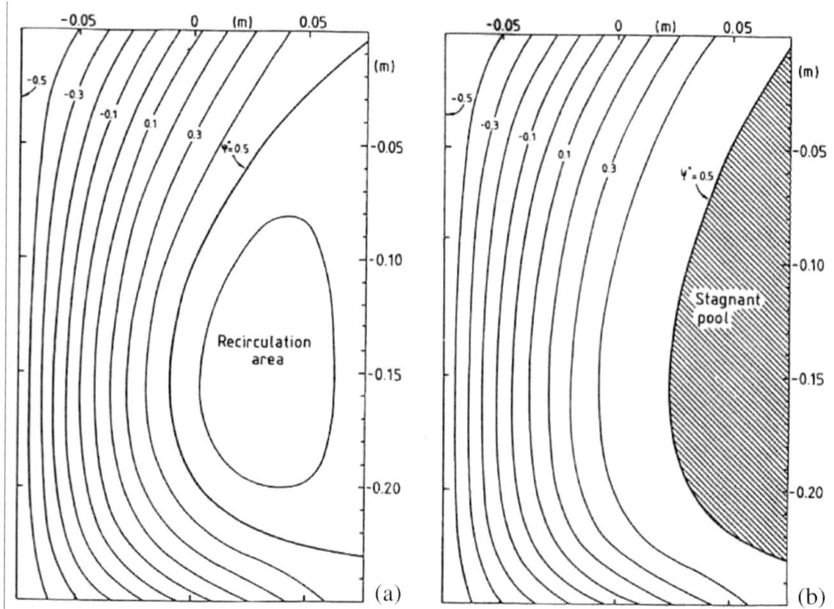

FIGURE 2.7.3. Comparison between gyres imbedded in the throughflow with uniform potential vorticity. The gyre can have the same uniform potential vorticity as the throughflow (left) or be stagnant (right). Each case is characterized by $w^* f/(2gD_\infty)^{1/2} = 0.39$. From Borenäs and Whitehead (1998).

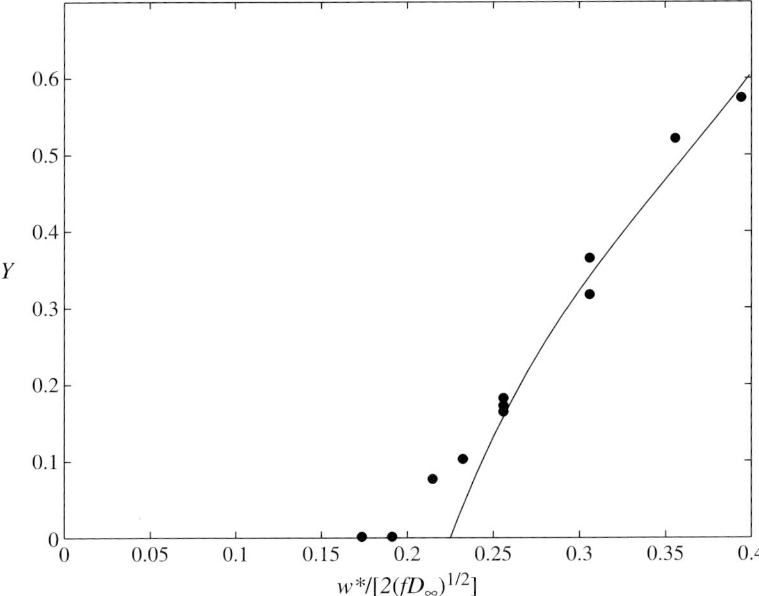

FIGURE 2.7.4. The gyre length as measured by the dimensionless distance between stagnation points $Y = (y_2 - y_1)f/[2(gD_\infty)^{1/2}]$ and plotted as a function of the dimensionless width $w^*f/[2(gD_\infty)^{1/2}]$. The curve is based on a theory with uniform potential vorticity upstream flow with $\psi_i = 1$ and with the experimental values $q = 0.25$ and $h_m^*/D_\infty = 3/4$. (From Borenäs and Whitehead, 1998).

by replacing the stagnation streamline $x = (w/2) - w_g$ by a such a wall. In this view, the gyre might choke the flow in same way as a true side wall contraction. One must be wary of this analogy: if the stagnations streamline is replaced by a wall, the value of $T = tanh[q^{1/2}w/2]$ is replaced by $T = tanh[q^{1/2}(w - w_g)]$, and the value of the corresponding Froude number (2.5.7) is altered. The value based on placing an artificial wall at $x = (w/2) - w_g$ is invalid since it does not account for the true physical characteristics of a Kelvin wave propagating through the flow. Such a wave would see the gyre edge as pliant, and not a rigid wall.

Exercises

(1) Suppose that the gyre is stagnant and that the exterior fluid (the throughflow) has uniform potential vorticity (as in Figure 2.6.3b). What is the matching condition along the separating streamline that allows solutions in the two regions to be joined?

(2) Show that the positions y_1 and y_2 of the stagnation points are independent of the potential vorticity distributions inside the gyre, provide that the potential vorticity of the exterior (noncirculating) fluid is the same constant.

2.8.   Nonrectangular Cross Sections

Up to this point we have dealt strictly with channels with rectangular cross sections. The only allowable variation of bottom elevation has been in the longitudinal (y) direction. Although this geometry lends mathematical convenience, it means that one must consider attached and detached flows separately. Were the two states dynamically similar, one might be content to put up with the implicit bookkeeping. The fact that there are significant differences raises some doubts concerning the artificial nature of rectangular geometry. For example, differences can be found in the dynamics of upstream disturbances; attached flow is controlled by Kelvin waves whereas detached flow is controlled by frontal waves. It has even been suggested that critical flow with respect to the latter can be difficult to achieve. A unifying theory taking into account the more realistic, rounded nature of natural straits would be quite advantageous. Such theory would allow a seamless merger between Kelvin and frontal wave dynamics. The simplest such model makes use of a channel with a parabolic cross section (Figure 2.8.1a). Borenäs and Lundberg (1986, 1988) investigated this geometry for the case of finite, uniform potential vorticity and later zero potential vorticity. The following discussion is based largely on their work.

Consider a channel with bottom elevation:

$$h^*(x, y) = h^*(0, y) + \alpha(y)x^{*^2},$$

nondimensionally

$$h(x, y) = h(0, y) + x^2/r(y). \tag{2.8.1}$$

In the usual manner, D is used as a depth scale and $(gD)^{1/2}/f$ as a length scale. The parameter $r(y) = f^2/g\alpha(y)$ can be interpreted as the ratio of the square of two length scales. The first is the half-width w_p of the level surface when the channel is filled evenly to a depth $d_p = \alpha\, w_p{}^2$ (Figure 2.8.1b). The second is a local Rossby radius of deformation $(gd_p)^{1/2}/f = (g\alpha)^{1/2}w_p/f$ based on this depth. Large values of r occur when the bottom curvature α is small compared to f^2/g. As suggested in Figure 2.8.1b this is equivalent to a small local deformation radius $(gd_p)^{1/2}/f$ in comparison to the resting half-width $w_p = (d_p/\alpha)^{1/2}$. By the same measure, a dynamically narrow channel occurs when the curvature is large compared to f^2/g (Figure 2.8.1c). That this measure of narrowness should depend only on the background parameters α, g, and f, and not fluid depth itself, is a special feature of the parabolic geometry and its uniform curvature.

The solution to (2.1.14) for the topographic profile (2.8.1) and for constant potential vorticity q can be written as

$$d(x, y) = \frac{1 + 2r^{-1}}{q \sinh\left(q^{1/2}(a+b)\right)} \left[\sinh\left(q^{1/2}(x - b)\right) - \sinh\left(q^{1/2}(x + a)\right) \right] + \frac{1 + 2r^{-1}}{q}$$

$$\tag{2.8.2a}$$

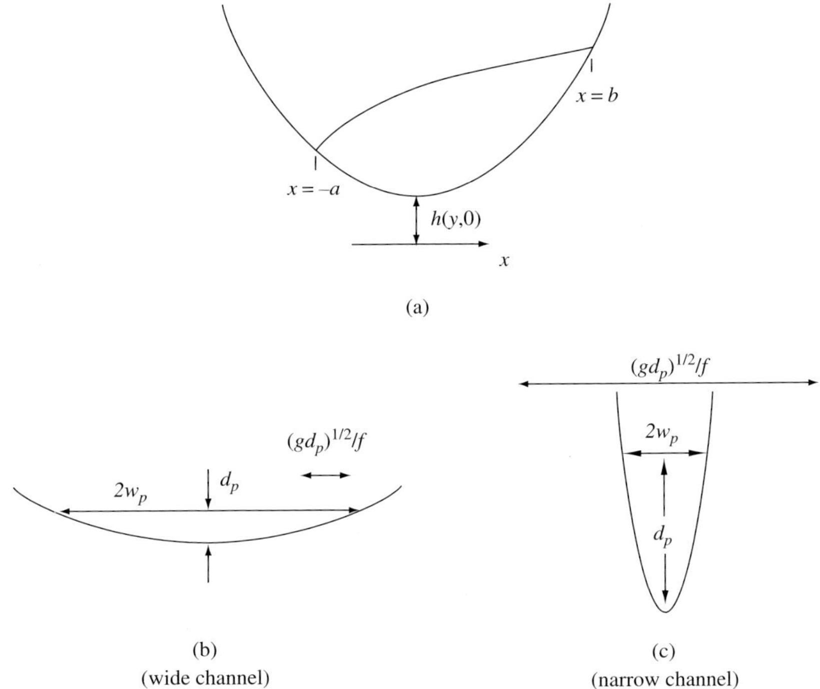

FIGURE 2.8.1. Definition sketch for flow in a parabolic channel (a). The wide and narrow limits of the parabolic channel (b) and (c).

The corresponding geostrophic velocity is

$$v(x, y) = \frac{1 + 2r^{-1}}{q^{1/2} \sinh \left(q^{1/2}(a + b) \right)} \left[\cosh \left(q^{1/2}(x - b) \right) - \cosh \left(q^{1/2}(x + a) \right) \right] + 2r^{-1}x.$$

$$(2.8.2b)$$

The surface or interface intersects the bottom at the two points $x = b$ and $x = -a$ (Figure 2.8.1a). The wetted width of the flow is therefore $a + b$.

In addition to the scales described above, the global deformation radius $(gD_\infty)^{1/2}/f$ is present but hidden in arguments like $q^{1/2}(x + a) = (x^* + a^*)f/(gD_\infty)^{1/2}$. As before, we might imagine that the potential depth D_∞ is set in an upstream reservoir. If the stream width is large in comparison to $(gD_\infty)^{1/2}/f$ at a particular section, the depth profile will have a boundary layer structure similar to that of the Gill (1977) model. If the range is small, arguments like $q^{1/2}(x - a)$ remain small, and the boundary layer structure is lost. The limiting case for the latter is the 'zero potential vorticity' limit, in which the fluid may be imagined to originate in a very deep, quiescent upstream basin.[7] It should be pointed out that the flow may still be 'wide' in the sense $r \gg 1$,

---

[7] To be self-consistent, the reservoir must have vertical sidewalls, else the depth would go to zero at the edges.

as in Figure 2.8.1b, while remaining narrow in the sense $q^{1/2}(a+b) \ll 1$. The Denmark Strait sill has an r value of 10–20, based on the average value of α. The value of $q^{1/2}(a+b)$ based on observations cited in Nikolopoulos *et al.* (2003) ($D_\infty = 600$m, $g = 4.8 \times 10^{-3}$m/s^2, and $a^* + b^* \cong 50$ km) is about 2.5.

The 'zero potential vorticity' case is the easiest to explore. The depth and velocity profiles may be obtained by taking the $q \to 0$ limit of (2.8.2), or simply by direct integration of (2.1.12) and (2.1.13) with $q = 0$:

$$d = \frac{1}{2}(1 + 2r^{-1})(a+x)(b-x) \tag{2.8.3}$$

The accompanying velocity profile has constant shear

$$v(x) = -\left[x + \frac{1}{2}(1 + 2r^{-1})(a-b)\right] \tag{2.8.4}$$

The Bernoulli function

$$\frac{v^2}{2} + d + h(0, y) + \frac{x^2}{r} = \frac{D_\infty}{D}$$

is uniform in the present limit. Substitution of (2.8.3) and (2.8.4) into this relation leads to

$$(1 + 2r^{-1})\left[\frac{(a-b)^2}{r} + \frac{(a+b)^2}{2}\right] = 4\Delta z, \tag{2.8.5}$$

where $\Delta z = D_\infty/D - h(0, y)$ is the elevation of stagnant water in the upstream basin above the deepest point of the parabolic bottom.

The volume flux is found with the help of (2.8.5) to be

$$Q = \int_{-a}^{b} d(x)v(x)dx = \frac{(a+b)^3(2+r)}{6r^2}\left[\Delta z\frac{r^2}{2+r} - \frac{r(a+b)^2}{8}\right]^{1/2} \tag{2.8.6}$$

and the right-hand side has the required form for a hydraulic functional in the single variable $(a+b)$. Setting the derivative of this expression with respect to $(a+b)$ to zero leads to the critical condition. It can be verified in the usual manner that critical flow must occur at the sill and h must therefore be evaluated at the corresponding position $y = y_s$. The resulting critical condition is

$$(a+b) = \sqrt{\frac{6r\Delta z}{(2+r)}},$$

or, with the help of (2.8.5),

$$6(a-b)^2 = r(a+b)^2 \tag{2.8.7}$$

The corresponding controlled flux is given by

$$Q = \frac{\Delta z^2}{2+r}\sqrt{\frac{3r}{2}}, \qquad (2.8.8)$$

or

$$Q^* = \frac{\Delta z^{*^2}}{2+r}\sqrt{\frac{3g}{2\alpha}}. \qquad (2.8.9)$$

This 'weir' formula can be compared with the case of a separated flow with rectangular cross section (2.4.15) with the result

$$\frac{Q^*_{\text{parabolic}}}{Q^*_{\text{rectangular}}} = \frac{2}{(2+r)}\sqrt{\frac{3r}{2}}. \qquad (2.8.10)$$

The comparison is meaningful for moderate or large values of r (wide channels) since the flow in the rectangular section is assumed to be separated. For large r it can be seen that the flux in the parabolic channel is less than the rectangular case by a factor proportional to $r^{-1/2}$. One of the reasons for this mismatch is that wide parabolic openings tend to favor reversals in velocity along the right edge, even when the flow is critical. In fact, it can be shown that flow reversals occur at the sill when $r > {}^2/_3$.

The wide channel or weak curvature case ($r \gg 1$) can be developed a bit further by noting that (2.8.3) reduces to

$$d = \frac{1}{2}(a+x)(b-x).$$

Such profiles tend to have flow and counterflow with positive velocity on the left and a return flow almost as great to the right (Figure 2.8.2). Since the velocity at the top of the profile is zero, the interface elevation there must equal that in the quiescent upstream reservoir. All possible solutions for a given reservoir interface elevation are therefore found by simply sliding a parabola with fixed curvature and fixed maximum elevation back and forth, as suggested in the figure. Upstream of the sill section, the profile must be centered slightly to the right of $x = 0$ in order to achieve positive Q. At the shallower sill section, the interface profile is obtained by sliding the parabola to the right and this results in a weaker counterflow. Downstream of the sill, the parabola is slid further to the right and the resulting supercritical flow is unidirectional.

The existence of a counterflow at a critical (or supercritical) section would appear to confound the notion of upstream influence. Such flows seem to be sensitive to downstream information despite the fact that no upstream wave propagation is possible. The situation may be made clearer by remembering that simple advection is quite different from propagation of mechanical information due to waves. One could place a drop of dye into a counterflow downstream

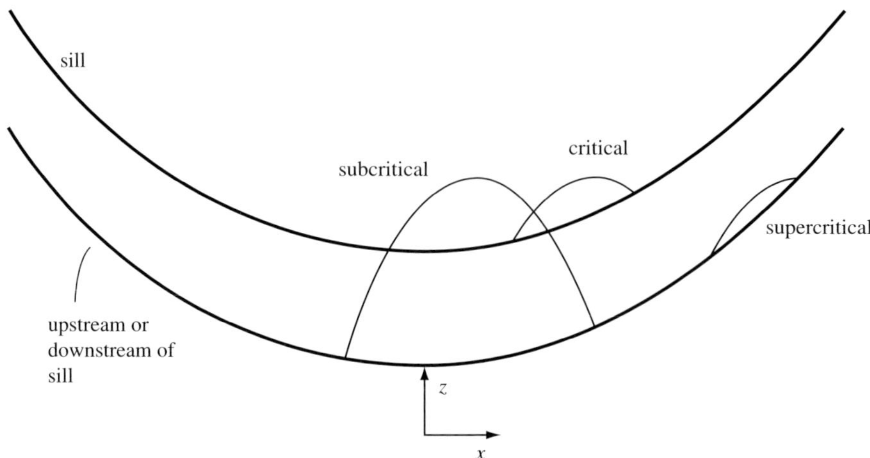

FIGURE 2.8.2. Example of zero potential vorticity flow in a wide parabolic channel at three sections. The upper thick curve represents the bottom at the sill section, whereas the lower thick curve represents the bottom at sections upstream and downstream of the sill. At the upstream section the subcritical solution is valid. Other solutions, including the critical solution at the sill and the supercritical solutions downstream, are obtained by sliding the parabola sideways. The apex of the parabola, where the velocity goes to zero, must remain at the same elevation.

of a controlling sill and follow its motion upstream and into the subcritical reaches of the current. However, the dye would not alter the transport or energy of the upstream flow, so there would be no real upstream influence. Rotating channel flows with countercurrents are just one example of physically realizable, geophysically relevant flows that can have velocity reversals at the critical sections. Another example is the two-layer exchange flow (Chapter 5).

So far the discussion has revealed an important difference between the rectangular and rounded cases. Critical flow in a rectangular section must be unidirectional, provided the potential vorticity is uniform. At a parabolic section of sufficiently low curvature, critical flow will experience a velocity reversal and reduced flux. Whether reversals actually occur at wide sills such as the Denmark Strait is not well-known; observations there suggest a stagnant region along the right edge (see Figure I.8a).

We will discuss only a few aspects of the case of constant, nonzero potential vorticity. To begin with, the characteristic speeds are given by:

$$c_{\pm} = \hat{v} \pm \left\{ \alpha^2 T^{-2}(\hat{w} - 2Tq^{-1/2}) \left[\hat{w} - 2Tq^{-1/2} + (T^2 - 1) \right. \right.$$
$$\left. \left. (\hat{w} - (1 + 2\alpha)T\alpha^{-1}q^{-1/2}) \right] \right\}^{1/2} . \tag{2.8.11}$$

(Pratt and Helfrich, 2005). Here $\hat{v} = \alpha(b - a)$, $T = \tanh(q^{1/2}\hat{w}/2)$ and $\hat{w} = a + b$. The corresponding Froude number is given by

$$F_p{}^2 = \frac{T^2(b-a)^2}{(\hat{w}-2Tq^{-1/2})\,[\hat{w}-2Tq^{-1/2}+(T^2-1)\,(\hat{w}-(1+2\alpha)T\alpha^{-1}q^{-1/2})]}$$

$$(2.8.12)$$

can be useful in assessing the hydraulic criticality of an observed flow, provided that the potential vorticity q can be estimated and the bottom shape can be reasonably fit to a parabola. Girton *et al.* (2006) discuss an example of application to the Faroe Bank Channel. Equation (2.8.12) can also be guessed directly from the condition for steady, critical flow (Borenäs and Lundberg, 1986). Finally, we note that long wave speeds, Froude numbers and critical conditions for zero potential vorticity flow across a section of arbitrary topography can be written down. The derivation arises in the consideration of the stability of such flows and is presented in Section 3.9.

The differences between the zero- and finite-potential vorticity cases is particularly evident when the parabola is wide ($r \gg 1$). As shown by (2.8.3), the zero potential vorticity profile occupies a width $b+a$ that is comparable to the Rossby radius based on the maximum depth within the profile. On the other hand, a flow with finite potential vorticity (see 2.8.2a) may be spread over a much larger width. The interior of the depth profile consists of a wide region having constant depth q^{-1}, the nondimensional potential depth. The free surface or interface therefore parallels the bottom, implying a broad geostrophic flow with local velocity proportional to the cross-channel bottom slope. Where this slope is negative, the velocity is also so. The depth is brought to zero at the edges by boundary layers with width equal to the potential-depth-based Rossby radius, nondimensionally $q^{-1/2}$. Negative flow often occurs in the right-hand boundary layer.

Killworth (1992) has argued that the picture of a broad flow with a sluggish interior, high-velocity boundary layers, and flow reversals (Figure 2.8.3a) is characteristic of wide channels with more general shapes and potential vorticity distributions. Some of the elements of his elaborate argument are as follows. The channel is considered dynamically wide in the sense that changes in h with x occur over a scale much greater than boundary width scale $q^{-1/2}$. For this definition to have meaning, $q(\psi)$ must remain nonzero across the breadth of the flow. Now consider an upstream region in which the flow is sluggish ($v \ll 1$), so that $B(\psi) \cong d+h$ and $q(\psi) \cong 1/d$. It follows that, $d = d(\psi)$, $h = h(\psi)$ and therefore $d = d(h)$, at least to a first approximation. The depth at any particular x in the interior region is therefore given by the potential depth $q^{-1}(\psi)$ for the value of ψ at that point. Since $d = d(h)$, the potential depth is determined by the local value of h. These features are characteristic of the *planetary geostrophic dynamics*, in which inertia is neglected but large variations in depth are allowed. In this limit, streamlines follow contours of constant h.

If streamlines originating in the sluggish region are followed downstream to the sill section, and if the topography remains gradually varying in x, then the streamlines will simply follow isobaths and the flow will remain sluggish. It is not possible, for example, for an isolated band of rapid geostrophic flow

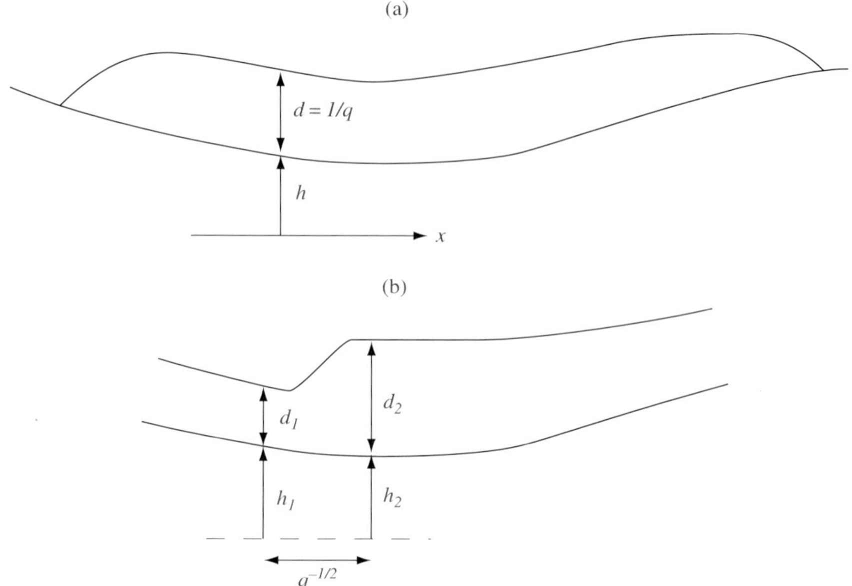

FIGURE 2.8.3. (a) Flow across a section in which the topography varies gradually with x. The interior depth at each point is equal to the potential depth for that particular streamline, and streamlines flow along contours of constant h. (b) A hypothetical band of flow with $v = O(1)$.

($v = O(1)$, Figure 2.8.3b) to arise in the interior of the stream. There the geostrophic relation would require an $O(1)$ depth change, from d_1 to d_2, over a distance $q^{-1/2}$, already assumed to be $O(1)$. However, the change in h across this distance is negligible for the assumed, gradually varying topography, and thus the relation $d = d(h)$ is clearly violated. It therefore would seem that rapid bands of flow can only occur at the edges. The flow in the right-hand boundary layer will tend to be negative, since it must bring the depth to zero over a short distance.

It is not hard to construct examples of geostrophic flow across a broad sill that varies rapidly in the interior. The above arguments point out the difficulty in achieving such a state as the result of evolution from a slow, gradually varying upstream state.

Exercises

(1) Compare the weir formula (2.8.9) to the case of attached, zero potential vorticity flow in a rectangular channel. Do the two formulae agree for $r \ll 1$? Should they?

(2) Prove that a velocity reversal at a critical section with parabolic geometry and $q = 0$ can only occur if $r > 2/3$.

2.9. Nonuniform Potential Vorticity

Our discussion of semigeostrophic models has largely been restricted to flows with uniform potential vorticity q^*. The only waves supported by such flows are the two Kelvin waves, or their frontal relatives. As noted in Section 2.1, the introduction of a potential vorticity gradient gives rise to a new restoring mechanism and a new class of waves that are nondispersive at long wave lengths. We discussed the case of topographic Rossby waves in a channel with a constant bottom slope $\partial h^*/\partial x^* = -S$ and a rigid upper boundary. The dispersion relation (2.1.30) governing a long wave propagating on a background state of rest can be generalized to include a uniform background velocity V, in which case the wave speed becomes

$$c^* = V + \left(\frac{dq^*}{dx^*}\right)\frac{w^{*2}D}{n^2\pi^2}, \quad n = 1, 2, 3, \cdots \tag{2.9.1}$$

where $\frac{dq^*}{dx^*} = -\frac{Sf}{D^2}$. For positive S, $\partial q^*/\partial x^* < 0$ and higher potential vorticity is found on the left-hand side (facing positive y^*) of the channel. In this case the propagation tendency of the waves is against the background flow. The latter is hydraulically critical ($c^* = 0$) when

$$V = \frac{Sfw^{*2}}{Dn^2\pi^2}. \tag{2.9.2}$$

In the opposite case ($\partial q^*/\partial x^* > 0$), all waves propagate towards positive y^*. Critical flow for this example therefore requires that the potential vorticity increase to the left of the flow direction.

Readers versed in the dynamics of large-scale flows in the ocean and atmosphere might choose to express (2.9.2) in the form

$$\frac{V}{\beta L^2} = 1,$$

where $\beta = Ddq^*/dx^*$ is a measure of the potential vorticity gradient and $L = w^*/n\pi$ is the horizontal length scale. The dimensionless parameter $V/\beta L^2$, sometimes called the *beta Froude number*, is generally interpreted as a measure of nonlinearlity of the flow field, values $\ll 1$ indicating linear dynamics. In the present context, the parameter is an indication of the importance of advection and its value must be O(1) for hydraulic effects to be possible. Various forms of the beta Froude number will arise throughout the remainder of the book in discussions of flows dominated by potential vorticity dynamics.

The presence of a potential vorticity gradient in combination with a free surface or interface leads to analytical difficulties in connection with the cross-stream structure equation (2.1.14). The difficulty can be described by first noting the connection between ψ and d implied by the geostrophic relation:

$$\frac{\partial\psi}{\partial x} = vd = \frac{1}{2}\frac{\partial d^2}{\partial x} + d\frac{\partial h}{\partial x}$$

If $\partial h/\partial x = 0$, integration of this equation from the channel side wall at $x = w/2$ to a point in the interior yields

$$\psi = Q + \tfrac{1}{2}(d^2 - d^2(\tfrac{1}{2}w, y)) \tag{2.9.3}$$

where $\psi = Q$ has been imposed at $y = w/2$. Equation (2.1.14) may now be written as

$$\frac{\partial^2 d}{\partial x^2} - q[Q - \tfrac{1}{2}(d^2 - d^2(\tfrac{1}{2}w, y))]d = -1 \tag{2.9.4}$$

If q is constant (2.9.4) reduces to the familiar linear equations that form the basis for models considered earlier. However, a nontrivial dependence of q on ψ introduces a nonlinearity that generally precludes analytical solutions for the cross-channel structure.

a. Stern's Criterion

Some progress can be made without actually knowing the particulars of the cross-stream structure. For example, Stern (1974) derives a generalized critical condition with no restriction on potential vorticity and with the requirements that the channel cross-section be rectangular ($\partial h/\partial x = 0$) and that the flow be unidirectional. A version of the proof, grounded in Stern's approach but simpler than his original, begins with the relation

$$v = \pm 2^{1/2}[B(Q + \tfrac{1}{2}(d^2 - d^2(\tfrac{1}{2}w, y))) - d - h]^{1/2}$$

which follows from the definition of the semigeostrophic Bernoulli function $B(\psi)$ and from (2.9.3). Assume that the velocity is positive, so that the '$+$' sign is appropriate. One then proceeds from a trivial relation thus:

$$w = \int\limits_{-w/2}^{w/2} dx = \int\limits_{0}^{Q} \frac{dx}{d\psi} d\psi = \int\limits_{0}^{Q} \frac{1}{vd} d\psi$$

$$= \int\limits_{0}^{Q} \frac{1}{2^{1/2}d[B(\psi) - d - h]} d\psi$$

$$= \int\limits_{[d^2(\tfrac{1}{2}w,y)-2Q]^{1/2}}^{d(\tfrac{1}{2}w,y)} \frac{\partial d}{2^{1/2}[B(Q + \tfrac{1}{2}(d^2 - d^2(\tfrac{1}{2}w, y))) - d - h]} \tag{2.9.5}$$

The use of d as an integration variable assumes a one-to-one correspondence between x and d, and this is guaranteed when v remains positive for $-w/2 \leq x \leq w/2$. The lower limit of integration is the left-wall depth expressed in terms of the flow rate and the right-wall depth. If $B(\psi)$ is known in advance, then the first and last of (2.9.5) can be combined to form the hydraulic functional

$$\mathcal{G}(d\tfrac{1}{2}w,\, y;\, w;\, Q) = \int\limits_{\left[d^2(\frac{1}{2}w,\, y)-2Q\right]^{1/2}}^{d(\frac{1}{2}w,\, y)} \frac{\partial d}{2^{1/2}[B(Q+\frac{1}{2}(d^2 - d^2(\frac{1}{2}w,\, y))) - d - h]} - w = 0$$

expressing a relationship between the single dependent variable $d(\tfrac{1}{2}w, y)$, the geometric variables w and h, and the parameter Q. A critical condition can thus be obtained by taking $\partial\mathcal{G}/\partial d(\tfrac{1}{2}w, y) = 0$. After use of Leibnitz's Rule and some integration by parts, one obtains the result

$$\int\limits_{\left[d^2(\frac{1}{2}w,\, y)-2Q\right]^{1/2}}^{d(\frac{1}{2}w,\, y)} \left(\frac{1}{dv^3} - \frac{1}{d^2v}\right)\partial d = 0.$$

Changing the integration variable from d to x (using $\partial d = v\partial x$) leads to Stern's result, which can be written in dimensional terms as

$$\int\limits_{-w^*/2}^{w^*/2} \frac{1}{v^{*2}d^*}\left(1 - \frac{v^{*2}}{gd^*}\right)dx^* = 0. \tag{2.9.6}$$

In essence, the local value of the Froude number $v^*/\sqrt{gd^*}$ must be 1 for some x^* across the channel in order for the flow to be critical. It is remarkable that this result does not depend on the Coriolis parameter f. It is also interesting that (2.9.6) appears to apply to potential vorticity waves as well as Kelvin and frontal waves. However, the restriction to unidirectional velocity profiles may disallow certain types of critical states, an issue that we will return to. As an aside, we note that the same reasoning that results in (2.9.6) can be used to estimate the speeds of certain long waves in a given flow. This subject is taken up in Exercise 3.

Stern's result can be used to define a type of generalized Froude number

$$F_d = \frac{\displaystyle\int\limits_{-w^*/2}^{w^*/2} \frac{1}{gd^{*2}}\,dx^*}{\displaystyle\int\limits_{-w^*/2}^{w^*/2} \frac{1}{d^*v^{*2}}\,dx^*} \tag{2.9.7}$$

having the property that $F_d = 1$ for critical flow and $F_d \to 0$ as $v^* \to 0$. The latter limit seems to imply that $F_d < 1$ for subcritical flow and that $F_d > 1$ for supercritical flow, but one should exercise caution in making this interpretation. Flows with nonuniform potential vorticity may admit to many wave modes and a particular value of F_d does not, in itself, indicate supercritical or subcritical conditions with respect to all possible waves. We only know that $F_d = 1$ indicates that one of the waves is arrested.

b. *The Solution of Pratt and Armi*

A detailed example of hydraulic effects in the presence of both gravitational and potential vorticity dynamics was worked out by Pratt and Armi (1987). In order to make the problem analytically tractable, they examined a nonrotating flow with the linear potential vorticity distribution

$$q^*(\psi^*) = q_o^* - a\psi^*, \tag{2.9.8}$$

in a channel with rectangular cross section. Although $f = 0$ this flow supports both gravity and potential vorticity waves and therefore contains some of the essential features we wish to investigate. Simplicity is provided by the fact that d^* is uniform across the channel, $d^* = d^*(y^*)$, so that the expression for potential vorticity reduces to

$$q^* = \frac{\partial v^*/\partial x^*}{d^*} \tag{2.9.9}$$

Differentiation with respect to x^* and use of (2.9.8) leads to the cross-stream structure equation

$$\frac{\partial^2 v^*}{\partial x^{*2}} + ad^{*2}v^* = 0. \tag{2.9.10}$$

There are two distinct cases to consider. When $a < 0$, $dq^*/d\psi^* > 0$ and the potential vorticity has higher values on the right side of the channel (where $\psi^* = Q^*/2$) then on the left side (where $\psi^* = -Q/2$), although the variation of q^* across the channel may not be monotonic. As suggested in Figure 2.9.1a, this setting would seem to favor potential vorticity wave propagation in the same direction as the overall transport. In this case the solutions to (2.9.10) will be exponential. If $a > 0$ the situation is as shown in Figure 2.9.1b, with generally higher potential vorticity on the left and possible upstream propagation of potential vorticity waves. Here the solutions to (2.9.10) are oscillatory.

Consider the case $a < 0$ first. The solution to (2.9.10) can be written as

$$v^* = \frac{\hat{v}^*\sinh(\alpha x^*)}{\sinh(\frac{1}{2}\alpha w^*)} + \frac{\overline{v}^*\cosh(\alpha x^*)}{\cosh(\frac{1}{2}\alpha w^*)}, \tag{2.9.11}$$

where

$$\alpha(y^*) = |a|^{1/2}\,d^*(y^*),$$

$$\overline{v}^*(y^*) = \tfrac{1}{2}[v^*(\tfrac{1}{2}w^*, y^*) + v^*(-\tfrac{1}{2}w^*, y^*)],$$

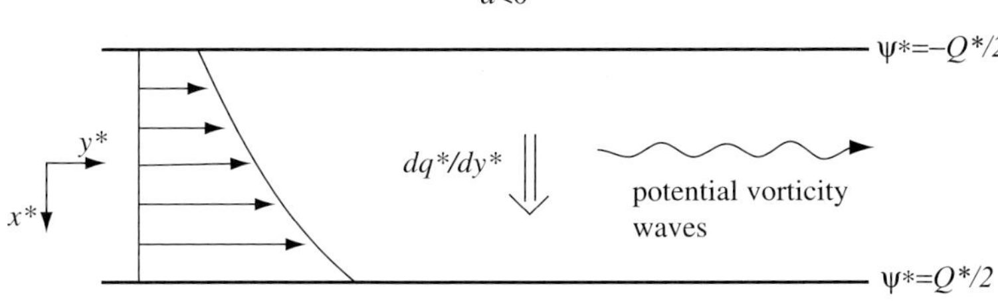

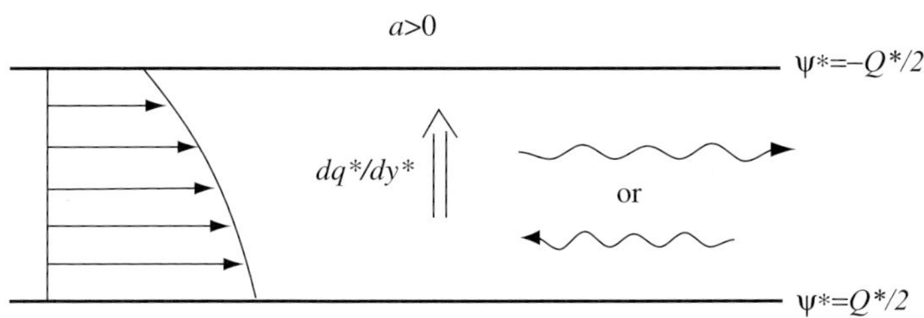

FIGURE 2.9.1. The direction of potential vorticity wave propagation, relative to background flow advection, for potential vorticity gradients of different signs. The channel is nonrotating and the potential vorticity gradient is determined entirely by the gradients in horizontal shear.

and

$$\hat{v}^*(y^*) = \tfrac{1}{2}[v^*(\tfrac{1}{2}w^*, y^*) - v^*(-\tfrac{1}{2}w^*, y^*)].$$

As in Gill's (1977) model the flow has a boundary layer structure, each layer here having thickness α^{-1}. However there are some important differences. One is that the decay scale depends only on the magnitude of the potential vorticity gradient $|a| = |dq^*/d\psi^*|$ and the depth d^*, and not on gravity[8]. Furthermore, the decay scale *depends on the dependent variable d* and is therefore a function of y, whereas Gill's decay scale (L_d) is universally constant.

The boundary conditions $\psi^*(\pm\tfrac{1}{2}w^*) = \pm\tfrac{1}{2}Q^*$ may be used to relate $\hat{v}^*$, $\bar{v}^*$ and d^* and form a hydraulic functional. The first step is to integrate the product of d^* and (2.9.11) across the channel, resulting in

$$\bar{v}^* = \frac{\alpha Q^* \coth(\tfrac{1}{2}\alpha w^*)}{2d^*}. \tag{2.9.12}$$

---

[8] The decay scale can also be written as $\left|\dfrac{v^*}{d^*(dq^*/dy^*)}\right|^{1/2}$ which may be compared with the thickness $\left(\dfrac{V}{\beta}\right)^{1/2}$ of inertial boundary currents on a beta-plane ocean (Charney, 1955). Here V is velocity scale and β is the planetary potential vorticity gradient.

Next, the potential vorticity equation (2.9.9) is applied at $x^* = w^*/2$, leading to $\partial v^*/\partial x = d^*(q_o^* - \frac{1}{2}aQ^*)$. The use of (2.9.11) to evaluate $\partial v^*/\partial x^*$ there results in

$$\hat{v}^* = d^* q_o^* \alpha^{-1} \tanh(\tfrac{1}{2}\alpha w^*). \qquad (2.9.13)$$

Finally, a functional relation of the required form is obtained by evaluating the Bernoulli equation along the right-hand wall:

$$\frac{1}{2}(\bar{v}^* + \hat{v}^*)^2 + d^* + h^* = B_R^*. \qquad (2.9.14)$$

Here B_R represents the right-wall value of the Bernoulli function. Substitution for $\hat{v}^*$ and $\bar{v}^*$ and nondimensionalization of the result leads to

$$\mathcal{G}(d; h, w) = \frac{1}{2}\left[\frac{1}{\tanh(\gamma d)} + \frac{\tanh(\gamma d)}{\Delta q}\right]^2 + d + h - B_R = 0 \qquad (2.9.15)$$

where $(d, h, B_R) = (d^*/D, h^*/D, B_R^*/gD)$ and

$$\gamma = \frac{1}{2}w^* |a|^{1/2} D, \qquad (2.9.16)$$

$$D = \frac{|a| Q^{*2}}{4g}, \qquad (2.9.17)$$

and

$$\Delta q = \frac{|a| Q^*}{2q_o^*}, \qquad (2.9.18)$$

all of which are nonnegative.

$\mathcal{G}(d; h, w)$ contains two parameters γ and Δq. The former is one-half the ratio of the channel width to the boundary layer width based on the scale depth D. It is a measure of the strength of potential vorticity effects over the cross-section of the flow. If $\gamma \ll 1$ potential vorticity effects are relatively weak. The other parameter Δq is a measure of the relative importance in the two terms q_o^* and $a\psi^*$ which comprise the potential vorticity. Specifically, Δq is the difference between the potential vorticity at the right and left walls normalized by their sum.

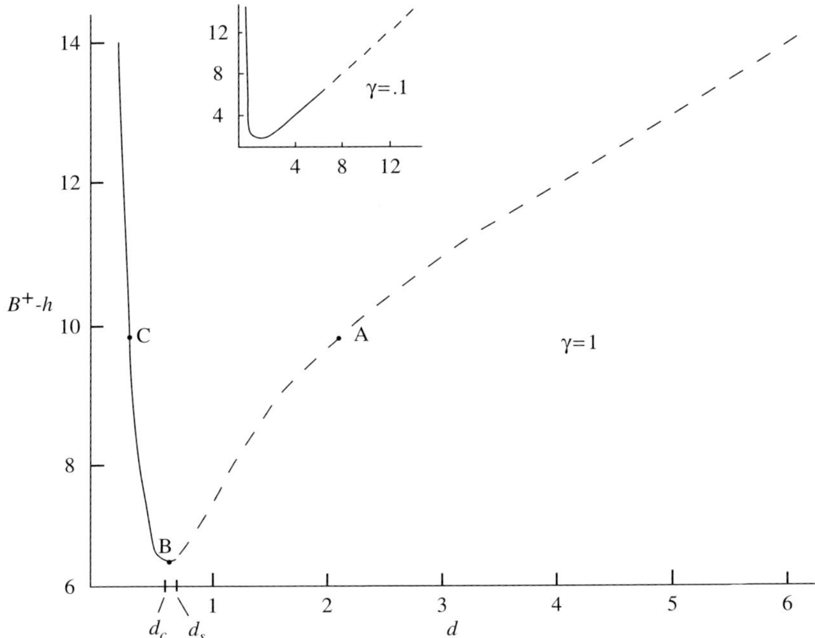

FIGURE 2.9.2. A plot of $B_R - h$ as a function of d for a channel of constant width, and with $a < 0$. The solution is based on equation (2.9.15) with $\Delta q = \gamma = 1$. The dashed section of curve corresponds to bidirectional flow. In the inset plot, γ has been reduced to 0.1.

The critical condition $\partial \mathcal{G}/\partial d = 0$ leads to

$$\gamma \sinh(\gamma d_c)\operatorname{sech}^3(\gamma d_c)[\coth^4(\gamma d_c) - \Delta q^{-2}] = 1 \qquad (2.9.19)$$

and the left-hand side of this expression decreases monotonically from positive ∞ to zero, indicating at the most a single root. A typical solution curve (Figure 2.9.2), based on (2.19.15) with the width w held constant, shows a single minimum in the value of $B_R{-}h$ plotted as a function of d with $\Delta q = 1$. Solutions are constructed in the usual way by following the curve as h changes. A hydraulic transition occurs if the maximum h coincides with the minimum of the curve. It can be shown that, in the limit of vanishing Δq and γ, (2.9.19) reduces to the result for one-dimensional flow: $\bar{v}^* = v^* = (gd^*)^{1/2}$. In this limit the left- and right-hand branches of the solution curve correspond respectively to supercritical and subcritical flows. We will assume that this characterization continues to hold for nonzero Δq and γ with the caveat that the actual wave speeds along the two branches have not been calculated.

There is nothing so far that dramatically distinguishes the character of the model from its one-dimensional counterpart. However, a closer look at the velocity structure reveals an important difference, namely that stagnation points with corresponding separating streamlines can exist on the left-hand wall. The required condition is $\hat{v}^* = \bar{v}^*$, or if (2.9.12) and (2.9.13) are used:

$$\Delta q = \tanh^2(\gamma d_s). \tag{2.9.20}$$

Here d_s denotes the value of d at the section of wall stagnation. The corresponding right-wall condition is obtained by reversing the sign of the right-hand term and cannot be satisfied for positive Δq. Hence, stagnation can occur only on the left wall. The use of (2.9.20) to substitute for Δq in the critical condition, (2.9.19) leads to

$$\gamma \frac{\sinh(\gamma d_c)}{\cosh^3(\gamma d_c)}[\coth^4(\gamma d_c) - \coth^4(\gamma d_s)] = 1, \tag{2.9.21}$$

and thus d_c must be $< d_s$ (the flow must be subcritical) for separation to occur. This *stagnation* separation should be distinguished from the rotation induced separation in which the wall depth vanishes. In Figure 2.9.2, subcritical solutions with $d > d_s$ are indicated by dashing. In this case, most of the subcritical curve has this property. Corresponding velocity profiles (Figure 2.9.3) will have reverse flow along the left-hand wall. The three sections correspond to points A, B, and C of the solution curve. Immediately upstream of the sill lies the stagnation point and beyond it a counterflow. At section A most of the channel contains recirculating fluid; only that passing close to the right wall reaches the sill.

We now turn to the more interesting case $a > 0$, which is favorable for potential vorticity wave propagation against the mean flow. The solution to (2.9.10) is

$$v^* = \hat{v}^* \frac{\sin(\alpha x^*)}{\sin(\alpha w^*/2)} + \bar{v}^* \frac{\cos(\alpha x^*)}{\cos(\alpha w^*/2)}, \tag{2.9.22}$$

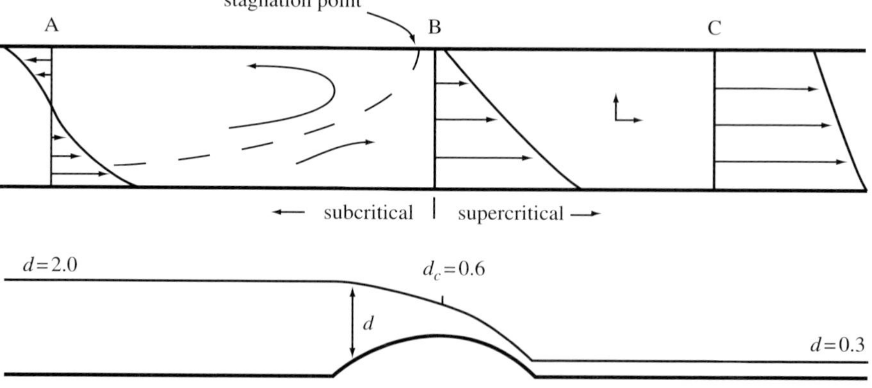

FIGURE 2.9.3. Plan view of a hydraulically controlled flow. Sections A, B and C correspond to the points indicated in Figure 2.9.2.

so that the velocity profile is oscillatory. Repetition of the earlier procedure leads to

$$\bar{v}^* = \frac{\alpha Q^* \cot(\alpha w^*/2)}{2d^*} \tag{2.9.23}$$

and

$$\hat{v}^* = \frac{q_o^* d^* \tan(\alpha w^*/2)}{\alpha}. \tag{2.9.24}$$

Substitution of (2.9.23) and (2.9.24) into the Bernoulli equation (2.9.14) and nondimensionalization gives

$$\mathcal{G}(d; h, w) = \frac{1}{2}[\cot(\gamma d) + \frac{\tan(\gamma d)}{\Delta q}]^2 + d + h - B_R = 0 \tag{2.9.25}$$

where γ and Δq are defined as before and are considered positive. Note that the squared term has the value $+\infty$ for $\gamma d = \frac{1}{2}n\pi$ $(n = 0, 1, 2, \cdots)$, suggesting that the solution 'curve' consists of a series of disconnected lobes. This is confirmed by a plot (Figure 2.9.4) showing $B_R - h$ as a function of d for $\gamma = \Delta q = 1$. Note that the minimum value of $B_R - h$ increases as the lobe number increases. For a given upstream state (here determined by B_R) and a given topographic elevation h, there may be more than two possible steady states. For example, the value $B_R - h = 10$ corresponds to 12 possible states. However, once a particular solution lobe is determined, perhaps on the basis of further information about the upstream state, then at the most two states are possible for any given h. Of course, a hydraulic jump or some other nonconservative feature might allow the solution to switch from one lobe to another, thereby allowing more possibilities.

Stagnation along the left wall is also possible and occurs when $\hat{v}^* = \bar{v}^*$, or

$$\Delta q = \tan^2(\gamma d_s). \tag{2.9.26}$$

As before, separation along the right wall is not possible for nonzero Δq.

At the minimum of each lobe the flow is critical and the corresponding depth d_c can be calculated from the condition $\partial \mathcal{G}/\partial d = 0$, which yields

$$\gamma \frac{\sin(\gamma d_c)}{\cos^3(\gamma d_c)}[\cot^4(\gamma d_c) - \cot^4(\gamma d_s)] = 1. \tag{2.9.27}$$

It is not difficult to show that $d_s > d_c$ within each lobe.

The most obvious qualitative difference between solutions corresponding to different lobes is in the number of zero crossings of the cross-channel profile of v. It can be shown that the solutions corresponding to lobe n have either n or $n - 1$ zero crossings, the greater number occurring for larger values of d. Thus the higher lobes correspond to intricate flows with multiple bands of fluid moving upstream and downstream. Figures 2.9.5 and 2.9.6 show examples taken from lobes #1 and #2.

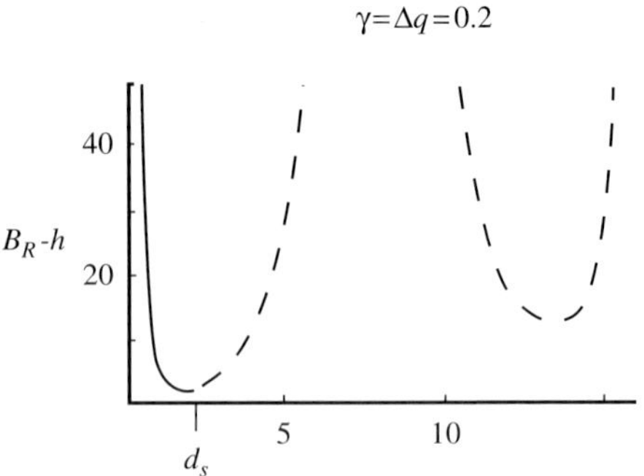

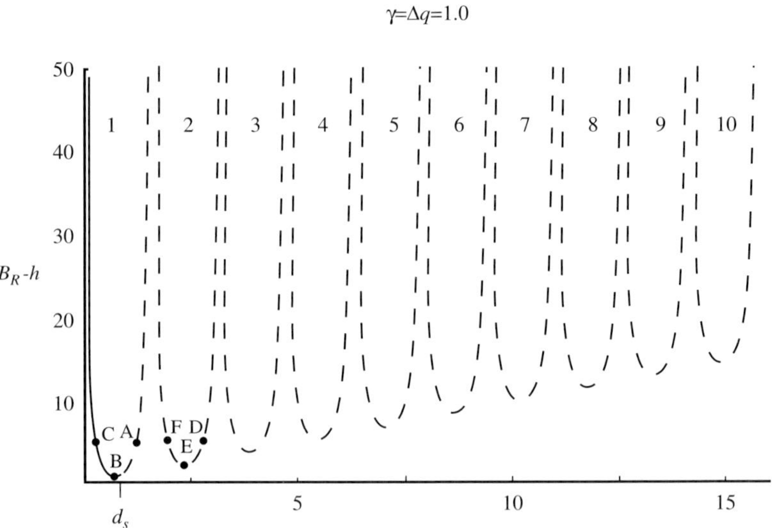

FIGURE 2.9.4. A plot of $B_R - h$ as a function of d for a channel of constant width, and with $a > 0$. The solution is based on equation (2.9.25) with $\gamma = \Delta q = 1$. The dashed section of curves corresponds to flows with velocity reversals. In the inset plot, γ and Δq have been reduced to 0.2.

There remains some mystery concerning solutions corresponding to different solution lobes. If the cross-channel solution is reduced by taking the limits γ and $\Delta q \to 0$, lobe #1 tends toward the solution curve for a one-dimensional, nonrotating flow (e.g. Figure 1.4.1). The inset of Figure 2.9.4 shows how this limit is approached: as γ and Δq are reduced, the depth range of the first lobe grows and the remaining lobes are pushed off to infinity. Controlled solutions belonging to the first lobe appear then to be governed by the dynamics of a shear-modified, long gravity wave. For the other solutions, it is evident that the

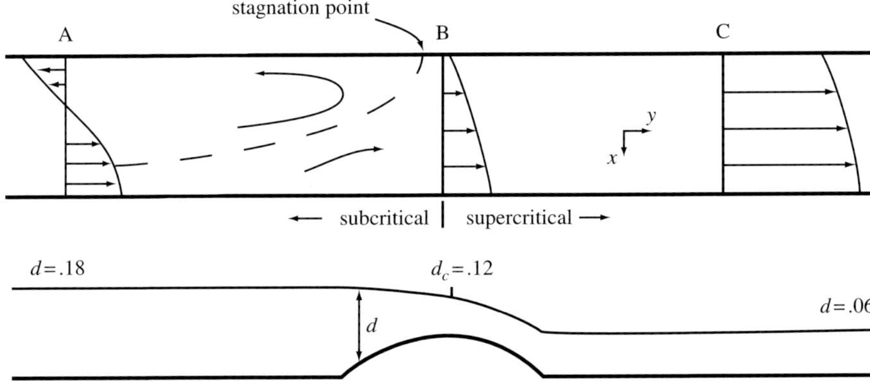

FIGURE 2.9.5. Plan view of a controlled solution based on Lobe 1 of the Figure 2.9.4 solution curve. Lettered sections match points in Figure 2.9.4.

change in depth across the sill is relatively small and becomes vanishingly so for the higher lobes. The change in the flow as it passes through a critical section is primarily one of horizontal structure. This idea can be formalized by calculation of the cross sectional enstrophy

$$e_n^* = \tfrac{1}{2} \int_{-w^*/2}^{w^*/2} (\partial v^*/\partial x^*)^2 \, dx = \tfrac{1}{2} d^{*2} \int_{-w^*/2}^{w^*/2} q^{*2} \, dx, \qquad (2.9.28)$$

a measure of the horizontal shear across a particular section. As explored in Exercise 3 it can be shown that the change in e_n^* caused by a small change in depth as the flow passes through a critical section increases as the lobe number becomes higher. This indicates that control of the flow corresponding to higher lobes primarily affects the horizontal shear and not the depth. Because of this

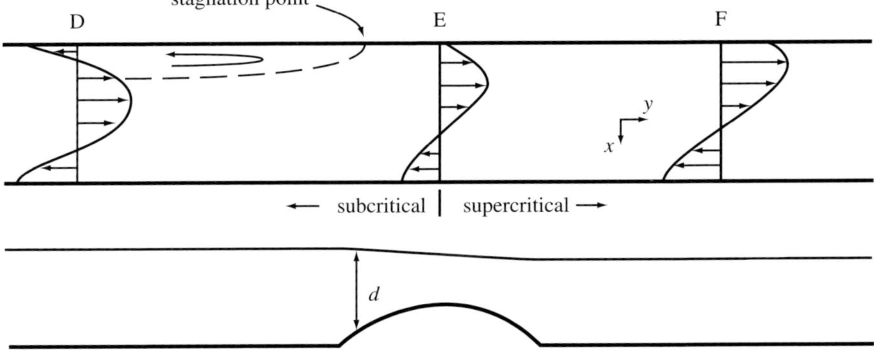

FIGURE 2.9.6. Same as for the previous figure, but now the solution is based on lobe 2 of the Figure 2.9.4 solution curve.

feature, and because the higher lobes owe their presence entirely to a finite potential vorticity gradient, it is evident that the corresponding solutions are controlled by a potential vorticity wave.

Further to the ongoing discussion, it can be shown that Stern's condition for criticality (2.9.6) succeeds in predicting the control condition for the first lobe, but fails for the remaining lobes. Failure is due to the fact that the higher lobe solutions all have velocity reversals, whereas the derivation of (2.9.6) assumes unidirectional flow. Flows with potential vorticity gradients may therefore experience a multiplicity of controlled configurations, not all of which obey Stern's criterion.

A final consideration, one that could render much of the above discussion academic, is stability. The most pertinent theorem for the present case is Fjortoft's necessary condition for instability (see Drazin and Reid, 1981), which does not strictly apply to our flow in general, but would be applicable if the flow were bounded by a rigid lid. Instability is possible when $dq^*/d\psi^* < 0$, or $a > 0$, the case permitting multiple solutions.

There remains uncertainty regarding the interpretation of the $a > 0$ solutions, how they are established, which branches of the higher lobes are supercritical and subcritical, and what their stability is. One of the difficulties is that the model allows a mix of potential vorticity and gravity wave dynamics. More recent investigations of hydraulic effects in the presence of potential vorticity gradients have utilized models that expunge gravity waves by placing a rigid lid on the surface. Also, piecewise constant (rather than continuous) distributions of q^* can reduce the number of wave modes to just one or two, further simplifying the problem and allowing the peculiar dynamics to be investigated in isolation. These models take us away from the topics and applications of the current chapter, but they are revisited in Chapter 6.

c.   Killworth's Solution

Abyssal flows that occur in deep ocean basins tend to be slow and nearly geostrophic, and perhaps not of the character envisioned by WLK and Gill in the upstream basins of their models. In an attempt to pose more realistic upstream conditions, Killworth (1992b) considered an inviscid model with a broad, geostrophically balanced upstream flow over a horizontal bottom. As it turns out, this assumption is sufficient to determine the potential vorticity of the flow, which turns out to be nonuniform.

The starting point is the assumption that the upstream velocity is nondimensionally small ($v \ll 1$). The expressions for the potential vorticity and Bernoulli functions then reduce to

$$q(\psi) = \frac{1 + \partial v / \partial x}{d} \cong \frac{1}{d}$$

and

$$B(\psi) = \frac{v^2}{2} + d + h \cong d,$$

assuming that $h = 0$ in the upstream reservoir. When combined with (2.9.3), these two relations yield

$$B(\psi) = \left[2(\psi - Q) + d^2(\tfrac{1}{2}w, -\infty) \right]^{1/2} \qquad (2.9.29)$$

and

$$q(\psi) \cong \left[2(\psi - Q) + d^2(\tfrac{1}{2}w, -\infty) \right]^{-1/2},$$

where $d(\tfrac{1}{2}w, -\infty)$ is the depth at the right wall of the reservoir.

If the flow drains into a narrow and/or shallow channel and develops O(1) velocities, it is constrained by the semigeostrophic equations. In particular, the flow must obey the integral constraint (2.9.5), or

$$w = \int_0^Q \frac{1}{2^{1/2}d[B(\psi) - d - h]} d\psi. \qquad (2.9.30)$$

With $B(\psi)$ given by (2.9.29), and d given in terms of ψ by (2.9.3), (2.9.30) forms an implicit relation between w, h, and the right-wall depth $d(w/2, y)$, the single remaining flow variable. The reader is reminded, however, that the derivation of this relation requires one-to-one relation between x and ψ, and thus flow reversals are not permitted.

Killworth (1992b) solved a version of (2.9.30) and obtained standard hydraulic curves relating $d(w/2, y)$ to either h or w. All such curves are similar to those shown in Figure 2.9.4 in having a single maximum or minimum, and the corresponding control is associated with Kelvin wave dynamics. The author speculates that solutions with potential vorticity wave controls may be possible, but the model would have to be extended to allow flow reversals. This is left as a project for an interested student.

d. *Summary*

The role of potential vorticity waves and controls in deep overflows and other gravity-driven flows remains imperfectly understood. If the potential vorticity gradient is single-signed, and if high values of q lie to the left, facing downstream, then the waves attempt to propagate against the current and hydraulically critical flow is possible. The implied critical control primarily affects the horizontal vorticity of the flow field, rather than the surface or interface height. Solutions with a potential vorticity wave control appear to be disconnected from solutions that exhibit control by a gravity or Kelvin wave, or by a potential vorticity wave with a different modal structure. If the Pratt and Armi (1987) model is any indication, it does not seem to be possible to combine two types of controls within the same conservative current system. The fact that deep-ocean overflows appear to exhibit gravitational control may disqualify potential vorticity controls. The latter may, however, act in broad ocean jets or strait flows that are not controlled with respect to gravity waves. All of these comments involve conjecture, begging further investigation.

Another feature suggested by this small body of work is that the presence of a potential vorticity control requires velocity reversals across the control section. This may be connected to the modal structure of the stationary wave, which is itself wiggly. The presence of velocity reversals means that certain analytical results, including Stern's critical condition and Killworth's model, both of which allow for nonuniform potential vorticity, do not allow for potential vorticity wave control. Both rely on an x-to-ψ coordinate transform, which requires a unidirectional flow.

A further cloud on the horizon is instability, which by Fjortort's theorem is favored by the same potential vorticity distributions that allow potential vorticity wave criticality.

Exercises

(1) Show that the velocity profile (2.9.22) can be written in the nondimensional form

$$v = \frac{\sin(\gamma dx)}{\Delta q \sin(\gamma d)} + \frac{\cos(\gamma dx)}{\cos(\gamma d)},$$

where $v = 2v^*/|a|^{1/2}q$, $d = d^*/D$, $x^* = x/(w^*/2)$ and γ is as defined above. Using this expression, calculate the nondimensional version e_n of the enstrophy e_n^* (first equality in 2.9.28). Then take the derivative of the result with respect to d and evaluate it at the critical depth. From the result, show that

$$(\partial e_n/\partial d)_{d=d_c} \sim d_c^2 (d_c \to \infty),$$

and therefore the change in enstrophy relative to a change in depth increases as the critical depth (and therefore the lobe number) increases.

(2) Using the methods of Part a of this section, show that the phase speed c^* of a long wave propagating along the (rectangular) channel is given by

$$\int_{-w^*/2}^{w^*/2} \frac{1}{(v^* - c^*)^2 \, d^*} \left(1 - \frac{(v^* - c)^2}{gd^*} \right) dx^* = 0,$$

provided that c^* does not lie in the range of the variation of v^*. Note for given $v^*(y^*)$ and $d^*(y^*)$, c^* will obey a quadratic equation. There are therefore only two such waves. Speculate on why the integral constraint does not capture the remaining waves.

2.10.   Transport Bounds

We have seen how difficult it is to calculate the volume flux Q of a hydraulically controlled, rotating flow when idealizations such as uniform potential vorticity and rectangular cross section are relaxed. Although calculations are still possible through numerical means, one might first ask whether any general statements about Q can be made without regard to the details of q and h. An approach developed by Killworth and McDonald (1993) and Killworth (1994) is to seek bounds on Q in terms of simple measures of the upstream flow and the channel geometry. Given some information about the available energy, one attempts to find the maximum Q that can be forced through a section of a channel with a given geometry. Although the bounds are formulated without reference to hydraulic control, the result bears a remarkable similarity to hydraulic laws developed in early sections.

The topographic cross section is arbitrary and it is only assumed that the bottom is wetted continuously across, so that the flow occurs in one coherent stream. In contrast to the situation in typical hydraulic models, $B(\psi)$ need not be conserved from one section to the next. However, it is most meaningful to imagine that all the streamlines that cross through the section originate in an upstream basin where the maximum B is denoted E. This maximum applies only to those basin streamlines that make their way to the sill section. If nonconservative processes are then limited to a quadratic bottom drag, $B(\psi)$ can only decrease along a particular ψ and the maximum B at any downstream section must be equal to or less than E. These ideas require some modification if the streamlines originate far downstream (as in Figure 2.9.4) or are part of a local closed gyre (Section 2.7). Although the section in question may lie anywhere, the tightest bound is obtained at the sill, meaning the section with the highest minimum bottom elevation across the flow, h_{min}. The smallest possible value that B (nondimensionally $v^2/2 + d + h$) can possibly have occurs when the depth d and velocity v are zero at $h = h_{min}$. It follows that

$$h_{\min} \leq B(\psi) \leq E \qquad (2.10.1)$$

In addition to geostrophy, the chief assumption made is that the potential vorticity of the flow is nonnegative.

Now consider a hypothetical flow at the sill section (Figure 2.10.1a). The layer thickness is assumed to go to zero at the edges $x = -a$ and $x = b$ of the stream, but the side walls could just as well be vertical. The surface or interface may have segments of negative slope indicating $v < 0$. The bound on Q is formulated by making a sequence of changes to the flow, each of which maintains or increases the original flux. This will lead to a simplified state for which a bound may be formulated.

The first step is to excise any segments of reverse flow along the side walls, so that the new edges of the current lie at $x = b'$ and $x = -a'$ (Frame b). A vertical wall now exists at $x = b'$. We next alter the bottom topography to the left of

$x = -a'$ such that it becomes flat and has the elevation h_{min} (Frame c). Over this flat portion we add a region of positive flow that brings the layer depth smoothly to zero at a point $x = -a''$. The width of the side region is arbitrary. None of the alterations thus far could decrease the volume flux. The flux of the altered flow is given by

$$\int_{-a}^{b} dv\,dx = \int_{-a}^{b} (z_s - h)v\,dx = \frac{1}{2}(z_s^2(b) - z_s^2(-a)) - \int_{-a}^{b} h\frac{\partial z_s}{\partial x}dx \geq Q \qquad (2.10.2)$$

where $z_s = d + h$.

We next eliminate each interior minimum in z_s by slicing off the top of the mound of water to its left (Frames c and d). The segment extending from $x = x_1$

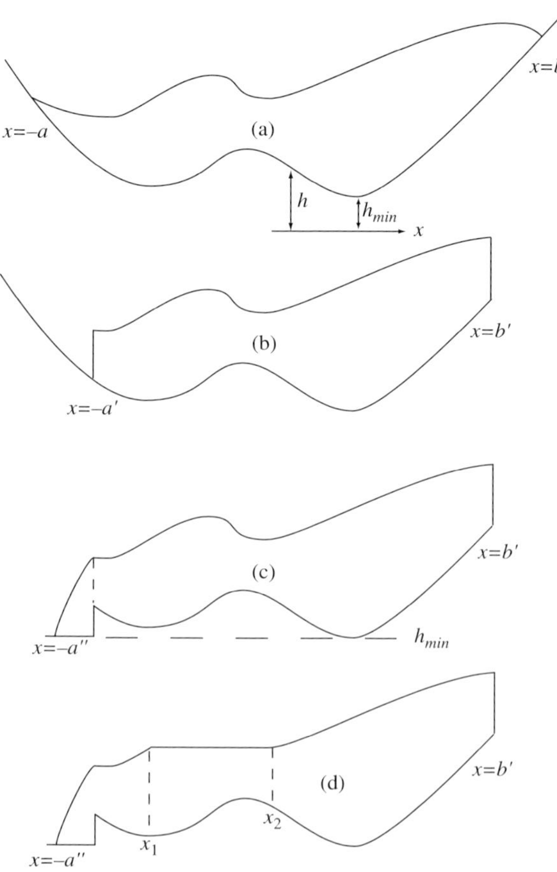

FIGURE 2.10.1. Series of surgical procedures used to alter a given flow (a) in order to produce a simpler flow (d) whose transport is known. The transport cannot be decreased in any step and thus the transport of (d) acts as a bound. (Based on a figure in Killworth and MacDonald, 1993).

to $x = x_2$ in the figure is therefore replaced by a quiescent region, and the same is done to the left of any remaining minimum. To prove that this operation cannot increase the flux note that for the Figure 2.10.1c. flow we have

$$z_s(x_2) = B(\psi(x_2)) \tag{2.10.3}$$

and

$$\frac{1}{2}v(x_1)^2 + z_s(x_1) = B(\psi(x_1)). \tag{2.10.4}$$

Since $z(x_1) = z(x_2)$,

$$B(\psi(x_1)) - B(\psi(x_2)) = \frac{1}{2}v(x_1)^2 > 0. \tag{2.10.5}$$

Finally, the previous assumption of positive potential vorticity q along with the relationship $dB/d\psi = q$ means that B must increase with ψ and thus

$$\psi(x_2) - \psi(x_1) \leq 0. \tag{2.10.6}$$

The flux to be removed must therefore be nonpositive.

The end result of this surgery is a water surface rising monotonically to the right, so the stream has positive or zero velocity everywhere across the channel with flux equal to or greater than the original. A bound on the altered flow can be formulated beginning using definition (2.10.2) of flux:

$$\frac{1}{2}(z_s^2(b') - z_s^2(-a'')) - \int_{-a''}^{b'} h \frac{\partial z_s}{\partial x} dx, \tag{2.10.7}$$

Since $\partial z_s/\partial x$ is nonnegative, the integral in the above expression cannot be less than

$$\int_{-a''}^{b'} h_{\min}(\partial z_s/\partial x)dx = h_{\min}(z_s(b') - z_s(-a'')) = h_{\min}(z_s(b') - h_{\min}), \tag{2.10.8}$$

The original flux Q is therefore bounded according to

$$Q \leq \frac{1}{2}(z_s^2(b') - z_s^2(-a'')) - \int_{-a''}^{b'} h \frac{\partial z_s}{\partial x} dx$$

$$\leq \frac{1}{2}(z_s^2(b') - h_{\min}^2) - h_{\min}(z_s(b') - h_{\min}) = \frac{1}{2}(z_s(b') - h_{\min})^2 \tag{2.10.9}$$

Now $z_s(b')$ cannot exceed the maximum value E of the Bernoulli function, and therefore $Q \leq \frac{1}{2}(E - h_{\min})^2$. Also, if we associate with E an equivalent surface elevation $h_{\min} + \Delta z_{\mathrm{E}}$, then the transport bound becomes $Q \leq \frac{1}{2}\Delta z_E^2$ or, in dimensional terms:

$$Q^* \leq \frac{g(\Delta z_E{}^*)^2}{2f}. \tag{2.10.10}$$

There are a number of examples, all with rectangular cross sections and all with separated sill flow, for which the right-hand side of (2.10.10) gives the exact flux. The first is the case of flow from an infinitely deep and quiescent basin across a sill (Section 2.4). Here $\Delta z_{\mathrm{E}}{}^*$ is just the reservoir head, Δz of (2.4.15), and is a constant over the upstream basin. We also argued in Section 2.6 that any separated sill flow that stagnates along the right wall is critical and that the corresponding flux is given by interpreting $\Delta z_{\mathrm{E}}{}^*$ as $\Delta z_R{}^*$, the upstream elevation along the right wall. If $q(= dB/d\psi)$ is nonnegative, and the reservoir flow is unidirectional, then $\Delta z_R{}^*$ does indeed represent the maximum upstream value of the Bernoulli function and (2.10.10) is exact. In both of these cases the flow is either positive or zero at the edges, so that no fluid need be excised from the end points (Figure 2.10.1a,b). Also, since the bottom is horizontal, the shaving off of mounds of fluid (Figure 2.10.1c) does not alter the volume flux. Therefore the sequence of steps taken to formulate the bound results in no decrease in transport. The cases serve notice that the bound is achievable.

The fact that (2.10.10) is achievable in two examples with rectangular cross-sections suggests that departures from this geometry might generally tend to reduce the flux. However, if the geometry is sufficiently irregular that the flow becomes divided into two or more streams, then the combined flux can exceed the bound, though (2.10.10) continues to hold for each individual stream. Whitehead (2003) presents an example. Simply put, the formation of multiple streams is similar to the existence of multiple openings through which fluid may drain from the basin.

Killworth and McDonald (1993) have shown that the bound can be extended to a fluid with N layers, each with its own uniform density, and all lying below a deep and inactive upper fluid. The volume flux Q_{n} in layer n is bounded according to

$$F_n \leq \frac{g_n}{2g}(E_n - h_{\min})^2, \tag{2.10.11}$$

where g_n is the reduced gravity and E_n is the maximum Bernoulli function for that layer, the latter defined with the same restriction as the single-layer case.

2.11.   Anatomy of An Overflow: The Faroese Channels

As outlined in the introduction, there are two major overflows that supply dense water from the Nordic Seas to the Atlantic Ocean. Each represents a substantial source of North Atlantic Deep Water. The Denmark Strait overflow, which has already been described, is traditionally regarded as an example of a rotating, hydraulically controlled flow. Dense fluid spills over the sill and forms a descending current that is banked against the Greenland slope. Strong interactions with shallower layers cloud comparisons with the simple models that we have explored. Some portions of the flow have been observed to be strongly barotropic, possibly due to interactions with the East Greenland Current. The latter generally lies to the east of the overflow, but also covers much of it with a layer of lighter, southward-flowing, low-salinity water. In addition, the descending outflow contains large, horizontal eddies whose expressions can be seen at the free surface. All of these factors make it difficult to think of the overflow as isolated and lying below motionless fluid.

In contrast, the deep current in the Faroese Channel system, situated between Iceland and Scotland, is more stable and less engaged with surface layers. The deep Norwegian Sea is drained to the south through the Faroe-Shetland Channel, which lies to the east of the Faroe Islands (Figure 2.11.1). After passing the Wyville-Thompson Ridge, the channel makes a sharp bend to the northwest and

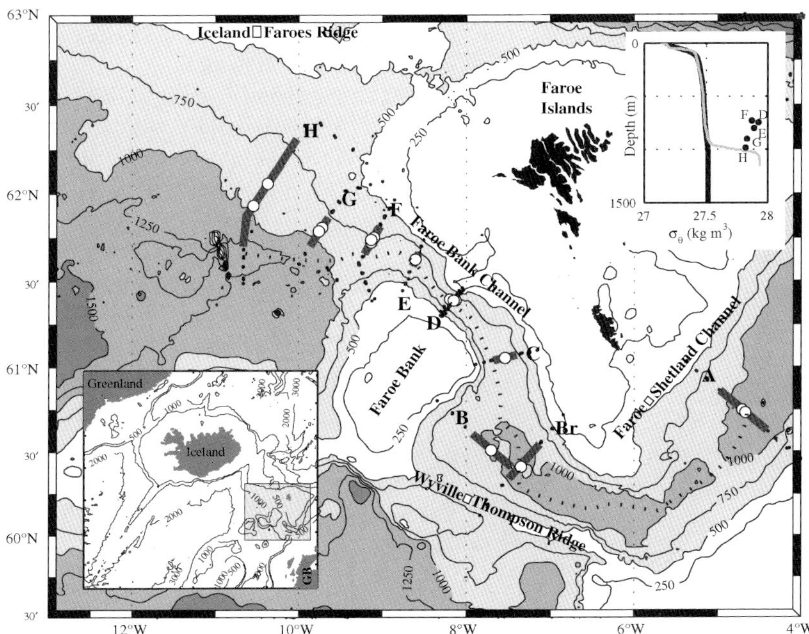

FIGURE 2.11.1. Bathymetric map of the Faroese Islands. The letters A-H indicate sections at which data were collected (Mauritzen et al., 2005). An open circle marks the center of mass anomaly of the dense overflow at the time the section was taken. Multiple circles on the same section indicate repeat measurements. The upper right inset shows a density profile (dark curve) of the background Atlantic Water along with dots indicating the mean overflow density for sections D–H. The path of the deepest part of the main channel system (the thalweg) is indicated by a dotted line. (From Girton et al., 2006).

becomes the Faroe Bank Channel. The most constricted section occurs near 'D' where the shallowest (850 m) topography and narrowest width roughly coincide. From this point, the channel bottom gradually descends and widens for 60km and then plummets into the Iceland Basin of the North Atlantic. A combination of intermediate and deep-water masses enters the Faroe-Shetland Channel from the Norwegian Sea. A small fraction of the volume flux leaks southward across the Wyville-Thompson Ridge (Hansen and Osterhus 2000), but the bulk continues into the Faroe Bank Channel and over the sill. It then spills down to about 4000 m in the Iceland Basin.

Though complex, this overflow provides one of the more clear-cut examples of a hydraulically controlled flow that is vividly influenced by Earth's rotation. The overflow has been the subject of a number of observational programs and our treatment relies on a 2000 survey (Mauritzen *et al.*, 2005; Girton *et al.*, 2006). We now describe the flow in more detail, using data collected across the lettered sections indicated in Figure 2.11.1. Some of the sections were repeated in order to gain a measure of time variability.

a. Hydrographic Properties

An along-channel section of surface referenced potential density (σ_θ) has been constructed along the axis of the deep current from the data at the individual cross sections (Figure 2.11.2). The isopycnal slopes suggest spilling of dense water as it flows from right to left out of the Norwegian Sea, over the sill (near D), and

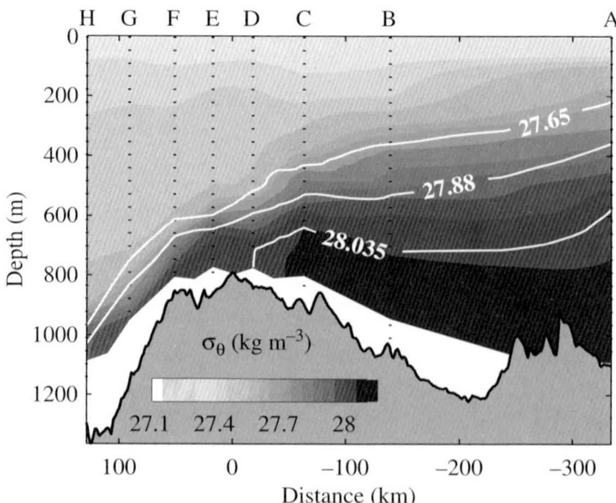

FIGURE 2.11.2. (a): An along-channel density (σ_θ) section based on the first sampling of data from Sections A-H. The section track is shown in Figure 2.11.1. The bathymetry along the deepest part (thalweg) is shaded gray while the bathymetry along the section track is shaded white. Selected density contours for layers described by Mauritzen *et al.* (2005) are shown in white. (From Girton *et al.*, 2006).

down into the Iceland Basin of the North Atlantic proper. Were the isopycnals associated with a broad-scale, nearly geostrophic flow, such as in a subtropical gyre, the along-channel isopycnal tilt would imply a 'thermal wind'; that is, a vertically sheared, geostrophic, cross-channel velocity. In the present channel setting, where the velocities are strong and cross-channel motion is restricted, the cross-channel velocity need not be geostrophically balanced. The expectation is that the along-channel isopycnal tilt is instead due to inertial acceleration, present in the hydraulic models we have already discussed, or to friction acting along the channel bottom. In either case the tilt need not imply cross-channel motion. In fact, the observations indicate flow primarily along the channel axis, though actual resolution of the transverse and longitudinal components is not accomplished. Downstream of the sill, where the tilts are strongest, the dense water experiences enhanced mixing and entrainment of overlying fluid. The resulting dilution of the overflow is suggested by the reduction or disappearance of the densities greater than $\sigma_\theta = 28.00$.

Potential temperature can be used as a proxy for density in this overflow and many past discussions have used the temperature structure to perform geostrophic estimates of volume flux and to define a hypothetical interface. In temperature sections A, D, F, and H (Figures 2.11.3–2.11.6) there is a strong isothermal, and therefore isopycnal, tilt across the channel. Based on the semigeostrophic scaling arguments introduced at the beginning of this chapter, the along-channel velocity should be in near geostrophic balance. Were the upper fluid motionless, which is never completely true, then a downward tilt toward the left would indicate a flow toward the Atlantic. In section A, which is 300 km upstream of the sill (at the far right in Figure 2.11.2), water colder than $4\,^\circ C$ occupies the upstream channel below about 400 meters, except near both sides where the isotherm tilts downward. Mauritzen *et al.* (2005) describe the $4\,^\circ C$ isotherm as

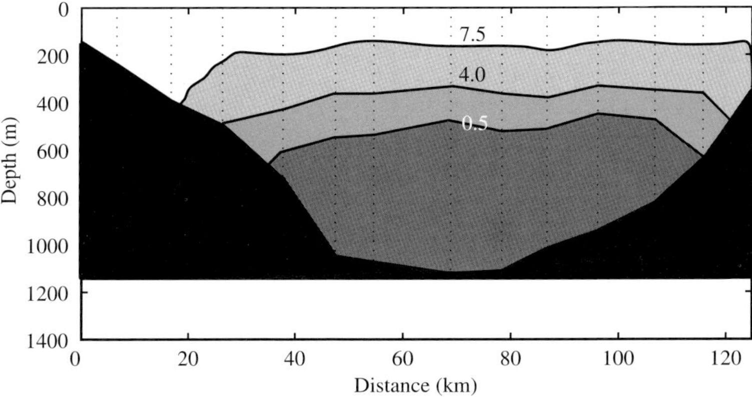

FIGURE 2.11.3. Potential temperature across Section "A" of Figure 2.11.1.

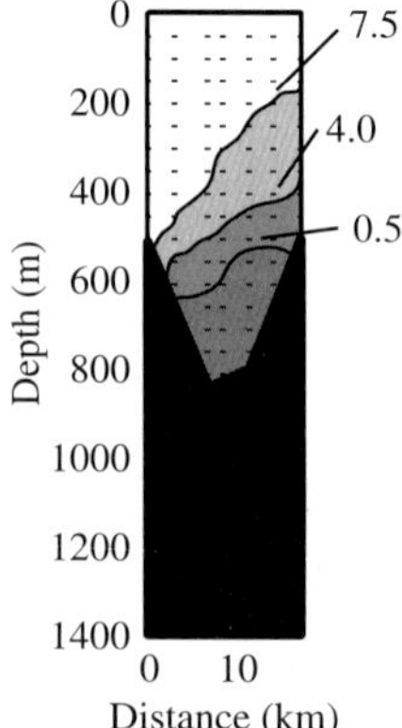

FIGURE 2.11.4. Potential temperature across Section "D", which lies close to the sill and the narrowest section. (From J. Price, private communication).

defining the upper temperature limit of the dense water that ultimately descends to great depths in the North Atlantic. (Some investigators use the $3\,°C$ surface as a boundary for calculating volume transport.) The tilt along the left side nominally indicates a current along the left side of the channel directed toward the sill, and thus toward the North Atlantic. The tilt along the right side appears to indicate a current directed away from the sill (toward the Norwegian Sea). Whether the implied countercurrent is a robust feature is not known. The two other isothermal surfaces (7.5 and $0.5\,°C$) also tilt downward in the same way on the left, and the $0.5\,°C$ surface does so on the right. This suggests that sublayers within the deeper fluid act with some vertical coherence.

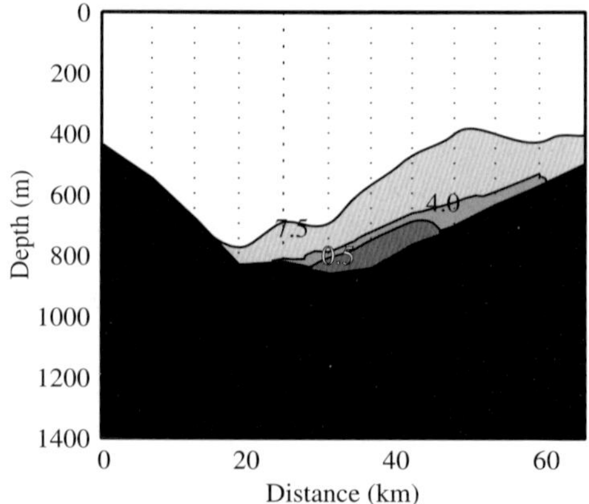

FIGURE 2.11.5. Potential temperature across Section "F". (From J. Price, private communication).

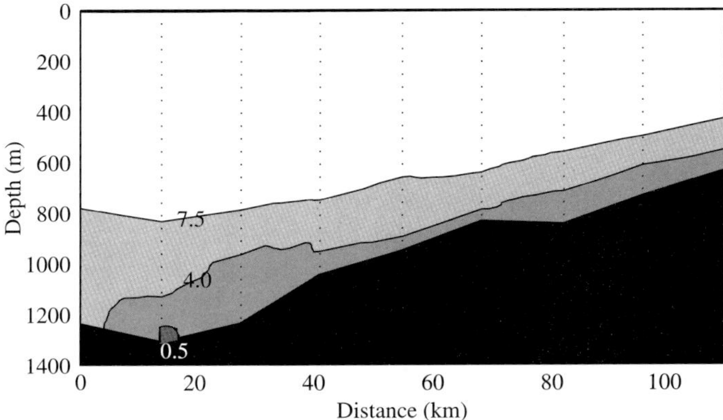

FIGURE 2.11.6. Potential temperature across Section "H" downstream of the sill, where the deep water has descended a few hundred meters.. (From J. Price, private communication).

It has become traditional to think of overflows as being separated into two regions with distinct dynamics. The first begins at the entrance to the Faroe-Shetland Channel, and ends at the sill (near section D). The flow between these two sections is often thought of as inviscid and conservative. The second portion extends downstream from the sill and contains the descending 'outflow' or plume, often marked by enhanced turbulence, mixing, spreading, and entrainment of overlying fluid. The controlling dynamics is strongly nonconservative and is often dominated by a balance between bottom drag, entrainment stresses, gravity, and rotation. Mauritzen *et al.* (2005) have calculated stresses in the water column in order to identify regions of enhanced mixing and drag. Although both processes are enhanced in the Faroe-Bank plume region, they are not necessarily negligible in the approach region and may, in fact, be large enough to significantly modify momentum and energy budgets. The dichotomy between an inviscid upstream region and a dissipative plume is therefore not as clear-cut as traditionally assumed.

The development of the descending plume is illustrated in temperature Sections D-H (Figures 2.11.4–2.11.6). The sill section (D) is also the narrowest and the isothermal tilt there suggests a unidirectional flow of water toward the Atlantic below 4 °C. Section E is slightly deeper, much wider, and lies where the along-axis slope suddenly becomes much steeper. The current itself is wider there and still apparently unidirectional. At section H, furthest downstream, the current has spread to an even greater width and continues to be unidirectional. The water in the layer is noticeably warmer; for example, there is only a small portion colder than 0.5 °C. A broader view showing both the upstream region and descending plume is given by the complete suite of sections (Figure 2.11.7). The panels, which proceed downstream from top to bottom, show the $\sigma_\theta = 27.65$ isopycnal, sometimes used to represent a hypothetical interface. The spreading of the outflow and its confinement to the right bank of the descending channel is

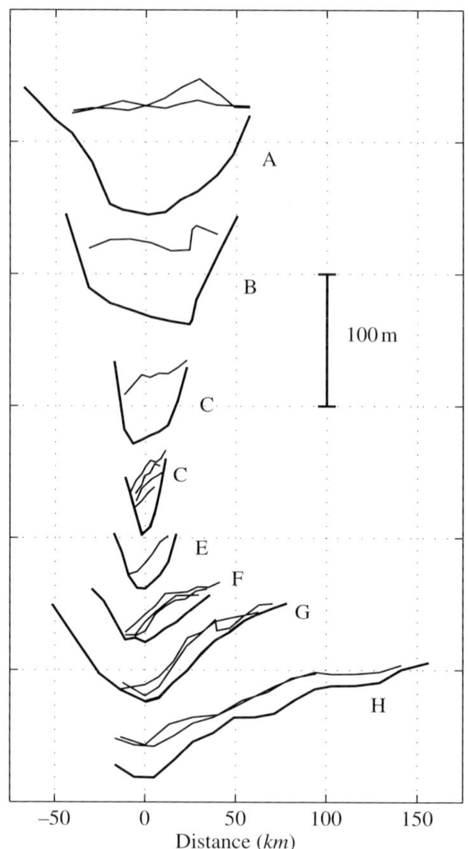

FIGURE 2.11.7. Stacked cross sections. The bathymetry (thick black lines) and the $\sigma_\theta =$ 27.65 isopycnal (thin black lines) across sections A-H. Multiple realizations of the isopycnal indicate repeat sections. ((From Girton *et al.*, 2006)).

evident. Multiple realizations of the interface corresponding to repeat measurements indicate a significant amount of time variability.

b.   Geostrophic and Direct Velocity

It can be useful for readers unfamiliar with physical oceanography to perform the simple exercise of estimating velocity of the flow using a geostrophic balance. We now do so at section D, leaving the remaining sections as homework exercises. Begin by thinking of the entire overflow as being contained in a single layer, with the 7.5°C isotherm representing the bounding interface. The density data in Figure 2.11.2 can be used to estimate the difference between the average density of the deep layer and that of the overlying fluid. The resulting relative change $\Delta\rho/\rho_o$ is approximately 5×10^{-4}. Using temperature as a proxy for density, the isopycnals corresponding to the 7.5°C isotherm has a $\delta = 200\,\mathrm{m}$

descent over a width of $w = 20\,\text{km}$. With a typical value of the Coriolis parameter rounded off to $f = 10^{-4}\,\text{s}^{-1}$, the geostrophic velocity is

$$v = \frac{g\Delta\rho\delta}{\rho_o f w}, \qquad (2.11.1)$$

or $0.67\,\text{ms}^{-1}$. The depth of the layer is about $D = 400\,\text{m}$ on average, which leads to a volume flux estimate of 4.0 Sv. for water colder than $7.5\,^{\circ}C$. The flux of water colder than $4\,^{\circ}C$, which was mentioned above to define the water reaching great depth in the Atlantic, could be estimated to be about half that number.

Overall, geostrophic estimates of velocity using the simple method described above range from 0.1 to $0.67\,\text{ms}^{-1}$. These values are somewhat smaller than those directly measured with a profiling current meter (Figure 2.11.8a). At locations B and C upstream of the sill, the greatest current meter speed is about 0.4 meters per second, whereas the most constricted section (D) has speeds approaching $1\,\text{ms}^{-1}$. In these three locations there is a surface current in the opposite direction and with speeds that can approach those of the deep flow. It is not clear that interactions with this shallow flow are negligible. The deep velocities at E–H are considerably larger than those of the overlying fluid. Speeds vary, with

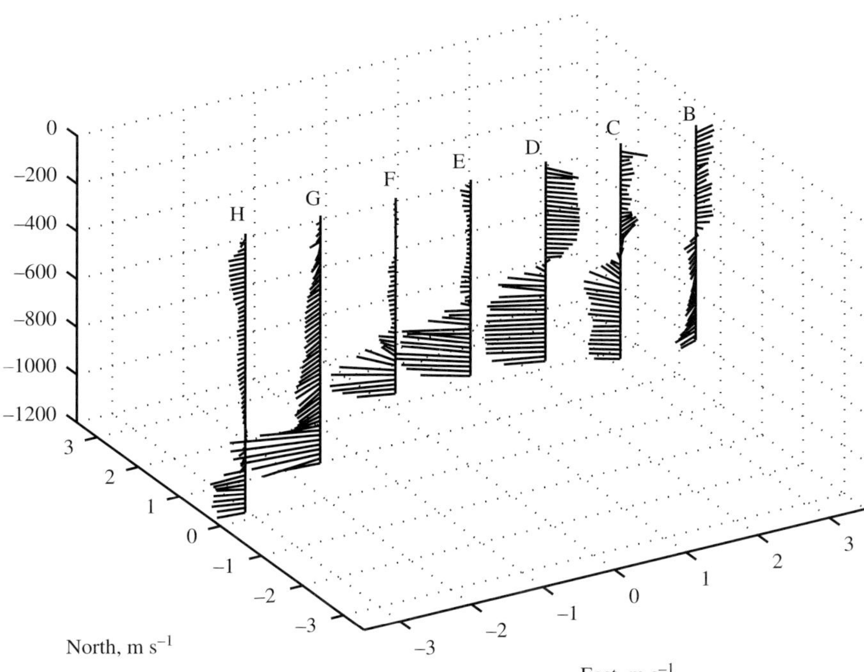

FIGURE 2.11.8. (a) Direct velocity measurements along the sections shown in Figure 2.11.1 and at 20m depth intervals. (From J. Price, private communication).

maximum values at F, G and H of $0.9\,\mathrm{ms}^{-1}$, $1.1\,\mathrm{ms}^{-1}$ and $0.5\,\mathrm{ms}^{-1}$ respectively. These sections also show how the deep current thins in the downstream direction.

The velocity measurements can be averaged over the depth of the dense outflow to produce a single-layer representation (Figure 2.11.8b). The plan view also gives the layer thickness over which the averaging was done and the value of the local Froude number (discussed below) based on that thickness and average velocity. There is an indication of reverse flow along the edges of the current upstream of the sill. In the downstream plume region, the thickest, highest-speed portion of the flow lies on its deepest (left) side. Over shallower regions of the slope, the velocities are smaller and less coherent.

Because density and velocity profiles were measured across each section, one can make a comparison between the measured speeds and geostrophic estimates. The geostrophic velocity at the middle of Section D (solid line in Figure 2.11.9) is plotted along with two directly measured profiles (dotted lines), and the average of the two (dashed line). The direct profiles contain more fine structure than the geostrophic profile, an artifact of smoothing of the temperature and salinity data. In addition, the direct velocity profiles show a bottom boundary layer. The geostrophic profile, which does not account for frictional effects, has no such feature.

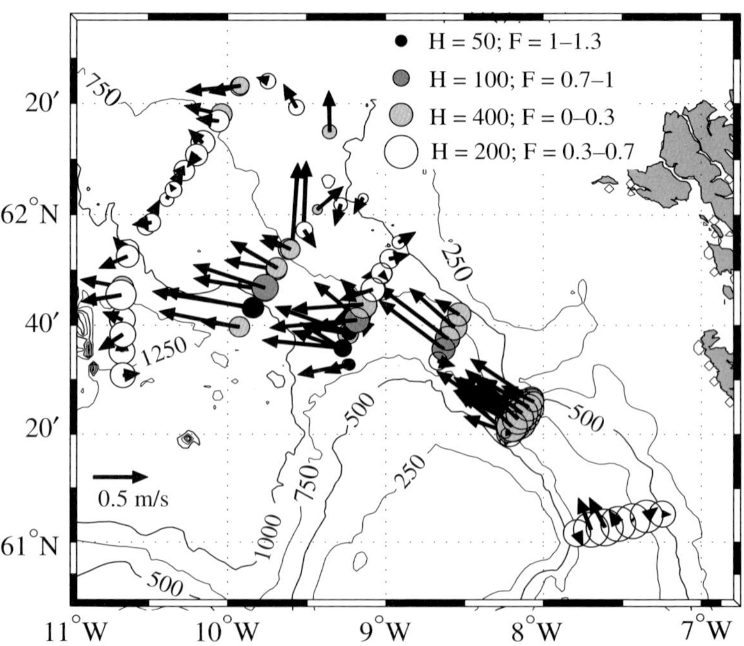

FIGURE 2.11.8. (b) The depth-average velocity below the $\sigma_\theta = 27.65$ isopycnal. The circle size indicates the thickness of the deep layer and the shading indicates the magnitude of the local Froude number. The data include repeat sections. (From Girton *et al.*, 2006).

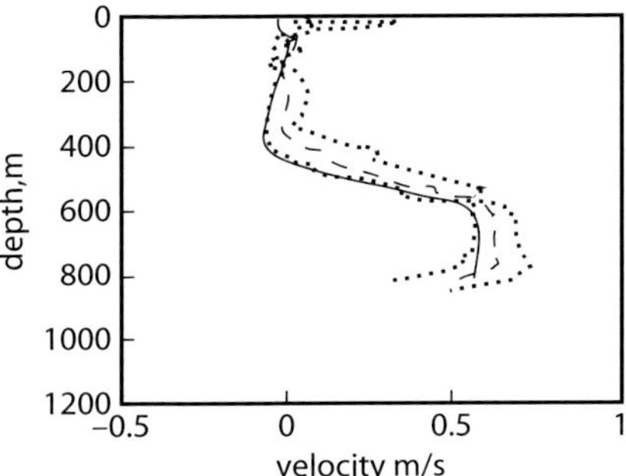

FIGURE 2.11.9. ADCP velocity data (dotted profiles) from two adjacent locations on Section D, their average (dashed line), and a geostrophic velocity profile from CTD data at the two locations (solid line). (From J. Price, private communication).

c. *Volume Flux*

Volume flux estimates made using the direct velocity data show variability from section to section (Figure 2.11.10), but with a general increase in flux downstream of the sill. Repeat sections, usually taken a few weeks apart, indicate considerable time variability as well, perhaps from the natural variability of the current itself, or from eddies, tides, or surface forcing. The estimates of flux for the deep overflow water are all positive, indicating a flow from the Greenland-Norwegian Sea toward the Atlantic.[9] The overall increase in transport from Section D to H is thought to be due to the turbulent entrainment of overlying

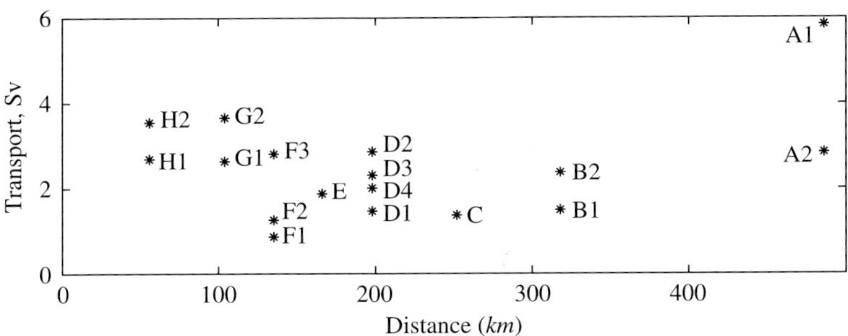

FIGURE 2.11.10. Volume flux estimates for water colder than $5\,^{\circ}C$ at the sections shown by letters in Figure 2.11.1. Duplicate estimates at the same passage locations are from data taken at different times. The flux estimates ignore counter flows. (From J. Price, private communication. Part b appears on following page.)

---

[9] There is one very large, and unexplained, estimate at section A. One possibility is that this estimate is correct, but there that is loss of water in the upstream channel from a flux of deep water over the Wyville-Thompson ridge, as mentioned earlier.

warmer water into the deep current. This view is supported by the observation that the overflow becomes warmer and less dense as it descends.

Sill flow toward the North Atlantic has been found every time measurements have been made. Borenäs and Lundberg (1988) estimated a geostrophic transport of 1.8 Sv below $3\,^{\circ}C$ with data collected in 1983 close to Section D. Although this value is in good agreement with present data, no clear picture of time-dependence was acquired. To detect changes in the flow over a few months duration, Saunders (1990) deployed an array of current meters in the same vacinity in 1987 and recovered them in 1988. Although many of the current meters were lost, velocity records of the cold overflow water with 363 days duration at depths of 492 and 693 m were recovered from one mooring. The current was found to persist all year with only a small seasonal fluctuation. The average volume flux of water colder than 3.0 degrees C was estimated to be 1.9 Sv.

More recent measurements of longer duration indicate a stronger seasonal variation along with a possible long-term trend. The current at the sill has been measured with upward looking, acoustic Doppler current profilers (ADCPs) since 1995. The corresponding transports (Figure 2.11.11) suggest a seasonal cycle in the water below $3\,^{\circ}C$ with maximum outflow during the fall. Hansen *et al.* (2001) used the same data to calibrate a relation between the transport and the upstream elevation $\Delta z_M{}^*$ (above the sill) of the $\sigma_t = 28.0$ surface. This elevation is found from hydrographic data monitored by Ocean Weather Ship-M, positioned in the eastern Norwegian Sea about 400 km upstream of the sill. The *ad hoc* relationship resembles a weir relation, with transport proportional to a power of $\Delta z_M{}^*$. A comparison between the calibrated relation and the measured transport appears in the figure. The weather ship has produced a temperature and salinity data set in the Norwegian Sea since 1948 and a time-history of the depth of $\sigma_t = 28.0$ can be extracted from this record. This history was used

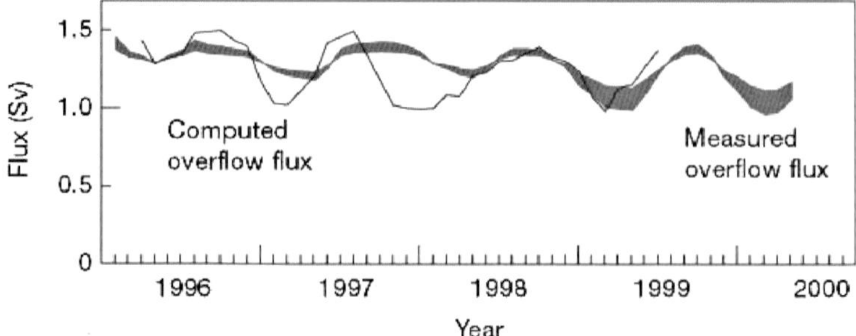

FIGURE 2.11.11. Estimated flux of overflow water colder than $0.3\,^{\circ}C$ through the Faroe Bank Channel from mid-1995 to late in 2000. The wide line is based on current meter data. The width indicates estimated error from uncertainties in depth of the $0.3\,^{\circ}C$ isotherm. The other line uses an empirical formula and data from Ocean Weather Ship-M to estimate flux. An eight-month time lag gives the best coherence with the current meter data; the result from the empirical formula has been displaced that much. (From Hansen *et al.*, 2001).

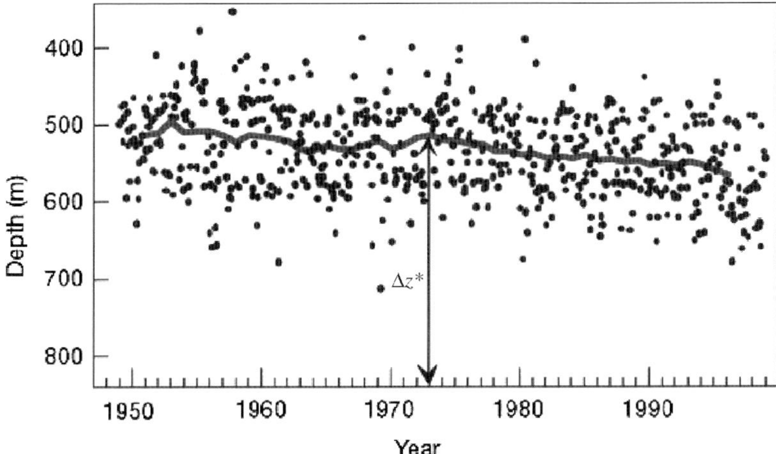

FIGURE 2.11.12. Depth of the density $\sigma_t = 28.0$ at Ocean Weather Ship-M from 1949 to 1999. The dots are a monthly average and the line is a five-year running mean. These data were used to infer a 20% decrease in flux over the 50-year interval. (From Hansen *et al.*, 2001).

by Hansen *et al.* (2001) in conjunction with the calibrated transport relation to estimate a history of transport over the period 1950–2000 (Figure 2.11.12). The value of Δz_M^* has diminished over this period and the corresponding decrease in transport is estimated to be about 20%. Continuation of such a trend would have important consequences for the meridional overturning cell in the Atlantic Ocean and therefore for climate. Some caution should be used in interpreting this result; for one thing, the Gill (1977) model calls into question the idea that transport can be monitored using a single upstream measurement. This and a set of related questions are taken up in Section 2.14.

d. *Potential Vorticity*

One of the key assumptions of the benchmark hydraulic models is that of uniform potential vorticity. Lake *et al.* (2005) estimated potential vorticity using three ADCPs deployed across the sill for 69 days. The relative vorticity is estimated using differences in velocities between neighboring profiles, while the interface position is estimated from the vertical shear. The result is two side-by-side potential vorticity time series. The contribution of the lateral shear is found to be roughly 1/4 as large as the Coriolis parameter f. The shear is negative, a result consistent with the layer being thinner at the sill than upstream. The magnitude of the potential vorticity was found to vary in time by a factor of two during the measurements. The average values at the two side-by-side locations differ by about 30%. The assumption of constant potential vorticity is therefore not obeyed in detail; it varies in space and time but may still be a useful first approximation.

e.   Hydraulic Criticality

The Faroe-Bank Channel inspired Borenäs and Lundberg's (1986) theory of uniform potential vorticity flow in a parabolic channel, discussed in Section 2.8. The authors followed this work with a (1988) report on the first large-scale observational study of the overflow. Among other findings, their critical condition appears to be satisfied, or nearly so, at the sill. Model estimates of volume flow range from 1.5 to 2.5 Sv., which compare well with their measured 1.5 to 1.9 Sv. By contrast, zero potential vorticity estimates for flow through a passage with a rectangular cross section give 2.1 to 3.4.

Girton *et al.* (2006) report on an extended effort to verify that the Faroe-Bank Channel flow becomes hydraulically critical and to determine the position of the critical section. The authors compute three independent indicators of flow criticality, the most general and reliable of which is the phase speed of the long-wave modes of the flow. The speeds are found by treating the observed flow at each section as a basic, steady parallel state and calculating the linear normal modes of this state. The numerical procedure is based on a method described by Pratt and Helfrich (2005) that uses an approximation of the actual bottom topography. In each case, two Kelvin-like modes are found along with a set of potential vorticity waves. The Kelvin modes can be recognized by the fact that their eigenfunction structures (Figure 2.11.13) indicate trapping to the left or right wall. For example, the wave structure shown in the upper left panel (mode #1) shows relatively large displacements of the interface along the left wall, while that of the lower right panel (mode #2) shows the largest displacements along the right wall. The cross-channel velocity of the wave is indicated by the displacements of the dark and light dashed lines, which are slight but still intensified along the left and right walls. The side-wall trapping, the prominence of vertical displacements of the interface, and the weakness of lateral displacements are characteristic of linear Kelvin waves and we therefore conclude that modes #1 and #2 in the figure are waves of this type. Both modes are similar to those found in a model with uniform potential vorticity (see Section 2.2). In contrast, modes #4 and #8 have relatively weak vertical displacement, relatively strong lateral displacements, and no evidence of side-wall trapping. These features are characteristic of the potential vorticity waves discussed on Section 2.1.

A section-by-section compilation (Figure 2.11.14) shows that the phase speeds of the potential vorticity modes (dashed lines) are bounded by the Kelvin wave speeds (solid lines). The right-wall Kelvin mode (upper solid line) always propagates in the downstream (positive) direction, as expected. The left-wall Kelvin mode (lower solid line) has a speed that is generally upstream (negative) indicating subcritical conditions with respect to that mode. However there is a single section (F), approximately 50 km downstream of the sill, where the wave speed goes to zero, or nearly so, and this suggests critical flow in the vicinity. Note that F lies where the bottom slope increases abruptly in Figure 2.11.2. The flow at the sill (D) appears by this measure substantially subcritical. The placement of the critical section on the downstream slope would be consistent with remarks made

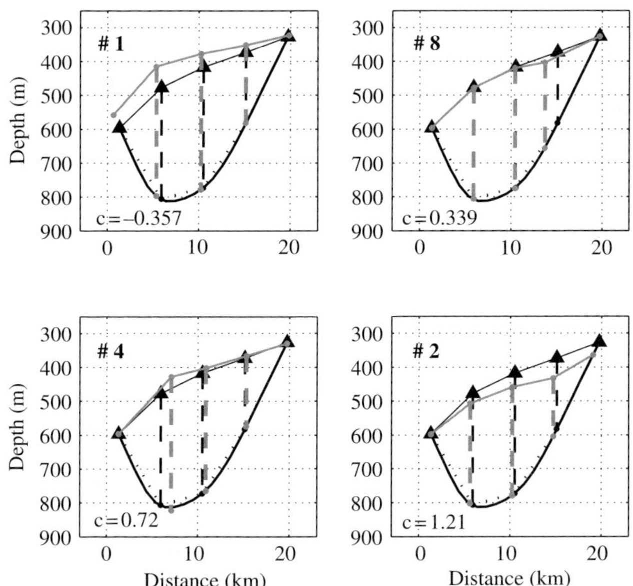

FIGURE 2.11.13. The linear eigenfunction structures for four of the lowest modes at Section D. Black lines and symbols indicate the interface of the observed 'background' flow. The solid gray line shows the change in the interface due to the presence of the mode. The magnitude (and sign) of this change is arbitrary and has been selected for visual convenience. The difference between the vertical solid and gray lines indicates the lateral excursions of the dense water due to the presence of the mode. Modes 1 and 2 resemble Kelvin waves, whereas 4 and 8 resemble potential vorticity waves. The phase speeds are given in m/s. (From Girton *et al.*, 2006).

in the first chapter concerning the effects of friction and entrainment, though a model that includes these and retains rotation has not been developed.

Although the real part of the phase speeds associated with the potential vorticity modes are generally positive (downstream), there is a case in which one of the speeds goes to zero (lowest dashed curve). The 'critical' section in question lies at B, approximately 140 km upstream of the sill. The ramifications and importance of a potential wave vorticity control are not well-understood, but the topic is revisited in Chapter 6. The present situation is further complicated by the fact that some of the modes have complex phase speeds (open circles), indicating instability. However, since overflows are driven by gravity, the Kelvin wave control would seem to be most relevant. As is generally the case, and suggested by the eigenfunction structures (Figure 2.11.13), the potential vorticity modes are manifested mainly in the lateral structure of the horizontal velocity, and less in the elevation of the interface.

The suggestion that the Kelvin wave critical section lies downstream of the sill and narrows is generally consistent with two other measures calculated by Girton *et al.* (2006). One involves the parabolic Froude number for uniform potential vorticity (eq. 2.8.12). The other is the distribution of the 'local' Froude number

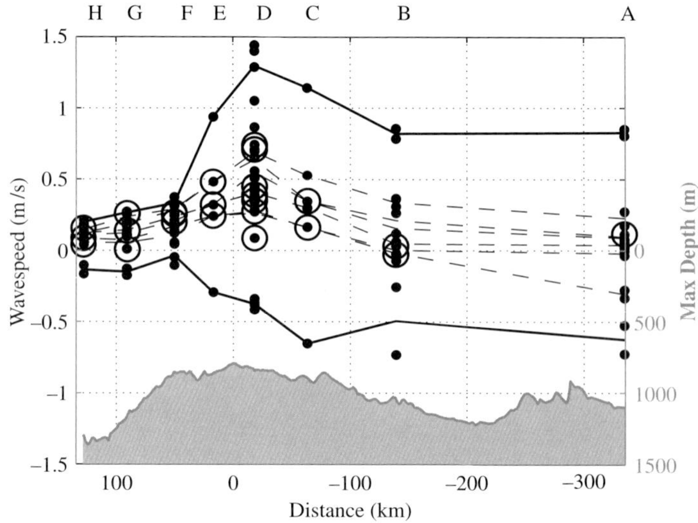

FIGURE 2.11.14. Phase speeds for the first eight long-wave channel modes of the flow, calculated at various sections. Curves indicate averages of values calculated for repeat sections, whereas dots indicate values for specific measurements. The upper curve gives the speed of a Kelvin wave propagating along the right edge (facing downstream) of the current. The lower curve gives the speed of the left-edge Kelvin wave. The intermediate curves correspond to potential vorticity modes. Open circles indicate that the phase speed is complex, though only the real part is plotted. Positive values indicate downstream propagation. (From Girton *et al.*, 2006).

$v/(g'd)^{1/2}$ across each section. One must exercise caution in interpreting the value of this last quantity at any particular point: hydraulic criticality implies the arrest of a Kelvin mode, or some other discrete cross-channel mode. The required conditions depend on the structure of the flow across the whole cross section and not just at a single point. Nevertheless the individual values of $v/(g'd)^{1/2}$ across a particular section may give some information as to whether critical flow is *possible*. There are two guiding pieces of information: First, a local region of flow over which $v/(g'd)^{1/2} > 1$ is one in which localized disturbances propagate downstream (see Appendix C or Section 4.3). Although hydraulics is more concerned with the propagation of cross-channel modes (that feel the side walls through satisfaction of boundary conditions), it is clear that a section of flow having $v/(g'd)^{1/2} > 1$ all the way across must be supercritical. The second piece of information concerns a conjecture that $v/(g'd)^{1/2}$ must equal one at some point across a section in order for that flow to be critical with respect to a normal mode. This result can be shown to hold for the simplified model flows that have been discussed thus far and, for example, is particularly clear in the formulation of Stern's critical condition (2.9.6). The latter holds for flow with arbitrary potential vorticity in a channel with a rectangular cross section and with unidirectional flow. The result has not been proven for cases in which the cross-section is nonrectangular or when velocity reversals exist.

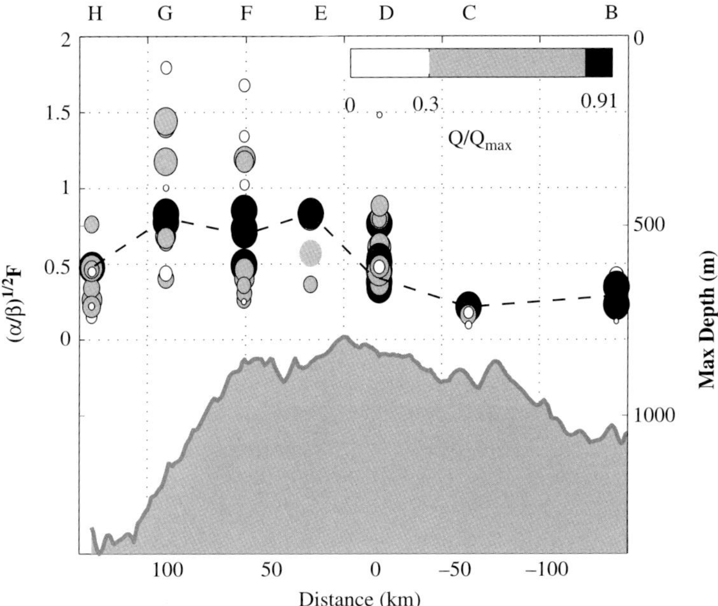

FIGURE 2.11.15. Local Froude numbers measured at various points across the various sections. A correction factor α/β has been applied to attempt to correct for the effects of vertical shear and continuous stratification (see Nielsen *et al.*, 2004). (The uncorrected values, which are smaller, are shown in Figure 2.11.8b). The shading in the circle and the circle size indicates the relative size of the transport velocity vd at the point at which the Froude number was measured. (From Girton *et al.*, 2006).

A compilation of local Froude numbers at all points of direct velocity measurement indicates that values exceed unity at only two sections, G and F (Figure 2.11.15). [The quantity plotted is actually a version of $v/(g'd)^{1/2}$ adjusted to compensate for the effects of vertical shear and continuous stratification.] It is at one of these sections that the left-wall Kelvin wave speed comes close to zero. There is no section over which the local Froude number is uniformly greater than zero. In fact, one of the striking aspects of the study is the lack of evidence at any section for a strongly supercritical flow.

f. Other Reading

A number of additional observational or data analysis projects have been completed as this book was being prepared. Duncan *et al.*, (2003) used three sections of velocity and density data at the sill and up to 60 km downstream. These data, taken over a five-day span, allowed estimates of frictional and mixing rates. Results include values of the bottom drag coefficient, the Von Karmen constant of the turbulence, the turbulent diffusivity, Richardson numbers and local Froude numbers. They found that the strongest mixing, characterized by

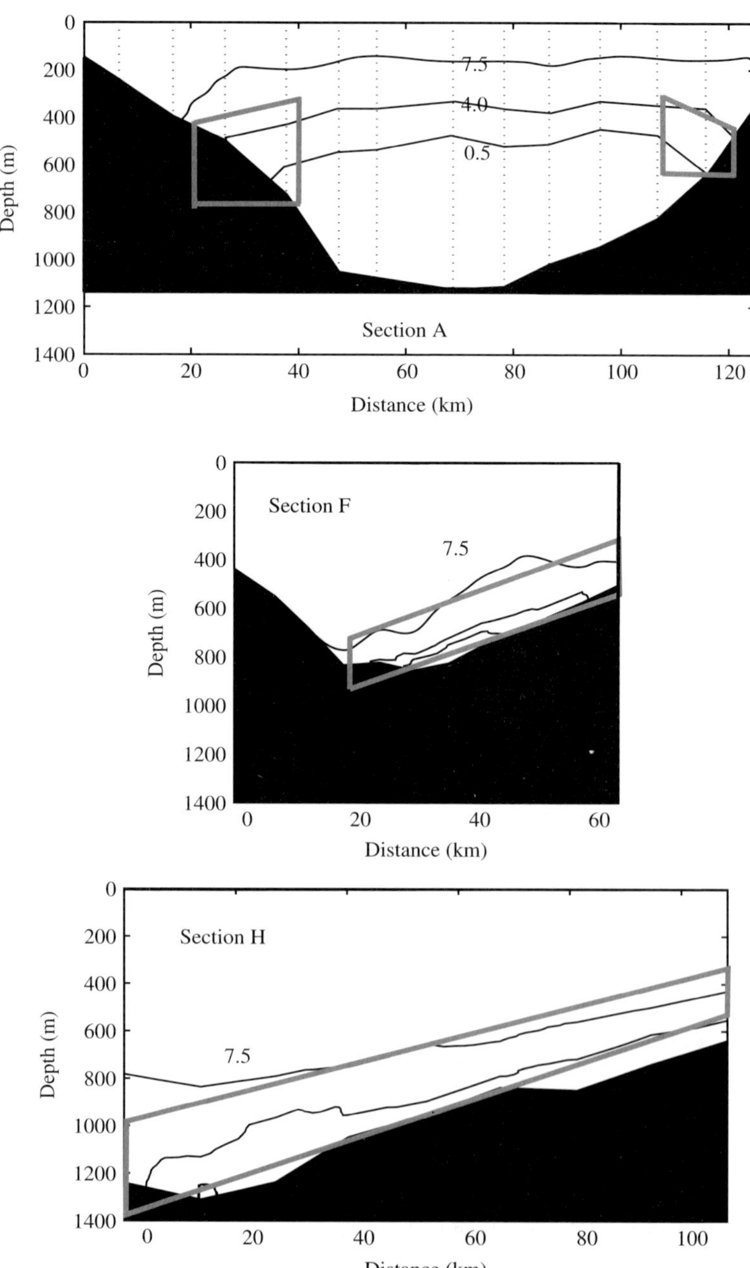

FIGURE 2.11.16. Sections A, F and H for use in homework exercise (based on J. Price, personal communication).

turbulent diffusivities up to $500\,\mathrm{cm}^2\ \mathrm{s}^{-1}$, is found 20 km downstream of the sill where Richardson numbers are small. At 60 km downstream, the diffusivities have decreased to $50\,\mathrm{cm}^2\ \mathrm{s}^{-1}$, and Richardson numbers are generally larger.

Readers who desire to learn more about the history of observations and models of the Faroese Channel system are referred to a review by Borenäs and Lundberg (2004). In addition, a presentation created by James F. Price containing many results of the Faroe Bank Channel Field Program is available at http://www.whoi.edu/science/PO/people/jprice/website/projects_overflows.html.

Exercises

(1) Estimate geostrophic velocity, layer depth, and volume flux at sections A, F, and H using the grey lines to represent the bounding interface, bottom and sides as shown in Figure 2.11.16. Note that Section A has left and right regions that are treated separately. You may use the approximations $f = 10^{-4}/\mathrm{s}$ and $\Delta\rho/\rho = 5 \times 10^{-4}$.

Our estimates are as follows:

Section A (on the left): $\delta = 100\,\mathrm{m}$, $w = 20\,\mathrm{km}$, $v = 0.25$ $D = 400\,\mathrm{ms}^{-1}$, $Q = 2.0$. On the right: $\delta = -100\,\mathrm{m}$, $w = 15\,\mathrm{km}$, $v = -0.34\,\mathrm{ms}^{-1}$, $D = 300\,\mathrm{m}$, $Q = -1.5\,\mathrm{Sv}$.

Section F: $\delta = 350\,\mathrm{m}$, $w = 45\,\mathrm{km}$, $v = 0.39\,\mathrm{ms}^{-1}$, $D = 200\mathrm{m}$, $Q = 3.5\,\mathrm{Sv}$.

Section H: $\delta = 600\,\mathrm{m}$, $w = 110\,\mathrm{km}$, and $v = 0.27\,\mathrm{m}\ \mathrm{s}^{-1}$. We picked $D = 400\,\mathrm{m}$ on the left, $D = 200\,\mathrm{m}$ on the right for the average $D = 300\,\mathrm{m}$, so $Q = 8.1\,\mathrm{Sv}$.

2.12.   Outflow Plumes

One of the most important aspects of deep ocean overflows is the mixing and water mass modification that occur as dense water spills over a sill and descends down the continental slope. Turbulent mixing and entrainment can lead to significant dilution and to increases in volume flux by 200% or more. The mixing may be due to bottom boundary layer turbulence or to interfacial instability or both. Bottom and interfacial drag may also be particularly important in determining the path that the flow takes. The portion of the overflow in which these processes are most active normally extends from the sill to some point tens or hundreds of kilometers downstream and is called the *plume* or *outflow plume*. It is a subregion of the *overflow*, meaning the entire hydraulically driven flow that begins at the upstream mouth of the channel and ends at the downstream extent of the plume. The overflows of the Denmark Strait, Faroe-Bank Channel, and Strait of Gibraltar all have distinctive plumes.

We will discuss two elementary models of outflow plumes. The first is a linear model that allows resolution of frictional (Ekman) boundary layers and the associated secondary circulations but ignores the effects of entrainment and inertia. The second model accounts for the latter two, but sacrifices resolution of Ekman layers and other cross-sectional variations of the flow. Both models are based on simplifications that disallow hydraulic effects such as subcritical-to-supercritical transitions.

a. Frictional Plumes in a Channel

Some insights into the structure of an outflow plume can be gained by comparing sections taken downstream of the Faroe-Bank Channel flow (Figures 2.11.4–2.11.6) showing the lateral spreading and thinning of the flow in the downstream direction. As the plume descends it becomes thinner and wider and is increasingly confined to the right-hand slope of the channel. These features have been reproduced in a laboratory model (Davies *et al.*, 2006, Figure 2.12.1) with a V-shaped channel that approximates the Faroe-Bank Channel bathymetry downstream of the sill. The observer faces upstream and the upper and lower frames show upstream and downstream sections of a particular realization of the plume. Variations in grey scale correspond to variations in density. Three main water masses can be distinguished and these consist of a quiescent overlying fluid (middle grey), a dense core region (dark) lying within the plume, and an intermediate interfacial region (light grey) consisting of a mixture of the two. A comparison of the upstream and downstream sections shows that the thickest portion of the core region has shifted from the bottom of the V-shaped channel to higher up on the slope, possibly indicating lateral motion from right to left in

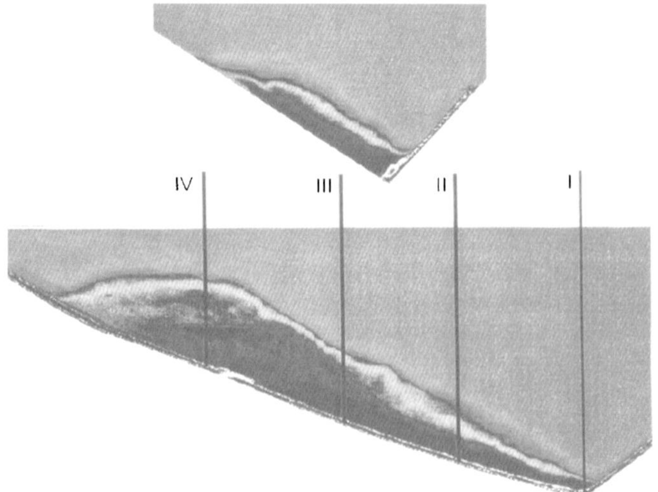

FIGURE 2.12.1. Dye images at two sections of a laboratory plume in a V-shape channel. Different shades of gray correspond to different densities, though there is no calibration. The viewer faces upstream. The lighter fluid is a mix between deeper (dark) and upper (middle gray) fluid. (From Davies *et al.*, 2006).

the figure. The lateral spreading and thinning of the plume in the downstream section is also evident.

It is difficult to configure laboratory experiments to match all the relevant nondimensional scale ratios that characterize the ocean application. In the above experiment it is possible to match quantities like the average Froude and Rossby numbers $V/(g'D)^{1/2}$ and V/fw, where V, D, and L are velocity, depth and width scales for the plume). More difficult to match is the Reynolds number $R_e = VD/\nu$, where ν is the kinematic viscosity of water. The magnitude of R_e potentially determines characteristics of the turbulence that lead to mixing and entrainment. Values for the present experiment lie in the range 1600–8000 whereas ocean values for the Faroe Bank channel are closer to $O(10^8)$. However the dependence of the flow on R_e considerably weakens after the threshold value $\cong 1000$ is exceeded (Davies *et al.*, 2002) and this criterion is approached in the experiment.

Deductive analytical models of plumes are generally not available due to the difficulty in dealing with the combined effects of friction, entrainment and nonlinear advection. However there are some very helpful models that rely partially on *ad hoc* assumptions. We will discuss two such models. The first ignores entrainment but includes some elementary Ekman layer effects. The second includes both frictional drag and entrainment but does so within the context of a 'streamtube' flow that has uniform properties across each section. The flow takes place on a uniformly tilting plane.

In the Davies *et al.*, 2006 model, the flow is confined to a V-shaped channel that has side slopes α and $-\alpha$ (Figure 2.12.2a, viewed from upstream) and that tilts with slope S along it axis. The flow grounds on the left and the right slopes at positions $x^* = x_L{}^*$ and $x^* = x_R{}^*$, as shown. If the slopes α and S are constant, the neglect of entrainment allows for a second simplifying assumption, namely that the flow is locally uniform in y^*. Changes in the flow along its path can then occur only as a result of changes in the bottom slopes (α or S), which could be allowed to vary gradually with y^*. Finally, it is assumed that the plume thickness d^* is much greater than the thickness of bottom or interfacial frictional layers and that the inviscid core region is in geostrophic balance in both directions. Thus the along-channel inertial effects fundamental to hydraulic behavior will be absent.

The geostrophic relations for the inviscid core are

$$u^* = \frac{g}{f}S, \tag{2.12.1}$$

for the cross-channel velocity component, and

$$v^* = \frac{g}{f}\frac{\partial}{\partial x^*}(d^* + h^*) = \frac{g}{f}\frac{\partial d^*}{\partial x^*} + \frac{g}{f}\begin{cases} \alpha \ (x^* > 0) \\ -\alpha \ (x^* < 0) \end{cases}, \tag{2.12.2}$$

for the along-channel component.

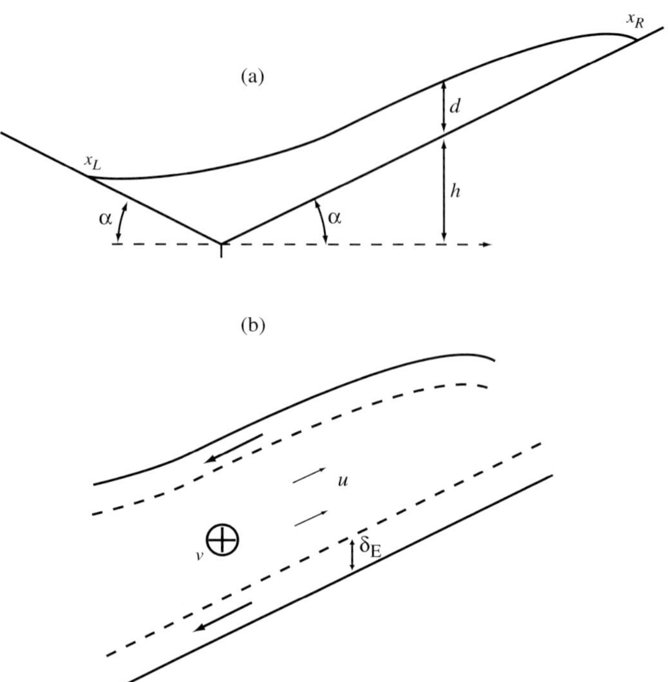

FIGURE 2.12.2. Definition sketch for Davies *et al.*, 2006 model.

The volume flux associated with the core region is

$$Q^* = \int_{x_L^*}^{x_R^*} v^* d^* (dx^*) = \frac{g}{f} \int_{x_L^*}^{x_R^*} d^* \left(\frac{\partial d^*}{\partial x^*} + \begin{cases} \alpha \; (x^* > 0) \\ -\alpha \; (x^* < 0) \end{cases} \right) dx^*$$

$$= \frac{\alpha g}{f} \left(\int_{x_L^*}^{0} (d^*) dx^* - \int_{0}^{x_R^*} (d^*) dx^* \right) \qquad (2.12.3)$$

in which $d^*(x_L^*) = d^*(x_R^*) = 0$ has been used. The flux in any frictional boundary layers at the top or bottom must be added to Q^* to get the total volume flux, but these contributions are small.

The effect of bottom friction on the current is to reduce the velocity to zero at the bottom. In a rotating flow, the reduction occurs within a boundary layer, the Ekman layer, which is discussed in most texts on geophysical fluid dynamics [e.g. Pedlosky (1987), Ch. 4]. For a laminar flow the Ekman layer thickness is $\delta_E = (2v/f)^{1/2}$, where v is the molecular viscosity. In a turbulent flow, v must be replaced by a much larger, hypothetical turbulent viscosity. One of the consequences of the Coriolis acceleration acting within the Ekman layer is that a volume transport is established transverse to the overlying geostrophic flow

(Figure 2.12.2b). If the overlying flow is directed primarily in the y-direction, which is the case here, then a transport in the negative x^* direction equal to $v^*\delta_E/2$ takes place in the boundary layer. Davies *et al.* (2006) assume that a similar frictional layer occurs at the upper interface and thus the total boundary layer transport is $v^*\delta_E$. If δ_E is much less than the fluid depth, which is the case assumed here, then the water column will consist of thin upper and lower Ekman layers, with combined transverse flux $v^*\delta_E$, separated by a relatively thick, geostrophic interior, referred to hereafter as the *core*, with transverse flux u^*d^*.

The transverse circulation in the interior and bottom Ekman layer has been visualized (E. Darelius, private communication) in an experiment with the same V-shape channel (Figure 2.12.3). Dye is injected at the left edge (facing upstream) of the flow, directly into the lower Ekman layer. As seen in the upper part of the figures, the dye crosses the channel to the viewer's right and then returns to the left edge by way of the geostrophic core region. Most of the dye then re-enters the bottom Ekman layer and the cycle repeats. The result is a spiral motion of the dye as it descends down the channel.

FIGURE 2.12.3. The spiral motion of a dye stream in a descending laboratory flow, facing upstream. The dye is introduced at the upper left, directly into the lower Ekman layer. It crosses the channel to the right in the bottom Ekman layer, returns to the left in the geostrophic core region, and repeats the cycle. (2006 photo by Elin Darelius).

For constant bottom slopes α and S, the flow is independent of y^* and the depth integrated mass flux in the x^*-direction is therefore nondivergent. This flux is clearly zero at the edges of the plume and the flux must therefore be zero everywhere else:

$$u^* d^* - v^* \delta_E = 0. \tag{2.12.4}$$

The use of (2.12.1) and (2.12.2) to substitute for the velocities leads to

$$\frac{\partial d^*}{\partial x^*} - \frac{S}{\delta_E} d^* = \begin{cases} -\alpha \ (x^* > 0) \\ \alpha \ (x^* < 0) \end{cases}. \tag{2.12.5}$$

The solutions in the two regions are given by

$$d^* = \begin{cases} Be^{Sx^*/\delta_E} + \alpha \delta_E/S & (x^* > 0) \\ Ae^{Sx^*/\delta_E} - \alpha \delta_E/S & (x^* < 0) \end{cases}. \tag{2.12.6}$$

The theory formally breaks down near the edges, where the depth vanishes[10]. The fact that the thickness is brought to zero monotonically means that the slope of the interface cannot be greater than the slope of the bottom.

Matching the two solutions across $x^* = 0$ leads to

$$A = B + 2\alpha \delta_E/S.$$

In addition, the position of the right-hand outcrop is obtained by setting $d^* = 0$ in the first of (2.12.6):

$$B = -\frac{\alpha \delta_E}{S} e^{-Sx_R^*/\delta_E}.$$

The full solution is therefore

$$d^* = \frac{\alpha \delta_E}{S} \begin{cases} 1 - e^{S(x^* - x_R^*)/\delta_E} & (x^* > 0) \\ [(2 - e^{-Sx_R^*/\delta_E})e^{Sx^*/\delta_E} - 1] & (x^* < 0) \end{cases} \tag{2.12.7}$$

The edge positions x_R^* and x_L^* are obtained by setting $d^* = 0$ in the second part of (2.12.7):

$$(2 - e^{-Sx_R^*/\delta_E})e^{Sx_L^*/\delta_E} = 1. \tag{2.12.8}$$

A second relation between x_R^* and x_L^* is provided by the formula (2.12.3) for the flux Q:

$$Q = \frac{\alpha^2 g \delta_E^2}{f S^2} \left[\frac{S}{\delta_E}(x_R^* - x_L^*) + 2(e^{-Sx_R^*/\delta_E} - 1) \right], \tag{2.12.9}$$

---

[10] The viscous edge regions are described by a alternative theory (Wåhlin and Walin, 2001).

in view of (2.12.7). If Q is specified, x_R^* can be computed though substitution of

$$x_L^* = -\frac{\delta_E}{S}\ln(2 - e^{-Sx_R^*/\delta_E}), \qquad (2.12.10)$$

which follows from (2.12.8), into (2.12.9). If x_R^* increases, so does $-x_L^*$ and the plume broadens as a whole.

Suppose that along-axis slope S decreases gradually in the y^*-direction. Then it is easily shown from (2.12.9) that x_R^* must increase in order that Q remain constant. The tendency of the plume to broaden as the slope decreases can be motivated using the following crude argument. Consider the flow at a point that lies on the right side $(x^* > 0)$ of the channel, but well away from the right edge. The interface elevation in this region is approximately uniform across the channel and v^* is proportional to α (see 2.12.2). If S is diminished, v^* therefore remains fixed, as does the lateral Ekman layer flux $v^*\delta_E$. The lateral velocity u^* is decreased, however, and d must increase, widening the flow, in order to maintain zero net volume transport across the channel (see 2.12.4).

We leave it as an exercise for the reader to show that the same broadening occurs when α decreases and S is kept constant. A decrease in α in the downstream direction is a crude representation of a widening of the channel that occurs in the Faroe-Bank Channel downstream of the sill. The broadening of the stream that is predicted by the model may account, at least in part, for that observed there and in the laboratory experiment.

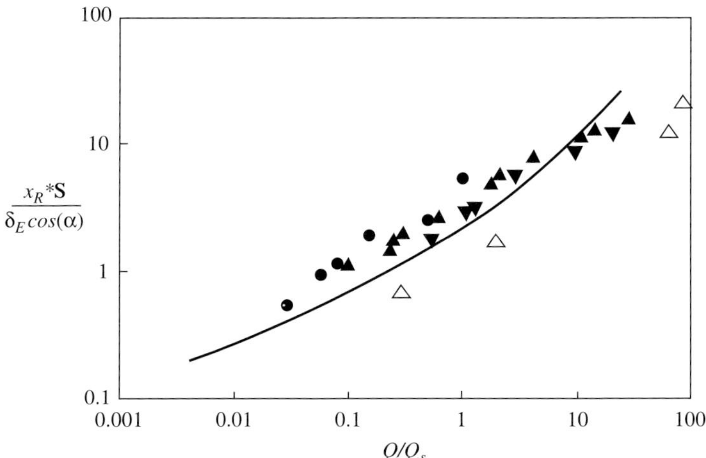

FIGURE 2.12.4. The dimensionless width $\frac{x_R^* S}{\delta_E \cos\alpha}$ of the right-hand $(x > 0)$ portion of the current as a function of dimensionless transport Q/Q_s, where $Q_s = \alpha^2 g \delta_E^2 / f S^2$. The curve shows the predicted width based on (2.12.9) and (2.12.10). The solid symbols represent data from the Davies *et al.* (2006) experiment with $[g_o'\ (\mathrm{cm\ s^{-2}}),\ f(\mathrm{s^{-1}})]$ given by ▲ [8.82,0.37]; ▼ [8.82,0.50]; and ● [23.50; 0.37]. The symbols △ represent Faroe-Bank Channel field data from Mauritzen *et al.* (2005). (From Davies *et al.*, 2006).

Equation (2.12.9) suggests $\frac{\alpha^2 g \delta_E{}^2}{fS^2}$ as a natural scale for Q. A plot of the corresponding nondimensional transport **vs** a nondimensional version of $x_R{}^*$ (Figure 2.12.4) shows that the right side of the plume broadens as the transport increases. This tendency is consistent with what is observed in the experiment (solid symbols), and to some extent the Faroe Bank Channel overflow (Δ symbols).

b. *Streamtube Models for Entraining Plumes*

The previous model is restricted in that it requires channel geometry and does not allow for entrainment. All major outflow plumes undergo entrainment, though it may only be significant over certain reaches. The plumes associated with outflows from the Mediterranean, the Denmark Strait, and the Weddell Sea's Filchner Ice Shelf ride along continental slopes and gradually descend. A traditional tool for simulating the combined effects of entrainment and bottom friction in these flows is the streamtube model, pioneered by Smith (1975) and improved by Killworth (1977), Price and Baringer (1994) and others. We will discuss the original version of this model and comment on later refinements.

The term 'streamtube' implies a coherent flow that can be characterized at any cross section by a few variables that define the bulk properties like average velocity. In the case of the Smith (1975) model the flow takes place on a flat surface with uniform slope S in the y-direction (Figure 2.12.5). Natural coordinates (s^*, n^*) are used to measure distance along and normal to the axis of the plume. The plume properties are assumed to vary gradually in the s^*-direction. Different versions of the model make different assumptions with regard to the variables used to characterize the flow and the way in which entrainment, friction and stratification are handled, but there is one far-reaching assumption made by all. The plume is supposed to be sufficiently thin so that the slope of the interface remains nearly equal to the slope of the bottom in any direction. Accordingly, the pressure gradient in the plume is assumed proportional to the

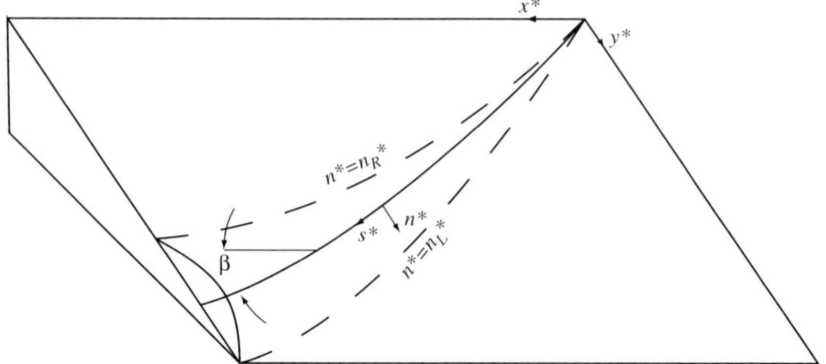

FIGURE 2.12.5. Definition sketch for streamtube model.

bottom slope; contributions from the gradient of the plume depth are neglected. The main consequence of this simplification is elimination of gravity wave dynamics from the model and thus the transmission of information upstream. As we will see, the steady plume equations can be integrated downstream beginning at some point where the flow properties are known without regard for conditions far downstream. This considerably simplifies the computational problem for ocean applications. However, the local Froude numbers in deep plumes are observed to consistently fall below unity, particularly far downstream of the source, suggesting that upstream wave propagation is possible. The importance of the consequences is not fully understood at the time of this writing.

The plume axis makes an angle β with respect to the x-axis and thus the position (X^*, Y^*) of the plume in the Cartesian frame is given by

$$\frac{dX^*}{ds^*} = \cos \beta \quad \text{and} \quad \frac{dY^*}{ds^*} = \sin \beta.$$

The along- and cross- axis velocities will be denoted by V^* and U^* respectively, with $|U^*| \ll |V^*|$, and it will be assumed that V^* and the plume density ρ are constant across the section of the plume at any particular s^*. The overlying fluid is stably stratified and has density $\rho_a(z)$, which can alternatively be expressed as a function of s^* for a given plume path.

Entrainment into the plume is represented as in Section 1.11 by a cross-interface, positive downwards, vertical velocity w_e^*. If A^* is the cross-sectional area of the plume, the volume flux A^*V^* must therefore obey

$$\frac{\partial}{\partial s^*}(A^*V^*) = \int_{n_R^*}^{n_L^*} w_e^* \, dn^*, \tag{2.12.11}$$

while the total mass flux ρA^*V^* is subject to

$$\frac{\partial}{\partial s^*}(\rho A^*V^*) = \int_{n_R^*}^{n_L^*} \rho_a w_e^* \, dn^* \cong \rho_a \int_{n_R^*}^{n_L^*} w_e^* \, dn^*. \tag{2.12.12}$$

Here n_R^* and n_L^* denote the grounding positions of the interface on the right and left edges of the plume, facing downstream.

The along-axis momentum balance for the plume is expressed in terms of the depth-integrated shallow water equations, the Cartesian versions of which are (2.1.17a,b). If at a particular section of the plume, the coordinates are aligned such that the former y^* axis coincides with the present s^*-direction, and the former x^* points in the minus n^* direction, then the along-axis momentum equation is

$$\frac{\partial}{\partial s^*}[V^{*2} d^* + \tfrac{1}{2} g' d^{*2}] + \frac{\partial}{\partial n^*}(U^*V^* d^*) - f U^* d^*$$
$$= g' d^* \sin \beta - (\tau_B - \tau_I)/\rho_o \tag{2.12.13}$$

where $g' = g(\rho - \rho_a)/\rho_o$. This result may also be compared with the depth-integrated momentum equation (1.9.10) for a one-dimensional flow subject to entrainment from above. Note that there is no source of momentum from the entrainment since the overlying fluid is assumed to be motionless ($v_1 = 0$ in 1.9.10). There are sinks of momentum from the frictional stresses at the bottom and interface, as represented by τ_B and τ_I. The Boussinesq approximation has been made, meaning that ρ is replaced by a constant reference density ρ_o where it multiplies inertial terms. It will be convenient to choose ρ_o as the density of the plume at some source point.

Following the assumption that the plume is so thin that the pressure gradient is due entirely to the bottom slope, the term proportional to d^{*^2} in (2.12.14) is neglected. An important consequence is that horizontal pressure gradients due to horizontal variations of density along the path of the plume are ignored. The remaining pressure gradient (first term on the right-hand side) is proportional to the along-axis component of the bottom slope $S \sin \beta$. In addition fU^*d^* is neglected on the basis that U^* is rendered small by the use of the natural coordinate system and the assumption of gradual variations in s^*. This last assumption is not consistent with the usual semigeostrophic approximation, where in spite of gradual variations the term fU^* must be retained. This and other nondeductive assumptions place the streamtube theory in the realm of *ad hoc* models.

Integration of the simplified version of (2.12.13) across the plume and use of $d^*(n_R^*) = d^*(n_L^*) = 0$ leads to

$$\frac{\partial}{\partial s^*}(A^*V^{*^2}) = g'SA^* \sin \beta - \int_{x_R^*}^{x_L^*} [(\tau_B + \tau_I)/\rho_o]dn, \qquad (2.12.14)$$

where $A^* = \int_{x_R^*}^{x_L^*} (d^*)dn$.

The cross-stream momentum equation is a form of the geostrophic relation, modified to include the effects of curvature along the s^*-axis. The equation is identical to the n^* momentum equation written in a cylindrical coordinate system with n^* as the radial variable (see Batchelor, 1967, Appendix 2) and with the advective terms ignored. (A similar problem will be considered in detail in Section 4.5.) The resulting momentum balance is

$$V^* \left(f + V^* \frac{d\beta}{ds^*} \right) = Sg' \cos \beta. \qquad (2.12.15)$$

The factor $d\beta/ds^*$ is the curvature of the plume axis and $V^*d\beta/ds^*$ can be thought of as an augmentation of the Coriolis acceleration.

A simple example that provides a reference for further analysis is a nonentraining, frictionless outflow that moves along isobaths ($\beta = 0$). Equation (2.12.15) gives the velocity of such a flow as

$$V^* = \frac{Sg'}{f}. \qquad (2.12.16)$$

This formula applies in more general settings as well. It can be shown (Exercise 3) that the right-hand side is the average velocity of a geostrophic current flowing along a constant slope, provided the interface elevation is properly taken into account in the calculation of the pressure. The same factor was also shown by Nof (1983) to be the speed of a geostrophically balanced, lens-like eddy propagating along a slope.

Let g_o' denote the value of reduced gravity at the source. The above result suggests $S g_o'/f$ as a scale for V^* and it is then natural to choose $L = S g_o'/f^2$ as a horizontal length scale. We also scale A^* with its upstream value $A_o{}^*$. The corresponding dimensionless forms of (2.12.11,12,14 and 15) are then

$$\frac{\partial}{\partial s}(AV) = E_n, \tag{2.12.17}$$

$$\frac{\partial}{\partial s}\left(\frac{\rho}{\rho_o} AV\right) = \frac{\rho_a}{\rho_o} E_n, \tag{2.12.18}$$

$$\frac{\partial}{\partial s}(AV^2) = \gamma\, A \sin\beta - F_\tau, \tag{2.12.19}$$

and

$$V^2 \frac{d\beta}{ds} = \gamma\cos\beta - V, \tag{2.12.20}$$

where $\gamma(s) = (\rho - \rho_a)/(\rho_o - \rho_{ao})$ and ρ_{ao} is the upstream value of ρ_a.

The expressions for entrainment and drag

$$E_n = \frac{1}{fA_o} \int_{n_R{}^*}^{n_L{}^*} w_e{}^* dn^* \tag{2.12.21}$$

and

$$F_\tau = \frac{1}{A_o g_o'} \int_{x_R{}^*}^{x_L{}^*} \left[(\tau_B + \tau_I)/\rho_o\right] dn \tag{2.12.22}$$

must be parameterized. The bottom and interfacial stresses are most commonly specified using a quadratic drag law of the form

$$F_\tau = \kappa V^2 \tag{2.12.23}$$

Entrainment is normally parameterized as described in Section 1.9 in terms of a Froude number. Many of the laboratory experiments or field studies that have been used to develop empirical formulae involve nonrotating flows in which the local Froude number $v/(g'd)^{1/2}$ is constant across the descending stream. Some of the corresponding data are shown in Figure 1.10.4. In the present case the value of $v/(g'd)^{1/2}$ varies across the stream and the parameterization is written in terms of a bulk Froude number characterizing the whole cross-section (Price and Barringer, 1994). For present purposes it is adequate to regard E_n as a function of V and A, and make no further specification.

The presence of frictional drag causes a flow that would otherwise follow isobaths to descend. Suppose that entrainment is ignored ($E_n = 0, \rho = \rho_o$), the overlying fluid is homogeneous ($\rho_a = \rho_{ao}$, and therefore $\gamma = 1$) and a uniform (in s) solution is sought. If the friction parameterization (2.12.23) is used, the velocity and angle of descent are found by setting $\partial/\partial s = 0$ in equations (2.12.19–20) and setting A to its upstream value of unity. Thus

$$\sin \beta = \kappa V^2 \qquad (2.12.24)$$

and

$$\cos \beta = V, \qquad (2.12.25)$$

These relations summarize a force balance (Figure 2.12.6) in which gravity attempts to pull the plume down the fall line, friction tends to retard the flow, and the Coriolis effect tends to accelerate the plume to the right of the flow direction. Only the frictional and gravitational forces act parallel to the plume axis and the velocity must have a down-slope component ($\beta > 0$) in order for the two to balance. In the direction normal to the axis the tendency of gravity to accelerate the fluid down the fall line is balanced by the Coriolis acceleration

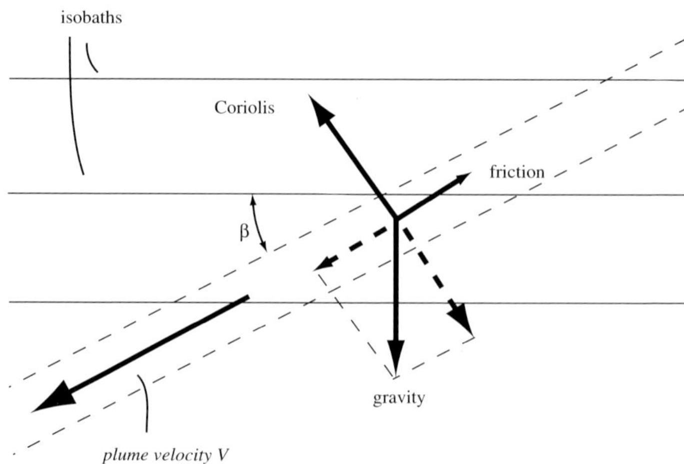

FIGURE 2.12.6. Plan view showing the equilibrium state of a descending, nonentraining plume in a homogeneous environment. The dashed arrows show the normal and tangential components of the gravitational force; these must be balanced by the Coriolis acceleration and the frictional drag vectors.

in the opposite direction. If V is eliminated between the above two relations, it follows that

$$\frac{\sin\beta}{\cos^2\beta} = \kappa, \qquad (2.12.26)$$

and thus the angle of descent increases as the friction coefficient increases. When perturbed from this parallel state, the plume executes stable meanders about its original path (see Exercise 4).

When entrainment is present, the flow can no longer be uniform in s, making simple solutions harder to come by. However, if the overlying fluid is homogeneous ($\rho_a = $ constant) then the problem can be simplified somewhat. To begin with, subtraction of the product of ρ_a/ρ_o and (1.12.17) from (1.12.18) yields

$$\frac{\partial}{\partial s}\left(\frac{\rho-\rho_a}{\rho_o}AV\right) = 0, \qquad (2.12.27)$$

showing that the buoyancy flux

$$B_f = [(\rho-\rho_a)/\rho_o]AV \qquad (2.12.28)$$

is conserved. We will assume that the plume at its point of origin ($s = 0$) flows parallel to the isobaths ($\beta = 0$), as it would if entrainment and friction were nil. The upstream values of the plume variables at this point are then $A(0) = V(0) = 1$, $\beta(0) = 0$, and $\rho(0) = \rho_o$. Downstream of this point we will track the evolution of the flow, proceeding on the assumption that entrainment and friction are finite but weak ($E_n \ll 1$ and $\kappa \ll 1$). The entrainment may vary with V and A, and it is assumed only that it retains the same general nondimensional size as the friction term, i.e. $E_n(V, A) = O(\kappa)$.

Next expand the dependent variables according to

$$A = 1 + \kappa A^{(1)} + \kappa^2 A^{(2)} + \cdots$$
$$V = 1 + \kappa V^{(1)} + \cdots \qquad (2.12.29)$$
$$\beta = \kappa\beta^{(1)} + \cdots$$
$$\rho = \rho_o + \kappa\rho^{(1)} + \cdots .$$

Substitution into (2.12.17,19,20 and 27) and retention of only $O(\kappa)$ terms leads to

$$\frac{d}{ds}\left(A^{(1)} + V^{(1)}\right) = \frac{E_n}{\kappa},$$

$$\frac{d}{ds}\left(A^{(1)} + 2V^{(1)}\right) = \beta^{(1)} - 1,$$

$$\frac{d\beta^{(1)}}{ds} = \tilde{\rho}^{(1)} - V^{(1)},$$

and

$$\tilde{\rho}^{(1)} = -(V^{(1)} + A^{(1)}).$$

where $\tilde{\rho}^{(1)} = \rho^{(1)}/(\rho_o - \rho_a)$.

The solutions satisfying the upstream conditions $\beta^{(1)} = V^{(1)} = A^{(1)} = 0$ are

$$\beta^{(1)} = 1 - \cos s,$$

$$V^{(1)} = -\sin s - \frac{E_n}{\kappa} s,$$

$$A^{(1)} = \sin s + \frac{2E_n}{\kappa} s,$$

and

$$\tilde{\rho}^{(1)} = -\frac{E_n}{\kappa} s.$$

Thus the combined influence of friction and entrainment causes the plume to turn downslope ($\beta^{(1)}$ becomes positive). The velocity V decreases but the area A increases at twice the rate, this in order for the volume flux to increase. The plume is diluted ($\tilde{\rho}^{(1)}$ becomes negative) in proportion to the entrainment rate but in inverse proportion to the drag coefficient. Along with these trends, the plume undergoes a meandering motion with wavelength 2π, dimensionally $2\pi S g'_o/f^2$, caused when the path overshoots the equilibrium angle $\beta^{(1)} = 1$ or $\beta = \kappa$. The linear solution remains valid only for distances of order κ^{-1} downstream of the point of origin; the secular growth associated with the terms proportional to s invalidate the asymptotic expansion further downstream.

Application of streamtube models to specific outflows have resulted in a number of refinements, including variable bottom slope and separate treatments of temperature and salinity. In his simulation of the Weddell Sea plume, which is observed to spill to the bottom of the slope, Killworth (1977) notes that the simulated flow will not reach the bottom without inclusion of the *thermobaric* effect, the increase with depth of the coefficient of thermal expansion. A less subtle process is entrainment, which is addressed by the Price and Barringer (1995) model. Since the entrainment velocity is parameterized by the Froude number, an explicit treatment of the plume width and depth, and not just the cross-section area, is required. Price and Barringer base their treatment on a spreading law in which the downstream rate of increase of the plume width is proportional to the bottom drag.

One of the most important factors determining the fate of ocean plumes is the density of the overlying water. It is well-known that the densest source waters come from the Mediterranean Sea, but the densest product waters (after entrainment) come from the high latitude overflows (Weddell Sea, FBC and Denmark Strait). For these applications, the least dense source waters tend to produce the densest product waters (Table 2.12.1). This is largely due to the fact that the density of the overlying water is greatest where the product water density is greatest.

TABLE 2.12.1. Densities (σ_θ) of the average source, product and overlying water for four major outflow plumes. (Data from Price and Baringer, 1995).

Location	source	product	ambient (overlying)
Filchner Ice Shelf (Weddell Sea)	27.93	27.89	27.82
Denmark Strait	28.04	27.92	27.72
Faroe Bank Channel	28.07	27.90	27.56
Mediterranean	28.95	27.70	27.06

Exercises

(1) Show that the theory that leads to (2.12.7) fails to provide a solution for the case in which the plume rides entirely over the positively sloping portion of bottom ($x^* > 0$), i.e. a solution with $x_{\mathrm{L}}^* > 0$.

(2) For the Davies *et al.* (2006) model, show that the plume broadens when the cross-channel slope α is decreased but S remains fixed.

(3) Show that the velocity defined by (2.12.16) is the average velocity of a geostrophic current flowing along a constant slope if the interface elevation is properly taken into account in the calculation of the pressure.

(4) *The meandering of a nonentraining plume.* Consider the Smith (1975) streamtube model for the case in which there is no entrainment and where the overlying density is uniform.

(a) Show that the two momentum equations (2.12.19 and 2.12.20) can be written for this case in the form

$$Q\frac{dV}{ds} = \frac{Q\sin\beta}{V} - \kappa V^2,$$

and

$$V^2\frac{d\beta}{ds} = \cos\beta - V,$$

where $Q = AV$ is the (now constant) volume flux.

(b) For a given Q, show that the fixed point (i.e. $\partial/\partial s = 0$) solutions, which are just the parallel flows discussed earlier, are given by (2.2.26) and

$$\kappa^2 V^6 Q^{-2} + V^2 - 1 = 0.$$

(c) By linearizing the momentum equations about this solution, show that small departures from the parallel state consists of meanders of the flow axis. Calculate the meander wavelength.

(d) Show that motions not restricted to small amplitude perturbations of the parallel state are described by the relationship

$$\frac{d\beta}{dV} = \frac{Q(\cos\beta - V)}{QV\sin\beta - \kappa V^4}.$$

Sketch some of the corresponding solutions in the (V, β) plane and show that they consist of periodic orbits that surround the fixed point calculated in (b).

2.13. Closed Upstream Basins with Forcing and Dissipation

The cornerstone models of rotating hydraulics make the assumption that the upstream reservoir is infinite in extent and that potential vorticity q^* is conserved. In reality, oceanic basins such as the Norwegian Sea, the Greenland Sea, and the Brazil Basin are finite and are subject to forcing and dissipation. Although the deep circulations in these regions have not been mapped out, the conventional view is that the flows are weak and in near geostrophic balance. Forcing and dissipation may be weak in a local sense, but they have a cumulative effect that can be significant over the broad expanse of the basin. These processes are instrumental in the determination of q^* and it is quite unlikely that this quantity will be uniform. In fact, direct observations of the velocity of the deep flow in the Faroe-Bank Channel (Lake *et al.*, 2005) suggest that q^* can be significantly nonuniform.

The main difficulty in trying to extend models like Gill's is one of tractability: the combination of forcing, dissipation and nonlinearity gives rise to formidable mathematical obstacles. However, one might expect nonlinear advection to be relatively unimportant in the upstream basin, where the flow is weak. In addition, one might expect forcing and dissipation to be of minor consequence within the strait that drains the basin, where advection is quite strong. The neglect of friction may not be supportable within the 'plume' region, downstream of the sill, where the relatively swift outflow descends into the downstream basin, but it may be valid upstream of the sill.

a. *Linear Model for the Basin*

Keeping these expectations in mind, we now develop an approach in which the slow, linear, dissipative circulation in an upstream basin is linked to an inertial outflow that takes place through a strait (Figure 2.13.1). The approach is presented in greater detail by Pratt and Llewellyn Smith (1997) and Pratt (1997). The basin topography is variable and the upstream flow may be fed by a variety of sources, including deep convection, lateral inflows through other straits, or dense fluid sliding down the continental slope. Since there is no reason

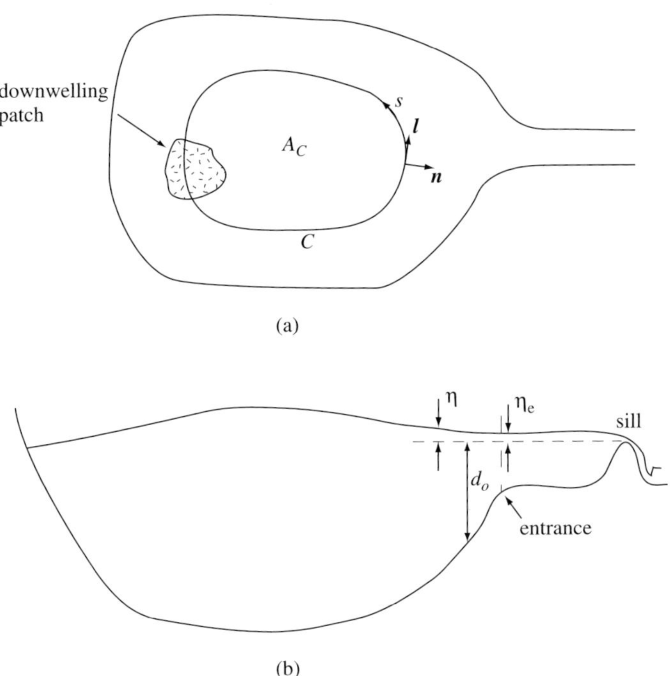

FIGURE 2.13.1. Definition sketch.

to expect the basin flow to take place in a preset direction, one must abandon
the semigeostrophic approximation and consider the full steady shallow water
equations:

$$(\mathbf{u}^* \cdot \nabla^*)\mathbf{u}^* + f\mathbf{k} \times \mathbf{u}^* = -g\nabla^*\eta^* - \frac{r_f \mathbf{u}^*}{d^*} \tag{2.13.1}$$

and

$$\nabla^* \cdot (\mathbf{u}^* d^*) = w_e^*. \tag{2.13.2}$$

where w_e^* is a positive downward entrainment velocity and r_f a drag coefficient.
The elevation η^* of the interface is measured relative to the sill and the depth
of basin below the sill is given by d_o^*:

$$d^*(x^*, y^*) = d_o^*(x^*, y^*) + \eta^*(x^*, y^*).$$

Note that the effects of entrainment have been included in the continuity equation
but not the momentum equation, an approximation that the reader can justify by
working through Exercise 1.

   Let N and D be scales for η^* and d_o^* and assume that $\varepsilon = N/D \ll 1$.
Then if $(gD)^{1/2}/f$ is a typical length scale for the basin flow, the geostrophic

relation suggests $\varepsilon(gD)^{1/2}$ as a velocity scale. The corresponding nondimensional versions of (2.13.1) and (2.13.2) are then given by

$$\varepsilon(\mathbf{u}\cdot\nabla)\mathbf{u}+\mathbf{k}\times\mathbf{u}=-\nabla\eta-\frac{r_f\mathbf{u}}{fD(d_o+\varepsilon\eta)}, \tag{2.13.3}$$

and

$$\nabla\cdot[\mathbf{u}(d_o+\varepsilon\eta)]=\frac{w_e{}^*}{fN}, \tag{2.13.4}$$

where $d_o(x,y)$ is the nondimensional basin depth below sill level. It is assumed that friction and entrainment are weak and this is formalized by replacing r_f/fD and $w_e{}^*/fN$ with εR_f and εw_e, where the nondimensional functions R_f and w_e are regarded as O(1).

The next step is to expand the dependent variables in power series in ε:

$$\mathbf{u}=\mathbf{u}^{(0)}+\varepsilon\mathbf{u}^{(1)}+\cdots \tag{2.13.5}$$

and

$$\eta=\eta^{(0)}+\varepsilon\eta^{(1)}+\cdots . \tag{2.13.6}$$

The lowest order approximations to (2.13.3) and (2.13.4) are just the geostrophic relation $\mathbf{k}\times\mathbf{u}^{(0)}=-\nabla\eta^{(0)}$ and $\nabla\cdot(\mathbf{u}^{(0)}d_o)=\mathbf{u}^{(0)}\cdot\nabla d_o=0$, which show that fluid circulates along contours of constant d_o. At the present order of approximation, these 'geostrophic' contours are equivalent to contours of constant f/d^*.

In order to determine the strength of the circulation on a particular contour, it is necessary to proceed to the next order of approximation:

$$\mathbf{k}\times\mathbf{u}^{(1)}+\nabla\eta^{(1)}=-(\mathbf{u}^{(0)}\cdot\nabla)\mathbf{u}^{(0)}-\frac{R_f\mathbf{u}^{(0)}}{d_o} \tag{2.13.7}$$

and

$$\nabla\cdot[\mathbf{u}^{(1)}d_o]=-\nabla\cdot[\mathbf{u}^{(0)}\eta^{(0)}]+w_e \tag{2.13.8}$$

Now consider a closed geostrophic contour C having unit normal and tangent vectors $\mathbf{n}$ and $\mathbf{l}$ and arclength coordinate s (Figure 2.13.1a). Integration of the tangential component of (2.13.7) about C and use of $\mathbf{u}^{(0)}\cdot\mathbf{n}=0$ leads to

$$d_o\oint_C\mathbf{u}^{(1)}\cdot\mathbf{n}\,ds=-R_f\oint_C\mathbf{u}^{(0)}\cdot\mathbf{l}\,ds. \tag{2.13.9}$$

Integration of (2.13.8) over the area A_C enclosed in C gives

$$d_o\oint_C\mathbf{u}^{(1)}\cdot\mathbf{n}\,ds=\iint_{A_C}w_e\,d\sigma. \tag{2.13.10}$$

Equation (2.13.9) is a form of Kelvin's theorem stating that the damping of circulation $\oint_C \mathbf{u}^{(0)} \cdot \mathbf{l}\,ds$ due to bottom friction is balanced by the input of circulation as the result of advection of planetary vorticity f (nondimensionally unity) across the contour by the normal velocity $\mathbf{u}^{(1)} \cdot \mathbf{n}$. Equation (2.13.10) relates this normal velocity to the influx of volume by the entrainment velocity acting over the area enclosed by the contour. Had our formulation taken the bottom Ekman layer into consideration, the cross contour flow would have been confined to this layer. In the present slab model, the velocity is evenly distributed over the layer depth.

Subtraction of the last two equations gives an expression for the average geostrophic speed about the contour

$$\oint_C \mathbf{u}^{(0)} \cdot \mathbf{l}\,ds = -R_f^{-1} \iint_{A_C} w_e\,d\sigma. \qquad (2.13.11)$$

The reader may wonder how the above steps were discovered. The roots of the procedure for obtaining (2.13.11) can be found in Greenspan (1968). Our asymptotic expansion yields a lowest order approximation (the geostrophic flow) that cannot be completely determined. The standard resolution to this problem of 'geostrophic degeneracy' is that one proceeds to the next order of approximation. In order to calculate the $O(\varepsilon)$ fields, a solvability condition must first be satisfied and it is this condition that determines the geostrophic fields. In quasigeostrophic theory (e.g. Pedlosky, 1987) the compatibility condition is the quasigeostrophic potential vorticity equation obtained from the $O(\varepsilon)$ equalities. Our problem differs from quasigeostrophic dynamics only in the allowance for large depth variations; the compatibility condition (2.13.11) is the same as that obtained by integrating the potential vorticity equation over the area enclosed in a geostrophic contour (see Exercise 2). Our use of the circulation integral is simply a shortcut to this procedure.

Consider an isolated patch of downwelling (Figure 2.13.1a) that is crossed by C. Equation (2.13.11) dictates that the average geostrophic velocity about C is proportional to the volume flux due to downwelling over the portion of the patch lying inside C. The geostrophic interface elevation $\eta^{(0)}$ is constant along C and its value can be determined from the relation $\mathbf{u}^{(0)} \cdot \mathbf{l} = \partial\eta^{(0)}/\partial n$. Let α be a parameter that identifies the contour ($C = C(\alpha)$). Then

$$\mathbf{u}^{(0)} \cdot \mathbf{l} = \frac{d\eta^{(0)}}{d\alpha} \frac{\partial\alpha}{\partial n}, \qquad (2.13.12)$$

and (2.13.11) becomes

$$\frac{d\eta^{(0)}}{d\alpha} \oint_C \frac{\partial\alpha}{\partial n}\,ds = -R_f^{-1} \iint_{A_C} w_e\,d\sigma. \qquad (2.13.13)$$

Since $\partial\alpha/\partial n$ is given by the geometry of the geostrophic contours, its integral can be determined as a function of α, as can the right-hand side of (2.13.13). The result is a first-order differential equation for $\eta^{(0)}$. Once the solution is

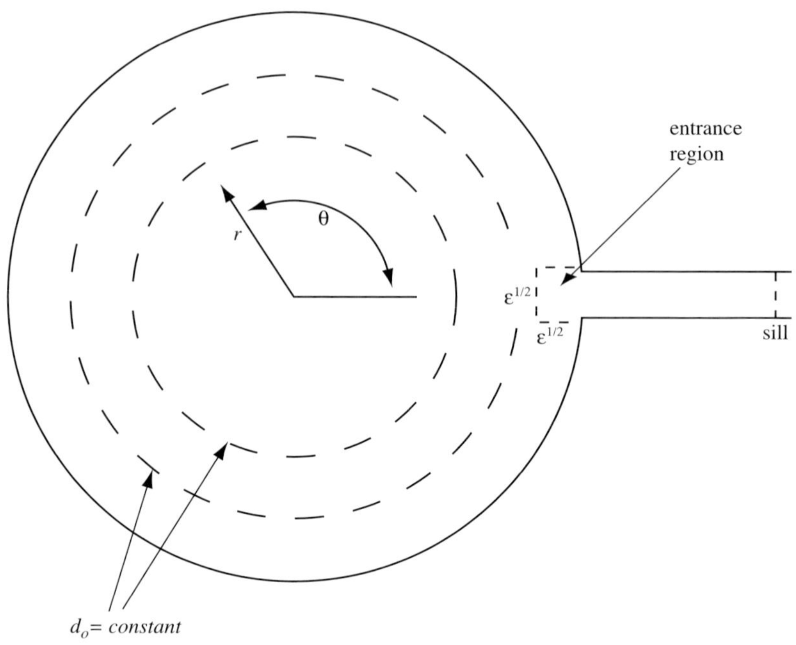

FIGURE 2.13.2. Basin with azimuthal symmetry.

obtained, the value of geostrophic velocity at any point in the basin can be calculated from (2.13.12).

As an example, consider a basin with azimuthal symmetry (Figure 2.13.2). All of the constant d_o contours are circular and the flow is fed by a central, uniform patch of downwelling:

$$w_e = \begin{cases} T/\pi r_o^2 & (r < r_o) \\ 0 & (r \geq r_o) \end{cases}. \tag{2.13.14}$$

The resulting geostrophic flow circulates around the circular geostrophic contours. Let u and v temporarily denote the radial and azimuthal velocity components, so that $u^{(0)} = 0$ and (2.13.11) gives

$$v^{(0)} = \frac{-T}{2\pi r R_f} \begin{cases} r^2/r_o^2 & (r < r_o) \\ 1 & (r \geq r_o) \end{cases}. \tag{2.13.15}$$

The geostrophic relation $v^{(0)} = \partial \eta^{(0)}/\partial r$ can then be integrated to obtain the interface elevation

$$\eta^{(0)} = \eta^{(0)}(a) - \frac{T}{2\pi r R_f} \begin{cases} \ln(r_o/a) + (r^2 - r_o^2)/2r_o^2 & (r < r_o) \\ \ln(r_o/a) & (r \geq r_o) \end{cases}, \tag{2.13.16}$$

where a is the basin radius. Finally, the radial velocity is determined from (2.13.10) as

$$u^{(1)} = \frac{T}{d_o(r)2\pi r} \begin{cases} r^2/r_o^2 & (r < r_o) \\ 1 & (r \geq r_o) \end{cases}.$$ (2.13.17)

In summary, specification of the transport T yields a basin state determined to within a constant, with no regard for boundary conditions or interactions with the strait and sill. The constant is $\eta^{(0)}(a)$, the interface elevation above the sill at the basin edge $r = a$. The leading order radial and azimuthal velocities are completely determined. If the basin is closed except for a single draining channel, then the normal component of the transport velocity $\mathbf{u}^{(1)}d_o$ must be zero at solid edges and must take on some finite distribution (yet to be determined) at the channel entrance. On the other hand, the flow might be fed by dense fluid sliding down the sloping walls of the basin or by inflow from a second strait. Then the correct boundary condition may involve the specification of the normal component of $\mathbf{u}d$ about the perimeter. Evaluation of (2.13.17) at the basin edge $r = a$ leads to $u^{(1)}(a)d_o(a) = T/2\pi a$, which generally satisfies neither of these conditions.

b. Diffusive Boundary Layer

A boundary layer is clearly needed to close the circulation and we therefore amend (2.13.5) and (2.13.6) so as to include boundary layer fields $\tilde{u}$, $\tilde{v}$ and $\tilde{\eta}$ that decay inwards from the edges of the basin:

$$u = \varepsilon u^{(1)}(r) + \varepsilon \tilde{u}(\xi, \theta) + \cdots,$$ (2.13.18)

$$v = v^{(0)}(r) + \varepsilon v^{(1)}(r, \theta) + (\varepsilon/\delta_b)\tilde{v}(\xi, \theta),$$ (2.13.19)

$$\eta = \eta^{(0)}(r) + \varepsilon \eta^{(1)}(r, \theta) + \varepsilon \tilde{\eta}^{(1)}(\xi, \theta) + \cdots$$ (2.13.20)

Here, δ_b represents the boundary layer thickness and $\xi = (a - r)/\delta_b$ is a stretched coordinate that varies by O(1) over this thickness. The size of the boundary layer correction $\varepsilon \tilde{u}(\xi, \theta)$ in (2.13.18) is dictated by the requirement that the O(ε) interior radial velocity must be brought to zero at $\xi = 0$. The correction $(\varepsilon/\delta_b)\tilde{v}(\xi, \theta)$ to the azimuthal velocity in (2.13.19) is determined by the requirement that the boundary layer must drain the O(ε) radial transport and carry it to the channel entrance within an O(δ_b) width. Since $(\varepsilon/\delta_b) \gg \varepsilon$, $\tilde{v}(\xi, \theta)$ enters the problem at a lower order than does dissipation, forcing and nonlinearity, this velocity will be geostrophically balanced. The normal derivative $\partial/\partial r = \delta_b^{-1}\partial/\partial\xi$ of the boundary layer correction for η must therefore be O(ε/δ_b), and the correction must itself be O(ε), as specified in (2.13.20).

The dynamics of the boundary layer can be determined through substitution of (2.13.18–20) into (2.13.3) and (2.13.4) and identification of the largest terms involving boundary layer fields. The thickness δ_b is then chosen in order to achieve a balance between these terms. This procedure is detailed in Pratt (1997), who finds that $\delta_b = \varepsilon^{1/2}$ and that the boundary layer is governed by:

$$\tilde{u}\left[\frac{\partial d_o}{\partial r}\right]_{r=a} + \frac{\tilde{v} d_o(a)}{a}\frac{\partial^2 \tilde{v}}{\partial\theta\partial\xi} + R_f\frac{\partial\tilde{v}}{\partial\xi} = 0. \tag{2.13.21}$$

The dynamical balance is one between various sources and sinks of vorticity. As a fluid column enters the boundary layer from the interior and moves up the sloping bottom towards the wall, negative vorticity is generated as a result of squashing of the column (first term). This effect is balanced by advection of vorticity along the wall (middle term) and dissipation of vorticity by bottom friction (final term). If the wall depth $d_o(a)$ is zero, or at least $\ll 1$, the middle term may be neglected. Under this condition, use of the geostrophic balance

$$\tilde{v} = -\frac{\partial\tilde{\eta}}{\partial\xi} \quad\text{and}\quad \tilde{u} = -\frac{1}{a}\frac{\partial\tilde{\eta}}{\partial\theta}$$

leads to a single equation for the normal boundary layer velocity:

$$\alpha^2\frac{\partial\tilde{u}}{\partial\theta} - \frac{\partial^2\tilde{u}}{\partial\xi^2} = 0, \tag{2.13.22}$$

where $\alpha^2 = -\frac{[\partial d_o/\partial r]_{r=a}}{aR_f}$.

Equation (2.13.22) is a diffusion equation with the time variable replaced by θ. The corresponding boundary layer on a straight coast is sometimes referred to as the "arrested topographic wave" (Csanady, 1978). The solution is also equivalent to the northern or southern boundary layer arising in a homogeneous Stommel circulation on a β-plane (Pedlosky, 1968). In the present setting, the sloping boundary at the basin edge provides a topographic β effect that makes the edge act like a Stommel northern boundary, with increasing θ equivalent to the westward direction.

c.   *Joining the Basin to the Strait*

In order to pose boundary conditions on (2.13.22) it is necessary to consider the conditions in the strait. In general, the fluid from outside has nonuniform potential vorticity and will have a complicated velocity distribution as it enters the strait. If this flow is hydraulically controlled at some point in the interior of the channel, it may be possible to relate the volume flux T to the interface elevation at the entrance by a weir relation. Although no general relation is available under conditions of nonuniform potential vorticity, the situation becomes considerably simplified if the layer thickness d over the sill is relatively small compared to d in the entrance region (Figure 2.13.1b). Fluid columns entering the strait must therefore be severely squashed as they pass over the sill, rendering $\partial v/\partial x \cong -f$, as assumed in the WLK model (Section 2.4). We will also assume that the strait width is much less than the nominal deformation radius (dimensionally

$w_s^* \ll (gD)^{1/2}/f$), so that the variation in η over the width is $\ll 1$. Under these conditions, the assumptions of the WLK theory hold provided that Δz^* in the weir formula [(2.4.10) for attached sill flow or (2.4.15) for separated sill flow] is equated with interface elevation η_e^* at the entrance to the strait, and not in the interior of the basin. If the basin source transport T^* is matched to the transport Q^* given by these formulae, and the various scaling factors are reconciled, one obtains

$$T = \begin{cases} (\frac{2}{3})^{3/2} w_s \left[\eta_e - \dfrac{w_s^{\,2}}{8} \right]^{3/2} & (w_s^2 < 2\eta_e) \\ \eta_e^2/2 & (w_s^2 \geq 2\eta_e) \end{cases}, \qquad (2.13.23)$$

where $w_s = w_s^* f/(gN)^{1/2}$. The strait width w_s^* scales with the deformation radius based on N, while the basin scale is generally assumed to be $\geq$ the deformation scale based on $D(\gg N)$. Since $\varepsilon = N/D$, the strait occupies a vanishingly small portion of the basin circumference as $\varepsilon \to 0$.

Returning to the question of boundary conditions, the value of u must be zero along the basin edge: $u(a, \theta) = \varepsilon(u^{(1)}(a) + \tilde{u}(0, \theta)) = 0$ for values of θ away from the entrance. Suppose that the entrance is centered at $\theta = 0$ and that the basin edge spans $-\pi \leq \theta < \pi$. As $\varepsilon \to 0$, the strait exists only within a vanishingly small interval about $\theta = 0$. The boundary condition there must be chosen to insure that the correct transport is accommodated. Thus

$$\tilde{u}(0, \theta) = -u^{(1)}(a) + \frac{T\delta(\theta)}{ad_o(a)} \qquad (2.13.24)$$

where $u^{(1)}(a) = T/2\pi a d_o(a)$ and where $\delta(\theta)$ denotes the Dirac delta function. Note that the integral of $d_o(a)u(a, \theta)$ across the strait entrance gives the correct transport:

$$\lim_{\theta_o \to 0} \int_{-\theta_o}^{\theta_o} d_o(a)[u^{(1)}(a) + \tilde{u}(0, \theta)]ad\theta = T \int_{-\theta_o}^{\theta_o} \delta(\theta)d\theta = T.$$

A general solution to (2.13.22) for the periodic geometry is

$$\tilde{u} = \mathrm{Re} \left\{ \sum_{i=0}^{n} A_n U_n(\xi)e^{in\theta} \right\}. \qquad (2.13.25)$$

with $U_n(\xi) = e^{-\alpha(1+i)(n/2)^{1/2}\xi}$. Application of the boundary condition (2.13.24) leads to

$$A_o = -u^{(1)}(a) + \frac{T}{2\pi a d_o(a)} = 0 \text{ and } A_n = \frac{T}{\pi a d_o(a)}(n \geq 1). \qquad (2.13.26)$$

One of the weaknesses of the above solution is that it does not resolve the flow near the entrance of the strait. The boundary layer approximation is lost within an $\varepsilon^{1/2} \times \varepsilon^{1/2}$ entrance region where the flow must turn the corner and enter the strait. Also, the depth $d_o(a)$ across the entrance has tacitly been assumed to match the constant depth about the perimeter of the basin, an assumption that leads to problems when $d_o(a)$ vanishes. In such cases, $d_o(a)$ needs to be replaced by the actual strait depth.

To close the problem completely, it only remains to evaluate the integration constant $\eta^{(o)}(a)$ in (2.13.16). Since the interface elevation changes by only $O(\varepsilon)$ across the boundary layer, and by extension, the $\varepsilon^{1/2} \times \varepsilon^{1/2}$ entrance region, we may approximate the elevation η_e just inside the entrance by $\eta^{(o)}(a)$. Equation (2.13.23) can then be inverted to obtain

$$
\eta^{(o)}(a) = \begin{cases} \dfrac{3}{2}\left(\dfrac{T}{w_s}\right)^{2/3} + \dfrac{w_s^2}{8} & (w_s^2 < 2\eta_e) \\[2ex] (2T)^{1/2} & (w_s^2 \geq 2\eta_e) \end{cases}
\tag{2.13.27}
$$

The complete solution for the basin interior is now given by (2.13.15–17) and (2.13.27). The horizontal circulation can be summarized by considering an element of fluid introduced into the middle of the basin as a result of the downwelling. The element circulates anticyclonically and slowly spirals outward until it reaches the basin edge, where it enters the boundary layer. The contribution to the azimuthal velocity from the boundary layer is weak in comparison to the $O(0)$ azimuthal velocity, and the element therefore continues to circulate anticyclonically in the boundary layer. The main impact of the boundary layer will be to allow the element to pass into the strait.

Since the total transport T out of the basin is specified, the effect of the sill is contained entirely in $\eta^{(0)}(a)$. If T changes, $\eta^{(0)}(a)$ is altered according to (2.13.27) and the interface elevation at the edge of the basin is raised or lowered. The overall circulation intensifies or diminishes uniformly and the interface in the basin becomes more or less domed [see (2.13.15) and (2.13.17)]. The circulation pattern is not altered, however. In a more realistic model, the downwelling velocity w_e might itself be altered by changes in the interface elevation and this would allow the circulation pattern to change.

If the fluid is introduced into the basin laterally, with $w_e = 0$ over the interior, then there is no interior circulation and the source water is transported entirely in boundary layers. Suppose that the basin is fed by an inflow at the opposite edge of the basin ($\theta = \pi$) from the draining strait. Then the boundary condition (2.13.24) is replaced by

$$
\tilde{u}(0, \theta) = \frac{T}{a d_o(a)}(\delta(\theta) - \delta(\theta - \pi)).
$$

and the coefficients in (2.13.25) become

$$
A_0 = 0 \quad \text{and} \quad A_n = \frac{2T}{a\pi d_o(a)}.
\tag{2.13.28}
$$

The inflow splits into two boundary layers that circle the basin and join at the draining strait (Figure 2.13.3). Note the overshoot of the cyclonic boundary

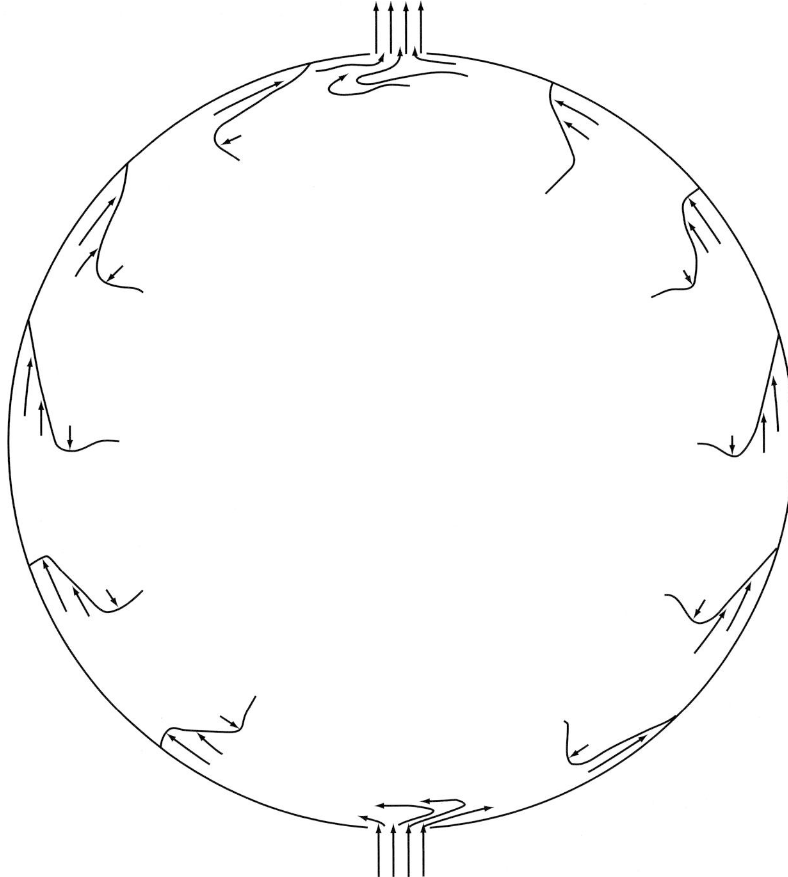

FIGURE 2.13.3. Example of a solution from the linear model in which fluid is fed into the parabolic basin through a strait in the side wall. The geometries of the entrance and exit are identical. The parameters are given by $T = 1$, $R_f = 0.2$, $a = 4.0$, and $w_s = 0.5$. (From Pratt, 1997).

layer, which causes the bulk of the flow to enter the strait along the 'left' wall of the basin.

A striking difference between the flows driven by downwelling and those driven by injection through the side walls lies in the way the fluid approaches the draining strait. In the first case the flow about the outer rim of the basin is anticyclonic and all the fluid approaches along the 'left' wall (facing into the draining strait). In the second case, the rim flow is split into two boundary layers carrying equal transports and the approach is from both walls (discounting the overshooting effect that diverts more fluid to the left wall immediately upstream of the exit). Some insight into the dynamical processes responsible for these differences can be gained by developing a circulation theorem for the rim flow.

To this end, consider the shallow water momentum equations in the vector form (2.1.15). If the tangential component of this equation is integrated about a circuit C_R that follows the basin perimeter and cuts across the entrance to any straits, one obtains

$$\frac{\partial}{\partial t^*} \oint_{C_R} \mathbf{u}^* \cdot \mathbf{l}\, ds = -\oint_{C_R} (\zeta^* + f)\mathbf{u}^* \cdot \mathbf{n}\, ds + \oint_{C_R} \mathbf{F}^* \cdot \mathbf{l}\, ds, \qquad (2.13.29)$$

where the integration direction is counterclockwise. The rate of change of circulation $\oint_{C_R} \mathbf{u}^* \cdot \mathbf{l}\, ds$, essentially the net swirl velocity about rim, is therefore equal to the flux of absolute vorticity $\zeta^* + f$ across the rim (due to inflows and outflows) plus the tangential component of forcing and dissipation along the rim.

In the cases considered above $\partial/\partial t^* = 0$, $\mathbf{F}^* = -r_f \mathbf{u}^*/d_o^*$, and $\zeta^* \ll f$. With a downwelling-driven flow drained by a single strait, (2.13.29) reduces to

$$r_f \oint_{C_R} \frac{\mathbf{u}^* \cdot \mathbf{l}}{d_o^*(a)}\, ds = -f \oint_{C_R} \mathbf{u}^* \cdot \mathbf{n}\, ds = -fT^*/d_o^*(a) < 0,$$

and therefore the net swirl velocity about the rim must be negative, as observed. For the case in which fluid is introduced through a source strait, we have

$$r_f \oint_{C_R} \frac{\mathbf{u}^* \cdot \mathbf{l}}{d_o^*(a)}\, ds = -fT^*/d_o^*(a) + fT^*/d_o^*(a) = 0.$$

The net swirl velocity in this case is zero, a property consistent with the presence of boundary layers on both left and right walls.

d.   Numerical Simulations and the Potential Vorticity of the Outflow

Numerical experiments based on the full shallow water equations (Helfrich and Pratt, 2003) have reproduced the overall circulation patterns anticipated by the linear theory. In the three simulations shown in Figure 2.13.4, fluid is introduced into a bowl-shaped basin at the same volume rate but in different geographic locations. When fluid is introduced through the back wall (Frame a) the inflow splits into two boundary layers that make their way to the draining strait. Introduction of fluid through uniform downwelling over the basin interior (Frame b) sets up a cyclonic circulation. The draining flow is now fed entirely along the 'left' wall. When the downwelling is concentrated in a localized region near the back wall (Frame c) the circulation is still predominantly anticyclonic but localized. Fluid is again fed into the strait along the left wall. These general patterns follow expectations based on the basin circulation integrals. An unanticipated finding is that the flow in the draining strait itself remains remarkably consistent as the basin circulation patterns vary. A comparison between the three cases (Figure 2.13.5) suggests that the flows at the sill, and in the channel immediately upstream, are nearly indistinguishable.

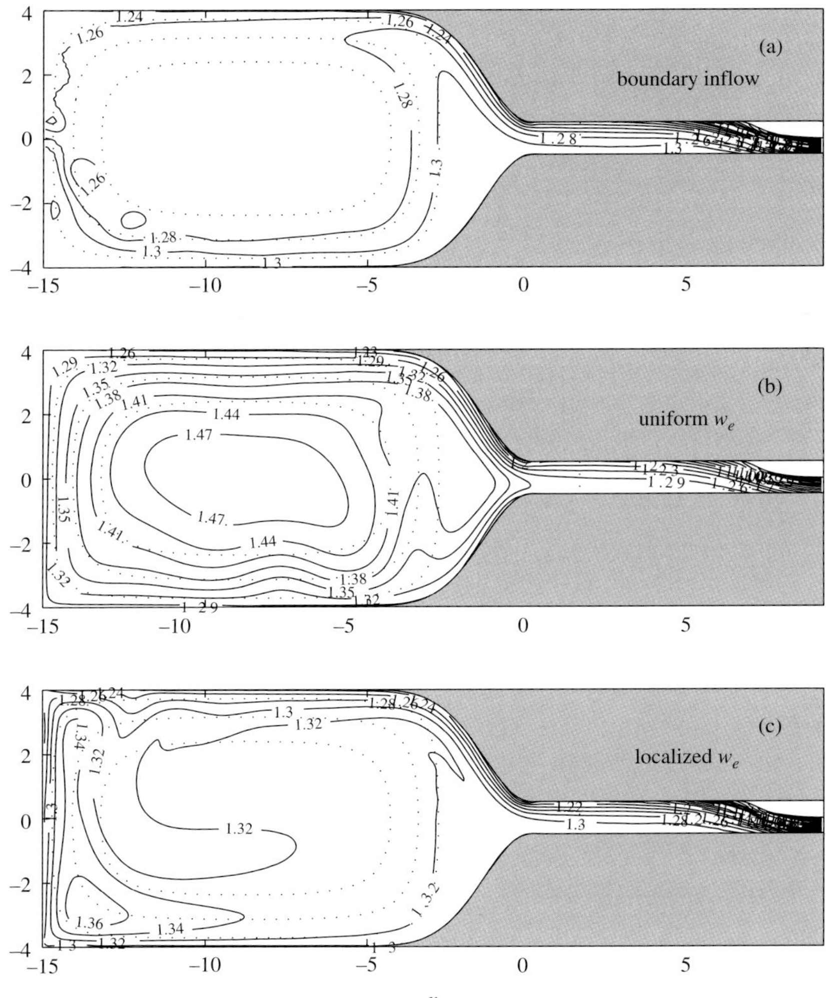

FIGURE 2.13.4. Three numerical simulations of a basin flow that is drained through a strait. The fluid is introduced into the basin (a) through the back wall (at $x = -15$), (b) through a downwards entrainment velocity w_e distributed uniformly over the basin, and (c) through a w_e concentrated near the back wall of the basin. The contours are ones of interface elevation. (From Helfrich and Pratt, 2003).

The linear model uses a weir formula based on 'zero potential vorticity' theory, but this approximation is not enforced in the numerical simulations. The potential vorticity (q) in the strait is self-determined and its value and distribution provide a basis for comparison with the cornerstone hydraulic models, most of which are based on uniform q. The observed potential vorticity distribution is nonuniform (Figure 2.13.6c), but the flow in the strait turns out to be qualitatively the same

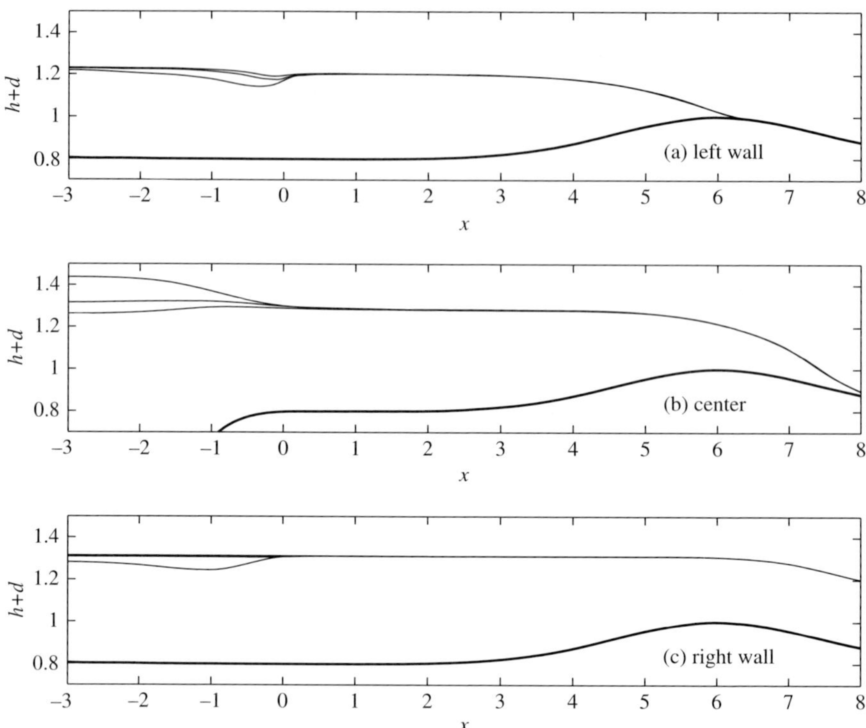

FIGURE 2.13.5. Side views of interface elevations along (a) the left wall, (b) along the centerline, and (c) along the right wall of the basin and channel for the three flows depicted in Figure 2.13.4. (Figure 6 of Helfrich and Pratt, 2003).

as that given by the Gill (1977) model for the same transport Q and with Gill's constant q replaced by the mean value $\overline{q}$ measured across the entrance to the strait. A comparison between two realizations (Figure 2.13.6, frames a and b vs. frames d and e) reveals only minor differences.

As suggested above, the flow in the strait, and the value of $\overline{q}$ in particular, are insensitive to the distribution of sources in the upstream basin. In fact $\overline{q}$ also tends to be quite insensitive to the value of the friction coefficient. As it turns out, the main factors controlling the potential vorticity are the sill width and the ratio of the sill elevation to the entrance elevation of the channel.[11] The potential vorticity selection can be therefore viewed as an aspect of the upstream influence due to the hydraulic control at the sill. The selection of $\overline{q}$ is clarified somewhat by consideration of the possible Gill solutions for a given strait geometry and transport Q. With Q and the sill geometry fixed, the Gill

---

[11] In the Helfrich and Pratt (2003) experiment, the entrance width is different than the sill width and their ratio provides a third geometric parameter that influences the observed value of $\overline{q}$.

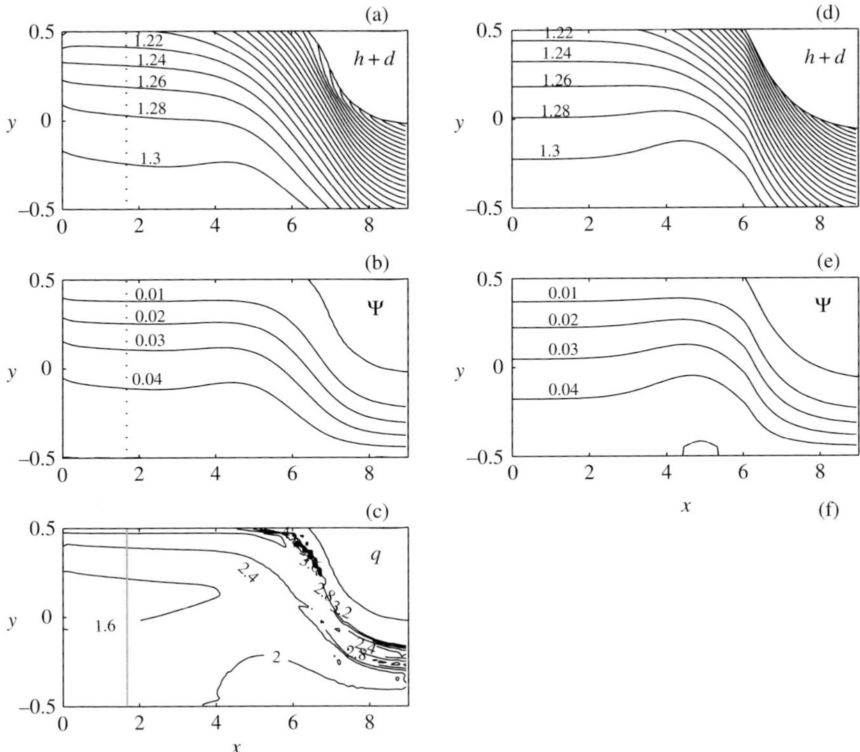

FIGURE 2.13.6. Comparison of the strait flows in plan view from the numerical experiments (a-c) and the Gill (1997) theory (d-e) based on the mean potential vorticity $\overline{q} = 1.78$ measured at the entrance (dotted line). The nondimensional parameters are given in both cases by $Q = .05$, $w_{\rm s} = 1$, and $R_{\rm f} = .01$. Also, the ratio of the sill elevation to the entrance elevation (both measured above the deepest point of the basin) is 0.8.

model still permits a range of steady, critically controlled solutions, each with its own q. The velocity and depth profile at the channel entrance is different in each case. An interesting quantity to focus on is the elevation $\Delta z_{\rm R}$ of the interface at the right wall. Helfrich and Pratt (2003) find that the observed $\overline{q}$ corresponds to a Gill solution for which $\Delta z_{\rm R}$ is maximized, or very nearly so, over the range of permissible solutions. Since the maximization occurs for fixed Q $[= (\Delta z_R^2 - \Delta z_L^2)/2]$, it follows that the left wall elevation $\Delta z_{\rm L}$ is also maximal.

In the linear model, the mean basin interface elevation is determined completely by the flux Q. If the latter is held fixed and the sill height is raised, the basin interface elevation is uniformly raised at the same rate. The same behavior is found in the numerical model, where a change in sill height simply causes the mean basin interface level to change an equal amount. Since $\Delta z_{\rm R}$ and $\Delta z_{\rm L}$ are maximal for all the possible Gill solutions with a particular Q, there is a strong suggestion, if not outright verification, that the basin has the maximum mean elevation over all such solutions. Of all the possible basin states corresponding to the various Gill solutions, the one realized apparently has maximal potential energy. The basin flow is highly subcritical, with kinetic energy dominated by potential energy, and a finding of maximal potential energy is tantamount to one of maximum energy.

e.   Upstream Monitoring

We have seen that changes in the location of the source have a profound effect on the circulation and the shape of the interface in the basin, but not in the strait. Transport formulae that are based on a single measurement of the upstream interface elevation, are therefore risky to use. In fact, Gill's (1977) transport relation (Section 2.5d) fails in the present experiments when the parameter ψ_i is measured in the interior of the upstream basin. Opportunities for monitoring the flow from the entrance of the channel are more promising. The numerical solutions, which all maximize the right-wall interface elevation Δz_R, tend to have relatively sluggish flow in that region (see Figure 2.13.6d or e). The Bernoulli function at this location is therefore nearly proportional to Δz_R. If it is also the case that the flow is separated at the sill, then by the arguments presented in Section 2.6, the dimensional transport is given by (2.6.7b), with the properly interpreted Δz_R^*.

Even when the sill flow is not separated, the robust nature of the strait flow means that monitoring is best done using quantities measured at the strait entrance rather than in the basin proper.

f.   'Westward' Intensification of the Approach Flow

The presence of sluggish flow near the entrance right wall (and rapid flow at the left wall) has also been observed in laboratory experiments by Whitehead and Salzig (2001) and is suggested by the linear theory for the basin flow (Figure 2.13.3). In the experiment (Figure 2.13.7) fluid is pumped into a deep, arc-shaped basin and it escapes through a broad, shallow channel. As a fluid column enters the channel it becomes squashed and acquires excess anticyclonic vorticity. There are two scenarios describing what happens next. In the first, which is consistent with traditional, inviscid hydraulic theory, the fluid simply continues into the channel and develops a strong shear. In the Gill (1977) model, for example, the shear would be confined to a boundary layer. The anticyclonic would favor the left wall boundary layer, and thus the flow would enter along that wall. In the second scenario, which is consistent with ideas about slow, nearly-geostrophic flow, streamlines tend to follow the isobaths, crossing them only to an extent allowed by friction. This is exactly what happens in the above linear model, where a strong swirling flow along the closed isobaths is accompanied by a weak flow towards shallower depths. The excess vorticity generated by the vortex squashing is dissipated by friction. In the entrance region, the isobaths

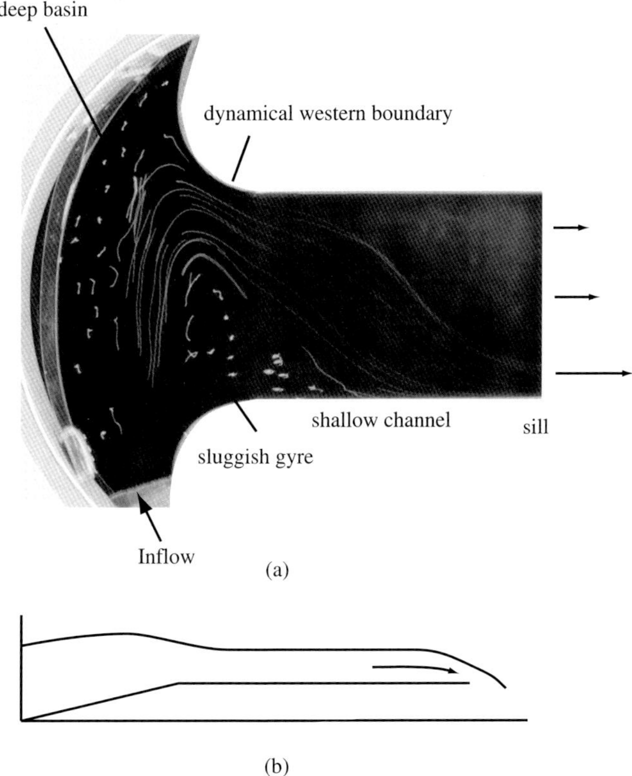

FIGURE 2.13.7. (a) Plan view of laboratory flow established by injecting fluid into a deep basin (left) and allowing it to drain through a shallow strait with a flat bottom (right). The width of the channel is roughly one deformation radius based on the elevation difference between the maximum surface height in the deep basin and the channel bottom. The streak lines are due to the motion of white floats. (From Whitehead and Salzig, 2001). (b) Side view showing the sloping bottom in the deep basin.

are not closed but, instead, intersect the left and right walls of the channel. These isobaths steer fluid towards the walls, where frictional boundary layers may exist. Once a fluid column has reached a frictional boundary layer, it is able to more easily cross isobaths and continue further into the channel. The question now is whether the left- or right-wall boundary layer is preferred.

In the earlier linear model, where the isobaths parallel the basin edge, the frictional boundary layer (or 'arrested topographic wave') is different than the frictional layer that exists near the entrance, where the isobaths intersect the sidewalls. It was first established by Stommel (1948) that such a layer can occur only where the ambient potential vorticity f/d_o^* increases in the direction with the wall on the left. For a broad ocean basin with constant depth d_o^*, and a Coriolis parameter f that increases to the north, the frictional layer must occur on the western boundary. In the present channel, where f is constant but d_o^*

decreases into the strait, the boundary layer must occur on the left wall. It is expected, then, that the flow entering the strait should be concentrated along this 'dynamical western boundary', a feature borne out by the laboratory and numerical experiments. This 'westward' intensification can also be motivated using a circulation integral, as explored in Exercise 5. The effect may account for the observation that the Denmark Strait overflow hugs the Iceland coastline, its dynamical western boundary, upstream of the sill (Jonsson and Valdimarsson, 2004). Support for this idea can also be seen in the inflow to the Barents Sea (Slagstad and McClimans, 2005).

Exercises

(1) The effects of entrainment have been included in the continuity equation (2.13.2) but not in the momentum equation (2.13.1). Using results obtained in Section 1.10, show that the conditions under which this assumption is valid are $w_e^*/fr_f \ll 1$ and $(g'd/v^2)(w_e^*/r_f) \ll 1$. (The second condition is clearly more stringent than the first for the conditions in the basin.)

(2) Derive the potential vorticity equation for the geostrophic flow component of the basin flow. Show that integration of this equation over the area A_C enclosed by a closed geostrophic contour yields the relation (2.13.11).

(3) Suppose that some of the constant-depth contours intersect the vertical side walls of the basin and are therefore not closed. How are the tangential and normal flow to the contour determined?

(4) Derive equation (2.13.29)

(5) (*Westward intensification of basin flow*). Consider the (northern hemisphere) case in which fluid is injected into a basin through one strait and drained through another, with no interior downwelling. Suppose that the source strait enters the north edge of the basin and the draining strait leaves the south edge. The basin is large enough for the beta effect to be important and therefore the value of f at the mouth of the source strait is larger than that at the mouth of the draining strait. Argue the flow in the basin will be concentrated in a western boundary layer. Show that the same effect occurs in an f-plane basin if the depth at the mouth of the source strait is greater than the depth at the entrance of the draining strait. Show that the same results hold for a southern hemisphere basin.

2.14. Comparisons Between Observed and Predicted Transports

We have described a number of deep straits and sills (Figure I.4) that act as potential sites of hydraulic control and, therefore, as choke points in the lower limbs of the ocean conveyor. At the time of this writing, comparisons between the observed features of these overflows and inviscid hydraulic models were based largely on volume transport (or flux). The lack of measurements with sufficient coverage to resolve boundary currents, and other features of the flow upstream of their sills, has precluded detailed comparisons with models such as Gill (1977). We do not, for example, have a good understanding of how well the reservoir states postulated by Whitehead, *et al.* (1974, hereafter WLK), Gill (1997), Killworth (1992) and Pratt (1997) agree with reality. Nevertheless, comparisons of observed and predicted transport are important, not only as a test of the models but also as a step towards the development of strategies for monitoring the ocean thermohaline circulation.

The most common transport (or 'weir') formula in current use is that due to WLK. As shown in Section 2.4, the volume flux across the sill is given by

$$Q_0^* = \frac{g'(\Delta z^*)^2}{2f} \qquad \text{if } w_c^* \geq \frac{\sqrt{2g'\Delta z^*}}{f}, \qquad (2.14.1a)$$

or

$$Q_0^* = \left(\frac{2}{3}\right)^{3/2} w_c^* \sqrt{g'} \left[\Delta z^* - \frac{w_c^{*2} f^2}{8g'}\right]^{3/2} \qquad \text{otherwise.} \qquad (2.14.1b)$$

The symbol Q_0^* for volume transport is used here as a reminder that the 'zero potential vorticity' approximation is in effect. Also, the reduced form of the gravitational acceleration $g' = g\Delta\rho/\rho$, $\Delta\rho = \rho - \rho_1$, is explicitly used to acknowledge application to an overflowing layer of density ρ that underlies an inactive layer of density ρ_1. The interface separating the two layers is usually chosen to correspond to a particular isopycnal. The transport then depends on the elevation difference Δz^* between the sill and the upstream interface. The geometry of the sill section is assumed to be rectangular and the upstream interface is assumed to be horizontal. In reality, the choices of the bounding isopycnal and its upstream level, the layer densities, and the elevation and width w_c^* at the sill section, require a number of *ad hoc* assumptions. Once Δz^* is estimated, a choice is made between the first and second formulae corresponding to separated and attached sill flow. We later describe a systematic method for estimating the parameters.

Several features make the WLK model a good starting point for comparison. First, it is based on the simplest of models and therefore requires the fewest parameters. More sophisticated models such as Gill's (1977) require additional upstream information that may or may not be available. Also, as explained in Section 2.6, the formula (2.14.1) is valid for a wide class of flows with arbitrary potential vorticity, provided that the flow is separated at the sill and that Δz^* is measured along the right wall of the upstream basin. The same formula also gives a bound on inviscid, rotating channel flow across a sill of arbitrary topography, provided that $g'\Delta z^*$ is interpreted as the maximum value of the Bernoulli function over streamlines in the upstream basin that cross the sill.

The WLK model formally depends on an assumption that is difficult to justify. In particular, fluid columns must able to make their way from a hypothetical deep and quiescent basin up over a shallow sill. It is well-known that slow, nearly geostrophic motion on an f-plane resists such changes: the fluid columns tend to move along, not across, isobaths. In the ocean, it is more likely the case that fluid passing over a deep sill originates from an upstream layer that lies at an intermediate depth and is suspended above the bottom. Fluid below this layer is blocked from passing across the sill. The layer thickness over the sill is not much greater than the upstream thickness and the zero potential vorticity approximation no longer holds. The Gill (1977) or Killworth (1992) models, in which the potential vorticity is finite, are now more appropriate.[12]

Several attempts have been made to use uniform potential vorticity models (such as that of Gill 1977) and to take into account sill geometries. Prediction of the flux requires that a represenative value of the potential vorticity of the overflow must be measured and this is not always possible. In the examples we shall cite, the potential vorticity is often unknown. However, it is still possible to estimate the importance of the effect of finite potential vorticity on the transport. To this end, consider the case in which the nondimensional potential vorticity q is equal to unity. For the theory developed in Section 2.6 this means

$$q = \frac{g'^{-1}B_R{}^* - h_c{}^*}{D_\infty},$$

where $B_R{}^*$ is the value of the Bernoulli function on the right wall and $h_c{}^*$ is the elevation of the sill above the flat bottom. Although a range of flux exists for any fixed q, one may construct a benchmark flux $Q_1{}^*$ for the case $q = 1$. Suppose that the sill height above the upstream bottom is zero ($h_c{}^* = 0$) and that all the current approaches the deep strait along the left wall ($B_R{}^* = g'D_\infty = g'\Delta Z_R{}^*$, say). Then the transport is given by (2.6.7) as

$$Q_1{}^* = \frac{g'(\Delta z_R{}^*)^2}{f}Q(1, w_c{}^*), \tag{2.14.2a}$$

and the function $Q(1, w_c{}^*)$ can be approximated as

$$Q(1, w_c) = 0.5 - 0.6331e^{-1.45w_c} + 0.1331e^{-2.9w_c} \tag{2.14.2b}$$

to within an error less than 1.3%. The dimensionless sill width is defined as $w_c = fw_c{}^*/(g'\Delta Z_R{}^*)^{1/2}$. $Q_1{}^*$ will be compared with $Q_0{}^*$ in order to gain a simple, if imperfect, measure of the sensitivity of the flux to the potential vorticity.

There are many reasons why formulae like (2.14.1 or 2.14.2) could fail. Among the most worrisome liabilities are the neglect of friction and time-dependence, and the restriction to rectangular geometry. A few recent studies have been able to account for more realistic sill geometries and these will be mentioned below. The effects of friction are much more difficult to deal with. The presence of

---

[12] A slight adjustment would have to be made in how the upstream conditions are viewed. In the Gill (1977) model, which assumes flow over a horizontal bottom, the value of g' would have to be changed to that relevant for a suspended layer.

a bottom Ekman layer and possible frictional layers along a sheared interface lead to energy dissipation and, as shown in Section 2.12, secondary circulations transverse to the channel axis. Johnson and Sanford (1992) report on observations suggesting such features in the Faroe-Bank Channel. The secondary circulations are demonstrated by Johnson and Ohlsen (1994) in a laboratory experiment. Development of a hydraulic transport relation that takes account of these effects has proved elusive.

The issue of time-dependence is also problematic. The hope of steady models is that actual time variations are slow enough to allow the model to be valid at any given instant. There are examples where this is clearly not the case. In fact, awareness of rapid fluctuations dates back to an early current meter deployment at the Denmark Strait (Worthington, 1969). Despite massive loss of instrumentation, one current meter recorded large bursts of overflow water, with velocities up to 1.4 m/s and with time scales as short as one day. Other overflows are strongly episodic and can switch on and off or meander back and forth across a moored instrument. MacCready et al. (1999) report that the deep flow through the Anagada-Jungern passage can behave this way, with about 10 episodes per year. Variation over longer time scales is also common. The flow approaching the Ceara Abyssal Plain was found to have a large annual signal with a much weaker interannual component (Hall et al., 1997 and Limeburner et al., 2005). We have already cited the apparent 50-year trend of decreasing transport in the Faroe-Bank Channel (Figure 2.11.12). In general, one can expect to see time dependence on a variety of scales due to internal waves, tides, mesoscale eddies, interactions with nearby currents, atmospheric forcing, and seasonal and longer scale changes in the surroundings.

For a model to be considered quasi-static, the time scale of variation must be much longer than the time required for a disturbance to propagate the length of the strait. This time is roughly the strait length divided by $(g'D)^{1/2}$. Two-day oscillations in the Denmark Strait do not meet this criterion; 1–2 month variability in Anagada-Jungfern passage probably does.

In cases where the dominant time variability is rapid, the standard practice is to compare the hydraulic prediction with some time-mean transport. The presence of a variety of time scales begs the question of how to measure the mean. We have identified ten locations (Table 2.14.1) having current meter data of one month or more, which is long enough to average out tides and storms. The overflows of the Faroe Bank Channel and the Denmark Strait, which contribute to the formation of North Atlantic Deep Water, are the most thoroughly observed. Five other lie in the path of the northward moving Antarctic Bottom Water in the Atlantic. Starting from the south, they are the Vema Channel, Ceara Abyssal Plain, Romanche Fracture Zone, Vema Gap, and Discovery Gap. The remaining straits include the Anagada-Jungfern Passage, composed of the Grappler Channel and Anagada Passage, which provide the deepest inflows into the Caribbean Basin. Also included is the Samoa Passage in the tropical Pacific, where Antarctic Bottom Water moves northward into the Pacific.

Flux estimates using (2.14.1) or (2.14.2) require the values Δz^*, g' and w_c^*, and Whitehead (1989b) has suggested a systematic method of computing

TABLE 2.14.1. Observed volume transports vs. predictions based on a rectangular sill section. All observed transports are based on current meter records of a month or more.

#	Name	g' 10^{-3} ms^{-2}	Δz^* m	f 10^{-4} s^{-1}	w_c^* km	w_c	Q_0^* Sv.	Q_1^* Sv.	$Q_{observed}^*$ Sv.
1	Jungfern Passage	0.45^a	165	0.45	10	1.65	0.136	0.12	0.085^b
2	Ceara Abyssal Plain	0.5^a	430	-0.1	700	15.0	4.62	4.53	2.1^c
3	Denmark Strait	3.0^a	580	1.3	350	34.5	3.88	3.8	2.9^d
4	Discovery Gap	0.1^a	600	0.87	80	28.4	0.21	0.2	0.21^e
5	Faroe Bank Channel	5.0^a	400	1.3	20	1.84	2.82	2.62	1.9^f
6	Grappler Passage	0.22^b	160	0.45	–	1.2	0.06		0.026
7	Romanche Fracture Zone	0.73^a	380	-0.02	20	0.08	2.15	2.09	0.66^g
8	Samoa passage	0.3^a	1050	-0.23	240	9.8	7.19	7.05	6.0^h
9	Vema Channel	1.0^a	1100	-0.7	446	29.8	8.64	8.62	4.0^i
10	Vema Gap	0.5^a	1000	0.28	9	0.36	3.35	3.03	2.1^j

Note: citations pertain to all information from that citation to the next. Also, negative values of f imply that $|f|$ should be used in the transport formula.
[a] Whitehead 2005
[b] MacCready et al. (1999)
[c] Hall et al. (1997)
[d] Dickson, Gmitrowicz and Watson (1990)
[e] Saunders (1987)
[f] Saunders (1990)
[g] Mercier and Speer (1998)
[h] Rudnick (1997)
[i] Hogg, Siedler, and Zenk (1999)
[j] McCartney et al. (1991)

these quantities. The method makes use of density profiles taken upstream and downstream of the strait in question. As an example, we use two profiles measured in the upstream and downstream basins of the Samoa Passage (Figure 2.14.1). Densities are similar at above 3950 m but differ below this depth, ostensibly as a result of the mixing and redistribution of density due to the overflow. Below the 'bifurcation' depth, the split extends to below the Samoa passage sill at approximately 5000 m. The tendency for the upstream and downstream profiles to split is observed for the ten straits under consideration, and is represented in Figure 2.14.2 by a generic profile pair. The 'interface' bounding the overflowing layer is selected to coincide with an isopycnal that lies at the bifurcation depth in the upstream basin. The bottom cross-section at the sill is plotted next to the profiles. The deepest point is selected to be the sill depth and Δz^* is chosen as the difference between the bifurcation depth and the sill depth. The value of $\Delta\rho$ is chosen as the difference in density between the two profiles at sill depth. For the Samoan Passage, we estimate values $\Delta z^* = 1050\,\mathrm{m}$ and $\Delta\rho = 3 \times 10^{-5}\,\mathrm{gm/cm}^3$. The channel width w_c^* is defined as the width of the cross-section at the bifurcation depth and is 240 km for the Samoan Passage. Finally, the local Coriolis parameter is given by $f = 2\Omega \sin \vartheta$ where ϑ is the latitude of the sill. The volume flux values based on (2.14.1) and (2.14.2), and

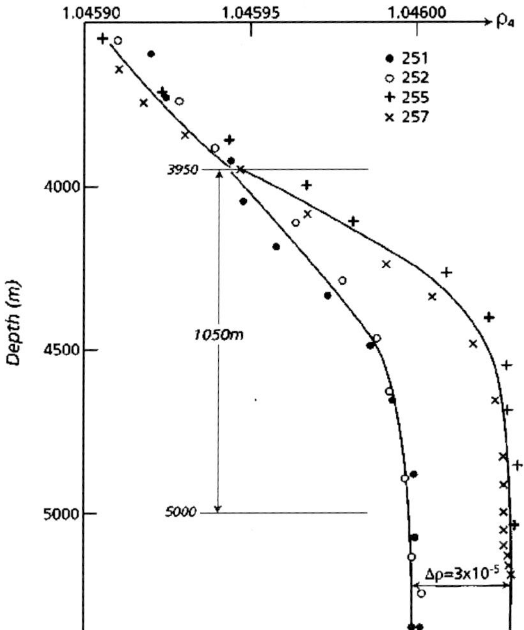

FIGURE 2.14.1. Density (gm/cm^3) corrected to 4000 m depth versus ocean depth at both ends of the Samoa Passage (From Whitehead, 1998).

using the 'bifurcation' method, for all ten examples are listed in Table 2.14.1. All except the Grappler passage correspond to values listed in Table 1 of Whitehead (2005). Original bifurcation figures are in Whitehead (1989) and Whitehead (1998) except for the Anagada and Grappler Passages.

Attempts to improve the accuracy of (2.14.1) have concentrated on more realistic topography and on the effect of nonzero potential vorticity. The outcomes (Table 2.14.2) suggest that the latter is not as important as the former. Realistic topography often, but not always, leads to a decrease in the predicted transport and this decrease can be substantial. This tendency is suggested in

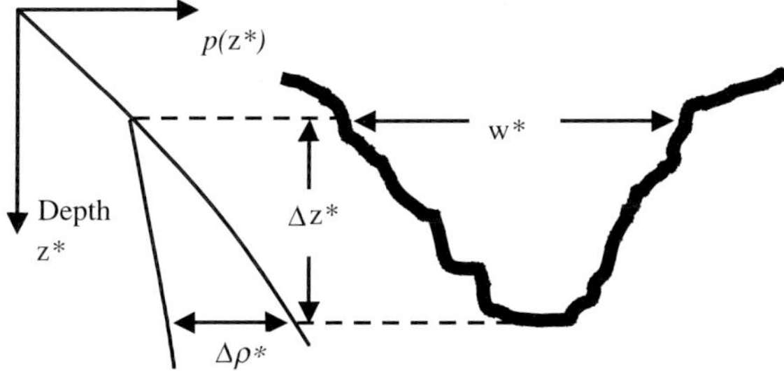

FIGURE 2.14.2. Ocean data needed to produce values of density, depth and width of a passage.

TABLE 2.14.2. Observed volume transport *vs* predictions based on a nonrectangular sill section. Observations of less than one month are included.

#& Name	Shape	$g^{*\prime}$ 10^{-3} ms^{-2}	Δz^* m	w_c^* km	w_c	Q^* Zero Sv	Q_1^* other Sv	$Q^*_{observed}$ Sv
1. Jungfern	Flat	0.28[a]	265	6	1.65	0.21		0.085
Passage	Flat	0.40[b]	100	5		0.04[d]		0.056
	Flat	0.40[b]	150	6		0.09		0.099
	Real	0.45[b]	165			0.079		0.072
	Real	0.40[b]	150			0.027		0.056
	Real	0.40				0.055		0.099
	RealX[c]	0.45[b]				0.085		0.099
2. Ceara	Flat	0.5[e]	430	700	15.00	4.6		1.1–2.1
Abyssal	Parabola					1.4	2.1	
Plain	V					1.7		
3. Denmark	Flat	3.0[e]	580	20		3.8		2.5
Strait	Parabola					0.5	0.6	
	V					0.7	0.96	
	Flat	4.78[f]	520	20	34.50	4.97		3.7
	Parabola					4.49		
	V					4.33		
	Real					2.32	2.45	
	RealX					4.47	4.35	
	Flat	3.82	370	20		2.01		2.1
	Parabola					1.85		
	V					1.73		
	Real					1.22	1.31	
	RealX					1.83	1.79	
5. Faroe Bank	Flat	5.0[e]	400	20	1.84	3.6		1.9
Channel	Parabola					0.5	0.53	
	V					0.7	0.86	
	Parabola	4.3[h]	350			1.5		1.4
	Parabola		450			2.5		1.9
9. Vema	Flat	1.0[e]	1540	446	29.80	16.4		4.1
Channel	Parabola					2.9	4.5	
	V					3.9	8.8	

Note: In the following, 'flat' refers to a rectangle, and 'real' implies a fit to the actual topography. Also note that the observed volume flux $Q^*_{observed}$ refers to the particular value used for comparison in the study cited. This value depends on the definition of the overflowing fluid and on the time and manner of measurement.

Citations pertain to all information from that citation to the next.

[a] MacCready *et al.* (1999)

[b] Borenäs and Nikolopoulos (2000)

[c] RealX- the abbreviation for real bottom topography with reverse flow excised

[d] Stalcup, Metcalf and Johnson (1975)

[e] Killworth (1992)

[f] Nikolopoulos, Borenas, Hietala, and Lundberg (2003)

[g] Girton *et al.* (2001)

[h] Borenäs and Lundberg (1988)

the work of Borenäs and Lundberg (1986, 1988) for flow in a channel of parabolic section. As discussed in Section 2.8, the predicted transport depends on the parameter $r = f^2/g'\alpha$, where α is the bottom curvature. Sections with large curvature act more like rectangles and departures from this shape therefore become more pronounced as r increases. For zero potential vorticity, it can be shown (Equation 2.8.10) that a reduction in transport relative to the rectangular case occurs in proportion to $r^{-1/2}$. The reduction is due largely to the presence of counterflow at the right side of the sill section, which occurs for $r > 2/3$. Among the notable case studies is Killworth (1992), who fit rectangular, V-shape, and parabolic cross sections to sills in four passages. For the Denmark Strait and Faroe-Bank Channel, a reduction by factor 4 or 5 in predicted flux is found for the parabola and V-shape relative to the rectangle. Finite potential vorticity effects are found to be much weaker. Borenäs and Nikolopoulos (2000) investigate the Jungfern Passage using a model that takes into consideration various shapes, including a close fit to the real sill topography. Reductions in flux relative to the rectangular case are by a factor of two or three. A small amount of reverse flow is also predicted at the sill for the real topography. The predicted transport is slightly raised when this counterflow is excised. Nikolopoulos *et al.* (2003) apply a similar technique to the Denmark Strait. For the actual sill geometry, a reduction in transport by a factor of two relative to a rectangle is found. However, the counterflow is much stronger and a flux value close to the rectangular case is restored when thecounterflow is excised. The effects of finite potential vorticity are again found to be moderate. These are among the studies summarized on Table 2.14.2. When comparing flux predictions, note that different authors may use different values of Δz^*, g', and w_c^* for the same location.

The flux values in both tables span nearly three orders of magnitude and require a log-log plot to show the entire range (Figure 2.14.3). The volume flux values from Table 2.14.1 for zero potential vorticity and rectangular sill geometry are shown by X-symbols. As expected, the corresponding values lie above the perfect fit diagonal. The values for the finite potential vorticity benchmark (Equation 2.14.2, open circles) lie slightly below the zero potential vorticity values. The greatest difference is approximately 10%, indicating that the value of upstream potential vorticity is not the greatest factor needed to bring the theory into agreement with measurements. In some cases with nonrectangular topography, the finite potential vorticity prediction exceeds that of the zero potential vorticity. The flux values over various bottom shapes (Table 2.14.2) are shown by assorted symbols. For a given strait, variations in the observed flux are indicated by horizontal scatter of like symbols. Variations in predicted fluxes for different geometric fits to the sill topography are indicated by the vertical scatter of different symbols with the same Q^*_{observed}. The latter is generally larger than the former. Values lying below the diagonal may contain reverse flow in the model, while those lying above have none, or have had the reverse flow excised. Overall, the steady component of flow is bounded by the predictions for flow over a flat bottom, and the influence of bottom shape introduces a wide range of variability in predictions.

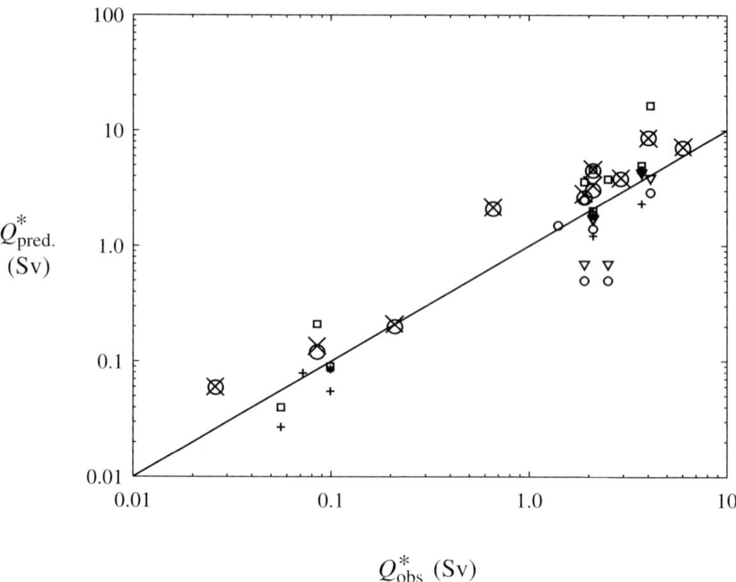

FIGURE 2.14.3. Comparison of predicted volume flux to observed volume flux based on values listed in Tables 2.14.1 and 2.14.2. The X-symbols are based on the WLK model (Equations 2.14.1). The large open circles are based on a uniform potential vorticity theory that assumes a sill height of zero and an approach flow entirely along the right wall of the channel (see Equation 2.14.2). The smaller symbols are based on theories that use a nonzero potential vorticity distribution and/or nonrectangular cross sections. Different symbols correspond to different geometries as follows: rectangular cross section (squares), parabola (small circle), V-shaped bottom (triangle), real bathymetry (plus), and real bathymetry with reversed flow excised (star).

Numerical models of overflow regions have received considerable development since about 1990. The numerical schemes attempt to resolve topographic features and eddies on a horizontal scale that is a fraction of the Rossby radius of deformation based on the local depth. Sigma coordinates[13] are often used because of their ability to handle regions with large topographic variations. The models resolve 60 or more levels and are based on primitive equation dynamics. The formulation typically includes parameterizations for both internal mixing and bottom drag.

The earliest studies focus on the dense overflow plume downstream of the sill (Jungclaus and Backhaus, 1994; Krauss and Käse, 1998; Shi *et al.*, 2001). The lateral scale is smaller than 10 km, allowing partial resolution of fronts and eddies. A second generation of models resolves the entire region around the Denmark Strait. In a fully three-dimensional computation by Käse and Oschlies (2000),

---

[13] 'Sigma' is defined as $(z_T^* - z^*)/(z_T^* - h^*)$, where z_T^* is the elevation of the sea surface. Thus the bottom corresponds to sigma $= 1$.

the computed volume flux agrees within a few tens of percent with (2.14.1). (The value of Δz^* is the depth, above the sill, of an isopycnal surface, averaged over a region approximately 85 km upstream of the sill.) Kösters (2004) compares a number of simple hydraulic estimates to the output of a slightly more elaborate numerical model. The model, driven by regional buoyancy forcing, has a lateral grid of 5 km and both idealized and real topography. The hydraulic criticality of the flow is evaluated using several forms of the Froude number, including (2.5.7) for the Gill model. The flow is judged to be critical approximately 80 km downstream of the sill. His volume flux comparisons with hydraulic predictions are very similar to those in Table 2.14.2. For example, the zero potential vorticity prediction for a rectangular sill is about double the model flux. Also, the method from Nikolopoulos *et al.* (2003), using a realistic bottom profile, yields predictions much smaller than the numerical model due to the presence (in the theory) of a return flow. The numerical models generally produce unidirectional flow at the critical section. If the reverse flow predicted by the theory is excised, the prediction comes within 30% of the numerical value. Consistent with the arguments of the previous section, upstream height values progressively closer to the sill produce better predictions.

We have seen that all predictions of the crudest zero potential vorticity theory tend to exceed ocean measurements for volume flux with ratios between one and three. Predictions for parabolic and realistic bottoms can extend below the observed values of flux. A rounded bottom profile sometimes leads to a prediction of return flow at the sill that produces a smaller flux, but excising the return flow increases flux toward flat-bottom values. Overall, it is clear that theory has produced a reasonable bound to observation, but that there is scope for much improvement. A number of aspects could be developed that might lead to improved agreement with observations. These include reconciliation of the issue of counterflow at the sill, and inclusion of time dependence, friction, and continuous stratification.

3
Time-Dependence and Shocks

None of the hydraulically driven oceanic or atmospheric flows discussed in the introduction to this book are steady, and very few are approximately so. A striking example of unsteadiness was discovered by Worthington (1969) who, in 1967, placed an array of 30 moored current meters in the Denmark Strait in order to measure the deep velocities of the overflow. When Worthington returned a year later he found that all but one of the instruments had been destroyed or could not be recovered. The surviving current meter showed a history of rapid velocity fluctuations ranging from near zero to 1.4 m/s, most likely due to the meandering of the edge of the jet-like overflow back and forth across the instrument. Worthington concluded "the currents associated with the overflow water were too strong on this occasion to be measured using existing mooring technology".

How applicable are the steady models of Chapters 1 and 2 to such flows? The answer to this question depends on how rapid the time fluctuations are. A rough measure of rapidity is the ratio of the free response time of the flow over the topographic feature to the characteristic time scale T_u of the unsteady motions. The response time is a measure of how long it would take the steady flow to adjust to a sudden change in conditions and is equal to the length L of the topographic feature divided by characteristic speed c_- of a wave propagating upstream. The importance of time-dependence is therefore measured by the parameter $L/(c_-T_u)$. If $L/(c_-T_u) \ll 1$, the response of the flow is instantaneous compared to the imposed fluctuation time scale and conditions are quasi-static. Local time derivatives in the equations of motion may be neglected (see Exercise 1) and the time variable simply becomes a parameter. If $L/(c_-T_u) = O(1)$ then the time dependence is dynamically important and local time derivatives should be retained. This last situation occurs in many oceanographically important straits where steady theory has traditionally been applied. For example, the Bab al Mandab (the strait connecting the Indian Ocean with the Red Sea) is about 200 km in length and is subject to tides with a dominant period (T_u) of about 12 hr. The typical speed c_- of the first internal gravity wave in the strait is 1 m/s or less[1] giving a travel time of at least 55 hr. With these numbers $L/(c_-T_u) \cong 4.6$ and the flow must be considered unsteady. Even if the length of the sill ($\cong 40$ km) is used in place of the total strait length, $L/(c_-T_u)$ remains O(1).

---

[1] The speeds of the lowest internal modes have been calculated by Pratt *et al.* (2000), taking into consideration the continuous stratification and shear.

The development of a general understanding of the hydraulics of flows undergoing continuous time variations has proven to be difficult and frustrating. The applicability of concepts such as hydraulic control and upstream influence in these situations is not well-established, even in the simplest cases. To gain ground, past investigators have concentrated on initial-value problems in which the evolution beginning with some simple initial state leads to the establishment of a steady flow. Such models help one to understand how steady solutions are established, how upstream influence is exercised, and why hydraulically controlled states are sometimes preferred over states that lack control. Initial-value experiments can also provide a first step towards understanding flows that are subject to continuously time-varying forcing, such as tidal forcing. One might think of continuous time variations as a sequence of initial value problems. The latter often provide a convenient setting for the study of rotating shocks, which tend to arise when sudden transitions are forced by suitable initial conditions.

In this chapter we shall consider two initial-value problems, both grounded in classical fluid mechanics. The first is a rotating version of the dam-break problem considered in Sections 1.2 and 1.3, carried out in channel geometry. Also referred to as "Rossby adjustment in a channel", this problem has received the attention of a number of authors. The second problem has received less attention but is directly related to the models of sill flows discussed in Chapter 2. It is a rotating version of Long's experiment, in which an obstacle is placed in the path of a steady stream in a rotating channel. We also discuss the special case in which the flow takes place on an infinite plane with no channel geometry. Hydraulic jumps, bores and other rotating shocks appear in all of these problems and several sections are devoted to a discussion of these objects. The chapter closes with a brief discussion of flow instability, one of the supposed sources of time variability.

Exercise

(1) Consider a shallow flow in a channel that contains an obstacle confined to $-L/2 < y < L/2$ but has a horizontal bottom elsewhere. The flow consists of a steady component that passes over the obstacle plus a small-amplitude, unsteady component with characteristic period T_u. The mean characteristic wave speeds in the vicinity of the obstacle are c_- and c_+. Show that the local time-derivatives in the shallow water equations become negligible over the obstacle when $L/(c_- T_u) \ll 1$. What about the flow away from the obstacle?

3.1.   Linear Rossby Adjustment and Geostrophic Control in a Channel

The Rossby adjustment problem in a channel belongs to a group of classical initial-value problems that includes dam-break and the lock exchange flows. The linear version of the problem was first solved by Gill (1976). Following the nondimensional notation and conventions introduced in Chapter 2, consider a rotating channel with sidewalls at $x = \pm w/2$ and a barrier at $y = 0$. As shown in Figure 3.1.1a, the barrier separates stagnant fluid with uniform depth $d = 1 + a$ in $(y < 0)$ from depth $1 - a$ in $(y > 0)$. At $t = 0$ the barrier is removed allowing the deeper fluid to spill forwards. If $a \ll 1$ the shallow water equations can be linearized and solved for the subsequent evolution.

Assuming $a \ll 1$, we first expand the velocity and depth fields as

$$u = au^{(o)} + a^2 u^{(1)} + \dots$$

$$v = av^{(o)} + a^2 v^{(1)} + \dots$$

$$d = 1 + a\eta^{(o)} + a^2 \eta^{(1)} + \dots$$

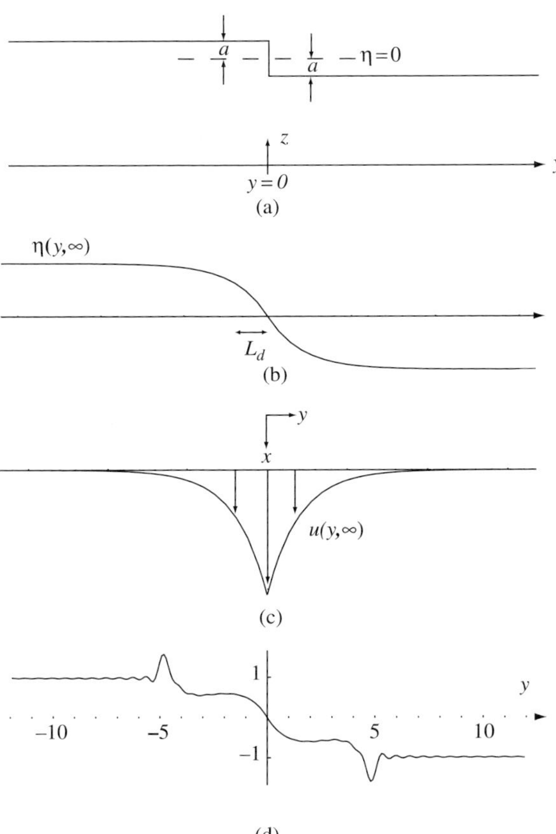

FIGURE 3.1.1. (a): Initial condition for Rossby adjustment problem. (b) and (c): Interface elevation and velocity for $t \to \infty$ state on infinite plane. (d): Radiation of Poincaré waves at $t = 5$ as computed from (3.1.21).

and substitute these expansions into the full shallow water equations. Because of the abrupt nature of the initial conditions, the semigeostrophic approximation will fail, at least in the vicinity of $y = 0$ for small t. We therefore set $\delta = W/L = 1$. With this minor change, the equations governing the lowest order fields are just the linear shallow water equations (2.1.20–2.1.22) with $\delta = 1$, or

$$\frac{\partial u}{\partial t} - v = -\frac{\partial \eta}{\partial x}, \tag{3.1.1}$$

$$\frac{\partial v}{\partial t} + u = -\frac{\partial \eta}{\partial y}, \tag{3.1.2}$$

and

$$\frac{\partial \eta}{\partial t} + \frac{\partial u}{\partial x} + \frac{\partial v}{\partial y} = 0 \tag{3.1.3}$$

Note that the $()^{(o)}$ notation has been dropped.

The initial conditions are given by

$$\eta(x, y, 0) = \eta_o = -\text{sgn}(y) \tag{3.1.4}$$

and $u(x, y, 0) = v(x, y, 0) = 0$. The boundary conditions are

$$u\left(\pm\frac{1}{2}w, y, t\right) = 0 \tag{3.1.5}$$

The linearized statement of potential vorticity conservation (see 2.1.23) for this case is

$$\frac{\partial v}{\partial x} - \frac{\partial u}{\partial y} - \eta = -\eta_o \tag{3.1.6}$$

The linearized potential vorticity $\frac{\partial v}{\partial x} - \frac{\partial u}{\partial y} - \eta$ is thus conserved at each (x, y), rather than following a fluid element. 'Forward' locations (with $y > 0$) thus maintain higher potential vorticity than locations with $y < 0$. In reality, a forward location will see an $O(a)$ change in potential vorticity if it is reached by fluid initially lying in $y < 0$. The consequences must be explored within the context of a nonlinear model and this is done in the following sections.

The variables u and v may be eliminated from (3.1.6), resulting in a single equation for η:

$$\frac{\partial^2 \eta}{\partial t^2} - \frac{\partial^2 \eta}{\partial x^2} - \frac{\partial^2 \eta}{\partial y^2} + \eta = \eta_o \tag{3.1.7}$$

(also see 2.1.24). It will be of interest to calculate the energy of solutions to this equation and compare the kinetic and potential contributions. An expression for the linearized energy may be obtained by taking $u \times (3.1.1) + v \times (3.1.2) + \eta \times (3.1.3)$, resulting in

$$\frac{\partial}{\partial t}\left(\frac{u^2 + v^2 + \eta^2}{2}\right) = -\nabla \cdot (\mathbf{u}\eta) \qquad (3.1.8)$$

The left-hand term is clearly the time rate of change of the kinetic plus potential energy at a point, whereas the right-hand term can be interpreted as the divergence of an energy flux.

a. The Rossby Adjustment in an Infinite Domain

The solution procedure for the problem described above is technically involved and it is helpful to first consider the special limit of an infinite domain ($-\infty < x < \infty$). This limiting case is the classical 'geostrophic adjustment' problem considered by Rossby (1938). Since the initial conditions are independent of y and since the governing equation contains no x-dependent forcing terms, it follows that $\partial/\partial x = 0$ for all time. This fundamental simplification rules out Kelvin waves.

Another simplification recognized by Rossby is that potential vorticity conservation allows the asymptotic ($t \to \infty$) solution to be predicted without the need to calculate the time evolution. Assuming this state to be steady, (3.1.7) reduces to

$$\frac{\partial^2 \eta_\infty}{\partial y^2} - \eta_\infty = -\eta_o = \mathrm{sgn}(y) \qquad (3.1.9)$$

Of course, $\eta_\infty = \eta_o$ is a possible solution to this equation, provided that one is willing to overlook the singularity at $y = 0$. However, we anticipate that the true asymptotic solution will be continuous in y at all points. [This expectation can be made rigorous by replacing the initial discontinuity in depth by an abrupt but continuous transition.] Granted the continuity of η_∞ at $y = 0$, it follows from integration of (3.1.9) over a small interval about $y = 0$ that $\partial \eta_\infty / \partial y$ must also be continuous. With these provisos, the solution to (3.1.9) becomes

$$\eta_\infty = \begin{cases} -1 + e^{-y} & (y > 0) \\ 1 - e^{y} & (y < 0) \end{cases}$$

while the corresponding geostrophic velocity is

$$u_\infty = -\frac{\partial \eta_\infty}{\partial y} = e^{-|y|}$$

As shown in Figures 3.1.1b,c this solution consists of a frontal region with a cusped jet centered at $y = 0$ and flowing in the positive-x direction. In contrast to the nonrotating version of this problem (Section 1.2), where the final flow is uniform and in the y-direction, the final flow here is parallel to the initial step. Another significant difference can be identified by computation of the total energy of the asymptotic state and comparing it with the initial energy. Integrating (3.1.8) over $0 < t < \infty$ and over a long interval $[-L, \ L]$ in the y-direction leads to

$$\frac{1}{2} \int_{-L}^{L} \left(\eta_o^2 - \eta_\infty^2 - u_\infty^2\right) dy = 1 = \int_0^\infty [v\eta]_{y=-L}^{y=L} dt \qquad (3.1.10)$$

The left-hand term is the difference between the energies (per unit x) of the initial and asymptotic states. By direct calculation this difference is unity (see Exercise 1). Thus the asymptotic state contains one unit of energy less than that of the initial state. This deficit is due to the radiation of energy towards $\pm\infty$ by Poincaré waves, as measured by the final term in (3.1.10). We next calculate these transient terms.

For reasons that will prove advantageous in our treatment of the channel geometry, we will solve for the transient part of the solution in terms of the variable u, rather than η. A single equation for the former may be obtained by eliminating v and η from (3.1.6) (also see Exercise 2 of Section 2.1). The result is

$$\frac{\partial^2 u}{\partial^2 t^2} - \frac{\partial^2 u}{\partial^2 y^2} - \frac{\partial^2 u}{\partial^2 x^2} + u = -\frac{\partial \eta_o}{\partial y} = 2\delta(y), \qquad (3.1.11)$$

and we again take $\partial/\partial x = 0$.

Next, let $u = u_T + u_\infty = u_T + e^{-|y|}$ where u_T denotes the transient part of the solution. It follows that

$$\frac{\partial^2 u_T}{\partial^2 t^2} - \frac{\partial^2 u_T}{\partial^2 y^2} + u_T = 0, \qquad (3.1.12)$$

$$u_T(y, 0) = -e^{-|y|}, \qquad (3.1.13)$$

and

$$\frac{\partial u_T}{\partial t}(y, 0) = 0, \qquad (3.1.14)$$

where the last relation follows from (3.1.1).

Since $u_T(y, 0)$ is even in y and since (3.1.12) has $\pm y$ symmetry, $u_T(y, \ t)$ must also be even. We therefore seek a solution in the form of a Fourier cosine integral:

$$u_T(y, t) = \int_0^\infty \hat{u}_T(l, t) \cos(ly) dl \qquad (3.1.15)$$

Taking the Fourier cosine transform $\hat{f}(l) = \frac{1}{\pi} \int_{-\infty}^\infty f(y) \cos(ly) dy$ of (3.1.12–14) leads to

$$\frac{\partial^2 \hat{u}_T}{\partial^2 t^2} + (1+l^2)\hat{u}_T = 0,\tag{3.1.16}$$

$$\hat{u}_T(l,0) = -\hat{u}_\infty(l) = -\frac{1}{\pi} \int_{-\infty}^{\infty} e^{-|y|}\cos(ly)\,dy$$

$$= -\frac{2}{\pi} \int_0^\infty Re\{e^{(il-1)y}\}\,dy = -\frac{2}{\pi(1+l^2)},\tag{3.1.17}$$

and

$$\frac{\partial \hat{u}_T(l,0)}{\partial t} = 0.\tag{3.1.18}$$

The solution to (3.1.16) subject to (3.1.17) and (3.1.18) is given by

$$\hat{u}_T = -\frac{2}{\pi(1+l^2)}\cos[\omega(l)t],$$

where $\omega(l) = (1+l^2)^{1/2}$ is the dispersion relation (2.1.27a) for Poincaré waves
with $w \to \infty$.

In summary,

$$u(y,t) = u_\infty + u_T = e^{-|y|} - \frac{2}{\pi} \int_0^\infty \frac{1}{1+l^2}\cos[\omega(l)t]\cos(ly)\,dl\tag{3.1.19}$$

Note that $\cos[\omega(l)t]\cos(ly) = \frac{1}{2}[\cos(ly+\omega t)+\cos(ly-\omega t)]$, so that the
integrand of (3.1.19) is a superposition of forward and backward propagating
Poincaré waves. An alternative form of (3.1.19) that will be of use later on is

$$u(y,t) = \begin{cases} \int_0^{(t^2-y^2)^{\frac{1}{2}}} (y^2+r^2)^{-\frac{1}{2}} J_o(r)r\,dr & (|y|<t), \\ 0 & (|y|>t) \end{cases}\tag{3.1.20}$$

(Cahn, 1945).

The free surface elevation follows from (3.1.6) with $\partial/\partial x = 0$. Thus

$$\eta(y,t) = \eta_o - \frac{\partial u}{\partial y}$$
$$= \begin{cases} -1+e^{-y} & (y>0) \\ 1-e^y & (y<0) \end{cases} - \frac{2}{\pi}\int_0^\infty \frac{l}{(1+l^2)^{1/2}}\cos[\omega(l)t]\sin(ly)\,dl\tag{3.1.21}$$

A sketch of the developing solution (Figure 3.1.1d) shows a wave front and a
Poincaré wake.

b.   The Channel Problem

With the channel geometry and boundary condition the asymptotic state η_∞ now can be expected to depend on both x and y. This state is still constrained by the principle of pointwise conservation of potential vorticity (3.1.7) and one might hope to find it by solving this equation. However, doing so would require as boundary conditions the values of η_∞ along the sidewalls, and there is no clear way of anticipating these values. One is therefore forced to consider the entire problem at once.

The condition of vanishing u on the channel sidewalls makes a solution in terms of u convenient and we therefore solve (3.1.11) directly. However, u is identically zero in linear Kelvin waves and so this approach will capture only that part of the solution due to Poincaré waves. The remaining portion of the solution will be addressed in due course.

The initial conditions:

$$u(x, y, 0) = 0$$

and

$$\frac{\partial u}{\partial t}(x, y, 0) = 0$$

remain the same as above. Since the initial conditions and the forcing term $(\partial \eta_o/\partial y)$ in (3.1.11) are x-independent, and since this equation has $\pm x$ symmetry, the solution will be an even function of x. Therefore a Fourier cosine series of the form

$$u(x, y, t) = \sum_{m=0}^{\infty} u_m(y, t) \cos(a_m x) \quad (a_m = (2m+1)\pi/w, \ \ m = 0, 1, 2, \ldots)$$

$$(3.1.22)$$

is appropriate. To find u_m, multiply (3.1.11) by $2w^{-1}\cos(a_m x)$ and integrate with respect to x over the width of the channel. After several integrations by parts and application of boundary conditions, one obtains

$$\frac{\partial^2 u_m}{\partial^2 t^2} - \frac{\partial^2 u_m}{\partial^2 y^2} + (1 + a_m^2)u_m = \gamma_m \frac{d\eta_o}{dy} = -2\gamma_m \delta(y)$$

$$= -2\gamma_m (1 + a_m^2)^{1/2} \delta[(1 + a_m^2)^{1/2} y], \qquad (3.1.23)$$

where $\gamma_m = \frac{4(-1)^m}{a_m w}$. The identity $\delta(y) = c\delta(cy)$ has been used in the final step.

Equation (3.1.23) is a forced equation for the nth Poincaré channel mode, with $n = 2m + 1$. Only odd-numbered modes are excited. The solution for $u_n(y, t)$ is closely related to the solution to (3.1.11) obtained in the case of an infinite domain ($\partial/\partial x = 0$). In particular, (3.1.11) can be transformed into (3.1.23) by replacing y by $(1 + a_m^2)^{1/2} y$, t by $(1 + a_m^2)^{1/2}t$, and the amplitude (-2) of the forcing term by $-2\gamma_m(1 + a_m^2)^{1/2}$. Since the solution to the $\partial/\partial x = 0$ version of

(3.1.11) obeys the same initial conditions $[u_n(y, 0) = 0]$ as are required here, the solution is the transformed version of (3.1.20):

$$u_m(y, t) = \gamma_m \begin{cases} \int_0^{(t^2-y^2)^{\frac{1}{2}}} (y^2 + s^2)^{-\frac{1}{2}} J_o[(1 + a_m^2)^{1/2}s]sds & (|y| < t) \\ 0 & (|y| > t) \end{cases} \qquad (3.1.24)$$

The corresponding solution for $\eta(x, y, t)$ requires several steps and these are navigated in Exercise 4. The result may be written as

$$\eta(x, y, t) = K(x, y, t) + P(x, y, t), \qquad (3.1.25)$$

$$K(x, y, t) = N(y, t)\frac{\cosh(x)}{\cosh(w/2)} + V(y, t)\frac{\sinh(x)}{\cosh(w/2)} - \text{sgn}(y)\left[1 - \frac{\cosh(x)}{\cosh(w/2)}\right], \qquad (3.1.26)$$

$$P(x, y, t) = -\sum_{m=0}^{\infty} (1 + a_m^2)^{-1}\left[\frac{\partial u_m}{\partial y}\cos(a_m x) + a_m\frac{\partial u_m}{\partial t}\sin(a_m x)\right], \qquad (3.1.27)$$

$$N(y, t) = -\frac{1}{2}\left[\text{sgn}(y - t) + \text{sgn}(y + t)\right], \qquad (3.1.28)$$

and

$$V(y, t) = \frac{1}{2}\left[-\text{sgn}(y - t) + \text{sgn}(y + t)\right]. \qquad (3.1.29)$$

The function $K(x, y, t)$ contains the Kelvin wave dynamics while $P(x, y, t)$ is a superposition of Poincaré modes. Consider the Kelvin waves first. The time-dependence is contained in the coefficients N and V which describe step functions propagating at speeds ± 1 (dimensionally $(gD)^{1/2}$, where D is the mean depth). Furthermore, N and V are exactly the surface height perturbation and velocity for the nonrotating version of this problem discussed in Section 1.2. For $w \gg 1$, K can be approximated as

$$K(y, t) \begin{cases} = -1 & (y > t) \\ \cong \left[2e^{(x-w/2)} - 1\right] & (0 < y < t) \\ \cong \left[-2e^{-(x+w/2)} + 1\right] & (-t < y < 0) \\ = 1 & (y < -t) \end{cases} \qquad (3.1.30)$$

The forward-moving Kelvin wave therefore consists of a step that propagates along the right wall and decays inward over one Rossby radius of deformation (here unity). This wave propagates into the resting fluid (with surface elevation $\eta = -1$) and its role is to raise the elevation along the right wall to that far upstream $(\eta = 1)$. The wave establishes a current along the right wall that extends from $y = 0$ to $y = t$. The velocity of this current is given by

$$v = \frac{\partial \eta}{\partial x} = 2e^{(x-w/2)}.$$

In addition (3.1.30) contains a backward-moving Kelvin wave trapped to the left wall. This wave establishes a boundary current that carries the same volume transport ($= 2$) as the right-wall current. The current extends from $y = 0$ to $y = -t$. The surface displacement along the left wall is reduced from 1 to the initial elevation (-1) of the downstream reservoir. Had we attempted to solve the problem using the semigeostrophic approximation, these Kelvin waves would have described the entire solution.

The Poincaré waves are contained in the function P. Since $u_m = 0$ at $x = \pm w/2$, P is zero there as well and the Poincaré waves do not affect the value of η along the walls. The Kelvin waves are therefore responsible for fixing the transport of the asymptotic solution. This transport is given by

$$Q = \int_{-w/2}^{w/2} v\,dx = \int_{-w/2}^{w/2} \frac{\partial \eta}{\partial x}\,dx = \eta_{w/2} - \eta_{-w/2} = 2\,\tanh(w/2) \qquad (3.1.31)$$

The dimensional equivalent is $Q^* = 2a^* gDf^{-1}\,\tanh[fw^*/2(gD)^{1/2}]$, where $a^* = Da$. As $w = fw^*/(gD)^{1/2} \to \infty$, $Q^* \to 2a^* gDf^{-1}$, also the transport of the boundary layers described above.

Despite the fact that Poincaré waves are inconsequential to the bulk transport, they are important in establishing the characteristics of the solution near $y = 0$. In particular, they set up a jet-like flow along $y = 0$ that carries fluid from the left-wall boundary layer to the right-wall boundary layer. This structure is particularly striking in the case $w \gg 1$, where (3.1.24) reduces to the cusped jet expression (3.1.13) obtained in the Rossby adjustment problem on an infinite plane. In this limit, the left-wall boundary current separates near $y = 0$, forming a jet that crosses the channel and forms the source for the right-wall boundary current. This crossing flow can be seen along with the forward and backward Kelvin waves in a picture of the developing flow for $w = 4$ (Figure 3.1.2). The Poincaré waves lag behind the Kelvin waves due to their slower group speeds.

Although the Poincaré waves transport no mass, they do transport energy. Calculation of the energy of the asymptotic state is somewhat involved and will

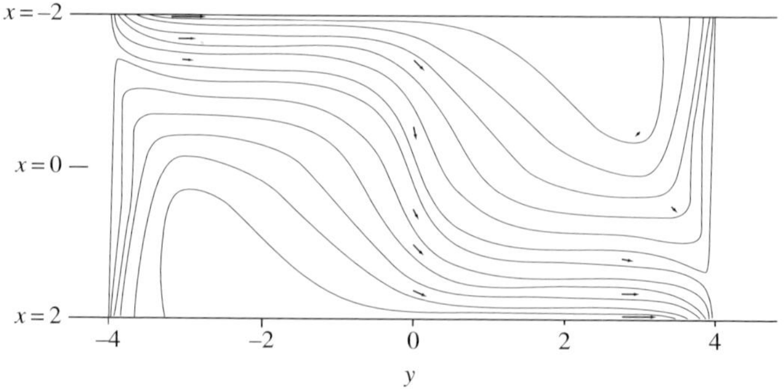

FIGURE 3.1.2. Contours of surface elevation for the linear Rossby adjustment problem in channel with $w = 4$ at $t = 4$. The contour values range from -0.9 to $+0.9$ in even increments as the channel is crossed. (Constructed from Figure 10.7 in Gill, 1982).

not be carried out here. However it can be shown that the difference between initial and asymptotic energies per unit width of channel tend to zero as $w \to 0$, the result obtained in Section 1.2 for the nonrotating version of the problem. Here all initial energy is converted to kinetic energy and there is no wave radiation. For finite w, energy radiation occurs and the asymptotic state contains a deficit. As $w \to \infty$, this deficit approaches the value unity, as we found for Rossby adjustment on an infinite plane.

c. *Geostrophic Control*

The crossing of the channel from the left to the right wall is the basis for the concept of 'geostrophic control' (Toulany and Garrett, 1984). Imagine two wide basins separated by a strait and suppose that a steady flow exists from basin A into basin B. Also assume that the dynamical setting (f-plane, horizontal bottom, homogeneous fluid, etc.) is the same as in the problem considered above. Then, if the flow is set up by the removal of a barrier that initially separates fluid of depth d_A^* in A from fluid of depth $d_B^* < d_A^*$ in B, we expect that a similar outcome will occur. The flow originating in A will be confined to a boundary current along the 'left' wall and this current will cross to the right wall within the strait and continue into basin B along that wall.

The principle of geostrophic control states that the dimensional transport between the basins is bounded by the difference in interior depths $d_A^* - d_B^*$ according to

$$Q^* \leq \frac{g}{2f}(d_A^{*2} - d_B^{*2}). \tag{3.1.32}$$

The depths d_A^* and d_B^* (or, equivalently, the layer thickness in a reduced gravity model) must be measured in interior regions, away from the boundary currents. The right-hand side of (3.1.32) is the geostrophic transport of a current flowing along a horizontal bottom with depths d_A^* and d_B^* on either side. The basic idea is that the throughflow cannot exceed the geostrophic transport of the current that crosses the channel. If the depth difference is relatively small, (3.1.32) can be approximated by the linearized formula

$$Q^* \cong \frac{gD}{f}(d_A^* - d_B^*), \tag{3.1.33}$$

where D is the mean depth.

In the linearized Rossby adjustment problem we can identify the channel sectors $y < 0$ and $y > 0$ as basins A and B. In the limit $w \to \infty$ the predicted transport $(2gDa^*f^{-1})$, is exactly given by the linearized version (3.1.33) of the bound (with $d_A^* - d_B^* = 2a^*$). In this case the geostrophic control bound gives the exact transport.

The concept of geostrophic control is most easily supported when the two basins are very wide and extend infinitely far from the separating strait. Problems arise when these conditions are not satisfied. If the upstream basin is finite, for example, the Kelvin wave that sets up the left-wall boundary current in the upstream basin would travel around the perimeter and return to the strait along the right wall, setting up a boundary flow there. Even when the basin has infinite area, ambiguities arise if its width is not large in comparison with the Rossby radius of deformation. An example is the channel adjustment problem with finite width. The exact transport in this case is $Q^* = (d_A{}^* - d_B{}^*)gDf^{-1}\tanh[fw^*/2(gD)^{1/2}]$, where $d_A{}^*$ and $d_B{}^*$ are to be interpreted as the *initial* surface elevations upstream and downstream of the barrier. Although it is true that this transport lies below the geostrophic control bound (3.1.33), it is the *steady* upstream values of $d_A{}^*$ and $d_B{}^*$, and not the initial values, that are typically given in practical applications. Alternatively, one might interpret $d_A{}^*$ and $d_B{}^*$ as characterizing the steady, upstream and downstream surface elevations of the final adjusted state. This interpretation is straightforward if $w \gg 1$, for then the two depths can be measured in the channel interior, away from boundary layers. In the case of moderate channel width, $d_A{}^*$ and $d_B{}^*$ cannot be defined unambiguously. Finally, one might also question whether the channel crossing calculated in the linear adjustment problem will persist if the full nonlinearities are retained. This topic is taken up in the next section, where we will also revisit the subject of geostrophic control.

In summary, we have explored the separate roles played by Kelvin and Poincaré waves in the linear adjustment problem for a rotating channel. Kelvin waves are responsible for setting up boundary currents that carry all of the volume flux. Although the Poincaré waves carry no volume flux they do transport energy. The associated energy radiation is relatively unimportant when the channel is narrow ($w \ll 1$): there is no room for Poincaré waves to be generated. Energy radiation increases as w becomes larger, and eventually half of the initial energy is radiated away. Poincaré waves also establish the jet-like flow that crosses the channel at the position of the initial barrier. The picture that emerges, with upstream flow confined to a left-wall boundary layer and downstream flow in a right-wall layer, both having widths equal to the Rossby radius of deformation, motivates the concept of geostrophic control.

Exercises

(1) Show by direct calculation that the nondimensional initial energy for the linear Rossby adjustment problem in an unbounded domain exceeds that of the final steady state by one unit.

(2) For the Rossby adjustment problem in an unbounded domain, calculate the final steady state without using linearization of the equations of motion. (*Hint: Use the x-momentum, written in Lagrangian form, to help find the final position of the front separating the two main water masses.*)

(3) In the limit $w \gg 1$, show that the velocity field set up by Poincaré waves in the vicinity of $y = 0$ is exactly the cusped jet found in the Rossby adjustment problem on an infinite plane.

(4) *Calculation of the surface displacement $\eta(x, y, t)$ given $u(x, y, t)$. Here we follow a procedure used by Gill (1976) which exploits the fact that $u(x, y, t)$ is an even function of x.*

(a) Show that (3.1.1–3.1.3 and 3.1.6) can be separated into even and odd parts as follows

$$\frac{\partial u}{\partial t} - v^{ev} = -\frac{\partial \eta^{odd}}{\partial x}, \quad v^{odd} = \frac{\partial \eta^{ev}}{\partial x}, \tag{3.1.34a, b}$$

$$\frac{\partial v^{ev}}{\partial t} + u = -\frac{\partial \eta^{ev}}{\partial y}, \quad \frac{\partial v^{odd}}{\partial t} = -\frac{\partial \eta^{odd}}{\partial y}, \tag{3.1.35a, b}$$

$$\frac{\partial \eta^{ev}}{\partial t} + \frac{\partial v^{ev}}{\partial y} = 0, \quad \frac{\partial \eta^{odd}}{\partial t} + \frac{\partial u}{\partial x} + \frac{\partial v^{odd}}{\partial y} = 0, \tag{3.1.36a, b}$$

$$\frac{\partial v^{odd}}{\partial x} - \frac{\partial u}{\partial y} - \eta^{ev} = -\eta_o, \quad \frac{\partial v^{ev}}{\partial x} = \eta^{odd}. \tag{3.1.37a, b}$$

where $\eta = \eta^{ev} + \eta^{odd}$, $\eta^{ev} = \frac{1}{2}[\eta(x, y, t) + \eta(-x, y, t)]$, $\eta^{odd} = \frac{1}{2}[\eta(x, y, t) - \eta(-x, y, t)]$, and similar decompositions apply to v.

(b) From the result in (a) show that

$$\frac{\partial^2 \eta^{ev}}{\partial x^2} - \eta^{ev} = \frac{\partial u}{\partial y} - \eta_o,$$

$$\frac{\partial^2 v^{ev}}{\partial x^2} - v^{ev} = -\frac{\partial u}{\partial t}.$$

Show that the general solutions to these equations can be expressed as

$$\eta^{ev}(x, y, t) = -\sum_{m=0}^{\infty} \frac{\gamma_m \operatorname{sgn}(y) + \partial u_m/\partial y}{(1 + a_m^2)} \cos(a_m x) + N(y, t)\frac{\cosh(x)}{\cosh(w/2)} \tag{3.1.38}$$

and

$$v^{ev}(x, y, t) = \sum_{m=0}^{\infty} \frac{\partial u_m/\partial t}{(1 + a_m^2)} \cos(a_m x) + V(y, t)\frac{\cosh(x)}{\cosh(w/2)}, \tag{3.1.39}$$

where again $\gamma_m = (2m+1)\pi/w$.

(c) Show that the functions N and v can be determined through substitution of these last solutions into (3.1.35a) and (3.1.36a) leading to

$$\frac{\partial V}{\partial t} + \frac{\partial N}{\partial y} = 0 \text{ and } \frac{\partial N}{\partial t} + \frac{\partial V}{\partial y} = 0. \tag{3.1.40}$$

(d) From the results of (b) show that an initial condition on V is $V(y, 0) = 0$ whereas $N(y, 0)$ is determined by the relation

$$-\operatorname{sgn}(y) = -A(x)\operatorname{sgn}(y) + N(y, 0)\frac{\cosh(x)}{\cosh(w/2)}$$

where

$$A(x) = \sum_{m=0}^{\infty} \frac{\gamma_m \cos(a_m x)}{(1 + a_m^2)}. \tag{3.1.41}$$

To help evaluate $A(x)$, use the above series to show that

$$\frac{d^2 A}{dx^2} - A = - \sum_{m=0}^{\infty} \gamma_m \cos(a_m x) = -1.$$

From this equation and from the boundary conditions $A(w/2) = A(-w/2) = 0$, which follow from (3.1.41), deduce that $A = 1 - \frac{\cosh(x)}{\cosh(w/2)}$ and therefore $N(y, 0) = -\text{sgn}(y)$. From (3.1.40) and the initial conditions just described, deduce the solutions (3.1.28) and (3.1.29).

(e) Finally, show using (3.1.37b) that

$$\eta^{odd} = - \sum_{m=0}^{\infty} \frac{a_m}{(1 + a_m^2)} \frac{\partial u_m}{\partial t} \sin(a_m x).$$

3.2.   Rossby Adjustment: Weakly Nonlinear Behavior

The nonlinear terms neglected in Gill's solution can be expected to remain small over a time of $O(a^{-1})$, where $2a$ is the dimensionless amplitude of the initial discontinuity in fluid depth, assumed $\ll 1$. We now discuss some new processes that arise after this time period is exceeded. One is the motion of the front that separates the regions of high and low potential vorticity and that initially lies at the position of the barrier. There is also a variety of nonlinear processes that act on the forward and backward Kelvin waves that establish the boundary currents. For $a \ll 1$ the Kelvin waves become well-separated from the potential vorticity front and the evolution of the two features may be treated separately. When a is $O(1)$, nonlinearities arise immediately after the barrier is removed and it become more difficult to treat specific processes in isolation. This topic is taken up in Section 3.3.

a.   Motion of the Potential Vorticity Front: Contour Dynamics

Since advection of linear potential vorticity $1 + \zeta - \eta$ is neglected in linear shallow water theory, $1 + \zeta - \eta$ at any (x, y) remains equal to its initial value. The steady flow that emerges as $t \to \infty$, sometimes referred to as the *wave adjusted state*, maintains $1 + \zeta - \eta = 1 - a$ upstream of $y = 0$ and $1 + \zeta - \eta = 1 + a$ downstream. As a fluid column crosses $y = 0$ its linear potential vorticity jumps from the first to the second of these values. Of course, the original shallow water equations require that the full potential vorticity $q = (1 + \zeta)/(1 + \eta)$ be conserved following the flow. Thus, the lower q of the upstream region would be

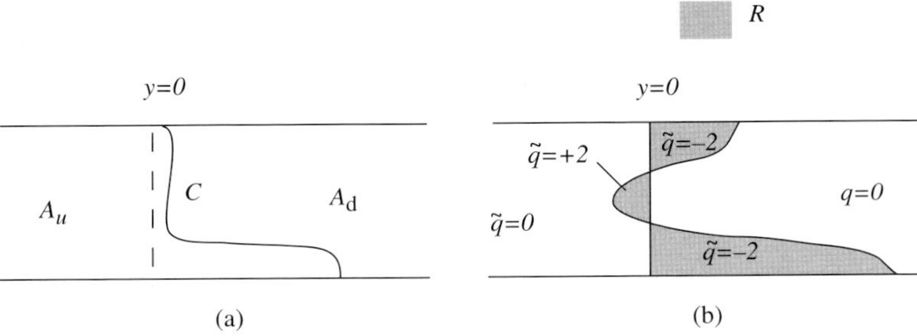

FIGURE 3.2.1. Definition sketches showing potential vorticity front (a) and potential vorticity anomalies (b).

carried downstream, leading to a departure from the linear steady state. The fluid at any t would therefore be divided into two bodies A_u and A_d having $q = 1 - a$ and $q = 1 + a$ (Figure 3.2.1a). The boundary C separating these two bodies is a *potential vorticity front*, a material contour across which q is discontinuous. C initially lies along $y = 0$ but later becomes convoluted. The potential vorticity distribution at any time is determined by the location of C.

The time required for the wave adjusted state to be established is roughly the time needed for a Kelvin wave to propagate a few deformation radii. In dimensionless terms both the Kelvin wave speed and the deformation radius are unity and therefore this time scale is O(1). However, the resulting fluid velocities are O(a), so that the time required to advect C one deformation radius is O(a^{-1}). For $a \ll 1$, C evolves very slowly and, in comparison, the wave adjusted state develops instantaneously. This scale separation was exploited by Hermann *et al.* (1989) who realized that for $a \to 0$ the wave adjusted steady state can be considered an initial condition for the calculation of C.

In order to compute the evolution of C, let

$$\eta = a[\eta_\infty(x, y) + \tilde{\eta}(x, y, \tau)] \tag{3.2.1}$$

where $\tau = a^{-1}t$. The first term on the right-hand side represents the wave adjusted state

$$\eta_\infty(x, y) = \frac{\sinh(x)}{\cosh(\frac{1}{2}w)} + \mathrm{sgn}(y)\left\{ \frac{\cosh(x)}{\cosh(\frac{1}{2}w)} - 1 \right.$$
$$\left. + \frac{4}{w}\sum_{n=1}^{\infty} \frac{(-1)^n}{a_n(1+a_n^2)} \cos(a_n x)e^{-(1+a_n^2)^{1/2}|y|} \right\}, \tag{3.2.2}$$

which can be obtained by taking of limit $t \to \infty$ in Gill's linear solution (3.1.25). This steady state consists of boundary layers on the left and right walls for $y \to -\infty$ and $+\infty$, and a crossing region about $y = 0$. The velocity field is geostrophically balanced:

$$u_\infty = -\frac{\partial \eta_\infty}{\partial y} \quad \text{and} \quad v_\infty = \frac{\partial \eta_\infty}{\partial x} \tag{3.2.3a, b}$$

The second term in (3.2.1) is a correction to the wave adjusted state that varies slowly in time and is determined by the requirement of potential vorticity conservation following fluid motion. The full velocity field is the sum of wave adjusted and transient parts.

$$u = u_\infty + \tilde{u}(x, y, \tau) \quad \text{and} \quad v = v_\infty + \tilde{v}(x, y, \tau). \tag{3.2.4a, b}$$

Substitution of (3.2.1) and (3.2.4) into the shallow water momentum equations (2.1.5 and 2.1.6 with $\delta = 1$ and $F = 0$) then shows that, to lowest order, the correction fields are also geostrophically balanced:

$$\tilde{u} = -\frac{\partial \tilde{\eta}}{\partial y} \quad \text{and} \quad \tilde{v} = \frac{\partial \tilde{\eta}}{\partial x}. \tag{3.2.5a, b}$$

The side wall boundary conditions $u(\pm w/2, y, t) = 0$ imply that $\tilde{u}(\pm w/2, y, \tau) = 0$ since $u_\infty(\pm w/2, y, \tau) = 0$ has already been imposed. Equation (3.2.5a) then requires $\frac{\partial \tilde{\eta}}{\partial y}(\pm w/2, y, \tau) = 0$. It is also necessary that the full solution approaches the wave adjusted solution ($\tilde{\eta} \to 0$) as $|y| \to \infty$ and therefore

$$\tilde{\eta}(\pm w/2, y, \tau) = 0. \tag{3.2.6}$$

Although the transient solution rearranges the velocity field, it does not alter the surface displacement along the sidewalls. The total transport $[2\tanh(w/2)]$ is therefore unaffected by the motion of the potential vorticity front.

In order to calculate the transient solution, one must go beyond the geostrophic approximation and consider higher order balances. First note that the potential vorticity itself can be written in terms of the present variables as

$$\frac{1+\zeta}{1+\eta} = \frac{1 + a\zeta_\infty + a\tilde{\zeta}}{1 + a\eta_\infty + a\tilde{\eta}} = 1 + a[\zeta_\infty - \eta_\infty + \tilde{\zeta} - \tilde{\eta}] + O(a^2).$$

The perturbation potential vorticity can therefore be partitioned into a wave adjusted part $q_\infty = \zeta_\infty - \eta_\infty$ and a transient part $\tilde{q} = \tilde{\zeta} - \tilde{\eta}$. By definition,

$$q_\infty = \text{sgn}(y). \tag{3.2.7}$$

Furthermore, the geostrophic relation for the transient velocities leads to $\tilde{\zeta} = \nabla^2 \tilde{\eta}$ and therefore

$$\tilde{q} = \nabla^2 \tilde{\eta} - \tilde{\eta}. \tag{3.2.8}$$

Substitution of the partitioned velocity and potential vorticity into the full shallow water potential vorticity equation and neglect of $O(a^3)$ terms results in

$$\left(\frac{\partial}{\partial \tau} + (u_\infty + \tilde{u})\frac{\partial}{\partial x} + (v_\infty + \tilde{v})\frac{\partial}{\partial y} \right)(q_\infty + \tilde{q}) = 0. \tag{3.2.9}$$

Thus $q_\infty + \tilde{q}$ $(= \mathrm{sgn}(y) + \tilde{q})$ is advected by the velocity field composed of the sum of the wave adjusted and transient velocities. A fluid column originating from $y < 0$ will initially have $q_\infty = -1$ and $\tilde{q} = 0$. Moreover, $\tilde{q}$ will remain zero as long as the column remains in $y < 0$. Upon crossing $y = 0$, q_∞ jumps to the value $+1$ and $\tilde{q}$ jumps to the value (-2) required to conserve $q_\infty + \tilde{q}$. Similarly, fluid that originates in $y > 0$ and crosses into $y < 0$ has $\tilde{q} = +2$. The situation is summarized in Figure 3.2.1b, which shows that $\tilde{q}$ is nonzero only within lobes of fluid that have crossed $y = 0$.

The transient solution can be computed using a method known as *contour dynamics* that was developed by Zabusky, *et al.* (1979) for two dimensional flows and extended for quasigeostrophic flows (the type under consideration) by Pratt and Stern (1986). Hermann *et al.* (1989) applied the method to the problem at hand in a way that differs only slightly from what is now described.

If the location of the contour C is known at a particular time τ_o, then the distribution of $\tilde{q}$ is known and $\tilde{\eta}(x, y, \tau_o)$ can be found by solving

$$\nabla^2 \tilde{\eta} - \tilde{\eta} = \tilde{q}(x, y, \tau_o). \tag{3.2.10}$$

subject to the boundary conditions (3.2.6). Note that $\tilde{q}(x, y, \tau_o)$ will be nonzero only within the shaded region R (Figure 3.2.1b).

The solution can be expressed in terms of the Green's function $G(x, y; \xi, \mu)$ defined by

$$\nabla^2 G - G = \delta(\xi)\delta(\mu) \text{ and } G = 0 \text{ at } (x = \pm w/2). \tag{3.2.11}$$

Thus

$$\tilde{\eta}(x, y, \tau_o) = \iint_R \tilde{q}(\xi, \mu, \tau_o)G(x, y; \xi, \mu)\,d\mu\,d\xi. \tag{3.2.12}$$

The (geostrophic) velocity fields $(\tilde{u}, \tilde{v})$ can be found by taking the gradient of $\tilde{\eta}$ and adding the results to the known (u_∞, v_∞). As discussed below, $(\tilde{u}, \tilde{v})$ can be expressed in terms of a contour integral around the edge ∂R of region R. Since the contour C is advected by this total velocity, the position of C at $\tau_o + \Delta \tau$ can be estimated. Then (3.2.12) can be applied to the new $\tilde{q}$ to determine the corresponding surface elevations and velocities. These steps are then repeated leading to an iterative procedure that can be implemented numerically. The nondimensional solution depends only on the channel width $w[= w^* f/(gD)^{1/2}]$.

A convenient and computationally efficient form of the Green's function is

$$G(x, y; \xi, \mu) = \frac{1}{2}\pi \sum_{n=-\infty}^{\infty} (\mathrm{K}_{1,n} + \mathrm{K}_{2,n}), \tag{3.2.13}$$

where

$$\mathrm{K}_{1,n} = -K_o\left\{\left[(x - \xi - 2nw)^2 + (y - \mu)^2\right]^{1/2}\right\},$$

$$\mathrm{K}_{2,n} = K_o\left\{\left[(x + \xi + w + 2nw)^2 + (y - \mu)^2\right]^{1/2}\right\},$$

and K_o denotes the modified Bessel function of zero order.

In advance of the actual computation, a certain amount of physical intuition can be gained by careful consideration of (3.2.13). First consider the term $K_{1,0}$:

$$K_{1,0} = -K_o \left\{ \left[(x-\xi)^2 + (y-\mu)^2 \right]^{1/2} \right\}$$

describing a cyclonic Helmholtz point vortex centered at $(x, y) = (\xi, \mu)$. At large distances from the center, the free surface displacement and associated counterclockwise swirl velocity decay exponentially. The decay scale is the Rossby radius of deformation, here unity. In an infinite domain, this term would comprise the entire Green's function. An isolated eddy composed of a patch of uniform potential anomaly $\tilde{q} = \tilde{q}_o$ would have an η field obtained by integrating $\tilde{q}_o$ times this Green's function over the area of the patch.

If a single boundary in the form of a wall at $y = w/2$ is present, the boundary condition can be satisfied by adding an image vortex to a hypothetical body of fluid lying inside the wall (Figure 3.2.2a). The image vortex is equal in

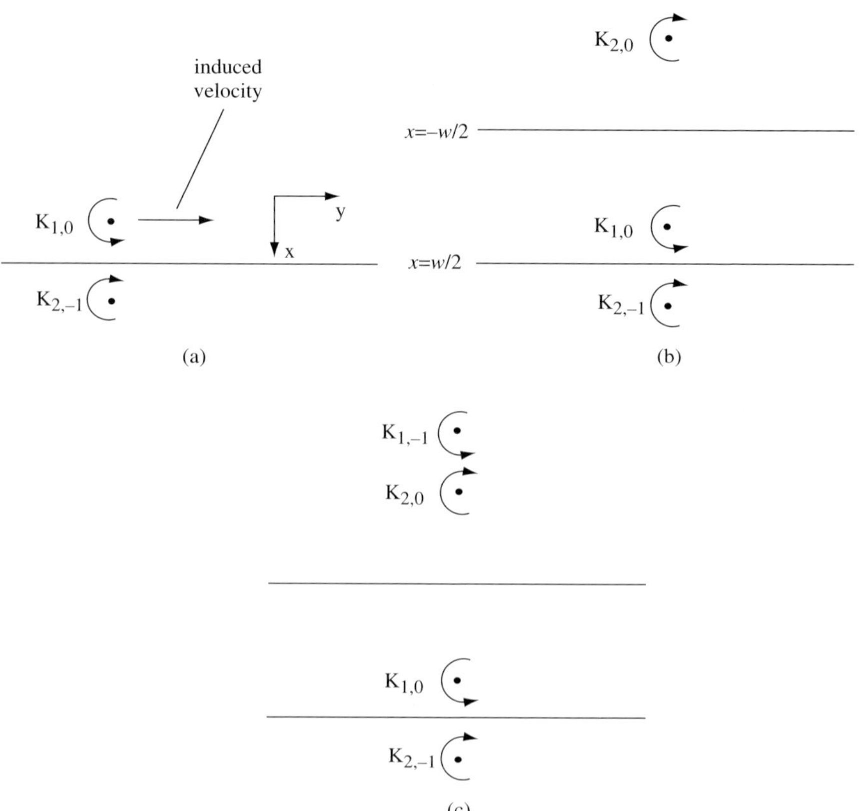

FIGURE 3.2.2. Point vorticies and images needed to satisfy the condition of no normal flow at the channel sidewalls.

strength but opposite in sign and is located an equal distance inside the wall. The x-velocity at the wall created by the image is equal and opposite to that due to the original vortex. The condition of zero normal flow is thereby satisfied. Note that the image for $K_{1,0}$ is $K_{2,-1}$ and that the velocity field created by the latter causes the original vortex to move parallel to the wall. An anticyclonic vortex will move towards negative y whereas a cyclonic vortex will move towards positive y. The motion is identical to that of a dipole (a pair of equal and opposite vorticies).

The boundary condition problem for the normal velocity become more difficult in the presence of two walls (Figure 3.2.2b). Beginning with the images $K_{1,0}$ and $K_{2,-1}$ one could add a third $K_{2,0}$ beyond the wall at $x = -w/2$. Alone, $K_{1,0}$ and $K_{2,0}$ would satisfy the boundary condition at $x = -w/2$. The original image vortex $K_{2,-1}$ gives rise to a small nonzero u at $x = -w/2$ but this vortex lies further from $x = -w/2$ than either $K_{1,0}$ or $K_{2,0}$. Since the velocity field of $K_{2,-1}$ decays away exponentially there is hope that the error in the boundary condition at $x = -w/2$ might not be too large. The same remarks can be made for the pair $K_{1,0}$ and $K_{2,-1}$ which alone would satisfy the boundary condition at $x = w/2$ if not for the presence of $K_{2,0}$.

In order to construct a Green's function that exactly satisfies the condition $u = 0$ on both walls, it is necessary to add further images at successively larger distances from the walls. For example, the contaminating effect of $K_{2,-1}$ on the boundary condition at $x = -w/2$ can be countered by adding its image $K_{1,-1}$ (Figure 3.2.2c). The contaminating effect of the $K_{2,0}$ on the boundary condition at $x = w/2$ can be countered by adding its image $K_{1,1}$. In general, $K_{1,n}$ corrects $K_{2,n-1}$ for $(n \geq 0)$ whereas $K_{1,n}$ corrects $K_{2,n}$ for $(n < 0)$. The series in (3.2.13) is constructed following this principle. As the reader might gather from an inspection of Figure 3.2.2c, the effect of adding all the extra images is to somewhat retard the dipole effect mentioned earlier. Thus the primary vortex $K_{1,0}$ does not move towards larger values of y as rapidly. In fact, a single vortex placed at the centerline $x = 0$ of the channel would not translate at all.

The geostrophic velocities for the transient solution can be obtained from (3.2.12) as

$$\nabla_{(x,y)} \tilde{\eta}(\xi, \mu, \tau_o) = (\tilde{v}, -\tilde{u}) = \iint_R \nabla[\tilde{q}(\xi, \mu, \tau_o) G(x, y; \xi, \mu)] d\xi d\mu \qquad (3.2.14)$$

where $\nabla_{(x,y)} = \frac{\partial}{\partial x} + \frac{\partial}{\partial y}$. The region R of anomalous potential vorticity is composed of several subregions or lobes, each of which contains fluid of uniform $\tilde{q}$. Consider a subregion R_o for which $\tilde{q}$ has the uniform value $\tilde{q}_o = -2$ (Figure 3.2.3a). The contribution to the integral in (3.2.14) from this subregion is

$$I_{R_o} = \tilde{q}_o \iint_{R_o} \nabla_{(x,y)} G(x, y; \xi, \mu) d\mu d\xi \qquad (3.2.15)$$

$$= \frac{\tilde{q}_o}{2\pi} \iint_{R_o} \sum_{n=-\infty}^{\infty} (\nabla_{(x,y)} K_{1,n} + \nabla_{(x,y)} K_{2,n}) d\mu d\xi$$

$$= -\frac{\tilde{q}_o}{2\pi} \iint\limits_{R_o} \sum_{n=-\infty}^{\infty} (\nabla_{(\xi,\mu)} K_{1,n} + \nabla_{(-\xi,\mu)} K_{2,n}) d\ \mu d\xi$$

$$= -\frac{\tilde{q}_o}{2\pi} \oint\limits_{\partial R_o} \sum_{n=-\infty}^{\infty} \left[K_{1,n}(d\mu, d\xi) + K_{2,n}(-d\mu, d\xi) \right]$$

The third step is made possible by the $(x \leftrightarrow -\xi, y \leftrightarrow -\mu)$ symmetry in $K_{1,n}$ and by the $(x \leftrightarrow \xi, y \leftrightarrow -\mu)$ symmetry in $K_{2,n}$. The final step follows from the application of Green's theorem. The contour integral is performed counterclockwise around the edge ∂R_o of R_o.

In the problem at hand, where $\tilde{q}_o$ equals $+2$ or -2, depending on the subregion, application of (3.2.15) over each subregion and summation of the results leads to

$$(\tilde{v}, -\tilde{u}) = \frac{1}{\pi} \oint\limits_{\partial R} [\ \sum_{n=-\infty}^{\infty} [K_{1,n}(d\mu, d\xi) + K_{2,n}(-d\mu, d\xi)]] \tag{3.2.16}$$

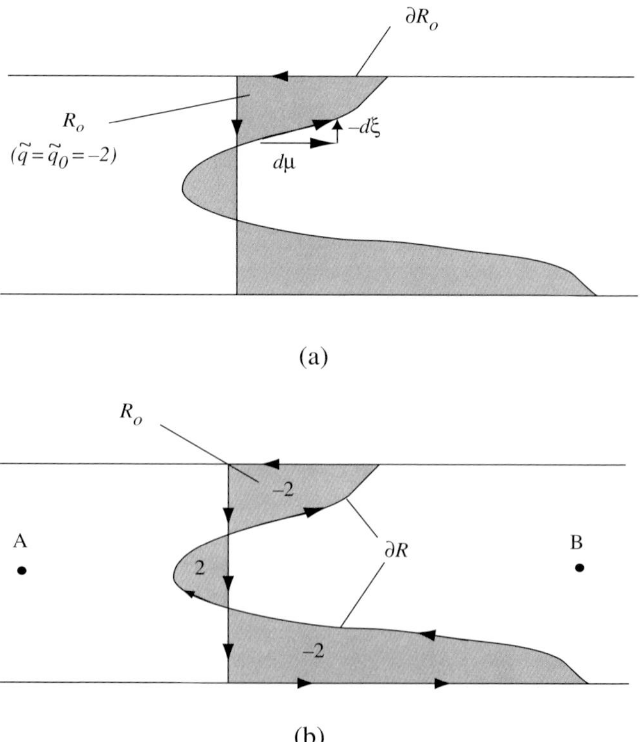

FIGURE 3.2.3. (a): Integration contour about the lobe R_o of anomalously low potential vorticity. (b): Integration contour for the entire region of anomalous potential vorticity.

where the integration circuit ∂R is shown in Figure 3.2.3b and the direction of integration is such as to keep higher $\tilde{q}$ values on the right. In normal practice the evolution of ∂R is calculated by seeding a group of material points along it and using (3.2.16) to follow the motion of each point. If $[x_n(t),\ y_n(t)]$ represents the coordinates of point n, then

$$dx_n/dt = u(x_n, y_n) \text{ and } dy_n/dt = v(x_n, y_n),$$

with u and v given by (3.2.16). These relations are integrated numerically over a small time increment for all the material points on ∂R and the new positions are used to update ∂R. In this way, the evolution of the potential vorticity front can be calculated without the need to explicitly consider any quantities measured away from the front.

Examples of the solutions reveal significant departures from the linear case, even when the channel is very wide (Figure 3.2.4). On the right-hand side of the channel, the front is carried rapidly downstream by the boundary layer flow and its leading edge has moved beyond the frame boundary by $\tau = 10$. More significantly, there is a tendency for the front to move towards positive y along the *left* wall. In the channel interior the front roughly maintains its original position. [The apparent movement of the interior front towards negative y is actually due to the fact that the plot is made in a frame of reference moving with the mean velocity $U = 2w^{-1}\tanh(w/2)$.]

The intrusion of low potential vorticity fluid along the left wall is due primarily to the image effect described above. The lobe of intruding fluid that is bounded

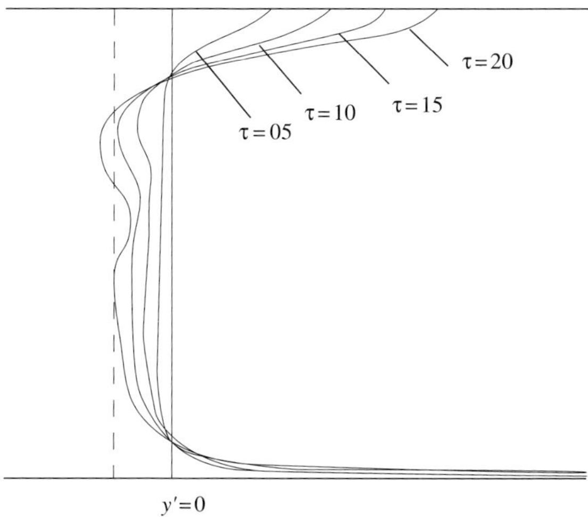

FIGURE 3.2.4. Evolution of the potential vorticity front for the case $w = 25$ (plan view). In order to visually separate the contour at different times, the results are plotted in a frame of reference y' that translates towards positive y at speed .08. The original position of the barrier ($y = 0$) at $\tau = 0$ is indicated by a dashed line. (From Hermann *et al.*, 1989).

by the potential vorticity front on the right, the wall $x = -w/2$ on the left, and $y = 0$ has potential vorticity anomaly $\tilde{q} = -2$ as suggested in Figure 3.2.3a. The vorticity of this blob is anticyclonic and must have a cyclonic image lobe on the other side of the left wall in order that the boundary condition $u = 0$ be satisfied. The tendency of the image lobe is to advect the anticyclonic fluid towards positive y.

Associated with the left-wall intrusion is an overshooting across $y = 0$ of the boundary current (Figure 3.2.5). Further downstream the boundary current veers away from the wall and reverses course. The current returns to $y = 0$ where it crosses the channel. As time progresses, the intrusion widens and the crossing route increasingly departs from $y = 0$. Hermann *et al.* (1989) speculate that eventually the crossing route will be swept downstream and that the final steady state at *any* fixed y will eventually be one with only a left-wall boundary layer. Confirmation is made for the case $w = 10$, where the downstream movement of the potential vorticity front is clear (Figure 3.2.6a). As the front moves away from the original position of the barrier, the cross-sectional profile of surface elevation there evolves to the point where only a left-wall boundary flow remains (Figure 3.2.6b).

The loss at $y = 0$ of the cross flow presents difficulties for the principle of geostrophic control. If we choose points A and B (Figure 3.2.3b) as our interior reference locations, then the total transport $2\tanh(w/2)$ is initially bounded by the value $\eta_A - \eta_B = 2$, as required. After the potential vorticity front travels beyond B, however, η_A and η_B become equal and the bound fails. Thus, geostrophic control applies after the wave adjusted flow is established but before the potential vorticity dynamics have affected the final adjustment.

The foregoing results suggest that geostrophic control might apply in systems where the time dependence is imposed by the tides or some other oscillatory forcing. The forcing period T must be longer than the wave adjustment time in order to allow the cross-channel flow to become established. The dimensional wave adjustment time is roughly the deformation radius $(gD)^{1/2}/f$ divided by the Kelvin wave speed $(gD)^{1/2}$. In addition, T must be much shorter than the advective time of the potential vorticity front, else the crossing flow will be carried away. The advective time is at least (D/a^*) times the wave adjustment

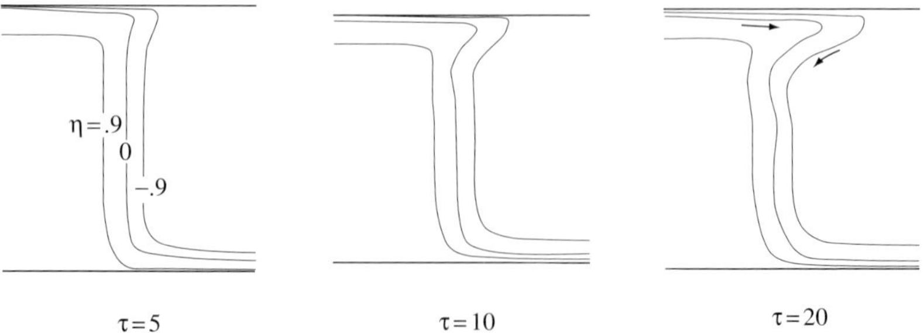

FIGURE 3.2.5. Evolution of the surface elevation (η) field for the case shown in Fig. 3.2.4. (From Hermann *et al.*, 1989).

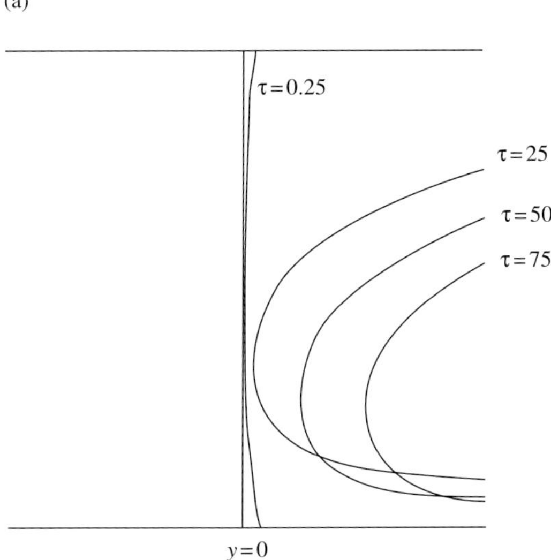

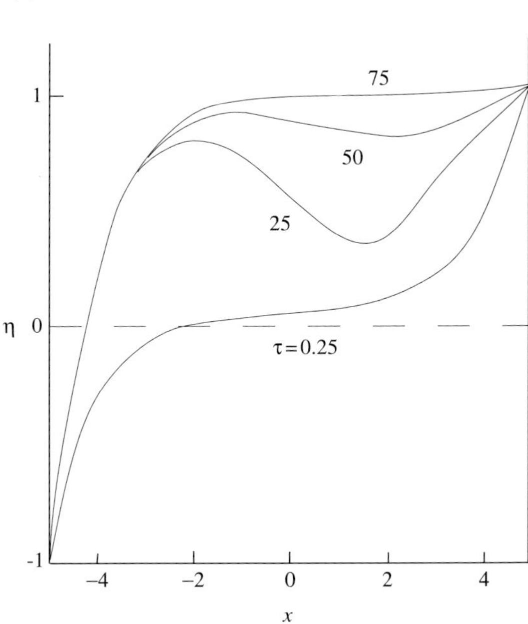

FIGURE 3.2.6. (a) The evolution of the potential vorticity front for the case $w = 10$. The portions of the curves near the side walls have been carried downstream and out of the frame. (b) The corresponding surface elevation (η) field at $y = 0$. (From Hermann *et al.*, 1989).

time, where a^* is now the (dimensional) tidal amplitude. Thus the principle of geostrophic control appears to require

$$1 \ll Tf \ll \frac{D}{a^*}. \tag{3.2.17}$$

(Pratt, 1991). As shown by the calculations of Herman *et al.* (1989) the upper bound in (3.2.17) may be overly conservative when the channel is much wider than the deformation radius.

Middleton and Viera (1991) and Hannah (1992) have assessed the validity of the geostrophic control in the Bass Strait, the channel separating Tasmania from Australia. When a low frequency (240 hr) wind and pressure forcing period is used for T, (3.2.17) is satisfied and geostrophic control holds, at least in their models. If one or both of the neighboring basins is effectively finite in extent, which is apparently not a problem for the Bass Strait, then the above arguments become complicated. A Kelvin wave can circle the basin and return to the strait. Wright (1987) has investigated models of this behavior and found that geostrophic control is more restricted in scope.

b. *Interactions between Kelvin and Poincaré Waves*

Given sufficient time, weak nonlinearities can also alter the character of the transients that set up the wave adjusted state. A numerical solution[2] obtained by Tomasson and Melville (1992) for $a = 0.15$ and $w = 2$ (Figure 3.2.7a) gives an overview. The solution is obtained by integrating the Boussinesq equations, an approximation to the full Euler equations permitting weak nonlinearity and weak nonhydrostatic effects. One of the most striking differences with Gill's linear solution is the lack of symmetry between the forward and backward moving waves. The forward waves, especially the region $75 < y < 210$, contain an abundance of smaller spatial scales, while the backward moving waves $(-200 < y < -75)$ remain relatively smooth. Another difference is that the leading edges of the waves (near $y = \pm 210$) exhibit curvature. An enlarged view of the forward waves (Figure 3.2.7b) shows this feature clearly. The leading edge of the advancing front is perpendicular to the right wall at the wall, but becomes increasingly oblique as one moves away from this wall. This aspect will be addressed further in Section 3.6.

A physical process that accounts for much of the new behavior is the resonant excitation of Poincaré waves by finite amplitude Kelvin waves. According to the linear solution, the removal of the barrier at $y = 0$ excites two Kelvin waves that move away along their respective walls. Poincaré waves are also generated but they are outrun by the Kelvin waves, which have larger group speeds. There is a tendency for the forward-propagating Kelvin wave to steepen and the backward-propagating Kelvin wave to rarefy, as described in Section 2.2. The smaller the

---

[2] To obtain this solution, the initial step in depth was slightly smoothed, so there is no distinct potential vorticity front.

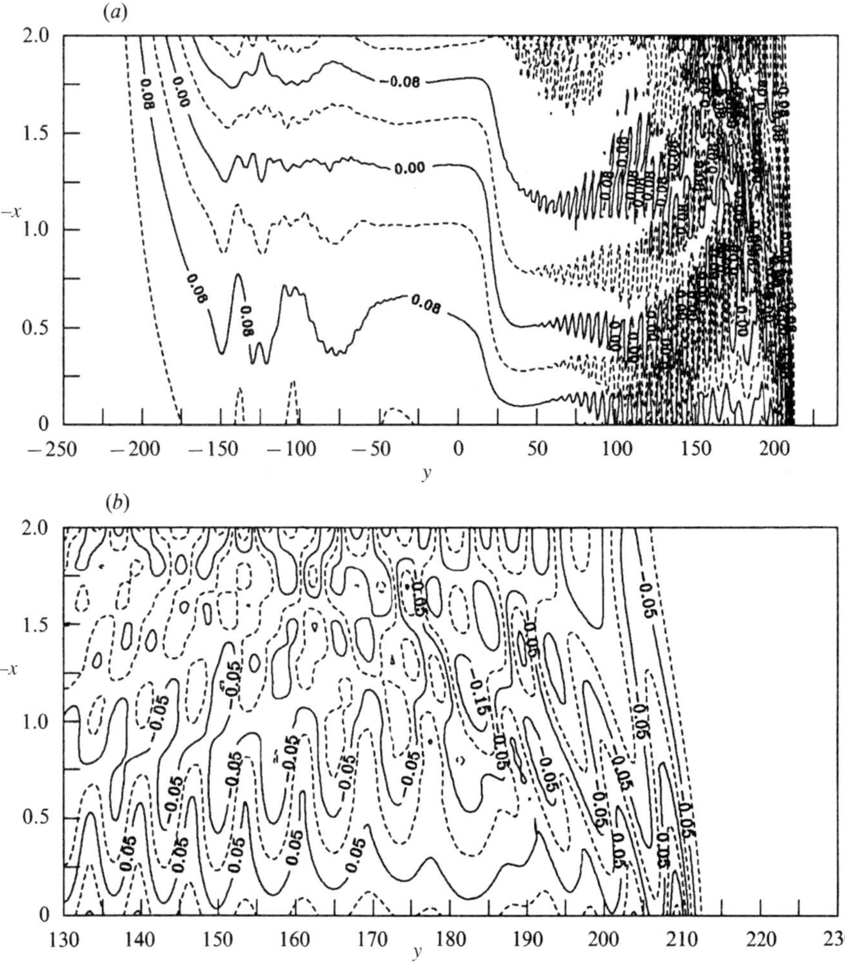

FIGURE 3.2.7. Surface elevation (η) field at $t = 200$ based on a numerical solution to the dam break problem with $a = .15$ and $w = 2$. Frame (b) shows the region near the edge of the forward wave front. (Tomasson and Melville, 1992, Figure 12).

step size a, the more slowly the steepening or rarefacation occurs. For sufficiently large a, the forward Kelvin wave may break, leading to the formation of a shock. This process is discussed in the next few sections. However, for moderate or small values of a, the steepening process may be arrested by dispersive effects due to nonhydrostatic accelerations. The equilibrated, finite-amplitude Kelvin wave propagates a bit more rapidly than its linear counterpart.

Now consider the linear dispersion relations for Poincaré and Kelvin waves in a channel geometry (Figure 3.2.8 and equations 2.1.27 and 2.1.29). The forward Kelvin wave has dimensional frequency $\omega^* = (gD)^{1/2}l^*$, as represented by the straight line. The effect of the nonlinear increase of speed for the forward Kelvin

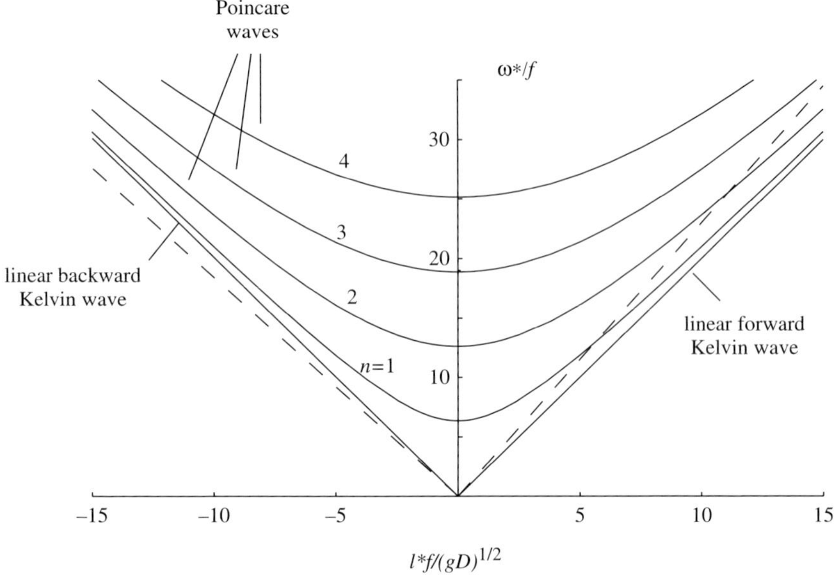

FIGURE 3.2.8. Dispersion relations for Poincaré and Kelvin waves. The dimensional frequency and along-channel wave number are denoted by ω^* and l^*. (Based on a figure from Tomasson and Melville, 1992).

wave can qualitatively be demonstrated by increasing the slope of this line thereby creating intersections with the dispersion curves for the Poincaré modes. A slight increase in slope leads to intersections only at high wave numbers. At the intersection points the phase speed of a Poincaré mode matches that of the Kelvin mode, a necessary condition for nonlinear interaction between the two. The Poincaré modes feed on energy contained in the Kelvin wave. Their presence accounts for some of the wave activity behind the leading edge of the forward Kelvin wave. The energy drain causes the Kelvin wave amplitude to gradually decay and the process of energy transfer is gradually attenuated. Since the backward Kelvin wave rarefies, the slope of the corresponding dispersion curve $\omega^* = -(gD)^{1/2}l^*$ decreases, moving it away from those of the Poincaré modes. The region to the rear of the backward advancing wave front is therefore relatively free of small-scale wave activity.

Exercises

(1) Find the wave adjusted steady state $\eta_\infty(x, y)$ using local conservation of linearized potential vorticity. That is, use the same approach as in the Rossby adjustment problem on an infinite plane, as discussed at the beginning of Section 3.1. First show that the mathematical problem is

$$\nabla^2 \eta_\infty - \eta_\infty = \text{sgn } y$$

subject to the boundary conditions

$$\eta_\infty = \pm \frac{1}{2} \tanh\left(\frac{1}{2}w\right) \quad \left(x = \pm\frac{1}{2}w\right)$$

and

$$\eta_\infty \to \pm\left[-1 + \frac{e^{\pm x}}{\cosh(\frac{1}{2}w)}\right] \quad (y \to \pm\infty)$$

(Note: These boundary conditions are deduced from the solution for K in the Kelvin wave part of the solution.) Then solve for the wave adjusted state.

(2) *Propagation tendency of a potential vorticity wedge.* As a crude model of the behavior near the leading edge of the left-wall intrusion, consider a semi-infinite wedge of fluid with $\tilde{q} = -2$ intruding into an ambient fluid with $\tilde{q} = 0$. The outside edge of the wedge forms an angle θ with the wall. Show that the velocity $v = v_L$ at the leading edge of the wedge is given by:

$$v_L = 1 - \cos(\theta) \tag{3.2.18}$$

(The identity $\int_0^\infty K_o(y)dy = \frac{\pi}{2}$ may prove helpful.)

By consideration of the image of the wedge, deduce the nose speed for a wedge of potential vorticity $\tilde{q} = 2$ propagating along a right wall:

$$v_R = \cos(\theta) - 1. \tag{3.2.19}$$

For $\theta < \pi/2$ note that $v_L > 0$ while $v_R < 0$. In the Rossby channel problem, however, a background velocity v_∞ exists along the right wall and this tends to advect the wedge towards positive y. The net result is that the wedge moves towards positive y at something less than the advective speed.

By a more complicated analysis (Hermann *et al.* 1989) it is also possible to demonstrate that the left wall wedge will steepen and the right wall wedge will rarify, as observed in the numerical solutions for large values of w.

3.3. Rossby Adjustment: Fully Nonlinear Case

We now come to the most difficult case of the channel Rossby adjustment: that in which the initial depth difference is moderate or large. Let

$$d(x, y, 0) = \begin{cases} 1 & (y<0) \\ d_o & (y>0) \end{cases}, \tag{3.3.1}$$

with $u(x, y, 0) = v(x, y, 0) = 0$ as before, so that $1 - d_o$ corresponds to the step height $2a$ of the previous sections. For O(1) values of $1 - d_o$ there is no longer a clear spatial separation between the Kelvin waves and the potential vorticity front. In the limiting case $d_o = 0$, the channel is initially dry downstream of the barrier and the leading edge of the advancing intrusion *is* the potential vorticity front. The introduction of a finite step size brings other new processes into play, including the separation of the fluid from the left wall and the formation of shocks. The results described below are largely due to Helfrich *et al.* (1999).

a. Case $d_o = 0$

The case of zero depth on the downstream side of the barrier is simplest to treat and will be considered first. Some of the general features of the flow can be anticipated. After the barrier is removed, we expect that the fluid will spill forward into $y > 0$ forming an intrusion (Figure 3.3.1). The linear solution suggests that the velocities along the right wall will be largest and therefore the leading edge or nose of the intrusion should follow that wall. We denote the position and speed of the nose by y_{nose} and c_{nose}. Behind the nose the width of the intrusion will gradually increase until contact with the left wall is made at $y = y_{\mathrm{sep}}$. The speed c_{sep} of the separation point is shown to be positive in Figure 3.3.1, and this will be confirmed later.

It is possible to approximate the solution for this case using semigeostrophic theory for uniform potential vorticity, the tools of which were developed in Sections 2.2 and 2.3. Central to the semigeostrophic approximation is the assumption of gradual variations in the y-direction, a condition that is clearly violated near $y = 0$ in the early stages of evolution. Nevertheless, there is some hope that, as time progresses, the changes in y will become gradual enough that semigeostrophic theory will capture the predominant features of the solution. We will therefore describe this theory and compare it to numerical solutions based on the full shallow water equations.

The semigeostrophic solution can be obtained using the method of characteristics in much the same way that the nonrotating dam-break problem was solved

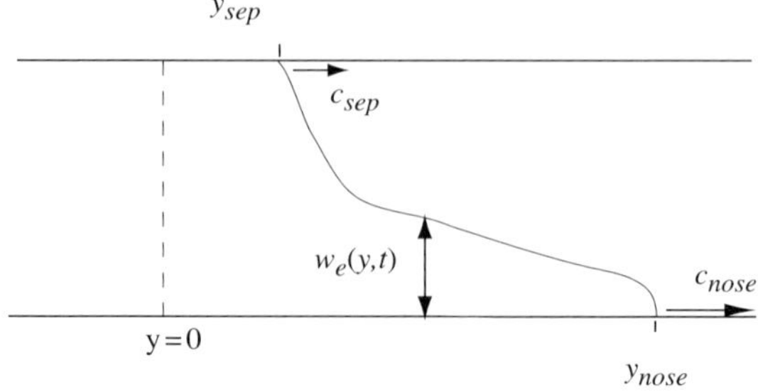

FIGURE 3.3.1. Schematic plan view of the intrusion following a dam-break in a rotating channel.

(see Section 1.3). Where the flow remains attached, the independent variables are $\bar{d}$ and $\hat{d}$, one-half the sum and difference of the wall depths (see 2.2.5 and 2.2.6). The initial conditions are therefore

$$\bar{d}(y,0) = \begin{cases} 1 & (y<0) \\ 0 & (y>0) \end{cases} \quad \text{and} \quad \hat{d}(y,\ 0) = 0 \tag{3.3.2}$$

Although the characteristic speeds $c_\pm$ and Riemann invariants $R_\pm$ are more complicated than in the nonrotating case, the central ideas involved in the construction of the characteristic curves remain the same. In particular, one must address the fact that there is an infinity of solutions that satisfy the discontinuous initial condition. Replacement of the discontinuity with a short interval of continuous depth change (as in Figure 1.3.2) resolves this problem. As was the case in the nonrotating version, the assumption of uniform R_- within this interval leads to the unsatisfactory conclusion that the leading edge of the intrusion will move towards *negative* y. We therefore take R_+ as being uniform as before. It follows from (2.2.23–2.2.24) that

$$R_+ = T^{-1}\hat{d} + \int^{\bar{d}} \alpha^{-1/2}[1 - T^2(1-\alpha)]^{1/2}\,d\alpha \tag{3.3.3}$$

$$= T^{-1}\hat{d} + \bar{d}^{1/2}\left[1 - T^2(1-\bar{d})\right]^{1/2} + \frac{1-T^2}{T}\log\left\{2\bar{d}^{1/2}T + 2[1 - T^2(1-\bar{d})]^{1/2}\right\}$$

$$= 1 + \frac{1-T^2}{T}\log(2T+2),$$

where $T = \tanh(w/2)$. The first two lines follow from the definition of R_+ for nonseparated flow. Also the value q of the potential vorticity has been set to its initial value of unity. The final step results from the evaluation of R_+ using the initial conditions (3.3.2). The result is a relationship between $\bar{d}$ and $\hat{d}$ that holds for $y < y_{\text{sep}}$.

In accordance with the ideas developed in Section 1.3, the uniformity of R_+ over the whole body of fluid implies that the '$-$' characteristic curves must have constant slope. The characteristic speeds are given by (2.2.22)

$$c_- = T^{-1}\hat{d} - \bar{d}^{1/2}[1 - T^2(1-\bar{d})]^{1/2}. \quad (y < y_{\text{sep}}) \tag{3.3.4}$$

Calculation of the values of c_- over the short interval replacing the step gives a value that increases from left to right (see Exercise 1), and thus the '$-$' characteristic curves fan out from the origin, as in Figure (1.3.2). Conservation of $R_-(\bar{d}, \hat{d})$ along each of the fanning curves in conjunction with the independent relationship (3.3.3) implies that $\bar{d}$ and $\hat{d}$ are individually conserved along each curve (just as v and d were conserved along the '$-$' characteristics of the nonrotating problem).

In the separated portion of the flow, a convenient choice of variables is $\bar{d}$ and the separated width w_e of the current. As shown in Section 2.3 the definition (3.3.4) of c_- remains valid provided we replace $\hat{d}$ by $\bar{d}$ and w by w_e. Thus

$$c_- = T_e^{-1}\bar{d} - \bar{d}^{1/2}[1 - T_e^2(1 - \bar{d})]^{1/2} \quad (y_{\text{sep}} < y < y_{\text{nose}}) \tag{3.3.5}$$

where $T_e = \tanh(w_e/2)$. An analytical expression for $R_+(\bar{d}, w_e)$ is not known and values must therefore be tabulated. The procedure for doing so [Stern et al., 1982; Kubokawa and Hanawa, 1984a; or Helfrich et al., 1999] is algebraically involved and will not be detailed here. Nonetheless, the method for obtaining the solution in the separated region is essentially the same as in the attached region. We set R_+ equal to its initial value (the final expression in 3.3.3) and thereby obtain an implied relationship between $\bar{d}$ and w_e that must hold throughout the separated region. The values of $\bar{d}$ and w_e along each '$-$' characteristic are then determined from this relationship in conjunction with (3.3.5).

Solutions to the unapproximated initial-value problem require a numerical algorithm for the full shallow water equations. The finite-difference code in the present case is summarized by Helfrich et al., 1999. It allows the fluid to vanish,[3] resulting in the formation of a free edge, either in the interior or through sidewall separation. The code also allows for the formation and maintenance of jumps and bores consisting of large changes in the depth and velocity over a few grid points. We have already shown that ideal hydraulic jumps in one-dimensional flows conserve volume flux and flow force. These ideas extend to two-dimensional jumps in rotating systems in a way that will be made clear in Section 3.5. The numerical code in question is written in a way that enforces conservation of the correct properties and, in particular, adds no mass or momentum to the flow. A key step is to base the code on the flux form (2.1.17) of the shallow water equations, which is essentially a conservation law for the quantities that need to be conserved across the jump.

The semigeostrophic solution (upper panel of Figure 3.3.2) can be compared with numerical solutions to the full 2-D shallow water equations (lower panel) for the case $w = 2.0$ (a channel width equal to two deformation radii based on the initial upstream depth). Both solutions agree qualitatively with each other and with the anticipated scenario. For example, the separated intrusion becomes increasingly narrow with time, in agreement with the fanning characteristic curves. The semigeostrophic theory does not capture the oscillations that are apparent in the lower panel of the figure and are associated with Poincare' waves. These waves are filtered by the semigeostrophic approximation. In the full numerical solution for $w = 4$ (Figure 3.3.3), the rarefying nature of the intrusion is apparent and the oscillations are more pronounced.

---

[3] The fluid depth is formally judged to vanish when d reaches a value of the order 10^{-5}. A thin sheet of fluid with this thickness is then maintained over those portions of the channel considered 'dry'.

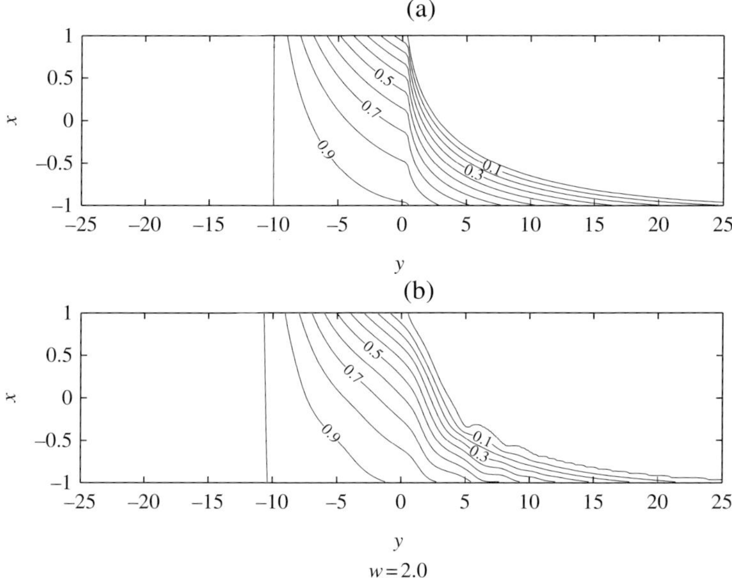

FIGURE 3.3.2. Contours of the depth d field at $t = 10$ for a channel with $w = 2.0$. (a) The semigeostrophic solution. (b) The numerical solution to the full shallow water equations. (From Helfrich *et al.*, 1999).

The value of c_{nose} can be determined from the semigeostrophic solution by noting that the nose moves along a '$-$' characteristic. Taking the simultaneous limit $\bar{d} \to 0$ and $w_e \to 0$ in (3.3.5) leads to $c_{nose} \to 2\bar{d}/w_e$. The values of $\bar{d}$ and w_e are also constrained by the requirement that $R_+(\bar{d}, w_e)$ is equal to its initial value, as specified by (3.3.3). However, $R_+(\bar{d}, w_e)$ is only known in tabulated form, and the limit of $2\bar{d}/w_e$ as $\bar{d}$ and w_e both approach zero must be ascertained numerically. The resulting c_{nose} depends on the value of T (Figure 3.3.4). The nose speed increases from its nonrotating value of 2 (dimensionally $2\sqrt{gD}$, where D is the initial depth behind the barrier) for $w = 0$ to the asymptotic value 3.8 for infinite w (or $T = 1$). The circles indicate the nose speed as given by the full numerical solution. Although the 'exact' result mirrors the behavior of the semigeostrophic result, the actual values are substantially less than those predicted by semigeostrophic theory.

The value of c_{sep} can be predicted by setting $c_- = c_{\text{sep}}$ and $\bar{d} = \hat{d}$ in (3.3.4), giving a relation between c_{sep} and the value of d at the separation point. A second relation for this $\bar{d}$ can be obtained by setting $\bar{d} = \hat{d}$ in (3.3.3). The relationship between c_{sep} and $T(w)$ is then obtained through elimination of $\bar{d}$ between the two, a procedure that must be carried out numerically. The prediction (lower solid curve in Figure 3.3.4) agrees well with the c_{sep} predicted by the full solution (squares). In the narrow channel limit ($T \to 0$), the nonrotating limit of equal c_{sep} and c_{nose} is obtained, implying that the free edge is directed perpendicular

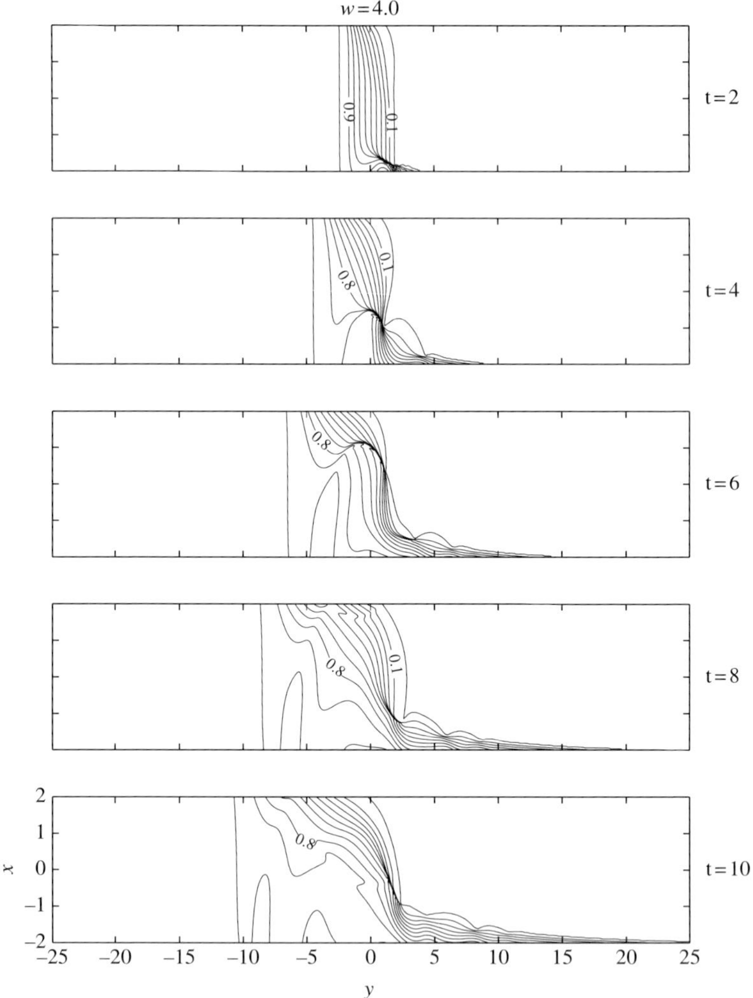

FIGURE 3.3.3. Full numerical solution for $d(x, y, t)$ at the indicated times for $w = 4.0$. The contour interval is 0.1. (From Helfrich *et al.*, 1999).

to the channel walls. In the other extreme ($T \to 1$), c_{sep} vanishes implying that the separation point remains pinned to the position of the initial barrier.

As in the nonrotating version of the dam-break problem, the steady state achieved for $t \to \infty$ is hydraulically critical. The corresponding values of $\bar{d}$ and $\hat{d}$ may be found by setting $c_- = 0$ in 3.3.4. A second relationship between these $\bar{d}$ and $\hat{d}$ is provided by (3.3.3). The volume transport $2\bar{d}\hat{d}$ of the resulting flow, shown by the solid curve in Figure 3.3.5, agrees quite well with the numerically determined values (diamonds). Both approach the nonrotating prediction of $(2/3)^3 w$ (dashed line) obtained in Section 1.3. For large w the

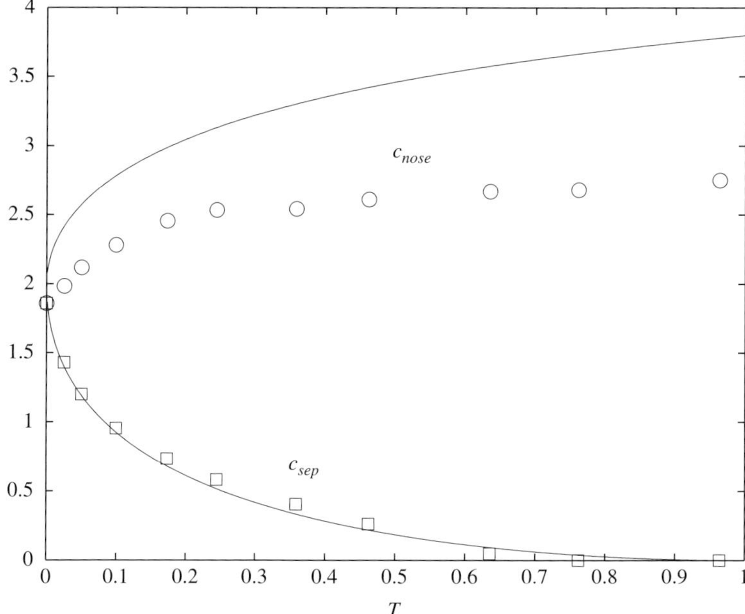

FIGURE 3.3.4. The intrusion nose speed c_{nose} and the separation point speed c_{sep} as functions of the width parameter T. The solid curve corresponds to theory and the symbols correspond to the numerical model results. (Based on Helfrich *et al.*, 1999, Figure 3).

predicted and observed transports approach the asymptotic value 0.5 (dimensionally $0.5gD^2/f$).

When the full dam break problem is carried out in the presence of a deep, overlying or underlying fluid, the behavior of the nose region is quite different. As discussed on Section 4.5, the leading edge becomes blunt and propagates at a much slower speed than that predicted above for the equivalent gD^2/f.

b. *The Case* $d_o > 0$

When the initial depth ahead of the barrier is nonzero, the flow will differ in many respects from what has just been discussed, or so one might anticipate. Separation of the fluid from the left wall will presumably not occur. The potential vorticity front formed between the initially shallow and deep fluid will be distinct from the leading edge of the Kelvin wave. Most importantly, the Kelvin wave may break and form a shock. As will be shown later in the chapter, potential vorticity is generally not conserved across such a shock and the foregoing theory, which was based on the assumption of globally uniform q, becomes more difficult to justify. All of these factors contribute to an understanding of the partial dam-break problem that is relatively poor. We will simply present some numerical solutions that show the major features of the flow.

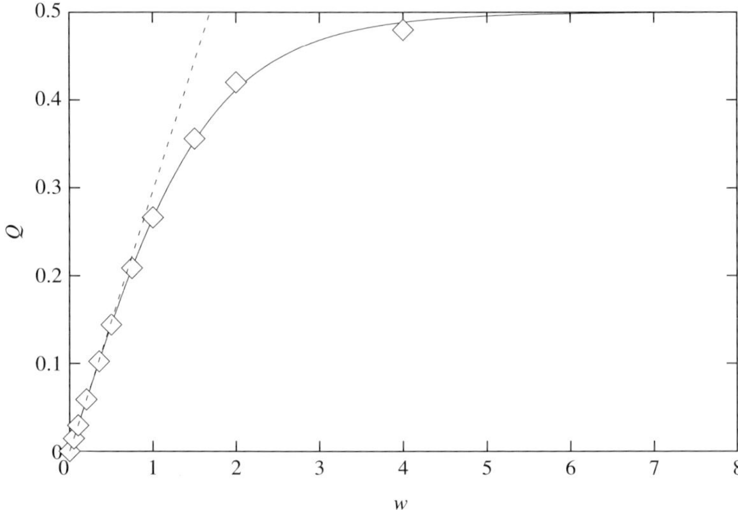

FIGURE 3.3.5. Final transport Q vs. w for the full dam-break ($d_o = 0$). The solid curve is from the semigeostrophic theory; it is also used as a scale factor in Figure 3.3.8b, where it is referred to as Q_o. The dashed line is based on the nonrotating theory and the diamonds correspond to the numerical solution. (From Helfrich *et al.*, 1999).

Based on simulations, there appear to be two main forms that the Kelvin wave shock may take. The first is favored by small values of d_o or large w. An example for the case $d_o = 0.1$ and $w = 4$ (Figure 3.3.6) shows a fully developed shock near $y = 15$ at $t = 12$. The leading edge consists of an abrupt change in depth, the amplitude of which decays away from the right wall. Virtually none of the disturbance is felt at the left wall. In addition, the leading edge curves backwards, forming an oblique angle with the right wall. The potential vorticity front, which is shown as a heavy, solid line, lies just to the rear of the shock along the right wall. As one moves away from the right wall, the position of the front lags further behind the shock. At the left wall, the front remains pinned near $y = 0$.

The second type of shock is reminiscent of the unstable Kelvin wave discussed in Section 3.2. It is favored by small or moderate $1 - d_o$ or w. In an example based on $d_o = .25$ and $w = 1$ (Figure 3.3.7), the potential vorticity front now lags well behind the shock at all x. In addition, the entire front moves forward and is no longer pinned to the left wall at $y = 0$. The shock, which can be seen at $t = 20$ near $y = 20$ is now felt across the entire channel width. The Poincaré waves just behind the front and along the left wall may have been generated resonantly by the Kelvin wave, as described in Section 3.2b. This generation mechanism is further suggested by the lack of similar oscillations near the backward-propagating, left-wall Kelvin wave (which is rarefying and which roughly occupies the interval $-20 < y < -5$ at $t = 20$). Both of the shocks described above will be revisited later in this chapter.

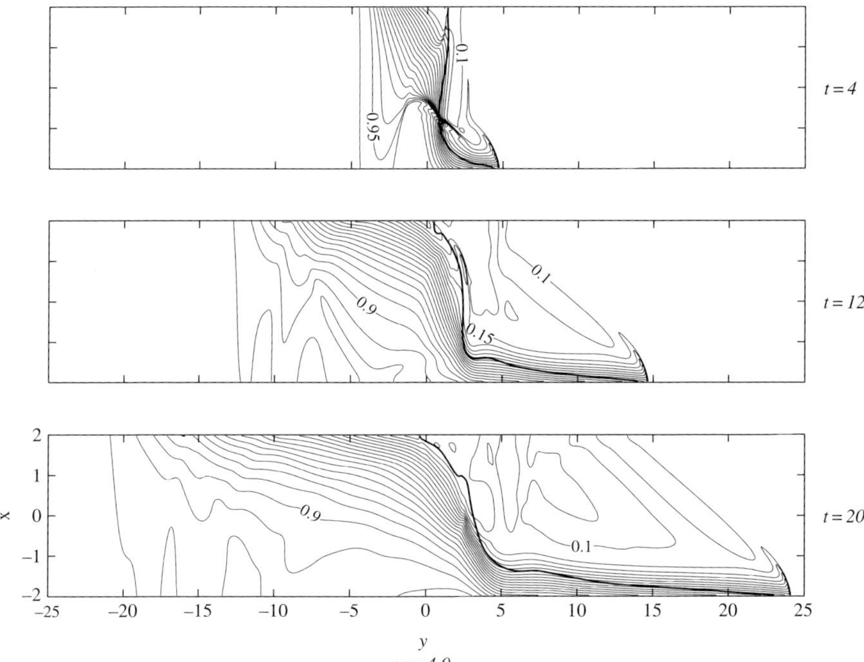

FIGURE 3.3.6. Full numerical solution for $d(x,\ y,\ t)$ to the partial dam-break problem with $w = 4.0$ and $d_o = 0.1$. The contour interval is 0.05. (From Helfrich *et al.*, 1999).

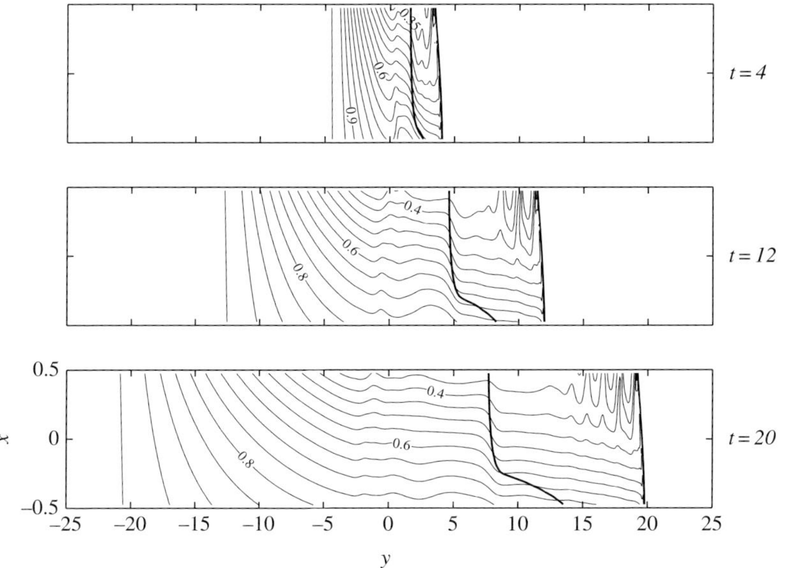

FIGURE 3.3.7. Same as Figure 3.3.6, except that $w = 1$ and $d_o = 0.5$. (From Helfrich *et al.*, 1999).

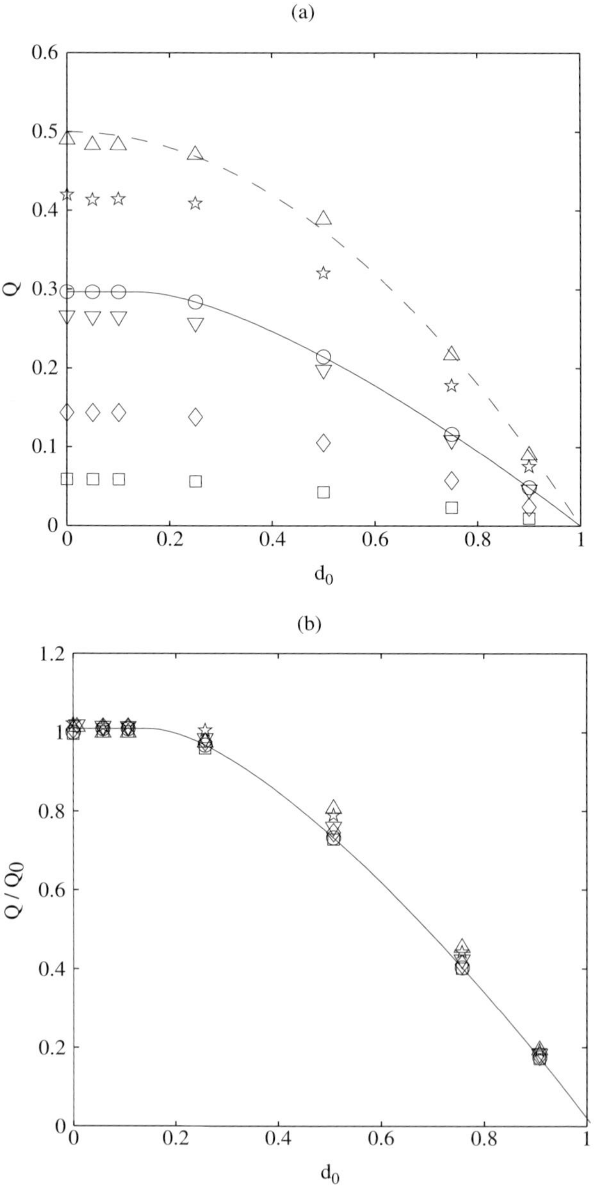

FIGURE 3.3.8. The final transport Q as a function of d_o and w. In (a) the solid line indicates the nonrotating theory for a channel of unit width, while the dashed line is the geostrophic transport $(1 - d_o^2)/2$ based on the initial difference in depths. In (b) the transport is normalized by the transport Q_o from the semigeostrophic theory for $d_o = 0$. In both (a) and (b) the numerical model transports are indicated by symbols according to $w = 0$ ($\bigcirc$), 0.2 ($\square$), 0.5 ($\diamond$), 1.0 (∇), 2.0 ($\star$), 4.0 ($\triangle$). (From Helfrich $et\ al.$, 1999).

The transport Q of the final steady state, measured at $y = 0$, shows that larger w and smaller d_0 lead to larger values of Q (Figure 3.3.8a). For comparison, the transport predicted for the nonrotating version of the dam-break (Stoker, 1957) in a channel of unit width is indicated by a solid line. In addition the 'geostrophic control' bound $(1 - d_o{}^2)/2$ based on the initial depth is shown as a dashed line. The calculated transport exceeds this bound only once, and then only slightly. However, it should be reemphasized that the loss of the crossing flow for all but large w obscures the interpretation of this result.

The increase in Q with w for any fixed d_o is similar to that of the semigeostrophic transport for $d_o = 0$ (solid line in Figure 3.3.5). When the measured Q values are normalized by the latter using common values of w, the data points for each d_o nearly collapse (Figure 3.3.8b). The solid curve in the same figure is the Stoker (1957) prediction for zero rotation, scaled as above.

Exercises

(1) Prove that the intersection of the solid line with the Q axis is Figure 3.3.8 lies at $Q = \left(\frac{2}{3}\right)^3 = .295\ldots$ (Hint: You will need to use a result from Section 1.3.)

3.4.   Adjustment to an Obstacle in a Rotating Channel

Although the process of Rossby adjustment provides valuable insight into the nature of transients in rotating channels, a further step is necessary to relate these transients to the establishment of hydraulically controlled states. The Long adjustment problem for nonrotating flow (Long, 1954 and 1970, Sections 1.6 and 1.7) provides a vehicle for doing so. In the original laboratory version of the experiment, an obstacle is towed at a fixed speed through a channel of shallow, resting fluid. Numerical versions of the experiment [e.g. Houghton and Kasahara (1968), Baines and Davies (1980)] place a fixed obstacle in the path of an initially steady, uniform flow, which is equivalent to the original set-up provided frictional effects are negligible. The outcome of the experiment for a single layer with a free surface depends on the Froude number $F_0 = v_0/\sqrt{gd_0}$ based on the initial depth d_0 and velocity v_0 of the moving stream and on a nondimensional obstacle height h_m/d_0 (Figure 1.7.2). For a given value of F_0 the outcome depends largely on whether h_m/d_0 exceeds a critical height (given by curve BAE). Beyond this height the obstacle partially blocks the approaching flow through the generation of a bore that moves upstream. The steady flow left behind has reduced volume transport and is hydraulically critical (hydraulically controlled) at the sill of the obstacle. For sufficiently large h_m/d_0 (given by the curve BC) the flow is completely blocked. Other boundaries can be calculated such as the curve AD, which separates flow having hydraulic jumps on the downslope of the obstacle from those that do not. Regime diagrams such as this figure and its generalizations in multi-layered flow are wonderful tools for developing knowledge and intuition about jumps, bores, upstream influence, hydraulic control, and hydraulics in general. They also give an indication of how high a sill must be in order to establish hydraulic control. More sophisticated versions of such models might indicate how high the sills in the abyssal ocean must be in order to alter the meridional overturning circulation.

Rotating versions of Long's experiment are quite difficult to carry out in the laboratory. Progress can and has been made using numerical models that are able to capture shocks. If the potential vorticity of the flow is uniform, predictions of the critical obstacle height and of certain aspects of the final steady solution can be made using the semigeostrophic theory developed in Sections 2.1–2.5. However, the extent to which features such as hydraulic jumps can be predicted is limited by the lack of a shock-joining theory for rotating jumps and bores, a subject explored in later sections of this chapter. The following discussion is based largely on the work of Pratt, 1983b and Pratt *et al.*, 2000.

a. *Initial Conditions*

In a rotating environment, the act of towing an obstacle along the channel at a fixed speed through an initially stationary fluid is no longer equivalent to introducing a stationary obstacle in a moving stream of the same speed. In the first case the free surface is horizontal; in the second it has a cross-stream, geostrophic tilt. We will consider the second version of the experiment since the upstream states seem more meaningful for ocean applications. The obstacle will therefore be introduced into a steady current that varies with x but is uniform in y. This current will be required to have uniform potential vorticity f/D_∞. For scaling purposes, the local depth D will be considered equal to the potential depth D_∞; the nondimensional potential vorticity is therefore given by $q = D/D_\infty = 1$. As was the case in the Rossby adjustment problem, the dimensionless decay scale of Kelvin waves across the channel will be unity. The choice of uniform (and nonzero) potential vorticity will naturally lead to comparisons with the Gill (1977) model for steady flow (Section 1.5).

If the potential vorticity remains at its initial value $q = 1$ throughout the adjustment and if the along-channel variations remain weak, than the cross-section profiles of depth and velocity are given by expressions developed in Sections 2.2 and 2.3 for semigeostrophic flow. For example, the profiles of d and v for nonseparated flow are described by (2.2.3) and (2.2.4). The average $\bar{B}$ of the semigeostrophic Bernoulli functions $v^2/2 + d + h$ on the two side walls is given by

$$\bar{B} = \frac{1}{2}[T^{-2}\hat{d}^2 + T^2(\bar{d}-1)^2] + \bar{d} + h \qquad (3.4.1)$$

where $T = tanh(w/2)$. The volume transport and Froude number are given by

$$Q = 2\bar{d}\hat{d}. \qquad (3.4.2)$$

and

$$F_d = \frac{\hat{d}}{\bar{d}^{1/2}T[1 - T^2(1 - \bar{d})]^{1/2}} \tag{3.4.3}$$

For detached flow the equations governing steady flow can be obtained by replacing $\hat{d}$ by $\bar{d}$ and T by $T_e = tanh(w_e/2)$ in these expressions.

To maintain continuity with Long's original experiment, we continue to use the Froude number of the initial flow, defined by (3.4.3), and the dimensionless obstacle height h_m/D_∞, to characterize the initial conditions. However, the presence of rotation brings two additional parameters into play and these may be selected in a variety of ways. An obvious choice is the channel width w (the dimensional channel width divided by $L_d = (gD_\infty)^{1/2}/f$), which determines the overall strength of rotation. As a fourth parameter, we will choose Gill's ψ_i, which determines relative amounts of volume flux contained in the right- and left-wall boundary layers of the initial flow. As a starting point, we will fix ψ_i by requiring that the total volume flux of the initial flow be contained in the left-wall boundary layer (i.e. $\psi_i = 1$). For this scenario to make sense, one could imagine that the channel broadens into a wide reservoir far upstream of where the initial-value experiment is to be performed. There the separation of the flow into left- and right-wall boundary layers is clear. A flow fed entirely by the left-hand boundary layer could have been set up as the result of a dam-break experiment in which motion is triggered by a Kelvin wave propagating upstream along the left wall.

The procedure for specializing the initial conditions to give zero approach flow along the right wall of the hypothetical reservoir is based on conservation of energy along that wall. Since the flow along the reservoir's right wall is stagnant, the value of the Bernoulli function there is unity (dimensionally gD_∞), and thus

$$\frac{v_0^2(w/2)}{2} + d_0(w/2) = 1,$$

where $()_0$ denotes the initial value. If (2.2.5–2.2.8) are used to write this relation in terms of $\bar{d}$ and $\hat{d}$ the result may be expressed in the nondimensional form:

$$\frac{[(\hat{d}_0/T) - T(1 - \bar{d}_0)]^2}{2} + \bar{d}_0 + \hat{d}_0 = 1. \tag{3.4.4}$$

To fix the initial conditions for given F_d and w the values of $\bar{d}_0$ and $\hat{d}_0$ must be computed. Once known, these last two quantities completely determine the initial depth and velocity profiles through (2.2.3) and (2.2.4). Equation (3.4.4) provides one equation for $\bar{d}_0$ and $\hat{d}_0$ while (3.4.3) provides a second. If $\hat{d}_0$ is eliminated between the two, the following relation for $\bar{d}_0$ is obtained:

$$\tfrac{1}{2}\{F_d\bar{d}_0^{1/2}[1 - T^2(1 - \bar{d}_0)]^{1/2} - T(1 - \bar{d}_0)\}^2 + \bar{d}_0 + F_dT\bar{d}_0^{1/2}[1 - T^2(1 - \bar{d}_0)]^{1/2} = 1. \tag{3.4.5}$$

One may also chose an initial flow that is separated from the left wall of the channel, in which case the above calculation will give $\bar{d}_0 < \hat{d}_0$. In this situation, the parameter T in (3.4.5) must be replaced by the *variable* $T_{e0} = \tanh(w_{e0}/2)$, where w_{e0} is the initial width of the separated current. The initial condition is now specified by the values of $\bar{d}_0$ (now equal to $\hat{d}_0$) and of T_{e0}. The two are related by

$$\bar{d}_0 = \frac{F_d^2(1 - T_{e0}^2)}{T_{e0}^{-2} - F_d^2 T_{e0}^2}, \tag{3.4.6}$$

which follows from (3.4.3). Substitution of this relation into (3.4.5) results, after some rearrangement, in

$$\frac{(F_d^2 - 1)^2}{2} + (1 - F_d^2 T_{e0}^4)(2F_d^2 - F_d^2 T_{e0}^2 - T_{e0}^{-2}) = 0. \tag{3.4.7}$$

The procedure is to first solve (3.4.7) for T_{e0} and then calculate the corresponding value of $\bar{d}_0$ from (3.4.6).

One consequence of the assumption that the volume flux in the initial flow is fed from the reservoir's left-hand boundary layer is that separated initial flow cannot be subcritical. The proof is the subject of Exercise 1.

b. *The Critical Obstacle Height*

It is anticipated that only values of h_m greater than some critical value h_c will lead to upstream influence: permanent alteration of the upstream flow. In Long's experiment the prediction of h_c follows from the consideration of a *steady* flow that passes over the obstacle and has the same volume flux Q and Bernoulli constant B as the initial state under consideration. There is a maximum h_m for which the upstream energy B is sufficient (at the given Q) to allow the fluid to surmount the crest. The corresponding sill flow is critical. If h_m exceeds the maximum allowable value, the values of B and/or Q must be altered in order to allow the flow to continue and this implies the generation of an upstream disturbance that alters the values of Q and B. Thus, the predicted h_c for given initial Q and B is that height for which these Q and B would, in a steady state, produce critical sill flow. An application of the same principles (with the upstream state now specified by F_d and w) results in a prediction of h_c in the rotating case.

For given initial values F_d and w, a unique initial flow with $\bar{d} = \bar{d}_0$ and $\hat{d} = \hat{d}_0$ is determined by the procedure laid out in Part *a* of this section. Consider a hypothetical steady flow with upstream values $\bar{d}_0$ and $\hat{d}_0$ that becomes critical ($\bar{d} = \bar{d}_c$ and $\hat{d} = \hat{d}_c$) at the crest ($h = h_c$) of the obstacle. Conservation of mass (3.4.2) requires that

$$\hat{d}_0 \bar{d}_0 = \hat{d}_c \bar{d}_c. \tag{3.4.8}$$

Together with the condition of criticality at the sill ($F_\mathrm{d} = 1$ in 3.4.3), (3.4.8) implies that

$$\bar{d}_c^4 + \frac{1 - T^2}{T^2}\bar{d}_c^3 - T^{-4}(\hat{d}_0\bar{d}_0)^2 = 0. \tag{3.4.9}$$

This equation determines $\bar{d}_c$ given the upstream/initial quantities $\hat{d}_0$ and $\bar{d}_0$. The value of $\hat{d}_c$ then follows from (3.4.8). Once $\bar{d}_c$ and $\hat{d}_c$ have been found it must be determined whether or not the flow at the sill is separated. If $\bar{d}_c \geq \hat{d}_c$ the flow is not separated and one may proceed to the next step, as described below. If $\bar{d}_c < \hat{d}_c$ the flow at the sill is separated from the left wall, and a revised procedure must be used (see Exercise 2). In either case the properties of the critical flow at the sill are known.

The critical sill height h_c can now be computed by equating the energy at the sill with that upstream. Employing the Bernoulli equation (3.4.1) with the computed values of $\hat{d}_c$ and $\bar{d}_c$ leads, in the case of nonseparated flow, to

$$h_c = 1 - \hat{d}_0\bar{d}_0 - \frac{1}{2}[T^2(\bar{d}_c - 1)^2 + (\hat{d}_c/T)^2 + 2\bar{d}_c]. \tag{3.4.10a}$$

When the sill flow is separated, this relation is replaced by

$$h_c = 1 - \hat{d}_0\bar{d}_0 - \frac{1}{2}\{T_{ec}^2[(\hat{d}_0\bar{d}_0)^{1/2} - 1]^2 + \frac{\hat{d}_0\bar{d}_0}{T_{ec}^2} + 2(\hat{d}_0\bar{d}_0)^{1/2}\} \tag{3.4.10b}$$

with

$$T_{ec}^2 = \left(\hat{d}_0\bar{d}_0\right)^{1/2} / \left[1 - \left(\hat{d}_0\bar{d}_0\right)^{1/2}\right] \tag{3.4.11}$$

(see Exercise 2).

The relationship between h_c and F_d is shown for a case of weak rotation ($w = 0.5$) by the curve CAE in Figure 3.4.1. The curve is composed of a number of segments indicating various states of separation. These states can be seen in the inset sketches showing plan views of the flow. To the left of CAE, there is no predicted upstream influence and the final flow upstream and downstream of the obstacle is identical to the initial flow. The final flow is altered directly over the obstacle but it does not become critical. To the right of CAE the predicted final upstream and downstream states have been altered by (unknown) transients. The predicted flow over the obstacle is critical at the sill and supercritical in the lee, possibly with some form of hydraulic jump. On CAE, the predicted flow is critical at the obstacle crest but the upstream flow is unaltered. Along the solid segment BA′, both the initial flow and the predicted sill flow are nonseparated. Along BC, which lies at the extreme lower right of the diagram, and is enlarged in an inset, the initial flow is attached but the predicted critical sill flow is separated. The predicted final flow thus separates from the left wall at some point slightly upstream of the sill. To the immediate left of BC, the upstream flow is attached and subcritical and the predicted flow over the obstacle is also subcritical but detached over a region extending upstream and downstream of the sill. If one begins at a point on BC and moves to the left, keeping F_d fixed, the region of separation shrinks. Where the curve BD is reached, the region has disappeared and the flow at the sill is marginally separated. The mathematical determination of this curve is discussed in Exercise 3.

The upper portion of Figure 3.4.1 corresponds to supercritical initial flow ($F_d > 1$). Part of the critical obstacle height curve consists of a segment A′E spanning a range of Froude numbers for which the initial flow is separated.

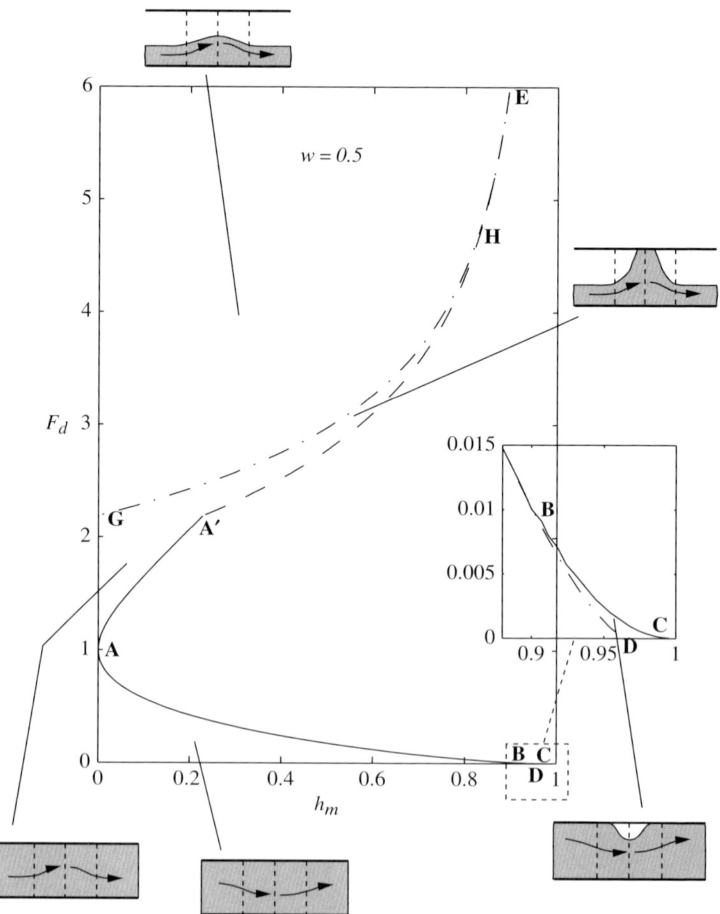

FIGURE 3.4.1. Regime diagram showing the predicted response in terms of the initial Froude number F_d and the obstacle height h_m, all for a channel of width $w = 0.5$. The curve CAE gives the critical obstacle height h_c, with different segments indicating different states of flow separation. The curves DB and GH indicate various states of flow separation for completely subcritical or supercritical flows. See the text for more details. (From Pratt et al., 2000).

Along subsegment A′H the predicted sill flow is critical and attached while along HE the sill flow is critical and separated. To the immediate left of A′H lies a wedge-shaped region A′HG in which the predicted final flow is supercritical everywhere, separated upstream and downstream of the obstacle, and attached near the sill. To the left of GE the predicted final flow is supercritical and separated everywhere.

An idea of the influence of rotation on the critical obstacle height can be gained by inspection of Figure 3.4.2, which shows the critical height curve CAE from the above weak rotation case ($w = 0.5$) plotted along with the $w = 2$ relation. For subcritical initial conditions rotation reduces the critical obstacle height whereas the reverse is true when the initial flow is supercritical. Note that the two curves merge when F_d is sufficiently large. Here the initial flow and the predicted sill flow are separated, implying that w is no longer a factor in determining h_m.

c. *Overview of the Temporal Evolution*

Numerical solutions to the full initial value problem show some similarities and important differences with what the semigeostrophic theory predicts. The runs are started at $t = 0$ with the flow specified in terms of F_d and w as described above. A Gaussian obstacle

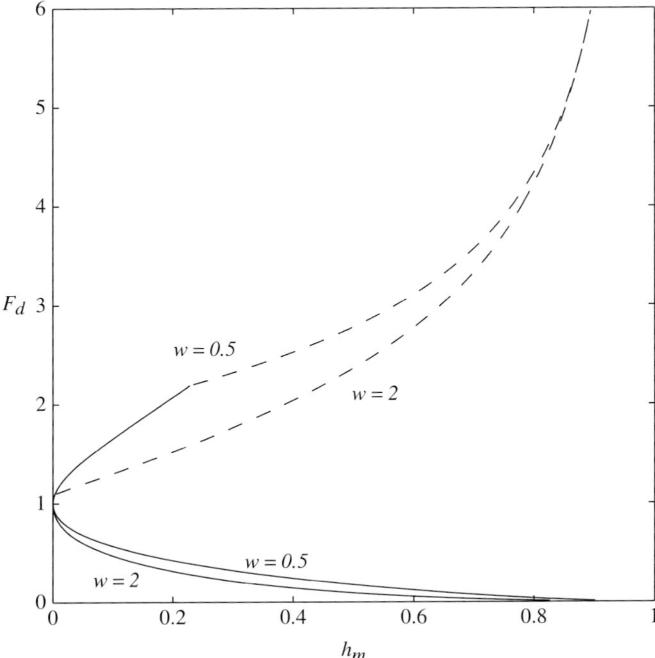

FIGURE 3.4.2. Curves of critical obstacle height $h_m = h_c$ as a function of F_d for $w = 0.5$ and $w = 2$. (From Pratt *et al.*, 2000).

$$h = h_o(t)\exp(-y^2/4)$$

is then quickly grown into the flow by increasing $h_0(t)$ linearly from zero to h_{m}. The shallow water equations are integrated using the shock-capturing code described by Helfrich *et al.* (1999).

The numerical results are summarized as part of regime diagrams (Figures 3.4.3–3.4.5) for channel widths $w = 0.5$, 2, and 4, respectively. The diagrams contain theoretical curves separating flows with and without control, and with and without separation, as computed from the semigeostophic model. The locations of numerical solutions are represented by symbols, with circles indicating solutions exhibiting a lack of permanent alteration of the original flow and the squares indicating cases exhibiting permanent upstream influence. The numerical results show versions of most of the features, including bores

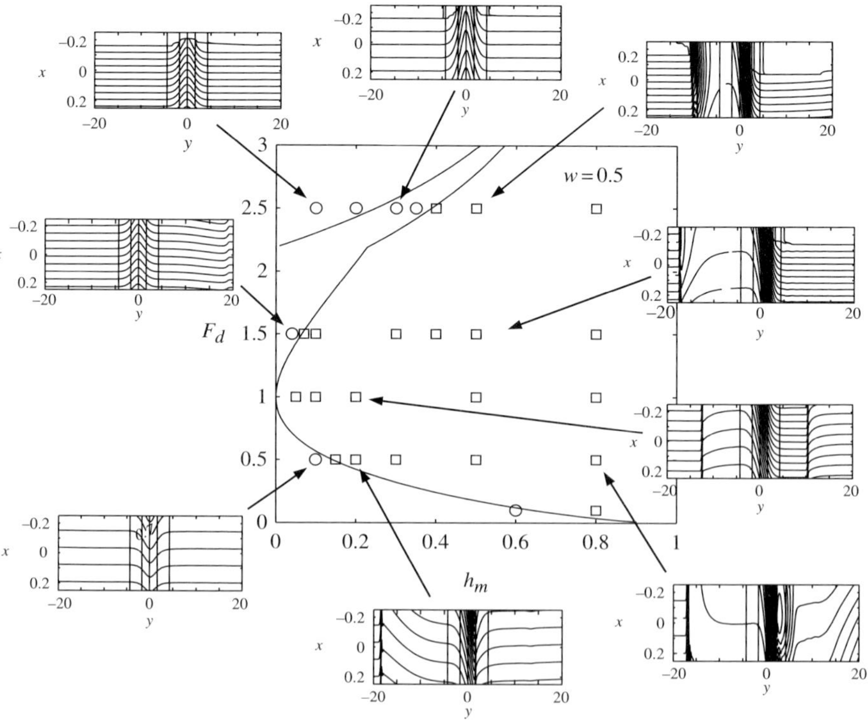

FIGURE 3.4.3. Summary of the numerical results for $w = 0.5$. The regime curves from the semigeostrophic theory are shown along with the locations of numerical runs. The circles indicate no permanent alteration of the original flow and the squares show cases of permanent upstream influence. Also shown are inset examples of the numerical results. The insets show contours of the free surface height $(d(x, y, t) + h(y))$. The shaded regions indicate those portions of the channel that are "dry" (defined by $d < 0.001$). The vertical lines are the 1, 0.5 and 0.001 times h_{m} contours of the bottom topography (From Pratt *et al.*, 2000).

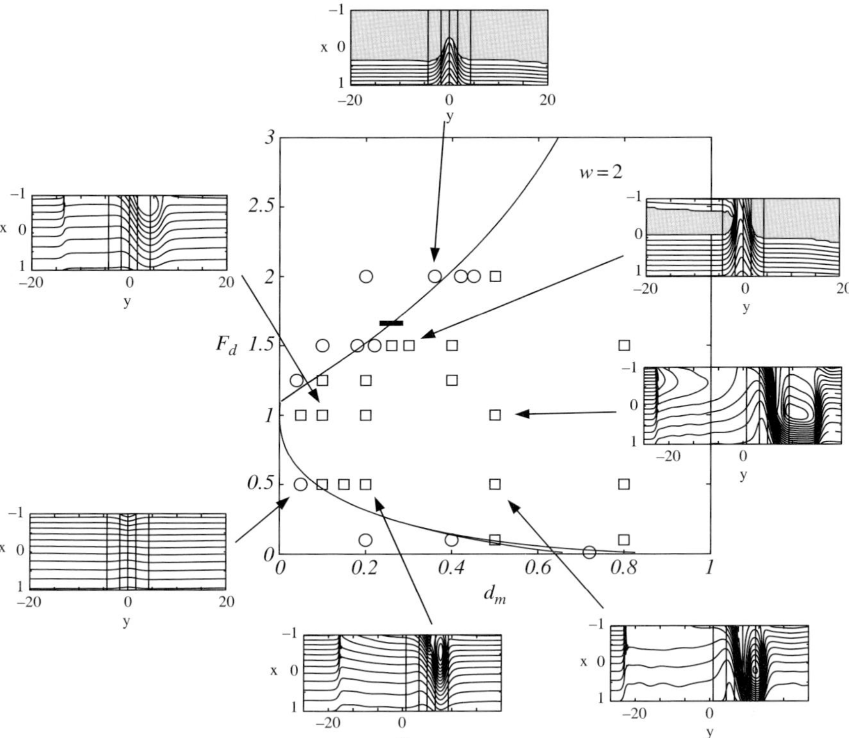

FIGURE 3.4.4. Same as Figure 3.4.3 except $w = 2$. The thick horizontal bar overlaid on the critical obstacle height curve indicates the value of F_d above which the predicted critical sill flow is separated from the left wall. (From Pratt *et al.*, 2000).

and jumps, which arise in Long's original experiments. They also reveal some features which are remarkable and unexpected. Since it is not possible to discuss each numerical run in detail, the reader is referred to the thumbnail insets in the three figures showing characteristic behavior found in different regions of the parameter space. These insets contain contours of the free-surface height, $d(x, y, t) + h(y)$, at later stages of the flow development. They illustrate the final steady flows over the topography and, in some cases, the structure of transient features. The gray-shaded regions indicate areas of the channel that are 'dry'.

The occurrence of upstream influence is indicated by asymmetry in the along-channel direction, relative to the sill, in the final state. Upstream influence is also indicated by a reduction in the transport at the sill crest compared to the initial transport. Perusal of the regime diagrams will show that the predicted critical obstacle height h_c generally overestimates the numerically determined value for subcritical initial flow ($F_d < 1$). The opposite is true for $F_d > 1$. The disagreement is minor for narrow widths but grows larger for wider values such as $w = 4$. In narrow channels the confinement provided by the wall suppresses cross-channel

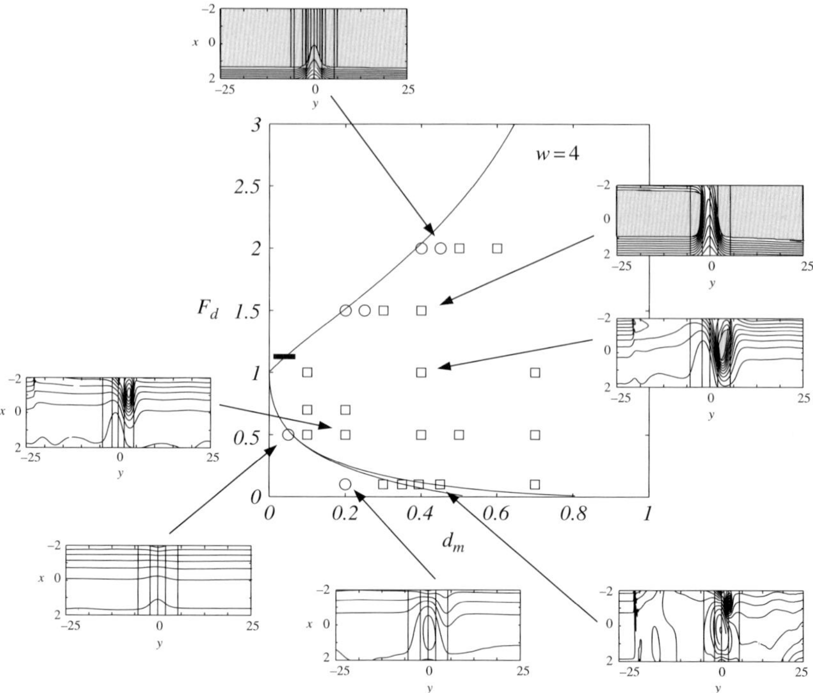

FIGURE 3.4.5. Same as Figure 3.4.3 except $w = 4$. (From Pratt et al., 2000).

accelerations and thus the along-channel flow should remain nearly geostrophic. For wider channels this effect is weakened and large cross-channel accelerations occur over the sill in the initial adjustment phase leading to departure from the semigeostrophic prediction. Nonconservation of potential vorticity could also affect the value of the critical height. The numerical model includes weak lateral viscosity and thus does not conserve potential vorticity following fluid parcels exactly. As we shall see later, fluid parcels that pass through shocks can suffer a change in potential vorticity due to dissipative processes within.

d.   The Case $w = 0.5$ (Weak Rotation)

We consider the flow evolution more carefully, beginning with the case $w = 0.5$ as summarized in (Figure 3.4.3). Despite the narrowness of this channel, rotation can be quite important. First consider some examples for which there is no predicted upstream influence ($h_{\mathrm{m}} < h_{\mathrm{c}}$) as illustrated by the insets on the left-hand side of the figure. Subcritical conditions give rise to an acceleration of the flow accompanied by a deflection of streamlines over the obstacle towards the right wall (e.g. $F_d = 0.5$, $h_m = 0.1$). The opposite occurs for supercritical initial conditions, as exemplified by the case ($F_d = 1.5$, $h_m = 0.04$). If F_d is large enough, the initial supercritical flow is separated and the corresponding final steady states may either be completely separated ($F_d = 2.5$, $h_m = 0.1$) or separated away from but attached near the sill ($F_d = 2.5$, $h_m = 0.3$). This last case is shown in greater detail in Figure 3.4.6. At $t = 10$ the disturbance generated by the introduction of the topography is evident immediately downstream of the sill. It consists of two waves that propagate downstream. The first is the faster Kelvin wave, centered at about $y = 15$ along the right wall (facing downstream). The second is the slower frontal wave (Section 2.3), centered at about $y = 8$ on the left free edge of the stream. By $t = 20$ the Kelvin wave has propagated out of the domain and the frontal wave (near $y = 17$) has steepened, nearly to the point where the stream width w_e is discontinuous. Both waves have propagated out of the visible domain by $t = 80$.

As noted above, upstream influence for the case $w = 0.5$ generally occurs where predicted. When the initial flow is attached, the disturbance that alters the upstream state takes the form of a 'Kelvin-wave' bore. For the case for $F_d = 0.5$

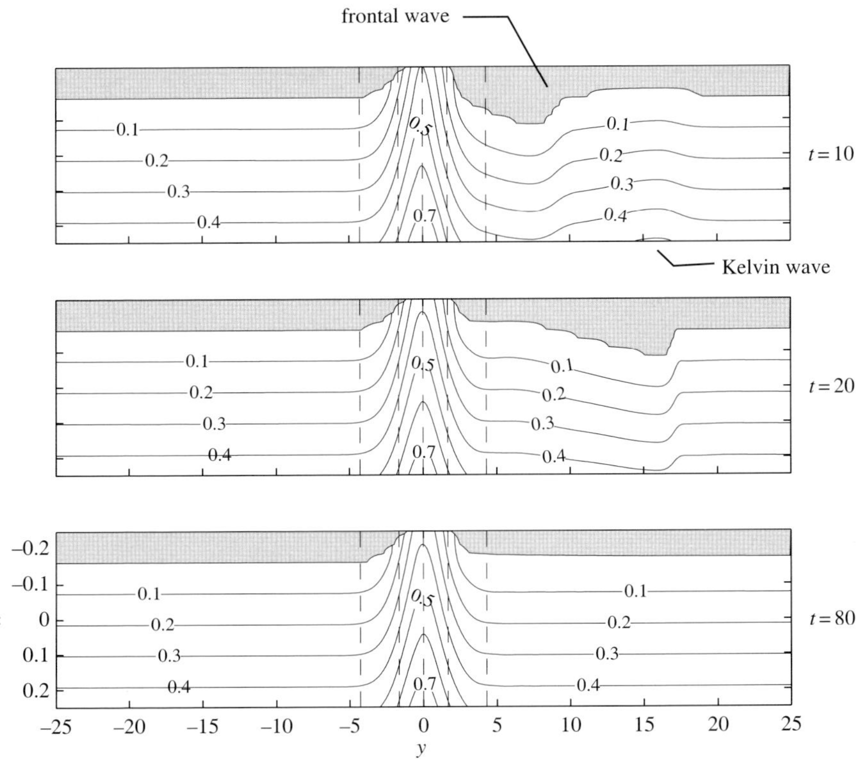

FIGURE 3.4.6. Numerical results for $F_d = 2.5$, $h_m = 0.3$ and $w = 0.5$. The panels show contours of the free-surface height ($d(x, y, t) + h(y)$) at the times indicated. The shaded regions indicate those portions of the channel that are "dry" (defined by d < 0.001). The dashed lines are the 1, 0.5 and 0.001 times h_m contours of the bottom topography. (From Pratt *et al.*, 2000).

and $h_m = 0.2$ (Figure 3.4.7), upstream and downstream propagating Kelvin waves are evident on each side of the topography at $t = 10$. The characteristic trapping of the Kelvin waves to the side walls is weakly apparent in this narrow channel. By $t = 30$ the downstream wave has left the domain, the upstream wave has steepened into a bore, and a hydraulic jump has formed on the downstream side of the obstacle. The jump remains over the downstream face of the obstacle in the final steady flow ($t = 50$). The Froude number (3.4.3) calculated from the numerical solution at $t = 30$ indicates a transition from subcritical to supercritical flow over the sill and a return to subcritical flow across the downstream jump (bottom panel of Figure 3.4.7). Also note how F_d decreases across the upstream

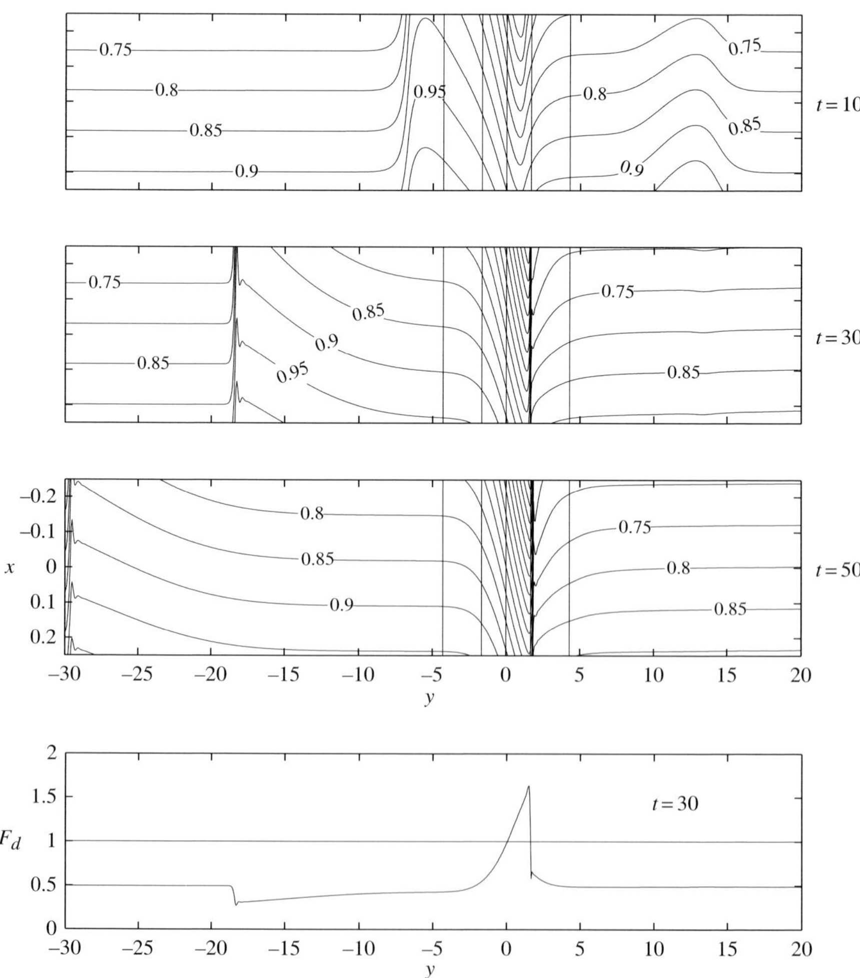

FIGURE 3.4.7. Same as Figure 3.4.6, except $F_d = 0.5$, $h_m = 0.2$ and $w = 0.5$. The bottom panel shows F_d as a function of y at $t = 30$. (From Pratt *et al.*, 2000).

bore. Generally speaking, the solution is similar to the nonrotating case. The most apparent sign of rotation is the deflection towards the right wall of the supercritical flow in the lee of the obstacle.

As in the nonrotating case, low values of the initial F_d favor stationary hydraulic jumps whereas higher values tend to cause the jumps to move downstream. The latter is illustrated by the inset in Figure 3.4.3 for $F_d = 1$ and $h_m = 0.2$, where the former hydraulic jump is shown as a discontinuity moving away from the topography in the downstream direction. With no rotation, the boundary separating regimes with and without jumps in the (h_m, F_d) plane can be constructed analytically using shock-joining theory. This boundary is given by the curve AD in Figure 1.7.2. A similar calculation is hindered in the rotating case due to the unavailability of a satisfactory shock-joining theory.

There are also a number of instances where the supercritical flow downstream of the sill separates from the left wall, a behavior that has important ramifications for downstream disturbances. A good example is the case $F_d = 1.5$ and $h_m = 0.5$ (Figure 3.4.8). At $t = 10$ the downstream-propagating Kelvin wave and upstream-propagating bore are evident. The flow approaching the sill is accelerated and veers toward the right wall downstream of the obstacle crest, leaving a small patch of dry channel near the left wall. The transition back to attached flow near $y = 7$ occurs as an abrupt expansion (located near $y = 16$ at $t = 30$). This transition

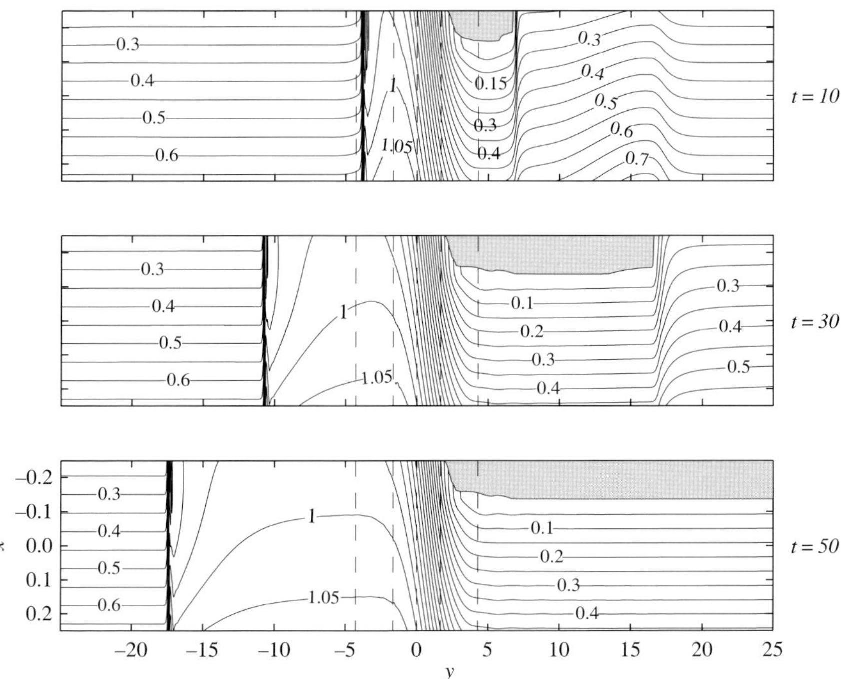

FIGURE 3.4.8. Same as Figure 3.4.6, except $F_d = 1.5$, $h_m = 0.5$ and $w = 0.5$. (From Pratt *et al.*, 2000).

is swept down the channel enlarging the dry region ($t = 30$) and ultimately leaving behind a detached supercritical flow in the lee of the topography ($t = 50$). The characteristic speed c_- has been calculated from (2.15) at points slightly upstream of and slightly downstream of the abrupt transition. On the upstream side, where the flow is separated and frontal wave dynamics apply, c_- is positive and greater than on the downstream side, where the flow is attached and Kelvin wave dynamics apply. Thus, linear disturbances generated just upstream of the transition overtake those generated just downstream, supporting the notion that the transition is a shock.

Flow separation in the lee of the obstacle is also observed for subcritical initial conditions and large values of h_m. In cases where hydraulic jumps occur, the usual abrupt change in depth is replaced by an abrupt change in the width of the stream. The jump is much like the transition in Figure 3.4.8, but with the feature stationary in the lee of the topography. An example (for $F_d = 0.5$, $h_m = 0.8$, Figure 3.4.9a) has a dry patch of bottom (shaded region in $1 < y < 2.3$) immediately downstream of the sill. This separated region terminates in a sudden

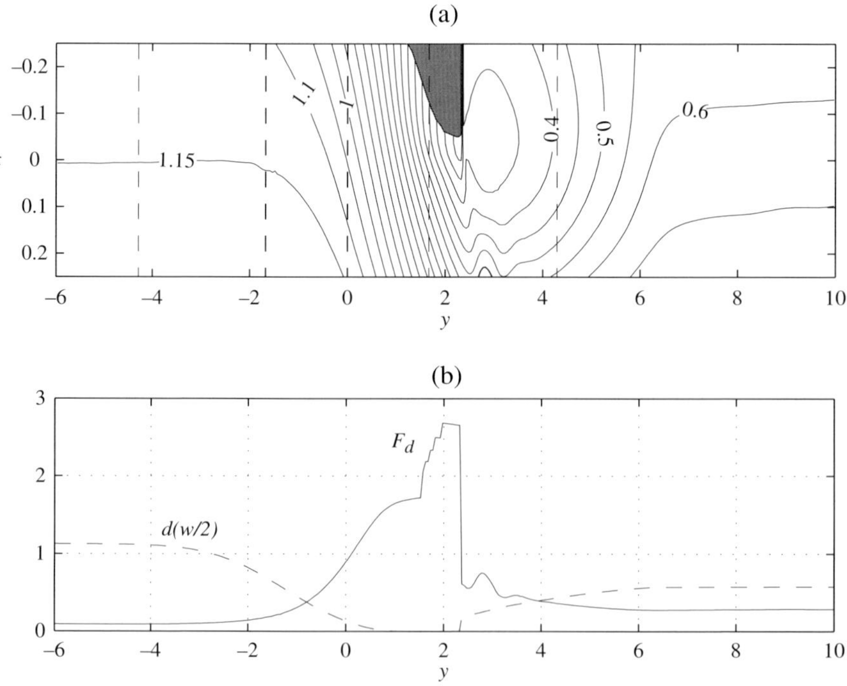

FIGURE 3.4.9. (a) Surface elevation contours for the steady flow that arises in the case $F_d = 0.5$, $h_m = 0.8$ and $w = 0.5$. A transverse hydraulic jump lies at $y \cong 2.3$. (b) Plot of F_d (solid line) and $d(-w/2, y)$ (dashed line) for the flow in (a). The transition from supercritical to subcritical flow near $y = 2.3$ coincides with the lateral expansion and reattachment of the flow to the left wall. (From Pratt et al., 2000).

expansion and reattachment of the flow. Downstream of the jump is a zone of cyclonic recirculation. A plot of F_d along with the left-wall depth (Figure 3.4.9b) shows that the transition from detached $[d(-w/2, y) = 0]$ to attached flow near $y = 2.3$ coincides with a supercritical to subcritical transition. This type of expansion is called a *transverse hydraulic jump* and its dynamics will be discussed later in this chapter.

When the initial flow is separated, upstream influence occurs in an unexpected manner. The leading portion of the upstream moving disturbance is a rarefying intrusion attached to the left wall as illustrated in the inset in Figure 3.4.3 for $F_d = 2.5$ and $h_m = 0.5$. In this example the intrusion is followed by a surge that leaves an attached flow upstream of the sill. The surge results in a rapid increase in depth, but the front is smooth and behaves like a rarefaction rather than a shock. We will return to this interesting situation below.

The foregoing examples show that rotation can lead to remarkable effects even when w is moderately small. These effects occur where high velocities are present, either due to supercritical initial conditions or because high velocities are induced in the lee of large obstacles. The high velocities lead to strong tilts in the free surface, sometimes resulting in separation of the flow. The Rossby radius of deformation based on the *local* depth becomes small in such cases and it is no surprise that inherently rotational features such as the transverse hydraulic jump arise under these conditions.

A final remark about the case $w = 0.5$ is that flows predicted along the curve BC of Figure 3.4.1 are not verified. The critical sill flow for such cases is predicted to be separated, whereas the numerical model produced attached sill flows. This behavior turns out to be quite general and will be revisited.

e.   *Cases $w = 2$ and $w = 4$ (Strong Rotation)*

The regime diagrams for the cases $w = 2$ *and* $w = 4$ (Figures 3.4.4 and 3.4.5) suggest that separation from the left wall occurs more readily. A region analogous to A′HG of Figure 3.4.1 exists for each case but is indistinguishably thin. The upper extent of the region is indicated by a horizontal bar in each figure. This bar marks that value of F_d above which the predicted critical sill flow is separated. For $w = 4$ nearly all supercritical initial states lie above this mark.

Strong rotation has consequences for the structure of the transients that occur as demonstrated by the subcritical case $F_d = 0.5$, $h_m = 0.2$ and $w = 2$ (Figure 3.4.10). Upstream influence is established as before by a 'Kelvin wave' bore and a stationary jump forms downstream. Both are trapped to within one deformation radius of the left wall. Downstream of the jump a region of cyclonic recirculation is generated. This region appears to expand in the downstream direction indefinitely. The structure of the bore and jump are evident in the topography of the free surface (Figure 3.4.11, which shows a similar case with right-to-left flow). When h_m is increased further, the lee flow tends

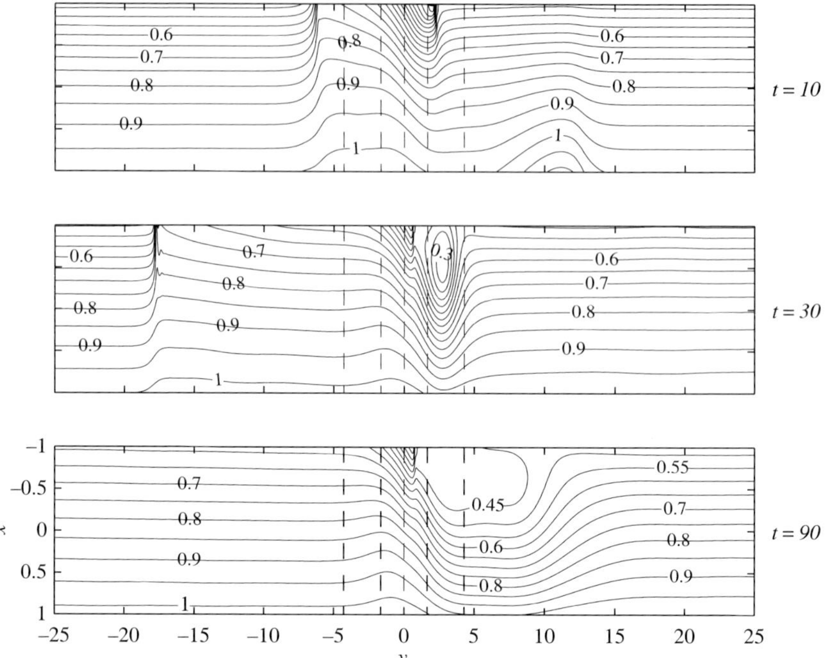

FIGURE 3.4.10. Same as Figure 3.4.6, except $F_\mathrm{d} = 0.5$, $h_\mathrm{m} = 0.2$ and $w = 2$. (From Pratt *et al.*, 2000).

to detach, then reattach over the topography to form a transverse jump, as discussed above.

For $F_d > 1$ the initial flow is separated for all Froude numbers save those close to unity. Upstream influence for these separated currents occurs in an unexpected manner, as demonstrated by the case $F_d = 1.5$, $h_m = 0.4$ and $w = 2$ (Figure 3.4.12). Although the initial flow is separated, the predicted critical flow at the sill is attached. Upstream influence occurs as the result of a splitting of the initial current over the topography ($t = 20$). A portion of the incident flow is diverted back towards negative y, forming a separated, rarefying intrusion along the left wall, while the rest continues over the topography ($t = 40$). The original current and upstream intrusion are narrow and do not contact each other. The final steady state upstream of the topography consists of two opposite, separated currents ($t = 50$). Remarkably, there is no upstream influence in the original current. However, the net flux towards the sill is reduced by the diversion of fluid into the left-wall intrusion.

Numerical calculations have also been carried out using F_d values sufficiently large that the predicted critical sill flow is separated. Such values lie above the horizontal bars in Figures 3.4.4 and 3.4.5. Significantly, these settings also result in left wall intrusions of the type just discussed and in

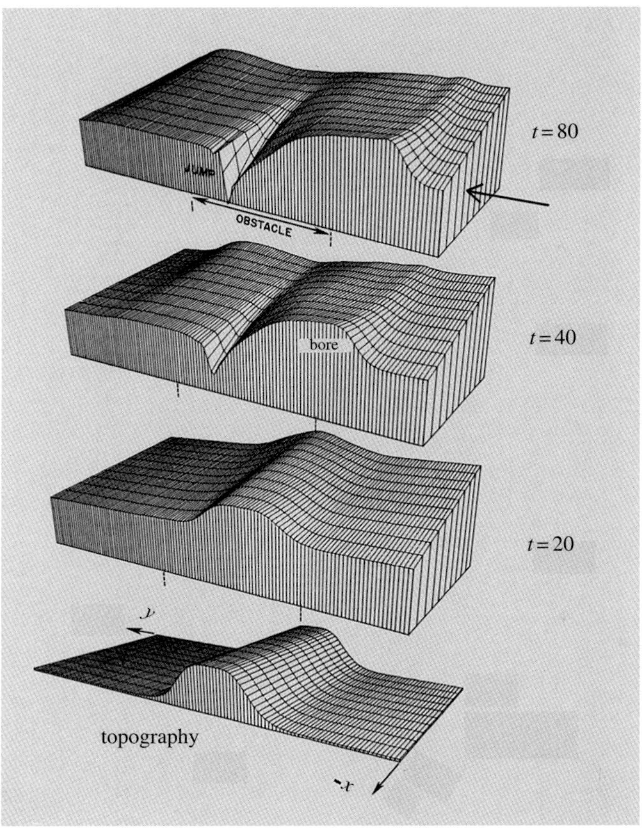

FIGURE 3.4.11. The top three frames show the free surface of the adjusting flow for the case $F_d = 0.41$, $h_m = .35$ and $w = 1.1$. The bottom frame shows the bottom topography and the flow is from right to left. (From Pratt, 1983b).

attached flow at the sill. In no case is the final sill flow observed to be critical and separated.

In the case of the widest channel considered ($w = 4$) the flow responds much as in the $w = 2$ case. One qualitative difference, evident for flows with $F_d < 0.5$, is the appearance of an anticyclonic recirculation cell over the sill. This feature occurs regardless of the presence of upstream influence, as illustrated by the two insets in Figure 3.4.5 for $F_d = 0.1$ and $h_m = 0.2$ and 0.45. In each case the velocity on the right wall at the sill crest $v(w/2, 0) < 0$. The recirculation cell occupies about three-quarters of the channel width, forcing the fluid that crosses the sill and continues downstream to do so in a narrow band adjacent to the left wall. The along-channel extent of the recirculation is comparable to the length of the topography. The existence of a counterflow at the sill is in violation of a theorem (Exercise 3 of Section 2.5) governing uniform-potential vorticity, semigeostrophic flow.

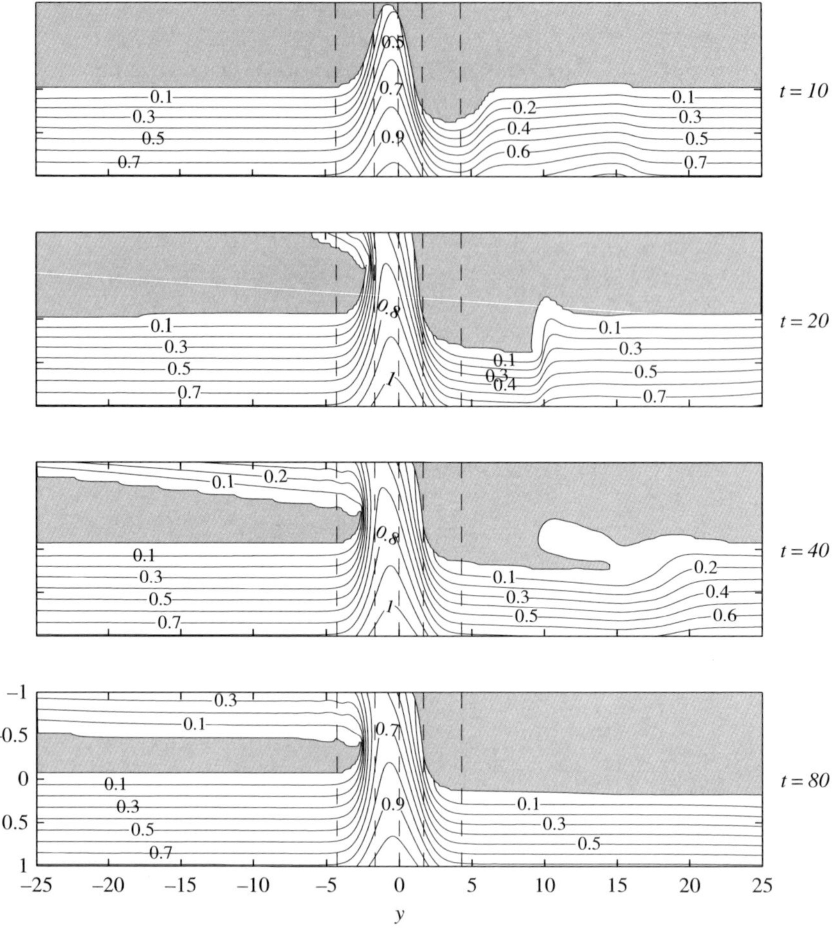

FIGURE 3.4.12. Same as Figure 3.4.6, except $F_d = 1.5$, $h_m = 0.4$ and $w = 2$. (From Pratt *et al.*, 2000).

f. The Lack of Hydraulic Control of Separated Flows

Many models of steady, hydraulically driven flow in rotating channels, including the studies of Whitehead *et al.*, 1974 and Gill 1977 (Sections 2.4 and 2.5) describe solutions that are hydraulically critical and separated at the controlling sill or narrows.[4] Whitehead, *et al.* (1974), Shen (1981) and Pratt (1987) attempted to reproduce such flows in the laboratory and were unsuccessful. The present numerical simulations also fail to produce such states, even where they are predicted. A possible explanation is the presence of an instability that acts when a separated flow is critical or subcritical. However, Paldor (1983) has shown

---

[4] Examples can be found in Gill's (1977) Figures 6 and 7, and 9d.

that separated currents of the type under discussion are stable, at least in the limit of zero potential vorticity, provided that the fluid depth along the right wall remains nonzero. So there does not yet appear to be a clear connection with inviscid instability. In general, the lack of hydraulic control of separated flows, both in numerical and laboratory experiments, remains a mystery.

g. *Breakdown of Semigeostrophic Theory*

Large channel widths provide room for significant cross-channel velocities to develop, leading to departures from semigeostrophic behavior. As an example, consider the hydraulically controlled flow shown in Figure 3.4.13a. The flow spills over the sill ($y = 0$) and veers across the channel. There is a weak hydraulic jump in the lee and a large region of cyclonic recirculation downstream. In principle, the semigeostrophic Froude number (3.4.3) for the flow should have value < 1 upstream and > 1 immediately downstream of $y = 0$. As shown in Frame b, however, F_d never exceeds unity, reaching a maximum value $\cong 0.95$ just upstream of the hydraulic jump. The value at the sill is considerably lower.

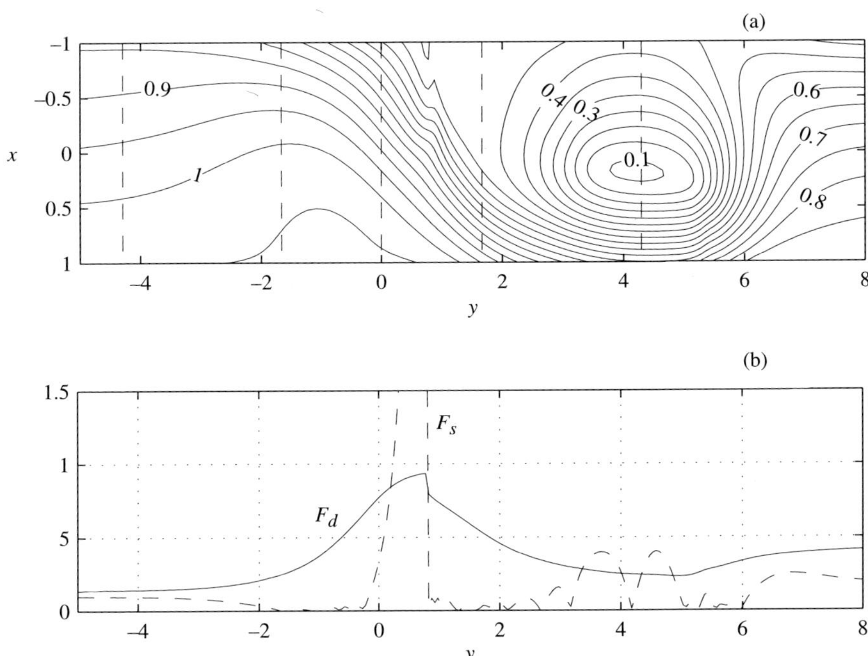

FIGURE 3.4.13. Details of the super- to subcritical transitions for the flow shown in the inset of Figure 3.4.4 with $F_d = 0.5$ and $h_m = 0.5$. (a) Contours of the free surface elevation $d + h$. (b) The Froude number F_d based on Equation 3.4.3 (solid line) and F_S based on Equation 3.4.12 (dashed line). The definition of F_S is invalid downstream of the jump (at $y \cong 1$) due to velocity reversals. (From Pratt *et al.*, 2000).

Apparently (3.4.3) is no longer a reliable measure of the hydraulicality of the flow. As shown in frame *a*, significant cross-channel velocities exist near and slightly downstream of the sill, suggesting that the failure of (3.4.3) may be due to a loss of semigeostrophy. This failure could also be due to potential vorticity nonuniformity that might have developed in the flow field as a result of the upstream shock.

In order to test the last hypothesis, consider the generalized Froude number (2.9.7) that was developed from Stern's (1974) critical condition for flow with nonuniform potential vorticity. In the current dimensionless variable set, the Froude number is defined as

$$F_S^2 = \frac{\int_{-w/2}^{w/2} d^{-2}dx}{\int_{-w/2}^{w/2} (v^2 d)^{-1}dx}, \tag{3.4.12}$$

and has value unity when the flow is hydraulically critical. This expression has meaning only when the flow is unidirectional and d is nonzero across the section in question. Although F_S does exceed unity on the downstream face of the obstacle (dashed curve in Figure 3.4.13b), its value ($\cong 0.4$) at the sill is even lower than the value of F_d there. This behavior suggests that the breakdown in our measure of the Froude number is due to the failure of the semigeostrophic approximation.

In addition to the failure of (3.4.12) to measure the true criticality of the flow, there are other indications of breakdown of the semigeostrophic approximation. As described earlier, the value of the critical obstacle height h_c predicted by semigeostrophic theory agrees well with the observed values for the narrowest channel ($w = 0.5$). As w increases, the agreement grows worse: the predicted h_c overestimates the actual h_c for subcritical initial flows and underestimates it for supercritical initial flows. More striking breakdowns in the semigeostrophic approximation occur within individual features. Perhaps the most dramatic is the grounding or separation of the flow ($d \to 0$) in the interior of the stream, as occurs at the bifurcation of the upstream flow ($t = 20, 40,$ and 80 in Figure 3.4.12 near $y = -2.5$) and at the detaching eddy (near $y = 14$ and $t = 40$ of the same figure). Such behavior is clearly in violation of the theorem proscribing the vanishing of d at a point where $\partial^2 d/\partial x^2 > 0$ in any semigeostrophic flow (see Section 2.2). Not surprisingly, semigeostrophic theory also fails in the vicinity of jumps, bores and other transients exhibiting rapid transitions in the y-direction. It is not necessary that w be large for such violations to occur, as evidenced by the presence of transverse jumps and bores when $w = 0.5$ (Figure 3.4.3).

h.   Upstream Recirculations

Although semigeostrophic theory admits solutions with closed recirculations, the location of the latter may be restricted by the assumed potential vorticity distribution. In Gill's (1977) uniform potential vorticity model, for example, it can be shown that the flow at any critical section must be unidirectional. Recirculations must therefore occur away from control sections. The laboratory simulations discussed in Section 2.6 contain recirculations, but in all cases the counterflow exists upstream of the sill crest. The potential vorticity distribution in these experiments is unknown. On the other hand, Section 2.9 makes it clear that flows with nonuniform potential vorticity may contain counterflow at a control section. The present numerical experiments contain examples in which recirculating fluid exists at the sill (Figure 3.4.14a). Inspection of the potential vorticity distribution across the sill confirms that it is nonuniform (Figure 3.4.14b). The boundaries of the recirculation (corresponding to $\psi = 0$) occur at the right wall $x = 2$ and at $x \cong -1$ in the cross section taken at the sill. Within these boundaries q is roughly constant, in agreement with conditions conjectured by Borenäs and Whitehead (1998). To the left of the recirculation the potential vorticity is much higher.

i.   *Concluding Remarks*

A review of the regime diagrams (Figures 3.4.3–3.4.5) suggests that most of the examples of hydraulically controlled flow can be placed in two broad classes. The first includes flows that remain attached to the left wall at every y. The

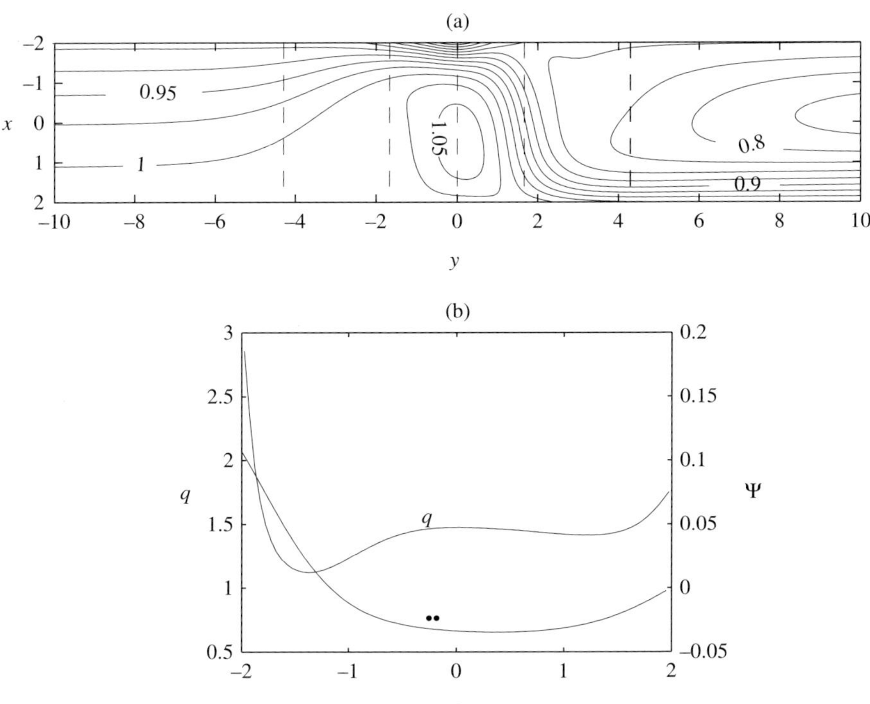

FIGURE 3.4.14. (a) Surface height contours for the case ($w = 4$, $F_d = 0.1$, and $h_c = 0.45$) in which a recirculation exists over the sill. (b) Potential vorticity and streamfunction profiles at the sill ($y = 0$). (From Pratt *et al.*, 2000).

time-dependent adjustment leading to the establishment of a controlled flow in this regime is similar to what takes place in Long's experiments, although the transients and the hydraulic jumps become trapped to the sidewalls. Energy dissipation due to jumps and upstream bores is then strongly localized near the left wall. One might collectively refer to these examples as the Kelvin-wave regime and note that it generally occurs for small-to-moderate F_d, h_m, and w. The second category includes flows that are separated from the left wall over some y. Significantly, the sill flow in all such cases remains attached. Further, all upstream disturbances and hydraulic jumps with separated upstream flow have attached downstream end states. Both Kelvin-wave and frontal-wave dynamics are important in these examples, which might collectively be referred to as a 'hybrid' regime. It is favored by large h_m, large F_d, and/or large w.

In no case is it possible to remove the left wall and achieve a hydraulically controlled flow. Even when the initial flow is separated and w is large, the critical sill flow becomes attached to the left wall. In addition, upstream influence for large w is transmitted in the form of an intrusion that travels along the left wall. These results imply that a 'coastal' version of the current, set up by moving the left wall to infinity, cannot be hydraulically controlled nor have a stationary hydraulic jump. One caveat should be mentioned: by restricting the initial conditions so as to require zero volume transport along the left-hand boundary of the hypothetical reservoir, all separated initial flows are supercritical. There is another family of separated but subcritical initial flows that could conceivably be subject to upstream influence without the aid of the left wall. This path has not been explored.

For some of the interesting features found in the simulations, no concrete oceanographic observations have been reported at the time of this writing. Such features include the Kelvin-wave hydraulic jump (Figure 3.4.11), the transverse hydraulic jump (Figure 3.4.9), and the bifurcation of the flow approaching the sill with resulting leakage back into the upstream part of the channel (Figure 3.4.12).

Exercises

(1) It has been assumed that the initial flow is fed entirely by a left-wall boundary layer. Show that such a flow, when separated, cannot be subcritical.

(2) *Calculation of the critical obstacle height for separated sill flow.* Show that h_c is given by (3.4.10b) when the sill flow is separated ($\bar{d}_c < \hat{d}_c$). Hint: by using $\bar{d}_c = \hat{d}_c = (\hat{d}_0 \bar{d}_0)^{1/2}$ and $T = T_{ec} = \tanh(w_{ec}/2)$, first show that (3.4.9) yields

$$T_{ec}^2 = \frac{\left(\hat{d}_0 \bar{d}_0\right)^{1/2}}{1 - \left(\hat{d}_0 \bar{d}_0\right)^{1/2}}$$

then combine this relation with the Bernoulli equation.

(3) *Determination of the curve BD in Figure 3.4.1.* Along segment BC of Figure 3.4.1 the initial flow is attached and the predicted critical sill flow is separated, as shown in the lower right inset. Reduction of the sill height with fixed F_d eventually results in reattachment of the flow at the sill, resulting in a state that is completely attached and subcritical (i.e. lower middle inset). The value of h_m at which reattachment occurs (i.e. $d_s(-w/2) = 0$ and $w = w_e$) will be denoted h_s. The value h_s of the sill height at which reattachment occurs can be found by replacing the critical condition in the steps leading to (3.4.10a) by the condition of marginal separation ($\bar{d} = \hat{d}$ and $T_e = T$) at the sill. Show that this procedure leads to

$$h_s = \tfrac{1}{2} T^2 [(\bar{d}_0 - 1)^2 - (\bar{d}_s - 1)^2] + \frac{\hat{d}_0^2 - \bar{d}_s^2}{2T^2} + \bar{d}_0 - \bar{d}_s,$$

where $\bar{d}_s = \hat{d}_s = (\bar{d}_0 \hat{d}_0)^{1/2}$.

(4) *Determination of the curve GH of Figure 3.4.1.* There exists a range of supercritical initial conditions for which the predicted final states are supercritical everywhere, separated upstream and downstream of the sill, and marginally separated at the sill. The obstacle height h_s for such solutions is smaller than the corresponding critical height and can be calculated by replacing the critical condition by the condition of marginal separation, this time in the steps leading to (3.4.10b). Show that this procedure yields

$$h_s = \tfrac{1}{2}(T_{e0}^2 - T^2)(\bar{d}_0 - 1)^2 + \frac{\hat{d}_0^2 - \bar{d}_s^2}{2T^2} + \tfrac{1}{2}\hat{d}_0^2(T_{e0}^{-2} - T^{-2}).$$

3.5.   Shock Joining

The reader of Sections 3.3 and 3.4 has seen a variety of shock waves, or 'shocks', composed of abrupt or discontinuous changes in the depth or width of the flow within which the semigeostrophic and/or hydrostatic approximations break down. Examples include the advancing Kelvin wave bores in the Rossby adjustment problem (Figures 3.3.6 and 3.3.7), the Kelvin wave hydraulic jump and upstream bore (Figure 3.4.11) and the transverse hydraulic jumps and bores (Figures 3.4.8, 3.4.9, and 3.4.12). We now take a closer look at these features by exploring the relationship between the flow immediately upstream and downstream of the abrupt transition. The problem of connecting these end states is known as *shock joining*. As a simple model, we will consider a hypothetical discontinuity in fluid depth occurring along a contour C (Figure 3.5.1). For the time being, it will be assumed that the fluid depth remains nonzero over C. Away from C the fluid motion is governed by the shallow water equations. It will be helpful to use a Cartesian coordinate system (n, s), placed such that n is aligned normal to and s parallel to C at the point P. The coordinate system remains fixed but C moves at speed $c^{(n)}$ in the n-direction.

If the system is one of reduced gravity, where the moving surface is an interface separating fluids of different densities, then the discontinuity may be associated with mixing of the two fluids. Closure of the shock joining problem then requires further assumptions or approximations. These difficulties have yet to be resolved in the current literature and will be avoided in the present discussion by limiting discussion to flows with a free surface.

A reader of Section 1.6 has seen two methods for obtaining the matching conditions across a shock. Both treat the shock as a discontinuity in d, v, etc. that exists in the presence of gradually varying topography. The approach that is most general, if not most popular, is to formulate the primitive conservation statements on mass and momentum over a control volume containing the discontinuity. Since the volume contains no sources of mass or momentum the conserved quantities are the volume flux and the flow 'force' (momentum flux plus pressure force). We will discuss the same procedure as applied to the shock of Figure 3.5.1.

The second approach is to integrate the shallow water equations over a small interval that contains the discontinuity. This method is generally less trustworthy because the equations themselves may not be valid within the region of rapid transition. Use of different forms of the shallow water equations yield different results. For example, integration of the common form (see 2.1.1 and 2.1.2) of the momentum equations yields the incorrect result that energy is conserved across the shock. The correct procedure is to write the equations so that they take the form of conservation laws for the quantites (in this case the volume flux and flow force) that are known to be preserved. The reasoning here is somewhat circular: one must know in advance which properties are to be conserved, and

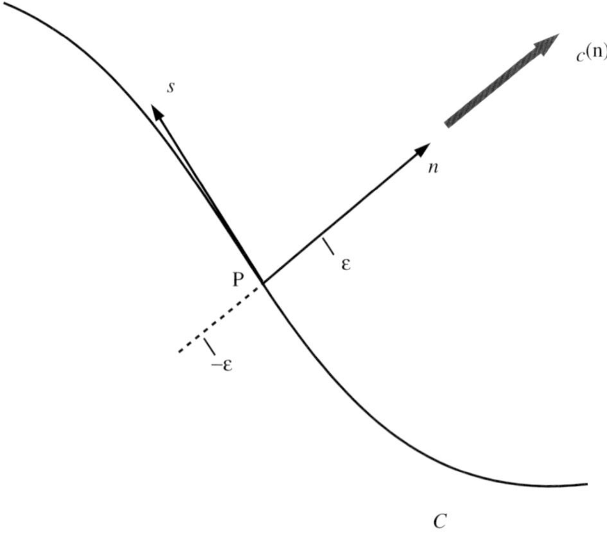

FIGURE 3.5.1. Definition sketch showing discontinuity in depth C that moves normal to itself at speed $c^{(n)}$ at the point P.

this knowledge derives from the fundamental reasoning behind the primitive control volume formulation! In fact, the desired 'flux' form of the momentum equations is that which follows directly from the control volume derivations.[5] Nevertheless, the approach is widespread and flexible: once the correct governing equations are known, they can be applied to a variety of shocks with differing structure and geometry. We will illustrate both methods, beginning with the primitive control volume approach. The following discussion is based largely on Pratt (1983b) and Schär and Smith (1993). Some of the basic ideas can be traced back to Crocco, as described by Batchelor (1967, Section 3.5).

a.   Shock Joining by Control Volume Analysis

Consider the force and mass budgets within a small box containing the shock, as shown in Figure 3.5.2a, b. The sides have length 2ε width $2l$ and the box extends from the bottom to the free surface. The box is fixed in space and is aligned so that its sides are parallel or perpendicular to n. It is assumed the velocity through edges of the box conform to the shallow water approximation and, in particular, is depth-independent, except possibly where the edges are intersected by the discontinuity.

The rate of change of n-momentum within the box must be balanced by the net flux of n-momentum into the box and the sum of the forces in the n-direction acting on the sides. One type of momentum flux is the normal flux $d(u^{(n)})^2$ across sides 1 and 2. Since $u^{(n)}$ is expected to be discontinuous across the shock, the difference in these normal fluxes remains finite as ε is decreased but decreases in proportion to l as l is decreased. Similarly, the depth-integrated pressure, nondimensionally $\frac{1}{2}d^2$, over side 1 is different from that over side 2, even as ε as decreased. All other forces and fluxes go to zero more rapidly as the box is shrunk. The tangential flux of normal momentum $(du^{(s)}u^{(n)})$ over sides 3 and 4 of the box are continuous in the s-direction and their difference decreases in proportion to εl as the box is shrunk. The Coriolis acceleration leads to a 'force' proportional to the integral of $du^{(s)}$ over the area of the box and is therefore proportional to εl. The same can be said for any contribution from bottom drag or topographic slope. Thus, as ε and l are decreased, the momentum budget reduces to

$$\frac{d}{dt}\iiint_v v\,dr \approx 2l\left[\tfrac{1}{2}(d_2^2 - d_1^2) + (d_2 u_2^{(n)^2} - d_1 u_1^{(n)^2})\right]$$

where V is the volume of the box. The left-hand integral reduces to $2lc^{(n)}(d_2 u_2^{(n)} - d_1 u_1^{(n)})$ as ε and l are reduced[6] and the matching conditions is thus

$$c^{(n)}\langle u^{(n)}d\rangle - \langle u^{(n)2}d + \tfrac{1}{2}d^2\rangle = 0, \tag{3.5.1}$$

---

[5] For example, the Section 1.10 control volume derivation (see Figure 1.10.3) leads directly to a flux form (Equation 1.10.4) of the momentum equations.
[6] A similar calculation was performed in connection with Equation 1.6.10.

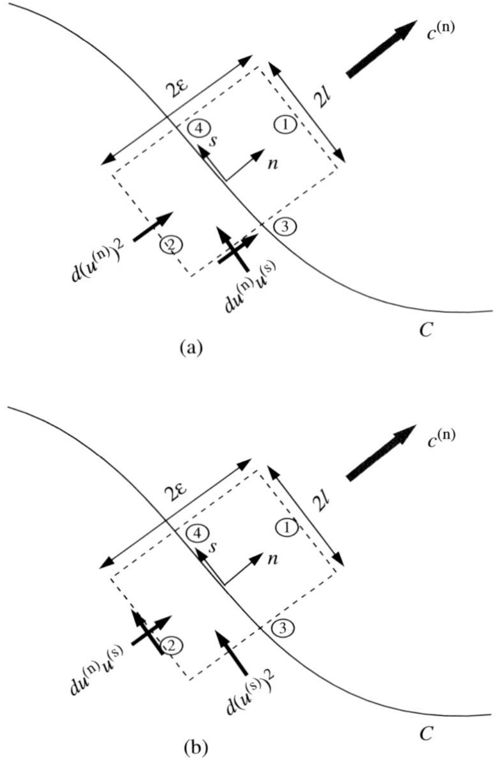

FIGURE 3.5.2. Control volumes (viewed from above) with (a) fluxes of momentum normal to the jump and (b) fluxes of momentum tangential to the jump.

where $\langle(\ )\rangle = \lim\limits_{\varepsilon\to 0}[(\ )_2 - (\ )_1]$. A similar treatment of the mass balance easily leads to

$$c^{(n)}\langle d\rangle - \langle u^{(n)}d\rangle = 0. \tag{3.5.2}$$

The tangential momentum balance (Figure 3.5.2b) is more subtle. Here the leading contribution comes from the difference in the normal flux of tangential momentum, proportional to the difference in $du^{(s)}u^{(n)}$ between sides 1 and 2. The flux $d\left(u^{(s)}\right)^2$ of tangential momentum and the pressure vary continuously between sides 3 and 4, and their difference leads to a negligible contribution as the box is shrunk. The same can be said for the contributions due to the Coriolis acceleration acting on the net normal velocity, the bottom drag, and topographic pressure. The result is that the change in net tangential momentum within the box, $c^{(n)}\langle u^{(s)}d\rangle$, is balanced by the difference in the normal flux of tangential momentum $\langle u^{(s)}u^{(n)}d\rangle$:

$$c^{(n)}\langle u^{(s)}d\rangle - \langle u^{(s)}u^{(n)}d\rangle = 0. \tag{3.5.3}$$

Together with (3.5.2), this result implies that the tangential velocity $u^{(s)}$ is conserved across the discontinuity:

$$\langle u^{(s)} \rangle = 0. \tag{3.5.4}$$

Note that (3.5.1) and (3.5.2) are identical to the conditions (1.6.4) and (1.6.5) governing one-dimensional shocks provided that the one-dimensional fluid velocity and shock speed are interpreted as $v^{(n)}$ and $c^{(n)}$. As a result, many of the properties of one-dimensional discontinuities apply locally to the two-dimensional, rotating discontinuities. For example, a stationary discontinuity requires that the local normal velocity of the upstream state be 'supercritical' $u_u^{(n)} > (d_u)^{1/2}$ (*cf.* Equation 1.6.7).

b. *Shock Joining using the Flux form of the Shallow Water Equations*

The correct matching conditions have been established as conservation laws for the normal fluxes of volume and tangential momentum, and for the normal component of flow force. It follows that the same conditions are derivable though integration of the differential form of these conservation laws, also the form that follows directly from the control volume analysis for a continuous flow. An interested reader might want to review the discussion in Section 1.10, in which a control volume derivation leads directly to a flux form (Equation 1.10.4) of the momentum equations. The two-dimensional form of these equations is given by (2.1.17), which can be written in the present coordinate system as

$$\frac{\partial (u^{(n)}d)}{\partial t} + \frac{\partial}{\partial n}(u^{(n)2}d + \frac{1}{2}d^2) + \frac{\partial (u^{(n)}u^{(s)}d)}{\partial s} = -d\frac{\partial h}{\partial n} + du^{(s)} + dF^{(n)} \tag{3.5.5a}$$

and

$$\frac{\partial (u^{(s)}d)}{\partial t} + \frac{\partial (u^{(n)}u^{(s)}d)}{\partial n} + \frac{\partial}{\partial s}(u^{(s)2}d + \frac{1}{2}d^2) = -d\frac{\partial h}{\partial x} - u^{(n)}d + dF^{(s)} \tag{3.5.5b}$$

To these we may add the continuity equation (2.1.7), expressed as

$$\frac{\partial d}{\partial t} + \frac{\partial}{\partial n}\left(u^{(n)}d\right) + \frac{\partial}{\partial s}\left(u^{(s)}d\right) = 0,$$

already in the desired form.

Integration of the last equation over a small interval $[-\varepsilon \le n \le \varepsilon]$ about the shock at this point results in

$$\int_{-\varepsilon}^{\varepsilon} \frac{\partial d}{\partial t}\,dn + [u^{(n)}d]_{n=\varepsilon} - [u^{(n)}d]_{n=-\varepsilon} + \int_{-\varepsilon}^{\varepsilon} d\frac{\partial u^{(s)}}{\partial s}\,dn = 0. \tag{3.5.6}$$

The first integral can be written as

$$\frac{\partial}{\partial t}\int_{-\varepsilon}^{\varepsilon}(d)\,dy = \frac{\partial}{\partial t}\int_{-\varepsilon}^{n_c(t)}(d)\,dy + \frac{\partial}{\partial t}\int_{n_c(t)}^{\varepsilon}(d)\,dy,$$

where $n_c(t)$ is the position of the discontinuity on the n-axis. If ε is reduced to zero, the right-hand side approaches $-c^{(n)}\langle d\rangle$, where $c^{(n)} = \partial n_c/\partial t$ and $\langle d\rangle$ is the change in d across the discontinuity, as defined earlier. Since the shock is parallel to the n-axis, the s-derivative in (3.5.6) is bounded along this integration path and the last integral in the same equation is made arbitrarily small by letting ε approach zero. The general constraint imposed by mass conservation thus reduces to (3.5.2).

We leave it as an exercise for the reader to show that a similar integration, applied to (3.5.5a,b), yields the correct conditions (3.5.1) and (3.5.3).

c. Consequences of the Shock Joining Conditions

If $u^{(s)} \neq 0$ then the change in $u^{(n)}$ required by (3.5.2) implies that the velocity vector $\boldsymbol{u} = (u^{(n)}, u^{(s)})$ must point in different directions on either side of a shock. Along a horizontal wall with free slip, the velocity vector is clearly aligned parallel to the wall regardless of whether a shock is present. These two facts can be reconciled only if C is aligned perpendicular to the wall at a point of contact, otherwise a flow into the boundary would be induced. In our slowly varying channel, where the walls are aligned in the y-direction, or nearly so, a shock must be aligned in the x-direction near the walls. One might now ask whether we can invoke the semigeostrophic approximation $v \gg u$ right up to the shock, which would force the shock to lie in the x-direction all across the channel. If so, one could start with a specified, geostrophically balanced $v(y)$ and $d(y)$ immediately upstream of a hydraulic jump and use (3.5.1) and (3.5.2) to compute $v(y)$ and $d(y)$ immediately downstream. However, since the shock-joining conditions do not depend on the Coriolis parameter, there is no guarantee that the downstream v will be geostrophically balanced; in general it will not be so. In summary, the semigeostrophic equations are not generally valid right up the shock, nor must the shock remain aligned with x away from the channel walls. Since rotational effects generally require a finite distance (the deformation radius) over which to act, we anticipate the existence of a transitional region around C within which the semigeostrophic far field flow adjusts to the (possibly) nongeostrophic flow at C.

This expectation is confirmed by the cross-stream momentum balance within the leading edge of the upstream-propagating 'Kelvin' bore of Figure 3.4.11. The momentum balance (Figure 3.5.3) is nearly geostrophic at $t = 20$, but becomes less so with time. The primary source of contamination is the development of strong, cross-channel accelerations within the steepening regions of the bore, an effect evidenced by the growth of the term $\partial u/\partial t$. By $t = 80$ the bore has steepened to the point where the depth changes occur over a fraction of a deformation radius $L_d(= (gD_\infty)^{1/2}/f)$. However, the ageostrophic region extends approximately 1/2 deformation radius upstream and downstream of the zone of rapid depth change.

Following the above remarks, one might expect a discontinuity in depth to occur within an ageostrophic region R that extends a distance $O(L_d)$ downstream and possibly upstream (Figure 3.5.4). The 'shock' might now be considered as whole region R with its imbedded discontinuity. R is joined upstream and downstream to semigeostrophic flows. It will be assumed that the flow in R is steady, but the same analysis can be carried out in the moving frame of a shock that translates at a steady speed c. The central problem of shock joining is to predict the downstream semigeostrophic end state given the upstream end state (and, in the case of a moving shock, the speed c). If the potential vorticity distribution $q(\psi)$ is preserved as the flow passes through R, then the shock joining problem is straightforward. For the $q(\psi)$ given by the known upstream condition, the downstream end state is found by solving the second order equation (2.2.2). The resulting profile of downstream depth, and the corresponding geostrophic velocity would then be known within two integration constants. These constants could be determined by two additional constraints, one being conservation of the total volume flux. A second constraint is provided by the conservation of

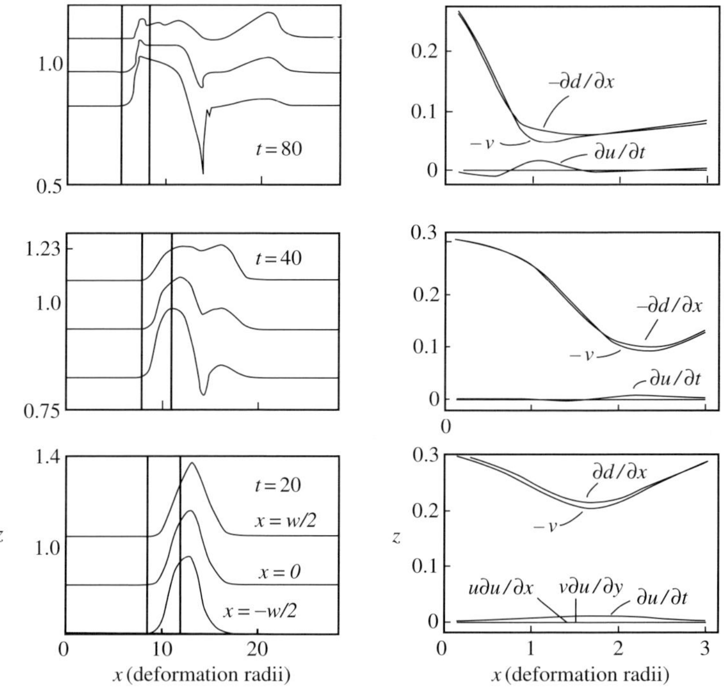

FIGURE 3.5.3. The frames on the left show the longitudinal sections of the surface elevation for the flow of Figure 3.4.11 at various times. The three sections in each frame are taken at the channel centerline and walls: $x=0$ and $x=\pm w/2$. The frames on the right show the terms in the y-momentum balance at the channel centerline over the interval indicated by vertical bars in the corresponding figure to the right. (From Pratt, 1983b).

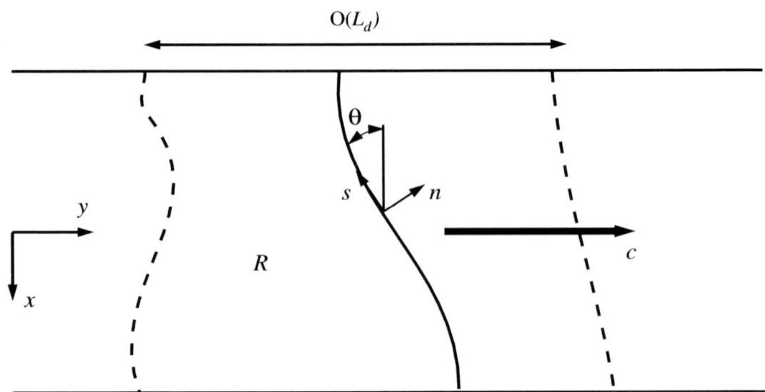

FIGURE 3.5.4. Idealized view of the ageostrophic region R and the imbedded depth discontinuity.

the total (width-integrated) flow force:[7] $\int_{-w/2}^{w/2}\left[v^2 d + \frac{1}{2}d^2\right]dx$. In summary, the conservation of volume flux, $q(\psi)$, and total flow force through R should be sufficient to close the problem.

Success of this procedure depends on potential vorticity conservation across the discontinuity, and we now ask whether this is consistent with (3.5.1, 3.5.2 and 3.5.4). Begin with the property that the Bernoulli function and potential vorticity are related by $q = dB/d\psi$, where ψ represents the streamfunction of the steady flow seen in the frame of reference moving with the steadily propagating shock. Since mass is conserved across the discontinuity, we have $\langle d\psi\rangle = 0$ and therefore

$$\langle q\rangle = \left\langle \frac{dB}{d\psi}\right\rangle = \frac{\langle dB\rangle}{d\psi}. \tag{3.5.7}$$

In addition, the jump in the value of B can be written in terms of the jump in depth using the previously derived relation (1.6.6) for energy dissipation, nondimensionally expressed as

$$\langle B\rangle = -\frac{\langle d\rangle^3}{4d_d d_u}. \tag{3.5.8}$$

Here d_u and d_d are the depths immediately upstream and downstream of the discontinuity at the point of interest. Thus

$$\langle q\rangle = -\frac{d}{d\psi}\frac{\langle d\rangle^3}{4d_d d_u} = \frac{1}{[u^{(n)}d]_u}\frac{d}{ds}\frac{\langle d\rangle^3}{4d_d d_u} \tag{3.5.9}$$

---

[7] The width-integrated flow force is conserved provided the horizontal component of bottom or side-wall pressure within R is not important. In a gradually varying channel, the length scale L of topographic and width variations is large compared to the length L_d of R and therefore the bottom and side-wall pressure alter the momentum flux through R by only an $O(L_d/L)$ amount.

where the final derivative is taken along the shock, as shown in Figure 3.5.4. The normal velocity $u^{(n)}$ is that seen in the moving frame. An observer facing the shock from upstream sees a positive normal velocity entering the shock, with ψ decreasing, and s increasing, from right to left. Potential vorticity is not conserved if the rate of energy dissipation varies with s. The dimensional version of (3.5.9) is obtained by multiplying its right-hand side by g and regarding all other variables as dimensional.

d. Geostrophic Shocks

Nof (1986) presents a special class of shocks that can be described analytically and for which the potential vorticity change can be calculated. The procedure is to look for a solution in which the channel flow is parallel ($v = 0$), and therefore geostrophic, right up to the discontinuity. The latter is assumed to be aligned in the x-direction so that C consists of a straight line perpendicular to the channel axis (Figure 3.5.5). Under the restrictions that both end states are parallel, and therefore geostrophically balanced, and that (3.5.1, 3.5.2, and 3.5.4) are satisfied at each x, a special class of upstream states can be found that permit stationary shocks with the assumed properties. As noted above, the upstream state must be 'locally supercritical' $v > d^{1/2}$ at each y. The results are classified in terms of two parameters: a Froude number $F_w = v_u|_{x=w/2}/d_u^{1/2}|_{x=w/2}$ and Rossby number $v_u(w/2)/w$, both based on right-wall values of the upstream flow. A set of examples of upstream and downstream depth profiles with fixed Rossby number are shown in Figure 3.5.6. Starting with the value $F_w = 1$, where there is no discontinuity, the jump $\langle d \rangle$ in depth across the shock tends to increase as F_w increases. In each case, $\langle d \rangle$ tends to increase from left-to-right and, according to (3.5.9), this is consistent with an increase in potential vorticity for the fluid passing through the discontinuity. The computed increases are shown

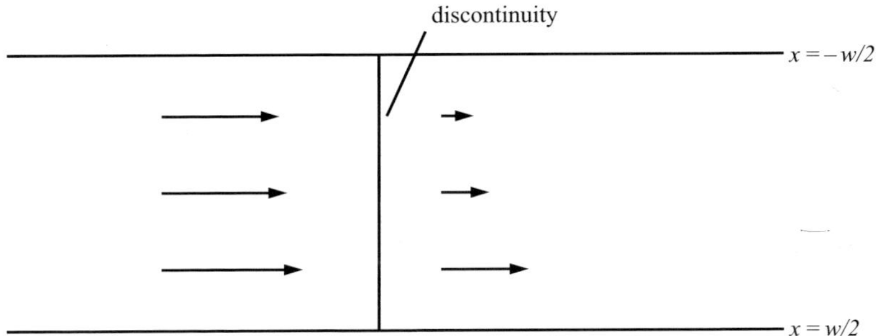

FIGURE 3.5.5. The shock hypothesized by Nof (1986). The depth discontinuity is perpendicular to the channel walls and the parallel, geostrophically balanced, upstream and downstream flows join directly to the discontinuity. (There is no adjustment region).

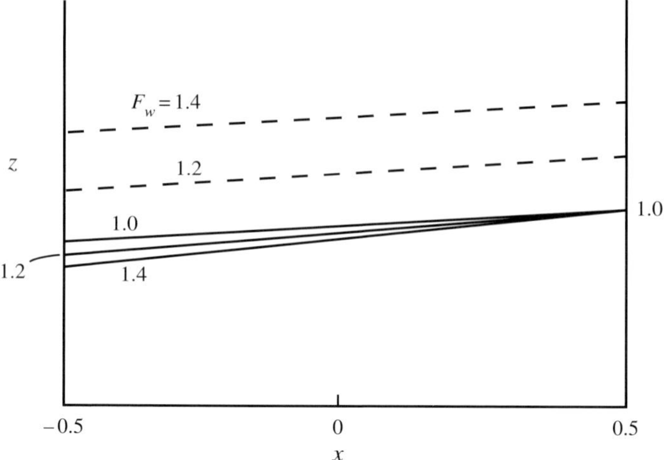

FIGURE 3.5.6. Upstream and downstream depth profiles for a shock of the type shown in 3.5.5. The governing upstream parameters are a Froude number $F_w = v_u(w/2)/d_u^{1/2}(w/2)$ and Rossby number $v_u(w/2)/w$, both based on values at the right channel wall $(x=w/2)$. The value of the latter for all plots shown is 0.2. (From Nof, 1986).

in Figure 3.5.7. Note that these changes can be O(1). Potential vorticity changes are also present in the various shocks discussed in Section 3.4.

e.   Vorticity Generation in Shocks

The nonconservation of potential vorticity across a shock can give rise to interesting downstream effects including jets and vortex streets. Consider a nonrotating jump in a channel with a rounded cross-section (Figure 3.5.7). This feature was modeled by Siddall *et al.* (2004) as part of a simulation of an ancient flood thought to have occurred in the Black Sea. The flow immediately upstream of the jump is parallel and uniform ($u=0$, $v=$ constant) and therefore $q_u = 0$. The jump consists of an abrupt, nearly uniform increase in the free surface elevation and thus the depth difference $[d_d(s) - d_s(s)]$ is constant. The differentiated term on the right-hand side of (3.5.9) is therefore controlled by the denominator, which decreases to the left and right of the channel center. The differentiated term therefore increases away from the channel center and it follows that $q_d > 0$ to the left and $q_d < 0$ to the right. With the neglect of f, q_d is proportional to the vorticity of the fluid downstream of the jump, the distribution of which is consistent with a jet-like velocity profile, as produced by a numerical simulation (Figure 3.5.9).

Vorticity production within a jump can be explored further by considering a helpful form of the vorticity equation:

$$\frac{\partial \zeta_a}{\partial t} + \nabla \cdot [\mathbf{u}\zeta_a + \mathbf{J}_n] = 0, \tag{3.5.10}$$

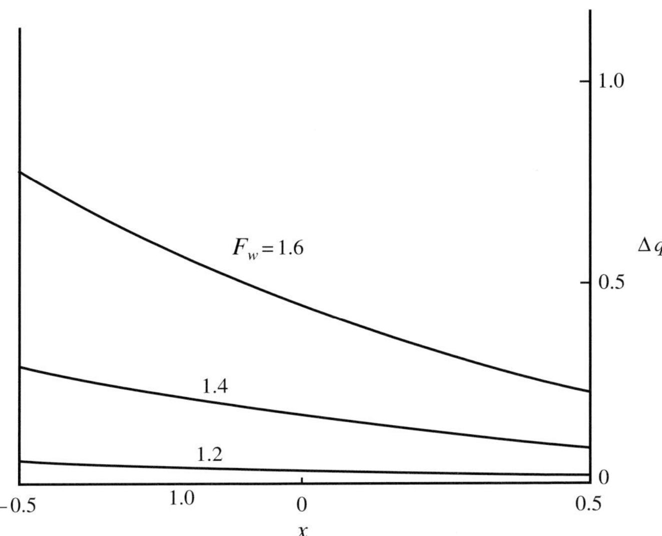

FIGURE 3.5.7. The change in potential vorticity across the shocks shown in Figure 3.5.6. (From Nof, 1986).

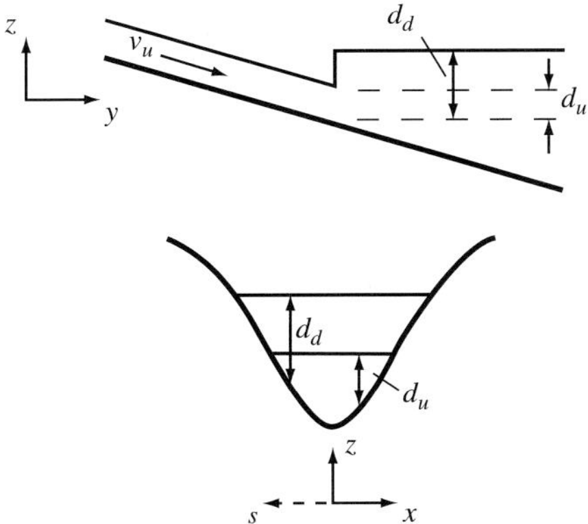

FIGURE 3.5.8. Schematic of a nonrotating hydraulic jump produced in a channel with a parabolic bottom. (From Siddall *et al.*, 2004).

(see Exercise 1 of Section 2.1). In this dimensionless form, $\zeta_a = 1 + \zeta$ is the absolute vorticity and $\mathbf{J}_n = \mathbf{k} \times \mathbf{F}$, where $\mathbf{F}$ contains the dissipation and horizontal body force. For the flows under consideration, the later is generally zero and we will think of $\mathbf{J}_n$ as arising only from dissipation. The vorticity flux vector $\mathbf{u}\zeta_a + \mathbf{J}_n$ is then composed of an advective part $\mathbf{u}\zeta_a$ plus a dissipative part.

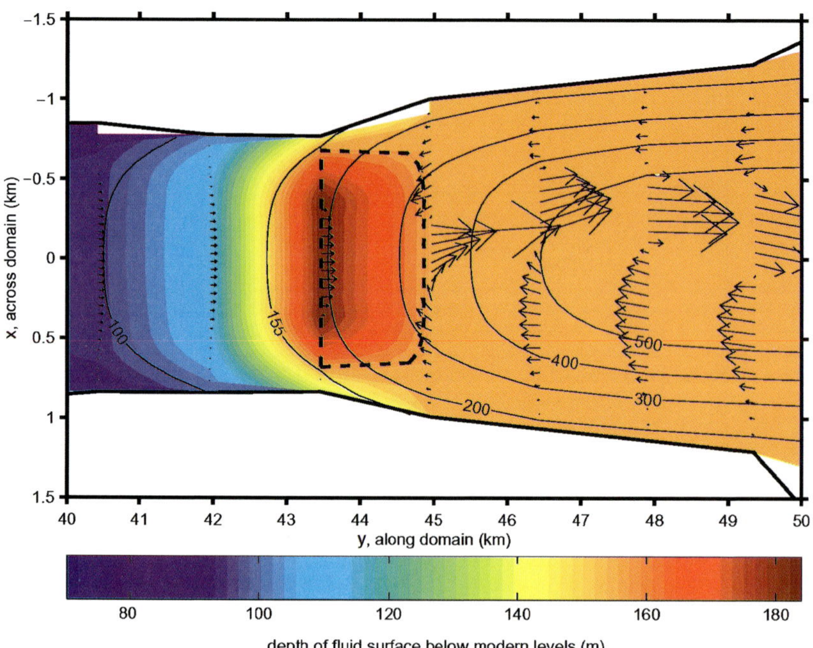

FIGURE 3.5.9. Plan view of the jump suggested in 3.5.8, as produced in a numerical simulation. The sudden change in depth occurs within the dashed area. The arrows indicate the depth-integrated velocity. (From Siddall, 2004).

Taking the cross-product of $\mathbf{k}$ with the dimensionless, steady version of (2.1.15) yields

$$\mathbf{k} \times \nabla B = \mathbf{u}\zeta_a + \mathbf{J}_n, \tag{3.5.11}$$

which shows that the Bernoulli function acts as a streamfunction for the vorticity flux (Schär and Smith, 1993).[8] Since $\mathbf{u}$ is parallel to streamlines, the derivative of B along them gives a contribution that is entirely due to dissipation. If the dissipation is zero, the vorticity flux is entirely due to advection and is proportional to the derivative of B in the cross-streamline direction. In the treatment of shocks we generally consider the dissipation to be negligible outside the region of rapid or discontinuous change.

A nice application of these ideas is to atmospheric wakes in the lee of islands and mountains (e.g., Smith *et al.*, 1997). For the islands in question, the effects of Earth's rotation are generally weak. The reduced airflow in the wake reduces the sea surface roughness, resulting in 'shadows' in the sea surface glint patterns (e.g., Figure 3.5.10). In an idealized view of the wake, the winds

---

[8] The inviscid form of (3.5.11) is related to a more general result obtained by Crocco (1937).

approaching the island are uniform and are confined to a shallow surface layer that obeys the reduced-gravity version of our shallow water equations. When the approach flow is subcritical and the island is not so high that it protrudes through the upper interface, the fluid spilling over the top can become supercritical and form a hydraulic jump (Figure 3.5.11). Regions of cyclonic and anticyclonic shear are also observed downstream of the jump and these are indicated in the figure. In some cases the vorticity is collected in a vortex street, a train of staggered eddies of alternating sign (Figure 3.5.12). If the approach flow is uniform and inviscid, the downstream vorticity must be generated by the jump.

FIGURE 3.5.10. Satellite photo showing sea surface glint around the Windward Islands. (NASA image S1998199160118).

The discontinuity in depth is largest at the center ($x = 0$) of the jump and (3.5.8) suggests that the loss in Bernoulli function should also be largest there. The flow immediately downstream of the jump should therefore have a minimum in B at $x = 0$ and B should increase as one moves along the jump in either direction (to the right of left, facing downstream). It is also assumed that B is conserved along streamlines ($\mathbf{J}_n = 0$) in the downstream region, changes having already taken place where the streamline passed through the jump. The y-component of (3.5.11) for the flow immediately downstream of the jump is $\partial B/\partial x = v\zeta_a$, where $v > 0$ and $\partial B/\partial x$ is > 0 for $x > 0$ and is < 0 for $x < 0$. The vorticity ζ_a, which is dominated by the relative vorticity ζ in these applications is therefore positive on the right-hand side of the wake (facing downstream) and negative on the left side. Since the approach flow has zero vorticity, the positive and negative vorticity must have been generated within the jump and could account for the vorticity in the alternating eddies. The time-dependent aspect of the alternating eddies requires an additional instability mechanism that is not explored here.

A complementary result can be found by applying (3.5.11) to the interior of the jump itself. To do so, it must be assumed that the rapid change in depth occurs over a small but finite distance and that (3.5.11) continues to hold within. Consider the component of this equation tangential to the jump. If one temporarily considers x to be the tangential direction, then this component is given by $-\partial B/\partial y = u\zeta_a + J_n^{(x)}$. Integration of this relation across the small interval ($-\varepsilon \leq y \leq \varepsilon$, say) of rapid depth change yields

$$\int_{-\varepsilon}^{\varepsilon} (u\zeta_a + J_n^{(x)})\,dy = -(B\big|_{x=\varepsilon} - B\big|_{x=-\varepsilon}) > 0.$$

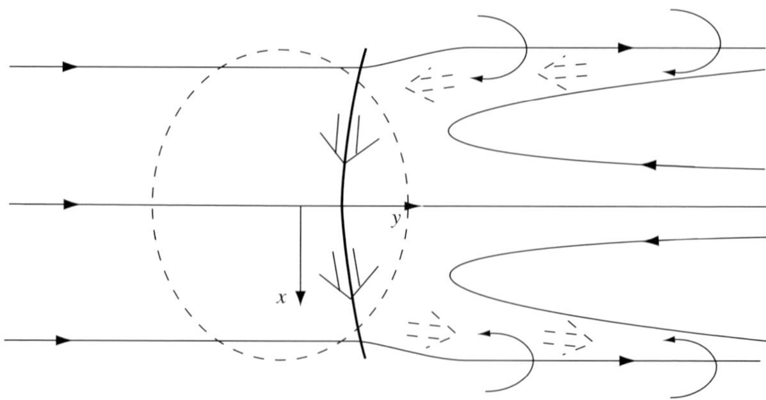

FIGURE 3.5.11. Idealized plan view of hydraulic jump and wake in the lee of an obstacle. The large arrows indicate vorticity fluxes. (Based on a figure from Schär and Smith, 1993).

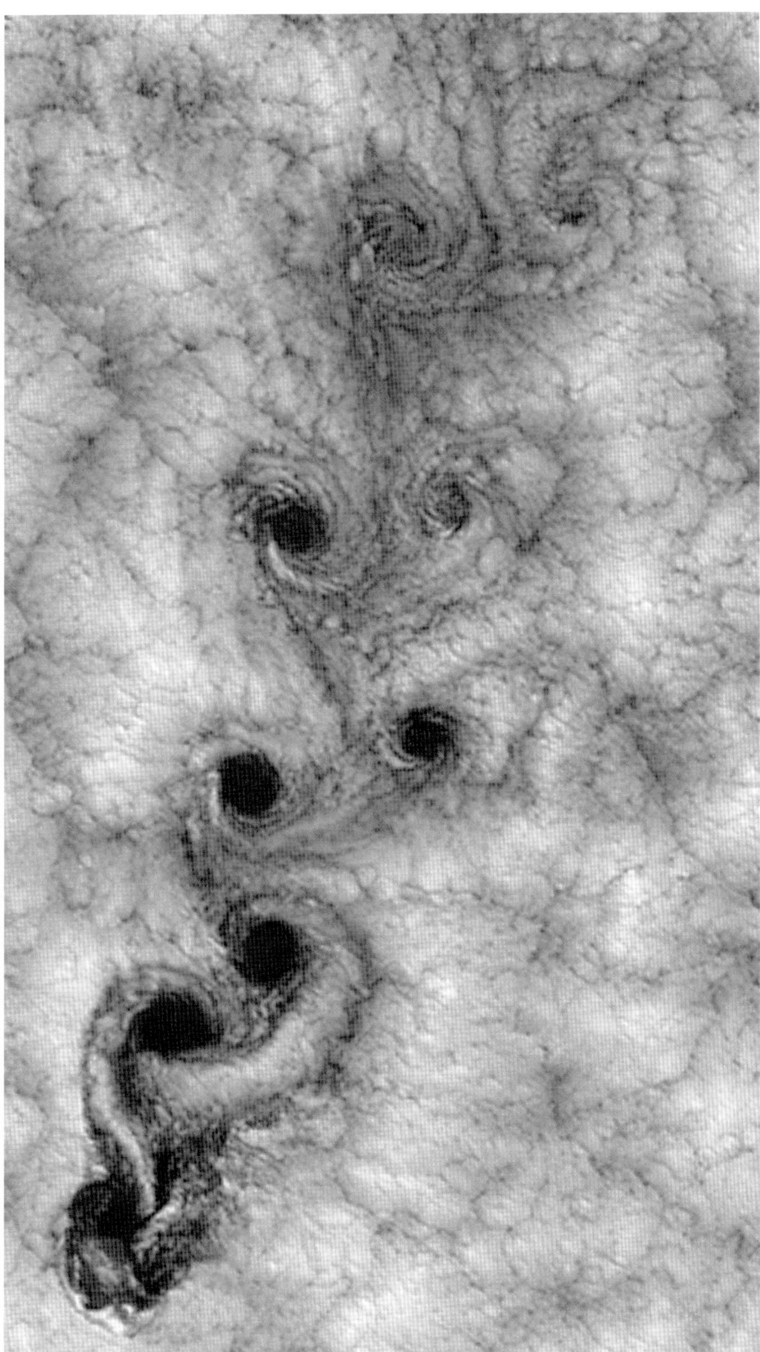

FIGURE 3.5.12. Landsat 7 image of a vortex street as apparent in the cloud cover off the Chilean coast near the Juan Fernandez Islands on September 15, 1999. (NASA image).

The left-hand term can be interpreted as a vorticity flux tangent to the jump (Figure 3.5.11), positive in the left-to-right direction (facing downstream). Its magnitude is zero at the extremities of the jump and therefore its divergence is positive over the left portion and negative over the right portion. A positive divergence is consistent with the generation of negative vorticity in the jump, whereas a convergent flux indicates a generation of positive vorticity. Both tendencies are in agreement with the vorticity carried away from the jump by the fluid.

Exercises

(1) Deduce the inviscid form of (3.5.11) directly from the relation $q = dB/d\psi$.
(2) For the nonrotating hydraulic jump shown in Figure 3.5.11, in which the depth is maximum at the centerline and the upstream velocity is uniform across the channel, use (3.5.11) to show that the downstream vorticity distribution is consistent with a jet.

3.6. A Kelvin Bore

A type of shock that can be treated with some success is the bore formed by a Kelvin wave propagating into a quiescent region of uniform depth. Examples arise in the nonlinear version of the Rossby adjustment problem when the channel is wider than several deformation radii. The downstream bore then consists of an abrupt change in depth that is trapped to the right channel wall (Figure 3.3.6). The near discontinuity in depth is aligned perpendicular to the wall at the contact point but is increasingly oblique away from the wall. In contrast with the bore observed in narrower channels (e.g. Figure 3.3.7) the present feature is felt only weakly at the left wall. The lack of influence of the left wall was exploited by Federov and Melville (1996) who developed a model describing the shape and speed of the bore.

Suppose that the discontinuity lies along a contour $y = Y(x, t)$ (Figure 3.6.1). It will be assumed that w is infinite so that the left wall is removed entirely. The fluid lying ahead ($y > Y$) is quiescent and the fluid lying immediately behind has velocity $u = u_o(x, t)$, $v = v_o(x, t)$, and depth $d = 1 + a(x, t)$. If the continuity equation (2.1.7) is integrated over a fixed interval $y_1 \leq y \leq y_2$ containing the shock, and if the interval is then reduced to zero, it follows that

$$\int_{y_1}^{y_2} \frac{\partial d}{\partial t} dy + \int_{y_1}^{y_2} \frac{\partial(ud)}{\partial x} dy - v_o(1 + a) = 0. \tag{3.6.1}$$

Since the integration interval is fixed, the derivatives may be taken outside of the integral and therefore

$$\int_{y_1}^{y_2} \frac{\partial d}{\partial t} dy = \frac{\partial}{\partial t} \int_{y_1}^{y_2} (d) dy = \frac{\partial}{\partial t} \left[\int_{y_1}^{Y} (d) dy + \int_{Y}^{y_2} (d) dy \right]$$

$$= \frac{\partial Y}{\partial t} [(1 + a) - 1] = a \frac{\partial Y}{\partial t}$$

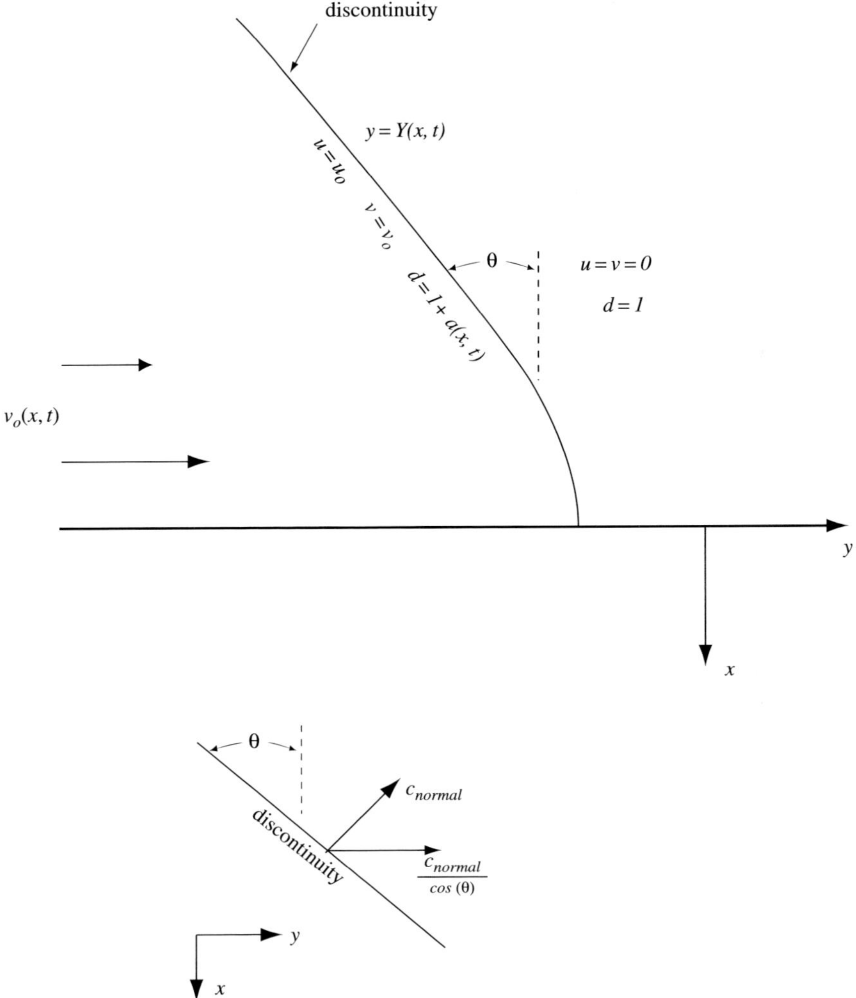

FIGURE 3.6.1. Definition sketch for a Kelvin wave bore propagating along a coastline.

After a similar treatment of its second term and division by $(1+a)$, (3.6.1) becomes

$$\frac{a}{1+a}\frac{\partial Y}{\partial t} + u_o\frac{\partial Y}{\partial x} - v_o = 0 \tag{3.6.2}$$

A second constraint follows from the continuity of the tangential velocity across the discontinuity (see 3.5.4). Since the fluid ahead of the discontinuity is quiescent the tangential velocity to the rear must also be zero:

$$v_o\frac{\partial Y}{\partial x} + u_o = 0 \tag{3.6.3}$$

A third constraint results from integration across the discontinuity of the flux form (2.1.17a) of the y-momentum equation. The resulting condition is

$$v_o \frac{\partial Y}{\partial t} - (v_o^2 + \frac{a(1 + \frac{1}{2}a)}{(1 + a)}) + u_o v_o \frac{\partial Y}{\partial x} = 0. \tag{3.6.4}$$

Use of (3.6.3) to eliminate u_o from (3.6.2) and (3.6.4) leads to

$$\frac{a}{1 + a} \frac{\partial Y}{\partial t} = v_o \left[1 + \left(\frac{\partial Y}{\partial x} \right)^2 \right] \tag{3.6.5}$$

and

$$v_o \frac{\partial Y}{\partial t} - \frac{a(1 + \frac{1}{2}a)}{(1 + a)} = v_o^2 \left[1 + \left(\frac{\partial Y}{\partial x} \right)^2 \right]. \tag{3.6.6}$$

Elimination of $\partial Y/\partial x$ between the last two equations gives

$$v_o = \frac{a(1 + \frac{1}{2}a)}{\partial Y/\partial t} \tag{3.6.7}$$

and substitution for v_o back into (3.6.5) yields

$$\left(\frac{\partial Y}{\partial t} \right)^2 = (1 + a)(1 + \frac{1}{2}a) \left[1 + \left(\frac{\partial Y}{\partial x} \right)^2 \right]. \tag{3.6.8}$$

Now suppose that the discontinuity propagates along the wall at a steady speed $\partial Y/\partial t = c$, so that

$$c^2 = (1 + a)(1 + \frac{1}{2}a) \left[1 + \left(\frac{\partial Y}{\partial x} \right)^2 \right]$$
$$= \frac{(1 + a)(1 + \frac{1}{2}a)}{\cos^2 \theta}, \tag{3.6.9}$$

where θ is the angle between the line of discontinuity and the normal to the wall. $\partial Y/\partial x$ must vanish at the point of contact in order to satisfy the condition of no normal flow, and it follows that

$$c^2 = (1 + a_o)(1 + \frac{1}{2}a_o) \tag{3.6.10}$$

where a_o is the value of a at the wall. This c is equivalent to the speed of a nonrotating, one-dimensional bore propagating into shallow water (see equation 1.6.7 with $v_u = 0$), based on the wall depth. Equation (3.6.9) can now be written as

$$\left(\frac{\partial Y}{\partial x} \right)^2 = \frac{(1 + a_o)(1 + \frac{1}{2}a_o)}{(1 + a)(1 + \frac{1}{2}a)} - 1. \tag{3.6.11}$$

The factor $(1+a)(1+\frac{1}{2}a)$ appearing in the last few equations can now be seen to have a simple interpretation. Consider a small segment of the discontinuity that is aligned at an angle θ and therefore faces the direction $(-\sin\theta, \cos\theta)$, as shown in the Figure 3.6.1 inset. Since the entire bore translates at speed $c = [(1+a)(1+\frac{1}{2}a)]^{1/2}/\cos(\theta)$, the segment in question moves in the y-direction at this speed. The speed of the front in the normal direction is therefore $[(1+a)(1+\frac{1}{2}a)]^{1/2}$. Equation (3.6.9) is just a statement of this geometrical consideration. At the wall, where y *is* the normal direction, c is given by (3.6.10). Since c is constant, the amplitude a of the discontinuity must diminish as θ increases.

A solution for Y as a function of x cannot be ascertained without a further assumption about the flow to the rear of the jump. Federov and Melville (1996) take the velocity component v_o to be geostrophic:

$$v_o = \frac{\partial a}{\partial x}, \tag{3.6.12}$$

implying that the nonsemigeostrophic region R described in the previous section is absent. This approximation is justified as long as the transverse velocity u_o remains $\ll v_o$, and this requires that the angle θ between the discontinuity and the x-axis remains small (cf. equation 3.6.3).

Substitution of (3.6.12) into (3.6.7) leads to

$$\frac{da}{dx} = \frac{a(1+\frac{1}{2}a)}{c}.$$

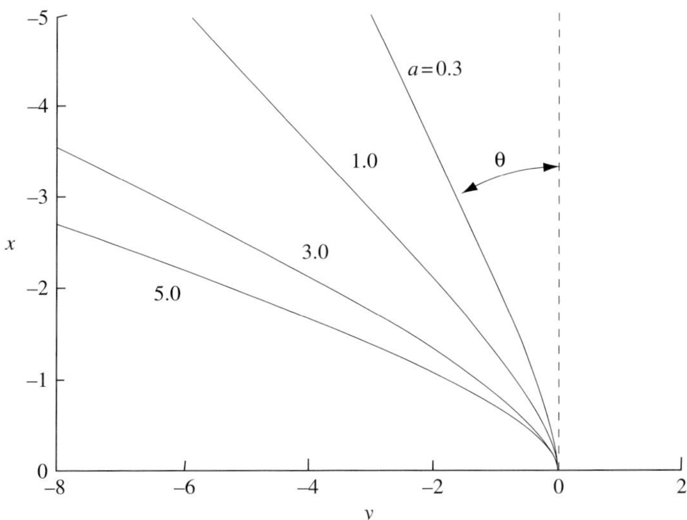

FIGURE 3.6.2. Solutions showing the path of the discontinuity $y = Y(x)$ in the moving frame of the bore for various values a_o. (Based on Federov and Melville, 1996, Figure 13).

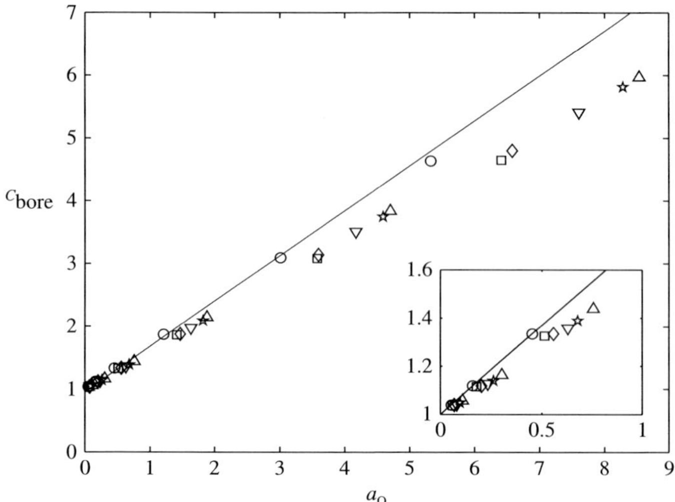

FIGURE 3.6.3. Comparison between the speed predicted by (3.6.10) (solid curves) and the speed observed by Helfrich *et al.* (1999) in their numerical calculations for various finite channel widths [$w = 0$ (O), 0.2 (), 0.5($\diamond$), 1.0(∇), 2.0($\star$) and 4.0(Δ)], where w has been scaled by the Rossby radius based on the depth $d_d{}^*$ in the quiescent region ahead of the bore. The bore speed c has been nondimensionalized by $(gd_d{}^*)^{1/2}$ and the bore amplitude a has been scaled by the depth $d_d{}^*$ ahead of the bore. (From Helfrich *et al.*, 1999).

If the right wall is temporarily assumed to lie at $x = 0$, the solution satisfying $a(0) = a_o$ is given by

$$a(x) = \frac{a_o e^{x/c}}{1 + \frac{1}{2}a_o(1 - e^{x/c})}.$$ (3.6.13)

Far from the wall ($x \to -\infty$) $a \to 0$ and, according to (3.6.11),

$$\left(\frac{\partial Y}{\partial x}\right)^2_\infty = \tan^2(\theta_\infty) = \frac{1}{2}a_o(3 + a_o).$$

The far field angle θ_∞ between the jump and the x-axis tends to zero as the amplitude a_o is reduced. When $a_o = 0.56$, θ_∞ is 45°.

The shape of the contour $y = Y$ can be found by substituting (3.6.13) into (3.6.11) and integrating that relation numerically. Examples of solutions for various a_o (Figure 3.6.2) show the curvature previously alluded to. The curvature and the angle θ increase as a_o does but Federov and Melville (1996) show that the geostrophic approximation for v_o remains good only as long as $a_o < 1$.

Although the theory assumes a coastal setting, verification of the predicted bore speed in channel geometry is best when the channel is narrow. In the simulations carried out by Helfrich *et al.* (1999), Poincaré wave radiation is observed where the bore contacts the left wall (e.g. Figure 3.3.7). The presence of such waves requires a nongeostrophic v and thus a violation of (3.6.12). A comparison (Figure 3.6.3) between the predicted bore speed and the speed measured from simulations shows that the observed speed is well predicted when the channel is narrow. Poincaré wave generation is minimal in such cases. As the channel width increases from moderate to large values in comparison with the deformation radius, the observed speed becomes moderately overestimated by (3.6.10), perhaps due to the radiation of Poincaré waves. The overestimation is also greatest for large values of the amplitude a_0 perhaps due to a failure of the geostrophic relation in the rear of the bore.

Exercises

1. Obtain (3.6.4) directly from (3.5.1) and (3.5.4).

3.7.   Shocks in Separated Flows

So far our discussion of shocks has concentrated on flows with finite layer thicknesses. However, the numerical simulations discussed in Section 3.4 reveal the existence of shocks in flows that are partially or completely separated from the left channel wall. The signature of these 'transverse' shocks is an abrupt change in the width of the stream that may propagate or remain stationary. We will begin the present discussion with the stationary version, an example of which is shown in Figure 3.7.1a.

An attempt to produce a hydraulic jump in a detached, laboratory flow was made by Pratt (1987). As suggested in Figure 3.7.2, fluid is pumped into the right-hand end of a channel, where it collects in a small reservoir and spills over an obstacle. Downstream of the sill, the flow becomes supercritical and undergoes a hydraulic jump. The subcritical flow downstream is withdrawn near the left-hand end of the channel. The procedure is to set up a steady state with no rotation, then spin up the channel to a steady rotation rate high enough to cause the supercritical flow to separate. An unavoidable difference between the laboratory flow and the numerical solution is that the 'global' deformation radius $(gD_\infty)^{1/2}/f$ based on the reservoir depth D_∞ is much greater than the channel width w^* in the laboratory, but comparable to w^* in the numerical experiment. This is due to the fact that the laboratory experiments are performed with free-surface flows and full gravity. A Kelvin wave in the laboratory channel is therefore felt across the whole width. Separation is still possible because the deformation radius based on the local depth scale D of the supercritical flow can be much smaller than w^*.

The qualitative features of the classical planar hydraulic jump depend primarily on the Froude number $V/(gD)^{1/2}$, where the velocity and depth scales V and D are normally based on the approach flow. Rotation leads to the addition of

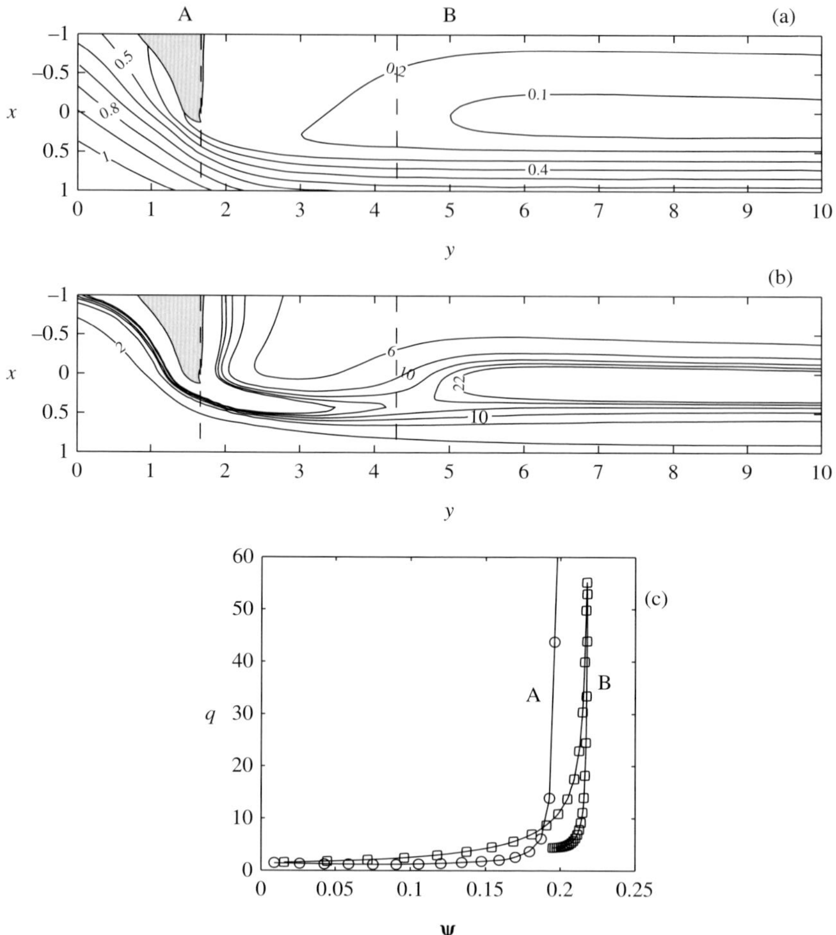

FIGURE 3.7.1. (a) Contours of surface elevation in the vicinity of a transverse hydraulic jump. The flow is left-to-right and is spilling down an obstacle whose crest lies at $y = 0$ and $w = 2$. The shaded region indicates dry channel bottom. (b) Potential vorticity distribution for the flow in (a). (c) Potential vorticity $q(\psi)$ distributions across sections A and B as marked in (a). From Pratt *et al.* (2000).

at least one dimensionless parameter and a natural choice is the ratio of the upstream width scale W to the 'local' Rossby radius of deformation $(gD)^{1/2}/f$ for the approach flow. If the approach flow is attached, W is just the dimensional channel width w^*. If the approach flow is detached then W is the separated current width w_e^* and V and D become related by the geostrophic condition $V = gD/fw_e^*$. It follows that $Wf/(gD)^{1/2} = (gD)^{1/2}/V$, and thus the upstream flow is characterized by a single parameter. However, the downstream end state in this case may be attached, implying that w_e/w is a relevant parameter for the jump as a whole. A third possibility in which both end states are detached is

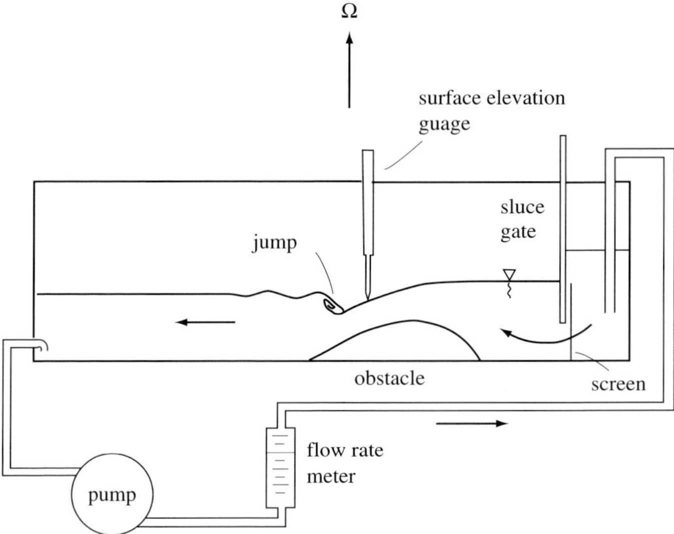

FIGURE 3.7.2. Laboratory apparatus for rotating hydraulic jump experiment. (From Pratt, 1987).

generally not observed for a stationary jump. In summary, important parameters include the upstream value of $V/(gD)^{1/2}$ along with[9] a second parameter

$$r = \begin{cases} w^* f/(gD)^{1/2} & \text{(completely attached)} \\ w^*/w_e^* & \text{(separated upstream, attached downstream)} \end{cases} \tag{3.7.1}$$

A representative sequence of experimental runs demonstrates the qualitative effects of increasing the rotation rate (Figure 3.7.3). The value of $V/(gD)^{1/2}$ is held within the range 7.1 ± 0.5 for all frames while r increases from 0.22 to 4.7. Frame (a) shows a case in which the rotation rate is small and the flow is indistinguishable from a nonrotating flow. The supercritical flow can be seen along with a hydraulic jump at the base of the obstacle. In Frame (b) the rotation rate has been increased to the point where some visual evidence of cross-channel variations in the flow field can be seen. In particular, the amplitude (depth change) of the jump is largest on the right side of the channel and waves along this edge have appeared downstream of the jump[10]. In Frames C and D, the

---

[9] Other parameters may be important as well. If the Rossby radius based on the potential depth D_∞ is comparable to w then the ratio of these lengths is relevant. If the potential vorticity of the approach flow varies, parameters measuring this variation may arise.

[10] This contrasts with some attached jumps in numerical models (e.g. Figure 3.4.11), which have maximum amplitude on the left side of the channel. However the differences may be due to the differences in the Kelvin wave decay scale $(gD_\infty)^{1/2}/f$, which is $< w$ in the numerical simulation and $\gg w$ in the laboratory experiment.

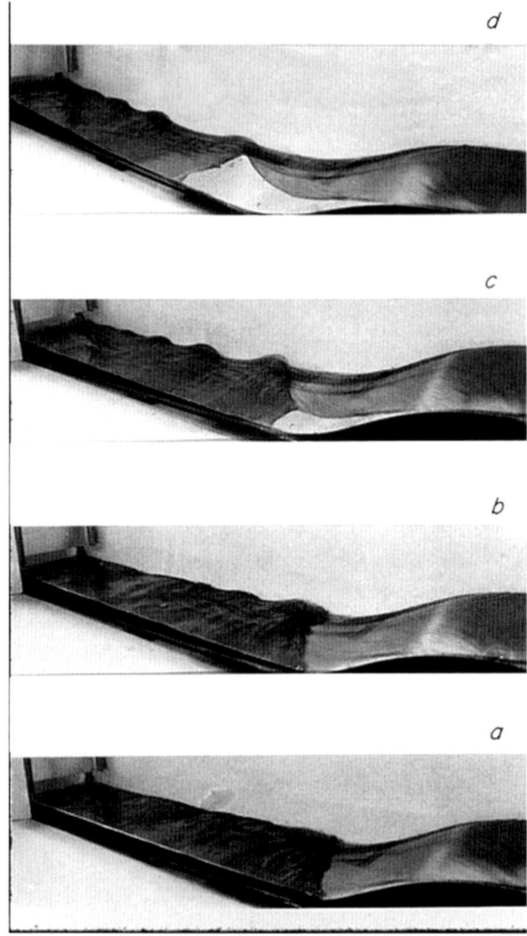

FIGURE 3.7.3. Photos of hydraulic jumps for (*a*) $r = 0.22$, (*b*) $r = 0.84$, (*c*) $r = 3.10$, (*d*) $r = 4.7$. The supercritical flow is spilling from right to left down an obstacle lying to the right in each photo. In (c) and (d) the supercritical flow has separated from the near wall. The Froude number $V/(gD)^{1/2}$ of the supercritical flow just upstream of the jump is 7.1 ± 0.5 in all cases. The value of V is estimated from the geostrophic relation as $g\Delta d^*/fw^*$, where Δd^* is the change in depth across the stream and w^* is the width (either separated or attached) of the flow. D is the average of the depths on the two sides of the stream. (From Pratt, 1987).

supercritical flow has separated and the hydraulic jump is manifested primarily as a discontinuity in the width of the stream. The conjugate subcritical flow remains attached to the left wall, even at the highest rotation rates. A broad, cyclonic recirculation (not visible in the photos) forms downstream of the zone of reattachment. The abrupt reattachment and the downstream recirculation are similar to that observed in the numerical simulations (Figure 3.7.1a, b).

As r increases, the upstream-to-downstream increase in depth becomes less abrupt. For separated upstream flow, the transition in depth is smooth and wave-like, and the region immediately downstream of the transition is observed to contain turbulent, horizontal eddies. There is no visual evidence of the vertical turbulence and mixing that characterizes nonrotating jumps. These observations suggest that potential vorticity may be approximately conserved across the jump at higher values of r, but a shock-joining model based on this assumption fails to accurately predict the downstream state. The numerical simulations clearly show that potential vorticity is altered (Figure 3.7.1b and c), much of the implied dissipation occurring downstream of the point of reattachment.

Another feature that complicates the discussion of dissipation is that the detached laboratory flow is bordered by a shallow but relatively wide, viscous region. As shown by a cross section of the supercritical flow (crosses in Figure 3.7.4) the fluid immediately adjacent to the right wall is strongly banked and is well-approximated by a zero potential vorticity profile (solid line) for the same volume flux and inviscid width. To the left lies a broad, shallow area where the fluid depth is close to the characteristic Ekman layer thickness $(2\nu/f)$ based on the kinematic viscosity ν.

Stationary shocks would appear to depend on the presence of a 'left' wall, even when the upstream flow is separated. Although this has not been proven in a general way, it is strongly suggested by the experiments described above.

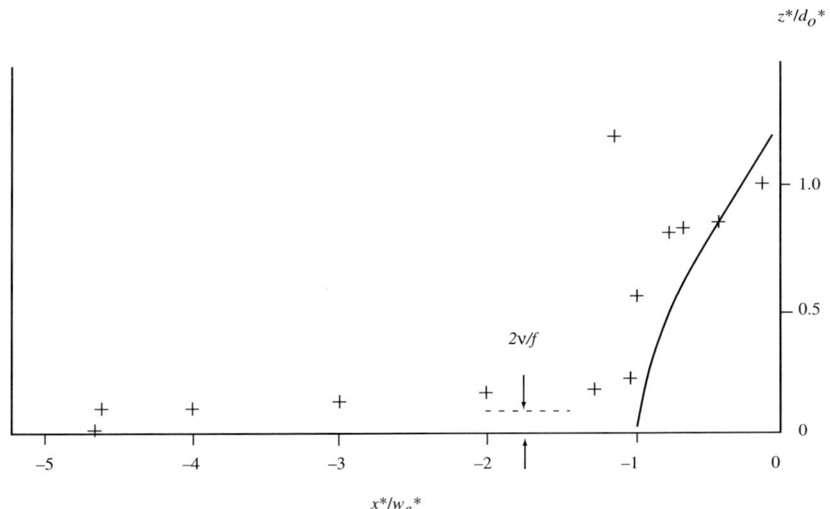

FIGURE 3.7.4. Cross section of the supercritical flow shown in Figure 3.7.3c. The crosses indicate measurements of the free surface elevation. The solid line shows the free surface profile of a zero potential vorticity flow with the same volume flow rate and same width. The width in this case is taken as the distance from the right wall at which the depth falls beneath the Ekman thickness $(2\nu/f)$, as shown by the dashed line. All depths are nondimensionalized by the observed wall depth and the cross-channel coordinate x by the observed inviscid width w_e^*. (Pratt, 1987, Fig. 5).

As shown by Nof (1984), this constraint is relaxed if the shocks are allowed to propagate. A possible version of such a feature consists of an expansion, both in width and depth, of a coastal current (Figure 3.7.5). The fluid is assumed to have zero potential vorticity and also have positive velocity, so that the current is supercritical (in view of the Section 3.2.3 discussion). A shock is postulated as a result of an increase in the transport, and therefore the wall depth, at some point far upstream. The resulting disturbance is imagined to steepen in the manner described in Section 2.3, eventually breaking and forming a steadily propagating discontinuity in depth that moves towards the observer.

Nof joins the end states of the shock using conservation of potential vorticity and width-integrated mass and momentum, even though the first constraint is not strictly justified. The calculation is further constrained by the requirement that the energy of fluid parcels cannot increase as they pass through the jump. This last condition rules out any solution for which fluid passes across the shock from deeper to shallower depths, meaning that the fluid velocity v in the moving frame of the shock must be < 0 for all x. The resulting theory gives a prediction of one end state given the other end state and the propagation speed. Or, the shock speed can be predicted from the knowledge of one end state and the change in wall depth. It is found that the shock speed is always greater than $(gd_{\rm d}{}^*)^{1/2}$ based on the wall depth $d_{\rm d}{}^*$ of the flow into which the shock propagates. Stationary shocks are therefore disallowed. As in nonrotating analogs of this shock, the propagation speed is greater than that of the linear wave propagating in the downstream region, but slower than the linear wave propagating down stream in the upstream region. This *must* be true to remain consistent with the steepening process that forms the shock in the first place.

The width is always increased by the passage of the shock (e.g. Figure 3.7.6) and this makes the solutions quite different from the transverse shocks found in the numerical simulations of Section 3.4. As exemplified by the feature shown at $y = 18$ at $t = 20$ in Figure 3.4.6, the numerically generated shock involve a

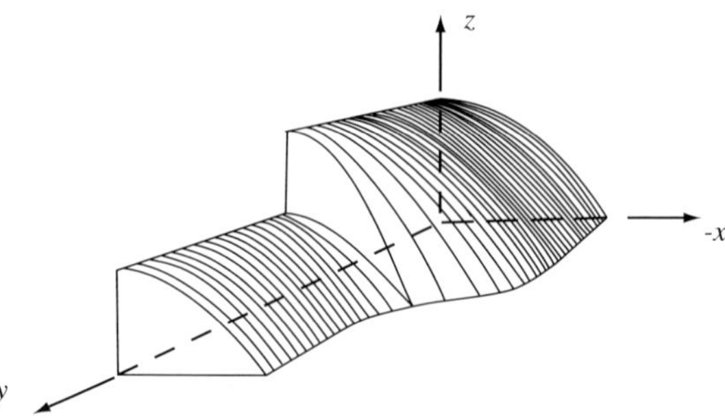

FIGURE 3.7.5. Sketch of hypothetical shock wave in a separated current. The observer faces upstream ($-y$). (Based on Nof, 1984, Figure 3a).

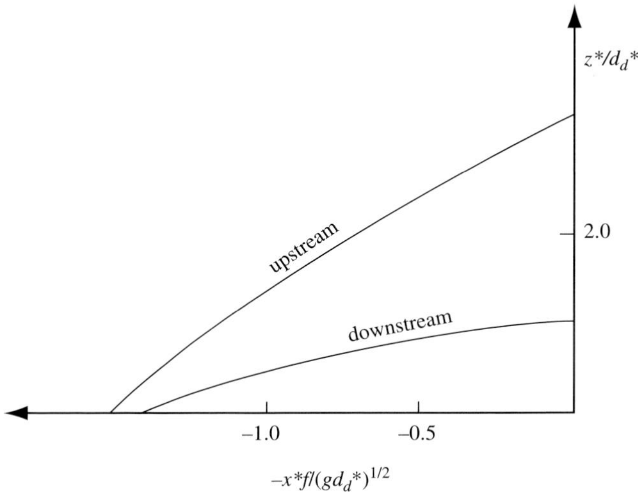

FIGURE 3.7.6. Example of the upstream and downstream depth profiles for a forward propagating shock in a separated current with zero potential vorticity. The Froude number $\overline{F} = 1$ and the downstream wall depth is 0.3 times the upstream wall depth. The Froude number is defined by $\overline{F} = (v_1{}^*(w^*/2) + v_1{}^*(w_e{}^*))/(2gd_d{}^*)^{1/2}$, where $v_1^*(w^*/2)$ and $v_1^*(w_e^*)$ are the values of v^* on the two edges of the downstream current (ahead of the shock) and $d_d{}^*$ is the downstream wall depth. (Based on Figure 10 of Nof, 1984).

decrease in width. Nof's solutions involve large depth changes along the wall that propagate at a speed comparable to $[gd^*(w^*/2)]^{1/2}$; both these features suggest Kelvin wave dynamics. The shocks in Figure 3.4.6 involve considerable changes in width accompanied by minor changes in depth, suggesting frontal wave dynamics. 'Frontal bores' are apparently not admitted in Nof's theory, perhaps due to the restriction to unidirectional velocity in both end states.

At the time of this writing, no direct observations of transverse shocks or jumps had been made in the ocean or atmosphere. Such features would occur internally and would possibly involve exchange of mass and momentum between layers, a process not accounted for in the above formulations. In the Denmark Strait, for example, the supposedly supercritical outflow gradually descends into the deep North Atlantic, gradually entraining overlying water as it does so. There is no evidence of a rapid, stationary change in the width of the flow. The suggestion that contact with the left channel wall is necessary for a stationary jump would mean that the jump would have to occur within the strait and not in the downstream basin. Observations in the Vema Channel (Hogg, 1983) reveal the type of rapid energy transformation that could be caused by a hydraulic jump, though there is no clear connection with the structural features of the jumps discussed above. In a classical jump one expects the fluid depth, and therefore the potential energy, to increase as the fluid passes through the jump. (Total energy is, of course, lost.) Hogg calculated the potential energy using hydrographic measurements taken along three different streamlines of the observed flow (solid curves in Figure 3.7.7). The streamlines are defined by intersections between potential density (σ_4) surfaces and the bottom. For the streamlines corresponding to potential density $\sigma_4 = 46.11$ and 46.13, the potential energy decreases and then increases as it would if an accelerating supercritical flow passed through a jump.

Exercises

(1) Consider a hypothetical stationary shock wave in which the upstream state is supercritical, the downstream state is subcritical, and both are detached and have zero potential vorticity. Show (as Nof, 1984 did) that such a feature cannot be stationary.

(2) Give a plausible reason that explains why Nof's (1984) shock solutions apparently do not include breaking frontal waves.

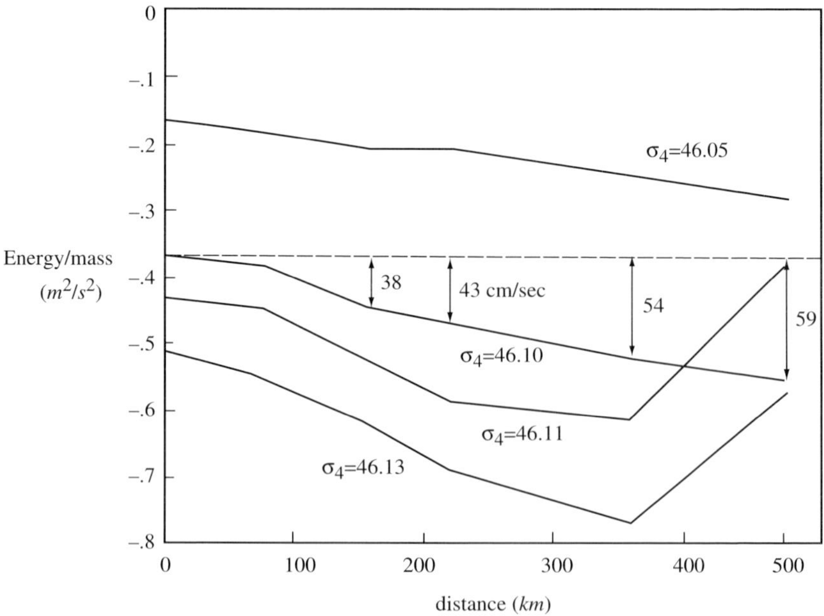

FIGURE 3.7.7. Effective potential energy along three streamlines of the deep Vema Channel overflow. The streamlines are defined as intersections between the indicated potential density (σ_4) surfaces and the bottom. The lower the value of σ_4, the closer the streamline is to the west wall. The flow is to the right in the figure and horizontal distance is measured downstream from the mouth, where the channel joins with the Argentine Basin. Speed arrows are those needed to make the total energy of the 46.10 surface uniform. (From Hogg, 1983).

3.8. Hydraulic Control in a Dispersive System: Flow Over an Infinite Ridge

For the most part, our discussion of hydraulics has been limited to cases where the geometry varies gradually in the direction of the channel or coastline. The linear waves permitted, being long in comparison to the lateral dimensions of the flow or conduit, have been nondispersive. Just what happens when this restriction is relaxed is a complex and unsettled matter. The majority of investigations have followed one of two approaches. In the first, the variations of the flow in the along-channel direction are considered long but finite, so that the waves are dispersive but only weakly so. Progress is also facilitated if the nonlinearity is assumed weak, implying that topographic variations must be small. The most interesting behavior occurs when the entire flow is close to a state of hydraulic criticality, so that waves are generated resonantly by the topography. The simplest theory is the one that results in the steady Kd.V. equation for a nonrotating, 1-d flow (Section 1.11). Grimshaw (1987) and others have extended this body of work to include weakly dispersive, hydraulically driven flows along coastlines. The second approach explores the opposite extreme in which the transverse length scale is essentially infinite. The flows are more idealized but analytically simpler than the nonlinear dispersive models, and waves of all lengths are permitted. The original Rossby adjustment problem (Section 3.1) is an example, though no topography is present. We will focus on another adjustment problem described by Baines and Leonard (1989). Although steady solutions can be readily calculated, the initial-value problem proves quite helpful in developing intuition about the flow. This accounts for the placement of the material in the present chapter on time-dependent flows.

Consider an infinite, horizontal plane with a uniform, shallow flow $v^* = v_o{}^*$, $d^* = d_o{}^*$, and $u^* = 0$. Since the layer depth is uniform, the velocity $v_o{}^*$ cannot be geostrophically balanced by a tilting free surface or interface. A geostrophic balance requires the presence of a uniform external pressure gradient $\frac{\partial p_o{}^*}{\partial x^*} = \rho f v_o{}^*$, perhaps transmitted by an overlying layer.[11] At $t = 0$, an isolated, uniform ridge $h = h(y)$ is placed in the path of the flow and the upstream effects in the resulting x^*-independent state are sought.

A fundamental departure from the models considered to this point is the lack of channel side walls and the Kelvin waves they support. Also, because the flow lacks a potential vorticity gradient, Rossby-type waves are absent. This leaves Poincaré, (inertia-gravity) waves as the only permissible linear transients and any upstream influence must be carried by them. However, we have demonstrated that such waves are inconsequential in establishing the volume transport in the channel adjustment

---

[11] For a system with a free surface, the imposed pressure gradient is rather artificial. However, the same governing equations hold if the shallow layer is imagined to be the lower of a two layer system, bounded above by a rigid lid. The pressure gradient is therefore imposed by the rigid lid and gives rise to a geostrophic v in both layers. If the upper layer is much thicker than the lower layer, the single-layer, reduced-gravity, shallow water equations will govern the latter. The reader who wishes to prove this might first consult Section 5.1.

problem (Section 3.1). The transport is generated solely in response to an upstream-propagating Kelvin wave, while the Poincaré waves act to establish the current that crosses the channel at the original position of the barrier. These considerations suggest that the ridge in the present problem will have no influence of the flow far upstream $(y = -\infty)$ and that the upstream state can therefore be specified for all time. This assumption is used in the steady theory presented below and is later tested as part of time-dependent numerical simulations.

The phase and group speeds (in the y^*-direction) of linear Poincaré waves are given in terms of the y^*-wave number l^* by

$$(c^* - v_o{}^*)^2 = g'd_o{}^* + \frac{f^2}{l^{*2}} \text{ and } (c_g^* - v_o{}^*)^2 = \frac{(g'd_o{}^*)^2}{(g'd_o{}^* + f^2/l^{*2})} \qquad \text{(3.8.1a, b)}$$

(see Exercise 2). When the wave length is large compared to the deformation radius $[l^*(gd_o{}^*)^{1/2}/f \ll 1]$, $c^* \to v_o{}^* \pm f/l^*$ and thus the waves remain dispersive. This behavior contrasts with the long-wave limit of nondispersion in a coastal or channel geometry. In the opposite limit $[l^* f/(gd_o{}^*)^{1/2} \gg 1]$ rotation becomes unimportant and the phase and group speeds approach the value $v_o{}^* \pm (g'd_o{}^*)^{1/2}$. Short waves are therefore nondispersive and are identical to the nonrotating, hydrostatic gravity waves discussed in Chapter 1. The approach flow will be called subcritical, critical, or supercritical with respect to these waves according to $F_o = v_o{}^*/(g'd_o{}^*)^{1/2} < 1, = 1, > 1$. Stationary waves $(c^* = 0)$ can exist if the flow is supercritical and the corresponding wavelength is given by

$$\lambda^* = 2\pi L_d(F_o^2 - 1)^{1/2}, \qquad \text{(3.8.2)}$$

where $L_d = (gd_o{}^*)^{1/2}/f$.

When scaled to the velocity and depth $v_o{}^*$ and $d_o{}^*$ of the approach flow, which is anticipated to remain fixed in time far upstream of the obstacle, the governing dimensionless shallow water equations become

$$\frac{\partial v}{\partial t} + F_o v \frac{\partial v}{\partial y} + u = -\frac{1}{F_o}\frac{\partial}{\partial y}(d+h), \qquad \text{(3.8.3)}$$

$$\frac{\partial u}{\partial t} + F_o v \frac{\partial u}{\partial y} - v = -1, \qquad \text{(3.8.4)}$$

and

$$\frac{\partial d}{\partial t} + F_o \frac{\partial (vd)}{\partial y} = 0. \qquad \text{(3.8.5)}$$

The constant factor on the right-hand side of (3.8.4) represents the externally imposed pressure gradient.

If (3.8.3–3.8.5) are linearized about the approach flow $v = 1 + \hat{v}$, $d = 1 + \hat{d}$, with $\hat{v}$, u, and $\hat{d}$ all $\ll 1$, it can be shown though elementary methods that the steady solution with $(u, \hat{v}, \hat{d}) \to 0$ as $y \to -\infty$ is given by

$$\hat{d} = -\frac{h}{1 - F_o^2} + \frac{1}{2(1 - F_o^2)^{3/2}} \left\{ \int_y^\infty h(\xi) e^{\frac{y - \xi}{(1 - F_o^2)^{1/2}}} \, d\xi + \int_{-\infty}^y h(\xi) e^{\frac{-(y - \xi)}{(1 - F_o^2)^{1/2}}} \, d\xi \right\} \quad (F_o < 1)$$

$$= -\frac{h}{1 - F_o^2} + \frac{1}{(F_o^2 - 1)^{3/2}} \int_{-\infty}^y h(\xi) \sin\left[\frac{y - \xi}{(F_o^2 - 1)^{1/2}} \right] d\xi \quad (F_o > 1) \qquad (3.8.6)$$

The term $-\frac{h}{1 - F_o^2}$ is just the linearized hydraulic solution that would exist if the flow were nonrotating and one-dimensional. It predicts that the free surface dips down over the obstacle for subcritical conditions and bulges up over the obstacle for supercritical conditions. However, this is not the whole story when dispersive effects are present. First, it can be seen that the flow is disturbed upstream and/or downstream of the obstacle. This response depends on the Froude number and on the shape of the obstacle. For $F_o < 1$ (Figure 3.8.1a) the flow has upstream/downstream symmetry and the disturbance is felt within a deformation radius of the obstacle. The free surface rises as the obstacle is approached from upstream and the corresponding pressure gradient gives rise to a transverse velocity $u < 0$ in this region. Over the obstacle the free surface descends, then ascends, and these slopes have bands of transverse velocity $u > 0$ and $u < 0$. For $F_o > 1$ there is no disturbance upstream of the obstacle and lee waves with lengths given by (3.8.2) exist downstream (Figure 3.8.1b).

To find steady solutions to the full problem, it is convenient to write the steady versions of (3.8.3–3.8.4) in the form

$$\frac{\partial d}{\partial y} = -\frac{F_o u + \partial h / \partial y}{(1 - F_o^2 / d^3)} \qquad (3.8.7)$$

$$\frac{\partial u}{\partial y} = \frac{(1 - d)}{F_o} \qquad (3.8.8)$$

The steady continuity equation ($vd = 1$) has been used to eliminate v in the first relation. The second equation makes use of the statement of uniform potential vorticity: $(1 - F_o \, du/dy)/d = 1$.

The vanishing of the denominator in (3.8.7) means that

$$\frac{F_o^{\,2}}{d^3} = \frac{F_o^{\,2} v^2}{d} = \frac{v^{*2}}{gd^*} = 1,$$

corresponding to the critical condition for nondispersive Poincaré waves. Again, the corresponding wave lengths are much shorter than L_d. For the critical flow to remain well-behaved, (3.8.7) requires that

$$F_o u = -\partial h / \partial y. \qquad (3.8.9)$$

(a)

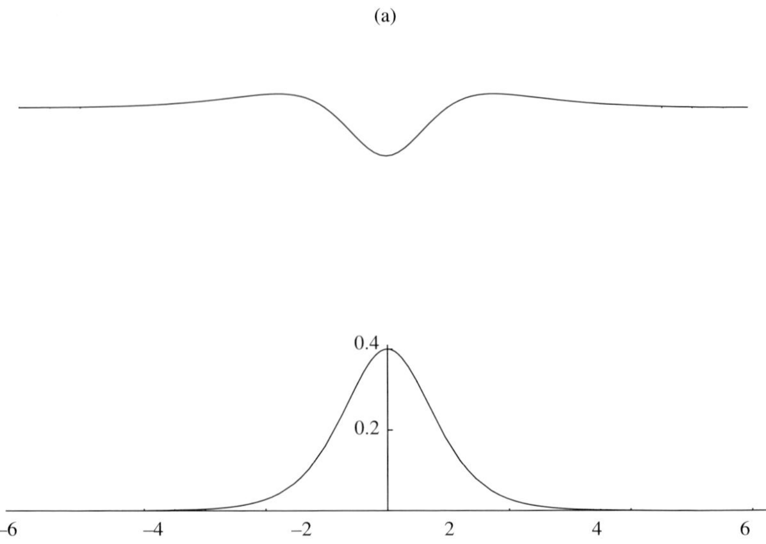

(b)

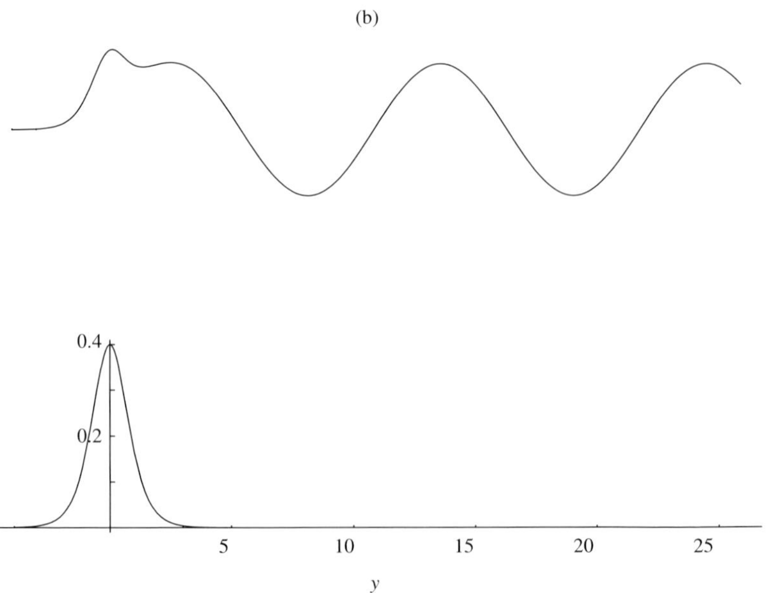

FIGURE 3.8.1. Sketches of linear solutions based on (3.8.6) for (a) $F_\mathrm{o} = 0.5$ and (b) $F_\mathrm{o} = 2.0$, both with $h(y) = 0.3\,sech^2(y)$. [Based on a figure from Baines and Leonard, 1989].

In the semigeostrophic case u is asymptotically small and the critical section must occur where $dh/dy = 0$. In the present case u may be large enough for the Coriolis acceleration to compete with the horizontal component of the bottom pressure force. Critical flow must occur where the two balance.

Numerical integrations of (3.8.7) and (3.8.8) reveal purely subcritical and supercritical solutions with properties qualitatively similar to the linearized solutions. Examples for a semicircular obstacle are shown in Figures 3.8.2a, c. A third solution exists in which a subcritical-to-supercritical transition occurs over the obstacle (Figure 3.8.2b). Here the upper surface rises as the flow approaches the obstacle, and is then drawn down and through a critical section on the upstream face. The transverse flow at the critical section is negative (into the figure) in accordance with (3.8.9). After its transition to supercritical flow the fluid descends, passes the sill, and a series of lee waves is excited. Lee waves also occur for the other solutions that are supercritical downstream (frames b–e), and are only partially seen in the figures. If the approach flow is supercritical ($F_\mathrm{o} > 1$), hydraulically critical states with upstream jumps are possible (frames d and e).

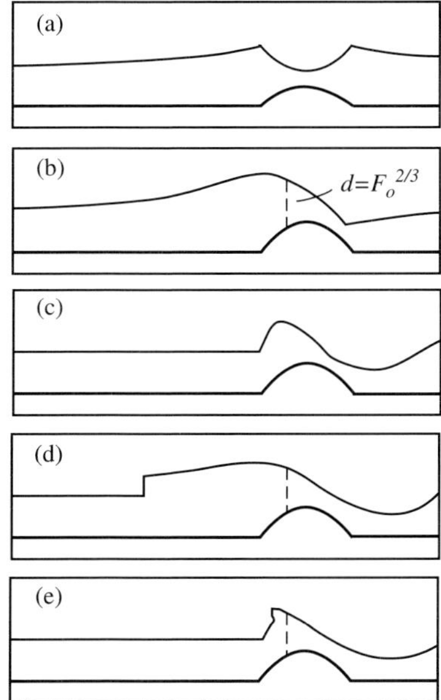

FIGURE 3.8.2. Possible steady, nonlinear solutions for flow over a semi-circular obstacle. (a) purely subcritical flow; (b) subcritical approach flow ($F_\mathrm{o} < 1$) with a hydraulic transition on the upstream face of the obstacle; (c) purely supercritical flow; (d and e) supercritical approach flow ($F_\mathrm{o} < 1$) with a stationary jump upstream of the obstacle and a hydraulic transition over the obstacle. Lee waves exist in cases (b)–(e). [Figure 5 of Baines and Leonard (1989)].

Suppose that one follows a fluid column that originates far upstream in the flow pictured in frame b of the Figure 3.8.2. As the obstacle is approached, the depth increases, the column is stretched, and its vorticity $(-\partial u/\partial y)$ must become positive in order that its potential vorticity be conserved. A transverse flow in the $-x$ direction is implied and the trajectory of the parcel turns to its left. When the upstream edge of the obstacle in encountered, the depth begins to decrease. The y-velocity v increases to conserve mass and the flow undergoes a transition to a supercritical state. The depth of the column decreases and eventually becomes less than its upstream value, implying a positive $\partial u/\partial y$. The transverse flow eventually becomes positive $(u > 0)$, and the trajectory sweeps back to the right. Downstream of the obstacle, a series of lee waves is encountered, with bands of transverse flow.

Returning now to consideration of the full, time-dependent adjustment problem, it is natural to ask how the solution shown in Figure 3.8.2b is established. Direct numerical integrations of (3.8.3–3.8.5) for $F_o < 1$ show that a bore is excited when the obstacle is sufficiently high (Figure 3.8.3), just as in a channel. The critical obstacle height depends on the topographic shape. As the bore moves upstream the pressure gradient associated with the rapid change in depth sets up a geostrophic transverse velocity. As the pressure gradient becomes increasingly opposed by the Coriolis acceleration due to the transverse flow, the bore decays and becomes smooth ($t = 4, 8, 20$ hr profiles in Figure 3.8.3). No permanent alteration of the flow occurs more than a distance $\cong L_d$ upstream of the obstacle. The main role of the bore is to establish the transition region in

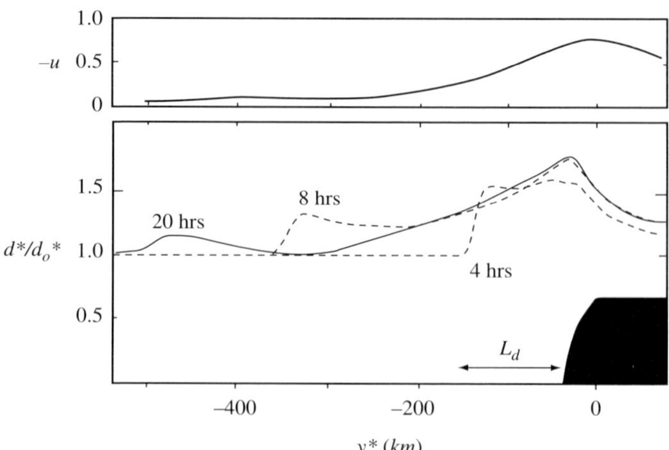

FIGURE 3.8.3. Evolution of the free surface as a result of adjustment to the introduction of a semi-infinite obstacle (shaded region), with $F_o = .85$ and $h_m^*/d_o^* = .667$. The curve in the upper frame shows the transverse velocity at 20hrs, while the lower set of curves shows the free surface at various times. [Based on Figure 3 of Baines and Leonard (1989)].

which the approaching flow deepens prior to meeting the upstream edge of the obstacle.

The above model illustrates a number of important departures from hydraulic theory due to wave dispersion. First, the flow away from the obstacle is influenced by the shape of the obstacle, not just its height. Second, lee waves can arise. Both of these features can be expected to arise in other problems when dispersion is present. A more important issue concerns the inability of the obstacle to alter the far upstream flow, regardless of the topographic height. This arises in the present model not so much because of the presence of dispersion, but because to the lack of channel walls. One might ask why the dispersive waves corresponding to the shortest wavelengths cannot alter the flow far upstream. The answer is that while these waves are too short to feel the effects of rotation over a wave period, they do feel rotation over the period (f^{-1}), roughly the time required to move a deformation radius. Rotation in this particular case is also associated with dispersion and thus the upstream influence is limited to a deformation radius from the topography.

Exercises

(1) The process of upstream influence due to an obstacle in a nonrotating channel has been described as occurring when the obstacle becomes so high that the upstream energy of the flow is insufficient to allow the flow to climb to the obstacle crest (Section 1.4). In the Baines and Leonard model, the obstacle can apparently be made arbitrarily high without necessitating any change in the upstream conditions. Show why this is possible by first deriving the expression of the Bernoulli function:

$$B(\psi) = F_o \frac{v^2}{2} + \frac{u^2}{2} + F_o^{-1}(d+h) - x(\psi),$$

where $x(\psi)$ is the x-position of the streamline in question. Also note that the flux vd per unit width is constant. As a reminder of the result for a channel, note that the terms involving u and x are absent, and that the maximum possible value of h for a given B is therefore finite. In the present case, however, an arbitrarily large value of h is permissible as it can be compensated for by an equally large $x(\psi)$. Explain this result in physical terms by thinking about the origin of the contribution $x(\psi)$ to the Bernoulli function.

(2) Derive the expressions (3.8.1a, b) for the phase and group speeds of the Poincaré waves by altering (2.1.27) to account for the presence of a mean flow $v^* = v_o^*$ and then specializing to the case of independence on the transverse coordinate x^*.

(3) Derive the equations governing the shape of nonlinear lee waves over a horizontal bottom in the context of the above model. (See Baines and Leonard 1989 for a solution).

3.9. Ageostrophic Instability

Our discussion to this point has largely avoided the question of stability. In fact, nearly all of the internal flows under discussion are unstable in some respect. The presence of the horizontal velocity discontinuity between the moving layer and the overlying fluid gives rise to interfacial instabilities. The wavelengths of the most unstable disturbances are finite and the disturbance pressure is therefore nonhydrostatic. The instabilities are avoided in traditional shallow water models with single layers because of the limitation to long wavelengths and the consequent hydrostatic approximation. It is natural to ask, however, whether the presence of the instabilities, and the mixing that they can cause, will wreck the idealization of the moving fluid as a single layer with uniform density. In cases where this length scale is small compared to the fluid depth, the instability may result in overturning and mixing that is limited to the vicinity of the interface. The sharp interface is replaced by a transition layer that may remain thin compared to the layer depth. The single-layer, reduced-gravity idealization may then still be appropriate for long-wave behavior. More on this point will follow in Chapter 5.

It is also reasonable to expect rotating-channel flows to be subject to instabilities that effect the horizontal structure. These include the well-documented barotropic instabilities that can arise in the presence of horizontal variations in velocity, and baroclinic instabilities that arise in rotating flows with horizontal variations in potential energy. Oceanic and atmospheric jets, boundary currents, and broad scale circulations are all subject to these instabilities. The theory for this subject has been developed most thoroughly within the quasigeostrophic approximation (e.g. Pedlosky, 1987). Hydraulically driven, rotating flows typically have strong horizontal shear and large variations in potential energy (interface elevation), and would therefore appear to be particularly vulnerable. Outflow plumes from the Mediterranean and the Denmark Strait are known to contain horizontal eddies that span the stream width and that could be attributed to instabilities. These flows are nonquasigeostrophic and a stability analysis requires that one abandon this approximation by allowing the horizontal velocity to be ageostrophic and the layer thickness to vary by large amounts, possibly vanishing at the edges.

At the time of this writing, the intersection between rotating hydraulics and ageostrophic instability is unclear. For example, the extent to which the steady flows of the Whitehead *et al.* (1974) and Gill (1977) models are unstable is not known. Nor is it understood how the presence of instabilities might alter these flows. For example, it is possible that the instabilities might act only in the supercritical portions of the flow and therefore have no upstream effects. Our inclusion of ageostrophic instability analysis is therefore made in the hope that other investigators will use the basic tools to answer some of these questions. Though it is not strictly necessary, the reader will benefit from some rudimentary knowledge of instability theory (e.g. Chapter 7 of Pedlosky 1986). The following development is based largely on the work of Griffiths *et al.* (1982), Ripa (1983), and Hayashi and Young (1987). We introduce some generalization in the bottom topography used by these authors.

a. Remarks on the Stability Problem and Review of Standard Conditions for Instability

We will be concerned with *linear* stability; that is the stability of an equilibrium *basic* state to infinitesimal disturbances. *Instability* means that it is possible to find an infinitesimal disturbance that will grow in time and lead to a permanent, finite departure from the basic state. We will also confine our discussion to flows that are inviscid and unforced, and therefore preserve their total energy and momentum. The growth of an unstable disturbance to the basic state must then occur without the benefit of any external forcing or dissipation. There certainly are classes of instabilities that act in nonconservative flows and that owe their existence to the presence of friction, but these will not be considered here.

In the traditional analysis of the barotropic stability of large-scale ocean currents and atmospheric winds, the basic state is parallel and zonal: nondimensionally $u = U(y)$. If the basic state has constant depth and takes place on an f-plane, stability is informed by Rayleigh's (1880) inflection point theorem. In particular, d^2U/dy^2 must change sign at some value of y for instability to be possible. Kuo (1949) showed that the β-plane extension of this result is that the potential vorticity gradient $\beta - d^2U/dy^2$ change sign. Charney and Stern (1962) extended this result further to include quasigeostrophic flows with continuous stratification. Instability requires that the horizontal gradient of potential vorticity (including the boundary contribution) must change sign at some point within the cross section. For a single layer with reduced gravity dynamics, this means that $\beta - d^2U/dy^2 + f^2U/(gD)$ must change sign. (Lipps, 1963).

The above necessary conditions can be strengthened by a result due to Fjøtorft (1950). His sufficient condition for stability of a barotropic flow is satisfied if a constant α can be found such that $(U - \alpha)(\beta - d^2U/dy^2) \leq 0$ for all y.[12] As an example, consider a 2D shear flow with $\beta = 0$ and suppose that d^2U/dy^2 changes sign at $y = y_o$. Rayleigh's inflection point theorem is therefore satisfied and the flow may be unstable. However, stability may still be demonstrated by choosing $\alpha = U(y_o)$, so that the Fjøtorft sufficient condition for stability becomes $[U(y) - U(y_o)](d^2U/dy^2) \geq 0$ for all y in the domain of interest. If it happens that the profile is such that $U(y) - U(y_o)$ and d^2U/dy^2 have the same sign, then the flow is stable. Fjøtorft's theorem is part of a sufficient condition for stability, developed below, that applies to shallow water flows.

In keeping with our convention for a rotating channel, we consider a steady basic state $v = V(x)$ and $d = D(x)$, whose stability is to be examined. The basic

---

[12] If the basic potential vorticity $\bar{q} = \beta - d^2U/dy^2$ is considered to be a function of the streamfunction, $\bar{q} = \bar{q}(\psi)$, then Fjøtorft's condition for stability is satisfied if a frame of reference, moving with constant speed c, can be found such that $d\bar{q}/dy \geq 0$.

flow is parallel, and therefore in geostrophic balance, and the channel cross section is arbitrary but uniform in y (Figure 3.9.1). The channel may contain vertical sidewalls $x = \pm w/2$, or the depth may vanish at one or both edges: $x = -a(y,\ t)$ and $x = b(y,\ t)$.

b.   Energy and Momentum in an Unstable Wave

Instability is traditionally defined and measured in terms of the growth in time of some positive definite quantity, usually a wave energy norm.[13] The wave draws on energy available in the mean (y-average) state due to horizontal shear or to

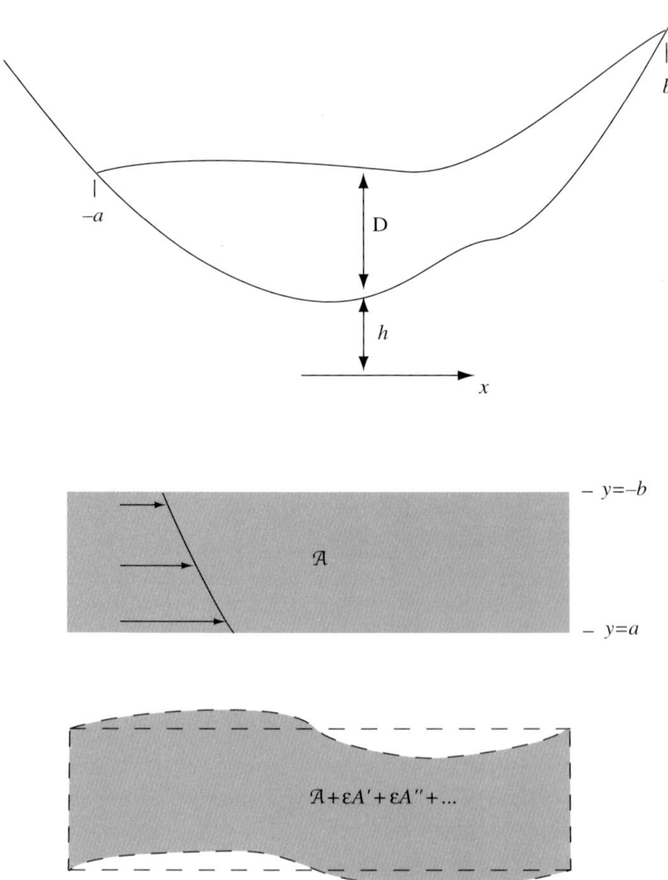

FIGURE 3.9.1. Cross section of the basic flow (top frame). The lower frames show plan views of the wetted areas of the basic flow and disturbed flow.

---

[13] Other norms are used, including enstrophy.

gradients in the elevation of the upper interface. As the wave energy grows, the energy associated with the mean diminishes. For the shallow water models used in hydraulics, in which Poincaré and Kelvin waves, and their relatives, are permitted, the energy associated with the wave is no longer positive definite. The notion that the wave draws energy from the mean flow must be reexamined. The sufficient conditions for quasigeostrophic stability are no longer valid; in fact, the instabilities that are most interesting from an energy perspective can occur when the potential vorticity gradient is zero.

The dimensionless shallow water energy equation is obtained from $ud \times (2.1.5) + vd \times (2.1.6) + d \times (2.1.7)$:

$$\frac{\partial}{\partial t} \frac{d(u^2 + v^2) + d^2 + 2dh}{2} = -\nabla \cdot (\mathbf{u}dB). \qquad (3.9.1)$$

The scaling introduced in Section 2.1, with $\delta = 1$, is in effect and the Bernoulli function B therefore takes its full two-dimensional form $\frac{1}{2}(u^2 + v^2) + d + h$. Suppose that the disturbed flow is periodic in y, or that the disturbance is isolated in y. Let A represent the horizontal region occupied by the fluid, the wetted area, over one wavelength (Figure 3.9.1). Integration of (3.9.1) over A and use of the side edge condition $ud = 0$, valid for vertical walls or for a free edge with vanishing depth, then yields

$$dE/dt = 0$$

where

$$E = \frac{1}{2} \iint_A [d(u^2 + v^2) + d^2 + 2dh] d\sigma \qquad (3.9.2)$$

and $d\sigma$ is the elemental area.

The total momentum of the flow over one period is also conserved, as can be shown from the following form of the momentum flux equation (see Exercise 1):

$$\frac{\partial}{\partial t}[d(v + x)] + \nabla \cdot [(v + x)vd] + \frac{\partial}{\partial y}\left(\frac{d^2}{2}\right) = -d\frac{\partial h}{\partial y}. \qquad (3.9.3)$$

With $\partial h/\partial y = 0$, integration over A yields

$$dM/dt = 0,$$

where

$$M = \iint_A d(v + x) d\sigma. \qquad (3.9.4)$$

We now separate the flow into a basic part $(V, \mathcal{D}, \mathcal{A})$ and a small perturbation. The amplitude of the perturbation is measured by the dimensionless parameter $\varepsilon \ll 1$. The flow field is formally represented as

$$v = \mathcal{V} + \varepsilon v' + \varepsilon^2 v'' + \dots,$$

$$u = \varepsilon u' + \varepsilon^2 u'' + \dots$$

$$d = \mathcal{D} + \varepsilon d' + \varepsilon^2 d'' + \dots \qquad (3.9.5)$$

$$A = \mathcal{A} + \varepsilon A' + \varepsilon^2 A'' + \dots$$

$$q = \mathcal{Q} + \varepsilon q' + \varepsilon^2 q'' + \dots.$$

The area perturbation $\varepsilon A' + \dots$ is due to lateral displacements of the free edges of the current and is zero when the fluid is bounded on both sides by vertical walls. If the edges are free, however, changes in the edge positions alter the horizontal area over which the flow exists (Figure 3.9.1).

Linear instability analysis determines the lowest order perturbation quantities like v', d', etc., which generally have a wave-like structure in y. We will refer to these lowest order quantities collectively as the *wave* field. The wave field can be considered as having no mean with respect to y. Such a mean can be shown to be time-independent and can therefore be disposed by redefining the basic flow. The entire perturbation field: $\varepsilon v' + \varepsilon^2 v'' + \dots, +\varepsilon d' + \varepsilon^2 d'' + \dots$, etc. will be referred to as the *disturbance*. The higher order contributions to the disturbance field, starting with $\varepsilon^2 v''$, etc., may have time-varying means with respect to y. Thus, if $\bar{v}$ represents the average of v over a spatial period in y, then

$$\bar{v} = \mathcal{V} + \varepsilon^2 \bar{v}^{(1)} + \cdots.$$

To the extent that higher order terms can be neglected, the total energy and momentum can now be decomposed into distinct parts associated with the wave and the mean. The latter can further be expressed as a sum of the basic state energy and the energy due to the mean of the disturbance. Substitution of the partitioned fields into the definition of E and neglect of $O(\varepsilon^3)$ terms leads to

$$E = E_b + E_w + E_m$$

where

$$E_b = \frac{1}{2} \iint_A \left[\mathcal{D}\mathcal{V}^2 + \mathcal{D}^2 + 2\mathcal{D}h \right] d\sigma,$$

$$E_w = \frac{\varepsilon^2}{2} \iint_A \left[\mathcal{D}(u'^2 + v'^2) + 2\mathcal{V}v'd' + d'^2 \right] d\sigma,$$

and

$$E_m = \frac{\varepsilon^2}{2} \iint_A \left[2\mathcal{D}\mathcal{V}v'' + (\mathcal{V}^2 + 2\mathcal{D} + 2h)d'' \right] d\sigma.$$

To avoid some unnecessary complexity we have temporarily assumed that the flow is bounded by rigid channel walls, and thus the disturbed area $\varepsilon A' + \dots$ is zero. If the effect of free edges is included, a second expression involving an integral over $\varepsilon A'$ is added to the final integral E_m. (See Hayashi and Young, 1987 for more details.)

The term E_b above is just the energy associated with the basic flow. The quantity E_w, sometimes called the *wave energy*, is the energy associated with the quadratic terms in the perturbation fields. The wave energy can be calculated from the solution to the linearized problem for u', v', etc. In two-dimensional or quasigeostrophic flow, the contribution to E_w from the term involving $\mathcal{V}v'd'$ is absent due to the fact that the depth perturbation is either zero or negligibly small. In this case E_w consists of a sum of nonnegative terms and is used as a measure of the size or growth of the perturbation. In the present shallow water setting, the term $\mathcal{V}v'd'$, and possibly the entire wave energy, can be negative. Finally, the term E_m is the contribution to the energy from the mean of the disturbance. The first order perturbations have no mean and thus E_m is composed of contributions from the means of the $O(\varepsilon^2)$ fields v'' and d''. The individual constituents cannot be calculated from the linearized problem, though as we will later see, the complete sum E_m can be.

For momentum,

$$M = M_b + M_w + M_m$$

where

$$M_b = \iint_A \mathcal{D}(\mathcal{V}+\mathrm{x})d\sigma,$$

$$M_w = \varepsilon^2 \iint_A (v'd')d\sigma,$$

$$M_m = \varepsilon^2 \iint_A [d''(\mathcal{V}+\mathrm{x}) + \mathcal{D}v'']d\sigma,$$

again neglecting terms of $O(\varepsilon^3)$ and assuming vertical side walls.

Another quantity of significance for stability analysis is the *disturbance energy*, defined as

$$E_d = E_m + E_w.$$

It is the sum of the wave energy and the energy associated with changes in the mean fields. It is also the difference $E - E_b$ between the energy of the actual flow and that of the basic flow. Since the total energy E is conserved, E_d is also be conserved. The disturbance energy of a growing wave that has sprung from an infinitesimal instability is zero. One way to think about this is to consider a disturbance observed to have finite but small amplitude of $O(\varepsilon)$. The individual terms that constitute E_d are $O(\varepsilon^2)$ and an uninformed observer might guess that E_d is also $O(\varepsilon^2)$. In fact, the disturbance can be traced back in time to when its amplitude is smaller. By retreating further in time, the disturbance amplitude, and therefore E_d, can be made arbitrarily small. The conserved disturbance energy is therefore essentially zero. The same remarks apply to the *disturbance momentum*, defined by $M_d = M_m + M_w$.

If, on the other hand, the observed disturbance has nonzero energy (or momentum) then it is clear that the disturbance, or some portion thereof, cannot have evolved as the result of an infinitesimal instability. A flow for which all possible disturbances alter the energy is therefore stable to infinitesimal perturbations.

A simple demonstration of the principle of zero-disturbance energy for an unstable system can be made with a pendulum (Figure 3.9.2a). First consider its stable equilibrium, with the arm and weight hanging straight down. A moderate perturbation sets the weight in periodic motion. Let a denote the maximum vertical displacement, relative to its equilibrium position, that the weight achieves during its swaying motion (frame a of the figure). The energy associated with the swaying motion is then proportional to a^2. This is also the disturbance energy: the difference between the total energy of the pendulum and its basic state energy. Note that all possible disturbances add energy to the system relative to the basic state.

Next consider the unstable equilibrium state, with the weight and arm suspended straight up (Figure 3.9.2b). A slight nudge sets the pendulum in motion and we consider a snapshot of that motion when the weight has undergone the same vertical displacement a as before. The total energy of the system at this point is the same as the basic state energy, or at least can be made to approach the

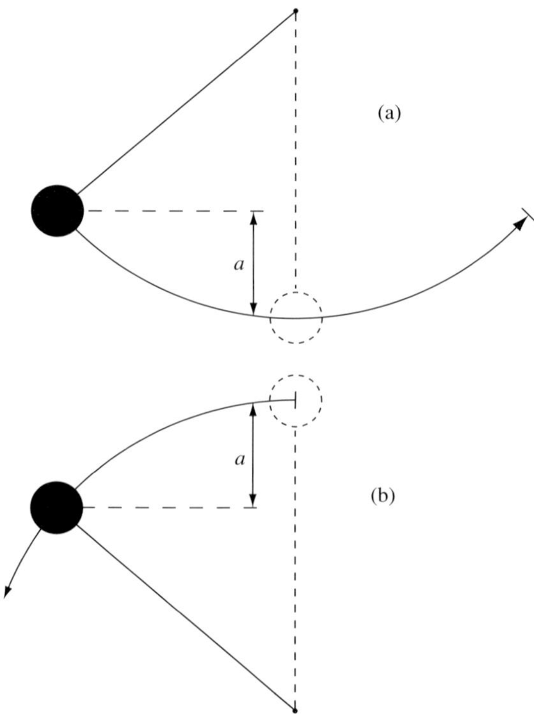

FIGURE 3.9.2. Periodic and amplifying disturbances of a simple pendulum.

basic energy by making the initial 'nudge' infinitesimally small. The disturbance energy is therefore essentially zero.

Another quantity of importance is the lateral displacement $x - x_\text{o} = \delta(y,\, t;\, y_o)$ of a fluid column away from it's original position x_o in the background state. Thus $\left(\frac{\partial}{\partial t} + v\frac{\partial}{\partial y}\right)\delta = u$ or, in linearized form,

$$\left(\frac{\partial}{\partial t} + \mathcal{V}(x_o)\frac{\partial}{\partial y}\right)\delta = u'(x_o, y, t). \tag{3.9.6}$$

c. Ripa's Theorem

A sufficient condition for stability (Ripa, 1983) can be formulated by making bounds based on the conservation laws for disturbance energy $E_w + E_m$ and momentum $M_w + M_m$. The 'wave' constituents E_w and M_w are composed largely of bound-friendly quadratic terms like v'^2. The terms that contribute to E_m and M_m are less so and require a bit more analysis. To this end we consider the linearized shallow water equations for the disturbance fields:

$$\left(\frac{\partial}{\partial t} + \mathcal{V}\frac{\partial}{\partial y}\right)u' - v' = -\frac{\partial d'}{\partial x} \tag{3.9.7a}$$

$$\left(\frac{\partial}{\partial t} + \mathcal{V}\frac{\partial}{\partial y}\right)v' + \mathcal{Q}\mathcal{D}u' = -\frac{\partial d'}{\partial y}, \tag{3.9.7b}$$

$$\left(\frac{\partial}{\partial t} + \mathcal{V}\frac{\partial}{\partial y}\right)d' + \frac{\partial(\mathcal{D}u')}{\partial x} + \frac{\partial(\mathcal{D}v')}{\partial y} = 0, \tag{3.9.7c}$$

and

$$\left(\frac{\partial}{\partial t} + \mathcal{V}\frac{\partial}{\partial y}\right)q' + u'\frac{\partial \mathcal{Q}}{\partial x} = 0, \tag{3.9.7d}$$

obtained through substitution of (3.9.5) into the unforced versions of (2.1.5–2.1.8) and neglect of $O(\varepsilon^2)$ terms. Here

$$\mathcal{Q} = \frac{1 + \partial \mathcal{V}/\partial x}{\mathcal{D}}$$

is the basic state potential vorticity and

$$q' = \mathcal{D}^{-1}\left(\frac{\partial v'}{\partial x} - \frac{\partial u'}{\partial y} - \mathcal{Q}d'\right)$$

is the perturbation potential vorticity.

It can be shown (see Exercise 2) that the above set leads to

$$\frac{\partial \overline{e_w}}{\partial t} - \mathcal{V}\mathcal{D}^2\overline{u'q'} = -\frac{\partial}{\partial x}\left(\mathcal{V}\mathcal{D}\overline{u'v'} + \mathcal{D}\overline{u'd'}\right) \tag{3.9.8}$$

$$\frac{\partial \overline{m_w}}{\partial t} - \mathcal{D}^2\overline{u'q'} = -\frac{\partial}{\partial x}\left(\mathcal{D}\overline{u'v'}\right) \tag{3.9.9}$$

where

$$e_w = \frac{1}{2}\mathcal{D}\left(u'^2 + v'^2\right) + \mathcal{V}v'd' + \frac{1}{2}d'^2 \tag{3.9.10}$$

and

$$m_w = v'd' \tag{3.9.11}$$

are the densities of the wave energy and wave momentum, and an overbar denotes an average in y over a spatial period.

Using the expression (3.9.6) for the linearized particle excursions, it follows from (3.9.7d) that

$$\left(\frac{\partial}{\partial t} + \mathcal{V}\frac{\partial}{\partial y}\right)\left[q' + \delta\frac{\partial \mathcal{Q}}{\partial x}\right] = 0. \tag{3.9.12}$$

The general solution to (3.9.12) can be written

$$q' = -\delta\frac{\partial \mathcal{Q}}{\partial x} + F(x, y - \mathcal{V}(x)t). \tag{3.9.13}$$

The first term on the right-hand side is the potential vorticity perturbation due to the transverse displacement of a fluid column in the basic state. It therefore results from a conservative rearrangement of the basic potential vorticity $\mathcal{Q}$. The second term reflects perturbations in q due to changes in the potential vorticity of fluid columns from their base values. These changes require some sort of external forcing. As shown by (3.9.13) the q' anomalies that result are passively advected by the basic velocity. If its initial spatial distribution is arranged advantageously, an isolated anomaly may temporarily amplify as a result of differential advection. According to linear theory, the disturbance will eventually decay, but its temporary growth might in practice lead to nonlinear effects that cause irreversible changes in the flow. The reader is referred to Farrell and Ioannou (1996) and references contained therein for further insight. The process described does not, however, qualify as instability according to our strict requirement that the disturbance is unforced.

If the forced contribution F to (3.9.13) is ignored, it follows that

$$\overline{u'q'} = -\frac{\partial}{\partial t}\overline{\frac{\delta^2}{2}\frac{\partial \mathcal{Q}}{\partial x}}.$$

The perturbation potential vorticity flux is therefore due to a motion, on average, of the fluid columns down the gradient of background potential vorticity. If (3.9.8) and (3.9.9) are integrated across the channel, and the above expression for $\overline{u'q'}$ is used, one finds

$$\frac{\partial}{\partial t} \iint_A \left(e_w + \mathcal{D}^2 \mathcal{V} \frac{\delta^2}{2} \frac{\partial \mathcal{Q}}{\partial x} \right) d\sigma = 0 \qquad (3.9.14a)$$

$$\frac{\partial}{\partial t} \iint_A \left(m_w + \mathcal{D}^2 \frac{\delta^2}{2} \frac{\partial \mathcal{Q}}{\partial x} \right) d\sigma = 0 \qquad (3.9.14b)$$

Comparison with the earlier energy decompositions suggests that the conserved integrals are the disturbance energy E_d and disturbance momentum M_d. Consequently, the integrals of the terms involving δ^2 *are* the mean energy and momentum, at least within a constant. If the potential vorticity gradient is zero, then the mean energy is identically zero and the disturbance energy E_d equals the wave energy E_w. A similar result holds for the disturbance and wave momentum. Instability is still possible as the growth in positive terms such as $\mathcal{D}u^2/2$ is compensated by the potentially negative term $\mathcal{V}v'd'$ in the wave energy. There is no exchange of energy between the growing wave and the mean flow. The mean flow may change, but the energy associated with that change is zero.

A sufficient condition for stability (Ripa, 1983) can be formulated as follows. Although e_w is not sign definite, it can be shown (Exercise 3) to be nonnegative provided that $\mathcal{V}^2/\mathcal{D} \leq 1$, for all y. That is, a flow for which the local Froude number, dimensionally $\mathcal{V}^*/(g\mathcal{D}^*)^{1/2}$, is everywhere ≤ 1, has nonnegative e_w. More generally, it can be shown that if a constant α can be found such that $(\mathcal{V} - \alpha)^2 \leq \mathcal{D}$, or

$$-\mathcal{D}^{1/2} \leq \mathcal{V} - \alpha \leq \mathcal{D}^{1/2}, \qquad (3.9.15a)$$

then $e_w - \alpha m_w \geq 0$. With this result in hand, we subtract the product of α and (3.9.14b) from (3.9.14a). A time integration of the result yields

$$\iint_A \left((e_w - \alpha m_w) + \mathcal{D}^2 (\mathcal{V} - \alpha) \frac{l^2}{2} \frac{\partial \mathcal{Q}}{\partial x} \right) d\sigma = \text{constant}.$$

Thus if a value of α can be found for which (3.9.15a) is satisfied, and if it is also the case that

$$(\mathcal{V} - \alpha) \frac{\partial \mathcal{Q}}{\partial x} \geq 0 \qquad (3.9.15b)$$

for each y, then the two grouped terms in the integrand are nonnegative. For an infinitesimal perturbation to the basic flow, the positive constant on the right-hand side is arbitrarily small. The integral of $e_w - \alpha m_w$, must then be bounded by an arbitrarily small positive constant, say $\hat{\varepsilon}^2$:

$$\hat{\varepsilon}^2 \geq \iint_A \left(\mathcal{D} \left(u'^2 + v'^2 \right) + 2(\mathcal{V} - \alpha)v'd' + d'^2 \right) d\sigma$$

$$\geq \iint_A \left((\mathcal{D}^{1/2}v' - d')^2 + \mathcal{D}u'^2 \right) d\sigma \qquad (3.9.16)$$

in view of the provision (3.9.15a). The transverse velocity u' must therefore be arbitrarily small, which rules out *shear* instability; that is, instability associated with the transverse motion of the fluid. An instability involving the growth of only v' and d' is still possible, but this would require $d' = \mathcal{D}^{1/2}v'$. This possibility can be eliminated by an argument explored in Exercise 4.

The two provisions in (3.9.15) therefore comprise a sufficient condition for stability: Ripa's Theorem. The first provision relates to gravity wave propagation while the second, which is identical to Fjøtorft's condition for stability, relates to potential vorticity wave propagation.

d. *Rotating Channel Flow with Uniform Potential Vorticity*

For the Whitehead et al. (1974) and Gill (1977) models, and other models of rotating channel flow with constant potential vorticity, the second requirement (3.9.15b) of Ripa's sufficient condition for stability is satisfied. The first requirement (3.9.15a) is essentially that a frame of reference $dy/dt = \alpha$ can be found such that all Froude numbers become less than one. A graphical interpretation of this condition can be obtained by plotting the profiles of $\pm \mathcal{D}^{1/2}$ and $\mathcal{V}$. The requirement is satisfied if one can shift the $\mathcal{V}$ profile up or down so that it fits between the curves for $\pm \mathcal{D}^{1/2}$ (Figure 3.9.3a). There is a range of states with uniform potential vorticity, in channels with rectangular cross sections, that satisfy this condition. However, this range has not been mapped out and it is not clear whether connections with the hydraulic and stability properties of the flow exist.

If the depth goes to zero at one or both edges of the channel (Figure 3.9.3b) then the condition is nearly impossible to satisfy. The value of α must be chosen as the velocity at the edge where the depth vanishes. Then if the depth vanishes at both edges, and the edge velocities differ, the condition cannot be satisfied. Thus, the majority of flows in the Borenäs and Lundberg (1986) theory for a parabolic cross section, and models with other rounded cross sections, generally do not satisfy the theorem and may be unstable.

e. *Modal Disturbances*

Let

$$
\begin{pmatrix} u' \\ v' \\ d' \end{pmatrix} = \mathrm{Re} \left[\begin{pmatrix} \hat{u}(x) \\ \hat{v}(x) \\ \hat{d}(x) \end{pmatrix} e^{il(y-ct)} \right] + O(\varepsilon),
$$

Substitution into (3.9.7a–c) then leads to

$$
[il(\mathcal{V} - c)\hat{u}] - \hat{v} = -\frac{d}{dx}\hat{d} \tag{3.9.17a}
$$

$$
il(\mathcal{V} - c)\hat{v} + \left(1 + \frac{d\mathcal{V}}{dx}\right)\hat{u} = -il\hat{d} \tag{3.9.17b}
$$

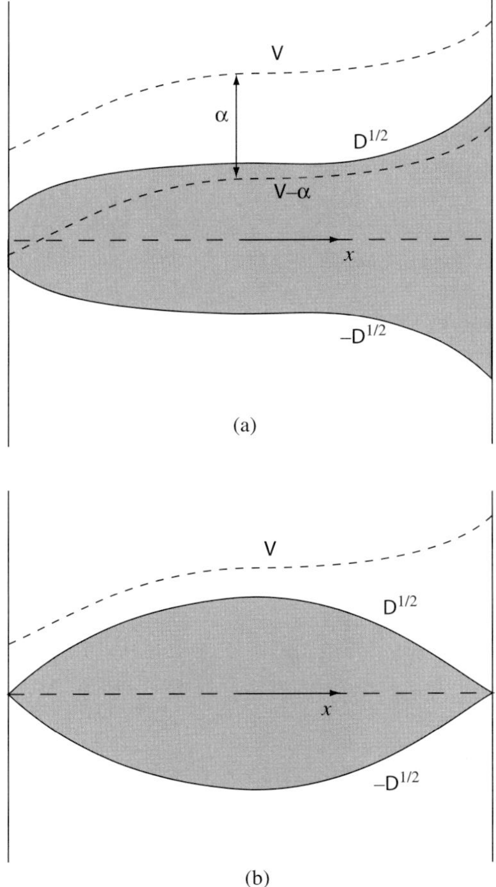

FIGURE 3.9.3. (a) Graphical representation of one of the two requirements (see 3.9.15a) of Ripa's Theorem. Stability requires that the velocity profile can be uniformally shifted up or down to fit entirely in the shaded area. (b) Same as (a) but for the case in which the layer thickness vanishes at the edges.

and

$$il(\mathcal{V} - c)\hat{d} + \frac{d}{dx}(\mathcal{D}\hat{u}) + il\mathcal{D}\hat{v} = 0. \qquad (3.9.17c)$$

Elimination of $\hat{u}$ and $\hat{v}$ in favor of $\hat{d}$ leads to

$$\frac{d}{dx}\left(\frac{\mathcal{D}}{R}\frac{d}{dx}\hat{d}\right) + \left[\frac{1}{\mathcal{V} - c}\frac{d}{dx}\left(\frac{\mathcal{D}}{R}\right) - l^2\frac{\mathcal{D}}{R} - 1\right]\hat{d} = 0 \qquad (3.9.18)$$

where

$$R = 1 + \frac{d\mathcal{V}}{dx} - l^2(\mathcal{V} - c)^2$$

The boundary conditions are

$$\mathcal{D}\hat{u} = 0 \text{ (edges of flow)}. \tag{3.9.19}$$

There are apparently no formal results informing solutions to the eigenvalue problem (3.9.18 and 3.9.19). However, numerical solutions in the long-wave limit generally reveal the presence of two Kelvin-like edge waves and an indeterminate number of potential vorticity waves. The latter are eliminated when the potential vorticity is uniform. The solutions presented in Figures 2.11.13 and 2.11.14 for the Faroe-Bank Channel comprise one example, although these were computed using a slightly different formulation. The phase speeds of the potential vorticity waves in this case are bounded above and below by the Kelvin waves speeds. Some of the potential vorticity waves are unstable. At finite wave lengths, a group of inertia-gravity (or Poincaré) waves is present as well. An example of the latter will be discussed below.

The analysis is substantially simplified in the case of zero potential vorticity ($\mathcal{Q} = 0$). Equation (3.9.17b) reduces to

$$(V - c)\hat{v} = -\hat{d}. \tag{3.9.20}$$

Also, the perturbation potential vorticity $q' = \mathcal{D}^{-1}(d\hat{v}/dx - il\hat{u} - \mathcal{Q}\hat{d})$ must vanish:

$$\frac{d\hat{v}}{dx} = il\hat{u}. \tag{3.9.21}$$

If these last two relations are used to eliminate $\hat{v}$ and $\hat{d}$ from (3.9.17c), one finds

$$\frac{d}{dx}\left(\mathcal{D}\frac{d\hat{v}}{dx}\right) - l^2\left[\mathcal{D} - (V - c)^2\right]\hat{v} = 0. \tag{3.9.22}$$

In view of (3.9.21) the boundary condition $\mathcal{D}\hat{u} = 0$ implies that $\mathcal{D}d\hat{v}/dx = 0$ at the edges. Integration of (3.9.22) across the flow then yields

$$l^2 \int \left[\mathcal{D} - (V - c)^2\right]\hat{v}dx = 0, \tag{3.9.23}$$

where the integration is understood to be across the width of the basic flow, whether or not vertical sidewalls are present.

Now let $c = c_r + ic_i$, so that $c_i > 0$ implies instability. The values of c_r and c_i can be bounded according to a 'semicircle' theorem, first derived by Howard (1961) in connection with stratified shear flow and extended by Hayashi and Young (1987) to an equatorial, shallow water flow. Multiply (3.9.22) by the complex conjugate of $\hat{v}$, integrate the result across the channel, and apply the boundary conditions to obtain

$$\int \left\{\left[\mathcal{D} - (V - c)^2\right]|\hat{v}|^2 + l^{-2}\mathcal{D}|d\hat{v}/dx|^2\right\} dx = 0.$$

The real and imaginary parts of this relation are

$$\int \left\{ \left[\mathcal{D} - (V - c_r)^2 + c_i^2 \right] |\hat{v}|^2 + l^{-2} \mathcal{D} |d\hat{v}/dx|^2 \right\} dx = 0 \qquad (3.9.24)$$

and

$$l^2 c_i \int (V - c_r) |\hat{v}|^2 dx = 0. \qquad (3.9.25)$$

Now let $V_{\min} \le V \le V_{\max}$ and suppose that $c_i > 0$. Then a series of inequalities (Exercise 5) leads to

$$\int \left\{ \left[c_r + \tfrac{1}{2}\left(V_{\max} + V_{\min}\right) \right]^2 + c_i^2 - \left[\tfrac{1}{2}\left(V_{\max} - V_{\min}\right) \right]^2 \right\} |\hat{v}|^2 \, dx$$

$$+ \int \mathcal{D} |\hat{v}|^2 + l^{-2} \int \mathcal{D} |d\hat{v}/dx|^2 \, dx \ge 0. \qquad (3.9.26)$$

The second and third integrals are nonnegative and instability therefore requires

$$\left[c_r + \tfrac{1}{2}\left(V_{\max} + V_{\min}\right) \right]^2 + c_i^2 \le \left[\tfrac{1}{2}\left(V_{\max} - V_{\min}\right) \right]^2. \qquad (3.9.27)$$

The complex phase speed of an unstable wave must therefore fall within the semicircle shown in Figure 3.9.4.

Of particular interest in hydraulics is the stability of long waves. Let $l \ll 1$ and write

$$\hat{v} = v_o + l v_1 + l^2 v_2 + \cdots \qquad (3.9.28)$$

For simplicity, we will normalize $\hat{v}$ such that its maximum value is unity.

We will now restrict attention to a current that vanishes at the two edges. Then the lowest order approximations to (3.9.22) and (3.9.23) are

$$\frac{d}{dx}\left(\mathcal{D} \frac{dv^{(0)}}{dx} \right) = 0$$

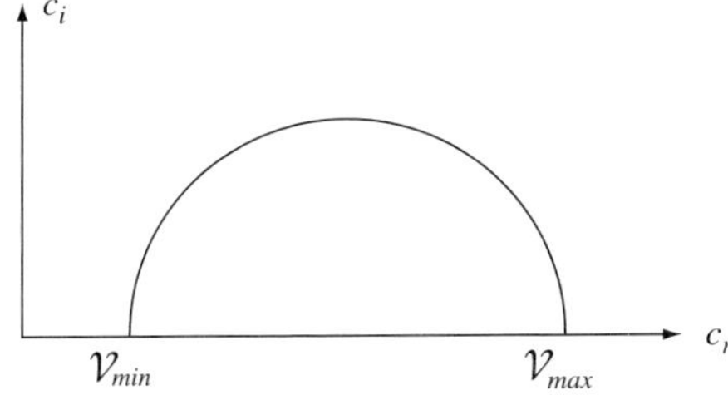

FIGURE 3.9.4. The semicircular region of the complex phase speed plane in which a growing wave must lie. (After Howard, 1961).

and

$$\int \left[\mathcal{D} - (V - c_o)^2 \right] v_0 = 0.$$

Integration of the first relation and enforcement of the boundary conditions lead to

$$v^{(0)} = \text{constant} = 1.$$

and the second relation then yields

$$c_0^2 - 2c_0 \langle V \rangle + \langle V^2 \rangle - \langle \mathcal{D} \rangle = 0.$$

The brackets denote a cross-channel average. The phase speeds of the two waves are given by

$$c_o = \langle V \rangle \pm \left[\langle V \rangle^2 - \langle V^2 \rangle + \langle \mathcal{D} \rangle \right]^{1/2}. \tag{3.9.29}$$

Long-wave instability occurs for $\langle V \rangle^2 - \langle V^2 \rangle + \langle \mathcal{D} \rangle < 0$.

For real c_o, (3.9.29) suggests the Froude number

$$F_o = \frac{\langle V \rangle}{[\langle V \rangle^2 - \langle V^2 \rangle + \langle \mathcal{D} \rangle]^{1/2}}. \tag{3.9.30}$$

The flow is hydraulically critical when $F_o = 1$. So far, all results hold for general bottom topography.

f. The GKS Instability

A example of a instability that acts in the presence of uniform potential vorticity, and therefore does not draw energy from the mean, was analyzed by Griffiths, Killworth and Stern (1982). As shown in Figure 3.9.5a, the basic flow rides over a constant bottom slope $dx/dx = S$ and has zero potential vorticity. [Paldor (1983) treated the special case $S = 0$.] The basic flow profile is computed from the geostrophic relation and from the zero-potential vorticity constraint $\partial V / \partial x = -1$. If basic current is positioned so that $x = 0$ lies midway between the two edges, and if the scale depth $\mathcal{D}$ is chosen as the centerline depth, the basic velocity and layer thickness are given by

$$V = S - x,$$

$$\mathcal{D} = 1 - \frac{x^2}{2}.$$

The edges of the current therefore lie at $x = \pm \sqrt{2}$.

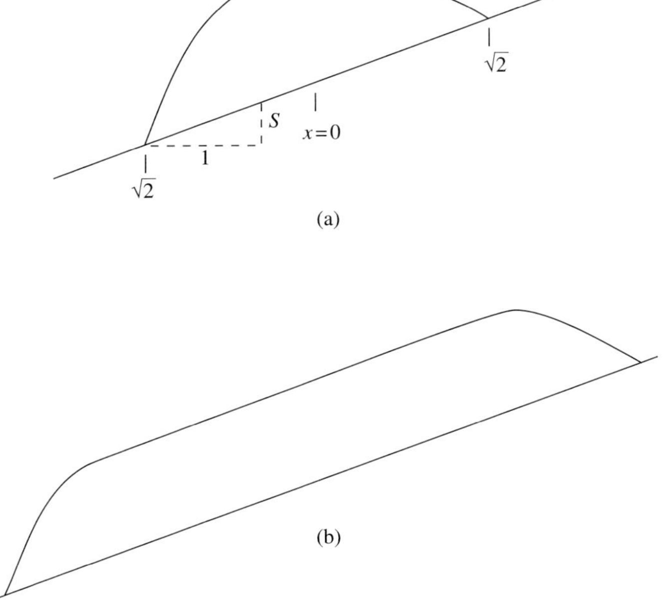

FIGURE 3.9.5. (a) The basic flow of the Griffiths, *et al.* (1982) stability model: a zero potential vorticity current over a sloping bottom. (b) Schematic view of a flow of uniform, nonzero, potential vorticity flow along a constant slope.

The speeds of the two long waves of the flow can be calculated from (3.9.29) using $\langle \mathcal{V} \rangle = S$, $\langle \mathcal{V}^2 \rangle = S^2 + 2/3$, and $\langle \mathcal{D} \rangle = 2/3$. Both waves have the same speed:

$$c_0 = S.$$

or $c_0^* = Sg'/f$.

If one attempts to calculate the next term in the wave number expansion (3.9.28), the eigenfunction is again found to be a constant. Our normalization requires this constant to be zero. It can then be shown that the integral determining the first correction c_1 to the wave speed is degenerate, and thus one must go to the next order of approximation. At $O(l^2)$ (3.9.22) and (3.9.23) give

$$\frac{d}{dx}\left(\mathcal{D}\frac{dv_2}{dx}\right) = -\left[\mathcal{D} - (\mathcal{V} - c_0)^2\right]$$

and

$$\int_{\sqrt{2}}^{\sqrt{2}} \left\{\left[\mathcal{D} - (\mathcal{V} - c_0)^2\right]v_2 - c_1^2 v_o\right\}dx = 0.$$

Substituting the solution to the first relation into the second leads, after a bit of algebra, to

$$c_1 = \pm \frac{2i}{\sqrt{15}}.$$

Waves with long, but finite, lengths are therefore unstable.

For the growing wave ($c_1 = +2i/\sqrt{15}$), it can also be shown that positions of the right and left edges edge of the current (at $t = 0$, say) are given by

$$\begin{pmatrix} b(y, t) \\ -a(y, t) \end{pmatrix} = \begin{pmatrix} \sqrt{2} \\ -\sqrt{2} \end{pmatrix} + \varepsilon \left[\begin{pmatrix} \cos(y) - \frac{2l}{\sqrt{30}} \sin(y) \\ \cos(y) + \frac{2l}{\sqrt{30}} \sin(y) \end{pmatrix} \right],$$

where ε is again a measure of the wave amplitude. Thus the original long wave ($l = 0$) has a meandering structure: it experiences displacements that are equal and in phase on either side of the flow. The lowest order correction introduces excursions that are equal but out of phase. This structure can be seen to some extent in the early stages of the instability as captured in a laboratory experiment (Figure 3.9.6).

Numerical solutions of the eigenvalue problem show that the central ingredients of the long-wave instability are preserved well into the range of finite l. As shown in Figure 3.9.7, the unstable wave continues to have $c_r = S$ and the growth rate lc_i increases with increasing l, reaching a maximum value of about .15 around $l = 0.8$. The most unstable wave therefore has a wavelength of about 8 deformation radii and will double in amplitude over several rotation periods. Both features are characteristic of the laboratory experiment (Figure 3.9.6), where the initial current width is about 3.5 deformation radii, the wave length is roughly twice that, and the instability reaches a large amplitude in eight rotation periods. Although the instability disappears when l exceeds a value $l_c \cong 1.1$, Hayashi and Young (1987) have shown that isolated bands of instability (the small lobes in Figure 3.9.7b) with smaller growth rates reappear at larger l. These weaker instabilities are shown as small lobes along the l axis.

The growth mechanism for the GKS instability is clarified by consideration of the phase speed curves shown Figure 3.9.7a. For l slightly greater than the cutoff value l_c, there are two neutral waves with phase speeds slightly greater and less than S. Analysis of the horizontal structure of these two shows that they are closely related to Kelvin waves: the faster is trapped to the right edge and the slower to the left edge of the flow. Where $l = l_c$ the values of c_r merge and the two wave resonate. Other bands of instability are similarly interpreted; they arise when the phase speeds of two neutral waves merge. In addition to the edge (Kelvin) waves, there is a family of inertia-gravity waves. The latter are closely related to the Poincaré waves discussed in section 2.1 and account for the additional dispersion curves of Figure 3.9.7a. Resonant interaction between an edge wave and an inertia-gravity wave, or between two inertia-gravity waves, accounts for the secondary bands of instability seen at higher wave numbers.

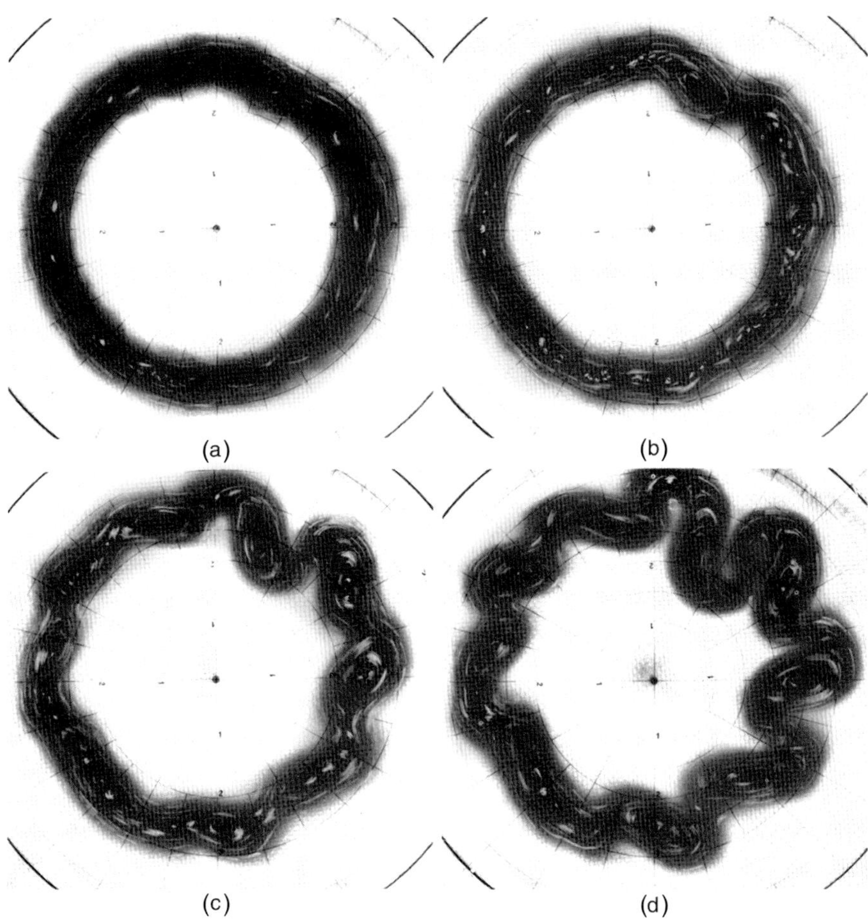

FIGURE 3.9.6. Streak photos showing the instability of a flow set up by introduction of an annular region of buoyant fluid at the upper boundary of a much deeper fluid. The initial width of the flow is approximately 3.4 deformation radii (based on the initial thickness of the buoyant layer). Photos *a–d* were taken at 2, 4, 6, and 8 revolutions following release of the fluid. (Figure 8 from Griffiths *et al.*, 1982).

Direct calculation of the disturbance energy E_d (also the wave energy for this case) for the waves shows that one member of a merging pair has negative and the other positive energy. In fact, it can be shown that the energy is opposite in sign to $c_r^{-1} dc_r/dl$ and thus the two members of any pair must have opposite signed E_d. For the unstable disturbance produced by the interaction between the two members, the disturbance energy is zero by definition. The potential vorticity gradient is zero for this flow and thus the mean energy E_m associated with the disturbance is also zero. The unstable pair does not draw on energy from the mean; instead, growth in the positive E_d of one member is offset by growth in the negative energy of the other. A similar result holds for the disturbance momentum.

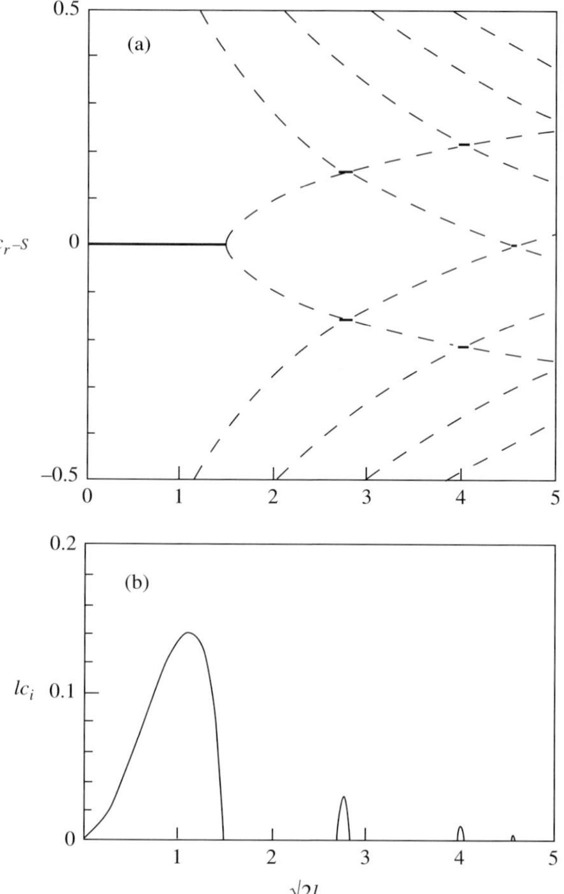

FIGURE 3.9.7. The phase speed (a) and growth rate (b) of instabilities of a zero potential vorticity current on a sloping bottom (from Hayashi and Young, 1987). The GLK instability corresponds to the band roughly spanning $0 < l < 1.1$. The $\sqrt{2}$ is due to a discrepancy between the present scaling and that of Hayashi and Young.

GKS have shown that the long-wave instability acts when the potential vorticity of the background flow is arbitrary. They compute the growth rates for several cases of uniform (nonzero) potential vorticity f/D_∞. The background flow for this last case (Figure 3.9.5b) is similar to that of the Gill (1977) model. In dimensionless terms, there is a central region with uniform depth D_∞, now moving at speed $g'S/f$, and flanked by boundary layers of dimensional thickness $(gD_\infty)^{1/2}/f$. When the width W of the whole current is wide compared to the latter, the modified Kelvin waves are trapped to the edges of the flow and the coupling is weak, as is the instability. When the width and deformation radius are comparable, the coupling is strong and, the system behaves more or less as the in the zero potential vorticity limit. Readers familiar with the classical Eady (1949) model of baroclinic instability will see similarities with the present problem. Both models involve edge waves that are separated by an interior region. (In the Eady problem the 'edges' are rigid, horizontal, upper and lower boundaries.) The tendency of the waves is to propagate in opposite directions, but the sheared background flow can, over a certain range, bring the two speeds into equality. The waves then couple and experience resonant growth. The effect weakens as the upper and lower boundaries are separated.

g.   Connections with Hydraulic Theory

The history of hydraulics, particularly with respect to flow criticality, is replete with tantalizing but vague connections to instability theory. However, it has proven difficult to make definitive statements about such connections. As an example, consider the Fjtørft sufficient condition for stability (also the second requirement (3.9.15b) of Ripa's condition for stability). It states that a necessary condition for *instability* of a single layer, quasigeostrophic flow is that $(V - \alpha)(\partial \mathcal{Q}/\partial x) < 0$. Thus the potential vorticity must increase in the direction to the left of the velocity seen in the moving frame. High potential vorticity on the left suggests that potential vorticity waves attempt to propagate counter the background flow, at least in simplified models. The rest frame $(\alpha = 0)$ version of this condition is also a necessary condition in the Pratt and Armi (1987) model for flow criticality with respect to potential vorticity waves (see Section 2.9). The first requirement (3.9.15a) of Ripa's theorem also appears to intersect with hydraulic theory in requiring the flow to be subcritical in a moving reference frame. Just how strong these connections are is not known.

Another connection between flow instability and hydraulic criticality is suggested by the physical mechanism of the GKS instability. Consider a steady flow that is evolving gradually in the y-directions and that becomes unstable to long waves downstream of some location y_0. If the instability results from the resonant coupling of two neutral long waves, then the corresponding wave speeds c_1 and c_2 must equal each other at y_0. The flow there must then be supercritical with respect to these waves, at least in the sense that information carried by the waves moves in one direction. It is also possible, though less likely, that the flow is critical, with $c_1 = c_2 = 0$. In any case, the flow cannot be subcritical with respect to the two waves. The importance of this property is the suggestion that long-wave instability may be confined to regions of supercritical flow in a wide range of applications. An example that will be reviewed in detail is two-layer flow in a nonrotating channel (see Section 5.2).

Exercises

(1) *Derivation of the equation for conservation of total momentum.* Begin with the flux form of the y-momentum equation with $dh/dy = 0$:

$$\frac{\partial}{\partial t}(vd) + \frac{\partial}{\partial y}\left(v^2 d + \tfrac{1}{2}d^2\right) + \frac{\partial}{\partial x}(uvd) + ud = 0.$$

Then write $ud = \frac{\partial}{\partial x}(xud) - x\frac{\partial}{\partial x}(ud)$, use (2.1.7) again, and integrate the result over the wetted area A to get (3.9.4).

(2) Derive the equation (3.9.8) for the wave energy density. One method follows this plan:

  (a) Begin by taking $u\mathcal{D} \times (3.9.7a) + v\mathcal{D} \times (3.9.7b) + d\mathcal{D} \times (3.9.7c)$, which should give the intermediate equation

$$\frac{\partial e_w}{\partial t} + \mathcal{V}\frac{\partial}{\partial x}(\mathcal{D}\overline{uv}) = -\frac{\partial}{\partial x}\left(\mathcal{V}\mathcal{D}\overline{uv} + \mathcal{D}\overline{ud}\right)$$

  (b) Then write out the definition of the potential vorticity flux $\overline{vq}$, rearrange some derivatives, and use the x-momentum equation to simplify. Substitution of the result for the second term on the left-hand side of the equation in (a) leads to the desired result.

(3) Show that the wave energy e_w is nonnegative provided that $\mathcal{V}^2/\mathcal{D} \leq 1$ for all x.

(4) *Completion of the proof of of Ripa's theorem.* Show that the relationship $\mathcal{D}^{1/2}v = d$ would prevent satisfaction of the both boundary conditions, whether free edges or vertical wall as present.

(5) *On the derivation of the semicircle theorem.* With $\mathcal{V}_{\min} \leq \mathcal{V} \leq \mathcal{V}_{\max}$ observe that

$$0 \geq \int (\mathcal{V} - \mathcal{V}_{\min})(\mathcal{V} - \mathcal{V}_{\max})|\hat{v}|^2\, dx$$

$$= \int \mathcal{V}^2|\hat{v}|^2\, dx - (\mathcal{V}_{\min} + \mathcal{V}_{\max})\int \mathcal{V}|\hat{v}|^2\, dx + \mathcal{V}_{\min}\mathcal{V}_{\max}\int \mathcal{V}|\hat{v}|^2\, dx$$

Next show using equations (3.9.24) and (3.9.25) that

$$\int \left[\mathcal{D} - \mathcal{V}^2 + c_r^2 + c_i^2\right]|\hat{v}|^2\, dx + l^{-2}\mathcal{D}\int |d\hat{v}/dx|^2\, dx = 0.$$

Using this last relation and (3.9.25) to substitute for the first two right-hand integrals in the first equation, obtain (3.9.26).

4
Coastal Applications

4.1. Curvature Effects

It was noted in Section 2.3 that a semigeostrophic channel flow that has become separated from the northern hemisphere left sidewall becomes immune to changes in the position of the right sidewall. As the position of the right wall changes the current moves with it, undergoing no other change in cross-sectional form. Only variations in bottom elevation influence the flow in a meaningful way. This aspect has been demonstrated under the usual conditions of gradually varying geometry, implying that the radius of curvature ρ^* of the wall or coastline is large compared to the characteristic width of the current. (This variable should not be confused with density.) As we discuss below, the effects of coastal curvature begin to become nontrivial once this restriction is relaxed. In order to make analytical progress, and thereby gain a better physical understanding, the ratio of the Rossby radius of deformation, though finite, must be kept small. Topographic effects continue to dominate in this limit if the flow contacts the bottom, but topography is irrelevant if the coastal flow takes place in a surface layer, insulated from the bottom by an inactive deeper layer. Sidewall curvature then provides the only forcing mechanism.

Consider a coast-following coordinate system in which s^* and n^* denote the along-shore and offshore directions, as shown in Figure 4.1.1. To motivate the equations of motion in the (n^*, s^*) system, first consider these equations in the more familiar cylindrical (r, θ) system (e.g. Batchelor, 1967, Appendix 2):

$$\frac{\partial u_\theta{}^*}{\partial t^*} + \frac{u_\theta{}^*}{r^*}\frac{\partial u_\theta{}^*}{\partial \theta} + u_r{}^*\frac{\partial u_\theta{}^*}{\partial r^*} + \frac{u_r{}^* u_\theta{}^*}{r^*} + f u_r{}^* = -\frac{g}{r^*}\frac{\partial(d^* + h^*)}{\partial \theta}$$

$$\frac{\partial u_r{}^*}{\partial t} + u_r{}^*\frac{\partial u_r{}^*}{\partial r^*} + \frac{u_\theta{}^*}{r^*}\frac{\partial u_r{}^*}{\partial \theta} - \frac{u_\theta{}^{*2}}{r^*} - f u_\theta{}^* = -g\frac{\partial(d^* + h^*)}{\partial r^*}$$

$$r^*\frac{\partial d^*}{\partial t^*} + \frac{\partial}{\partial \theta}(u_\theta{}^* d^*) + \frac{\partial}{\partial r^*}[r^* u_r{}^* d^*] = 0.$$

Here $u_r{}^*$ and $u_\theta{}^*$ denote the radial and azimuthal velocity and θ increases (and $u_\theta{}^*$ is positive) in the counterclockwise direction. Topographic forcing (terms with h^*) are relevant when the current runs along the bottom and will be retained for completeness. However these will be ignored in our discussion of surface currents.

Now consider a particular location (s^*-value) along the coastline. The radius of curvature $\rho^*(s^*)$ is considered positive if the coast curves to the right in the direction of increasing s. Position the cylindrical coordinates so that the constant-r^* circles are locally tangent to the coastline at the location in question, as shown in Figure 4.1.1. The origin $(r^* = 0)$ is positioned a distance $\rho^*(s^*)$ from the coast and therefore $r^* = \rho^* + n^*$ and $\partial s^* = -\rho^* \partial \theta$. Associating u_r^* and $-u_\theta^*$ with the off-shore and along-shore velocity components u^* and v^* then leads to

$$\frac{\partial v^*}{\partial t^*} + \frac{\rho^*}{\rho^* + n^*} v^* \frac{\partial v^*}{\partial s^*} + u^* \frac{\partial v^*}{\partial n^*} + \frac{u^* v^*}{\rho^* + n^*} - f u^* = -g \frac{\rho^*}{\rho^* + n^*} \frac{\partial (d^* + h^*)}{\partial s^*} \quad (4.1.1)$$

$$\frac{\partial u^*}{\partial t^*} + u^* \frac{\partial u^*}{\partial n^*} + \frac{\rho^*}{\rho^* + n^*} v^* \frac{\partial u^*}{\partial s^*} - \frac{v^{*2}}{\rho^* + n^*} + f v^* = -g \frac{\partial (d^* + h^*)}{\partial n^*} \quad (4.1.2)$$

$$\frac{\rho^* + n^*}{\rho^*} \frac{\partial d^*}{\partial t^*} + \frac{\partial}{\partial s^*} (v^* d^*) + \frac{\partial}{\partial n^*} \left[\frac{\rho^* + n^*}{\rho^*} u^* d^* \right] = 0 \quad (4.1.3)$$

When the coastline curves to the left in the positive s^*-direction, so that $\rho^* < 0$, the origin of the local cylindrical system lies *offshore* at $n^* = \rho^*$. The corresponding singularity appearing in (4.1.1–4.1.3) is avoided if the upper layer outcrops at a value of $n^* < \rho^*$, or if the fluid at $n^* = \rho^*$ is stagnant.

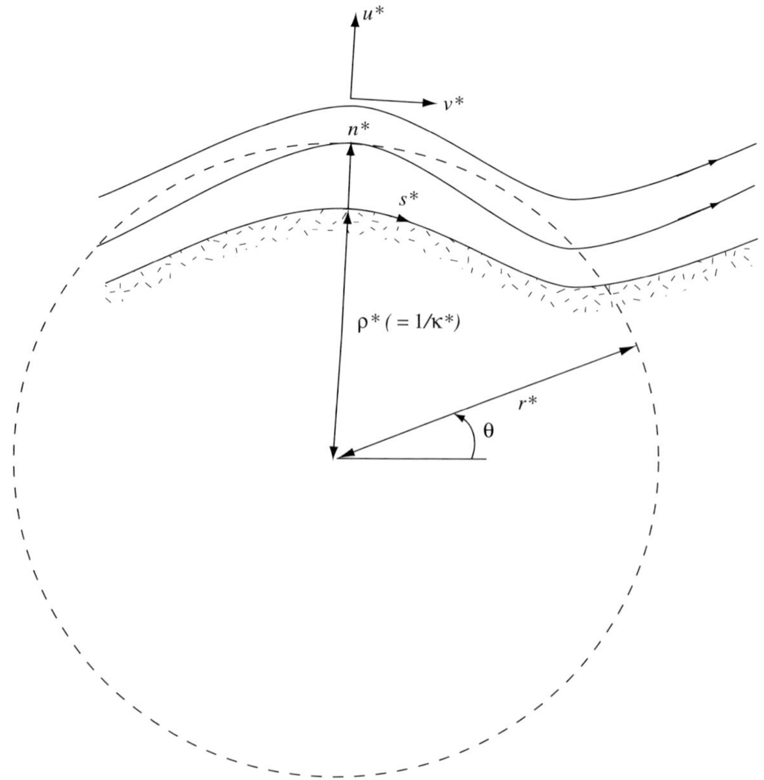

FIGURE 4.1.1. Curvilinear coordinate system.

Conservation of potential vorticity in the new coordinates can be expressed as

$$\left[\frac{\partial}{\partial t^*} + u^*\frac{\partial}{\partial n^*} + \frac{\rho^*}{n^*+\rho^*}v^*\frac{\partial}{\partial s^*}\right]q^* = 0 \qquad (4.1.4)$$

where

$$q^* = \frac{f + \dfrac{\rho^*}{\rho^*+n^*}\dfrac{\partial u^*}{\partial s^*} - \dfrac{\partial v^*}{\partial n^*} - \dfrac{v^*}{\rho^*+n^*}}{d^*}.$$

Steady flow can be described in terms of a stream function ψ^* such that

$$\frac{\partial \psi^*}{\partial n^*} = v^*d^* \quad \text{and} \quad \frac{\partial \psi^*}{\partial s^*} = -\frac{\rho^*+n^*}{\rho^*}u^*d^*. \qquad (4.1.5)$$

as suggested by (4.1.3). Conservation of the Bernoulli energy and the potential vorticity along streamlines then take the forms:

$$\frac{u^{*2}+v^{*2}}{2} + g(d^*+h^*) = B^*(\psi^*) \qquad (4.1.6)$$

and

$$q^*(\psi^*) = \frac{f}{D_\infty(\psi^*)}, \qquad (4.1.7)$$

where D_∞ denotes the potential depth.

We will examine a current with width w_e^* and with net positive transport in the positive s^*-direction. The bottom elevation is constant with n^* but may vary with s^*. We will view the flow as a surface current in which the bounding lower interface may outcrop off shore or may join to a motionless offshore region (Figures 4.1.2a, b). Treatment of a flow with the wall to the left (Figure 4.1.2c) will come later. Let ρ_o^* denote the characteristic radius of curvature of the

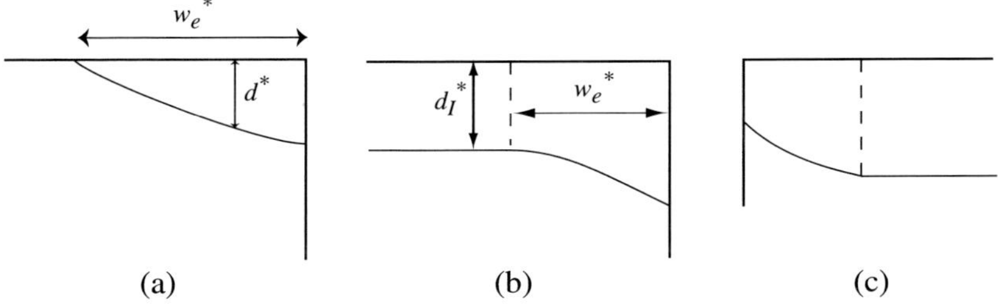

FIGURE 4.1.2. Upper layer geometry for (a): surface current with the wall to the right of positive flux and an outcropping interface; (b): a similar situation, but with the offshore edge joined to a finite depth, quiescent region by a free streamline; (c) a current running with the wall to the left, facing downstream.

coast, L the scale of s^*-variation of the topography, and w the characteristic value of w_e^*. Then the above equations can be rendered dimensionless through use of the scales (W, L) for $(n^*,\ s^*)$ and $(V,\ U)$ for $(v^*,\ u^*)$, with the latter related by $U = VW/L$. In accordance with usual scaling relations (Section 2.1), $W = (gD)^{1/2}/f$ and $V = (gD)^{1/2}$. The nondimensional forms of (4.1.1–4.1.3) are then

$$\frac{\partial v}{\partial t} + \frac{\rho}{\rho + \frac{W}{\rho_o^*}n}v\frac{\partial v}{\partial s} + u\frac{\partial v}{\partial n} + \frac{W}{\rho_o^*}\frac{uv}{(\rho + \frac{W}{\rho_o^*}n)} - u = -g\frac{\rho}{\rho + \frac{W}{\rho_o^*}n}\frac{\partial(d+h)}{\partial s},$$

$$(4.1.8)$$

$$\frac{W^2}{L^2}\left[\frac{\partial u}{\partial t} + u\frac{\partial u}{\partial n} + \frac{\rho}{\rho + \frac{W}{\rho_o^*}n}v\frac{\partial u}{\partial s}\right] - \frac{W}{\rho_o^*}\frac{v^2}{(\rho + \frac{W}{\rho_o^*}n)} + v = -\frac{\partial(d+h)}{\partial n}, \quad (4.1.9)$$

and

$$\frac{\rho + \frac{W}{\rho_o^*}n}{\rho}\frac{\partial d}{\partial t} + \frac{\partial}{\partial s}(vd) + \frac{\partial}{\partial n}\left[\frac{\rho + \frac{W}{\rho_o^*}n}{\rho}ud\right] = 0. \qquad (4.1.10)$$

There are two adjustable parameters; the aspect ratios W/L and W/ρ_o^*. If the geometry is gradually varying in the sense that $W/L \ll 1$, and W/ρ_o^* is also $\ll 1$, then (4.1.9) will, to leading order, reduce to the geostrophic balance $v = -\partial(d+h)/\partial n$ and all coefficients involving curvature will drop out of the remaining equations. Hence curvature effects disappear from the leading order equations in the limit of small W/L and W/ρ_o^*, even though ρ_o^* and L might be comparable. This result would appear to be formal justification of the earlier remarks concerning the insensitivity of the flow to wall curvature.

There is one exception to the remark just made. If the flow moves near the critical speed ($c_- = 0$) it becomes sensitive to gradual changes in curvature. As a demonstration, consider the lowest order approximation to (4.1.6–4.1.8) when $W/L = 0$, $0 < W/\rho_o^* \ll 1$ and when the fluid has uniform potential vorticity. To the lowest order the resulting equations are the same as those governing the separated channel flow discussed in Section 2.3. One of the two characteristic forms of these equations is

$$c_-\frac{dR_-}{ds} = O(W/\rho_o^*) \qquad (4.1.11)$$

where c_- is the characteristic speed and R_- is the Riemann invariant (e.g. 2.3.18 and 2.3.19). The right-hand side contains the numerically small curvature terms. If $c_- = O(1)$, then dR_-/ds must be $O(W/\rho_o^*)$ implying that the current experiences only slight changes in response to the curvature. On the other hand, a flow that is nearly critical in the sense that $|c_-| = O(W/\rho_o^*)$ will allow dR_-/ds to be $O(1)$ and is therefore sensitive to weak curvature.

One way to include curvature effects in a mathematically simplified setting is to assume $\rho_o^* \approx L$, with $W/L \ll 1$. Neglecting terms of $O(W/L)^2$ or higher in (4.1.7) leads to an equation in which advection is neglected but centrifugal acceleration is retained. In addition, the local radius of curvature $\rho(s) + n$ is approximated by its value at the coast $\rho(s)$. A common form of the offshore momentum equation that incorporates these approximations is

$$-\frac{v^{*2}}{\rho^*} + fv^* = -g^* \frac{\partial(d^* + h^*)}{\partial n^*}. \tag{4.1.12}$$

When applied to the potential vorticity equation, the same assumptions lead to

$$\frac{f - \dfrac{v^*}{\rho^*} - \dfrac{\partial v^*}{\partial n^*}}{d^*} = \frac{f}{D_\infty}, \tag{4.1.13}$$

also valid to $O(W^2/L^2)$. These dimensional forms are unfortunate in that they encourage the belief that centrifugal acceleration v^{*2}/ρ^* can be as large as the Coriolis acceleration fv^*. Equation (4.1.9) clearly shows that W/ρ_o^* would have to be $O(1)$ in such cases. The operative along-shore length scale L would then be ρ_o^* and thus $W/L = O(1)$, suggesting that the advective terms in (4.1.9) are no longer negligible. One is then obligated to solve the full shallow water equations. We will proceed with (4.1.12 and 4.1.13) with the caveat that their validity depends on the curvature terms remaining small compared to the remaining terms.

Most investigations of curvature have assumed that the potential vorticity is uniform ($D_\infty = constant$) and that the flow can be traced back into a region where the wall curvature $\kappa = 1/\rho$ is zero. In this upstream region the cross-sectional velocity and depth profiles are given by the semigeostrophic solutions (e.g. 2.2.3 and 2.2.4). If the upstream region in question is a reservoir bounded by two sidewalls, the flow is contained in geostrophic boundary currents. If the upstream geometry is coastal, a single boundary current is present.

In a seminal investigation, Röed (1980) considered flow originating from a wide reservoir. Through an unspecified process the reservoir outflow is imagined to separate from the left reservoir wall and become concentrated in a right-wall boundary current of the type shown in Figure 4.1.2a. Given the local value of ρ^* at a particular downstream location, one seeks a solution that preserves the potential vorticity, volume transport, and energy of the reservoir flow. Let gD_r represent the value of the Bernoulli function B^* along the right wall (facing downstream), where the streamfunction ψ^* is taken as zero. Then the relation $Q^* = dB^*/d\psi^*$ implies $B^*(\psi^*) = gD_r + f\psi^*/D_\infty$. The value of ψ^* along the left wall in the reservoir is Q^*,[1] the total volume transport, and ψ^* must also take on this value along the free edge $n^* = w_e^*$ of the separated current. The solution at a downstream section where ρ^* is nonzero may be obtained by first guessing the value w_e^* and then solving the pair of first order ODE's (4.1.12) and (4.1.13),

---

[1] In the present coordinate system, ψ increases from right to left as seen by an observer facing downstream.

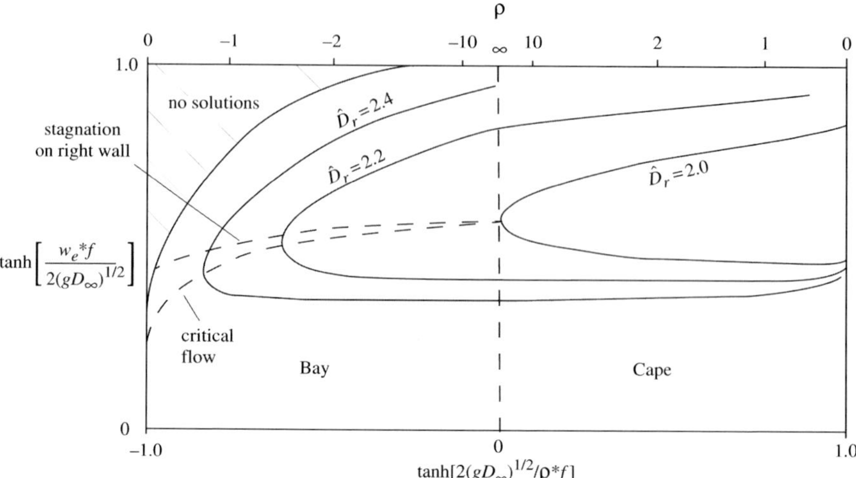

FIGURE 4.1.3. Solution curves for coastal current with uniform potential vorticity. The lower dashed line indicates critical flow and the upper dashed line indicates stagnation along the right wall. The upper branches of the (solid) solution curves correspond to subcritical flow. $\hat{D}_r = D_r(2g/fQ)^{1/2}$ is conserved along each solution curve and $D_\infty(2g/fQ^*)^{1/2} = 4$ for all curves. (From Röed, 1980).

or their dimensional versions, numerically.[2] The integration is started at the free edge $n^* = w_e{}^*$ of the flow using the conditions $d^*(w_e{}^*) = 0$ and

$$v^*(w_e{}^*) = (2B^*(Q^*))^{1/2} = (2(gD_r + fQ^*/D_\infty))^{1/2} ,$$

which follows from (4.1.6) with $v^{*2} \gg u^{*2}$. The integration is then carried to the wall ($n^* = 0$), where the condition $\psi^* = 0$ is checked. If ψ^* is nonzero at the wall, the value of $w_e{}^*$ is adjusted and the procedure is repeated until $\psi^* = 0$ is obtained there. More than one acceptable value of $w_e{}^*$ is generally possible.

By implementing this iterative method for various values of ρ^*, one can generate a sequence of cross sections, all with the same Q^*, D_∞, D_r (and therefore $B^*(\psi^*)$). Figure 4.1.3 contains a dimensionless graph showing solution curves obtained in this way. Each solid curve gives the stream width, represented by $\tanh\left[w_e{}^*f/2(gD_\infty)^{1/2}\right]$, as a function of the wall curvature, represented by $\tanh\left[2(gD_\infty)^{1/2}/\rho^*f\right]$. The dimensionless value of the wall energy $\hat{D}_r = D_r(2g/fQ^*)^{1/2}$ is conserved along each solution curve and $D_\infty(2g/fQ^*)^{1/2} = 4$ for all curves. Each curve has an upper and lower branch and direct calculation of the speed of the frontal wave that propagates on the free edge indicates that the upper branch is subcritical and the lower branch supercritical.[3] The lower

---

[2] Röed actually solved these equations with full variable curvature (ρ replaced by $\rho + n$).
[3] Since the curvature is assumed small, the characteristic speeds are approximately given by semigeostrophic theory (see Equation 2.2.22 with $\hat{d} = d$ and $w = w_e$).

dashed line corresponds to critical flow, as indicated by the merger of subcritical and supercritical solution branches. Just above it lies a second dashed line that marks solutions with zero velocity at the wall. Above this curve the solutions have reverse flow near the wall. This condition, which cannot occur when the flow is supercritical, is closely related to the stagnation condition discussed in connection with upstream gyres (Section 2.7). Although the plot extends over the whole range $-\infty < 2(gD_\infty)^{1/2}/\rho^* f < \infty$, the equations used to derive the solutions are formally valid only for $\left|2(gD_\infty)^{1/2}/\rho^* f\right| \ll 1$.

The solution curves of Figure 4.1.3 have several notable features. First, the subcritical branches show that the stream width decreases and approaches a critical state as the curvature decreases. A subcritical current originating upstream along a straight wall ($\rho^* \to \infty$) will therefore narrow and become less subcritical if the wall bends to the left (facing downstream). The same current becomes wider and more subcritical if the wall bends to the right. If the wall bends to the left and its (negative) curvature becomes sufficiently strong, the flow will undergo a subcritical-to-supercritical transition. The transition takes place at the point of maximum negative curvature. Downstream, the flow will become supercritical and will continue to narrow as the wall becomes less curved. If this supercritical flow then moves into a stretch of positive coastline curvature it can either narrow or widen depending on the particular value of $\hat{D}_r$. Thus, there is no simple rule governing the widening or narrowing of a supercritical current as the coastline curvature varies. The reader will also note that the dependence of the width on curvature is generally weak when $\left|(gD_\infty)^{1/2}/\rho^* f\right| \ll 1$. This behavior is consistent with the earlier finding that curvature effects are weak in the long-wave limit. An exception to this rule occurs when $\hat{D}_r = 2.0$. The corresponding current is exactly critical along the upstream section of straight coastline and will experience a rapid widening or narrowing upon encountering slight finite ρ^*. This is just an example of the sensitivity of a critical flow to its geometric constraints, anticipated by (4.1.11).

Röed also describes solutions that completely separate from the wall. The solutions arise for positive values of ρ^* that are O(1) and therefore outside the formal range of validity of the theory. The general problem of separation from a coast is difficult and of great oceanographic importance. The Gulf Stream, the Kuroshio, and the Mediterranean inflow are just three of many examples of boundary flows that experience separation. In the first two cases the separation is from a 'left-hand' boundary and almost certainly involves the variation in f with latitude. The Gibraltar inflow separates from a 'right-hand' boundary (the Moroccan coast) at a sharp corner that marks the beginning of the Alboran Sea. The latter contains the anticyclonic Alboran Gyre. To compare features like this with the (Röed, 1980) model, it should first be noted that the model permits two types of separation. In the first, the active layer remains in contact with the wall but a stagnation point forms there. This type of separation is demonstrated by Whitehead and Miller (1979) in a laboratory experiment based on the Strait of Gibraltar and Alboran Sea geometry (Figure 4.1.4). In terms of the Röed theory, the value of ρ^* required for separation is indicated

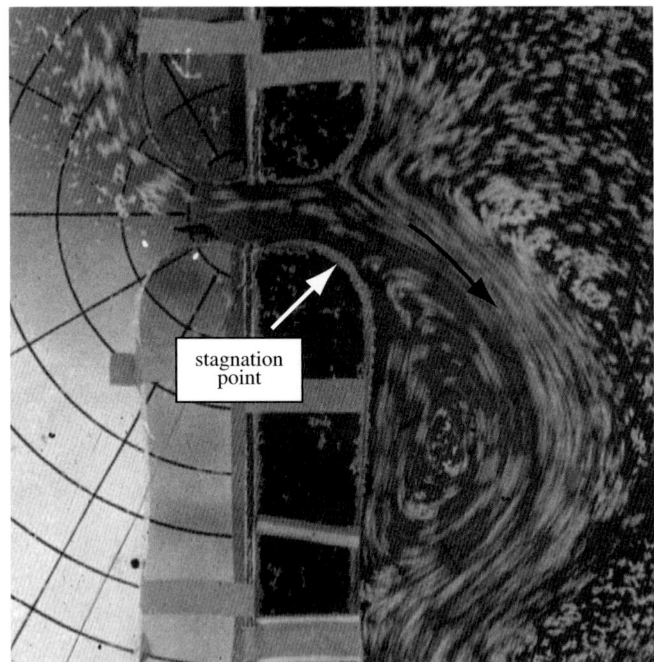

FIGURE 4.1.4. Flow separation in a two-layer lock-exchange flow. The (clear) surface layer enters the gap from the left reservoir and separates from the boundary at the indicated stagnation point. The separated flow continues in an anticyclonic arc, forming a gyre. The denser layer is dyed black and extends to the surface to the right of the gyre. The experiment is described more fully in Whitehead and Miller, 1979.

by the upper dashed line in the left half of Figure 4.1.3, where the flow is slightly subcritical. The Moroccan coast line has positive curvature where separation occurs, whereas the model separation in question requires negative curvature.

The second type of separation involves the detachment of the entire upper layer from the coast and the surfacing of the underlying fluid. It may seem surprising that the flow can outcrop on both sides and still maintain a positive flux, but this is made possible by centrifugal acceleration. If the product of (4.1.12) and d^* is integrated across the width of the stream, the transport can be shown to obey

$$Q^* = \frac{g'd^{*2}(0)}{2f} + \frac{1}{\rho^*}\int_0^{w_e^*} d^*v^{*2}dn, \qquad (4.1.14)$$

and thus a positive Q may be maintained by a positive ρ^* even when the wall depth $d^*(0)$ vanishes. As mentioned earlier, some of Röed's solutions undergo this type of separation for sufficiently large and positive curvature, though the locations in parameter space are not given. Klinger (1994) revisited this issue using essentially the same model and found that the radius of curvature required is roughly equal to the inertial radius $\bar{v}_u^*/f$ based on the average velocity of the upstream flow (measured where the wall curvature is zero). It is tacitly assumed that the upstream velocity profile is unidirectional, and therefore far from being separated from the left wall. If the upstream flow is nearly separated (and therefore bidirectional), the flow may easily separate for values $\rho^* \gg \bar{v}_u^*/f$. Klinger also explores a configuration in which the lower layer does not have an offshore outcrop (Figure 4.1.2b). Here the moving portion of the current is separated from a stagnant offshore region by a free streamline. The potential vorticity of the flow is again constant but the separation condition is found to be insensitive to its value. Despite the finite offshore depth, the wall depth may again go to zero causing the flow to separate. The separation condition over much of the parameter space of the solution is $\rho^* < 0.9\bar{v}_u^* w_{eu}^*/(g'd_I^*)^{1/2}$, where w_{eu}^* is the upstream current width and d_I^* is the interior depth. If the upstream width w_{eu}^* scales with the deformation radius $(g'd_I^*)^{1/2}/f$, then the criterion is nearly the same as for the first case. Again, this condition may violate the assumption of small wall curvature that underpins the model.

A similar technique can be used to explore the case with the wall to the left of positive Q^* (Figure 4.1.2c). Ou and de Ruijter (1986) use a model that is similar to Klinger's, but with the wall to the left. The potential vorticity of the moving fluid is constant and the flow is joined to a stagnant offshore region that has lower potential vorticity. The hydraulically relevant wave is now a Kelvin wave that attempts to propagate upstream. Its speed is approximated by $-(g/D_\infty)^{1/2}$ times the wall depth (*cf.* Equation 2.2.26) provided the potential vorticity front lies more than a distance $(gD_\infty)^{1/2}/f$ offshore. Under this condition the flow remains subcritical as long as the wall depth is finite. If d^* vanishes at the wall, leading to separation of the current, the flow is close to the critical speed.[4] The criticality of the separated current downstream depends upon the environment in which it flows: upstream propagation of long waves may or may not be permitted. Curving of the wall to the left of the direction of flow encourages broadening of the boundary current and separation of the flow, whereas negative curvature has the opposite effect. Ou and de Ruijter also take into account variations in the value of f along the wall and the resulting model is sufficiently complicated that no simple criterion for separation is written down. However, unlike the case of unidirectional flow with the wall on the right, the flow may separate at moderate curvatures.

Laboratory and numerical models allow one to escape the restriction of weak curvature (Figure 4.1.4). These studies traditionally seek local criteria for separation as derived from length scales that characterized the flow at a particular location. The scales include the local radius of curvature, the Rossby Radius of deformation $(g'D)^{1/2}/f$ based on a local upper-layer thickness scale D, and the inertial radius U/f based on the local velocity scale U. The ratio of the last two is a Froude number $F = U/(g'D)^{1/2}$. Many of the experimental flows are set up by a dam-break or lock exchange, and this tends to make F

---

[4] This property is valid as long as the radius of curvature remains large compared to the current width.

close to unity. In such cases the separation criterion is roughly $(g'D)^{1/2}/f\rho^* \geq 1$ (e.g. Whitehead and Miller, 1979). But since $(g'D)^{1/2}/f$ is roughly equal to U/f the criterion could also have been written as $U/f\rho^* \geq 1$. One study that allows a range of Froude numbers (Bormans and Garrett, 1989b) suggests that the latter is more general. The connection between the experiments and the theory described earlier is difficult to establish, not only because $U/f\rho^* \geq 1$ violates the underlying assumptions of the models but also because the models stress nonlocal (upstream) separation criteria, such as a dependence on D_∞.

Other factors cloud the picture, suggesting that more than two dimensionless parameters are relevant. Numerical experiments with no-slip boundary conditions produce separation more readily than those with free-slip conditions. Also, separation is sometimes found to be sensitive to the other properties such as the vorticity distribution in the flow. If the vertical wall is replaced by a sloping bottom or continental shelf the separation condition is altered and the tendencies that occur in response to wall curvature can actually be reversed, as shown in Section 4.2. In the end, flow separation may be sensitive to a whole array of physical circumstances that simple models have difficulty assimilating.

Exercises

(1) It was argued in connection with equation (4.1.14) that positive curvature will allow a current of the type shown in Figure 4.1.2a to maintain a positive flux even when depth along the right wall vanishes. Prove that this is also true for a current of the type shown in Figure 4.1.2b.

(2) Show that (4.1.12) and (4.1.13) can be solved analytically for the case of zero potential vorticity. Derive the resulting depth and velocity profiles assuming geometry of the form shown in Figure 4.1.2a. For given values of energy gD_r and flux Q what is the condition for separation of the entire upper layer from the right wall.

4.2.   Coastal Upwelling Fronts and Jets

When the wind blows along a Northern Hemisphere coastline such that the coast lies to the left of the downwind direction, offshore Ekman transport is created at the surface. Water moves offshore and is replaced by deeper fluid that upwells and creates colder surface temperatures at the coast. This circulation is of biological importance because it lifts large amounts of nutrient-rich seawater into the photic zone and, through photosynthesis, provides a basis for many of the world's fisheries. A view of the resulting state as it would appear in a two-layer idealization is shown in Figure 4.2.1. The continental shelf is represented by a sloping region over which the total depth increases from zero to D_o at the shelf break. At the shelf break ($x^* = w^*$) is a vertical wall that represents the continental slope. Offshore of this point the depth is infinite and the lower layer is inactive. The interface profile (I) shows the state that might occur before the

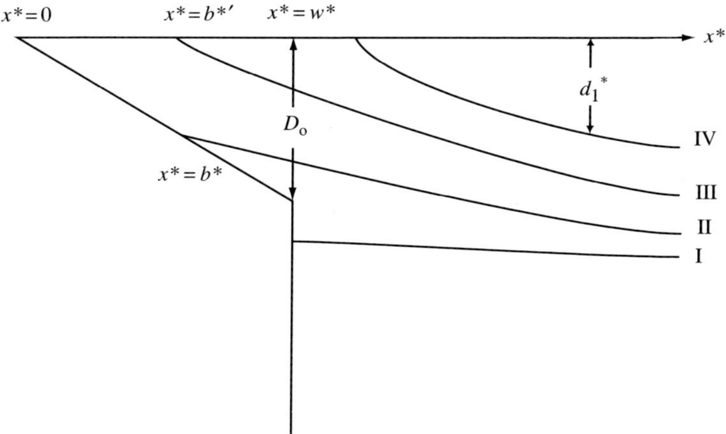

FIGURE 4.2.1. Side view of simplified continental shelf.

upwelling event. As a result of the upwelling, lower layer fluid is brought up onto the shelf causing the interface to ground on the bottom (II) or to outcrop at the surface (III or IV). The sloping interface implies a cross-shelf pressure gradient and the latter tends to be balanced by a geostrophic, along-shore flow. In the Northern Hemisphere the upper layer flow runs with the coast on its left. Jet-like flows are observed along the northwest American coastline and along other coasts that experience upwelling.

Once an along-shore current is established, it will experience topographic interactions due to capes, canyons, and other irregularities in the coastline. As shown in Figure 4.2.2, a southward flowing jet along the Oregon and Californian coastline passes several promontories, including Capes Blanco and Mendocino. The cool (lighter) areas in the lees of these features represent deeper fluid that has welled up to the surface. In the context of Figure 1, these pools could be created when the interface evolves from profiles I or II, for which the interface grounds on the shelf, to (III) or (IV), where it outcrops at the surface. A number of investigators have attempted to explain these and other aspects of along-shore evolution using a hydraulic theory for the coastal jet. The descriptions below are based primarily on the work of Gill and Schumann (1979), who applied such a model to the Agulhas Current, and on Dale and Barth (2001), who applied a closely related model to Cape Blanco. A simplified version of the model was used by Stommel (1960, Chapter 8) to simulate the Gulf Stream along the eastern United States coastline.

The story just told tacitly ignores frictional effects, even though Ekman layer dynamics are essential to the upwelling. Nevertheless, it will be assumed that once the along-shore flow is set up, friction will not contribute significantly to evolution over limited regions of strong topographic variation. At the same time, we invoke the usual assumption of gradual along-shore variations in the coastal geometry, meaning that $w^*(y)$ varies on a scale large compared to w^*

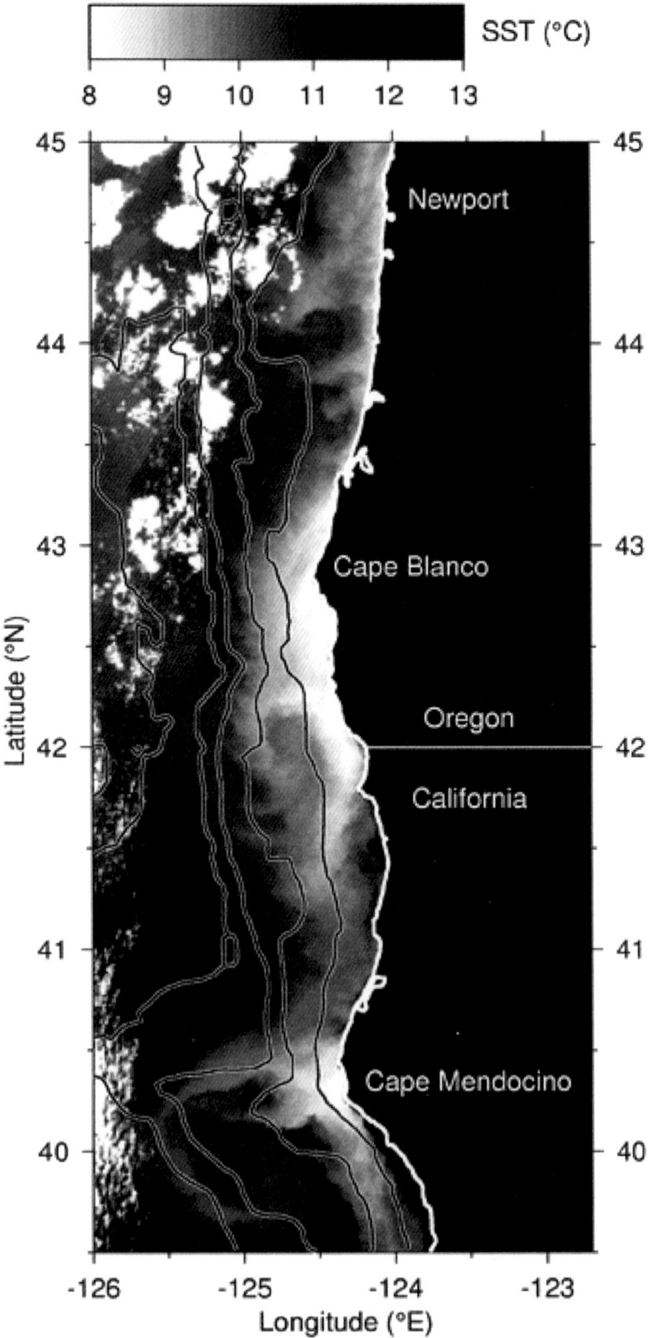

FIGURE 4.2.2. Sea surface temperature from May 18th 1995 at 21:00 UCT for a region of the US west coast. Isobaths are shown at 200 m (approximates the shelf edge), 1000 m, 2000 m and 3000 m. (From Dale and Barth, 2001).

itself. The sea surface will be treated as a rigid lid and the shelf break depth D_o will be considered fixed. Capes are then represented as a narrowing of the shelf (a decrease in w^*), which is consistent with the bathymetry of the Oregon coast.

The dynamics of the upper layer involve interactions with the lower layer, a process that has not been explored thus far. Although a detailed development of this subject takes place in Section 5.1, the uninitiated reader should be able to follow this section without assistance. The top and bottom layers will be numbered '1' and '2' respectively and, in accordance with the semigeostrophic approximation, the along-shore velocity components v_1^* and v_2^* will be considered geostrophic. The upper layer is assumed to be capped by a rigid lid and the pressure there is denoted by p_T^*. The geostrophic relations for the two layers are then given by

$$f v_1^* = \frac{1}{\rho_1} \frac{\partial p_T^*}{\partial x^*} \tag{4.2.1}$$

and

$$f v_2^* = \frac{1}{\rho_1} \frac{\partial p_T^*}{\partial x^*} - g' \frac{\partial d_1^*}{\partial x^*}$$

where d_1^* is the upper layer thickness and g' is the reduced gravity. The pressure gradient term on the right-hand side of the second equation follows from the hydrostatic relation. Subtraction of the second equation from the first results in the *thermal wind* relation

$$f(v_1^* - v_2^*) = g' \frac{\partial d_1^*}{\partial x^*}. \tag{4.2.2}$$

The semigeostrophic potential vorticity of the upper layer will be considered constant:

$$\frac{f + \dfrac{\partial v_1^*}{\partial x^*}}{d_1^*} = \frac{f}{D_{1\infty}}, \tag{4.2.3}$$

even though there is little in the way of observation or deduction to justify the constancy of $D_{1\infty}$. The assumption is made purely for convenience. As for the lower layer potential vorticity:

$$\frac{f + \dfrac{\partial v_2^*}{\partial x^*}}{d_2^*} = \frac{f}{D_{2\infty}}, \tag{4.2.4}$$

it will be sufficient to assume that $D_{2\infty} \gg D_{1\infty}$, as suggested by Figure 1, even though $D_{2\infty}$ need not be constant.

Let $D_{1\infty}$, $(g'D_{1\infty})^{1/2}$, $\rho_1 g' D_{1\infty}$, $(g'D_{1\infty})^{1/2}/f$ and f^{-1} serve as scales for depth, along-shore velocity, rigid lid pressure, horizontal length, and time. Then (4.2.1)–(4.2.4) become

$$v_1 = \frac{\partial p_T}{\partial x} \tag{4.2.5}$$

$$v_1 - v_2 = \frac{\partial d_1}{\partial x}, \tag{4.2.6}$$

$$1 + \frac{\partial v_1}{\partial x} = d_1. \tag{4.2.7}$$

and

$$1 + \frac{\partial v_2}{\partial x} = d_2 \frac{D_{1\infty}}{D_{2\infty}} << 1. \tag{4.2.8}$$

Note that the approximation in (4.2.8) is valid only over the shelf, where lower layer fluid has risen up and where $D_{1\infty}$ is a legitimate scale for the lower layer depth. Offshore of the shelf break $v_2 = 0$, and (4.2.6) and (4.2.7) then yield

$$\frac{\partial^2 d_1}{\partial x^2} - d_1 = -1. \tag{4.2.9}$$

With the requirement that d_1 remain bounded as $x \to \infty$, the solution takes the form

$$d_1 = p_T = 1 + \hat{d}(y, t)e^{-(x-w)}. \tag{4.2.10}$$

where the coefficient $\hat{d}$ is to be determined by matching (4.2.10) to the inshore solution.

Over the shelf, the solutions depend on the configuration (I), (II), (III), or (IV) of Figure 4.2.1. Where the upper layer occupies the whole water column, the velocity is computed from (4.2.7) with d_1 equal to the specified shelf depth. Where both layers are present, the solution is obtained by differentiating (4.2.6) with respect to x and using (4.2.7) and (4.2.8) to eliminate the derivatives of v_1 and v_2. The resulting equation for the upper layer depth is

$$\frac{\partial^2 d_1}{\partial x^2} - d_1 = 0. \tag{4.2.11}$$

Marching through the mathematics for all four configurations is rather tedious and we therefore limit a detailed discussion to case (II): the most difficult and most interesting of the four. Although the primary focus is on steady flow, the retention of time-dependence is not burdensome. Certain aspects of the nonlinear frontal or Kelvin waves that arise are explored in the exercises at the end of the section. Hereafter we will assume that the shelf break depth $d_o (= D_o/D_{1\infty})$ is constant, so that

$$d_1 + d_2 = xd_o/w(y). \tag{4.2.12}$$

For Case (II), let $b(y, t)$ denote the x-position at which the interface grounds. Then the cross-sectional structure of the flow in the various regions is given as follows. For the inshore region ($0 \leq x \leq b$):

$$v_1(x, y, t) = v_o + \frac{d_o x^2}{2w} - x \qquad (4.2.13)$$

and

$$p_T(x, y, t) = p_o + v_o x + \frac{d_o x^3}{6w} - \frac{x^2}{2}, \qquad (4.2.14)$$

where

$$v_o = v_1(0, y, t) = e^{b-w} - \frac{d_o b}{w}\left(1 + \frac{b}{2}\right) + w \qquad (4.2.15)$$

and

$$p_o = p_T(0, y, t) = -\tfrac{1}{2}w^2 - b e^{b-w} + \frac{d_o b}{w}\left(1 + b + \frac{b^2}{3}\right). \qquad (4.2.16)$$

The latter follows from the nondimensional geostrophic relation $v_1 = \partial p_T / \partial x$ for the upper layer.

For the shelf region occupied by both layers ($b \leq x \leq w$):

$$d_1(x, y, t) = \tfrac{1}{2}e^{x-w}(1 - e^{2(b-x)}) + \frac{d_o b}{w}e^{b-x}, \qquad (4.2.17)$$

$$v_1(x, y, t) = \tfrac{1}{2}e^{x-w}(1 + e^{2(b-x)}) - \frac{d_o b}{w}e^{b-w} + w - x, \qquad (4.2.18)$$

and

$$p_T(x, y, t) = d_1(x, y, t) - \tfrac{1}{2}(x - w)^2. \qquad (4.2.19)$$

Finally, the offshore region ($x \geq w$) has

$$v_1(x, y, t) = \tfrac{1}{2}e^{w-x}(1 + e^{2(b-w)}) - \frac{d_o b}{w}e^{b-x} \qquad (4.2.20)$$

and

$$p_T(x, y, t) = d_1(x, y, t) = 1 - v_1(x, y, t). \qquad (4.2.21)$$

The coefficients in the above expressions have been chosen such that p_T and v_1 are continuous across the boundaries $x = b$ and $x = w$ of the three regions. These requirements ensure that d_1 will also be continuous.

The y-momentum equation for the upper layer can be used to calculate the behavior of the along-shore current in y and t. A convenient form to use is

$$\frac{\partial v_1}{\partial t} + \left(1 + \frac{\partial v_1}{\partial x}\right)u_1 = -\frac{\partial}{\partial y}\left(p_T + \frac{v_1^2}{2}\right). \qquad (4.2.22)$$

The second term, which is just $d_1 u_1$ times the upper layer potential vorticity, goes to zero at the coastline ($x = 0$). Substitution of the expressions for v_1 and p_T (see 4.2.13–4.2.16) into this equation and evaluation of the result at $x = 0$ leads to

$$\frac{\partial v_o}{\partial t} + \frac{\partial}{\partial y}(B_o) = 0, \tag{4.2.23}$$

where B_o is the upper layer Bernoulli function at the coast:

$$B_o = \tfrac{1}{2}v_o^2(b, w) + p_o(b, w), \tag{4.2.24}$$

with v_o and p_o given by (4.2.15) and (4.2.16).

Since all the time dependence is contained in the variable $b(y, t)$, (4.2.23) can be written in the form

$$\frac{\partial v_o}{\partial b}\frac{\partial b}{\partial t} + \frac{\partial B_o}{\partial b}\frac{\partial b}{\partial y} = -\frac{\partial B_o}{\partial w}\frac{\partial w}{\partial y},$$

or

$$\frac{\partial b}{\partial t} + c\frac{\partial b}{\partial y} = -\frac{\partial B_o}{\partial w}\frac{\partial w}{\partial y}\left(\frac{\partial v_o}{\partial b}\right)^{-1}, \tag{4.2.25}$$

where

$$c(b, w) = \frac{\partial B_o}{\partial b}\left(\frac{\partial v_o}{\partial b}\right)^{-1}. \tag{4.2.26}$$

The signal that propagates at speed $c(b, w)$ is essentially an internal Kelvin wave that has been modified by the presence of the current and the sloping bottom. Since the model admits just one wave, the concepts of subcritical or supercritical flow need to be rethought. The usual practice is to call the flow *supercritical* when the wave propagates in the same direction as the upper layer transport. The nominal direction of upper layer transport will be positive (with the northern hemisphere coastline on the left) and therefore supercritical and subcritical flow is here characterized by $c > 0$ and $c < 0$. In the latter case the wave propagates with the coast on its right; that is, in the usual sense for northern hemisphere Kelvin waves. There will also be some cases with negative upper layer transport, making the classification of the flow less straightforward.

We are now in a position to discuss case II steady solutions and their hydraulic properties. However, it will be helpful to first describe the properties of cases (I), (III), and (IV), which will be stated without proof. The reader who has mastered Sections 2.2 and 2.3 will not be surprised by most of what comes next. For case (I), the interface grounds along the vertical wall and the wave dynamics are similar to those of a Kelvin wave that propagates along the left wall of a wide channel (Figure 2.2.3). Most importantly, the wave speed is negative regardless of the direction of the upper layer flux (cf. Equation 2.2.26). Case (I) flows are therefore always subcritical. In Case (IV) the interface outcrops at a position offshore of the shelf break and the wave dynamics are identical to those that occur when the Figure 2.2.3 flow separates from the left wall of the channel. The characteristic wave speed is zero and the flow is exactly critical. The flow no longer feels the coastal topography and is essentially unforced. Case (III) is more difficult to describe in terms of previous results, but it can be shown that the wave speed is always positive and thus the flow is always supercritical (see Exercise 3). Transitions between subcritical and supercritical flow can only occur when the interface grounds on the shelf (Case II).

Under conditions of steady flow, B_o is a constant prescribed by upstream conditions. Solutions in this case could, in principle, be computed from (4.2.24). A useful alternative to this relation can be derived by first noting that

$$\partial B_1/\partial x = v_1 \partial v_1/\partial x + \partial p_T/\partial x = v_1(\partial v_1/\partial x + 1) = v_1 d_1.$$

The upper layer volume transport may therefore be written as

$$Q_1 = \int_0^\infty (v_1 d_1)\,dx = B_\infty - B_o(y, t)$$

where B_∞ is the value of the upper layer Bernoulli function at $x \to \infty$ and is equal to unity. Equation (4.2.24) can therefore be expressed as

$$\tfrac{1}{2}v_o^2(b, w) + p_o(b, w) = 1 - Q_1. \tag{4.2.27}$$

A set of steady solution curves for various values of Q_1, showing the interface outcrop position as a function of the shelf width appears in Figure 4.2.3. In order to present all cases with a single figure, the composite variable

$$\alpha = \begin{cases} b' \text{ (case III)} \\ -b \text{ (case II)} \\ -wd_1^+/d_o \text{ (case I)} \end{cases} \tag{4.2.28}$$

has been introduced. The quantity d_1^+ represents the upper layer depth just offshore of the shelf break. Recall that b or b' denotes the position of the inshore edge of the interface over the shelf (Figure 4.2.1). The shelf edge depth $d_o = 2$ in all cases, and this means that the upper layer thickness far offshore ($= 1$) is less.

Critical states in Figure 4.2.3 lie along the dashed curve that passes through the minima of the family of $Q = $ constant curves. Subcritical flows lie to the left and correspond to instances of cases (I) or (II); supercritical flows lie to the right and correspond to (III) or (IV). The slope of all the curves for case (III) equals unity ($w = \alpha + constant$) and thus a solution with a surface outcrop over the shelf maintains a fixed distance $w-\alpha$ from the shelf edge. Case (IV) occurs to the right of the line $\alpha = w$ (or $Q = 0.5$). There the interface outcrops at the surface offshore of the shelf break and the flow is immune to topographic variations.

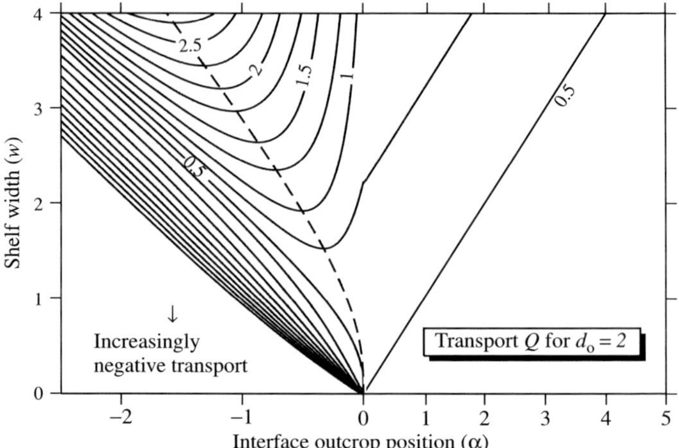

FIGURE 4.2.3. Solution curves relating the shelf width w and position of interface outcrop α (see Equation 4.2.28) for various values of the upper layer transport Q_1. The shelf break depth $d_o = 2$ in all cases. The figure is based on solutions to (4.2.27) for case II and the relations that govern steady flows in the remaining cases as they appear in Dale and Barth (2001). (From Dale and Barth, 2001).

A flow that is subcritical upstream may undergo a transition to supercritical flow due to a narrowing of the shelf. This evolution can be traced by following one of the constant Q curves from the left-hand portion of the figure, through the minimum in W, and onto the right-hand branch. As this occurs, the inshore termination of the interface continuously rises along the shelf, possibly striking the surface and exposing the cold lower layer. There are also some features that complicate this traditional picture; one is that some of the curves have negative Q. The latter terminate at the origin and can therefore be joined with a supercritical solution branch only if the shelf width w goes to zero. Another complication is that for other values of d_o the solutions cannot always be continued smoothly through the subcritical regime. Dale and Barth (2001) should be consulted for further details.

An example of a critically controlled solution appears in Figure 4.2.4, which is calculated using a more general model containing an approximation to the Cape Blanco topography. The dashed line in Figure 4.2.4b shows the position at which the interface grounds over the shelf and this curve turns solid where the grounding becomes a surface outcrop. Lower layer fluid is exposed along the coast south of this transition. Whether this accounts for the observed behavior of the front near Cape Blanco is unsettled; other explanations such as local enhancement of the winds by the Cape have also been put forward (e.g. Samelson *et al.*, 2002).

Further results on time-dependent features of upwelling fronts can be found in the literature. Gill and Schumann (1979) discuss the nonlinear properties of the coastal trapped waves that arise in all three cases. Some of these properties are

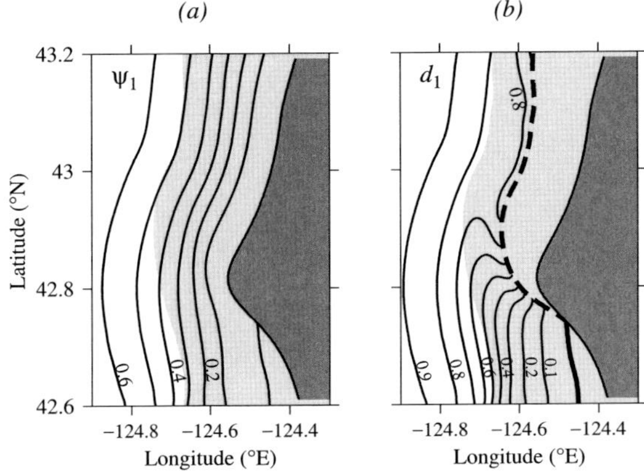

FIGURE 4.2.4. Fields of (a) streamfunction and (b) upper layer thickness d_1 for approximated Cape Blanco topography. The solution shown is critically controlled and has $Q = 0.701$. A dashed bold line shows the position $x = b$ where the interface grounds on the bed, and this turns to a solid bold line where the interface outcrops at the surface. (From Dale and Barth 2001).

drawn out in Exercise 3 below. Dale and Barth (2001) describe some initial-value experiments that demonstrate blocking of the upstream flow by the cape.

Since the model discussed above is constrained to have uniform potential vorticity in each of the two layers, potential vorticity waves have been expunged. Among this group is the continental shelf wave which, in the absence of a background flow, propagates in the same direction as a Kelvin wave. Coastal flows with potential vorticity gradients may exhibit hydraulic behavior, though field examples have yet to be clearly identified. Hughes (1985a, b; 1986a, b; 1987) describes a variety of models, some of which will be touched on in Section 6.2. Particularly relevant to the present discussion is the (1985b) model, which allows for potential vorticity and Kelvin wave dynamics and shows that hydraulic transitions with respect to both are possible.

Exercises

(1) Calculate the lower layer velocity over the shelf for Case (II).
(2) Calculate the characteristic wave speed for Case II explicitly. Consider an initial condition in which b increases monotonically from one constant value to another. Discuss the direction of propagation of the resulting wave and the tendency to steepen or rarefy.
(3) *The dynamics of Case* III.

(a) Show that the upper-layer, cross-shelf structure for case III is given by:

$$v_1(x, y, t) = \tfrac{1}{2} e^{w-x}(1 + e^{2(b'-w)}) \text{ and}$$

$$p_T(x) = d_1(x) = 1 - v_1(x)$$

for $x > W$, and by

$$v_1(x, y, t) = \tfrac{1}{2}e^{w-x}(1 + e^{2(b'-w)}) + w - x,$$

$$p_T(x, y, t) = \tfrac{1}{2}[e^{w-x}(1 - e^{2(b'-w)}) - (x - w)^2] \text{ and}$$

$$d_1(x, y, t) = \tfrac{1}{2}e^{w-x}(1 - e^{2(b'-w)})$$

for $b' \leq x \leq w$. Note that b' is the position of the surface outcrop.

(b) To find the evolution of the flow in x and t, consider the y-momentum equation applied at the outcrop $x = b'$. A particularly convenient form of this equation is obtained by applying it *along* the outcrop, so that derivatives in y and t are taken *after* x is set to b' in the expressions appearing in (a). To achieve this form first show that

$$\left(\frac{\partial}{\partial t}v(x, y, t)\right)_{x=b} = \frac{\partial}{\partial t}v(b'(y, t), y, t) - \left(\frac{\partial v(x, y, t)}{\partial x}\right)_{x=b} \frac{\partial b'(y, t)}{\partial t}$$

for any variable v. Show that a similar expression holds for y-derivatives. Now use expressions like this to replace local y- and t- derivatives in the upper layer y-momentum equation by derivatives of quantities first evaluated at $x = b'$. With the help of the geostrophic relation for v_1 and the potential vorticity definition, you should obtain

$$\frac{\partial}{\partial t}\left(v_1(b'(y, t), y, t) + b'(y, t)\right)$$

$$+ \frac{\partial}{\partial y}\left(p_T(b'(y, t), y, t) + \frac{v_1^2(b'(y, t), y, t)}{2}\right) = 0.$$

Finally, apply this equation to the expressions derived in part (a) to obtain an evolution equation for b'. Identify the characteristic wave speed and show that it is nonnegative.

(c) Show that for steady flow, $b' - w$ remains conserved.

4.3.   Oblique Shocks and Expansion Fans: The Supercritical Marine Layer

The marine layer is a relatively dense and well-mixed layer of moist air that lies above the sea surface and is often capped by a strong inversion in temperature and humidity. In the North Pacific the layer can extend all the way from California to Hawaii and its thickness can increase over that distance from around 600m or less to 2000m. The physical properties of the layer are particularly well observed along the Northern California coast (e.g. Dorman 1985, 1987; Winant et al., 1988; Dorman et al., 1999; Edwards, et al., 2001). During the summer upwelling season, the North Pacific High drives equatorward winds along the coastline. The winds are intensified to the west of the 1000 m high coastal mountain range, an effect that extends 100 km or so offshore. Wind speeds near the inversion level can reach values of up to 30 m/s. and internal Froude numbers can exceed unity. During such periods of ostensibly supercritical flow, irregularities in the coastline can produce dramatic changes in the wind speed and layer thickness. In one configuration (Figure I.2a, b) the winds accelerate and the layer thins as it passes Point Arena, where the coastline abruptly veers to the southeast. Speeds of 20 m/s are reached and the layer thickness decreases from 600 m to 300 m. The contours of constant wind speed, which are roughly perpendicular to the coastline near Point Arena, become more oblique as Stewarts Point is approached. Similar behavior has been observed along Peru's coastline by Freeman (1950), who likens the acceleration and thinning with an *expansion fan*, a phenomena well documented for supersonic flow by aeronautical engineers. The fan is sometimes marked by clearing as the high-speed air descends and warms (Figure 4.3.1). Between Stewarts Point and Bodega Bay, where the coast veers slightly southward, the wind speed diminishes and the layer thickens in what has been described as an *oblique hydraulic jump*. Different visualizations of marine layer jumps (Figures 4.3.2 and 4.3.3) show the abrupt and sometimes wavy character of the transition.

The subsidence associated with the Pacific high-pressure system creates a particularly sharp interface between the cold and moist marine layer and the

FIGURE 4.3.1. Aircraft photo, facing to the North, showing Cape Mendocino. The area of clear air corresponds to an expansion fan in the lee of the Cape. (Enhanced version of photo by Dr. Clive Dorman).

overlying warm and dry air. It is therefore natural to treat the entire layer as a 'slab' and to use the shallow water equations as a model. Expansion fans and oblique hydraulic jumps are not admitted in the long-wave limit of these equations and we must therefore allow full freedom in the two horizontal dimensions. If the flow is assumed steady and supercritical, the method of characteristics can be used to obtain solutions. A complete derivation of the characteristic form of the steady shallow water equations appears in Appendix B. The present section contains a nonrigorous discussion of characteristic curves, oblique jumps and expansion fans; a formal discussion appears in Section 4.4. We will ignore the effects of rotation since this considerably simplifies the discussion of characteristics and still allows for a description of the basic phenomena. The neglect of rotation will, however, preclude any discussion of the decay of features in the offshore direction.

It should also be noted that the supercritical mode of the marine layer is just one of several observed configurations. Another is the 'gravity current' mode, in which the layer moves northward along the coast with a distinct leading edge (Figure I.2c). This type of flow is discussed in Section 4.5.

The method of characteristics for the steady shallow water equations in two dimensions has origins in the theory of gas dynamics and compressible flow

FIGURE 4.3.2. Possible hydraulic jump near Point Arena, looking southeast. (Photograph by Dr. John Baine).

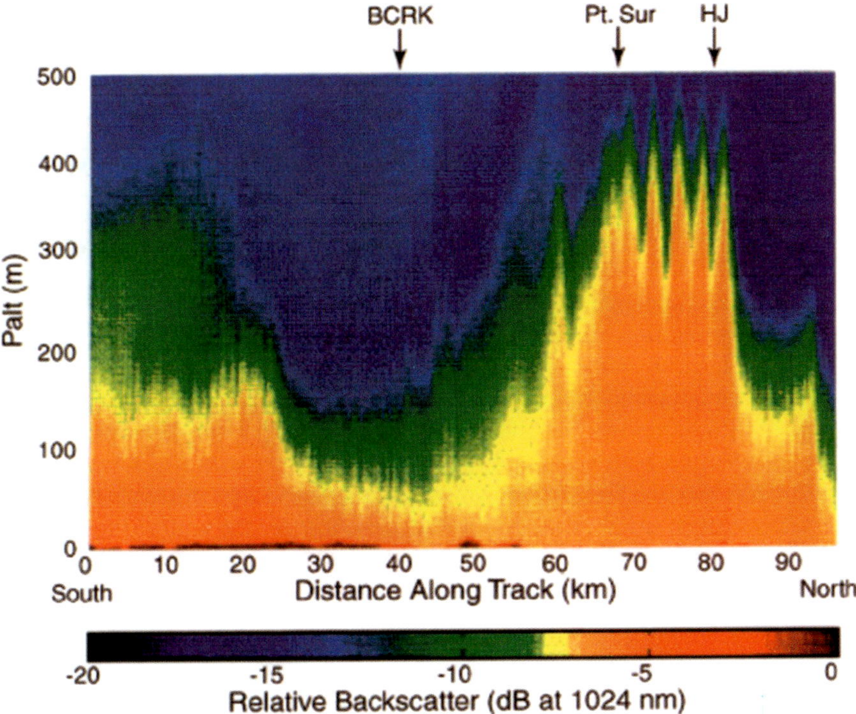

FIGURE 4.3.3. Image of a hydraulic jump near Point Sur based on LIDAR, a laser device that points upward. Air density variations cause the light to reflect back, similar to radar. The bottom of the air temperature inversion causes strong backscatter and is indicated by the yellow-green boundary. (From Dorman *et al.*, 1999).

(Courant and Friedrichs, 1976). The methodology can be applied in regions of the flow field where

$$F = \frac{(u^{*2} + v^{*2})^{1/2}}{gd^*} > 1. \tag{4.3.1}$$

F is clearly a local Froude number based on the full flow speed. The region over which (4.3.1) holds is sometimes called *supercritical*. This usage differs from that of our previous long-wave applications, where the entire cross section of a gradually varying flow is judged supercritical or subcritical depending on whether a long normal mode could propagate in one or two directions. The appropriate Froude number in those cases depends on the flow across the whole cross section and is aware of the boundary conditions. The Froude number defined in (4.3.1) is relevant to free, locally generated disturbances.

Where (4.3.1) holds, the influence of a localized forcing is limited to a downstream subregion of the flow field. The governing equations in this case are hyperbolic and information is carried downstream along characteristic curves. To be more precise, consider a uniform southward current with velocity v_o^* and

depth $d_o{}^*$ (Figure 4.3.4) such that $F = v_o{}^*/(gd_o{}^*)^{1/2} > 1$. A localized disturbance to the flow introduced at point p will spread out in a widening circle as
it is advected downstream. The radius of the circle will grow at rate $(gd_o{}^*)^{1/2}$
while the center of the circle will move southward at speed $v_o{}^*$. The disturbance
will therefore spread over a wedge or 'cone' of influence that spans the angle
$2A$, where

$$A = \sin^{-1}(F^{-1}).\tag{4.3.2}$$

The angle A and the edges of the cone are analogous to the Mach angle and Mach
lines of supersonic flow. In shallow water theory, A is referred to as the Froude
angle. If $F < 1$, the disturbance circle spreads upstream and downstream, carrying
the influence to all parts of the flow field. The steady shallow water equations in
this case are elliptic and the method of characteristics is no longer appropriate.

A related feature distinguishing two-dimensional flows with $F > 1$ from those
with $F < 1$ is that the former can support a stationary, free disturbance, while
the latter cannot. It is left as an exercise to show that for the uniform southward
flow considered above, a small-amplitude, stationary disturbance with horizontal
structure $e^{i(k^*x^*+l^*y^*)}$ can exist provided

$$\frac{l^*}{(l^{*2}+k^{*2})^{1/2}} = \sin(A) = \frac{1}{F} < 1.\tag{4.3.3}$$

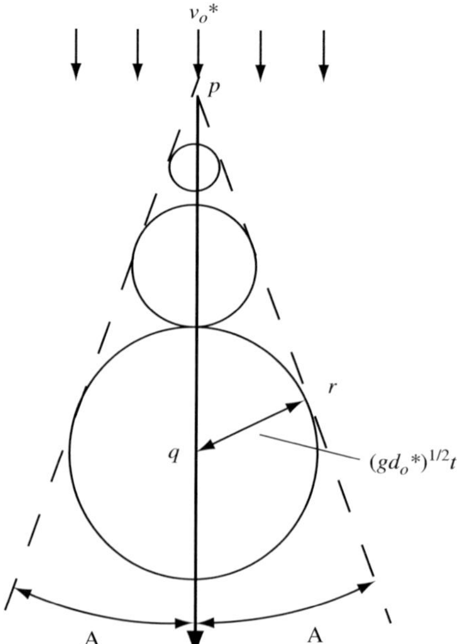

FIGURE 4.3.4. Wedge of influence and the Froude angle A.

There are two groups of waves (corresponding to $\pm k^*$), each with crests and troughs tilted at the Froude angle with respect to the background flow direction (Figure 4.3.5a). We denote the corresponding lines of constant phase by C_+ and C_- and note that they are aligned at the same angles as the edges of the wedge of influence in Figure 4.3.4. In both cases the alignment is such that the normal component of velocity equals the intrinsic propagation speed $(gd^*)^{1/2}$ of a gravity wave. As F approaches unity from above, the dashed and solid lines become perpendicular to the background flow. The flow is now one-dimensional and hydraulically critical in the sense explained in Chapter 1. For $F < 1$ the stationary disturbances cease to exist. In the next section, we will show that the Froude lines are also characteristic curves for the steady flow.

It can also be shown (Exercise 1) that disturbance energy propagates along the characteristic curves in the downstream sense. Stationary disturbances generated by coastline irregularities to the east of the flow should therefore be carried away from the coast along the C_- lines. Suppose that the coastline veers away from the upstream flow direction (Figure 4.3.5b) and that the background flow adjusts so as to run parallel to the coast with a new velocity and depth. The new Froude angle A_1 between the disturbance phase lines and the coast will depend on the new value $F = F_1$, which cannot be calculated without further analysis. However, we have already seen that a supercritical channel flow accelerates and shoals (Section 1.4) when the channel widens. F_1 might therefore be expected to exceed its upstream value F_0 and (4.3.2) then implies $A_1 < A_0$. One can infer an expansion fan, a family of fanning wave crests and troughs, in the intervening region (Figure 4.3.5b).

Where the coastline turns back into the flow (Figure 4.3.5c), one might expect the Froude number to decrease and A to increase, giving rise to intersecting crests and troughs, and perhaps a shock. A simple model that allows prediction of the angle β of the oblique shock is sketched in Figure 4.3.6. The coastline is assumed to turn into the upstream flow at an angle α and the flow upstream and downstream of this point is assumed to be locally parallel to the coast. The matching conditions across the shock were developed in Section 3.5.2. For example, equations (3.5.2) and (3.5.4) expressing the continuity of normal flux and tangential velocity lead to

$$v_o^* d_o^* \sin\beta = v_1^* d_1^* \sin(\beta - \alpha)$$

and

$$v_o^* \cos\beta = v_1^* \cos(\beta - \alpha),$$

so that

$$\frac{d_o^*}{d_1^*} = \frac{\tan(\beta - \alpha)}{\tan(\beta)}. \tag{4.3.4}$$

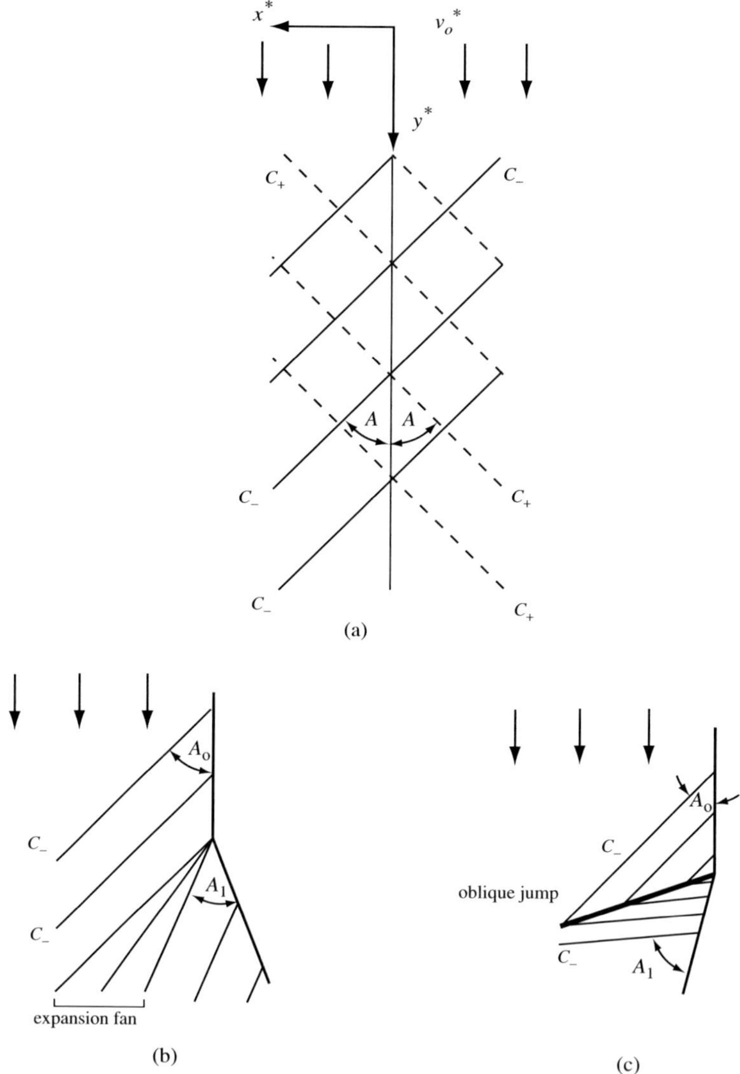

FIGURE 4.3.5. (a) Cross-waves in a supercritical flow. The crests and troughs are characteristic curves. (b) Expansion fan caused when the coastline veers away from the upstream flow. (c) Oblique hydraulic jump caused the by the coastline veering into the flow.

A third constraint based on the above relations plus the balance of flow force across the jump (see Equation 3.5.1 with $c^{(n)} = 0$) is

$$\frac{d_o^*}{d_1^*} = \frac{2}{-1 + \sqrt{1 + 8\dfrac{v_o^{*2} \sin^2 \beta}{g d_o^*}}} \tag{4.3.5}$$

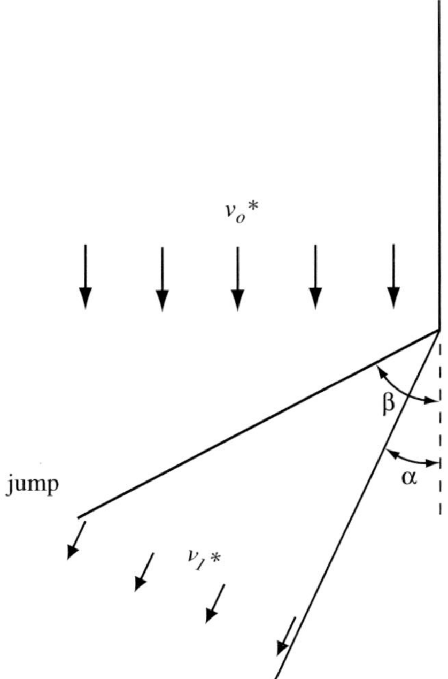

FIGURE 4.3.6. Oblique hydraulic jump at a corner.

(see Exercise 2). Eliminating $d_o{}^*/d_1{}^*$ between (6) and (7) gives

$$\tan(\beta - \alpha) = \frac{\tan\beta}{-1 + \sqrt{1 + 8\dfrac{v_o{}^{*2}\sin^2\beta}{gd_o{}^*}}}, \tag{4.3.6}$$

allowing determination of the jump angle given α and the upstream flow.

The above discussion covers only the most basic elements of shocks and expansion fans, making assumptions regarding the acceleration and thinning of the layer as it passes inward and outward bends in the coast. The results can be put on a firmer footing using the method of characteristics and this is done in the next section. The steady, characteristics-based, shallow water model is ultimately limited in its ability to address questions concerning time-dependence, three-dimensionality, and subcriticality. To learn more about modeling efforts that address these questions, one can refer to Samelson (1992), Rogerson *et al.* (1999), Burk *et al.* (1999) and Edwards *et al.* (2001). At the time of this writing, modern observational references include the Edwards *et al.*, paper, Perlin *et al.* (2004), and references contained therein.

Exercises

(1) For 2-dimensional plane waves in a uniform flow with velocity $(0, -v_o^*)$, derive the dispersion relation

$$\omega^* = -l^* v_o^* \pm (g d_o^*)^{1/2} (k^{*2} + l^{*2})^{1/2}$$

and deduce the condition (4.3.3) that the waves be stationary. For stationary disturbances, show that the group velocity is

$$\mathbf{c_g} = \frac{(g d^*)^{1/2} k^*}{l^* (k^{*2} + l^{*2})^{1/2}} (\pm l^* \mathbf{i} - k^* \mathbf{j})$$

and that energy therefore propagates along the lines of constant phase (characteristic curves) C_+ and C_-, and in the downstream direction of these lines.

(2) Derive equation (4.3.5). (Hint: show that Equation 1.6.8 holds for the oblique jump if F_u is interpreted as the upstream Froude number based on the normal component of velocity.)

4.4.   Expansion Fans and Compressions: Formal Theory

The ideas presented in the previous section can be formalized using the method of characteristics for steady, 2D, shallow flow. This methodology has been used widely in the field of aerodynamics to describe supersonic flow (Courant and Friedrichs 1976).

   A simple reinterpretation of variables in the governing equations leads to solutions of the shallow water equations, with or without rotation. The irrotational case is particularly simple and leads to descriptions of the marine layer expansion fans and compressions that are elegant and that capture most of the important physical mechanisms. The methodology can be extended to account for rotation but the governing equations for this case (Appendix C) are less transparent.

a.   Summary of the Method of Characteristics

The essential ideas underlying expansion and compression waves generated by flow along a coast can be illustrated through consideration of an irrotational, shallow flow. The governing characteristic equations are developed in Appendix B and we give only a brief recount of the central ideas here. We begin by attempting to cast the steady shallow water equations in a standard quasilinear form (see B1) with two dependent variables. The full shallow water equations, which are normally written in terms of the three variables u, v, and d, may be expressed in terms of just u and v through use of the Bernoulli equation:

$$\frac{u^2 + v^2}{2} + d = d_o. \tag{4.4.1}$$

All variables are now nondimensional, with length, depth, and velocity scales L, D, and $(gD)^{1/2}$. Although the Bernoulli function d_o is normally depends on the streamfunction, it is here rendered constant by the assumption of an irrotational velocity field:

$$\frac{\partial v}{\partial x} - \frac{\partial u}{\partial y} = 0. \tag{4.4.2}$$

If the gradient of (4.4.1) is taken and the continuity equation:

$$d\nabla \cdot \mathbf{u} + \mathbf{u} \cdot \nabla d = 0,$$

is used to eliminate ∇d from the result, one obtains

$$(d - u^2)\frac{\partial u}{\partial x} - uv\left(\frac{\partial u}{\partial y} + \frac{\partial v}{\partial x}\right) + (d - v^2)\frac{\partial v}{\partial y} = 0. \tag{4.4.3}$$

Together, (4.4.2) and (4.4.3) constitute the required form of two equations in the two unknowns u and v, d being considered a function of these variables through (4.4.1).

To achieve the characteristic form, (4.4.2) and (4.4.3) must be linearly combined to form an expression in which all derivatives are expressed in a single direction. This procedure can be carried out successfully provided that the local Froude number $F = (u^2 + v^2)^{1/2}/d$ exceeds unity within the region of interest. Under this condition there are two characteristic directions, and the slopes of the corresponding characteristic curves C_+ and C_- are given by

$$\left(\frac{dy}{dx}\right)_{\pm} = \tan(\theta \pm A), \tag{4.4.4}$$

where θ is the inclination of the velocity vector $\mathbf{u}$ with respect to the x-axis and A is the Froude angle defined by

$$d^{1/2} = \pm |\mathbf{u}| \sin A \tag{4.4.5}$$

(see 4.3.2). As sketched in Figure 4.4.1a, the characteristic curves C_+ and C_- at a point P are aligned at angles $\pm A$ with respect to the local velocity vector or streamline. The wedge formed between C_+ and C_- defines the region of downstream influence for P.

Let α and β serve as parameters that vary along the two characteristic curves. Then (4.4.4) implies

$$\cos(\theta + A)\frac{\partial y}{\partial \alpha} = \sin(\theta + A)\frac{\partial x}{\partial \alpha} \quad \text{along } C_+, \tag{4.4.6a}$$

and

$$\cos(\theta - A)\frac{\partial y}{\partial \beta} = \sin(\theta - A)\frac{\partial x}{\partial \beta} \quad \text{along } C_-. \tag{4.4.6b}$$

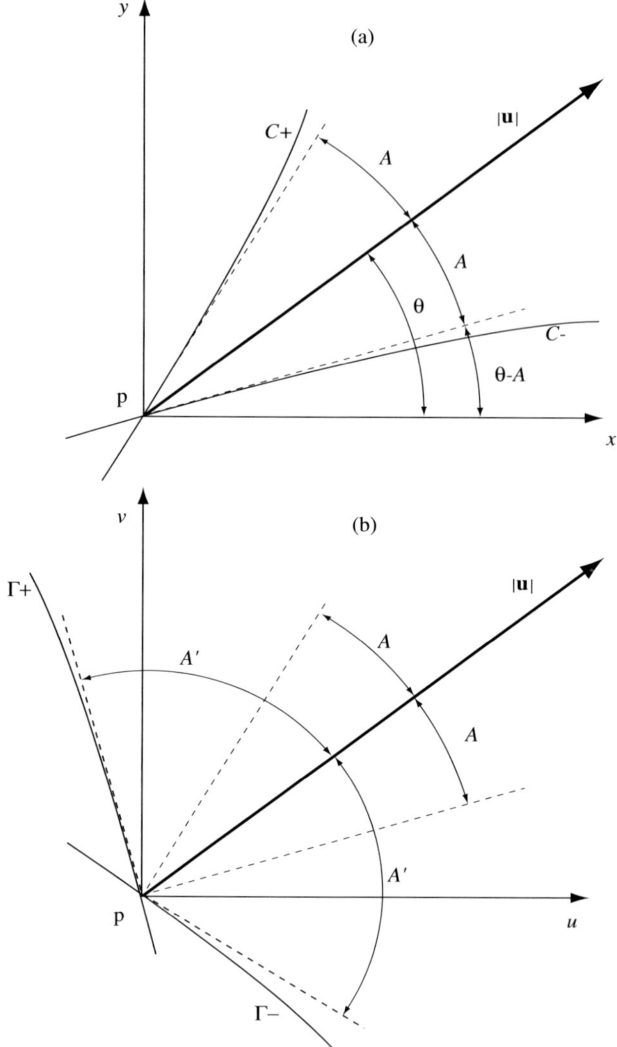

FIGURE 4.4.1. (a) The wedge of influence for a disturbance generated at the origin in the (x, y)-plane lies between the C_+ and C_- characteristics, which are inclined at the Froude angle A with respect to the velocity vector. In the hodograph (b) the wedge of influence lies between the images of the characteristics Γ_+ and Γ_-, which are inclined at angle A' with respect to the velocity and lie at right angles to C_- and C_+ respectively.

As shown in Appendix B, the characteristic equations governing the evolution of the flow along these curves are given by

$$\sin(\theta - A)\frac{\partial v}{\partial \alpha} = -\cos(\theta - A)\frac{\partial u}{\partial \alpha} \text{ along } C_+, \qquad (4.4.7a)$$

and

$$\sin(\theta + A)\frac{\partial v}{\partial \beta} = -\cos(\theta + A)\frac{\partial u}{\partial \beta} \text{ along } C_-, \qquad (4.4.7b)$$

If the flow field consists of linearized disturbances to a known background state, then θ and A are known in advance and (4.4.6) can be solved independently to determine the characteristic curves. Equation (4.4.7a, 4.4.7b) can then be integrated along these curves, beginning from a boundary at which u and v are known, in order to obtain a solution. In more general circumstances, the four equations must be solved simultaneously.

b.   The Hodograph for 2d, Irrotational Flow

A helpful alternative to the physical plane representation of characteristics is the $(u,\ v)$ plane, or *hodograph*. As suggested in Figure 4.4.1b, the characteristic curves C_+ and C_- have images Γ_+ and Γ_- determined by (4.4.7). Comparing (4.4.6) to (4.4.7), it is apparent that the tangent to C_+ is normal to the tangent to Γ_-, and vice versa, when the two directions are represented in the same space. The relationship between A and the angle A' in the $(u,\ v)$ plane between characteristics and streamlines is thus

$$A' = 90° - A. \qquad (4.4.8)$$

It follows from (4.4.5) that

$$d^{1/2} = |(u, v)| \cos(A'). \qquad (4.4.9)$$

Hodograph characteristics are inclined at the angle ω with respect to the u-axis, where

$$\omega = \theta + A' \text{ for } \Gamma_+, \qquad (4.4.10a)$$

and

$$\omega = \theta - A' \text{ for } \Gamma_-. \qquad (4.4.10b)$$

An advantage of the hodograph for two-dimensional, irrotational flow is that the general forms of the characteristic curves can be determined and represented graphically, without regard to the particular geometry or boundary conditions. To determine these forms, it is helpful to introduce the new variables $u^{(n)}$ and $u^{(t)}$ representing the projection of $\mathbf{u}$ normal and tangent to the hodograph characteristic in question. We use the convention that positive $u^{(n)}$ lies to the left of positive $u^{(t)}$. The following analysis applies to either Γ_+ or Γ_-, with ω defined by the corresponding (4.4.10a) or (4.4.10b). The tangential component of the velocity is given by

$$u^{(t)} = u \cos \omega + v \sin \omega = d^{1/2}, \qquad (4.4.11)$$

where the final equality follows from (4.4.9). The normal component is given by

$$u^{(n)} = v \cos \omega - u \sin \omega.$$

Rearrangement of these relations leads to

$$u = u^{(t)} \cos \omega - u^{(n)} \sin \omega \qquad (4.4.12a)$$

and

$$v = u^{(t)} \sin \omega + u^{(n)} \cos \omega. \qquad (4.4.12b)$$

If we now treat ω as a parameter along the Γ_+ or Γ_- curve in question, then differentiation of the last two relations leads to

$$\frac{du}{d\omega} = \frac{du^{(t)}}{d\omega} \cos \omega - \frac{du^{(n)}}{d\omega} \sin \omega - u^{(t)} \sin \omega - u^{(n)} \cos \omega \qquad (4.4.13a)$$

and

$$\frac{dv}{d\omega} = \frac{du^{(t)}}{d\omega} \sin \omega + \frac{du^{(n)}}{d\omega} \cos \omega + u^{(t)} \cos \omega - u^{(n)} \sin \omega. \qquad (4.4.13b)$$

The combination $\cos(\omega) \times (4.4.13b) - \sin(\omega) \times (4.4.13a)$ leads to

$$\frac{du^{(n)}}{d\omega} = -u^{(t)}, \qquad (4.4.14)$$

after use of (4.4.7) with (4.4.10) to eliminate the terms on the left-hand side.

A second equation for $u^{(n)}$ and $u^{(t)}$ can be found from (4.4.11), which allows Bernoulli's relation (4.4.1) to be expressed as

$$\frac{u^{(n)2} + u^{(t)2}}{2} + u^{(t)2} = d_o. \qquad (4.4.15)$$

Differentiation of the latter and use of (4.4.14) leads to

$$\frac{du^{(t)}}{d\omega} = \frac{1}{3} u^{(n)}. \qquad (4.4.16)$$

The solutions to (4.4.14) and (4.4.16) can be written as

$$u^{(n)} = -(2d_o)^{1/2} \sin\left[(\omega - \omega_o)/\sqrt{3}\right]$$

and

$$u^{(t)} = \left(\frac{2d_o}{3}\right)^{1/2} \cos\left[(\omega - \omega_o)/\sqrt{3}\right],$$

where ω_o is an arbitrary constant. The condition that $u^{(n)} = 0$ when $u^{(t)} = (2d_o/3)^{1/2}$, which follows from (4.4.15), has been imposed. Since Γ_+ may range from lying parallel to the velocity vector ($u^{(n)} = 0$, $u^{(t)} > 0$) to lying perpendicular and to the left of the velocity ($u^{(t)} = 0$, $u^{(n)} < 0$), $\omega - \omega_o$ can vary over $[0, \sqrt{3}\pi/2]$. Similarly, $\omega - \omega_o$ varies over $[0, -\sqrt{3}\pi/2]$ for Γ_-.

In view of (4.4.12) the solutions for u and v are given by

$$\frac{u}{(2d_o/3)^{1/2}} = \cos\left[(\omega - \omega_o)/\sqrt{3}\right]\cos\omega + \sqrt{3}\sin\left[(\omega - \omega_o)/\sqrt{3}\right]\sin\omega,$$

$$\text{(4.4.17a)}$$

and

$$\frac{v}{(2d_o/3)^{1/2}} = \cos\left[(\omega - \omega_o)/\sqrt{3}\right]\sin\omega - \sqrt{3}\sin\left[(\omega - \omega_o)/\sqrt{3}\right]\cos\omega.$$

$$\text{(4.4.17b)}$$

When $\omega = \omega_o$ the above pair give $v/u = \tan\omega_o$, confirming that the hodograph characteristics are aligned with the velocity vector ($A' = 0$). According to (4.4.8) and (4.4.9) this condition requires the Froude number to be unity and ω_o is therefore the orientation of a particular characteristic under conditions of criticality. If ω is increased from its critical value ω_o, the hodograph characteristic veers to the left of the velocity vector. The resulting curve should therefore be identified with Γ_+. Γ_- is generated by decreasing ω below ω_o.

Since the local Froude number must exceed unity ($u^2 + v^2 > d$) it follows from (4.4.1) that the hodograph characteristics must lie outside the 'critical' circle $u^2 + v^2 = 2d_o/3$, the equivalent of the 'sonic' circle in aerodynamics. An outer bound on the range of u and v is the 'separation circle' $u^2 + v^2 = 2d_o$ obtained by setting $d = 0$ in (4.4.1). Analogous to the 'cavitation circle' in aerodynamics, this bound indicates the flow speed that would occur when the layer depth vanishes, exhausting the available potential energy.

The curves defined by (4.4.17a, b) are epicycloids lying between the critical and separation circles (Figure 4.4.2). These curves can be constructed graphically by considering a point p fixed to the perimeter of the small circle that fits between the bounding circles. If the small circle is rolled around the circumference of the critical (inner) circle, the point p traces out an epicycloid. Figure 4.4.2a shows the curves generated when p initially lies along the critical circle. Rolling the small circle counterclockwise causes p to move to q; rolling the circle counterclockwise causes p to move to r. The associated epicycloids Γ_- and Γ_+ are both tangent to (u, v) when the latter touch the critical circle and the direction of the two curves at this point is ω_o (Figure 4.4.2b). As one moves from p along Γ_+, the tangent angle ω increases as does the orientation θ of the velocity vector. Since the physical plane characteristic curve C_- is perpendicular to Γ_+, its angle of inclination $\theta - A$ also increases. The family composed of all possible Γ_- and Γ_+ can be generated by varying the position of the starting point p, or equivalently the angle ω_o, around the critical circle.

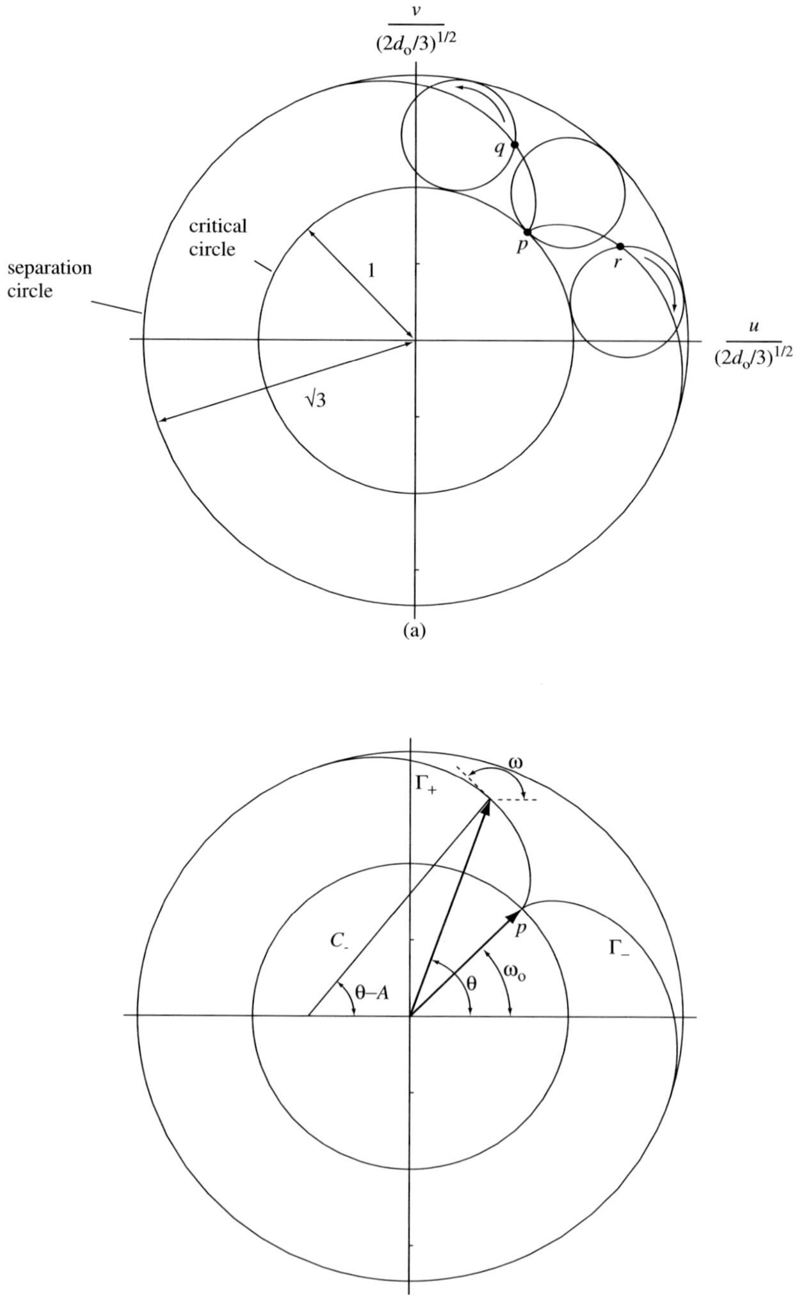

FIGURE 4.4.2. (a) The critical and separation circles and the epicycloids generated by rolling the small circle on the critical circle. (b) The hodograph characteristics Γ_+ and Γ_- for a particular ω_o.

c. *Riemann Invariants and Simple Waves*

Now consider the region $\mathcal{R}$ of physical space over which the flow is to be calculated. As an example, we take $\mathcal{R}$ as the area lying downstream of the open boundary $\mathcal{B}$ and to the west of the irregular coastline (Figure 4.4.3a). The flow crossing $\mathcal{B}$ is assumed to be uniform and southward: $d = d_o$, $u = 0$ and $v = v_o$. The C_+ or C_- curves at $\mathcal{B}$ are all inclined at the angle A or $-A$, $A = \sin^{-1}(d_o^{1/2}/v_o)$, relative to the velocity vector. The hodograph image Γ_+ of a particular C_+ curve crossing $\mathcal{B}$ can be found by drawing the velocity vector $\mathbf{u_o} = (0,\ v_o)$ in the hodograph (Figure 4.4.3b). The desired Γ_+ curve is the one touched by the tip of this vector and is sketched in bold. The uniformity of the flow crossing $\mathcal{B}$ implies that this particular Γ_+ corresponds to *all* the C_+ curves entering $\mathcal{R}$ across $\mathcal{B}$. A relationship between u and v, say $R_+(u,\ v) = $ constant, can be constructed by tracing the values along this curve. This relationship must hold over all of $\mathcal{R}$ covered by the C_+ curves originating from the upstream boundary. (In many cases the coverage of $\mathcal{R}$ by these curves is only partial, as when the downstream flow contains shocks.)

The function R_+ is a version of the Riemann invariant discussed in earlier sections in connection with time-dependent flows. Those discussions also alluded to the *simple wave*, a flow region for which one of the Riemann invariants is constant. In the present setting a simple wave corresponds to a region of flow for which all C_+ (or C_-) characteristics correspond to a single Γ_+ (or Γ_-). Since the particular relation between u and v along the unique characteristic holds for the entire simple wave, the individual values of u and v (and therefore d) must be constant along all characteristics of the opposite sign. The slopes of each such characteristic must therefore be constant. In the above example, where all possible u and v values lie along the bold Γ_+ curve (Figure 4.4.3b) the C_- curves must have constant slope. The latter must also lie normal to the bold Γ_+ curve when the two are plotted in the same space.

We now consider the effect of the coastline variation suggested in Figure 4.4.3a. The boundary condition of no normal flow and free slip implies that the inclination θ of the velocity vector at the coast is that of the coast itself. Since the possible range in velocity components u and v is restricted to the bold Γ_+ curve in Figure 4.4.3b, the complete velocity vector $\boldsymbol{u}$ can be found at each point on the coastline from the local angle θ. If one follows the coastline southward (Figure 4.4.3a) from the upstream boundary (point o) to point m, the value of θ increases from $3\pi/2$ to a slightly larger value. As θ increases, the velocity vector at the coast can be found by tracing along the bold Γ_+ curve in Figure 4.4.3b from o to m. It is clear that the flow speed increases as the point m is reached and that the tilt of the C_- curves, which are perpendicular to Γ_+, has also increased. The corresponding region of diverging C_- curves, or *expansion fan*, is shown in the upper frame. Further downstream, the coastline bends back southward and the above process is reversed. The result is a set of converging C_- curves that form a shock. The matching conditions appropriate to a shock formed at a simple corner were discussed in the previous section. The Riemann invariant relation between u and v is lost where the C_+ curves

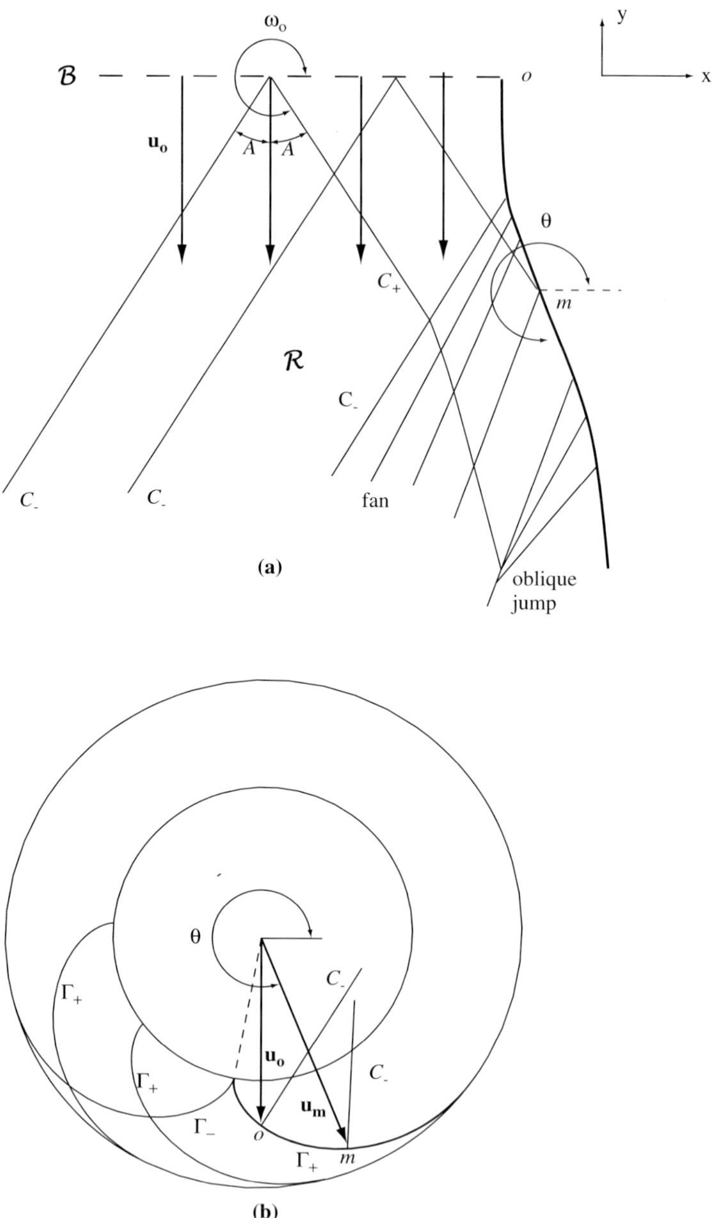

FIGURE 4.4.3. (a) Schematic view of the characteristics produced by flow along an irregular coastline. (b): The corresponding hodograph.

cross the dissipative shock. In fact, the dissipation may lead to the generation of vorticity that would invalidate the assumption of a constant Bernoulli function in the downstream region.

The method of characteristics may be used to compute rotating flows, or nonrotating flows with vorticity, but elegant graphical solutions are (apparently) no longer possible. Three characteristic directions and curves must be considered, two of which are defined by (4.4.4) and the third of which are the streamlines. The characteristic equations that must be integrated along these curves to compute the flow are developed in Appendix C and the reader is referred to Garvine (1987) for an application.

4.5. Rotating Gravity Currents

When relatively fresh river or estuary water discharges into the open ocean, it tends to turn to the (Northern Hemisphere) right and form a coastal surface flow or 'gravity current' (e.g. Munchow and Garvine 1993; Rennie, *et al.*, 1999). The outflow can be modulated, and sometimes blocked, by upwelling-favorable winds blowing across the mouth of the estuary. This is the case when northeastward winds blow across the mouth of Chesapeake Bay. When the winds relax or change direction, the brackish surface layer that normally resides in the bay is released. It exits and flows southwestward in a gravity current or plume along the Virginia and N. Carolina coasts (Figure 4.5.1). The leading edge of the current forms a blunt nose that can sometimes be seen at the free surface from the shoreline (Figure 4.5.2). A similar phenomenon occurs in the now familiar California coastal atmospheric marine layer (Beardsley *et al.*, 1987; Dorman 1987). At the beginning of the event shown in Figure I.2c, the winds were from the north and the marine layer had gathered offshore of the Southern California Bight. At the time the image was taken, the marine layer had surged northward in response to a wind reversal. The leading edge can be seen near Point Arena, where it stalled, formed an eddy, and resumed its northward travel.

The nonrotating gravity current has been studied extensively (Simson, 1997) and many of the ideas developed in this body of work form the basis for models with rotation. Much of our direct knowledge about rotating gravity currents is based on laboratory experiments, including Stern *et al.*, 1982, Griffiths and Hopfinger 1983, Kubokawa and Hanawa 1984b, and more recently, Helfrich and Mullarney 2005. In these experiments, a homogeneous layer floating on an ambient fluid of slightly greater density and held in a reservoir is released and allowed to flow into a rotating channel or annulus. The situation is similar in some respects to the full dam-break problem considered in Section 3.3, but turned upside down. The contact with the free surface avoids some of the frictional complications that would occur if the intrusion rubbed against the bottom. The fluid seeks out the right-hand wall of the channel and forms a boundary current, but unlike the thin nose found in the single-layer version of the problem (Figure 3.3.3), the two-layer gravity current forms a blunt nose. This feature is evident in Figure 4.5.2 and in a sequence of realizations of a laboratory current (Figure 4.5.3). The upper image in each pair is a plan view,

FIGURE 4.5.1. Synthetic aperture radar image showing a coastal gravity current flowing south out of the mouth of Chesapeake Bay. (From Donato and Marmorino, 2002).

while the lower image is a side view created by a mirror reflection. Lateral and vertical detrainment of the (dyed, fresh) fluid in the current into the (clear, saline) ambient fluid can be observed, particularly in the early stages. In the four laboratory experiments cited, the nose is observed to propagate at the speed

$$c_b^{\,*} = \beta(g'd_b^{\,*})^{1/2}, \tag{4.5.1}$$

where $d_b^{\,*}$ is the upper layer depth at the wall, just upstream of the head (Figure 4.5.4) and β ranges over 1.0–1.3. The width $w_b^{\,*}$ of the current behind the head is more difficult to define due to the presence of eddies and billows around the outer edge. Nevertheless all investigations show that, regardless of definition,

$$w_b^{\,*} = \beta_w \sqrt{g'd_b^{\,*}}/f,$$

with $0.5 < \beta_w < 0.8$. The values of $c_b^{\,*}$, $d_b^{\,*}$ and $w_b^{\,*}$ tend to decrease gradually with time. In some experiments the leading edge stagnates, creating an expanding gyre to the rear. The traditional view (Figure 4.5.4) is that the gravity current consists of a blunt nose followed by a relatively thick 'head' region, a thinner

FIGURE 4.5.2. Photo of nose of Chesapeake Bay plume near Duck, NC, March 1991. (Photo by William Birkemeier, US Army Corps of Engineers).

'neck', and a long and gradually thickening rear portion that joins to the reservoir. In some cases the neck and head can be distinguished only in the early stages of the experiment. If drifting particles are placed in the flow it is observed that the head is fed from the rear by a relatively laminar current near the wall. Upon reaching the blunt nose, some of this fluid is diverted offshore where it reverses direction and moves upstream relative to the nose, possibly becoming detrained. The basic elements of this circulation are shown in a numerical simulation of the current at an early stage (Figure 4.5.5). The positive flow that feeds the head lies between the (dashed) $v^* = c_b^*$ contour and the wall. The velocity vectors are plotted in a frame of reference translating at speed c_b^*, so the positive flow feeding the head appears weak. Retrograde motion is observed along the offshore portion of the intrusion.

Various attempts have been made to predict c_b^* in terms of the properties of the flow just upstream of the head. Although the turbulent character of the current makes it difficult to find properties that are conserved between the nose and upstream, this problem can be circumvented by restricting attention to the leading edge of the nose and along the wall, where the unsteadiness and turbulence is observed to be minimal. Consider a side view of an idealized version of the current in a frame of reference translating at speed c_b^* (Figure 4.5.6). The tip s of the nose is a stagnation point and the denser fluid approaches from the far right a at speed c_b^*. The Bernoulli function evaluated at the free surface (where the pressure is considered zero) is

$$B^* = \frac{u^{*2} + v_m^{*2}}{2} - fc_b^* x^* + g\eta^*, \tag{4.5.2}$$

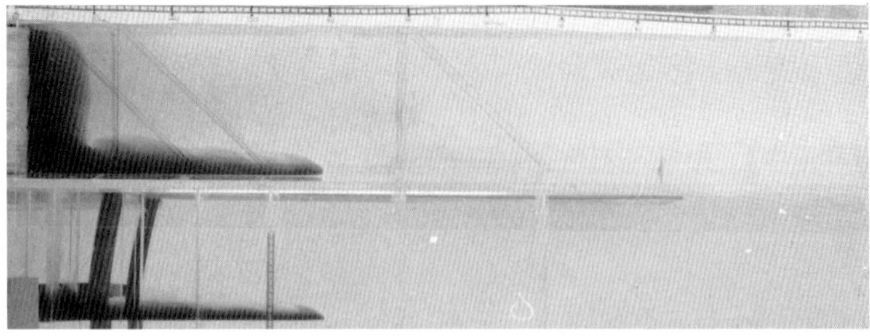

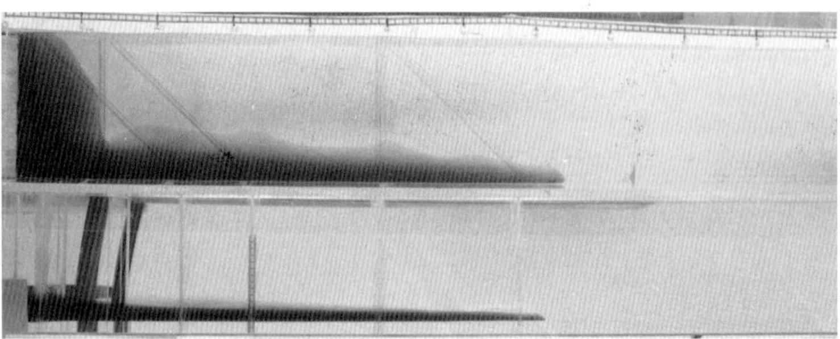

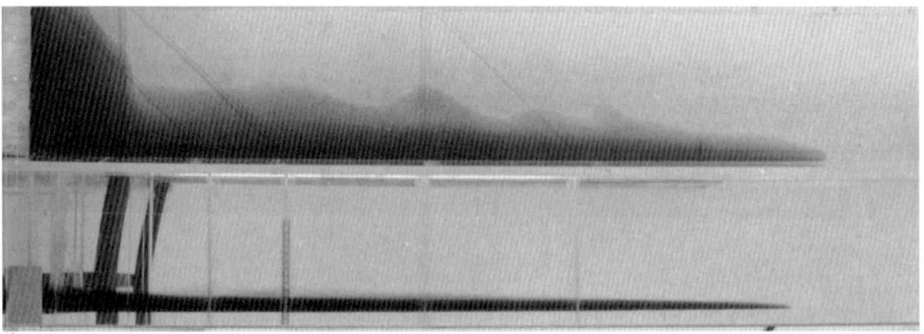

FIGURE 4.5.3. Advancing gravity current as seen in plan and (upside-down) side views (upper and lower half of each frame) for successive times. The side view in each case is a mirror reflection. The photos are based on the laboratory experiments performed by Stern, *et al.*, 1982.

where $v_m{}^* = v^* - c_b{}^*$ is the y-velocity seen in the moving frame. (The factor $fc_b{}^*x^*$ accounts for the moving reference frame and g represents full gravity.) Assuming that B^* is conserved along the wall streamline that connects a to s, (4.5.2) reduces to $g\eta_s{}^* = c_b{}^{*2}/2$, where $\eta_s{}^*$ is the free surface elevation at the stagnation point and $\eta_a{}^*$ has been set to zero. It is further assumed that the wall flow remains steady between the nose and an upstream location b where the

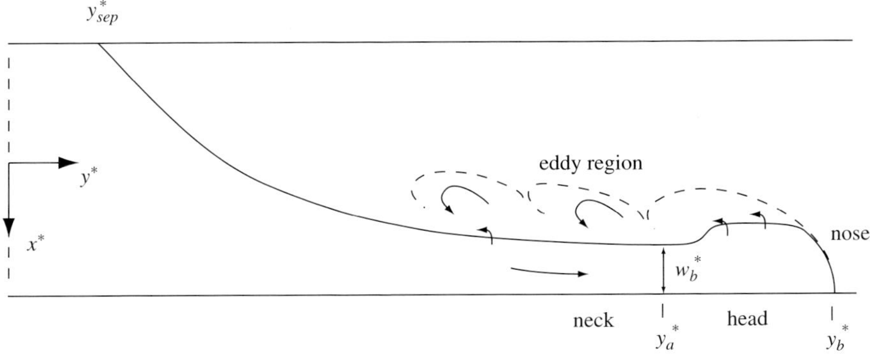

FIGURE 4.5.4. Traditional view of a gravity current in a rotating channel.

wall depth and velocity are approximately uniform. Conservation of B^* between s and b leads to $\frac{1}{2}(v_b^* - c_b^*)^2 + g\eta_b^* = g\eta_s^* = \frac{1}{2}c_b^{*2}$ if the previous relation is used. To this point, no restriction on the dynamics of the denser fluid has been made. However, if this denser fluid is considered dynamically inactive, then the relation $g\eta^* = g'd^*$ can be invoked, leading to $\frac{1}{2}(v_b^* - c_b^*)^2 + g'd_b^* = \frac{1}{2}c_b^{*2}$, or

$$c_b^* = \tfrac{1}{2}v_b^* + g'd_b^*/v_b^*. \tag{4.5.3}$$

By minimizing c_b^* over positive v_b^* it can easily be shown that $c_b^* \geq (2g'd_{-\infty}^*)^{1/2}$. The lower bound is actually achieved in the nonrotating version of the dam-break, where v and d become independent of x. If no detrainment

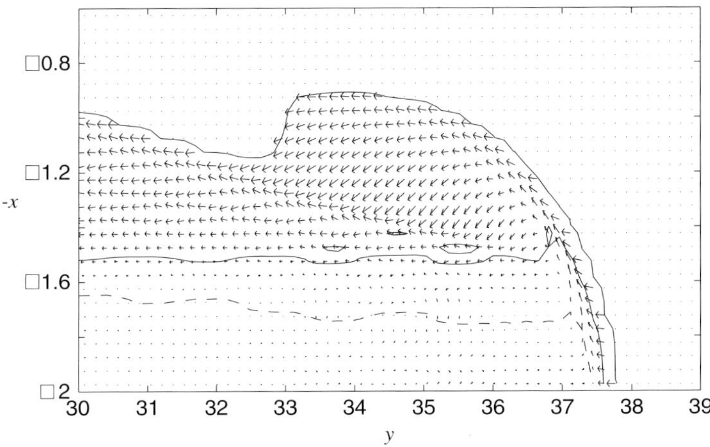

FIGURE 4.5.5. Numerical solution showing the horizontal circulation in the head of a gravity current as seen by an observed moving with the speed c_b. The dashed curve corresponds to zero along-shore velocity in the moving frame. (From Helfrich and Mullarney, 2005).

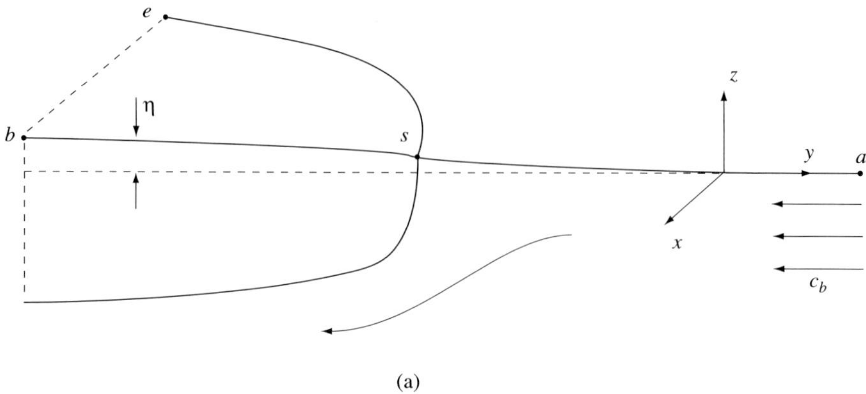

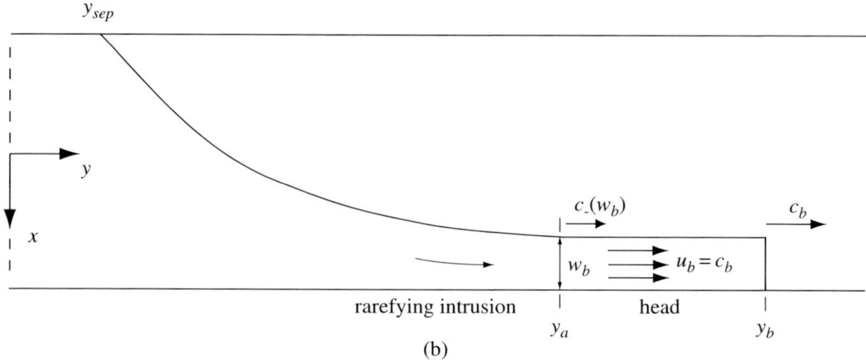

FIGURE 4.5.6. Idealization of gravity current nose region.

into the lower layer occurs and the plume remains in a steady state, conservation of mass requires that v_b must equal c_b. Equation (4.5.3) then reduces to the lower bound

$$c_b = (2g'd_b)^{1/2} \qquad (4.5.4)$$

(Benjamin, 1968). In this case $\beta = 2^{1/2}$. Thus, rotation generally increases the speed of the bore if the latter is scaled by $(g'd_b{}^*)^{1/2}$, a prediction that is in general agreement with laboratory observations. However, as already mentioned, the experimental values of β tend to fall below $2^{1/2}$. This failure has been attributed to various causes, including the presence of friction (Martin and Lane-Serff, 2005; Helfrich and Mullarney, 2006) and the participation of the lower layer (Benjamin, 1968; Klemp *et al.*, 1994 and 1997; Hacker and Linden, 2002; Martin *et al.*, 2005.)

If the entire gravity current head is regarded as steady, the volume flux at any section of the current must be zero. Applying this restriction at the section $y^* = y_b{}^*$ that coincides with point b in Figure 4.5.6a leads to

$$\int_{-w_b{}^*}^{0} (d^* v_m{}^*) dx^* = \int_{-w_b{}^*}^{0} [d^*(v^* - c_b{}^*)] dx^* = 0$$

or

$$Q_b{}^* = c_b{}^* A_b{}^*, \qquad (4.5.5)$$

where $Q_b{}^* = \int_{-w_b{}^*}^{0} (d^* v^*)_{y^*=y_b{}^*} dx^*$ is the rest frame volume transport at b and $A_b{}^* = \int_{-w_b{}^*}^{0} (d^*)_{y^*=y_b{}^*} dx^*$ is the cross section area there. By assumption, the flow is uniform at y_b and therefore geostrophically balanced. Thus

$$g' d_b{}^{*2} = f c_b{}^* A_b{}^*. \qquad (4.5.6)$$

The relation (4.5.6) adds the new variable $A_b{}^*$ to the mix and further information is required in order to close the problem for $c_b{}^*$. One approach is to assume that the current has uniform potential vorticity, which occurs if the source reservoir has constant depth and friction and entrainment are absent. The uniform potential vorticity depth profile (2.3.1) allows one to write $A_b{}^*$ in terms of $v_b{}^*$ and $d_b{}^*$, thus closing the system (4.5.4) and (4.5.6). Kubokawa and Hanawa 1984b and later Helfrich and Mullarney used equivalent approaches and found $c_b{}^*$ to be of the form (4.5.1), but with β only marginally greater than $2^{1/2}$. The corresponding $\beta_w = 0.78$.

Yet another theory for the nose speed is due to Nof (1987) with later refinements by Hacker and Linden (2002), who added a third constraint of momentum conservation and applied it to a model in which the gravity current has no flow relative to the front (i.e. $v^*(x^*, y^*) = c_b{}^*$) The resulting $\beta = 2^{1/2}$ is identical to the result for no rotation and $\beta_w = 2^{-1/2}$.

Attention to this point has been focused on the local properties of the gravity current near its leading edge. It still remains to relate $v_b{}^*$ and $d_b{}^*$ to the reservoir conditions. Assuming that the flow to the rear of the head varies gradually with y^*, it is reasonable to apply semigeostrophic theory and use the method of characteristics to link the head to the reservoir. One way of proceeding is to calculate the solution to a dam-break problem as in Section 3.3. Although one could in principle perform this calculation for a full two-layer system, the solution is complicated by the presence of a lower layer potential vorticity front that initially lies at the position of the barrier (where the lower layer thickness is discontinuous). When the barrier is removed this front is overridden by the upper layer gravity current and must be accounted for. The problem is avoided if the lower layer is considered infinitely deep, for then the solution proceeds as described in Section 3.3. As before, the problem is considerably simplified if one of the Riemann invariants R_+ or R_- is constant for the initial condition as a whole. Stern *et al.* (1982) opted for constant R_- since this produces a steepening flow that results in the formation of a blunt nose. The requirement that the nose be blunt leads to the identification of a unique value of R_- (see Exercise 1) and allows closure of the problem. The resulting current evolves into a uniform flow with width corresponding to $\beta_w = 0.42$. At the leading edge of the current is a shock that is interpreted as the nose. Energy conservation in the form of (4.5.3) is assumed to hold across the shock and this is sufficient to determine the nose speed coefficient $\beta = 1.57$. The solution is elegant in that the detrainment rate (32%) can be predicted. Kubokawa and Hanawa (1984b) altered this approach by relaxing the requirement that R_- be consistent with a smooth, blunt nose. The missing constraint is instead provided by a requirement of conservation of volume transport (4.5.6) across the nose. The resulting solution therefore has no detrainment.

Both of the solutions are subject to the objection raised in Sections 1.3 and 3.3, namely that negative v occurs at the position of the barrier at the instant of its removal. The alternative is to consider R_+ uniform, as is done in the traditional dam break. If this approach were to be carried to its logical conclusion, the result would be a rarefying intrusion with a thin leading edge (as in Figure 3.3.3). The blunt nose that is actually observed in the two-layer system might, however, be explainable as a local feature, created by processes that tend to hinder the leading edge. Clarification and guidance can be gained from a peculiar version of the dam-break problem with zero rotation. Suppose that instead of being removed altogether, the initial barrier is moved horizontally at a fixed speed $< 2(g'd_o{}^*)^{1/2}$ away from the reservoir. Then as shown by Stoker 1957 (also see Exercise 5 of Section 1.3) the flow near the moving barrier consists of a slab-like region with constant depth and velocity. This region extends upstream from the barrier and joins with a second, rarefying region. The structure of the second region is the same as in the classical dam-break. Abbott (1961) and Garvine (1981) used this piecewise continuous solution, interpreting the slab region as a model of the head and taking $(2g'd_b{}^*)^{1/2}$ as the barrier speed. Helfrich and Mullarney (2005) have taken a similar view of the rotating gravity current in a channel (Figure 4.5.6b). The head consists of a translating slab with width $w_b{}^*$ and velocity c_b^*. The head is joined to a rarefying feeder current that extends from the rear of the head back into the reservoir. The feeder current is just a truncated version of the rarefying intrusion shown in Figure 3.3.2a. It becomes attached to the left sidewall at an upstream point $y_{\text{sep}}{}^*$. At the point of transition $y_a{}^*(t^*)$ between the head and feeder current, the volume transport and width are required to be continuous. Continuity of transport implies that the head suffers no detrainment. Under the constraint of uniform R_+, each width value w_e within the separated portion of the rarefying intrusion travels at a characteristic speed $c_-{}^*$ that depends only on the local width. The transition point $y_a{}^*(t^*)$ therefore travels at the characteristic speed $c_-{}^*(w_b{}^*)$.

In order to complete the solution, a separate model for the nose speed $c_b{}^*$ must be used. The model takes the form (1) with the value of β given empirically or by one of the above theories. It turns out that $dy_a{}^*/dt^* < c_b{}^*$ regardless of this choice and the transition point therefore recedes relative to the nose. The value of w_b itself is determined by the requirement of continuity of volume transport. As shown by Helfrich and Mullarney (2005), the general procedure can be carried out for nonseparated gravity currents in a channel, as generally occurs for dimensionless channel width $w = w^*/(g'd_{1\infty}{}^*)^{1/2}$ less than about 0.5 in the experiments. The resulting model allows properties like $c_b{}^*$, $c_{\mathrm{sep}}{}^*$, $w_b{}^*$ and $h_b{}^*$ to be related to w^*.

A comparison is made (Figure 4.5.7) between the predicted values, all based on a nominal empirical value $\beta = 1.2$, and data from the Stern *et al.* (1982) and Helfrich and Mullarney (2005) experiments. In browsing through this figure the reader will find significant discrepancies between the two sets of data, despite the similarity between the two experiments. Some of these differences may be due to varying definitions of variables or the way they are measured. For example, the

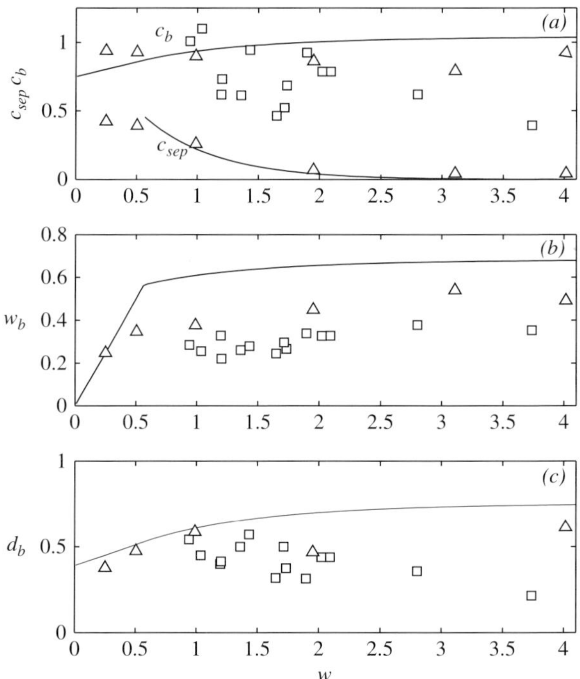

FIGURE 4.5.7. Comparison of solutions for a semigeostrophic gravity current with uniform Reimann invariant R_+ with data from various laboratory and numerical simulations. The speeds shown have been nondimensionalized using $(g'd_{1\infty})^{1/2}$, where $d_{1\infty}$ is the initial depth in the reservoir for the upper layer. The widths w and w_b have been scaled by $(g'd_{1\infty})^{1/2}/f$ and d_b by $d_{1\infty}$. The theory, which is shown by solid curves in each frame, is based on a nose speed of the form (1) with the nominal empirical value $\beta = 1.2$. The theory and some of the experiments extend into the range in which the gravity current does not separate from the left channel wall (roughly $w < 0.5$). Triangles and squares represent data from the laboratory experiments of Helfich and Mullarney (2005) and Stern *et al.* (1982), respectively. In (a) $c_b{}^*$ and $c_{\mathrm{sep}}{}^*$ are both shown, though Stern *et al.* (1982) did not measure the latter. Frames (b) and (c) show comparisons based on the nose width and depth. [From Helfrich and Mullarney (2005), Figure 15].

experimental values of $c_b{}^*$ are generally observed to suffer a slow decrease with time, probably due to frictional effects, and discrepancies in measured values of $c_b{}^*$ may be due to the time at which the measurements were taken. The theory in this case is not completely pure since it is based on an empirical β. Nevertheless, the comparison indicates some success in the prediction of values or trends for certain quantities such as $c_b{}^*$, $c_{\mathrm{sep}}{}^*$, $w_b{}^*$; the prediction for $d_b{}^*$ is less successful.

The foremost shortcoming in the theory of rotating gravity currents is failure to adequately address the entrainment problem. Only the theory of Stern *et al.* (1982) offers a prediction of entrainment. At the time of this writing there exist no laboratory measurements of entrainment against which this theory can be compared. It may also seem odd to the reader that surface gravity current doctrine emphasizes detrainment of fluid into the ambient fluid whereas the literature on descending plumes (Section 2.12) emphasized entrainment of ambient fluid. To some extent, the difference between entrainment and detrainment is based on how the current and ambient fluid are defined. But, as pointed out by McClimans (1994), different turbulent regimes naturally lead to different categorizations. For example, a surface gravity current that has a nonturbulent and undiluted core region and whose turbulence exists only near the outer edge, may naturally be regarded as detraining. On the other hand, a descending plume that experiences turbulence over its entire cross section, with consequent dilution of density, may naturally be regarded as entraining.

Exercises

(1) *The gravity current as a steepening bore.* Following Stern *et al.* (1982), consider the curves of constant Riemann invariant for separated, zero-potential vorticity flow (Figure 2.5). (They considered finite but constant potential vorticity and the diagram for this case is similar to Figure 2.5.) As a model of a gravity current we seek a solution that has R_- =constant and that allows a blunt 'nose' (i.e. permits w_e to go to zero while $\partial w_e/\partial y$ remains finite). Of all the candidate R_- =constant curves in Figure 2.5 that have this property, show that only the curve that intersects the origin is consistent with a blunt nose. (Hint: one approach is to use equations (2.3.19).

(2) *A bound on the gravity current width.*

    (a) Suppose that the gravity current is considered steady in a frame of reference moving with the nose speed c_b and that the width approaches a uniform value w_b upstream of the nose. By applying the Bernoulli equation between the nose (point s in Figure 4.5.6a) and an upstream point on the outer edge (e in the same figure) show that $w_b{}^* \le c_b{}^*/2f$. (This result was first obtained by Stern *et al.* 1982.)

    (b) Show that the result is invalidated if an energy loss from s to p is permitted.

    (c) Show that the bound is equivalent to $\beta_w \le \beta/2$.

5
Two-Layer Flows in Rotating Channels

The exchange flow between a marginal sea or estuary and the open ocean is often approximated using two-layer stratification. Two-layer models are most convincing when the interfacial region separating the upper and lower fluids is relatively thin. The exchange flow in the Strait of Gibraltar exhibits this behavior, at least at certain locations and times (Figure I.9). The vertical density and velocity profiles taken near the Camerinal Sill show a relatively sharp transition between slab-like upper and lower water masses. Elsewhere, the Gibraltar interface can be thicker and can contribute significantly to the overall mass budget for the strait. The Bab al Mandab (BAM) exchange flow experiences variations throughout the water column that are quite continuous (Figures 1.10.7 and 5.0.1). Under such conditions, a two-layer model might still give guidance provided that motions over the water column are associated with the lowest internal mode of the stratified shear flow.

Rotational effects are often ignored in applications such as Gibraltar and the BAM, where the narrowest widths are about the same or less than the internal Rossby radius of deformation based on the local depth scale. However valid this assumption is, it certainly fails where the strait broadens into the neighboring marginal sea or ocean. For most deep-ocean overflows rotation is paramount but exchange dynamics are less important. However, the overflow itself is often composed of fluid drawn from an intermediate water mass in the upstream basin, with a weaker contribution from deep waters. The deep and intermediate water masses may exhibit independent behavior that might be captured by treating the two as separate homogeneous layers. The flow of Antarctic Bottom Water through the Vema Channel (Figure 5.0.2) provides an example. Upstream of the sill (right-hand section), the deep isopycnals slope downwards from left to right, or west to east. Slightly downstream of the sill section (middle frame), the slopes in the deepest (dark shaded) water are reversed and slope upwards. Isopycnals on the right-hand side of the Channel become pinched together as a result. Further downstream the deep isotherms regain their original slope (left-hand frame).

The early sections in this chapter provide a review of two-layer hydraulic phenomena in nonrotating systems. Much of this subject is covered in Baines (1995); our review concentrates on maximal exchange, overmixing, and other concepts that arise in the consideration of ocean exchange flow and that have not been emphasized in other texts. The final sections explore the difficult and nascent field of two-layer hydraulics with rotation.

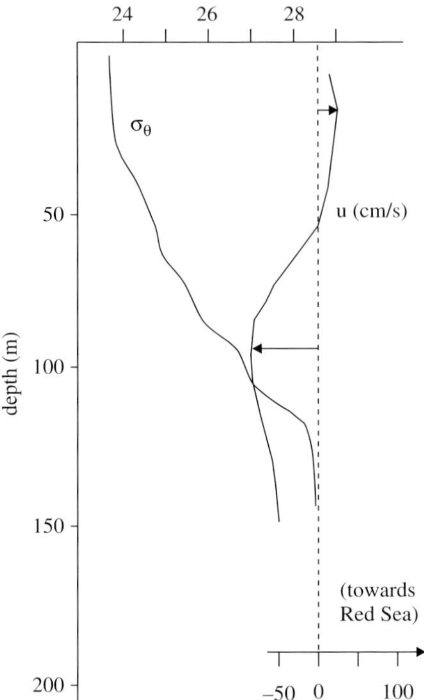

FIGURE 5.0.1. April 1996 CTD cast at the Bab al Mandab sill along with a March 1996 average velocity profile (from ADCP). (From Pratt *et al.*, 1999).

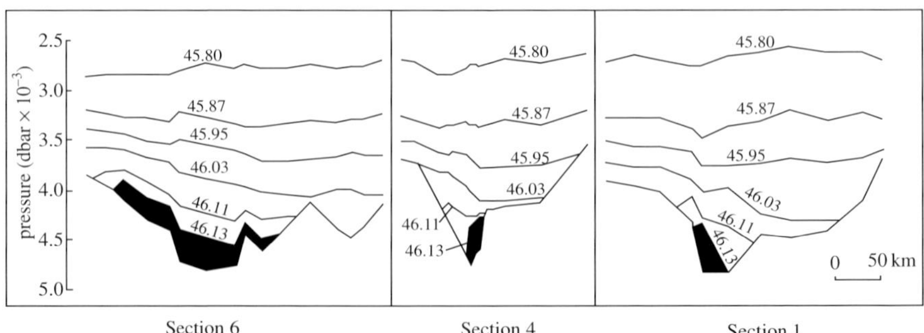

FIGURE 5.0.2. Three cross sections of the Vema Channel showing depths of selected potential density (σ_4) surfaces. Sections 1 is upstream of the sill, Section 4 is close to the sill, and Section 6 is downstream of the sill. (From Hogg, 1983).

5.1.   Formulation of Two-Layer, Semigeostrophic Models

The channel is laid out as before, with a rectangular cross-section and width and bottom elevations w^* and h^* (Figure 5.1.1). Two homogeneous layers of fluid are now present and we follow the oceanographic convention in numbering the top and bottom layers 1 and 2 respectively. The density ρ_2 of the bottom layer is only slightly greater than ρ_1.

In formulating the governing equations, we will employ a number of standard approximations. The first involves the treatment of the upper boundary of the two-fluid system. If this boundary were a free surface, overlain by a vacuum or by a substantially less dense fluid such as air, then free surface gravity waves would exist. In nearly all oceanographic applications the propagation speeds of these waves are much greater than the typical fluid velocities. In the Denmark Strait overflow, for instance, typical peak velocities are about 1m/s whereas the speeds of long, free surface gravity waves are up to 100 times larger. The Froude number $F_d = v^*/\sqrt{gd^*}$ based on free-surface dynamics is therefore $\ll 1$. Our previous experience with homogeneous flows suggests that it is unlikely that bottom topography (or width variations) will cause significant departures of the free-surface

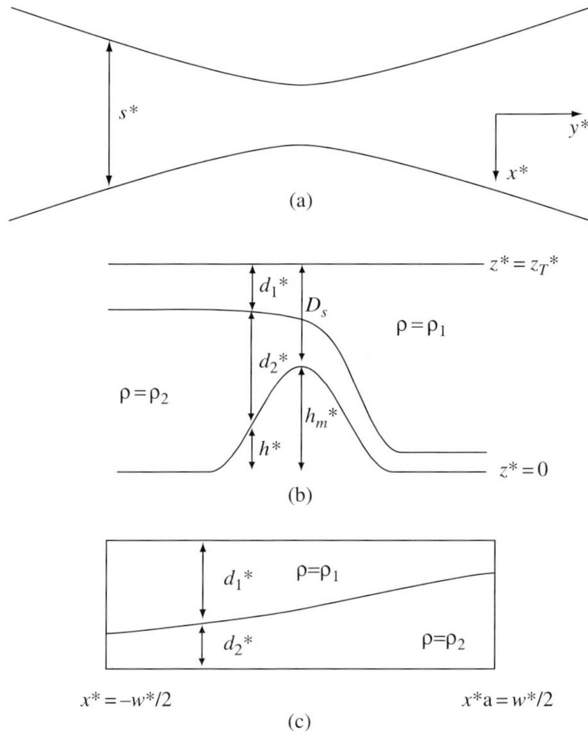

FIGURE 5.1.1. Definition Sketches.

elevation from a horizontal plane. In fact, the free-surface elevation $z_T{}^*$ can be shown to obey

$$\frac{dz_T{}^*}{dy^*} \approx F_d^2 \left(\frac{d^*}{w^*}\frac{dw^*}{dy^*} - \frac{dh^*}{dy^*} \right),$$

obtained by rearrangement of Equation (1.4.3) with $F_d \ll 1$. Thus, when F_d is small, the departure in the free surface elevation is smaller by a factor F_d^2 than the variations in w or h. For the Denmark Strait, F_d^2 is about 10^{-4}. On the other hand, the speeds of the *internal* waves that propagate on the interface between the two layers are much smaller and the associated Froude numbers much larger. One might expect, then, that the typical vertical excursions of the interface will be much greater than those of the free surface. Since the latter now give a negligible contribution to variations in the upper layer depth, we simply regard the upper surface as rigid and horizontal. If $z^* = z_T{}^*$ denotes the constant elevation of this surface, h^* the bottom elevation, and $d_1{}^*$ and $d_2{}^*$ the thicknesses of the two layers, then

$$z_T{}^* = h^* + d_2{}^* + d_1{}^*.$$

The rigid lid approximation is explored more formally in Exercise 1.

Under the assumption that variations in the horizontal are gradual on the scale of the depth, the pressures in each layer will be hydrostatic. Therefore

$$p_1{}^* = p_T{}^* + \rho_1 g(z_T{}^* - z^*) \tag{5.1.1}$$

and

$$\begin{aligned}
p_2{}^* &= p_T{}^* + \rho_1 g(z_T{}^* - h^* - d_2{}^*) + \rho_2 g(h^* + d_2{}^* - z^*) \\
&= p_T{}^* + g(\rho_2 - \rho_1)(h^* + d_2{}^*) + g(\rho_1 z_T{}^* - \rho_2 z^*),
\end{aligned} \tag{5.1.2}$$

where $p_T{}^*(x^*, y^*, t^*)$ denotes the pressure at the rigid upper lid.

There are two other standard assumptions. The first is that the channel geometry varies only gradually along its axis, suggesting that the along-channel velocity $v_i{}^*$ is geostrophically balanced. The formal arguments leading to this 'semigeostrophic' approximation are essentially those laid out in Chapter 2. The second assumption is that the density difference between the two layers is relatively small: $\Delta\rho/\bar{\rho} = (\rho_2 - \rho_1)/[\frac{1}{2}(\rho_2 + \rho_1)] \ll 1$. This is the basis for the Boussinesq approximation, in which the actual density ρ_1 or ρ_2 is replaced by a representative value such as the average $\bar{\rho}$, except where they are multiplied by g. The reasoning here is that g is much larger than accelerations of the fluid itself, and its product with the small $\Delta\rho/\bar{\rho}$ is nonnegligible. The semigeostrophic equations governing the inviscid, Boussinesq, two-fluid system are then

$$fv_1{}^* = \frac{1}{\bar{\rho}}\frac{\partial p_T{}^*}{\partial x^*} \tag{5.1.3}$$

$$\frac{\partial v_1{}^*}{\partial t^*} + u_1{}^*\frac{\partial v_1{}^*}{\partial x^*} + v_1{}^*\frac{\partial v_1{}^*}{\partial y^*} + fu_1{}^* = -\frac{1}{\bar{\rho}}\frac{\partial p_T{}^*}{\partial y^*} \tag{5.1.4}$$

$$f v_2{}^* = \frac{1}{\bar\rho} \frac{\partial p_T{}^*}{\partial x^*} + g' \frac{\partial d_2{}^*}{\partial x^*} \tag{5.1.5}$$

$$\frac{\partial v_2{}^*}{\partial t^*} + u_2{}^* \frac{\partial v_2{}^*}{\partial x^*} + v_2{}^* \frac{\partial v_2{}^*}{\partial y^*} + f u_2{}^* = -\frac{1}{\bar\rho} \frac{\partial p_T{}^*}{\partial y^*} - g' \left(\frac{\partial d_2{}^*}{\partial y^*} + \frac{\partial h^*}{\partial y^*} \right) \tag{5.1.6}$$

where $g' = \Delta\rho g / \bar\rho$ is the reduced gravity.

The equation of mass conservation within layer i is

$$\frac{\partial d_i{}^*}{\partial t^*} + \frac{\partial (u_i{}^* d_i{}^*)}{\partial x^*} + \frac{\partial (v_i{}^* d_i{}^*)}{\partial y^*} = 0 \tag{5.1.7}$$

If (5.1.5) is subtracted from (5.1.3) the result is the *thermal wind* relation for the along-channel velocity component:

$$f(v_1{}^* - v_2{}^*) = -g' \frac{\partial d_2{}^*}{\partial x^*}. \tag{5.1.8}$$

The difference in velocities between the two layers is thus proportional to the cross-channel slope of the interface.

The semigeostrophic potential vorticity within layer i is defined by

$$q_i{}^* = \frac{f + \dfrac{\partial v_i{}^*}{\partial x^*}}{d_i{}^*}, \tag{5.1.9}$$

and conservation of this property following the fluid motion,

$$\left(\frac{\partial}{\partial t^*} + u_i{}^* \frac{\partial}{\partial x^*} + v_i{}^* \frac{\partial}{\partial y^*} \right) q_i{}^* = \frac{d_i{}^* q_i{}^*}{dt^*} = 0,$$

may be shown in the same manner as for a homogeneous fluid.

In the event the potential vorticity is uniform within each layer, it is convenient to write

$$q_i{}^* = \frac{f}{D_{i\infty}} \tag{5.1.10}$$

where $D_{i\infty}$ represents the potential depth of layer i. Using the above definition in (5.1.9) and combining the two results with (5.1.8) leads to an equation for the cross-channel structure of the flow

$$\frac{\partial^2 d_2{}^*(x^*, y^*, t^*)}{\partial x^{*2}} - L_I^{-2} d_2{}^*(x^*, y^*, t^*) = -\frac{f^2(z_T{}^* - h^*)}{g' D_{1\infty}}, \tag{5.1.11}$$

where

$$L_I = \left[\frac{g' D_{1\infty} D_{2\infty}}{f^2 (D_{1\infty} + D_{2\infty})} \right]^{1/2} \tag{5.1.12}$$

is the internal Rossby radius of deformation. Equation (5.1.11), which is similar to the cross-channel structure equation (2.2.2) governing the single-layer case, shows that the interface will have a boundary layer structure with e-folding scale L_I when the channel width is $\gg L_I$. Through the thermal wind relation this structure will be imposed on the shear velocity $v_1^* - v_2^*$. However, v_1^* and v_2^* need not individually decay away from the sidewalls.

When the flow is steady, individual transport stream functions ψ_1^* and ψ_2^* can be defined such that

$$v_i^* d_i^* = \frac{\partial \psi_i^*}{\partial x^*}, \quad \text{and} \quad u_i^* d_i^* = -\frac{\partial \psi_i^*}{\partial y^*}. \tag{5.1.13}$$

The semigeostrophic Bernoulli functions for each layer are conserved along their respective streamlines:

$$B_1^*(\psi_1^*) = \frac{v_1^{*2}}{2} + \frac{p_T^*}{\bar{\rho}} \tag{5.1.14}$$

and

$$B_2^*(\psi_2^*) = \frac{v_2^{*2}}{2} + \frac{p_T^*}{\bar{\rho}} + g'(d_2^* + h^*). \tag{5.1.15}$$

We leave it as an exercise to show

$$\frac{dB_i^*}{d\psi_i^*} = q_i^*. \tag{5.1.16}$$

In most problems it is convenient to eliminate the rigid lid pressure and work with quantities that govern the internal structure of the flow. For example, subtraction of (5.1.15) from (5.1.14) eliminates p_T^*, leaving

$$\Delta B^*(\psi_1^*, \psi_2^*) = B_2^*(\psi_2^*) - B_1^*(\psi_1^*) = \frac{v_2^{*2} - v_1^{*2}}{2} + g'(d_2^* + h^*). \tag{5.1.17}$$

The quantity ΔB is sometimes referred to as the internal energy (per unit mass). It should be stressed that ΔB^* is not conserved following ψ_1^* or ψ_2^* unless the two streamlines coincide, such as at a vertical boundary that extends through both layers.

Exercises

(1) Reformulate equations (5.1.3–5.1.7) to allow for a free upper surface (at which the pressure may assumed to be zero). Through inspection of these equations, formulate velocity, length and time scales based on the *internal* dynamics of the flow (i.e. use g' rather than g). Under this scaling, show that the contribution to d_1^* from a typical displacement of the interface is much greater than the contribution from a typical displacement of the free surface. Deduce that the free surface displacement can be neglected in the continuity equation for the upper layer, so that the upper surface can effectively be treated as rigid.

(2) Show that the Bernoulli functions as defined by (5.1.14) and (5.1.15) are indeed conserved along streamlines of the respective layers, provided that the flow is steady.

(3) Prove (5.1.16).

(4) Using the expression for the linear wave speed of an internal disturbance in a nonrotating, two-layer system (see 5.2.3) show that the two-layer Rossby radius of deformation (5.1.12) may be interpreted as the distance that such a wave will travel in a period $2\pi/f$.

5.2. Basic Theory for a nonrotating Channel

There are several articles that deserve special mention in the annals of two-layer hydraulics, the earliest being Stommel and Farmer's (1952, 1953) model of estuary dynamics. Many of the distinctive properties of these flows, including the possibility of two control sections, were identified by Wood (1968, 1970) in his laboratory simulations of lock exchange between basins and selective withdrawal from stratified reservoirs. The steady theory was unified and extended in a series of articles by L. Armi and D. Farmer, including Armi (1986), Armi and Farmer (1986, 1987, 1988) and Farmer and Armi (1986), who were interested in the Strait of Gibraltar and other oceanographic examples of exchange flow. This work forms the foundation for our summary and their fingerprints are on much of what follows. A slightly different view is provided by Long's (1954) towing experiments and subsequent investigations of initial-value problems by various authors (Baines, 1995 and references contained therein). This literature gives considerable insight into how two-layer flows are set up.

The governing equations are the x^*-independent, $f = 0$ versions of (5.1.4, 5.1.6, and 5.1.7). These equations can be put into characteristic form [Baines (1995) pp. 98–99] using the methods laid out in Appendix B. The characteristic speeds are given by

$$c_\pm{}^* = \frac{v_1{}^* d_2{}^* + v_2{}^* d_1{}^*}{d_1{}^* + d_2{}^*} \pm \left\{ \frac{g' d_1{}^* d_2{}^*}{d_1{}^* + d_2{}^*} \left[1 - R_b^{-1}\right] \right\}^{1/2}. \tag{5.2.1}$$

For an evolving flow containing disturbances of arbitrary amplitude, we may regard $c_+{}^*$ or $c_-{}^*$ as the local and instantaneous speed of a signal propagating forward or backward with respect to the advective speed defined by the first expression on the right-hand side. Although no linearization has been made, we can also regard $c_+{}^*$ and $c_-{}^*$ as the speeds of small-amplitude, long waves propagating on a steady and uniform background flow with depth and velocity $d_i{}^*$ and $v_i{}^*$. Note that these speeds are real only so long as

$$R_b = \frac{g'(d_1{}^* + d_2{}^*)}{(v_1{}^* - v_2{}^*)^2} \geq 1. \tag{5.2.2}$$

Thus, if the magnitude of the *shear velocity* $v_1{}^* - v_2{}^*$ is large enough, $c_\pm{}^*$ become imaginary, corresponding to long-wave Kelvin-Helmholtz instability of the background flow. The parameter R_b is a discrete (or 'bulk') form of the *Richardson number* $R_i = [g\rho^{-1}\partial\rho/\partial z^*]/(\partial v^*/\partial z^*)^2$ for continuously stratified shear flow.

The possibility of instability is an important departure from the behavior of the single-layer case considered in the first chapter. It is natural to ask whether traditional properties such as hydraulic control and upstream influence remain meaningful when part or all of the flow is unstable. The answer to this question is largely unknown at the time of this writing. For many of the two-layer flows encountered in nature or in the laboratory, the primary instabilities occur in supercritical regions away from control sections. The associated disturbances propagate away from the control section(s) and conditions there remain steady.

There is another aspect of the stability issue that bears consideration. An analysis (e.g. Turner 1973, Section 4.1) of the inviscid, two-layer system with respect to an arbitrarily short (nonhydrostatic) disturbances shows that the flow is *always* unstable provided that $v_1^* \neq v_2^*$. In a two-layer system with infinite layer depths, for example, all sinusoidal interfacial waves with lengths less than $\pi |v_1^* - v_2^*|/g'$ are unstable. The resulting mixing can destroy the sharp interface and create an intermediate transitional layer. Wilkinson and Wood (1983) present a laboratory demonstration using a hydraulically driven, two-layer system. If the shear is weak, unstable waves have small scales and the intermediate layer remains thin. Its thickness d_I^* can be estimated using the hypothesis that the layer will grow until the mean flow becomes stable. A necessary condition for instability of a thin, laminar, intermediate layer is that the bulk Richardson number $g^* d_I^*/(v_1^* - v_2^*)^2$ based on d_I^* falls beneath 1/4. Empirical evidence (e.g. Thorpe, 1973; Koop and Browand, 1979) suggests a transitional value closer to 0.3, and thus the expected layer thickness is

$$d_I^* \cong 0.3(v_1^* - v_2^*)^2/g'.$$

As long as d_I^* remains much less than d_1^* and d_2^*, the presence of the intermediate layer may to a first approximation be disregarded and the two-layer protocol adopted.

Some of the important differences between single- and two-layer hydraulics may be anticipated from an examination of the formula for the long-wave phase speed. If the background flow is at rest, (5.2.1) reduces to

$$c_\pm^* = \pm\left(\frac{g' d_1^* d_2^*}{d_1^* + d_2^*}\right)^{1/2}. \tag{5.2.3}$$

When the lower layer is relatively thin ($d_2^* \ll d_1^*$), $c_\pm^*$ reduces to the value $\pm\sqrt{g' d_2^*}$ for a single layer under reduced gravity. A corresponding result for the upper layer is obtained by taking $d_2^* \ll d_1^*$. If the total depth $d_2^* + d_1^*$ is held constant while the interface is varied from the top-to-bottom boundary, $|c_\pm^*|$ vary from zero to their maximum values at mid-depth ($d_1^* = d_2^*$), then back to zero. This is quite different from the case of a single layer, in which $|c_\pm^*|$ increases monotonically as the lower layer depth increases.

From (5.2.1) it can be shown that

$$c_+{}^* c_-{}^* = \frac{g' d_1{}^* d_2{}^*}{d_1{}^* + d_2{}^*} \left(\frac{v_1{}^{*2}}{g' d_1{}^*} + \frac{v_2{}^{*2}}{g' d_2{}^*} - 1 \right), \tag{5.2.4}$$

and thus at least one of the characteristic speeds is zero if the sum of the squares of the layer Froude numbers,

$$F_1 = \frac{v_1{}^*}{(g' d_1{}^*)^{1/2}} \quad \text{and} \quad F_2 = \frac{v_2{}^*}{(g' d_2{}^*)^{1/2}}, \tag{5.2.5}$$

is unity. This result makes it convenient to define a *composite* Froude number (Stommel and Farmer, 1952) as

$$G^2 = F_1{}^2 + F_2{}^2. \tag{5.2.6}$$

Critical flow corresponds to $G^2 = 1$, implying that one or both of $c_\pm{}^*$ is zero. If $G^2 < 1$ then (5.2.4) indicates that the product of $c_+{}^*$ and $c_-{}^*$ is < 0, implying that the two internal gravity waves propagate in opposite directions. This type of flow is considered *subcritical* since information can move in both directions. Similarly, $G^2 > 1$ implies that both waves propagate in the same direction and the flow is *supercritical*. These definitions avoid reference to 'upstream' or 'downstream', a tacit acknowledgement that two layers may flow in opposite directions. Thus, supercritical flow may have both waves moving in the $+y^*$ direction or in the $-y^*$ direction. It is not meaningful to talk about the criticality of an individual layer unless the other layer is inactive. For example, it is not meaningful to state that layer 1 is 'critical' when $F_1 = 1$, unless $F_2 \ll 1$. (However, it can be stated with certainty that the two-layer flow is supercritical if either F_1 or F_2 is > 1.)

Imagine a flow that is evolving in the y^*-direction due to changes in the channel geometry and suppose that this flow changes from stable to unstable at a particular y^*. Since $R_b = 1$ at that section (5.2.1) requires that $c_+{}^* = c_-{}^*$ there. Thus the flow must first be critical or supercritical before it can become unstable with respect to long waves. This is a special case of the connection, discussed at the end of Section 3.9, between long-wave instability and critical/supercritical flow.

The volume transport within a layer is

$$Q_i{}^* = v_i{}^* d_i{}^* w^* \tag{5.2.7}$$

and both $Q_1{}^*$ and $Q_2{}^*$ are constants for steady flow. If $Q_1{}^*$ and $Q_2{}^*$ have opposite signs we have an *exchange flow*. *Pure exchange flow* occurs when the net or *barotropic* transport

$$Q^* = Q_1{}^* + Q_2{}^* \tag{5.2.8}$$

is zero. Another quantity that will prove useful is the transport ratio:

$$Q_r = \frac{Q_1{}^*}{Q_2{}^*}. \tag{5.2.9}$$

The time-dependent continuity equation for a particular layer, which may be obtained by integrating (5.1.7) across the channel, is

$$w\frac{\partial d_i^*}{\partial t^*} + \frac{\partial Q_i^*}{\partial y^*} = 0.$$

An important constraint on the barotropic transport can be formulated by adding together the time-dependent continuity equations for each layer. Noting that $d_1^* + d_2^*$ depends only on y:

$$\frac{\partial(d_1^* + d_2^*)}{\partial t^*} = -w^{*-1}\frac{\partial Q^*}{\partial y^*} = 0.$$

The total transport Q^* is therefore a function of t^* only. It follows that Q^* is constant in time if this is so at any section.

Steady solutions are normally calculated using the internal Bernoulli equation (5.1.17). In thinking about the various solutions, it often helps to imagine that the channel is connected to an infinitely wide basin where the layer depths $d_{1\infty}^*$ and $d_{2\infty}^*$ are nonzero and where the flow is therefore quiescent. If $h^* = 0$ in this basin then

$$\Delta B^* = g' d_{2\infty}^*. \tag{5.2.10}$$

If a hydraulic jump occurs within the channel, the value of ΔB^* will generally change across the jump.

At this stage, the mathematical problem for the steady two-layer flow involves four variables (the depth and velocity in each layer) governed by two continuity equations (5.2.7), the internal Bernoulli equation and the geometric constraint

$$d_1^*(y^*) + d_2^*(y^*) + h^*(y^*) = z_T^* \tag{5.2.11}$$

resulting from the rigid-lid assumption. It is possible to reduce the algebra to a single equation for one of the layer thicknesses and sketch solution curves analogous to that shown in Figure 1.4.1. Another approach is to reduce the algebra to two equations in two variables and sketch solution curves in the two-dimensional space of these variables. The choice of method is largely one of personal preference. Our preference is for the second approach, as developed by Armi (1986) using the layer Froude numbers as the dependent variables. Following his formulation, the layer depths and velocities may be written in terms of F_1 and F_2 using

$$d_i^* = \frac{Q_i^{*2/3}}{g'^{\,1/3}F_i^{2/3}w^{*2/3}} \quad \text{and} \quad v_i^* = \left(\frac{Q_i^* g'}{w^*}\right)^{1/3} F_i^{2/3}. \tag{5.2.12a, b}$$

Making these substitutions and using (5.2.11) allows (5.1.17) to be written in the form

$$\mathcal{G}_1(F_1, F_2; h^*, w^*) = \left(\frac{g' Q_1^*}{w^*}\right)^{2/3}\left(\frac{1}{2}F_1^{4/3} - \frac{1}{2}Q_r^{-2/3}F_2^{4/3} + F_1^{-2/3}\right)$$

$$+ \Delta B^* - g' z_T^* = 0. \tag{5.2.13}$$

Furthermore, (5.2.11) itself can be rewritten as

$$\mathcal{G}_2(F_1, F_2; h^*, w^*) = Q_r^{2/3} F_1^{-2/3} + F_2^{-2/3} - (z_T^* - h^*)g'^{1/3} w^{*2/3} Q_2^{*-2/3} = 0.$$
(5.2.14)

Using the two-variable generalization of Gill's approach, the critical condition may be calculated using (1.5.9), which leads to

$$\frac{\partial \mathcal{G}_1}{\partial F_1} \frac{\partial \mathcal{G}_2}{\partial F_2} - \frac{\partial \mathcal{G}_1}{\partial F_2} \frac{\partial \mathcal{G}_2}{\partial F_1} = 0.$$
(5.2.15)

The reader may wish to verify that application to (5.2.13) and (5.2.14) yields the result $G^2 = 1$, the condition for stationary disturbances derived from the wave speed formula.

The regularity condition that must hold at a critical section can be obtained by applying (1.5.11), which leads to

$$\frac{\partial \mathcal{G}_1}{\partial \gamma_i} \left(\frac{\partial \mathcal{G}_2}{\partial y} \right)_{\gamma_1, \gamma_2} - \frac{\partial \mathcal{G}_2}{\partial \gamma_i} \left(\frac{\partial \mathcal{G}_1}{\partial y} \right)_{\gamma_1, \gamma_2} = 0 \ (i = 1 \ or \ i = 2),$$
(5.2.16)

with $\gamma_1 = F_1^{2/3}$ and $\gamma_2 = F_2^{2/3}$, or any other set of suitably defined functions and variables. Exercise 2 guides the reader through a choice that minimizes the algebraic manipulations. The resulting condition is

$$[v_2^{*2}(y_c) - v_1^{*2}(y_c^*)]\frac{dw^*}{dy_c^*} - g'w^*(y_c^*)F_2^2 \frac{dh^*}{dy_c^*} = 0,$$
(5.2.17)

where y_c^* denotes the position of the critical section. If w^* is constant, critical sections must occur at a point where $dh^*/dy^* = 0$. In our previous, single-layer examples such points were generally restricted to sills. Later we will show that two-layer critical flow can also occur on a level part of the channel away from an obstacle. If h^* is constant but w^* varies, then critical flow can occur as before where $\partial w^*/\partial y^* = 0$, as at a narrows, or where $v_1^{*2} = v_2^{*2}$. The latter possibility was first identified by Wood (1970) and the corresponding control section is called a *virtual control*. If the flow is unidirectional ($v_1^* v_2^* > 0$) the shear velocity ($v_1^* - v_2^*$) is zero at such a control. A novel aspect of the virtual control is that it can occur where the channel width is changing, and we will later show that w^* must, in fact, be decreasing in the flow direction. The position y_c^* of the virtual control depends on the flow itself and is not locked to a particular width.

An advantage of the Froude number plane representation is that critical flow lies along the diagonal line $F_1^2 + F_2^2 = 1$ (Figure 5.2.1). In the triangular region to the lower left of the diagonal the flow is subcritical. Above, the flow is supercritical. Some of the flow states in the supercritical range may be unstable with respect to long waves. The condition for stability (5.2.2) can be written in terms of the layer Froude numbers using (5.2.12) and the resulting threshold curve

$$[Q_r^{1/3} F_1^{2/3} - F_2^{2/3}]^2 - Q_r^{2/3} F_1^{-2/3} - F_2^{-2/3} = 0$$
(5.2.18)

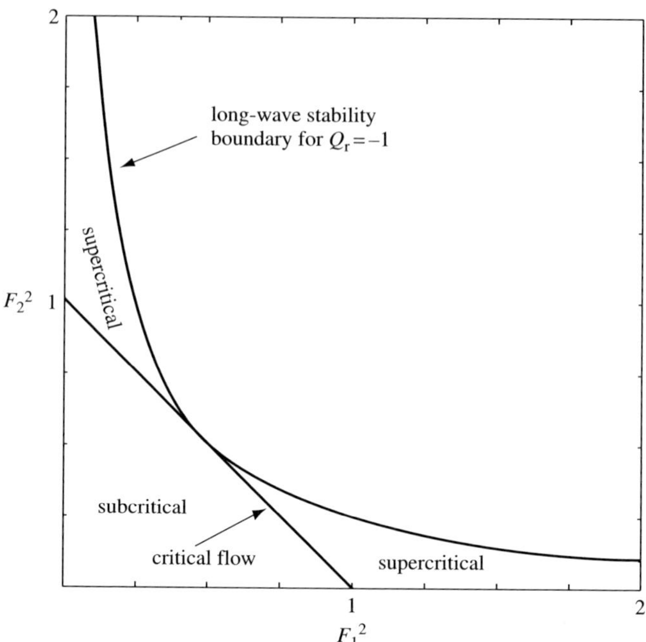

FIGURE 5.2.1. The critical diagonal and the long-wave stability threshold in the Froude number plane. (From Armi, 1986).

is plotted in Figure (5.2.1) for $Q_r = -1$ (pure exchange flow). The threshold curve for $Q_r = 1$ lies well above the critical diagonal and out of the range of the plot. An exchange flow state corresponding to any point lying above the (5.2.18) curve is formally unstable, though it remains to be seen whether such states are members of realizable solutions for reasonable upstream conditions.

The Froude number plane is not the only vehicle for representing solutions to two-layer flow. A reader seeking alternatives may wish to consult Dalziel (1991) or Baines (1995).

Exercises

(1) Show that application of (5.2.15) to (5.2.13) and (5.2.14) leads to the critical condition $G^2 = 1$. (Hint: Notice that F_1, F_2, and w^* only enter these relations in 2/3 power or 4/3 powers.)

(2) Derive the regularity condition (5.2.17) as follows:

(a) Use the layer velocities $v_1{}^*$ and $v_2{}^*$ as independent variables and define functions $\mathcal{G}_1$ and $\mathcal{G}_2$ written solely in terms of these variables (and the geometric variables). This can be accomplished using equations (5.1.17), (5.2.7) and (5.2.11).

(b) Obtain (5.2.17) by evaluation of (1.5.11) and use of the functions defined in (a) and the two-layer critical condition.

(3) Show that if critical flow occurs at a sill (where $dh^*/dy^*=0$ and $d^2h^*/dy^{*2} < 0$), then a supercritical-to-subcritical transition (or vice versa) must occur. That is, the flow cannot remain subcritical on either side of the sill.

5.3.   Flow over an Obstacle

We now consider the Froude number plane representation of solutions for flow over topography in a channel of constant width. Continuing to follow Armi (1986), it is helpful to rewrite the energy equation (5.2.13) in the normalized form

$$\tfrac{1}{2}F_1^{4/3} - \tfrac{1}{2}Q_r^{-2/3}F_2^{4/3} + F_1^{-2/3} = \frac{g'z_T^* - \Delta B^*}{(g'Q_1^*/w^*)^{2/3}}. \tag{5.3.1}$$

One interpretation of the quantity on the right-hand side follows by imagining that the straight channel is connected to an infinitely wide, quiescent basin as described above. Use of (5.2.10) and (5.2.11) then leads to

$$\frac{g'z_T^* - \Delta B^*}{(g'Q_1^*/w^*)^{2/3}} = \frac{g'd_{1\infty}^*}{(g'Q_1^*/w^*)^{2/3}} = d_{1\infty}.$$

The parameter $d_{1\infty}$ is the dimensionless upper layer thickness in the hypothetical wide basin. It may also be regarded as a measure of the potential energy in the basin, smaller $d_{1\infty}$ being associated with higher interface values and therefore higher potential energy.

In some applications, the transport ratio Q_r may be regarded as fixed. For example, some ocean straits are constrained to carry a net volume flux that is close to zero, so that Q_r has a value close to -1. Let us assume that Q_r is constant. Then there is a family of solutions to (5.3.1), each member having a particular upstream state as indicated by the value of $d_{1\infty}$. These solutions can be represented as a family of curves plotted in the Froude number plane (e.g. Figure 5.3.1a). The case shown has $|Q_r| = 1$ and the $d_{1\infty} =$ constant solutions are represented by the thicker curves. In the absence of hydraulic jumps or of other sources of dissipation, a solution must follow one of these curves. Some of the curves intersect the critical flow diagonal, raising the possibility that corresponding solutions can be critically controlled. Froude number diagrams for other values of Q_r have similar qualitative aspects (Armi, 1986) and we can therefore discuss most of the general features of the solutions using the one figure. Note that Q_r and Q_1^* enter (5.3.1) to 2/3 powers and therefore a solution curve valid for a combination (Q_r, Q_1^*) is also valid for $(-Q_r, Q_1^*)$, $(Q_r, -Q_1^*)$, or $(-Q_r, -Q_1^*)$. The direction of flow in a given layer for a particular solution is therefore arbitrary. Each curve yields four possible solutions corresponding

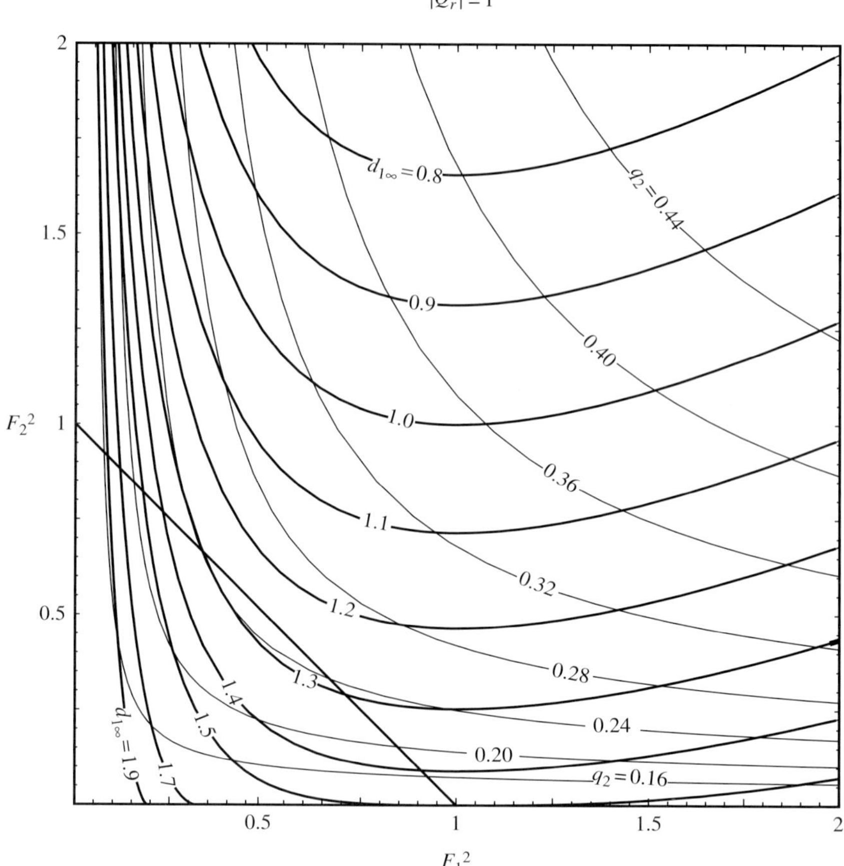

FIGURE 5.3.1a. The Froude number plane showing solution curves for flow over a variable bottom in a channel with constant width and $|Q_r| = 1$. Contours of constant internal energy $d_{1\infty}$ are represented by thick lines. Continuous solutions must lie along these contours. The thin contours represent constant q_2. For a fixed layer flux $Q_2{}^*$, larger values of the topographic height h^* correspond to smaller q_2. (From Armi, 1986).

to different directions of flow in the two layers. However, not all possibilities may be realizable: the stability of the flow and its ability to form hydraulic jumps does depend on the directions of layer transports. An obvious example is a unidirectional flow that is stable according to (5.2.2) but becomes unstable due to the increased interfacial shear that is created when the direction of one of the layers is reversed. More subtle examples arise when a change in direction of a layer flux gives rise to the shock-forming instability (Figure 1.4.4).

Now suppose that the value of $d_{1\infty}$ is given along with the topographic function $h^*(y^*)$. In order to construct a solution one needs to know how to move along the appropriate curve of constant $d_{1\infty}$ as h^* varies. This link between the solution and the topography is provided by (5.2.14), which can be cast in terms of the Froude numbers as

$$Q_r^{2/3} F_1^{-2/3} + F_2^{-2/3} = q_2^{-2/3}, \tag{5.3.2}$$

where

$$q_2 = \frac{Q_2^*}{[z_T^* - h^*(y^*)]^{3/2} g'^{1/2} w^*}.$$

The thin contours drawn in Figure 5.3.1 are ones of constant q_2. Since Q_2^*, g', and w^* remain fixed for a particular steady solution, changes in q_2 with y^* are entirely due to changes in h^*. Increases in h^* lead to increases in q_2 and inspection of Figure 5.3.1a shows that higher h^* are generally found by moving away from the origin.

a. Flow from a Deep Basin

One important class of solutions describes flow originating from an infinitely deep upstream basin that has the same width as the channel. Note that at least one of the layer depths must be infinite (and the corresponding velocity zero) in the basin and therefore the solution curve must begin along the horizontal ($F_2^2 = 0$) or vertical ($F_1^2 = 0$) axes of Figure 5.3.1. Inspection of the figure shows that the only possibilities originate from the horizontal axis. These solutions have $F_2 = 0$ in the basin, meaning that the lower layer is infinitely deep and (therefore) stagnant. The reverse situation, a stagnant upstream upper layer with a moving lower layer, is not possible. This asymmetry between the behavior of the upper and lower layers is due to the fact that the topography contacts only the lower layer. Although the formal solutions allow the direction of flow within each layer to be arbitrary, let us assume that the lower layer flow is out of the deep basin. The upper layer flow may then be in either direction, unless otherwise noted. We will continue to refer to the basin as 'upstream', even though the upper layer may flow into it.

Now suppose that the value of $d_{1\infty}$ is known to be 1.7, so that the solution must lie along the thick curve with that value. Keep in mind that $d_{1\infty}$ is not the actual upper layer depth in the deep basin, but rather the upper layer depth in a hypothetical reservoir that has infinite width and is therefore quiescent. (This reservoir might be imagined to lie upstream of the deep basin.) The flow state in the deep basin lies where the $d_{1\infty} = 1.7$ curve intersects the F_1^2 axis and is clearly subcritical. An observer moving from the basin into the channel will see an increase in h and must therefore move upwards along the '1.7' curve to higher contour values of q_2. If the sill is reached before the critical diagonal is encountered then the solution at points downstream is found by retracing the '1.7' curve back to the F_1^2 axis. In this way a completely subcritical solution is obtained. The value of F_1^2 is minimal at the sill, meaning that the upper layer depth reaches a maximum (see 5.2.12a). Figure 5.3.1b shows this situation schematically, with the '1.7' solution curve traced over a circuit *aba* and the corresponding subcritical solution (inset) experiencing an interfacial dip over the obstacle.

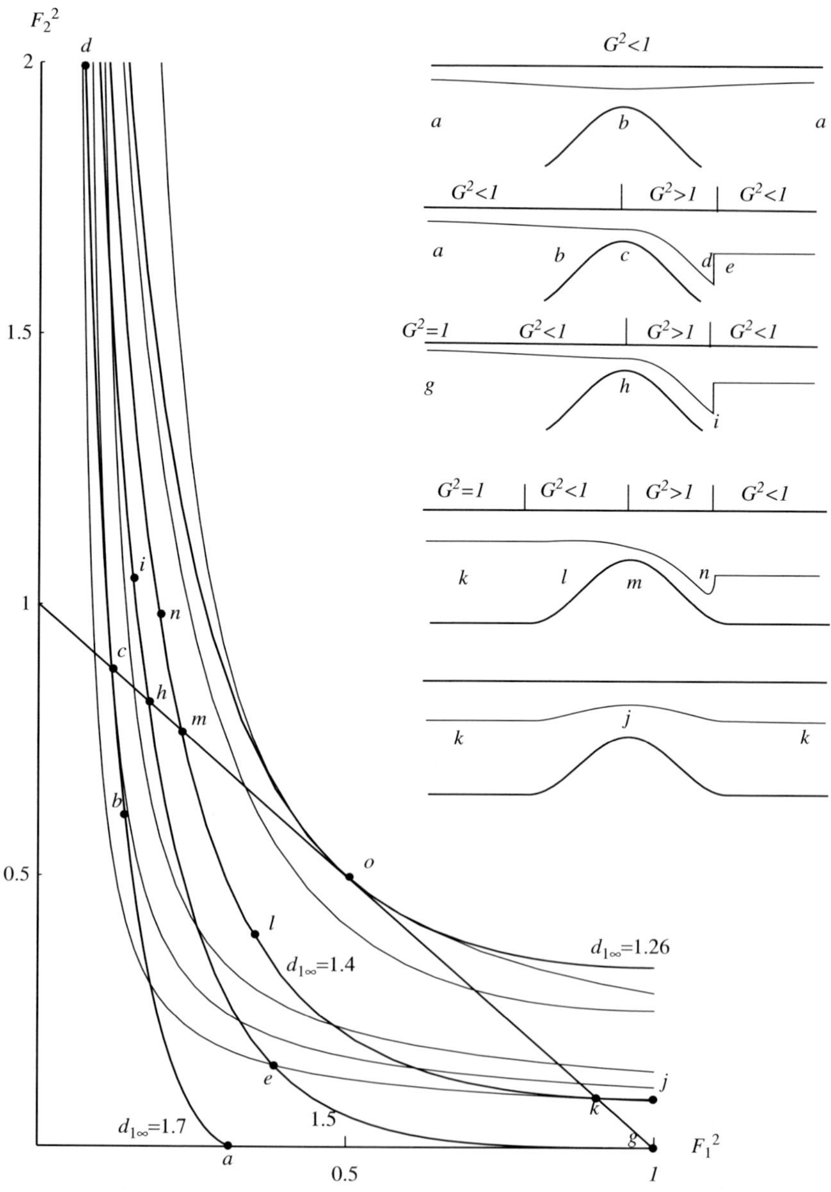

FIGURE 5.3.1b. A portion of the Froude number plane in (a) with examples of various solutions sketched in the insets.

If the sill height is increased to the point where the sill is encountered at the crossing with the critical diagonal, then a transition to supercritical flow is possible. Note that the thick and thin contours make grazing contact with each other along the diagonal, a property that has two important implications. First, the value of the topographic parameter q_2 at that point is the maximum that occurs along the thick curve. That is, the topographic elevation is the highest that can be achieved along that solution curve. The second implication is that the solution may be followed beyond the sill either by continuing upward along the '1.7' curve into the supercritical region or by retracing downward into the subcritical region. This same dilemma arises in the treatment of single-layer flows and it can be shown by similar arguments (see Exercise 3 of the previous section) that the correct option is to continue into supercritical space. The circuit is something like *abcd* in Figure 5.3.1b and the interface profile resembles the free surface profile for a hydraulically controlled, single-layer flow (second inset). It is natural to ask what would happen if the sill height is increased even further and we will return to this point shortly.

b. *Internal Hydraulic Jumps*

Given the similarity with the single-layer case, one might expect a hydraulic jump to arise in the supercritical part of the flow. The problem of shock joining in two layers is more difficult than for the single-layer case due to several factors. First, a transfer of horizontal momentum between the two layers can occur as the result of pressure forces on the steeply sloping interface within a jump. These forces exist in the region where nonhydrostatic effects are expected to be greatest, making a calculation of the pressure force problematic. The difficulty is avoided in single layers due to the fact that the pressure is essentially zero at the free surface. Second, entrainment of one layer into the other or creation of masses of intermediate density can occur as the result of mixing. These transformations complicate the mass, and perhaps the momentum balances. In some cases interfacial instability and mixing occur broadly and cause the transition from supercritical-to-subcritical flow to occur without any roller or other abrupt feature. An example of this limiting case is shown in the top frame of Figure 1.6.5.

One situation that allows simplification occurs when the two fluids are immiscible, so that Q_1^* and Q_2^* are conserved across the jump. If the jump occurs over a small interval in y^*, so that h^* is the same on either side, then the conjugate states must lie along the same constant$-q_2$ curve. As an example, suppose that a hydraulic jump occurs at point d in Figure 5.3.1b. The jump must return the supercritical flow to a subcritical state and must do so along the thin curve passing through d. It must therefore connect with another constant energy curve, perhaps at point e. Determination of the correct energy curve is quite difficult, however. The jump should cause an overall loss of total energy and it is not obvious what this means for ΔB, the difference between the upper and lower layer Bernoulli functions.[1] There have been a number of attempts to come to grips with these problems and the reader is referred to Jiang and Smith (2001a, b), Holland *et al.* (2002), and references contained therein for more information.

c.   Maximal and Submaximal Exchange

We now rejoin the discussion of the hydraulically controlled solution *abcd* (or a jump-containing variant like *abcde*) and ask what happens if the maximum value of q_2 is increased. This increase could occur as a result of raising the sill or of increasing the value of $Q_2{}^*$, both with $d_{1\infty}{}^*$ fixed. Since the new q_2 is higher than the maximum value possible along the $d_{1\infty} = 1.7$ curve, a time-dependent adjustment of the values of $d_{1\infty}$ and/or Q_r must occur. The adjustment involves the generation of a disturbance that propagates upstream and alters the conditions in the deep basin. If the adjustment leaves Q_r unchanged,[2] the new sill flow is found by following the critical diagonal from point c in Figure 5.3.1b down and to the right until the thin curve with the new value of q_2 is encountered. The solution now lies along the (thick) solution curve that intersects this point, and it can be seen that the corresponding $d_{1\infty}$ is lower than before. In cases where Q_r is altered, one would have to predict the new value and then consult the Froude number plane diagram appropriate for that value. Determination of Q_r generally requires analysis of the upstream disturbance. For purposes of illustration, we will proceed on the assumption that Q_r remains fixed.

The new solution curve intersects the lower axis at larger values of $F_1{}^2$ than before and thus the composite Froude number G^2 of the upstream flow is greater. The upper layer in the basin now has a higher velocity and smaller thickness. As the sill height is increased, one moves to solutions with lower values of $d_{1\infty}$ and with larger values of F_1 in the basin. Eventually the value $d_{1\infty} = 1.5$ is reached and it can be seen that the corresponding energy curve has an intersection with the $F_1{}^2$-axis at $F_1{}^2 = 1$. The flow in the basin is now critical. Since the basin is infinitely deep, the lower layer remains at rest and the upper layer moves at speed $(g'd_1{}^*)^{1/2}$. The value of G^2 is unity both in the basin (point g in Figure 5.3.1b) and at the sill (point h). The intervening flow is subcritical and the flow downstream of the sill is supercritical, perhaps with a hydraulic jump. The situation is represented by the solution *ghi* sketched in the inset.

To this point we have not distinguished between cases of unidirectional flow ($Q_r = 1$) and exchange flow ($Q_r = -1$). The solutions discussed can take either form. However, when the flow in the upstream basin becomes critical, important

---

[1] The simplest approach [suggested by Armi (1986)] is to assume that the energy loss in the jump is negligible, so that the conjugate states lie on the same energy curve.

[2] Cases where Q_r remains fixed can occur in systems with a closed upstream basin with a specified source of volume, often zero. The adjustment to a change in sill height then involves the original upstream disturbance, plus a set of reflected and re-reflected disturbances that pass information about the closed nature of the basin back into the strait.

differences arise in how the two situations should be interpreted. We will concentrate on the case of exchange and revisit unidirectional flow in Part f of this section. The upper layer now moves from the strait into the 'upstream' basin, right to left in the figures, and flows at the critical speed in the basin. A uniform critical flow of this type is typically vulnerable to frictional and dispersive effects and would be difficult to establish in the laboratory. However, a more robust version of the solution can be set up with a slight modification in the geometry of the deep basin. Suppose that the basin is made somewhat wider than the straight section of channel over which the topographic changes occur. Then the upper layer, which moves toward the left (or $-y^*$) direction, passes the sill and enters a subcritical stretch of flow over which the bottom drops away and the lower layer deepens. The upper layer accelerates to the critical speed and then exits into the 'upstream' basin, which is now wider. Since this layer has become effectively disengaged with the (motionless) lower, it behaves like a single layer and follows the behavior outlined in Chapter 1. In particular, the upper layer becomes supercritical after it exits into the wider basin. The supercritical flow generally terminates in a hydraulic jump. The critical section (point g) in the figure is now known as an *exit control*. Propagation of information from the basin into the channel is blocked by this control and by the region of supercritical flow.

The solution with both a sill control and an exit control has been obtained by allowing the value of $d_{1\infty} = \frac{d_{1\infty}^*}{(g'Q_1^*/w^*)^{2/3}}$ to decrease until the upper layer in the basin becomes critical. It can be shown (Exercise 6) that the corresponding $d_{1\infty}^*$ is approximately half the fluid depth over the sill. Since $d_{1\infty}^*$ is a measure of the internal energy of the flow, the decrease in $d_{1\infty}$ can be accomplished by holding the energy constant and increasing $|Q_1^*|$. The threshold state $d_{1\infty} = 1.5$ may therefore be regarded as having the maximum possible upper layer transport for the given available internal energy. As Figure 5.3.1a shows, this value cannot be exceeded by any solution that connects smoothly to a deep upstream basin. There are solutions with larger $|Q_1^*|$ (i.e., the ones with $d_{1\infty} < 1.5$) but none intersect the lower axis.

For flows with only a sill control ($d_{1\infty} > 1.5$) the upper layer remains relatively inactive and the behavior of the lower layer is similar to that of a single layer. For example, it can be shown that the layer Froude numbers at the sill fall in the ranges $0.8 < F_2^2 < 1$ and $F_1^2 < 0.2$. Thus the lower-layer Froude number is close to the critical value ($= 1$) for a single layer whereas the upper-layer Froude number is well into the subcritical range of a single layer. The wave arrested at the sill is dynamically similar to a wave propagating in an environment in which the upper layer is inactive. In contrast, the solution for $d_{1\infty} = 1.5$ involves the engagement of both layers. The exit control takes place where the lower layer is inactive and the sill control takes place where the upper layer is relatively inactive.

For exchange flows it is common to refer to the solution with both a sill control and an approach control as being *maximal*. It has the largest $|Q_1^*|$, and therefore the largest exchange transport $|Q_1^*-Q_2^*|$, of all the solutions that can be smoothly connected to a deep basin. The maximization assumes that Q_r

remains fixed. Maximal flow is distinguished by the property that information is allowed to enter the strait from neither the upstream nor the downstream basin. Exchange solutions with just sill controls ($d_{1\infty} > 1.5$) are called *submaximal.* Such flows block downstream information from entering the upstream basin, but not vice versa.

d. Basins with Finite Depth

For an upstream basin of finite depth there exists a similar family of sill-controlled solutions, each having a single control at the sill, and a limiting solution with two controls. The previously considered solution curves with constant $d_{1\infty}$ are still in play, but the possible upstream states now lie at finite F_2 and not along the abscissa of the Froude number plane. Suppose that $d_{1\infty} = 1.7$, so that the solution lies along the previously considered thick curve in Figure 5.3.1b. Then a solution with a sill control corresponds to something like *bcd*. What makes cases like this more difficult is the exercise of fixing the parameter $d_{1\infty}$ and identifying the upstream state b on the Froude number plane. Even if the transport ratio Q_r, the channel width w^*, the basin depth $z_T{}^*$ and the sill depth $D_s = z_T{}^* - h_m{}^*$ are known, and g' and $d_{1\infty}{}^*$ are measured, an algebraic process is still required to locate the solution on the Froude number plane. This problem is explored further in the exercises.

Notwithstanding this technical issue, one may proceed by decreasing the value of $d_{1\infty}$ as before and browsing through the continuum of solutions with sill controls. A limiting solution with two controls will eventually be obtained, this time with $d_{1\infty} < 1.5$. The limiting process can be implemented by fixing the topography and the value of $d_{1\infty}{}^*$ and increasing the layer fluxes. An example of the limiting case is shown by the curve segment *klmn* in Figure 5.3.1b. The upstream flow in the uniform, finite-depth section of channel (k in the figure) is critical. Once the bottom begins to shoal, the flow becomes subcritical (l). It then passes through a sill control (m) and becomes supercritical (n). A profile of the solution is sketched in the inset. For exchange flow, the flux is again maximal over all solutions with the same topography and same Q_r.

The limiting solution curve that determines the maximal solution for a given finite upstream depth is not easy to locate. However the curve and its $d_{1\infty}$ value can be calculated and shown to depend on the ratio of the sill depth D_s to the upstream depth $z_T{}^*$. By applying the definition of q_2 at the upper left intersection of the energy curve with the critical diagonal (i.e. at the sill control) it follows that

$$|Q_2{}^*| = q_2(D_s/z_T{}^*)g'^{1/2}w^*D_s^{3/2}. \tag{5.3.3}$$

The function $q_2(D_s/z_T{}^*)$ is simply q_2 at the upper left intersection point and the calculation of its dependence on $D_s/z_T{}^*$ is described in Exercise 5. For the case of an infinitely deep upstream basin ($D_s/z_T{}^* \to 0$), q_2 is given by 0.208, whereas $q_2 = 0.25$ for the point labeled o. Thus the range of variation is quite narrow. As $D_s/z_T{}^*$ increases so does the associated q_2 and thus the maximal flux for fixed D_s and g' increases as the upstream depth decreases.

Although (5.3.3) bears similarity to the single-layer weir formula (1.4.12), it is more constrained. It is no longer necessary to have knowledge of an upstream interface elevation or the like; the only state variable that needs to be measured is the reduced gravity g'. The insensitivity of the flux to upstream conditions is consistent with the existence of critical or supercritical upstream flow, which blocks mechanical information generated in the upstream basin from reaching the sill. The relevance of g' is consistent with the fact that density is advected by the flow and information about the density difference $\Delta\rho$ can pass right through the control section. The value of g' has been regarded fixed throughout this discussion, but one would wish to eventually relax this constraint by allowing $\Delta\rho$ to vary, say, in response to changes in forcing and/or mixing in the upstream basin. This topic will be pursued in Sections 5.5 and 5.6.

If the sill elevation $h_m{}^*$ is decreased to zero, so that $D_s/z_T{}^* = 1$, the upstream and sill controls merge. The coalescence point *lies* at o (5.3.1b) where the critical diagonal makes grazing contact with the curve $d_{1\infty} = 1.26$. It can be shown (see Exercise 2) that both c_+ and c_- are zero in this solution, which will emerge as an important type of flow through a contraction. It is left as an exercise to show that corresponding lower layer transport is given by

$$|Q_2{}^*| = (1+|Q_r|^{1/2})^{-2} g'^{1/2} w^* D_s{}^{3/2}, \tag{5.3.4}$$

where D_s is now just the depth in the uniform channel. Larger values of $|Q_2|$ correspond to (supercritical) solution curves lying entirely above the critical diagonal. These solutions do not connect directly to any geophysically relevant reservoir state, nor is it possible to connect the solutions to subcritical flow by hydraulic jumps along curves of constant q_2. Therefore (5.3.4) apparently gives an upper bound on $|Q_2|$ for relevant flow (i.e. flows that become subcritical somewhere upstream).

e. *Other Constraints*

In most cases of geophysical or engineering interest, geometrical variables like w^*, $h_m{}^*$, and $z_T{}^*$ are known in advance and $d_{1\infty}{}^*$ can be estimated from hydrographic data. In addition, a relation between $Q_1{}^*$ and $Q_2{}^*$ can often be stipulated, such as when the strait connects with a closed basin with known evaporation E and precipitation P. (The flow rates are then constrained by $Q_1{}^* - Q_2{}^* = \iint_{A_s}(E-P)dA$, where A_s is the surface area of the basin.) These constraints are still insufficient to determine the parameters $d_{1\infty}$, q_2, and Q_r required to fix the solution and the individual values of $Q_1{}^*$ and $Q_2{}^*$. To do so, one must assume that the solution is critical at the sill, and perhaps in the approach, and use these conditions to close the problem.

As an example, consider the case where the upstream basin is infinitely deep and it is suspected that an approach (or exit) control *and* a sill control occur $(d_{1\infty} = 1.5)$. For exchange flow, this would mean that the exchange transport is maximal. Further assume that the downstream basin is closed and has $\iint_{A_s} (E - P)dA = 0$, so that $Q_r = -1$. We have already shown that $Q_2^* = -Q_1^* = .208g'^{1/2}w^*D_s^{3/2}$ under these conditions. If $|Q_r| \neq 1$, then a more general version of the last relation can be used (see Exercise 3). If only a sill control exists, the flux is less constrained and it becomes necessary to measure the upstream interface level in order to close the problem.

f. Experiments on Two-Layer Sill Flows

Laboratory and numerical experiments have proven valuable in determining whether the flows discussed above are realizable and in demonstrating how they can be established. By no means have all of the theoretically possible steady behavior been investigated. We discuss two revealing experiments, the first dealing with unidirectional flows and the second with exchange flows. A review of the work on unidirectional flows will help illustrate some of the differences with the exchange flows discussed above. The experiments were originally performed by Long (1954, 1970) who towed an obstacle through a two-fluid system initially in a state of rest. Extensions have been carried out by Houghton and Isaacson (1970), Baines (1984, 1987), and others. One setting for numerical computation of the flow has two layers initially moving from left to right with equal velocity and in a uniform channel $(h^* = 0)$. Consider the case where this initial flow is subcritical and where the lower layer is much thinner than the upper layer. At $t^* = 0$ an obstacle of height h_m^* is placed in the path of the flow. Since the upper layer is relatively deep, the adjustment for moderate h_m^*/z_T^* is similar to that for a single-layer flow. If $h_m^*/z_T^* \ll 1$, the flow remains subcritical and there is no upstream influence. As h_m^*/z_T^* is increased, a critical value will be reached above which upstream influence occurs. The critical value is that required to establish critical flow over the sill for the upstream conditions given by the initial flow. The steady solution that develops over the obstacle will resemble solution bcd (Figure 5.3.1b) qualitatively. A slight increase in h_m^*/z_T^* past the critical value will result in the excitation of a bore that permanently alters the upstream flow by deepening the lower layer and decreasing the lower layer transport. Further incremental increases in h_m^*/z_T^* will have a similar effect. As long as the upper layer remains relatively inactive during this process, the linear wave speed $(c_-^* \cong v_2^* - (g'd_2^*)^{1/2} < 0)$ of the upstream flow increases in magnitude. As the obstacle height increases, it is possible for the lower layer to become completely blocked as a result of this process and further increases in h_m^*/z_T^* will cause the obstacle to protrude into the upper layer. In this case, additional upstream changes are prevented.[3] Up to this point the evolution is similar to that found in the single-layer version of Long's experiment (Section 1.6).

---

[3] Additional increases in the obstacle height will only impede the upper layer flow if frictional or nonhydrostatic effects come into play.

If the lower layer remains unblocked, increases in $h_m{}^*/z_T{}^*$ eventually lead to effects that are special to two-layer systems. To understand these changes, one should recall that growth of the obstacle does not alter the total volume transport $Q_1{}^* + Q_2{}^*$. Thus, the decrease in $Q_2{}^*$ is compensated by an increase in $Q_1{}^*$. In addition, the upstream thickening of the lower layer results in a thinning of the upper layer. Both effects tend to bring the initially inactive upper layer into play upstream of the obstacle and a consequence is that the growth in $-c_-{}^*$ is reversed. The bore achieves its maximum amplitude where $-c_-{}^*$ is maximum; the upstream disturbance beyond this threshold consists of a bore followed by a rarefaction. At some $h_m{}^*/z_T{}^*$ this trend causes $c_-{}^*$ to be reduced to zero: the upstream flow becomes critical. The flow over the obstacle now resembles the solution *klmn* of Figure 5.3.1b, qualitatively, with an upstream control and a sill control. The upstream critical section is called an *approach control*.

Although the shape of the interface and the distribution of layer Froude numbers in configuration *klmn* are the same as for the previously considered maximal exchange solution, there are some important differences. For one thing, the fact that the total volume transport remains fixed at its initial value makes it less meaningful to talk about maximal flux. (Maximal *exchange* on the other hand can be defined even when $Q_1{}^* + Q_2{}^*$ is constrained to be zero.) Another difference can be seen by imagining, as we did earlier, that the channel broadens at some upstream location. The flow therein becomes supercritical, as before, but now the direction of wave speed propagation is *towards* the sill. If one follows this supercritical flow as it leaves the broad basin and enters the narrower portion of channel, it becomes critical and then subcritical. This arrangement focuses wave energy towards the approach control section and therefore gives rise to a shock-forming instability.

Once the solution *klmn* is established, a slight increase $h_m{}^*/z_T{}^*$ leads to interesting changes in the flow that may not be completely described by hydrostatic theory. In order to understand these changes, it is helpful to remember that the upstream propagation speed has already been reduced to zero by the rarefaction triggered by previous adjustment. A new rarefaction caused by a further increase in $h_m{}^*/z_T{}^*$ would therefore be unable to propagate upstream. Numerical simulations with hydrostatic models have shown, in fact, that such an increase causes the flow over the obstacle to revert to a supercritical, symmetrical state, while the approach control is maintained. The flow near the obstacle now resembles solution *kjk*, with an approach control but no sill control. Beyond this point, increases in $h_m{}^*/z_T{}^*$ lead to no further upstream influence. Laboratory experiments give a somewhat different picture for the flow downstream of the sill. Here a nonhydrostatic, and possibly dissipative, feature known as a *supercritical leap* may form (Lawrence, 1993; Zhu and Lawrence, 1998). The 'leap' is a smooth transition from one supercritical state with a deep lower layer to another supercritical state with a shallower lower layer. This transition occurs on the downstream face of the obstacle and can be followed by a hydraulic jump.

The initial value problem has also been investigated for cases in which the initial lower layer depth is not small. The sequence of events that takes place may be different from what is described above and the reader is referred to Baines (1995, Chapter 3) for a thorough discussion. A fundamental point to keep in mind is that the formation of the upstream control, the central departure from single-layer hydraulics, occurs because $-c_-^*$ has a maximum value in the upstream flow at an intermediate interface level.

Turning now to the case of exchange flow, we review an experiment (Zhu and Lawrence, 2000) that shows how maximal and submaximal states can be established. As shown in Figures 5.3.2(a, b), the channel contains an isolated obstacle and opens abruptly at either end into wide reservoirs. The right and left reservoirs are initially filled to the top with fluids of slightly different densities, the left reservoir containing the denser fluid. A barrier that sits atop a sill separates the two fluids. The barrier is removed and the two fluids are

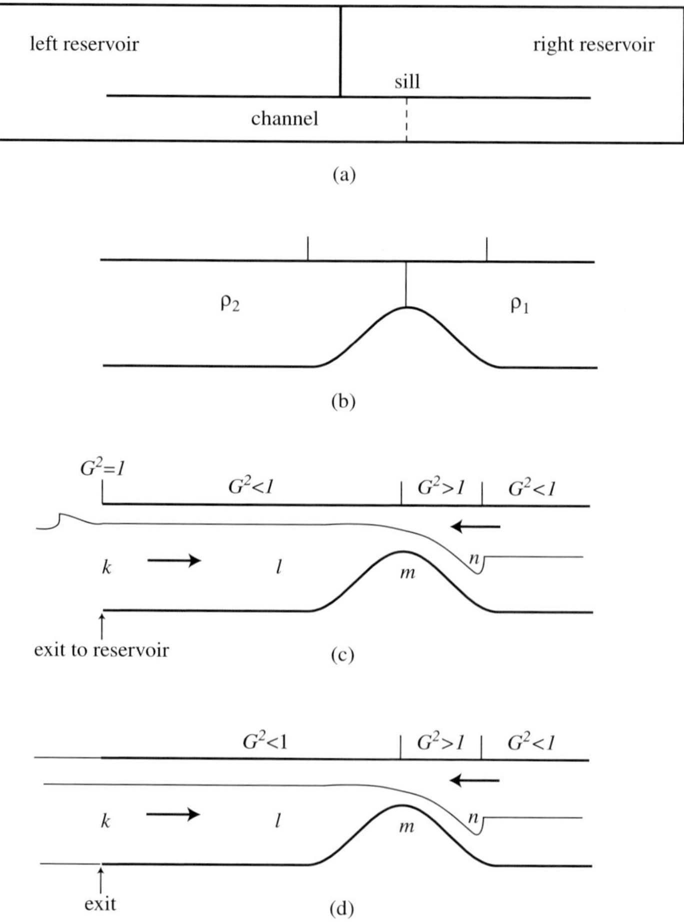

FIGURE 5.3.2. The experimental setup used by Zhu and Lawrence (2000) to simulate a lock exchange.

allowed to displace each other. After an initial period of transient activity, the flow within the channel settles into a nearly steady state. The layer velocities in the left reservoir are relatively weak and the upper-layer depth therefore approximates $d_{1\infty}{}^*$. Initially, this depth is relatively small (Figure 5.3.2c), but it gradually increases as lower-layer fluid is drained out of the reservoir. An exit control occurs near the left end of the channel (point k) and the flow immediately to the right is subcritical. To the left there is a brief span of supercritical flow. Ideally, this flow would be joined to the reservoir flow by a hydraulic jump. In the experiment, the supercritical flow actually enters the reservoir as a concentrated jet that gradually disperses. At the sill the subcritical flow passes through a second control and becomes supercritical. A hydraulic jump occurs on the right slope of the obstacle and the flow thereafter is subcritical. From the left end of the channel to the hydraulic jump the interface resembles the solution $klmn$ of Figure 5.3.1b. The transition from the left end of the channel into the left reservoir cannot be traced in this figure but is discussed below. While in this configuration, $|Q_1{}^* - Q_2{}^*|$ remains fixed at its maximal value, the determination of which is described in Exercise 4.

As the left reservoir loses lower-layer fluid, the interface there falls and approaches the level at the entrance (point k) of the channel. At the same time, conditions in the channel between the exit control and the sill control remain steady; the supercritical end states insulate that part of the flow from the two reservoirs. However, as the reservoir interface drops the region of supercritical flow diminishes and eventually disappears. The exit control becomes 'flooded', the flow there becomes subcritical, and the exchange becomes submaximal and dependent on the upstream interface elevation. This elevation continues to decrease and with it the exchange flux.

Exercises

(1) For arbitrary Q_r, which constant energy curve makes grazing contact with the critical diagonal in Figure 5.3.1a?

(2) For the solution designated by the point o in Figure 5.3.1b, prove that under conditions of pure exchange flow, $c_+{}^* = c_-{}^* = 0$.

(3) Consider the case of flow over an obstacle with the lower layer originating from an infinitely deep basin. If the flow has an exit control and a sill control, show that $d_{1\infty} = 1.5$ regardless of the value of Q_r. Further show that the transport in the lower layer is given the generalized weir formula:

$$Q_2{}^* = [Q_r{}^{2/3}F_{1c}^{-2/3} + (1 - F_{1c}^2)^{-1/3}]^{-3/2} g'^{1/2} w^* D_s{}^{3/2}$$

where F_{1c} is determined from

$$F_{1c}^{4/3} - Q_r^{-2/3}(1 - F_{1c}^2)^{2/3} + 2F_{1c}^{-2/3} = 3.$$

(4) In the experiment of Zhu and Lawrence (2000), described in part f, a maximal exchange flow was observed. The values of w^*, g', $z_T{}^*$, and $h_m{}^*$ are set by the geometry and by the initial conditions, and it is also known, due to the closed geometry of the channel and reservoir system, that $Q_r = -1$. To predict the maximal value of $Q_2{}^*$:

(a) Show that

$$F_{1e}^{4/3} - (1 - F_{1e}^2)^{2/3} + 2F_{1e}^{-2/3} = F_{1s}^{4/3} - (1 - F_{1s}^2)^{2/3} + 2F_{1s}^{-2/3},$$

where the subscripts e and s correspond to exit and sill. (Hint: Use energy conservation between the exit and sill along with the critical condition at both locations.)

(b) Further, show using volume flow rate continuity that

$$Q_2 = g'^{1/2} w^* (z_T{}^* - h_m{}^*)^{3/2} [F_{1s}^{-2/3} + (1 - F_{1s}^2)^{-1/3}]^{-3/2}$$

$$= g'^{1/2} w^* (z_T{}^*)^{3/2} [F_{1e}^{-2/3} + (1 - F_{1e}^2)^{-1/3}]^{-3/2}.$$

This gives three equations for the unknowns F_{1e}^2, F_{1s}^2, and Q_2 in terms of the known w^*, $z_T{}^*$, etc.

(5) *Calculation of the coefficient $q_2(D_s/z_T{}^*)$ in equation 5.3.3.* Consider a solution for flow in a channel with constant width and with $|Q_r| = 1$. The flow has two control points corresponding, say, to points k and m in Figure 5.3.1b. Show that the values of the lower layer Froude numbers at k and m can be computed from the relations:

$$\frac{Q_r{}^{2/3} F_{1m}^{-2/3} + (1 - F_{1m}^2)^{-1/3}}{Q_r{}^{2/3} F_{1k}^{-2/3} + (1 - F_{1k}^2)^{-1/3}} = \frac{D_s}{z_T{}^*}$$

and

$$\tfrac{1}{2} F_{1m}^{4/3} - \tfrac{1}{2} Q_r{}^{-2/3} (1 - F_{1m}^2)^{2/3} + F_{1m}^{-2/3} = \tfrac{1}{2} F_{1k}^{4/3} - \tfrac{1}{2} Q_r{}^{-2/3} (1 - F_{1k}^2)^{2/3} + F_{1k}^{-2/3}.$$

Here $z_T{}^*$ is the total depth upstream of the obstacle (where $h^* = 0$). Once F_{1m} has been calculated from these relations, F_{2m} follows from the critical condition $G^2 = 1$. Then q_2 follows from (5.3.2).

(6) For the maximal solution with a deep upstream basin and with $|Q_r| = 1$, show that $d_{1\infty}{}^*$ is 0.53 times the depth over the sill. That is, the interface in the hypothetical wide upstream basin lies about half the sill depth.

(7) Prove the result 5.3.4.

5.4. Flow through a Pure Contraction

If the bottom remains horizontal ($h^* = $ constant), and the flow is choked only by contractions in the width of the rectangular channel, a new type of control condition can come into play. Solutions can still be represented in the Armi (1986) Froude number plane and Figure 5.4.1a shows an example with $|Q_r| = 1$. The thinner contours continue to represent constant $|Q_2^*|/[(z_T^* - h^*)^{3/2} g'^{1/2} w^*]$, except that w^* rather than h^* is considered as varying from one contour to the next. Decreasing values of w^* generally lead one away from the origin. The form (5.3.1) of the energy equation is no longer convenient for constructing solution curves since w^* appears as a scale factor. A more helpful form

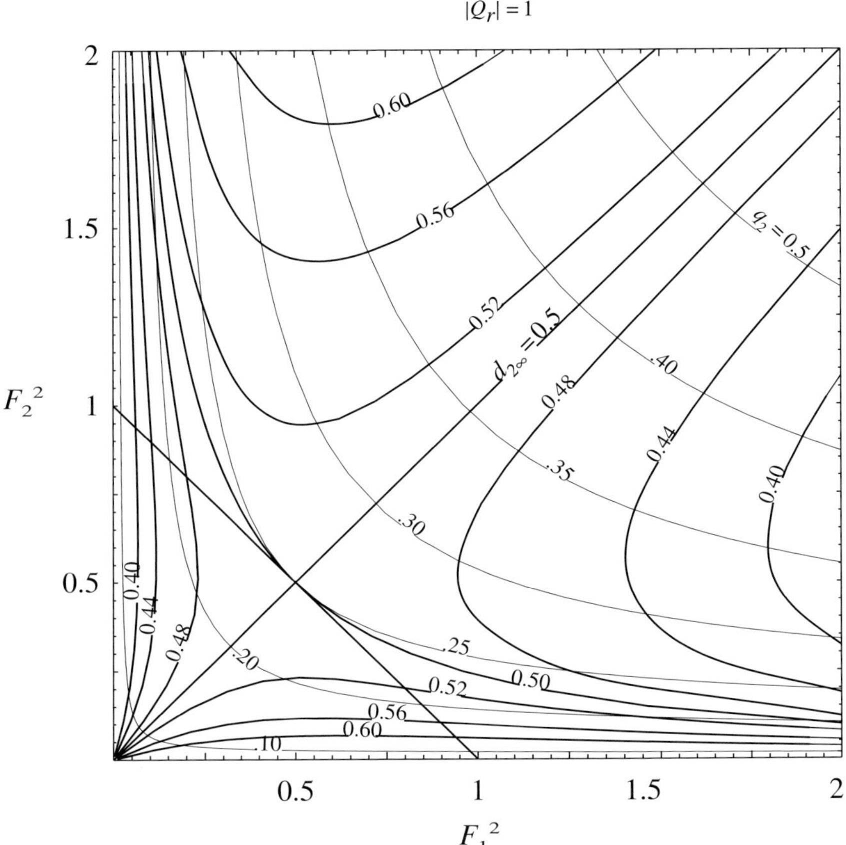

FIGURE 5.4.1a. The Froude number plane for flow through a pure contraction with $|Q_r| = 1$. Solutions must lie along the thick curves, which have constant $d_{2\infty}$. The thin curves are of constant q_2 and are the same as in Figure 5.3.1, but now the larger values of this parameter are associated with narrower widths. (From Armi, 1986).

$$\frac{\frac{1}{2}F_2^{4/3} - \frac{1}{2}Q_r^{2/3}F_1^{4/3} + F_2^{-2/3}}{Q_r^{2/3}F_1^{-2/3} + F_2^{-2/3}} = \frac{\Delta B^*}{g'z_T{}^*} = d_{2\infty} \qquad (5.4.1)$$

is obtained by setting $h^* = 0$ in the definition of q_2 and using (5.3.2) to eliminate w^* from (5.3.1). The internal energy is now represented by $d_{2\infty}$ which, in view of (5.2.10), is the dimensionless interface elevation in the hypothetical wide, quiescent basin. The thick curves in Figure 5.4.1 are contours of constant $d_{2\infty}$. Exchange flows and unidirectional flows having the same values of $|Q_r|$ and $|Q_2{}^*|$ are again represented by the same diagram, though differences in stability and jump-forming capabilities render some combinations unrealizable. In contrast to the case of variable topography, both layers feel the geometric variations directly. There is now a symmetry between the behavior of the upper and lower layers. For the solution curves shown in the figure, all of which have $|Q_r| = 1$, this means that a solution with a particular $d_{2\infty}$ has a counter part in which the layers are reversed. That is, $F_1{}^2$ and $F_2{}^2$ are interchanged and $d_{2\infty}$ is replaced by $1 - d_{2\infty}$. More generally, for a given Q_r there is a comparable solution with flow rate ratio $1/Q_r$ in which the two layers are interchanged (see Exercise 2).

a.   Submaximal Flow from a Wide Basin

Figure 5.4.1a represents solutions for which the volume flow rates in the two layers have equal magnitude. There is a family of constant energy curves that emanate from the origin $(F_1{}^2 = F_2{}^2 = 0)$ and represent flows originating from an infinitely wide, quiescent basin. Let us first restrict attention to unidirectional flow. All of the curves beginning at the origin intersect the critical diagonal, indicating the presence of a critical section for sufficiently small minimum width $w_m{}^*$. For all but one of these curves, the contours of constant width make grazing contact with the solution curves along the critical diagonal. Critical flow for these solutions occurs at the narrowest section. Continuation past this section leads to supercritical flow, possibly with a hydraulic jump. If the upper layer thickness is greater than the lower layer thickness in the basin ($d_{2\infty} < 0.5$) then the lower layer is thinned and accelerated, and the upper layer is thickened and decelerated, through the contraction. An example is given by the curve *afm* of Figure 5.4.1b. The opposite is true when $d_{2\infty} > 0.5$ as indicated by curve *ain*. The behavior of the thinner layer in each case is similar to single layer flow through a contraction.

b.   Self-Similar Flow

Of the Figure 5.4.1a curves originating from the origin, there is one that does not cross the critical diagonal at a point of minimum width. This 'similarity' solution is given by the straight line $F_1{}^2 = F_2{}^2$ and corresponds to equal basin layer thicknesses ($d_{2\infty} = 0.5$). Since $|Q_r| = 1$ this solution is characterized by equal layer depth and speed at each section. If the flow is unidirectional then

$v_1{}^* = v_2{}^*$ and the fluid behaves as if it were homogeneous, entirely bereft of internal dynamics. Beginning at the origin, one can trace a solution from the upstream basin through the narrowest section. For relatively large values of the narrowest width (in particular, $q_2 < 0.25$) the flow will remain subcritical and will resemble something like the trace *ala* in Figure 5.4.1b. As $w_m{}^*$ is decreased (and the corresponding q_2 increased), the trace may just reach the critical diagonal and will therefore become critical at the narrowest section, remaining subcritical elsewhere. A further decrease in $w_m{}^*$ causes the solution to cross the critical diagonal and become supercritical. Where the diagonal is crossed (point *b*) the solution curve is normal to the curves of constant q_2. In other words, critical flow occurs not at the minimum width but at a point of diminishing width: $\partial w^*/\partial y^* < 0$. The existence of this *virtual control* is consistent with the regularity condition (5.2.17) and the fact that $v_1{}^{*^2}(y_c) - v_2{}^{*^2}(y_c{}^*) = 0$. Downstream of the virtual control the flow becomes supercritical and remains so as the narrowest section, say *c* in Figure 5.4.1b, is passed. In principle, the solution then retreats back

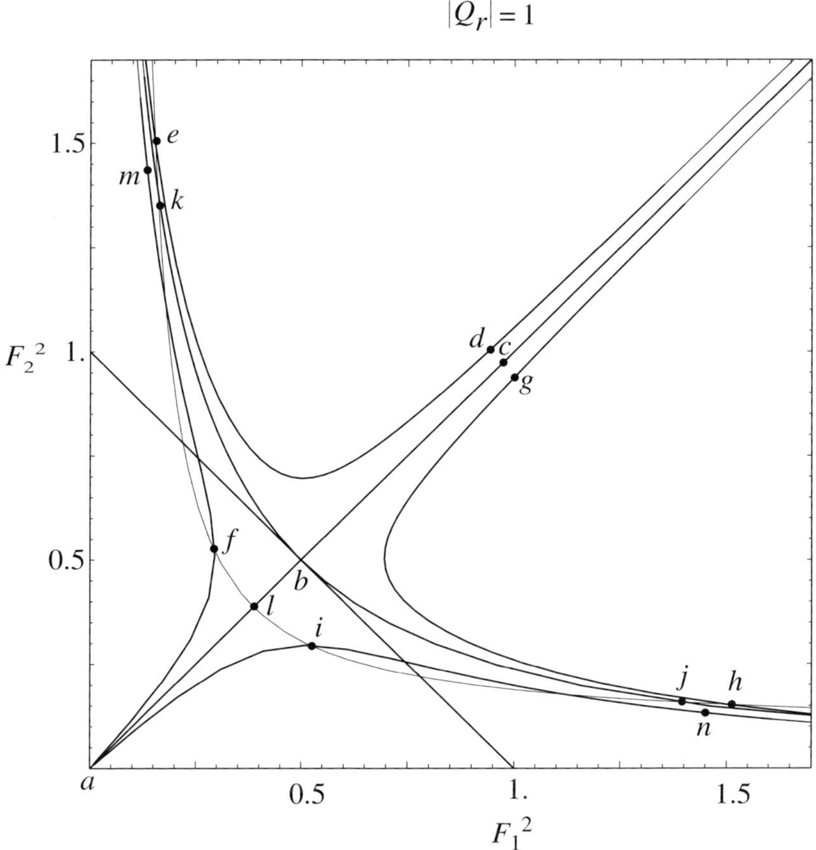

FIGURE 5.4.1b. Examples of solutions for the previous figure, as described in the text.

along the same line towards the origin, passing through another virtual control. The idealized path is something like *abcba* in Figure 5.4.1b. In reality, a slight amount of dissipation will cause the flow to move off of the supercritical portion of the similarity solution and onto one of the other supercritical solutions (with $d_{2\infty} \neq 0.5$). The downstream flow may eventually be returned to a subcritical state by a hydraulic jump. The circuit traced is therefore something like *abcdefa* or *abcghia*, the choice influenced by downstream conditions.

A similar barotropic solution exists for each value of Q_r (Exercise 4). Since the velocities in each layer are equal, Q_1^* and Q_2^* are each proportional to the net transport $Q^* (= Q_1^* + Q_2^*)$. We proved in Section 5.2 that the latter remains independent of time if it is so far upstream and thus Q_1^* and Q_2^* must also remain independent of time. Thus the similarity solution is not subject to blocking as the result of upstream influence. The minimum width may be made arbitrarily small without affecting the layer transports. In graphical terms, the act of making w_m^* small and q_2 large simply forces the solution flow at the narrowest section (point *c*) to extend farther from the origin. There is no nothing that forces Q_1^* or Q_2^* to change.

c. *Laboratory Examples of Unidirectional Flow*

Armi (1986) has produced examples of these solutions in a laboratory channel with a width contraction (Figure 5.4.2). The two layers are pumped from right to left at fixed values of Q_1^* and Q_2^* such that $Q_r = 1$. The channel narrows to a minimum width midway through and widens again at the left end. Here, there is nothing like a quiescent reservoir and the flow is varied by changing the net transport $Q^* (= Q_1^* + Q_2^*)$ and by altering the downstream conditions. For smaller values of Q^*, Armi finds hydraulically controlled flows that resemble the solutions with unequal layer depths ($d_{2\infty} < 0.5$ or $d_{2\infty} > 0.5$) as described above. Examples are given in Figures 5.4.2a, b and the corresponding solution traces are something like *afm* or *ain* in Figure 5.4.1b. In both cases the flow is subcritical until it reaches the narrowest section, where it undergoes a transition to supercritical flow. The particular solution arising from a specified $Q^* (= 2Q_2^*$ in the experiment) is predicted by calculating the value of $q_2 (= Q_2^* / (z_T^{*3/2} g'^{1/2} w_m^*))$ at the narrowest section and finding the intersection of the corresponding $q_2 =$ constant curve with the critical diagonal in Figure 5.4.1a. For $q_2 < .25$ there will be two such intersections and therefore one must choose between two solutions, one having $d_{2\infty} < 0.5$ and the other $d_{2\infty} > 0.5$. In the experiment, the choice is forced by downstream conditions that influence the initial evolution by which the steady flow is set up.

If Q^* is increased, the value of q_2 at the narrowest section increases, forcing the intersection with the critical diagonal to move closer to the midpoint $F_1^2 = F_2^2 = 1/2$. The value of $d_{2\infty}$ for the corresponding solutions approaches 0.5, meaning that the layer depths become equal. At the value $q_2 = 0.25$ the similarity solution is obtained and the layer depths become equal, at least in principle, at all points along the channel. A further increase in Q^* forces a self-similar flow with a virtual

FIGURE 5.4.2. Side views of unidirectional, two-layer flows through a contraction. Frames (a) and (b) show flows with a control section at the narrowest section, which lies approximately at the numeral '2'. Frame (c) shows a self-similar flow with a virtual control. At the upstream (right) entrance, the layer depths and velocities are equal and continue to be so as the channel converges and the narrowest section is passed. The virtual control occurs somewhere to the right of the narrowest section but is not distinguished by any visual property of the interface. A small amount of mixing is observed in the downstream end of the channel. (From photos appearing in Armi, 1986).

control. As discussed above, the predicted flow enters the channel in a subcritical state and passes through a virtual control, becoming supercritical, on its way to the narrowest section. It continues to flow at a supercritical speed through the narrows. A laboratory realization is shown in Figure 5.4.2c. The flow wanders a bit from its predicted self-similar state downstream of the narrows, probably due to frictional effects. In principle, Q^* can be increased without limit, not surprising when one considers that the flow is behaving as if the density were uniform.

A virtual control clearly operates in a different way, and has different implications for the upstream flow, than a standard narrows or sill control. For example, the flows shown in frames a and b of the figure experience upstream effects, manifested in the interface height, in response to changes in the minimum width w_m^*. The same in not true of the flow with the virtual control (frame c), which is supercritical through the narrows. As discussed above, changes in the value of w_m^* produces no upstream effect whatsoever. Instead, the virtual control acts to maintain the barotropic character of the flow. Its appearance coincides with the disappearance of shear between the two layers. Through the regularity condition, it requires that $v_1^* = v_2^*$, and thus establishes a shear-free state at the position of the control. Since the sidewall forcing acts equally on each layer, the shear-free state extends upstream and downstream from the position of the virtual control. In essence, the virtual control has driven all internal dynamics out of the flow, which now acts like a single, homogeneous layer.

d. Lock Exchange Flow

Under conditions of pure exchange ($Q_r = -1$) similar versions of most of the above solutions can be found. One that is not observed is the exchange version of the similarity solution, which now has $v_1^* = -v_2^*$ and is unstable upon entry into the supercritical region. However another solution comes into play: the one indicated by the energy curve $d_{2\infty} = 0.5$ that makes grazing contact with the critical diagonal in Figures 5.4.1a or b. This solution can be imagined to occur between two wide basins, one in which the top layer is very thin and the other in which the lower layer is very thin. This situation is difficult to realize when the flow is unidirectional (see Exercise 3), however it can readily be established for an exchange flow. The traditional method of doing so is to perform a 'lock exchange' experiment of the type suggested in Figure 5.3.2, but with a pure width contraction in place of an obstacle. Removal of a barrier placed at the contraction allows the fluids to move in opposite directions, displacing each other above and below, eventually resulting in a steady solution of the type just described. The flow is critical at the narrowest section, where *both* c_-^* and c_+^* vanish, and becomes supercritical on either side. Hydraulic jumps typically arise in these supercritical extensions, so that the complete solution circuit is something like $aijbkfa$ in Figure 5.4.1b. The direction of wave propagation in the supercritical regions is always away from the narrowest section and thus the flow there is insulated from small disturbances generated in the neighboring basins.

Asymmetric exchange solutions corresponding to solutions like *ain* or *afm* in Figure 5.4.1b can also be established in the laboratory. One approach (e.g. Lane-Serff *et al.*, 2000) to perform a 'partial' lock exchange experiment in which the barrier holding back the denser fluid extends from the bottom only part way up to the rigid lid. The dense fluid is filled only to the top of the barrier and the barrier itself is positioned at the narrowest section. After release, the steady exchange flow that is established has a lower layer that is thinner overall than the upper layer. The flow state corresponds to one of the solutions with $d_{2\infty} < 0.5$, of which *afm* is an example. Cases with $d_{2\infty} > 0.5$ may be established by positioning the initial barrier to extend from the rigid lid partially down to the bottom and filling the less dense fluid to this lower level. The complete range of exchange states is sketched in Figure 5.4.3.

The 'full' lock exchange solution achieves the maximum value of $q_2 (= 0.25)$ of any of the realizable exchange solutions. This solution therefore reaches the maximal flux

$$Q_2{}^* = .25 g'^{1/2} w_m{}^* D_s{}^{3/2}, \qquad (5.4.2)$$

for fixed minimum width $w_m{}^*$ over all possible internal energy levels. The formula follows from use of the definition of q_2, or from setting $|Q_r| = 1$ in (5.3.4). This solution is characterized by a double hydraulic control in the sense that both internal waves are frozen at the narrows. Stommel and Farmer (1952) identified this state and verified it experimentally.[4] Their analysis and their later (1953) application to estuary dynamics (Section 4.5) deserve special mention in the annals of hydraulics as one of the first applications of hydraulic theory to oceanographically relevant flows. Both layers are engaged: the upper layer being more so in one reservoir, the second in the other, and both being active at the narrowest section. The submaximal solutions ($d_{2\infty} \neq 0.5$) are characterized by having only one wave frozen at the narrowest section, by having a smaller $Q_2{}^*$ for the same $w_m{}^*$, D_s and g', and by being dominated by the dynamics of one of the layers.

e.   *Unequal Layer Fluxes*

Froude number diagrams for $|Q_r| \neq 1$ show similar features with a few twists. The case $|Q_r| = 0.5$ is shown in Figure 5.4.4a. Under conditions of exchange ($Q_r = -0.5$), the flow contains a barotropic component $Q_1{}^* + Q_2{}^*$, here equal to $Q_1{}^*/2$. A similarity solution with the virtual control exists and corresponds to the straight contour with $d_{2\infty} = 2/3$. For general $|Q_r|$, the corresponding value of $d_{2\infty}$ is given by $(|Q_r| + 1)^{-1}$ and the contour itself by $|Q_r| F_1{}^2 = F_2{}^2$ (see Exercise 4). However, the former 'lock exchange' solution, which occurs along the curved energy contour with $d_{2\infty} = 2/3$, now has two intersections with the critical diagonal. The lower right intersection corresponds to a virtual control and the upper left intersection to a

---

[4] However, it was not recognized as the maximal limit of a continuum of other controlled solutions until the work of Armi (1986) and Farmer and Armi (1986).

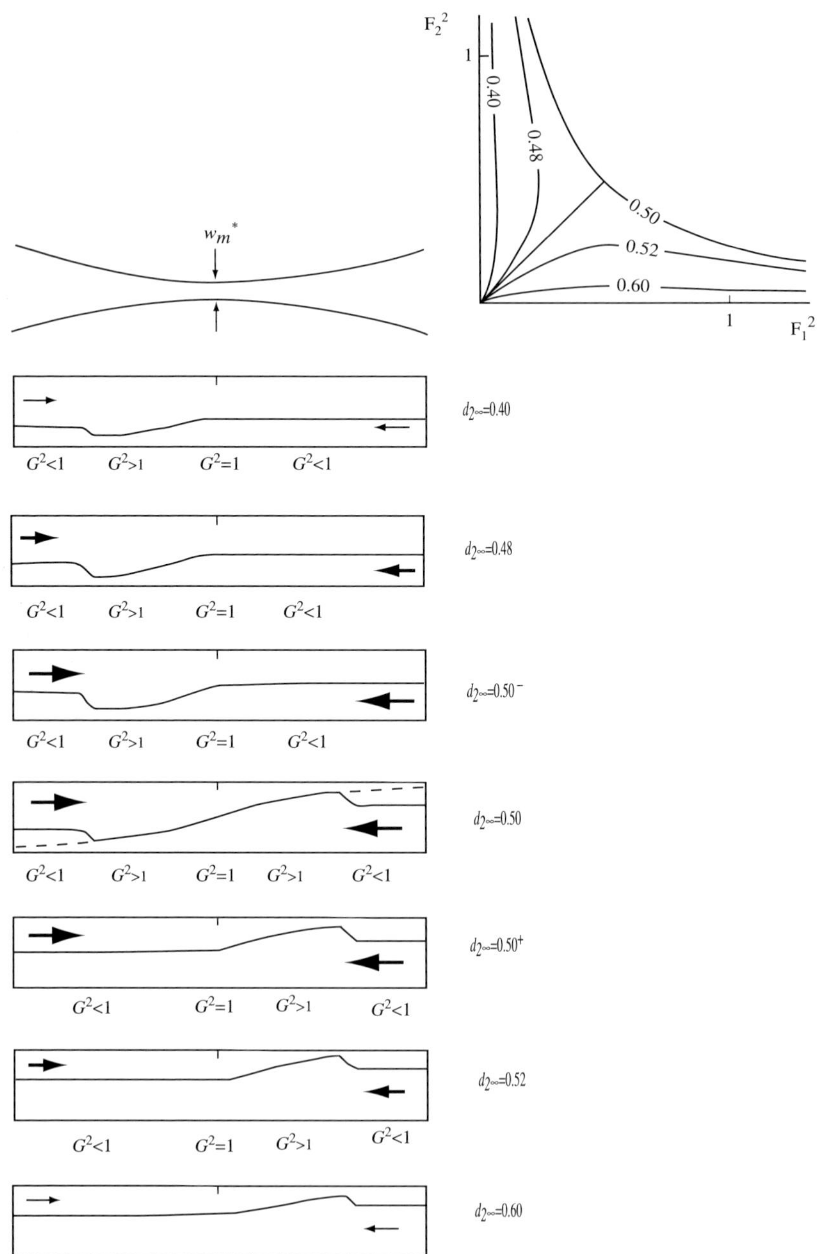

FIGURE 5.4.3. A sequence of steady solutions for two-layer exchange through a pure contraction, as described in the text. (Based a figure in Armi and Farmer, 1986).

$$|Q_r| = 0.5$$

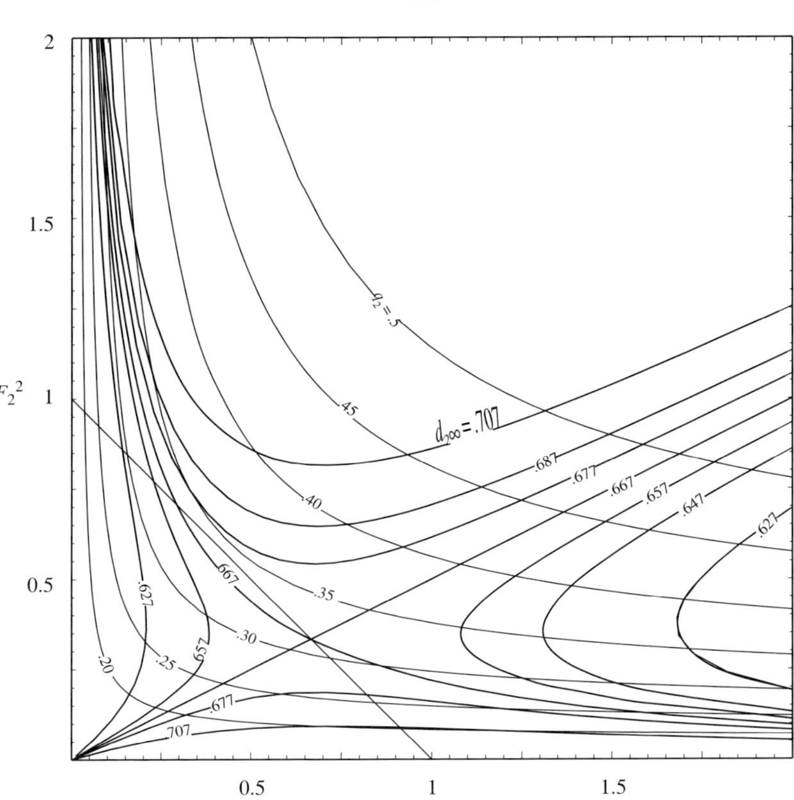

FIGURE 5.4.4a. Froude number plane for flow through a pure contraction with $|Q_r| = 0.5$. (Based on a figure in Armi and Farmer, 1986.)

narrows control. It can be shown that the virtual control lies on the side of the narrows from which the barotropic component of the flow originates. Also, there is a group of solutions with $d_{2\infty}$ slightly greater than 2/3 that intersect the critical diagonal twice and that go off into supercritical space at either end. Since both F_1^2 and F_2^2 go to infinity following the right-hand branch of these curves, the corresponding solutions cannot be smoothly connected to an infinitely wide reservoir.

The physical separation of the two control section in the presence of barotropic flow was first recognized by Wood (1970), who also coined the term 'virtual control'. In an exchange flow, the virtual control clearly operates in a different manner that its unidirectional counterpart. The compatibility condition only requires that the flow speeds in the two layers be equal. In the limiting case of zero barotropic flow studied by Stommel and Farmer (1952) the virtual control is hidden by the fact that the two controls occur together.

If the flow is unidirectional and originates from a wide reservoir, the range of possible behavior can be illustrated, as before, by imagining a series of experiments in which the value of $q_2 (= Q_2^*/(z_T^{*3/2} g'^{1/2} w_m^*))$ is gradually increased by increasing Q_2^*. We continue to assume that the flow is critical at the narrows. Beginning along a solution curve for which $d_{2\infty} > 2/3$, we move through a succession of flows with relatively deep lower layers. These solutions have upper layers that are relatively active and that are accelerated through the contraction. However, the transport in the lower layer is twice that in the upper layer and the dynamics of this layer are more easily brought into play. As Q_2^* is raised the similarity solution is realized when $d_{2\infty}$ reaches the value 2/3. Here the lower layer depth remains twice the upper layer depth along the entire solution curve. For further increases in Q_2^* the solution remains along the similarity curve and develops a virtual control upstream of the narrows. As before, the flow becomes supercritical through the narrowest section and, in the expanding section of channel, tends to wander off of the $d_{2\infty} = 2/3$ curve. Possible outcomes with weakly dissipative jumps are illustrated by the paths *abefgha* or *abeijha* in Figure 5.4.4b.

If we begin instead with a solution for which $d_{2\infty} < 2/3$, the approach to the similarity solution is just slightly different. We move through a series of solutions like *ahm* in which the lower layer is most active. As Q_2^* is raised, a solution traced by the curve *abcd* is approached from the left. The subcritical flow from the reservoir is nearly self-similar as it enters the contracting section and nearly passes through a virtual control there. The solution now lies just to the left of point *b* in Figure 5.4.4b. The solution then veers away from the similarity curve and passes through a narrows control (point *c*), after which it becomes supercritical. A further increase in Q_2^* gives rise to the similarity solution with a virtual control.

Under conditions of exchange, a similarly modified sequence of solutions exists. As Q_2 is increased from low values the limiting form is no longer the similarity solution (which is again unstable) but rather the full lock exchange solution. This solution has supercritical flow extending into the two reservoirs and subcritical flow between the two controls. A hydraulic jump or some other source of dissipation is required to join the supercritical flows to the quiescent reservoirs. An example is given by the circuit *alkbcdha* in Figure 5.4.4b. A procedure for obtaining a weir formula for this case is explored in Exercise 8.

Although it is beyond the scope of this text, generalizations of the concepts of virtual control and maximal exchange can be made for continuously stratified fluids. One place the interested reader might begin is Killworth (1992a) and the references contained therein.

Exercises

(1) By free hand, sketch the qualitative features of the solutions corresponding to the following circuits:

   a) *afm* (Figure 5.4.1b)
   b) *ain* (Figure 5.4.1b)

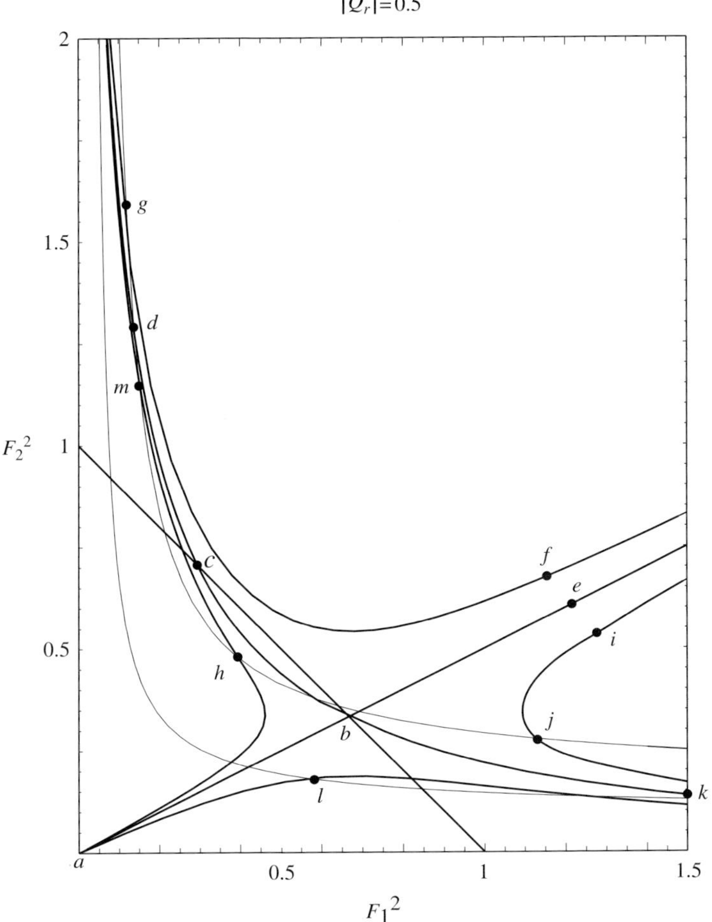

FIGURE 5.4.4b. Examples of solution traces based on the curves shown in (a).

c) *abcba* (Figure 5.4.1b)
d) *jbk* (Figure 5.4.1b)
e) *aijbkfa* (Figure 5.4.1b)
f) *kbcd* (Figure 5.4.4b)
g) *alkbcdha* (Figure 5.4.4b)

The sketches should be the style of the Figure 5.3.1b insets, with control sections and stretches of subcritical and supercritical flow labeled.

(2) For flow through a contraction with constant h^*, show that for each Q_r there is another solution with reciprocal flow rate ratio $(1/Q_r)$ in which the two layers are interchanged.

(3) Consider the following flows, each of which has at least one critical section. Remark on the stability of the hydraulic transition at the critical section(s) in

each case. (Refer to the shock-forming instability, not Kelvin-Helmholtz instability.)

(a) The solution *kjk* in Figure 5.4.1b.
(b) A solution of the type *abcd* in Figure 5.4.4b, with the lower layer entering the deep basin and the upper layer exiting the basin.
(c) The 'lock exchange' solution with $Q_r = 0.5$. In other words, the solution with both a virtual and narrows control lying along the $d_{1\infty} = .667$ curve in Figure 5.4.4b, but now with unidirectional flow.

(4) Prove that a barotropic ($v_1 = v_2$) solution exists for arbitrary Q_r. Show that the solution is represented in the Froude number plane by $|Q_r|F_1^2 = F_2^2$ and that the corresponding value of $d_{2\infty}$ is $(1 + |Q_r|)^{-1}$.

(5) Show that where the similarity solution of Exercise (4) intersects the critical diagonal, a second solution with the same $d_{2\infty}$ must exist.

(6) The two solutions implied in Exercise 5 must both satisfy (5.4.1) with $d_{1\infty} = (1 + |Q_r|)^{-1}$. With this setting show, in fact, that (5.4.1) can be written in the form

$$\left(F_2^{2/3} - |Q_r^{1/3}| F_1^{2/3}\right)\left[\tfrac{1}{2}F_1^{2/3}F_2^{2/3}(1 + |Q_r|)\left(F_2^{2/3} + |Q_r^{1/3}| F_1^{2/3}\right) - Q_r^{2/3}\right] = 0.$$
$$(5.4.3)$$

Show that satisfaction of this relation by setting the leading expression to zero gives the barotropic (similarity) solution. Setting the longer expression to zero gives the pure lock exchange solution.

(7) For $Q_r = -1$, show using equation (5.4.3) that the curve defining the lock exchange solution is identical to the curve (5.2.18) defining the long-wave stability threshold. (Hint: Note that Q_r is negative.) This result was first shown by Lawrence (1990).

(8) If the definition of q_2 is applied at the narrows, an implicit 'weir' formula is obtained:

$$Q_2{}^* = (q_2)_{narrows} w^* g'^{1/2} D_s{}^{3/2}.$$

Here $(q_2)_{narrows}$ is the value of q_2 corresponding to point c in Figure 5.4.4b, or the equivalent figure for the value of Q_r in question. Using a result obtained in Exercise 6, write out a procedure for calculating $(q_2)_{narrows}$ in terms of Q_r for the maximal lock exchange solution (e.g. *kbcd* of Figure 5.4.4b). (The algebra may be too complex to obtain a closed form solution.)

5.5.   Overmixing and Maximal Exchange in Estuaries

As discussed in the previous section, two-layer exchange flows exhibit a range of critically controlled steady states. Given certain restraints imposed by the upstream conditions, there generally exists a family of 'submaximal' solutions in which one of the layers acts more or less like a single-layer (reduced-gravity) flow while the other layer remains relatively passive. There is a single section of critical flow and the wave that is arrested is the one that attempts to propagate in the upstream direction of the 'active' layer. For a pure sill geometry, only the lower layer can be the relatively 'active' one. For a pure contraction, either layer may be the relatively active one. There is also a particular solution that is characterized by the presence of two critical sections and is a limiting case of the above solutions. One control often acts where the upper layer is active while the other acts where the lower layer is active. Such controls arrest wave propagation opposite to the direction of flow in the active layer. In the example of flow from a deep basin over a pure sill, the 'lower layer control' lies at the sill while the 'upper layer control' lies in one of the neighboring basins. In the case of a pure contraction with pure exchange, both critical states coincide at the narrowest section and both layers are active. The theory for these idealized geometries has been extended to include situations where the sill and the narrowest width occur at different sections. Farmer and Armi (1986) have shown that the maximal solution in this case has one control section at the narrows and the second at the sill.

For a steady exchange flow with fixed reduced gravity g' and flux ratio Q_r the solution with two critical sections has maximal exchange transport. The control sections for the maximal solution are insulated from the far field by stretches of supercritical flow that extend into the reservoir, terminating in hydraulic jumps. Linear wave propagation is permitted into, but not out of, the end basins. In this way, the flow at the control sections (particularly the exchange transport) is immune to mechanical changes that occur in the end basins. For example, a slight change in the interface level that is forced in one of the basins will generate an internal wave that will spread over the basin, but not into the strait. At the same time, changes in material properties such as density, which are advected by the flow, are not restricted in the same way. A forced change in layer density, and thus g', in an upstream basin, will be carried through the strait regardless of the state of hydraulic criticality. The value of g' ultimately depends on how the flow is forced and how the two layers are mixed.

Although Long's (1954) experiments, their descendants, and other initial-value experiments are helpful in developing intuition about maximal and submaximal flows, it is usually difficult to extrapolate the results to particular geophysical settings. For example, oceanographically relevant exchange flows often originate from an upstream basin or estuary that has finite extent and is subject to forcing, dissipation and mixing. The upstream conditions are therefore quite different from those envisioned by Long. Usual forcing mechanisms include cooling, evaporation, and precipitation over the basin surface, inflows and outflows from other straits or rivers, and mechanical forcing due to winds and tides. Estuaries are fed by a source of fresh runoff water that floats above the denser, saline ocean water and flows out into the ocean proper. Turbulence generated by tides, winds and internal instabilities can lead to mixing of the two water masses and an increased salinity of the upper layer. The export of salt that occurs where the upper layer exits must be balanced by an inflow of deeper, saltier water, and an exchange flow is set up. Semi-enclosed seas having strong evaporation or cooling can act as 'inverse estuaries', where the exchange flow is reversed. Two of the most widely studied examples are the Red Sea and Mediterranean Sea, which experience excessive evaporation and relatively little fresh water input from rivers. The combination of evaporation and surface cooling causes the surface waters to sink and eventually flow out into the ocean proper through the connecting passages, here the Bab al Mandab and the Strait of Gibraltar. Relatively fresh water is drawn in at the surface of these straits, resulting in exchange flows. Whether the latter are maximal or submaximal is a question that has excited a great deal of debate.

Under conditions of steady flow in a semi-enclosed closed basin with observable air-sea fluxes it is easy to write down a number of constraints on the overall exchange flow. For example, the net volume transport out of the basin must be balanced by river runoff, precipitation, and evaporation:

$$\iint_{A_s} (E - P)dA = -(Q_R^* + Q_1^* + Q_2^*), \qquad (5.5.1)$$

where $E - P$ represents the volume flux per unit surface area due to evaporation minus precipitation, A_s is the surface area of the basin, and $-Q_R^*$ is the volume inflow due to river runoff. If there are differences in the concentration of a chemical tracer between the inflow and outflow and if the sources and sinks of this tracer in the basin can be quantified, then a similar conservation law can be written down. For example, the input of salt due to river runoff in the Red Sea and Mediterranean Sea is negligible and thus the total influx of salt must be approximately zero:

$$Q_1^* S_1 + Q_2^* S_2 = 0. \qquad (5.5.2)$$

Equations (5.5.1) and (5.5.2) can be rearranged to yield Knudsen's relations

$$Q_1^* = \frac{S_2 \left[\iint_{A_s} (E - P)dA + Q_R^* \right]}{S_1 - S_2} \qquad (5.5.3a)$$

and

$$Q_2^* = \frac{S_1 \left[\iint_{A_s} (E - P)dA + Q_R^* \right]}{S_2 - S_1}, \qquad (5.5.3b)$$

named after the Danish chemist and oceanographer. The salinity of the inflowing layer (either S_1 or S_2) is equal to the salinity of the ocean water that is drawn in and can nominally be regarded as known. We will also assume that the values of $E - P$ and Q_R^* are known, even though the uncertainties in the measurement of these fluxes may be significant. If the salinity of the outflowing layer can be measured, then (5.5.3) can be used to calculate Q_1^* and Q_2^*.

The above approach was used by Nielsen (1912) to estimate the volume fluxes in the Strait of Gibraltar. Although they may provide a practical means for estimating layer fluxes, equations (5.5.3a, b) beg the question of what determines the salinity of the outflowing layer (or, equivalently, $S_2 - S_1$). A theory that provides an answer is based on the idea of *overmixing*, first proposed by Stommel and Farmer (1953). Their ideas were formulated in the context of an estuarine circulation, where $E - P$ is neglected, S_2 is regarded as fixed at an open-ocean value, and mixing between the upper and lower layers in the estuary interior is regarded as imposed independently of the mean circulation itself. One may begin by imagining an unmixed state in which the river discharge $Q_R{}^*$ produces a fresh layer of water ($S_1 = 0$) that passes through the surface of the estuary and exits at the mouth. If mixing with the lower saline layer is initiated, perhaps as a result of winds or tides, S_1 is increased and $S_2 - S_1$ is decreased. Equations (5.5.3a, b) then show that $Q_1{}^*$ and $Q_2{}^*$ increase: the estuary acquires a weak inflow of salty ocean water and an increased outflow of brackish surface water. If the mixing is increased further, the salinity difference between the layers continues to decrease and a stronger exchange circulation is induced. This process may not, however, continue unabated. Eventually the exchange at the mouth of the estuary must reach a maximal value permitted by hydraulic constraints and mixing beyond this threshold should have no further effect.

These ideas can be cast in quantitative form by requiring that the flow at the mouth of the estuary be hydraulically controlled, even when the exchange flow is weak. Then

$$\frac{v_{1c}{}^{*2}}{g'd_{1c}{}^*} + \frac{v_{2c}{}^{*2}}{g'd_{2c}{}^*} = \frac{Q_1{}^{*2}}{g'w_m^{*2}d_{1c}^{*3}} + \frac{Q_2{}^{*2}}{g'w_m^{*2}d_{2c}^{*2}} = 1, \tag{5.5.4}$$

where the subscript 'c' refers to the quantities measured at the mouth. For the time being, we will assume that the mouth consists of a pure contraction, with minimum width $w_m{}^*$, and with no sill or other topographic variation.

The density difference between the two layers is due primarily to the salinity difference and thus

$$\rho_2 - \rho_1 = \beta(S_2 - S_1),$$

where $\beta(= 0.77 \times 10^{-3} \text{ g cm}^{-3} \text{ ppt}^{-1})$ is the coefficient of expansion of water due to salinity. In terms of the reduced gravity:

$$g' = g\frac{\beta(S_2 - S_1)}{\rho_o}. \tag{5.5.5}$$

If the depth at the mouth of the estuary is D_s, then $d_{1c}{}^* + d_{2c}{}^* = D_s$ or

$$d_{1c} + d_{2c} = 1, \tag{5.5.6}$$

where $d_{nc} = d_{nc}{}^*/D_s$.

Substitution of the (5.5.3) layer transports into (5.5.4) leads to

$$\frac{S_2^2}{d_{1c}^3} + \frac{S_1^2}{(1 - d_{1c})^3} = \frac{g\beta w_m^{*2}D_s{}^3(S_2 - S_1)^3}{\rho_o Q_R{}^{*2}}, \tag{5.5.7}$$

after use of (5.5.5) and (5.5.6). Further discussion of this relation can be simplified if it is assumed that $(S_2 - S_1)/S_2 \ll 1$, implying that $Q_1{}^* \cong -Q_2{}^*$ and therefore $Q_R{}^* \ll Q_1{}^*$. With this simplification, (5.5.7) can be written as

$$\frac{1}{d_{1c}^3} + \frac{1}{(1 - d_{1c})^3} = (\Delta s)^3, \tag{5.5.8}$$

where

$$(\Delta s)^3 = \frac{g\beta w_m^{*2} D_s{}^3 (S_2 - S_1)^3}{\rho_o Q_R{}^{*2} S_2^2}. \tag{5.5.9}$$

The relationship between the nondimensional salinity difference Δs and d_{1c} takes the form of a curve with two vertical branches and single minimum (Figure 5.5.1). For a given value of the nondimensional salinity difference Δs, and provided $(\Delta s)^3 > 16$, there are two roots d_{1a} and d_{1b}. Let us assume for the time being that the left branch of the curve gives the appropriate root. Begin at the state $d_{1c} = d_{1a}$ and imagine that the mixing increases while $Q_R{}^*$ is held fixed. Then $S_2 - S_1$ should decrease, lowering the value of Δs, and the solution for d_{1c} is found by descent along the left branch of the solution curve. The minimum possible value of Δs lies at the base of the curve, where $d_{1c} = 1/2$. The corresponding salinity difference

$$(S_2 - S_1) = \left(\frac{16\rho_o Q_R{}^{*2} S_2^2}{g\beta w_m^{*2} D_s{}^3} \right)^{1/3} \tag{5.5.10}$$

is the minimum possible, corresponding to the largest $Q_1{}^* (= -Q_2{}^*)$, for the estuary. A further increase in the intensity of mixing in the estuary can apparently not alter these values and the resulting state is therefore 'overmixed'. It is not clear what this term implies for the interior state of the estuary itself, but some clues are provided by laboratory experiments to be presented here and in the next section. The analysis can also be carried out using the unapproximated version (5.5.7) of the governing relation and this leads to a skewed version of the Figure 5.5.1 curve (see Exercise 1).

In the overmixed limit, the interface depth at the estuary mouth lies at mid-depth and this corresponds to a state of maximal hydraulic exchange for flow through a pure contraction, as discussed in Section 5.4. Thus the state represented by the minimum of the Figure 5.5.1 curve represents a dynamically consistent state of maximal exchange in which the mouth, where the flow is critical, is insulated from both the ocean *and* the estuary by finite regions of supercritical flow. Other solutions lying along the left branch of the curve are hydraulically controlled, but submaximal. In these cases, supercritical flow exists only outside the estuary mouth.

The situation in which the mouth contains a sill is another matter. Let $z_T{}^*$ represent the depth, taken as constant, in the estuary interior, so that $D_s/z_T{}^* < 1$ when the mouth contains a sill. As discussed in Section 5.3, the corresponding maximal exchange solutions have unequal layer depths over the sill. When the sill is very shallow $(D_s/z_T{}^* \ll 1)$, $d_{1c} = 0.625$ and $d_{2c} = 0.375$ so that the interface lies below mid-depth. As the sill height is reduced, the interface rises eventually to mid-depth.

The corresponding range of d_{1c} values is indicated by the thicker segment of the curve in Figure 5.5.1. The limiting state of maximal exchange, and thus overmixing, in the presence of a sill therefore lies above the bottom of the curve and on the right branch. For a submaximal flow the interface at the sill lies below its level for maximal exchange. The corresponding 'undermixed' states lie along the right-hand branch of the curve. For these states supercritical flow extends from the mouth some distance *into* the estuary. If no sill is present, the choice between left and right branches depends on how the flow is established; the laboratory experiment described next selects the left branch.

The Stommel–Farmer hypothesis of an approach towards maximal estuary exchange and overmixing under conditions of controlled mixing has been investigated in a number of laboratory experiments. Similar experiments geared towards inverse estuaries will be discussed in the next section. One method of controlling the mixing rate is to introduce fresh water into a salty laboratory estuary in the form of a turbulent plume of adjustable depth, and hence variable mixing. In Timmermans (1998), a small basin representing an estuary is connected to a salt-water reservoir by a narrows (Figure 5.5.2). The estuary receives a steady flux of fresh water through a small submerged tube at adjustable depth. The fresh influx forms an ascending turbulent plume that entrains salty water as it rises to the surface. Brackish plume water accumulates at the surface and exits horizontally through the narrows while salty water enters beneath to supply salt to the plume. As the depth over which the plume rises increases, so does the total amount of entrainment. The net upstream mixing in the experiment, thought by Stommel and Farmer to be controlled by

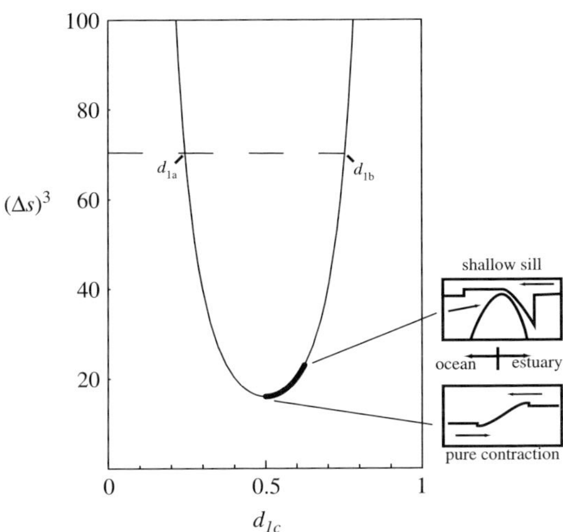

FIGURE 5.5.1. The dimensionless salinity difference Δs as a function of the dimensionless upper layer thickness d_{1c} at the mouth of an estuary, according to equation (5.5.8). The thickened portion of the curve shows the location of maximal exchange for a range of sill heights.

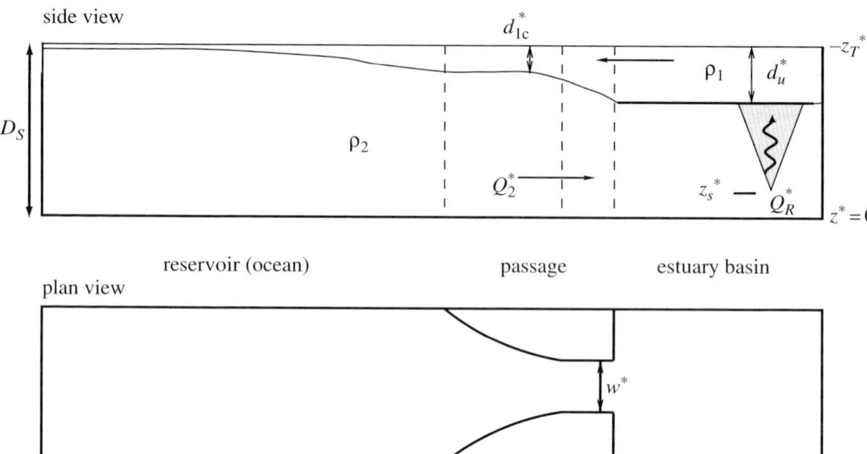

FIGURE 5.5.2. Sketch of the laboratory reservoir, fresh water source, and passage used to simulate an estuarine flow with partial mixing. The arrows indicate directions of flow.

the tides or winds, can therefore be varied by injecting the fresh water at different depths.

Suppose that the plume is fed at elevation z_S^* and volume rate Q_R^* and that it ascends a height $z_T^* - d_u^* - z_S^*$ in order to the reach the base of the upper layer (Figure 5.5.2). The entrainment into the plume, and the corresponding value of g' at its top, can be estimated (Turner, 1973) using a theory for a self-similar plume rising through a quiescent fluid. The theory, which is based on the assumption that the source is weak ($Q_R^* \ll Q_2^*$) yields

$$g' = 8.33 \left[(g\beta S_2 Q_R^*)^2 (z_T^* - d_u^* - z_S^*)^{-5} \right]^{1/3} \qquad (5.5.11)$$

where the leading coefficient is determined empirically.

This information may be used to predict the state of the exchange flow as a function of the source elevation z_S^*. To do so, one must equate the internal energy (Bernoulli function) in the basin near the source to that at the narrowest section. If the approximation of zero net exchange is made, it follows that

$$\left\{ \frac{1}{d_{1c}^2} - \frac{1}{(1-d_{1c})^2} \right\} = \frac{2g\beta w^{*2} D_p^3 (S_2 - S_1)^3}{S_2^2 Q_R^{*2}} (d_{1c} - d_u) \qquad (5.5.12)$$

where d_{1c} and d_u are the upper layer depths at and upstream of the sill, nondimensionalized by the total depth D_s.

Equations (5.5.8, 5.5.11 and 5.5.12) can be used to calculate the state variables (g', d_u, d_{1c}, etc.) as functions of the mixing parameter, z_S^*, or equivalently $d_S = (z_T^* - z_S^*)/D_S$. For a given d_S, the exact location along the curve of possible solutions (Figure 5.5.1) can be found. Timmermans (1998) verified that lowering the source (increasing z_R) causes the solution to tend towards the overmixed limit

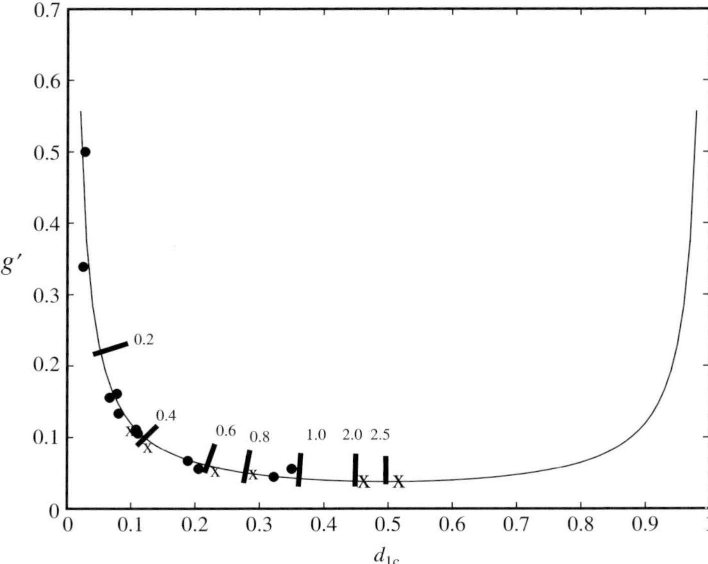

FIGURE 5.5.3. The curve shows the predicted value of g' as a function of the dimensionless, critical upper layer depth d_{1c} at the narrowest section. The hash marks on the curve indicate where a solution with the indicated value of d_S should lie. The symbols indicate data points from the experiment of Timmermans (1998). Points indicate forcing by a single plume while crosses indicate six plumes.

and that data track the predicted curve quite well (Figure 5.5.3). However, even when the plume source is positioned at the bottom of the tank ($d_S = 1$) the total entrainment is insufficient to reach the limit of overmixing, here the minimum of the curve. (The threshold d_S predicted by the theory is about 2.5.) This limitation can be overcome by adding more plumes and the 'x' symbols, representing experiments with 6 plumes, reach the threshold of overmixing.

One of the great mysteries raised by the hypothesis of overmixing concerns the state of the flow that occurs when this limit is exceeded. The basic premise is that the salinity difference between the two layers decreases as mixing increases, and that the exchange flow must increase to satisfy the overall salt budget. But what then happens when the exchange reaches its maximal value? A further increase in mixing would seem to require a further decrease in the salinity difference, leading to a violation of the salt budget. What happens under these conditions is not generally understood and undoubtedly depends on the way the flow is set up and driven. The laboratory experiments described below and in the next section provide some insight.

Using an inverted version of the previous experiment, Whitehead *et al.* (2003) attempted to exceed the overmixed condition with a single plume and to provide some insight into the corresponding upstream state. The reservoir now contains fresh water and salt water is pumped in through a tube elevated above the bottom the

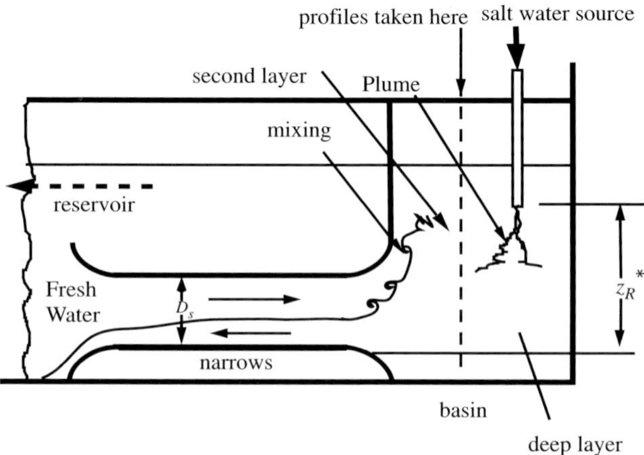

FIGURE 5.5.4. Sketch of the Whitehead et al. (2003) laboratory setup.

'estuary' basin (Figure 5.5.4). The full-depth narrows of the previous experiment is replaced with a submerged and shallower passage, similar to an upright experiment with a shallow sill. The mixing parameter d_S is now replaced by z_R $(= z_R^*/D_s)$ the dimensionless source elevation above the bottom of the narrows. The new config-uration allows a wider range of forcing.

The dyed salt plume appears on the far right in a photo (Figure 5.5.5). The salt-water layer appears black in the right basin and grey in the narrows because the basin is wider. The run shown is thought to exceed the limit of overmixing. The fresh upper layer flows into the basin from left to right and accelerates as it passes through the narrowest section and into the right basin. In the classical view, this flow would develop a hydraulic jump somewhere near the entrance to the basin. However, the region where this jump is expected shows billows (Figure 5.5.6). The latter cause the clear, fresh water entering the chamber to mix with the salty water, resulting in a brackish (grey) layer that extends into the basin up to the level of the tube source. The presumed maximal exchange flow should also have a hydraulic

FIGURE 5.5.5. Photograph of an experiment with $z_R = 3$, thought to be overmixed.

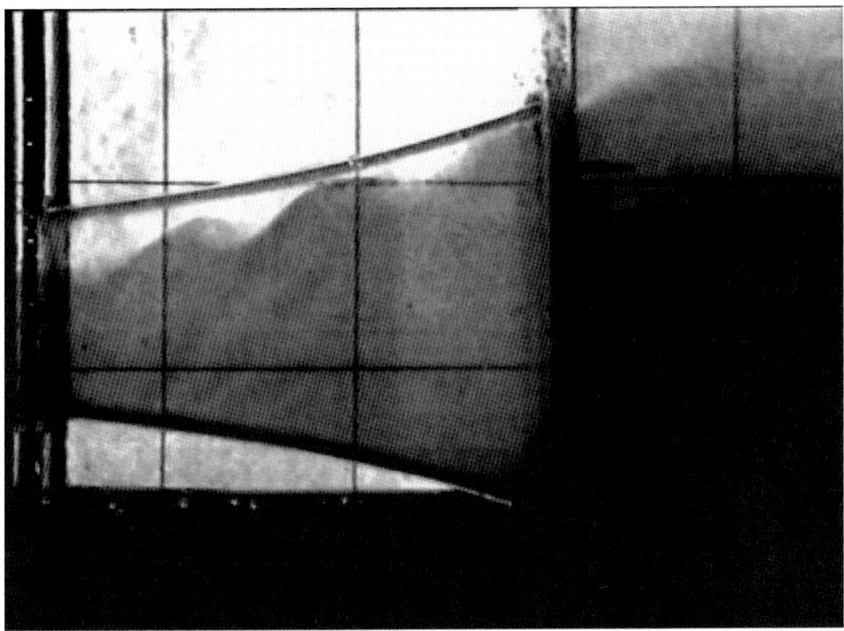

FIGURE 5.5.6. Close-up of the flared region between the passage and the right basin where clear water flows up and into the chamber with developing billows. The experiment is the same as shown in Figure 5.5.6.

jump at the left end of the channel, and while this feature may have been present, it was not documented.

The approach to and beyond the limit of overmixing can been seen in a set of salinity profiles taken in the right basin (Figure 5.5.7). The values have been normalized by the lower layer salinity predicted under conditions of maximal exchange for sill depth of the experiment. The source pumping rate and salinity are the same for each experiment but the mixing parameter z_R is varied. The value of z_R, now the elevation of the plume source, is labeled with each profile. The profiles show the top layer (salinity $S_1 = 0$), a homogeneous bottom layer, and an intermediate middle layer that may be partially stratified. It can be seen that as z_R is increased (the source is raised) from 1.5 to 2.5, the salinity in the bottom layer decreases as expected. Further increases in z_R cause the bottom salinity to cluster around a value approximately 1.25 times the predicted value for overmixing. There is an unexplained minimum at $z_R = 3.0$. Although the predicted limiting bottom salinity (or g' value) for this experiment is not reached, the convergence for values $z_R > 2.5$ suggests that the exchange flow is close to or has exceeded the limit of overmixing. The discrepancy may be due to the presence of frictional effects that have not been taken into account.

We now return to the conceptual question, raised earlier, regarding how the 'overmixed' flow conspires to keep g' at a relatively fixed value while the elevation

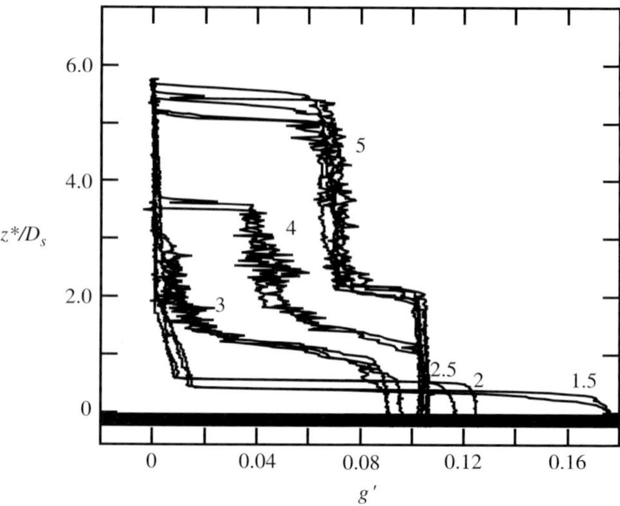

FIGURE 5.5.7. Density profiles for 14 experiments, measured at the location shown in Figure 5.5.4. The profiles are labeled with the nondimensional source elevation z_R.

of the plume source, and presumably the mixing, increases. An answer is suggested by two changes in the density distribution of the basin flow. One is a deepening of the lower layer and the other is the increased salinity of the overlying fluid (as evidenced by an increase in overall density). This second effect is due to the billows and other interfacial instabilities in and around the narrows, which cause the salty bottom layer to become entrained in the fresh layer entering from the reservoir. Now the total amount of salt in the estuary basin must remain constant, and thus the source salt flux must equal the salt flux through the narrows into the left reservoir. When the basin flow is undermixed, freshwater from the reservoir enters the basin and becomes entrained into the salty plume. As the plume mixing is increased and more fresh water is entrained, the plume is increasingly diluted, the density difference between layers decreases, and the exchange flow through the narrows intensifies. Once maximal exchange conditions in the narrows are reached, the amount of freshwater that can be drawn in from the reservoir cannot be increased. If one then attempts to increase the mixing further (by raising the source), the system responds in a way that limits the entrainment of fresh water. It does so by increasing the depth of the lower layer (thus limiting the vertical height over which mixing can occur) and by creating a mechanism by which salt is detrained into the incoming fresh water (thus increasing the salinity of the water that is entrained into the plume). In this respect the term 'overmixing' is misleading. Although the overall mechanical level of turbulence in the basin may increase, the actual net mixing between the fresh and salty layers remains fixed.

The overmixing hypothesis is by no means the only model of estuary dynamics. In fact, it is difficult to find examples of estuaries that clearly have maximal exchange at the mouth. The presence of strong time-dependence due to tides can cloud the interpretation. A reader interested in digging deeper into this field can consult Hetland and Geyer (2004), references contained therein, and also the textbook of Dyer (1997).

We end this section with a bit of speculation that some readers may wish to turn into careful research. The Black Sea acts like a giant estuary, with a relatively fresh surface layer fed by rivers and precipitation, and a deep, saline bottom layer. The Sea is connected to the Mediterranean by the Bosporus, which contains a two-layer flow that exchanges fresh surface water for saline Mediterranean water. The lower layer of Mediterranean water begins its journey at a salinity of about 38 psu, passes through the Bosporus, and descends in a turbulent plume into the Black Sea. Entrainment with the fresher (17 psu) water leads to dilution of this plume. The resulting water mass (about 22 psu) spreads throughout the deep Black Sea basin. The deep and shallow layers are separated by a pycnocline with a base at about 150 m depth, well below the 40 m deep Bosporus.

There are two features that suggest that the Black Sea could be overmixed. One is the relatively deep pycnocline, similar to that in the inverted experiment (Figure 5.5.7). In that experiment, the interface or pycnocline in the right basin is much shallower than the passage. The second suggestive piece of evidence is that multiple sections of critical flow have been observed in the Bosporus (Gregg and Özsoy, 2002), which could be consistent with maximal exchange.

Exercises

(1) By rearranging the primitive version (5.5.7) of the relation governing estuary flow, show that

$$\frac{1}{d_{1c}^3} + \frac{(1 - \tilde{\Delta}s)^2}{(1 - d_{1c})^3} = \gamma(\tilde{\Delta}s)^3,$$

where $\tilde{\Delta}s = (S_2 - S_1)/S_2$ and $\gamma = g\beta w_m D_s{}^3 S_2/\rho_o Q_R{}^{*2}$. For a (positive) γ of your choice, sketch the curve of $\tilde{\Delta}s$ vs. d_{1c} over $0 < d_{1c} < 1$ and note that the result is an asymmetrical version of the Figure 5.5.1 curve. Show that the minimum value of $\tilde{\Delta}s$ lies where

$$\frac{1}{d_{1c}^2} = \frac{(1 - \tilde{\Delta}s)}{(1 - d_{1c})^2}.$$

Deduce that this minimum must occur in $1/2 \le d_{1c} < 1$ and thus the interface must, in the overmixed limit, lie below mid-depth. Note that the minimum value of $\tilde{\Delta}s$ itself can be obtained by eliminating d_1 between the last two equations and solving the resulting polynomial.

(2) How would the original theory of Stommel and Farmer be modified to fit the experimental conditions suggested in Figure 5.5.4?

5.6.   Overmixing in Inverse Estuaries

Ideas about overmixing and maximal flow have also been applied to inverse estuaries, most notably the Mediterranean and Red Seas. The overturning circulation and exchange flow is driven primarily by excessive evaporation that exists in the arid climates of these marginal seas. Historical estimates of $E - P$ range from 0.5 to 1.0 m/year in the Mediterranean and roughly twice that amount in the Red Sea. The contribution Q_R^* from river runoff is minor in both cases and can be neglected in Knudsen's relations (5.5.3a, b). Evaporation concentrates salt in the surface waters, thereby increasing its density and leading to sinking. In some cases the evaporation preconditions the water for deep convection in response to atmospheric cooling events that take place in winter. In the two-layer idealization, upper layer water is mixed down into the lower layer and the exchange flow at the mouth, in this case the Strait of Gibraltar or the Bab al Mandab, is reversed relative to a normal estuary. The upper layer salinity S_1 at the mouth is considered to be observable since it is this water that is drawn in from the outside ocean.

It should be kept in mind that the intensity of the mixing and deep convection in an inverse estuary is, in principle, a function of the $E - P$ and/or the surface thermal forcing. It is therefore more difficult to think of mixing as an independent parameter in the way that estuary mixing might be considered independent of Q_R^*. Moreover, intensification of thermohaline forcing at the surface presumably causes the density of the product water to increase. That is, stronger forcing ostensibly leads to more mixing but also to an *increase* in the value of g'. In this situation it is unclear whether stronger surface forcing moves the system towards or away from an overmixed state.

In the Mediterranean and the Red Seas, the exact sequence of events by which the outflow waters are produced and the geographical distribution of such events are not fully understood. In the Mediterranean, deep convection events have been observed in the Gulf of Lyon and the northern Aegean Sea. The Red Sea is not as well observed but it is known that the bulk of the outflow comes from an intermediate water mass that lies only 100–300 m below the surface. In one view of the overturning circulation (Philips, 1969; see Exercise 1) the sinking occurs continuously over the whole area of the sea. Sporadic deep convection events in the northern reaches of the Red Sea (Woelk and Quadfasel, 1996) produce a deeper water mass that contributes a lesser amount to the outflow. The dynamics of the overturning circulation are further complicated during the summer monsoon by the appearance of a three-layer exchange flow in the Bab al Mandab.

The overmixing hypothesis was first applied to the Mediterranean Sea by Bryden and Stommel (1984). As in the original estuary formulation, the density difference between the two layers is due primarily to the salinity difference, and thus (5.5.5) continues to hold. The arguments leading to the Figure 5.5.1 curve can be repeated with river runoff Q_R^* replaced by $\iint_{A_s} (E - P)dA$ and S_1, rather than S_2, considered known. The approach to the limit of overmixing occurs as a result of the descent along the left or right branch of the curve until the maximal flux permitted by the mouth geometry is reached. For an estuary, this descent occurs as a result of holding Q_R^* fixed and supposing that $S_2 - S_1$ decreases as a direct result of increased mixing. This scenario is more problematic for an inverse estuary where increased mixing due to increased $\iint_{A_s} (E - P)dA$ might actually *increase* $S_2 - S_1$. Regardless of how it is achieved, the minimum possible salinity difference corresponding to the overmixed and maximal exchange conditions is found to be

$$(S_2 - S_1) = \left(\frac{\rho_2 S_1^2}{q_2^2(D_s/z_T^*)g\beta w_m^{*2}D_s^3} \right)^{1/3} \left(\iint_{A_s} (E - P)dA \right)^{2/3}. \qquad (5.6.1)$$

As discussed in Sections 5.3 and 5.4, the coefficient $q_2(D_s/z_T^*)$ ranges between 0.208 and 0.25 as the geometry changes from a sill with infinitely deep upstream basin $(D_s/z_T^* = 0)$ to a pure contraction $(D_s/z_T^* = 1)$. It has been assumed that the sill lies at the point of minimum width in the mouth and that there is no barotropic flow.

Bryden and Stommel employ the pure contraction limit of (5.6.1) along with $E - P = 0.95\,\text{m/year}$ to obtain $S_2 - S_1 = 1.7\%$, not far from the observed value 2.1%. This estimate has been refined by later investigators who considered physically separated sill and narrows and weak barotropic inflow. In the case of Gibraltar, the shallowest sill lies adjacent to Camarinal (Figure 5.6.1) while the narrowest section lies at Tarifa. Through the work of Farmer and Armi (1986) it is now understood that maximal flow under these conditions tends to have one control at the narrows (where the upper layer is most active) and the other at the sill (where the lower layer is most active). Further refinements can be introduced through the use of nonrectangular cross-sections to more accurately represent the true topography (Bormans and Garrett, 1989a; Dalziel, 1991). The most refined model to date is due to Bryden and Kinder (1991), who also describe this history in more detail. Their predictions based on maximal exchange are $S_2 - S_1 = 2.0\%o$, $Q_2^* = 0.88 \times 10^6\,\text{m}^3/\text{s}$, and $Q_1^* = -0.92 \times 10^6\,\text{m}^3/\text{s}$ as compared with contemporary observations $S_2 - S_1 = 2.1\%o$, $Q_2^* = (0.67 \pm 0.04) \times 10^6\,\text{m}^3/\text{s}$, and $Q_1^* = -(.78 \pm 0.17) \times 10^6\,\text{m}^3/\text{s}$ (Tsimplis and Bryden, 2000).

The debate continues as to whether the Gibraltar exchange flow is maximal or slightly submaximal (Garrett, 2004). Although a hydraulic jump is observed at the western end of the strait, it is less clear whether such a jump occurs at the eastern end. In addition, the flow is strongly modulated by the tides, which can cause the observed jump to break up and travel away as a train of internal waves. A comprehensive documentation of the flow over a tidal phase appears in Armi and Farmer (1988). Despite these complications, the reasonable agreement with observations has made the Mediterranean Sea/Gibraltar system the most promising large-scale application of the theory of overmixing.

In the Bab al Mandab, the exchange flow does not appear to be maximal. The sill and narrowest section occur at different locations and the flow at the latter appears to be subcritical (Pratt *et al.*, 1999). The exchange flow therefore lacks the double critical section characteristics of maximal exchange. The Red Sea therefore does not appear to be overmixed, or even close to such a state, despite the fact that it is subject

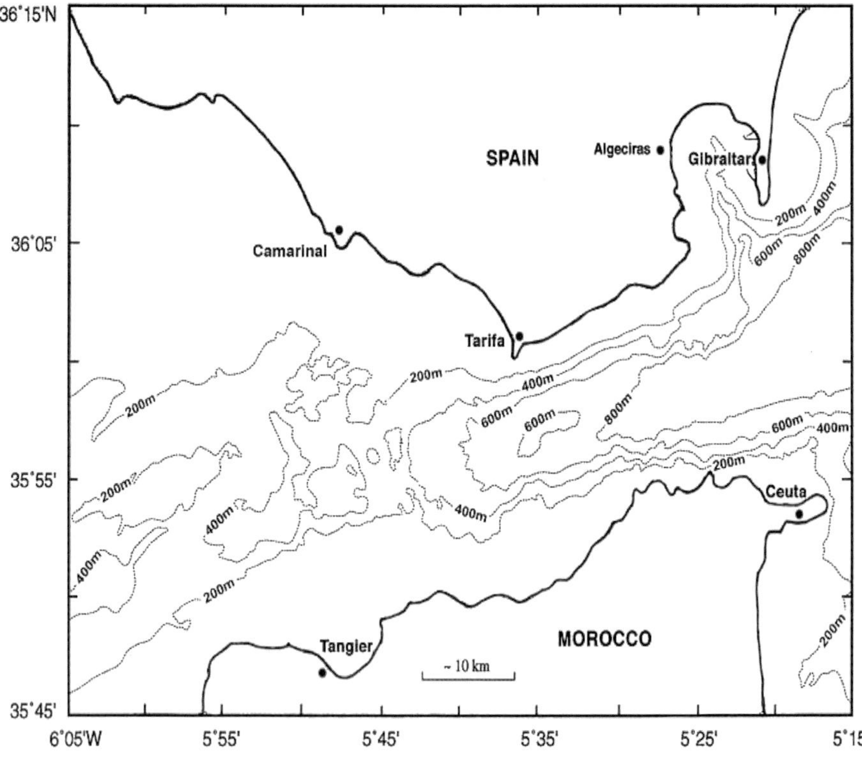

FIGURE 5.6.1. Bathymetry of the Strait of Gibraltar.

to fierce evaporation and winter cooling and that the sill depth is about the same as at Gibraltar. This apparent incongruity could be explained by the failure of one of the many assumptions leading to the hypothesis of overmixing. One of the most fundamental is that an increase in the sea surface forcing pushes the system towards maximal exchange. Intuitively, one would expect Q_2^* to increase as the rate of convection-generating buoyancy loss to the atmosphere is increased. However, the turbulence generated by vertical convection and the associated eddy stresses must also increase as the convection becomes more active. The resultant eddy momentum stresses may actually retard the exchange flows, making it difficult to predict how rapidly Q_2^* will actually increase. In summary, it is not obvious that an increase in $E - P$ must cause the solution to *descend* along the Figure 5.5.1 solution curve (or its equivalent for an inverse estuary).

In an attempt to investigate the competition between forcing and eddy damping Finnigan and Ivey (1999, 2000) performed the laboratory simulation suggested in Figure 5.6.2. The 'marginal sea' is really a section of strait channel that lies to the left of an obstacle. The flow there is driven by cooling at the upper surface, which produces vertical convection[5]. The convection does not penetrate completely to the

---

[5] The true system is an inverted version of what is shown in Figures 5.1 and 5.2, with the obstacle suspended from the top lid and the left hand basin heated from below.

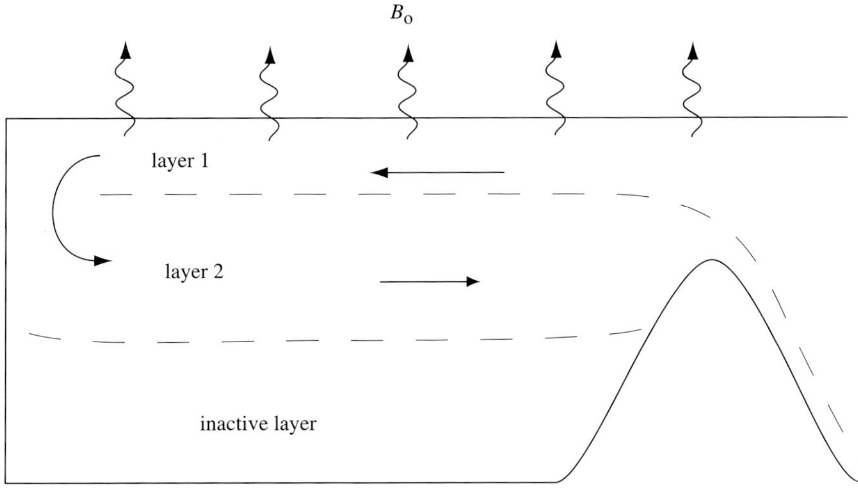

FIGURE 5.6.2. Sketches of the flow forcing and resulting two-layer regime in the experiments of Finnigan and Ivey (2000). The lowest layer is relatively inactive.

bottom, but does give rise to something like a two-layer exchange flow with most of the sinking occurring at the far left end of the basin (Figure 5.6.3). The two active layers overlie a third, inactive layer. In the area between the sinking and the sill, there is some return flow or recirculation of layer 2 fluid back into layer 1. The lower layer spills over the sill and is hydraulically controlled in all cases.

Since the experimental flow is driven entirely by thermal effects, a thermodynamic balance is needed to replace the statement of conservation of salt. The new constraint may be derived from the density equation, which for a continuously stratified fluid is

$$\frac{\partial \rho}{\partial t^*} + \nabla_3^* \cdot (\mathbf{u}^* \rho) = \nabla_3^* \cdot (\kappa \nabla_3^* \rho). \tag{5.6.2}$$

Here κ is a turbulent diffusivity and $\nabla_3^* \cdot$ represents the three-dimensional divergence operator. Relative differences in density throughout the domain are typically small and it has become traditional to work with buoyancy $b = -g \frac{\rho - \rho_o}{\rho_o}$ in place of density. Here ρ_o denotes a constant reference density. If the condition of incompressibility $\nabla_3 \cdot \mathbf{u}^* = 0$ is used, b may be used in place of ρ in (5.6.2) and the corresponding steady form is

$$\nabla_3 \cdot (\mathbf{u}^* b) = \nabla_3 \cdot (\kappa \nabla_3 b).$$

Integration of this relation over the volume of the marginal sea and use of the divergence theorem gives a relation between the advection of b into the basin at the mouth and the turbulent diffusion of b through the bounding surfaces. The former is given by the integral of $v^* b$ over the entrance of the strait and equals $(Q_1^* b_1 + Q_2^* b_2)$ for a two-layer exchange flow. In the case of pure exchange $(Q_1^* = -Q_2^*)$ the net flux is

$$Q_2^*(b_2 - b_1) = -Q_2^* g(\rho_2 - \rho_1)/\rho_o = -Q_2^* g'. \tag{5.6.3}$$

(a)

(b)

(c)

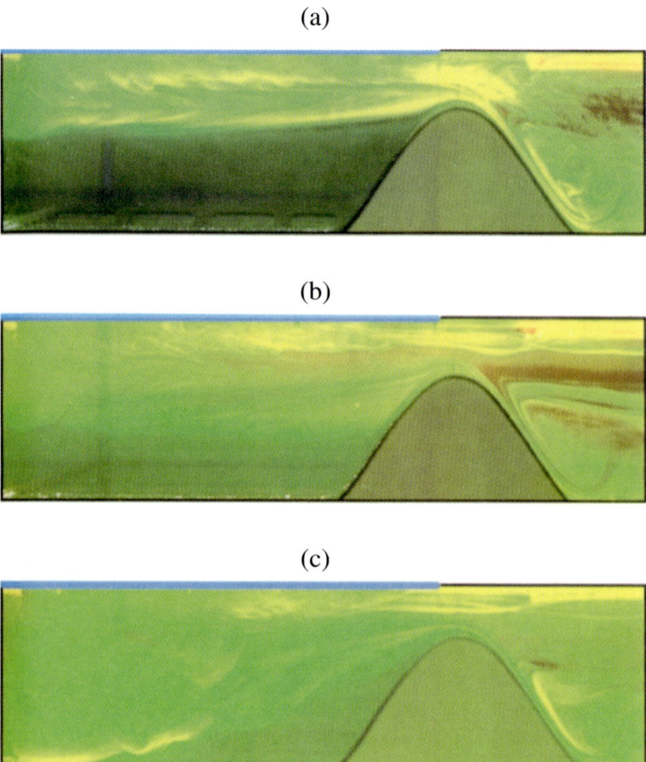

FIGURE 5.6.3. Photographs of three realizations of the Finnigan and Ivey (2000) experiment with progressively larger surface buoyancy flux (applied where the surface is marked by a thickened line). From top to bottom, the frames correspond to Experiments 2, 3, and 4 in Table 1.

The diffusive flux is presumably dominated by fluxes at the air-sea interface $z^* = z_T^*$ and is given in terms of the rates of evaporation E and sensible heat gain H_s by

$$\kappa \frac{\partial b}{\partial z^*}\bigg|_{z^*=z_T^*} = -B_o = \frac{g}{\rho_o}\left[(H_s - \rho_s L_H E)\frac{\alpha_v}{c_p} - \rho_s E S_s\right] \qquad (5.6.4)$$

(e.g. Phillips, 1966). The effects of precipitation are ignored. Here S_s is the surface salinity and L_H, α_v and c_P are the latent heat of evaporation, the coefficient of thermal expansion and the specific heat under constant pressure, all of sea water. B_o has units $[L^2/T^3]$ and positive values are associated with a buoyancy loss in the surface waters. If B_o is uniform over the surface area A_s of the basin then the buoyancy budget is given by

$$B_o A_s = g' Q_2^*. \qquad (5.6.5)$$

If (5.6.5) is used to replace $Q_2^*(= -Q_1^*)$ in the critical condition (5.5.4), the result may be written.

$$g' = (B_o A_s)^{2/3} w^{*-2/3} D_s^{-1} \left(\frac{1}{d_{1c}{}^3} + \frac{1}{(1-d_{1c})^3} \right)^{1/3}. \tag{5.6.6}$$

This relation is kin to (5.5.8), but now the effect of surface buoyancy flux $B_o A_s$ replaces that of river runoff Q_R^*. Using (5.6.5) this relation can also be written in terms of the lower layer volume flux:

$$Q_2* = (B_o A_s)^{1/3} w^{*2/3} D_s \left(\frac{1}{d_{1c}{}^3} + \frac{1}{(1-d_{1c})^3} \right)^{-1/3}. \tag{5.6.7}$$

If one were to follow the thread of the previous section, the buoyancy flux would be considered fixed and the value of g' would be minimized (or Q_2^* maximized) over d_{1c} to get a maximal, overmixed state. The overmixed state would be limited by the maximal Q_2^* permitted by the hydraulic constraints for the sill height in question. For a high sill, the maximal Q_2^* occurs when $d_{1c} = .625$ whereas $d_{1c} = 0.5$ for a pure contraction with no sill. The Finnigan and Ivey experiment utilizes a relatively high sill, and thus the former value is appropriate. Substitution d_{1c} into (5.6.7) yields

$$(Q_2^*)_{\text{max}} = 0.351(B_o A_s)^{1/3} w^{*2/3} D_s. \tag{5.6.8}$$

A measure of proximity to this overmixed state for the experimental setup is therefore

$$\left[\frac{Q_2^*}{(Q_2^*)_{\text{max}}} \right] = 2.849 \left(\frac{1}{(d_{1c})^3} + \frac{1}{(1-d_{1c})^3} \right)^{-1/3}. \tag{5.6.9}$$

Maximal exchange occurs when $d_{1c} = .625$; submaximal exchange occurs for larger values, or deeper interface levels.

Now consider a sequence of experiments in which the exchange is initially submaximal and the surface buoyancy flux (and therefore $g'Q_2^*$) is gradually increased. It is possible, though unlikely, that g' could stay fixed during this process. The increased buoyancy flux would then have to be carried by an elevated Q_2^* and, as suggested by (5.6.9), the interface would shallow and the system would evolve towards a maximal state. At the other extreme, Q_2^* might remain constant and the extra buoyany flux would be carried by a larger g'. In this situation, the flow would be pushed *away* from the maximal limit: Q_2^* would stay the same while $(Q_2^*)_{\text{max}}$ would increase. More generally, (5.6.7) suggests that Q_2^* must increase more rapidly than $(B_s A_s)^{1/3}$ in order to move towards the maximal state. According to (5.6.6), g' must then increase no more rapidly than $(B_s A_s)^{2/3}$.

Finnigan and Ivey cite four experiments with successively larger B_o (Table 5.6.1). B_o is lowest for experiment 1 and increases by factors of about 2 or 3 for experiments 2, 3 and 4. The value of $Q_2^*/(Q_2^*)_{\text{max}}$ increases slightly from Experiment 1 to 2, but it decreases for the remaining runs. Although the value of Q_2^* increases slightly

TABLE 5.6.1. (data from the four experiments of Finnigan and Ivey, 2000)

Experiment	$B_o \times 10^6$ (m^2 s^{-3})	Q_2^*/w^* (m^2 s^{-1})	g' (m s^{-2})
1	0.4	0.39	0.0036
2	0.9	0.56	0.0029
3	3.0	0.82	0.0068
4	6.0	1.02	0.0080

from 2 to 3 to 4, this trend is compensated for by increases in g'. The experiments therefore indicate that the increase in mixing with B_o does not overcome the increase in g'. There are some practical limits on how hard the experiment can be driven.

The inability of Q_2^* to increase sufficiently rapidly to approach a maximal state is attributed to a damping effect associated with the increase in turbulence as the experiment is driven harder. Some evidence of this can be seen in the Figure 5.6.3 photos, in which the thermal forcing increases from top frame to bottom. Another factor is the appearance, with stronger forcing, of a closed circulation cell near the left end of the tank. The two-layer model, in which all surface fluid is converted into lower layer fluid that exits the basin, no longer described the actual situation.

One aspect of this experiment that is unrealistic for applications such as the Mediterranean is the limited extent of the upstream reservoir. Suppose that the local surface buoyancy flux B_o is kept constant and the area A_s of the marginal sea is increased. If the local convection and overturning processes remain unchanged then g' should also remain the same and the increased buoyancy flux $g'Q_2^*$ should be contained largely in the flux Q_2^*. Grimm and Maxworthy (1999) describe an experiment in which A_s is effectively larger than in the Finnigan and Ivey experiment. (Also, the thermal forcing is replaced by an injection of a brine mixture at the surface.) Both submaximal and maximal (overmixed) exchange flows are observed depending on the strength of the forcing. Nevertheless, it is clear that there is more to overmixing and maximal exchange than perhaps met the eyes of the originators of this idea.

Exercises

(1) [*The Phillips(1966) Red Sea model*]: As a simple model of deep convection consider a quarter plane ($z^* < 0$ and $y^* > 0$) of initially homogeneous fluid. At the upper boundary $z = 0$ the fluid is subjected to a uniform buoyancy flux B_o due to cooling and evaporation. This forcing gives rise to turbulent convection, setting up a hypothetical circulation in which surface fluid approaching the sidewall ($y^* = 0$) is drawn down. The effects of viscosity are to be ignored and there is no buoyancy flux thorough the sidewall. Present a coherent argument, purely on dimensional grounds, that

$$v^* \propto (B_o y^*)^{1/3} f(z^*/y^*),$$

where v represents the time-mean y-velocity of the statistically steady flow that eventually occurs. Also, f is an unknown universal function.

Next suppose that the inflow and outflow must cross a sill of depth D_s that lies a distance L from the vertical wall. A plausible generalization of the above relation is then

$$v^* \propto (B_o y^*)^{1/3} f(z^*/y^*) g(z^*/D_s).$$

If the flow is now imagined to occur in a channel of width w^*, compute the exchange transport and comment on any similarities with (5.5.12).

5.7. Maximal and Submaximal Exchange between Two Deep Basins with Rotation

The exploration of two-layer hydraulic phenomena in rotating channels is quite limited in comparison to the nonrotating case. Progress to date has largely been limited to three special cases, the subjects of this and the following two sections. The first case is pure exchange flow over a sill separating two infinitely deep basins. The theory relies on the zero potential vorticity approximation in both layers, at least in the vicinity of the sill. The next section covers the case of the pure lock exchange through a contraction with no sill. By the nature of the initial value problem, the potential vorticity in this case is uniform and finite. The third special case (Section 5.9) is the only one that involves a net barotropic flow. The novel feature is a scale mismatch between the baroclinic and barotropic components of the flow. The latter is confined to sidewall boundary layers while the former is felt all across the channel. Even in a wide channel the effects of forcing by one sidewall can be transmitted into the baroclinic boundary layer on the opposite wall through the barotropic flow. As this is written, little is known regarding hydraulic jumps, bores, and other transient features of these flows. Also very little is known regarding their stability. The following discussions of purely steady models only scratch the surface of what is undoubtedly a rich body of physical behavior.

We begin with what might seem to be a simple calculation–that of pure exchange over a sill between two deep basins. This problem, which is a natural extension of the Whitehead *et al.* (1974, Section 2.4) model for single-layer flow, turns out to be anything but simple. The same authors consider certain aspects of the two-layer problem and our discussion is based on their work as well as that of Dalziel (1988, 1990), Riemenschneider *et al.* (2005), and Timmermans and Pratt (2005).

Consider a rectangular channel that has uniform width and that separates two relatively wide and deep basins with horizontal bottoms. An exchange flow between the basins could be established as a result of a lock exchange experiment in which the basins are filled with fluids of densities ρ_1 and ρ_2 and are separated by a barrier that sits atop the sill (Figure 5.7.1a). If the barrier is removed the two layers can be expected to displace each other and flow into the opposite reservoirs. As usual, the basin that initially contains the denser fluid will be referred to as 'upstream' (Figure 5.7.1b) and the direction of its outflow will be considered positive. The layer depths in the initially quiescent reservoirs are the potential depths $D_{1\infty}$ and $D_{2\infty}$. The potential vorticity of the layers within the channel will therefore be $f/D_{1\infty}$ and $f/D_{2\infty}$, provided that the initially shallow water has been entirely washed out of the strait and that dissipation does not intervene. If the sill depth $D_s \ll D_{1\infty}$ and $\ll D_{2\infty}$, then the definition (5.1.9) of semigeostrophic potential vorticity implies that

$$\left(f + \frac{\partial v_n^*}{\partial x^*} \right) / f = O\left(\frac{D_s}{D_{n\infty}} \right) \ll 1$$

in the vicinity of the sill. Therefore

$$\frac{\partial v_n^*}{\partial x^*} \cong -f, \tag{5.7.1}$$

as was the case in the single-layer analog.

As discussed in Chapter 2, models based on (5.7.1) are often referred to under the title *zero potential vorticity*. It is true that the relative vorticity exactly equals $-f$ when the potential vorticity of the flow is exactly zero. But (5.7.1) should in the present model be regarded as an approximation, valid only where the layer

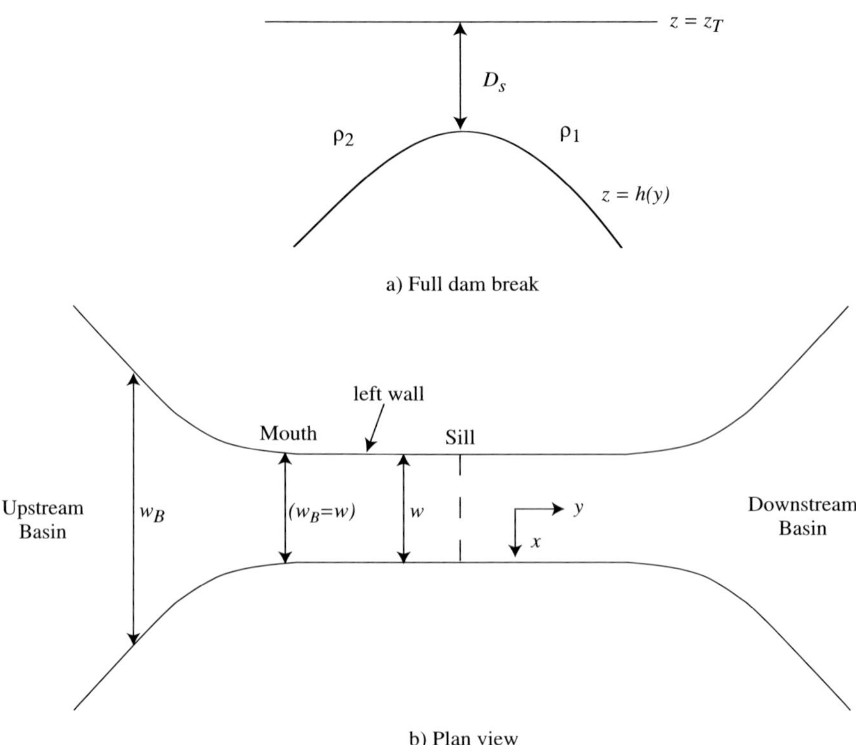

FIGURE 5.7.1. Initial condition for the lock-exchange problem (a). Plan view of the basin and strait geometry (b). (From Timmermans and Pratt, 2005).

depth $d_n{}^*$ is small compared to its potential depth $D_{n\infty}$. The dimensional value of the potential vorticity need not be zero. Equation (5.7.1) and the derived relations (5.7.2–5.7.4) below can be expected to hold near the sill, but can fail as either of the deep reservoirs is approached. We will analyze the sill region first and return to the basin states later.

In terms of the nondimensional variables,

$$x = \frac{x^* f}{\sqrt{g' D_s}}, \quad y = \frac{y^*}{L}, \quad z = \frac{z^*}{D_s}, \quad v_n = \frac{v_n^*}{\sqrt{g' D_s}}, \quad \text{and } u_n = \frac{fL}{g' D_s} u_n^*,$$

the thermal wind relation (5.1.8) is

$$v_2 - v_1 = \frac{\partial d_2}{\partial x}, \tag{5.7.2}$$

while (5.7.1) becomes

$$\frac{\partial v_n}{\partial x} = -1. \tag{5.7.3}$$

In addition, the dimensionless form of Crocco's relation (5.1.16) is $dB_n/d\psi_n = D_S/D_{n\infty}$. The Bernoulli functions B_1 and B_2 are therefore uniform where $D_S/D_{n\infty} \ll 1$, and it follows that the 'internal' Bernoulli function

$$\Delta B = B_2 - B_1 = \tfrac{1}{2}\left(v_2^2 - v_1^2\right) + d_2 + h \tag{5.7.4}$$

(see 5.1.17) is uniform wherever both layers are shallow with respect to their respective potential depths.

Equations (5.7.2) and (5.7.3) provide that the slope of the interface across the channel is constant. Some of the geometrically possible cross sections are shown in Figure 5.7.2. *Attached flow* will refer to a state in which the layer depths remain finite across the channel (Frame a). For a lock exchange flow, the lower layer velocity will generally be positive and the upper layer velocity negative, so the interface will slope from left to right, as in Frame a. Other possible states are *singly detached* (Frames b and c) and *doubly detached* (Frame d). It will be convenient to redefine the location of the origin $x = 0$ depending on the state of separation. The various cases of attachment and detachment are an artifact of the rectangular channel cross section. Natural straits have a smoothly varying topography, but this introduces difficulties more serious than the bookkeeping required to keep track of the various modes of separation.

a.   *Critical Conditions at the Sill*

With the channel spanning $-w/2 < x < w/2$ the solutions to (5.7.2) and (5.7.3) for attached flow can be written as

$$v_i(x, y) = -x + \bar{v}_n(y), \tag{5.7.5}$$

$$d_1(x, y) = \left(d(y) - \bar{d}_2(y)\right) - \left(\bar{v}_2(y) - \bar{v}_1(y)\right) x, \tag{5.7.6}$$

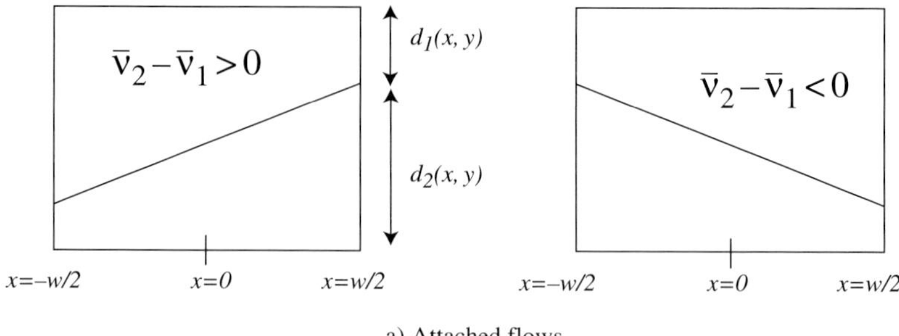

a) Attached flows

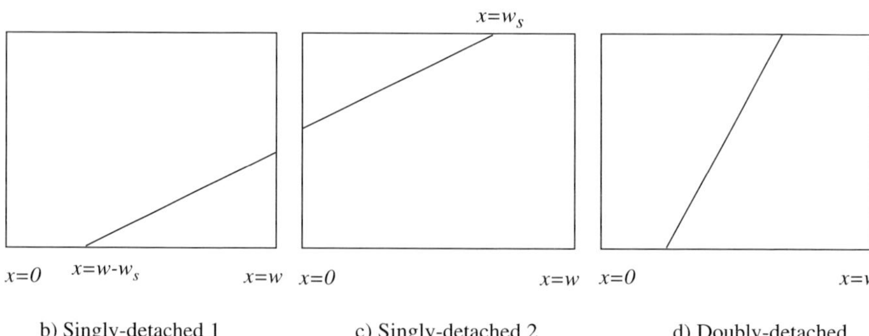

b) Singly-detached 1   c) Singly-detached 2   d) Doubly-detached

FIGURE 5.7.2. Definition sketches for the flow cross section (a) and the various possible separation states with a positive interface tilt (b–d). (From Timmermans and Pratt, 2005).

and

$$d_2(x, y) = \bar{d}_2(y) + (\bar{v}_2(y) - \bar{v}_1(y)) \, x. \tag{5.7.7}$$

where $d = d_1 + d_2$ is the channel depth, $d = 1$ at the sill, and an overbar denotes the value of the quantity at the channel center, $x = 0$.

The volume fluxes $Q_n = Q_n^* f / g' D_S^2$ in layers $n = 1, 2$ can be calculated using (5.7.5)–(5.7.7) as

$$Q_1 = \int_{-w/2}^{w/2} v_1 d_1 \, dx = \bar{v}_1 \bar{d}_1 w + (\bar{v}_2 - \bar{v}_1) \frac{w^3}{12} \tag{5.7.8}$$

and

$$Q_2 = \int_{-w/2}^{w/2} v_2 d_2 \, dx = \bar{v}_2 \bar{d}_2 w - (\bar{v}_2 - \bar{v}_1) \frac{w^3}{12}. \tag{5.7.9}$$

The dimensional versions are given by

$$Q_{1,2}^* = \bar{v}_{1,2}^* \bar{d}_{1,2}^* w^* \pm (\bar{v}_2^* - \bar{v}_1^*) \frac{w^{*3} f^2}{12 g'} \tag{5.7.10}$$

In the absence of rotation, the first term on the right would give the layer transport. With rotation the velocity in each layer decreases as x increases. If $\bar{v}_2 - \bar{v}_1 > 0$, the interface slope is positive and the lower layer is thicker on the right-hand side of the channel (facing in positive y direction). This thicker portion has smaller (perhaps even negative) velocities compared to the velocities in the thinner left-hand side. Hence the positive interface tilt reduces the transport in the lower layer. A similar effect occurs in the upper layer and thus rotation reduces the net exchange $Q_2 - Q_1$ relative to the value based on the average layer velocity and depth. This trend is reminiscent of the tendency of rotation to reduce transports in single-layer overflows. It will be shown later, however, that the tendency is reversed, at least in theory, when the two-layer flow becomes doubly detached from the sidewalls.

Attention is now restricted to pure exchange flow

$$Q = -Q_1 = Q_2 > 0. \tag{5.7.11}$$

Use of this relation with (5.7.8) and (5.7.9) leads to

$$\bar{v}_1 \bar{d}_1 = \bar{v}_1 \left(d - \bar{d}_2\right) = -\bar{v}_2 \bar{d}_2 \tag{5.7.12}$$

and it follows from (5.7.8) that

$$\bar{v}_2 = \frac{(Q/w)\left(d - \bar{d}_2\right)}{\bar{d}_2\left(d - \bar{d}_2\right) - w^2 d/12} \tag{5.7.13}$$

and

$$\bar{v}_1 = \frac{-(Q/w)\,\bar{d}_2}{\bar{d}_2\left(d - \bar{d}_2\right) - w^2 d/12}. \tag{5.7.14}$$

A function $\mathcal{G}$ relating a the single variable $\bar{d}_2$ to the geometric parameters can be found by substituting the last two relations for the velocities in (5.7.4):

$$\mathcal{G}\left(\bar{d}_2; d, w\right) = \frac{Q^2 d\left(d - 2\bar{d}_2\right)}{2w^2\left(\bar{d}_2(d - \bar{d}_2) - w^2 d/12\right)^2} + \bar{d}_2 + h - \Delta B = 0 \tag{5.7.15}$$

Application of the critical condition $\partial \mathcal{G}/\partial \bar{d}_2$ leads to

$$\frac{Q^2}{w^2} = \frac{\left(\bar{d}_2\left(d - \bar{d}_2\right) - w^2 d/12\right)^3}{d\left(\bar{d}_2\left(d - \bar{d}_2\right) - w^2 d/12 + \left(d - 2\bar{d}_2\right)^2\right)}, \tag{5.7.16}$$

which can also be written as

$$\frac{\bar{v}_1^2 \bar{d}_2 + \bar{v}_2^2 \bar{d}_1 - w^2 \left(\bar{v}_2 - \bar{v}_1\right)^2 /12}{\left(\bar{d}_1 \bar{d}_2 - w^2 d/12\right)} = 1. \tag{5.7.17a}$$

The left side of (5.7.17a) can be viewed as the square of a composite Froude number that characterizes the hydraulic state as subcritical, critical or supercritical for values $< 1, = 1$ or > 1. In dimensional terms, the composite Froude number is given by

$$G_r^{\,2} = \frac{\bar{v}_1^{\,*2}\bar{d}_2^{\,*} + \bar{v}_2^{\,*2}\bar{d}_1^{\,*} - \dfrac{w^{*2}f^2}{12g'}(\bar{v}_2^{\,*} - \bar{v}_1^{\,*})^2}{g'\left(\bar{d}_1^{\,*}\bar{d}_2^{\,*} - \dfrac{w^{*2}f^2}{12g'}d^*\right)}. \tag{5.7.17b}$$

In the limit of weak rotation $\left(w^{*2}f^2/g'd_n^{\,*} \to 0\right)$ it reduces to the familiar $G^2 = \frac{\bar{v}_1^2}{d_1} + \frac{\bar{v}_2^2}{d_2} = \frac{\bar{v}_1^{\,*2}}{g'd_1} + \frac{\bar{v}_2^{\,*2}}{g'd_2}$ for nonrotating flow.

It can be shown (Exercise 5) that the characteristic speeds for the two-layer system under conditions of attachment, and allowing for net barotropic flow, are given by

$$c_{\pm} = \frac{\bar{v}_2\bar{d}_1 + \bar{v}_1\bar{d}_2}{d} \pm \left(1 - \frac{(\bar{v}_2 - \bar{v}_1)^2}{d}\right)^{1/2}\left(\frac{\bar{d}_1\bar{d}_2}{d} - \frac{w^2}{12}\right)^{1/2}. \tag{5.7.18}$$

Aside from the factor of $w^2/12$, this formula is identical to its nonrotating counterpart (5.2.1). The critical condition (5.7.17) can be obtained by setting $c_- = 0$ and using $\bar{v}_1\bar{d}_1 + \bar{v}_2\bar{d}_2 = 0$.

The regularity condition (1.5.4) can be applied to (5.7.15) to determine further restrictions on the location of a section of hydraulic control. Attention is confined to the channel portion of the domain, for which $w = $ constant. After use of (5.7.16) and some lengthy algebra, the condition reduces to

$$\left[(d - \bar{d}_2)^2 - \frac{w^2 d}{12}\right]\frac{dh}{dy} = 0 \tag{5.7.19}$$

The control section can lie where $dh/dy = 0$ (as at the sill) or at a virtual control $(\bar{d}_2 = \bar{d}_{2v})$, where

$$\bar{d}_{2v} = d - \frac{wd^{1/2}}{2\sqrt{3}}. \tag{5.7.20}$$

In the limit of weak rotation $(w \to 0)$, satisfaction of (5.7.19) requires $\bar{d}_{2v} \to d$. The virtual control in this case occurs in the deep reservoir, when the upper layer is thin and the lower layer is infinitely deep and inactive. Rotation allows the virtual control to occur in the shallow reaches of the channel.

For attached flow, the volume flux for a particular critical sill state is obtained by setting $d = 1$ in (5.7.16). The flux depends on the mean lower layer thickness $\bar{d}_2 = \bar{d}_{2c}$ and the channel width w (Figure 5.7.3.). In order to ascertain the effect of rotation on the flux, it is best to plot the flux per unit width $Q/w(= Q^*/\left(w^*g'^{1/2}D_s^{3/2}\right)$ since f does not appear in the scaling for this quantity. For fixed $\bar{d}_{2c}$, the transport per unit width is reduced as the strength of rotation $(w = w^*f/\sqrt{g'D_s})$ is increased. The

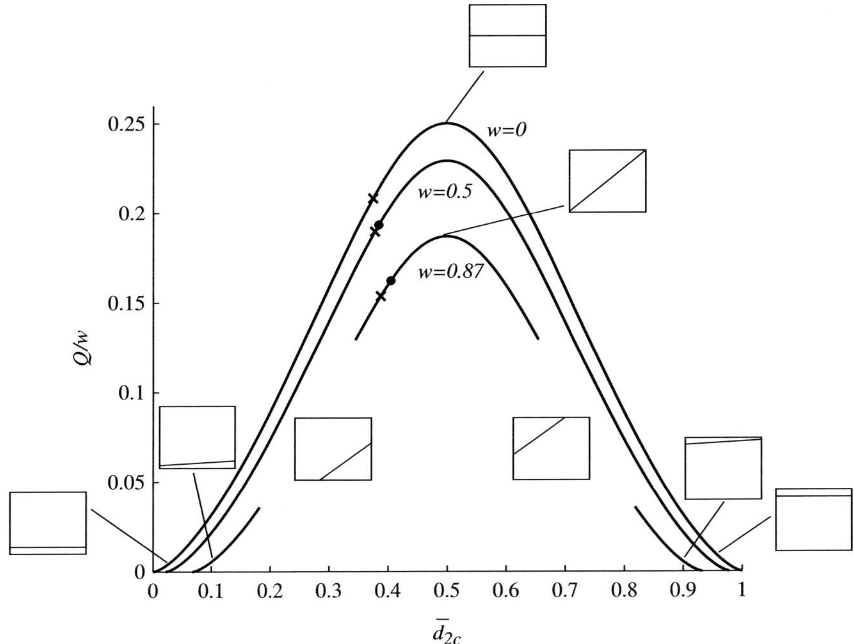

$$\overline{d}_{2c}$$

FIGURE 5.7.3. The nondimensional exchange transport per unit width as a function of the mean lower layer depth. The flow is assumed to occur at the sill section ($d = 1$) and is critical and attached to both sidewalls. Gaps in the curve for $w = 0.87$ correspond to separated flow, for which Figure 5.7.4 should be consulted. The dots and crosses indicate maximal exchange values calculated under two different assumptions as described in the text. (From Timmermans and Pratt, 2005).

gaps in the $w = 0.87$ curve correspond to $\overline{d}_2$ values over which the flow becomes detached from one of the sidewalls. For $w \leq 0.866$, the flow at the sill is always attached to both sidewalls.

According to Figure 5.7.3, the maximum possible Q/w for any w occurs when $\overline{d}_{2c} = 1/2$. The corresponding critical states are symmetrical in that the interface passes through mid-depth at the channel center ($x = 0$) and the cross-sectional areas occupied by the two layers are identical. According to (5.7.16) the flux is given by

$$Q/w = \frac{1}{4}\left(1 - \frac{w^2}{3}\right) \tag{5.7.21}$$

or $Q^* = \frac{w^* g'^{1/2} D^{3/2}}{4}\left[1 - \frac{w^{*2} f^2}{3 g' D}\right]$. In the limit of weak rotation ($w \to 0$), Q reduces to its maximal value $w/4$ for lock exchange through a pure contraction. The opposite extreme occurs when the channel is sufficiently wide to cause separation, which first occurs when the interface strikes the upper right and lower left corners. The reader may wish to verify that this marginal state requires $d = w^2$, or $w^* = (g'd^*)^{1/2}/f$

(see Exercise 4). Since $d = 1$ at the sill, double separation of the critical flow occurs when $w = 1$, implying

$$Q = \frac{1}{6}, \tag{5.7.22}$$

or $Q^* = \frac{1}{6}\frac{g'D^2}{f}$.

In an analysis that assumes symmetry between layers, Whitehead *et al.* (1974) derived this and the former expression. They also speculated that (5.7.22) would continue to hold after the flow becomes detached from both sidewalls. The basic notion is that the detached flow, which would resemble that of Frame d of Figure 5.7.2, would be quiescent in the flange regions occupied by only one of the layers. This idea is inconsistent with the maintenance of zero potential vorticity throughout each layer. Also, as we will eventually see, symmetrical critical states at the sill cannot be smoothly connected to a deep upstream basin. As in the nonrotating case, the lower layer is relatively thin at the sill and the flux is less than that given above. The flow is inherently asymmetrical due to the fact that the topography contacts only the lower layer. The maximum attainable exchange flux is therefore less than for $\overline{d}_{2c} = 1/2$. Symmetrical states would appear to be appropriate for a lock exchange flow carried out in the pure contraction. However, a difficulty with this scenerio is that the vortex squashing mechanism responsible for generating 'zero' potential vorticity would be lacking. It is therefore difficult to justify the corresponding flux formula, including the '1/6th' law, by any dynamically consistent argument. Despite these difficulties (5.7.21 and 5.5.22) have been shown to be relevant, either as a bound on, or an approximation of, laboratory exchange through a contraction. This matter will be discussed in Section 5.8.

The interface may detach from one or both of the sidewalls in a variety of ways. The three possibilities relevant to the lock exchange experiment are shown by the lower panels in Figure 5.7.2. An analysis of the type just described can be performed for each case, though the algebra is a bit more involved. A case with single separation is worked through in Exercise 3 and the case of double separation can be found in Timmermans and Pratt (2005). The results can be used to extend Figure 5.7.3 to cases of separated sill flow (Figure 5.7.4), in which Q/w is plotted as a function of the separated width: w_s for single separation and w_b for double separation. For example, associated with the sill width $w = 2.0$ is a range of critical solutions beginning with a singly separated case of zero width ($w_s = 0$) and zero flux. As w_s/w is increased the flux increases and double separation occurs at $w_s/w \cong .34$. Beyond this point the (dashed) curve for singly attached flow becomes solid, signifying double separation. The horizontal axis must now be interpreted as w_b/w (insets). For narrow channel widths (specifically $w < 1.01$) double separation does not occur. By symmetry, w_s may be interpreted in Figure 5.7.4 as the lower layer width for lower layer detachment or the upper layer width for upper layer detachment.

The curves in Figure 5.7.4 suggest, not surprisingly, that double separation occurs more readily as w increases. In addition, the maximum possible Q/w increases as rotation increases. It will be shown later that the maximal Q/w that can be linked

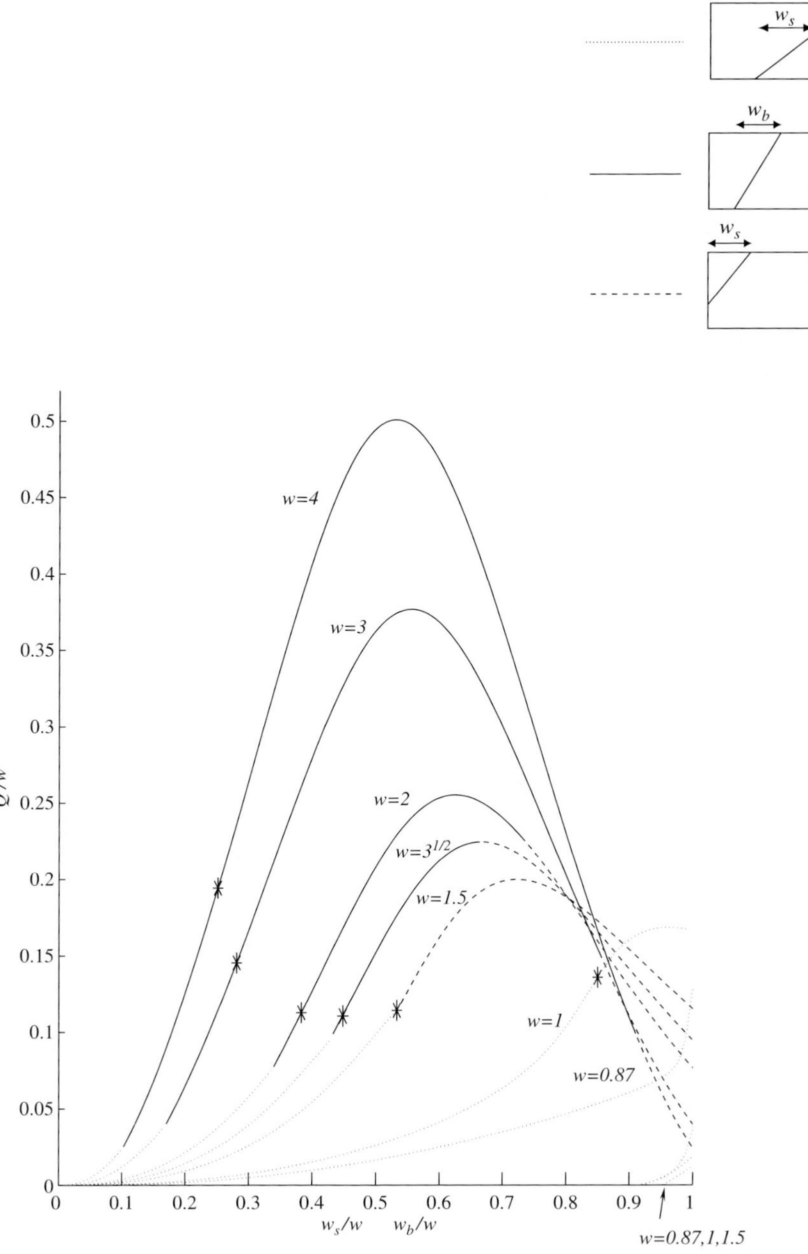

FIGURE 5.7.4. The flow rate per unit width as a function of the separated width of the current at the sill (w_s/w for single detachment and w_b/w for double detachment) for various w. By symmetry, w_s may be interpreted as the lower layer width for lower layer detachment, or the upper layer width for upper layer detachment. (From Timmermans and Pratt, 2005).

to a deep upstream basin (indicated by stars in the figure) actually decreases with increasing w for singly separated flows. However this trend is reversed when double separation occurs. Another subtlety is that the relationship between Q and w_s is not always unique for a given w. There are two sets of curves for $0.866 \leq w \geq 1.720$, as represented by $w = 0.87$, 1.0, and 1.5. The second set of curves appear in the lower right of Figure 5.7.4 and is associated with relatively small values of Q/w. The significance of these solutions is not well understood.

b.   *The Froude Number Plane*

Attention has thus far been confined to the sill, where the flow is assumed to be critical. In order to extend the solution upstream and downstream, and thereby describe the solution for the channel as a whole, one can use conservation of layer fluxes and internal energy. In the absence of background rotation, the solutions are conveniently represented in the layer Froude number plane (e.g. Figure 5.3.1). As shown by Reimenschneider *et al.* (2005), a similar approach is possible here, though under more restricted circumstances.

The governing relationships for the case of attached flow can be derived by using (5.7.12) to write the lower layer transport relation (5.7.9) and the Bernoulli equation (5.7.4) in the forms

$$Q_2 = Q = \frac{\bar{v}_2 \bar{v}_1 wd}{(\bar{v}_1 - \bar{v}_2)} + (\bar{v}_2 - \bar{v}_1)\frac{w^3}{12}$$

and

$$\Delta B - z_T = \frac{\bar{v}_1^2 - \bar{v}_2^2}{2} + \frac{\bar{v}_1 d}{\bar{v}_1 - \bar{v}_2} - d.$$

Division of these two by factors of d^2 and d respectively, leads to

$$\frac{Q}{d^2} = \frac{\bar{F}_1 \bar{F}_2}{\bar{F}_1 + \bar{F}_2}\left(\frac{w}{d^{1/2}}\right) - \frac{\bar{F}_1 + \bar{F}_2}{12}\left(\frac{w}{d^{1/2}}\right)^3 \tag{5.7.23}$$

and

$$\frac{\Delta B - z_T}{d} = \frac{\bar{F}_2^2 - \bar{F}_1^2}{2} - \frac{\bar{F}_2}{\bar{F}_1 + \bar{F}_2}, \tag{5.7.24}$$

where

$$\bar{F}_n = \frac{\bar{v}_n}{d^{1/2}} = \frac{\bar{v}_n^*}{(g'd^*)^{1/2}}. \tag{5.7.25}$$

$\bar{F}_1$, $\bar{F}_2$ are pseudo Froude numbers based on the total depth and layer velocities at mid-channel.

In the Froude number plane representation for nonrotating flow, a form of the energy equation (e.g. 5.3.1 or 5.4.1) was used to construct solution curves in the space of the layer Froude numbers F_1 and F_2. For a given range of bottom elevation or channel width, one traces out a solution by moving along the appropriate constant-energy curve. Control points occur where the solution curve intersects the critical diagonal $F_1{}^2 + F_2{}^2 = 1$. The range of bottom elevation (or width) is specified by contours that cross the energy curves and are based on mass conservation. A similar construction in terms of the pseudo Froude numbers $\overline{F}_1$, $\overline{F}_2$ is possible, but there are several complicating factors. One is that that the energy $(\Delta B - z_T)$ depends not only on $\overline{F}_1$ and $\overline{F}_2$ but also d. A more suitable quantity on which to base solution contours is $Q^{1/2}/(\Delta B - z_T)$, which depends on $\overline{F}_1$, $\overline{F}_2$ and on the ratio of the channel width to the 'local' radius of deformation:

$$\frac{w}{d^{1/2}} = \frac{w^*}{(g'd^*)^{1/2}}.$$

If the topography is chosen such that this ratio remains constant, then $Q^{1/2}/(\Delta B - z_T)$ depends only on the Froude numbers and contours of this function represent solutions in the Froude number plane (Figure 5.7.5). The channel geometry must therefore be one in which width varies in proportion to the square root of depth.

A second complication with the Froude number plane is in the representation of various regions with flow separation (Figure 5.7.5 inset). Where single or double detachment occurs the governing equations must be reformulated. It can be shown that $Q^{1/2}/(\Delta B - z_T)$ continues to depend only on $\overline{F}_1$, $\overline{F}_2$ and $w/d^{1/2}$. The corresponding solution curves can therefore be extended and are shown in the main part of the figure. $\overline{F}_1$ and $\overline{F}_2$ continue to be defined in terms of the velocities at $x = 0$. In cases of extreme separation, the upper (or lower) layer may not exist at midchannel and the corresponding $\overline{v}_1$ (or $\overline{v}_2$) will cease to be physically meaningful. In such cases, $\overline{F}_1$ (or $\overline{F}_2$) is defined by extrapolation to $(x = 0)$ using the mathematical form of v_1 (or v_2) valid in the nonseparated region. Although the Froude numbers so defined may have little physical meaning, they continue to act as a formal representation of the flow state.

In order to trace solutions along such curves for a given range in d, contours of constant Q/d^2 can be plotted. Equation (5.7.23) can be used for this purpose in the attached region, while modified relations hold in the other regions. The resulting contours are shown as dashed curves in Figure 5.7.5. For the case shown, the value of d is infinite at the origin, corresponding to a deep upstream (or downstream) reservoir. If Q is regarded as fixed, then shallower depths are found by moving away from the origin.

We may regard $Q^{1/2}/(\Delta B - z_T)$ and Q as Gill-type functions that depend on $\overline{F}_1$ and $\overline{F}_2$ as well as the geometric variables d and $w/d^{1/2}$. Critical flows therefore occur where

$$J_{\overline{F}_1,\overline{F}_2}[Q^{1/2}/(\Delta B - z_T), Q/d^2] = 0.$$

There are two geometrically distinct situations where this condition is satisfied. The first (indicated by stars in Figure 5.7.5) lie where the (solid) solution curves

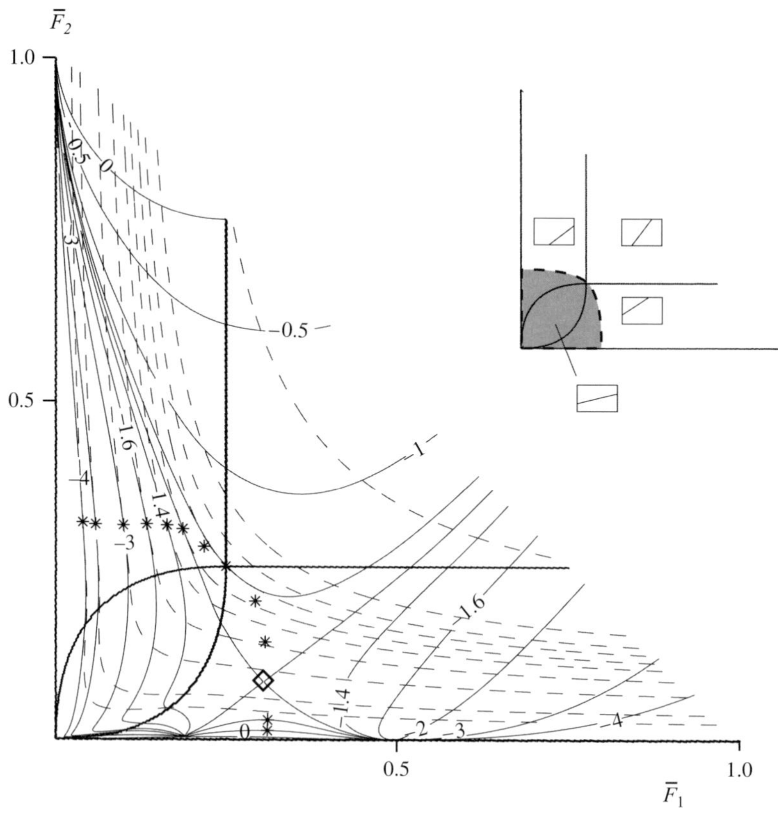

FIGURE 5.7.5. Pseudo Froude number plane for two-layer flow with background rotation, zero potential vorticity, and contained in a self-similar channel geometry: $w/d^2 = 1$. The inset shows the (dashed) critical curve and the (sold) boundaries demarking various regions of separation. The shaded region of the inset corresponds to subcritical flow. The layer Froude numbers are defined as $F_n = \bar{v}_n/d^{1/2}$, where d is the total channel depth and $\bar{v}_n$ is the centerline velocity. If the layer in question is separated and does not exist at the channel centerline, $\bar{v}_n$ is defined by formally extending the velocity profile for that layer to the centerline. The solid curves in the main part of the diagram are of constant $Q^{1/2}/(\Delta B - z_T)$, while the dashed curves are of constant Q/d^2. To prevent too much clutter, the contours on the dashed lines are not labeled. However, the value of Q/d^2 increases as one moves away from the origin, corresponding to shallower depths. Solutions are traced along solid lines. Stars indicate sill controls while the square indicates a virtual control. (A rotated version of Figure 5.7.9 from Reimenschneider *et al.*, 2005).

and (dashed) topographic curves make grazing contact. These correspond to sill controls. The second (indicated by a single square) occurs where two solution curves cross. The Jacobian vanishes here because $Q^{1/2}/(\Delta B - z_T)$ is locally constant. Since the bottom elevation changes as one follows a solution curve through the intersection point, this point is a virtual control.

The overall geometry of the cases shown in Figure 5.7.5 is similar in some respects to the Froude number planes for nonrotating flow through a contraction, particularly Figure 5.4.1. However there are also some important differences. To start with, there is nothing so simple as the critical diagonal; critical flows now occur along the arc indicated by a heavy dashed line in the inset. Flow states lying below this curve are subcritical. Also, where there was a single family of submaximal solutions with sill controls, there are now two. The first begins in the subcritical region at a point lying along the $\overline{F}_1$ axis and ends at a point on the $\overline{F}_2$ axis. The second begins at the former point and ends at another point on the $\overline{F}_1$-axis. Each member of this second family passes through a sill control.

Of the submaximal solutions, the ones beginning on the $\overline{F}_1$-axis and ending on the $\overline{F}_2$ axis are most consistent with physical expectations. An example corresponding to the -1.4-contour appears in Figure 5.7.6. At the upstream end, which corresponds to the left ends of the panels, the flow is subcritical, the lower layer is very deep and the upper layer is separated (panel *a*). The lower layer mid-channel velocity $\overline{v}_2$ is nearly zero and bands of positive and negative flow exist on the left and right walls (panel *b*). As the channel narrows and shoals the flow becomes attached and the lower layer velocity becomes unidirectional. As the sill is passed, the lower layer separates and becomes strongly trapped to the right wall. The upper layer becomes very deep and develops the same swirl velocity with backflow that characterizes the upstream lower layer (panel *c*). The second group of submaximal solutions has deep lower layers at both ends of the channel and does appear as relevant to the lock exchange problem. These solutions correspond to the set of small-flux solutions that appear in the lower right corner of Figure 5.7.4.

There is also a single solution with properties vaguely similar to the nonrotating maximal solution for flow through a pure contraction (curve *jbk* of Figure 5.4.1b). The corresponding solution curve begins in Figure 5.7.5 along the $\overline{F}_1$-axis near $\overline{F}_1 = 0.5$ and continues through a virtual control (square), becoming subcritical and then passing through a sill control (star) and ending on the $\overline{F}_2$-axis. The latter termination coincides with the downstream termination of the previously discussed submaximal solutions. There, the lower layer is thin, separated and unidirectional, while the upper layer is deep and has bidirectional flow. It can also be shown that the upstream state has a relatively deep, swirling lower layer and a separated and unidirectional upper layer. Both end states would almost certainly terminate in hydraulic jumps before the basin depth became infinite. The maximal nature of the flux for this solution is confirmed by the fact that the dashed contour lying at the sill control indicates a flux than is larger than for the previously considered solutions.

It is possible to construct other solutions that are more difficult to connect to deep upstream basins. For example, the solution curve that begins along the $\overline{F}_1$ axis near $\overline{F}_1 = 0.2$ and passes through the virtual control is similar to the curve *abc* of Figure 5.4.1b, which was disqualified under conditions of exchange flow for reasons of long-wave instability. However, the general level of numerical

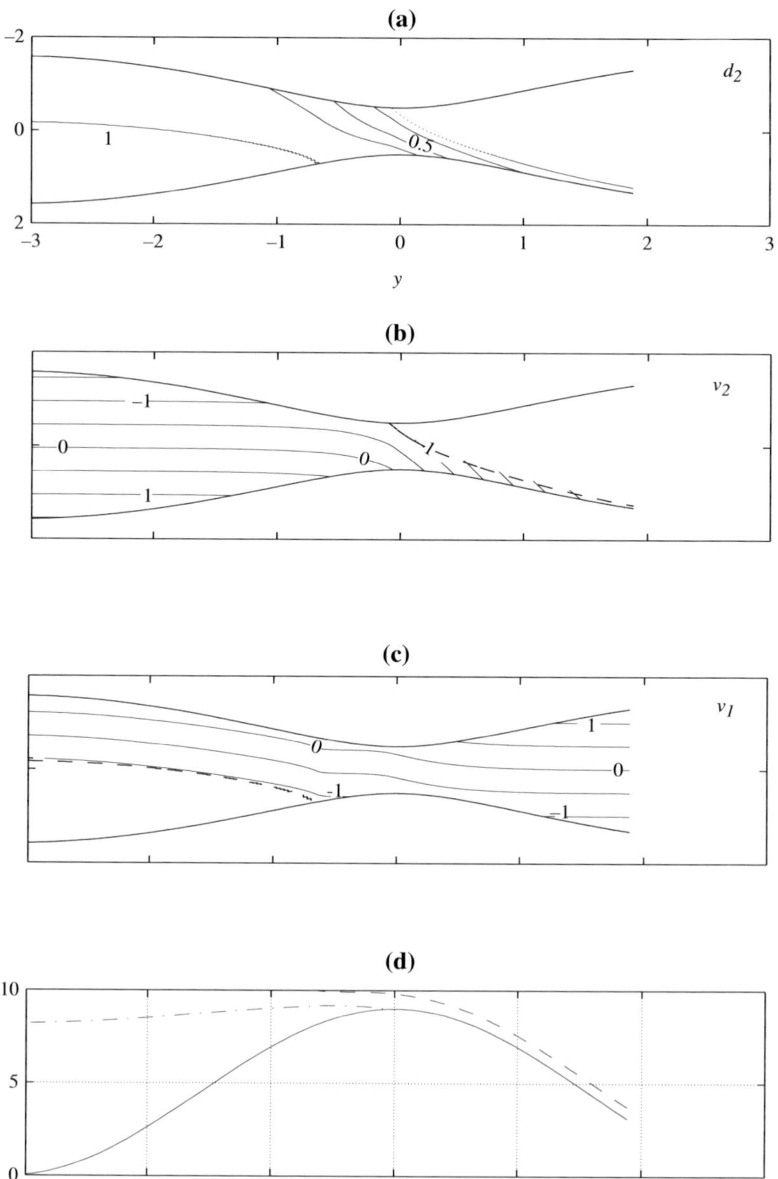

FIGURE 5.7.6. An example of submaximal flow based on the $Q^{1/2}/(\Delta B - z_T) = -1.4$ contour of Figure 5.7.5. The top three panels show d_2, v_2, v_1, respectively. The dashed curves in the middle two figures show where the edge of the layer in question contacts the bottom or top lid. The bottom panel is a side view showing the intersection of the interface with the left wall (dashed-dotted curve) and the right wall (dashed curve.) (Constructed from Reimenschneider *et al.*, 2005, Figure 12).

and laboratory verification of any of the solutions is poor. We also note that unlike the previous Froude number plane representation, in which the direction of the velocity in any particular layer is arbitrary, the present solution curves apply only to exchange flow.

c. *Connecting the Sill Flow to the Upstream Basin*

The solutions represented in the Froude number plane predict that strong recirculations will develop within at least one of the layers as the upstream or downstream basin is approached. This property derives from the global enforcement of the 'zero potential vorticity' condition $\partial v_n^*/\partial x^* = -f$. While this approximation will hold near the sill, or wherever else the layer thickness is much less than its potential depth $D_{n\infty}$, it becomes untrustworthy where the layer thickness becomes as large as $D_{n\infty}$. In the initial-value experiment suggested in Figure 5.7.1a, $D_{n\infty}$ is just the initial depth of layer n in the basin of origin. After the exchange flow has been initiated, the upstream depth of either layer can be expected to decrease a relatively small amount, and vorticity thus generated should also be small compared with f. In other words, the layers are expected to remain relatively quiescent in their basins of origin.

Let us now consider a model that incorporates this view and thereby acts as an alternative to the model with global zero potential vorticity. For simplicity, assume that the width w of the strait separating the two basins is constant and that all changes in total depth d occur within the strait (Figure 5.7.1b). The upstream basin formally begins at the channel *mouth*, where the channel starts to widen. We denote the width in this widening region $w_B(y)$ and note that the depth d there is already very large in comparison to D_s. The upper layer in this basin is expected to be relatively shallow and therefore subject to the zero potential vorticity approximation. The corresponding velocity and depth profiles are therefore determined by the single-layer expressions for zero potential vorticity: (2.2.29 and 2.2.30) for attached flow or (2.3.11 and 2.3.12) for separated flow. In the present context it is convenient to rewrite the expressions for attached flow as

$$v_1(x, y) = \frac{2\hat{d}(y)}{w_B} - x \qquad (5.7.26)$$

$$d_1(x, y) = \overline{d}(y) + \frac{2\hat{d}(y)}{w_B(y)} - x \qquad (5.7.27)$$

where $\overline{d}$ and $\hat{d}$ are one half the sum and difference of the depths at the side walls $x = \pm w_B/2$ (see 2.2.5 and 2.2.6). The associated flux and internal Bernoulli functions are given by

$$Q = -2\hat{d}\overline{d} \qquad (5.7.28)$$

and

$$\Delta B = -\frac{2\hat{d}^2}{w_B{}^2} - \frac{w_B{}^2}{8} - \overline{d} + z_T \tag{5.7.29}$$

where v_o and d_o are the velocity and depth at the side wall, here $x = -w_B/2$.

If the upper layer is detached, it is convenient to redefine $x = 0$ as lying at the left wall. The corresponding profiles are given by

$$v(x, y) = v_o(y) - x \tag{5.7.30}$$

and

$$d(x, y) = v_o(y)x - \frac{x^2}{2} + d_o(y), \tag{5.7.31}$$

where v_o and d_o are the left-wall velocity and depth. The flux and Bernoulli function for this case are given by

$$Q = d_o^2/2 \tag{5.7.32}$$

and

$$\Delta B = -\frac{v_o^2}{2} + z_T - d_o, \tag{5.7.33}$$

and the latter is obtained by evaluating (5.7.4) at the left wall. The separated current width w_e can be related to Q by

$$v_o = \frac{w_e}{2} - \frac{d_o}{w_e} = \frac{w_e}{2} - \frac{(2Q)^{1/2}}{w_e}, \tag{5.7.34}$$

which follows from setting $d = 0$ at $x = w_e$.

At this stage we have written down expressions for the upper layer velocity and depth that are valid in the vicinity of the mouth and points upstream. The derivation is based on the hypothesis that the lower layer is deep and quiescent and the upper layer is thin. In order to smoothly connect the flow at the mouth ($w = w_B$) to one of the previously described critical sill states, for which both layers are active, it is necessary that the two states have equal values of ΔB and Q. Since the mouth flow may be attached or detached, and the sill flow may lie in one of four configurations, a good deal of bookkeeping is required to check through all the possibilities. The general procedure is to begin with one of the critical sill flows represented in Figure 5.7.3 or 5.7.4, then check whether there is a mouth state, either attached or detached, with the same ΔB and Q. Proceeding thus will lead to one of three possibilities. First, there are no mouth states and therefore no smooth connection to the upstream basin. Examples are the aforementioned sill flows that lie at the maxima of the Figure 5.7.3 curves. The second possibility is that there are two mouth states, one subcritical and one supercritical. In this case, prior experience suggests selection of the subcritical root as the appropriate upstream state. The resulting solution can be considered submaximal. Finally, there exists a single, hydraulically critical, mouth state. The solution associated with the final possibility will have maximal flux and will have two controls, one at the mouth and one at the sill.

As an example, consider the case $w = 0.5$. The possible critical sill states are all attached and are specified along the middle curve in Figure 5.7.3. Begin at the lower left extremity of the curve, where both $\overline{d}_c$ and Q/w are small. For each location on this portion of the curve there are two possible mouth states, one supercritical and one subcritical. As one proceeds to the right, the value of Q/w increases and the two mouth states converge, eventually merging to a critical state. This state, which is indicated by a cross in the figure, is the maximal state for the w in question. For higher values of $\overline{d}_c$ there are no physically meaningful mouth states, and thus the whole right-hand portion of the curve is irrelevant to the problem at hand. The maximal Q/w for each case is always less than the maximum of the curve in question.

An interesting point of departure from the solutions with globally zero potential vorticity is in the location of upstream control. For the solutions under discussion, the upstream control lies at the mouth of the strait. For the Froude number plane solutions based on the global enforcement of zero potential vorticity, the upstream (virtual) control lies within the strait (see 5.7.20). Both results must answer to criticism; in the first case the upstream control is imposed by hypothesis, in the second the control may lie where the zero potential vorticity approximation fails. However, it turns out that the maximal fluxes predicted in either case differ by a negligible amount. The dots on the curves of Figure 5.7.3 indicate the maximal flow rate obtained when the zero potential vorticity relations are enforced globally. They indicate Q/w values that lie only slightly above the maximal fluxes obtained with a mouth control. Moreover, no virtual controls within the channel are found once the sill flow becomes separated.

A similar set of calculations can be carried out for separated sill flows. An overall view of the maximal solutions thus obtained for various values of w appears in Figure 5.7.7. It can be shown that the maximal Q/w for a given w is always less than the maximum value permitted by the critical condition alone. It is also generally true that the mean lower layer depth at the sill is less than the mean upper layer depth, and thus solutions such as shown in Figure 5.7.2c are ruled out. The maximal value of Q/w (Figure 5.7.8) decreases as the strength of rotation (the value of w) increases, provided that w is less than about 1.6. However this trend is reversed at higher values of w, corresponding to the transition between singly separated sill flow (dots) and doubly separated sill flow (crosses). The increase in the exchange transport with increasing rotation contrasts with the usual tendency in single-layer hydraulics for the flux to decrease with rotation.[6] The new trend is due in part to the zero potential vorticity model, which has no constraining boundary layer structure. In addition, the tendency for rotation to

---

[6] The trend for the maximal flux to increase when w exceeds 1–2 deformation radii is also predicted in the Reimenschneider et al., 2005 model.

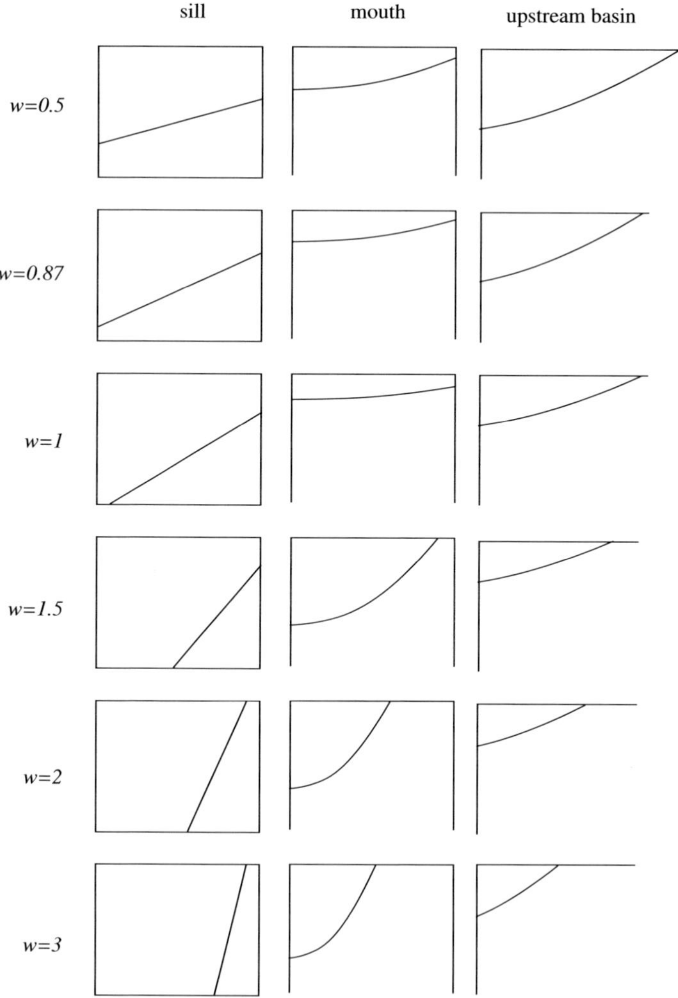

FIGURE 5.7.7. Cross sections of maximal exchange configurations at the sill, mouth and in the upstream basin. (From Timmermans and Pratt, 2005, Fig. 17).

squash a separated layer against its right-hand wall has different consequences in single- and two-layer flows. In the former, the effect is to constrict the cross section of the layer and thereby diminish the transport. In the two-layer setting, the squashing of one layer against a wall relieves the other layer. For comparison, the flux predicted by the Whitehead *et al.* (1974) model for symmetrical upper and lower layers (Equations 5.7.21 and 5.7.22) is indicated by a dashed line in Figure 5.7.8. The values overestimate the maximal exchange values at low and moderate rotation, but underestimates it at high rotation.

Laboratory and numerical experiments (e.g. Whitehead and Miller, 1979; Dalziel, 1988; Whitehead and Hunkins, 1992; Rabe *et al.*, 2007) with two-layer, rotating exchange flows have so far failed to reproduce double detachment of

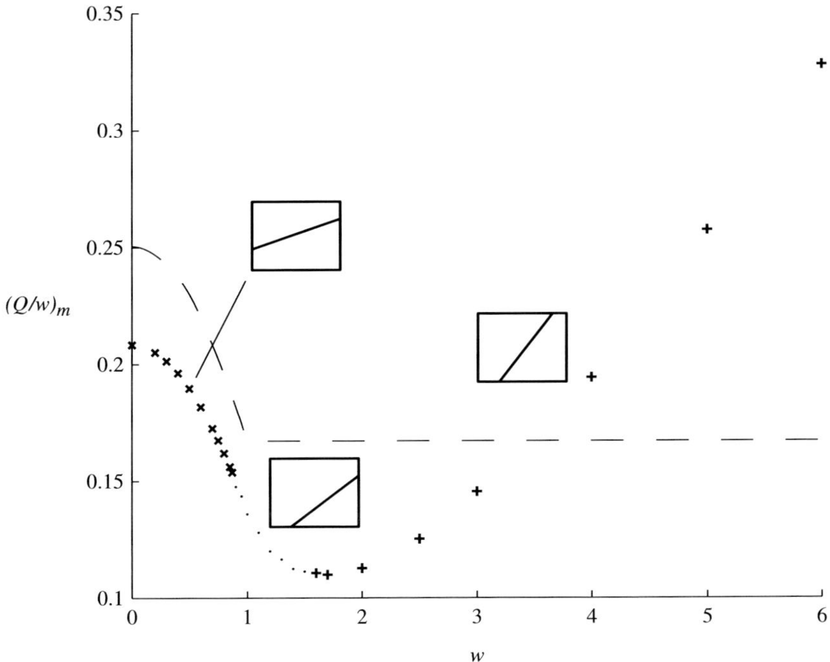

FIGURE 5.7.8. The nondimensional maximal volume flux per unit width $Q_{max}/w (= Q^*_{max}/w^* g'^{1/2} D_s^{1/2})$ for pure exchange flow. The geometry is shown in Figure 5.7.1 and the lower layer is assumed to be inactive in its deep upstream basin. The $\times$, $\bullet$ and $+$ correspond to attached, singly separated, and doubly separated sill flow, as shown by the insets. The dashed curve indicates the Q/w as given by the Whitehead et al. (1974) formula (5.7.21) for attached flow with symmetric upper and lower layers. The connecting dashed line gives the value $Q/w = 1/6$ given by the same authors for symmetrical, detached flow, assuming that regions occupied by only one of the layers are quiescent. (Based on Timmermans and Pratt, 2005, Figure 16).

the interface. Where double detachment is predicted ($w >$ about 1.5) the flow is instead observed to become unstable. The result is a time-dependent flow field marked by the presence of eddies and with mean features quite different from that predicted by the zero potential vorticity theory. The increase in maximal flux predicted beyond $w = 1.5$ has not been reproduced.

Is it possible to relate Q to some well-defined and easily measured property of the flow in the upstream basin? To investigate this question further it will be helpful to have a better understanding of the behavior of the upper layer in the basin. Any mouth state can be formally extended into the basin by allowing the width w_B to gradually increase from its value w at the mouth and requiring that Q and ΔB be conserved. In all cases, the upper layer will separate at sufficiently large w_B if it is not already separated at the mouth. Once separated, the flow will continue, unaltered, into the basin until some other process intervenes. The velocity and depth profiles of the separated upper layer are given by (5.7.30)

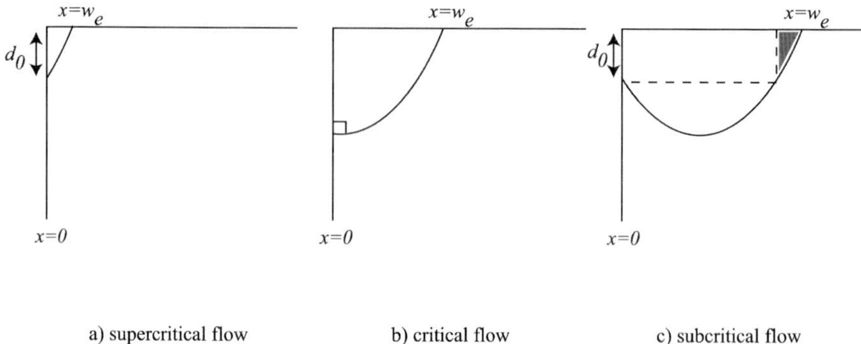

FIGURE 5.7.9. Cross section of the separated flow in the upstream basin. (From Timmermans and Pratt, 2005, Figure 5.7.7).

and (5.7.31). As explained in Section 2.3, the criticality of the flow may be identified by the presence or lack of velocity reversals. If the upper layer depth decreases monotonically away from the left wall (Figure 5.7.9a), so that v_1 is everywhere < 0, then the flow is supercritical. The presence of a depth maximum and a corresponding velocity reversal (Figure 5.7.9c) implies subcritical flow. In this case there will be a band of reverse flow along the left wall carrying fluid towards the mouth. If the maximum depth occurs at the wall (Figure 5.7.9b) then v_1 is zero there and the flow is critical.

If one begins by selecting a sill flow that is submaximal, then for that Q and w there are two possible mouth states, one supercritical and one subcritical. The usual stability considerations require that we select the subcritical root. This mouth state may be separated or nonseparated; in the latter case separation will occur within the wide basin. An example based on the case $w = 0.5$ (Figure 5.7.10a) shows an attached mouth flow that becomes separated in the basin. The anticipated band of counterflow is present along the wall. For the maximal state the mouth flow is critical. Stability considerations now require a transition to supercritical flow as the basin is approached (Figure 5.7.10b). The separated upstream state is distinguished from the submaximal case in that the current is narrower and contains no velocity reversals. Of course, the supercritical flow could pass through a hydraulic jump and lose these distinguishing characteristics.

The separated upstream width w_e is a clearly defined property of the basin flow and, as such, is a potentially convenient property on which to base a weir relation. An implicit weir formula for the case of attached sill flow can be developed by equating ΔB at the sill (see 5.7.15 with $d = 1$ and $h = z_T - 1$) with (5.7.33). With the help of (5.7.34) this equality can be written as

$$\left(\frac{w_e}{2} + \frac{(2Q)^{1/2}}{w_e} \right)^2 = 2 - 2\overline{d}_{2c} - \frac{Q^2\left(1 - 2\overline{d}_{2c}\right)}{w^2(\overline{d}_{2c}(1 - \overline{d}_{2c}) - w^2/12)^2}, \quad (5.7.35)$$

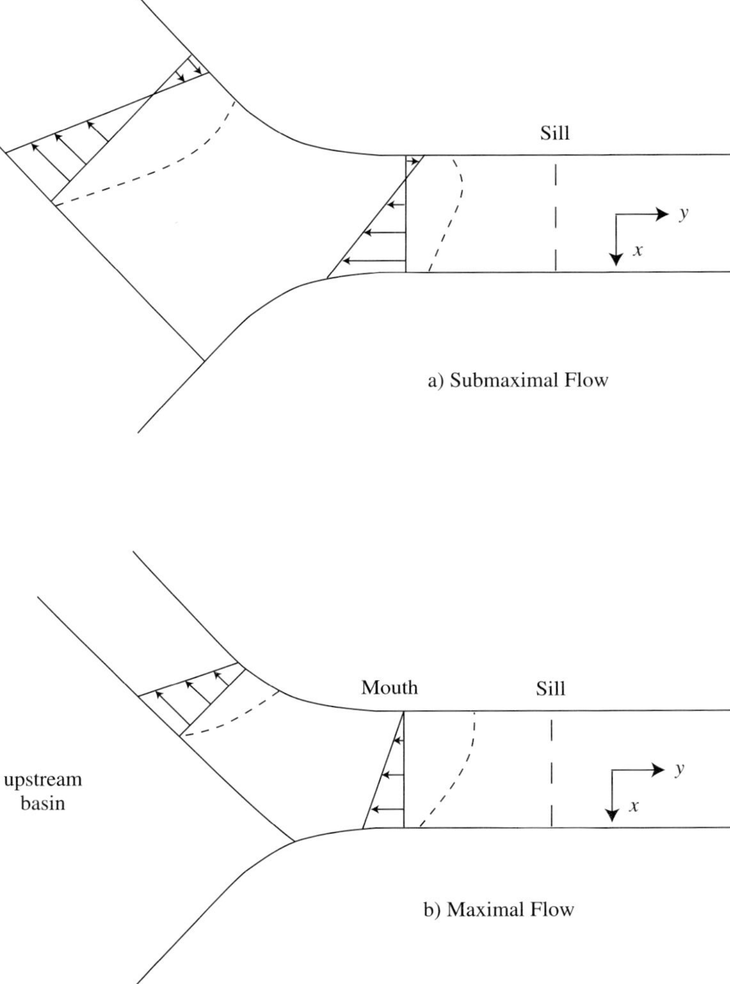

FIGURE 5.7.10. Plan views of the flow upstream of the sill. The upper layer velocity (solid) and depth profiles (dashed) are shown at the mouth (or entrance) of the strait and in the upstream basin after detachment has occurred. In both cases the sill width w and mouth width w_{m} are 0.5. Frame (a) shows a submaximal case with subcritical flow at the mouth and in the basin. Frame (b) shows the maximal flow with critical flow at the mouth and supercritical flow in the basin. (From Fig. 10 of Timmermans and Pratt, 2005).

where $\overline{d}_{2c}$ is determined by (5.7.16) as applied at the sill:

$$\frac{Q^2}{w^2} = \frac{\left(\overline{d}_{2c}\left(1-\overline{d}_{2c}\right) - w^2/12\right)^3}{\left(\overline{d}_{2c}\left(1-\overline{d}_{2c}\right) - w^2/12 + \left(1-2\overline{d}_{2c}\right)^2\right)}. \tag{5.7.36}$$

Together, (5.7.35) and (5.7.36) provide a relationship between Q/w, w, and w_e. Satellite or aircraft measurements of separated current width w_e might thereby provide an estimate of the flux, provide that the width is clearly defined.

d.   The Strait of Gibraltar Revisited

Conventional wisdom dictates that rotation is not a major influence in dynamically narrow straits such as Gibraltar and the Bab al Mandab. In the case of Gibraltar (Figure 5.6.1) the value of $(g'D_s)^{1/2}/f$ based on the average sill depth $D_s = 200\,\mathrm{m}$, and $g' = 0.02\,\mathrm{ms}^{-2}$ is 23 km, which is significantly greater than the width $w^* \cong 13\,\mathrm{km}$ at Tarifa Narrows. However, this gross estimate fails to account for the behavior of the upper layer at the eastern end of the strait, where it shallows and accelerates. Acoustic images and CTD sections have shown that this layer can separate from the northern coast at some point between Tarifa and Gibraltar. Corroboration of this phenomenon can be found in photographs of the strait from space (e.g. Figure 5.7.11), which suggest detachment of the (lighter) surface layer and outcropping of the (darker) lower layer near the northeastern corner of the strait.

We have already documented the tendency of maximal flows to produce an upstream separated width w_e that is significantly smaller than that for submaximal flow. The observed separation width ($w_e^* = 15 \pm 1\,\mathrm{km}$) is therefore a potential discriminator. Although the idealized theory does not account for barotropic flow, nor for the geometric complexities of the Strait, it is still instructive to compare the predicted maximal and submaximal values of w_e with observations.

FIGURE 5.7.11. October 1984 space shuttle photograph of the Strait showing a (dark) area thought to be the lower layer outcrop south of Gibraltar. The strait width at the narrowest section is about 13 km, while that at the separation point is 15± 1 km. (NASA, LBJ Space Center Photo S-17-34-080).

To do so, fix the width of the idealized, constant width channel as $w = .57$, which corresponds to the above estimate $w^* f/(g'D_s)^{1/2} = 13/23$. As shown by Timmermans and Pratt (2005) the corresponding $w_e{}^*$ for maximal flow is 15 km, within the range of uncertainty of the observed value, whereas submaximal solutions would have $w_e{}^* > 32$ km. The predicted volume flux for the maximal case is $Q^* = 0.92 \pm 0.03$, which lies at the upper end of the estimate 0.78 ± 0.17 for the upper layer flux obtained by Tsimplis and Bryden (2000). A number of refinements are called for here, but the results suggest that the flow was maximal at the time of the observations.

Exercises

1) Show that the 'zero potential vorticity' limit $D_s \ll D_{i\infty}$ is equivalent to the limit in which the global Rossby deformation radius (5.1.12) is large compared to the 'local' deformation radius $(g'D_s)^{1/2}/f$.

2) *Barotropic 'similarity' solution for flow through a pure contraction.*
   Suppose that for two-layer, zero potential vorticity flow, the layer fluxes Q_1 and Q_2, both > 0, are specified. Then show that equation (5.7.15) admits a barotropic solution for which the interface is horizontal ($\bar{v}_1 = \bar{v}_2$) and the layer depths are constant in x and y. Show that the values of $\bar{v}_1$, $\bar{v}_2$, and the layer depths are determined from

$$\bar{v}_1(y) = \bar{v}_2(y) = \frac{Q_1 + Q_2}{w(y)d} \quad \text{and} \quad d_2 = d - d_1 = \bar{d}_2 = \frac{Q_2 d}{Q_1 + Q_2}.$$

3) *Critical conditions for singly-detached sill flow.*

   (a) Consider a singly-detached exchange flow where the interface between the two layers intersects either $z = z_T$ or $z = 0$. For a positive interface slope $(\bar{v}_2 - \bar{v}_1 \geq 0)$, show that the flow detaches from the right wall $(x = w/2)$ when

$$\bar{d}_2 \geq d - (\bar{v}_2 - \bar{v}_1)\frac{w}{2},$$

   and that it detaches from the left wall $(x = -w/2)$ when

$$\bar{d}_2 \leq (\bar{v}_2 - \bar{v}_1)\frac{w}{2}.$$

   (b) With the origin $x = 0$ positioned as the left wall (as in Figure 5.7.2b), show that the velocity and depth profiles for the case of lower layer separation are given by

$$\bar{v}_i(x, y) = w - w_s - x + \hat{v}(y)$$

$$d_1(x, y) = \begin{cases} d(y) - \hat{v}_-(x + w_s - w) & (x \geq w - w_s) \\ d(y) & (x < w - w_s) \end{cases} \qquad (5.7.37)$$

and

$$d_2(x, y) = \begin{cases} \hat{v}_-(x + w_s - w) & (x \geq w - w_s) \\ 0 & (x < w - w_s) \end{cases}$$

where the $\hat{}$ over a variable implies its value at $x = w - w_s$, the point where the interface intersects the bottom of the channel, and $\hat{v}_\pm = \hat{v}_2 \pm \hat{v}_1$.

(c) Show that the volume fluxes in the two layers are given by

$$Q_2 = w_s^2 (\hat{v}_2 - \hat{v}_1) \left(\frac{\hat{v}_2}{2} - \frac{w_s}{3} \right) \tag{5.7.38}$$

and

$$Q_1 = dw \left(\hat{v}_1 - w_s + \frac{w}{2} \right) - w_s^2 (\hat{v}_2 - \hat{v}_1) \left(\frac{\hat{v}_1}{2} - \frac{w_s}{3} \right). \tag{5.7.39}$$

(d) By evaluating the Bernoulli function where the interface intersects the bottom, show that

$$\Delta B = \frac{\hat{v}_+ \hat{v}_-}{2} + z_T - d, \tag{5.7.40}$$

where $\hat{v}_\pm = \hat{v}_2 \pm \hat{v}_1$. Assuming the net transport to be zero ($Q = -Q_1 = Q_2 > 0$), show that (5.7.38), (5.7.39), and (5.7.40) can be written as

$$G_1 \left(\hat{v}_-, \hat{v}_+, w_s; d, w \right) = w_s^2 \hat{v}_-^2 - dw(\hat{v}_- - \hat{v}_+ - w + 2w_s) = 0, \tag{5.7.41}$$

$$G_2 \left(\hat{v}_-, \hat{v}_+, w_s \right) = w_s^2 \hat{v}_- \left(\frac{\hat{v}_+ + \hat{v}_-}{4} - \frac{w_s}{3} \right) - Q = 0, \tag{5.7.42}$$

and

$$G_3 \left(\hat{v}_-, \hat{v}_+; d \right) = \hat{v}_+ \hat{v}_- - 2 \left(\Delta B - z_T + d \right) = 0. \tag{5.7.43}$$

(e) By interpreting (5.7.41–5.7.43) as three functions in the three variables $\hat{v}_-$, $\hat{v}_+$ and w_s, show that the critical conditions is.

$$\hat{v}_- w_s \left\{ wd \left(3\hat{v}_+^2 - 6\hat{v}_+ w_s + 4w_s^2 \right) - 6\hat{v}_- wd \left(-\hat{v}_+ + 2w_s \right) \right.$$
$$\left. + \hat{v}_-^2 \left(3wd + 2w_s^2 \left(-3\hat{v}_+ + 4w_s \right) \right) \right\} = 0. \tag{5.7.44}$$

4. Show that a symmetric state of marginal separation, in which the interface contacts the upper right and lower left corners, occurs when $d = (\bar{v}_2 - \bar{v}_1)w$ and $\bar{v}_2 = -\bar{v}_1$. Show that this state is critical when $d = w^2$.

5. *Characteristic speeds under conditions of attachment.* By observing that the profiles (5.7.5–5.7.7) are valid for time-dependent flow, show that the equation for conservation of y-momentum is given by

$$\frac{\partial \hat{v}}{\partial t} + \left[\hat{v}\left(1 - \frac{2\overline{d}_2}{d}\right) + v_b\right]\frac{\partial \hat{v}}{\partial y} + \left(1 - \frac{\hat{v}^2}{d}\right)\frac{\partial \overline{d}_2}{\partial y} = 0,$$

where $\hat{v} = \overline{v}_2 - \overline{v}_1$ denotes the shear velocity and $v_b = (\overline{v}_1\overline{d}_1 + \overline{v}_2\overline{d}_2)/d$ denotes the (constant) barotropic velocity. Further show that the continuity equation for either of the layers may be written as

$$\frac{\partial \overline{d}_2}{\partial t} + \left[(d - \overline{d}_2)\frac{\overline{d}_2}{d} + \frac{w^2}{12}\right]\frac{\partial \hat{v}}{\partial y} + \left[\hat{v}(1 - \frac{2\overline{d}_2}{d}) + v_b\right]\frac{\partial \overline{d}_2}{\partial y} = 0.$$

From these two expression deduce the characteristic speeds $c_{\pm}$ and show that they can put in the simplified form given by (5.7.18).

5.8. Maximal Exchange Trough a Pure Contraction with Rotation

One of the fundamental problems of two-layer hydraulics is that of full lock exchange in a channel with a horizontal bottom. We have covered the case of zero background rotation and described the maximal exchange that is set up by the removal of the full-depth barrier. If the exchange occurs through a narrows, the flow is critical there and is joined to the end reservoirs by supercritical flows. The solution in the vicinity of the narrows is represented by the curve *jbk* of Figure 5.4.1b. In the rotating version of this problem we again imagine a barrier separating fluids of slightly different densities, ρ_1 and ρ_2, positioned at the narrowest section of a channel that is rectangular in cross section. The channel has constant depth D and is capped by the usual rigid lid. When the barrier is removed the two fluids begin to move in opposite directions, thrusting under and over each other as before. With Northern Hemisphere rotation, each layer veers to its right as it intrudes. The potential depth $D_{n\infty}$ of layer n is its initial depth D, whereas the average depth of each layer is at the narrows after adjustment $D/2$. The severe vortex squashing that is inherent in rotating flow over a shallow sill is therefore lacking and use of the zero potential vorticity approximation, which depends on this process, is no longer valid. The nondimensional potential vorticity, while uniform in each layer, must now be considered nonzero.

Two approaches have historically been used to predict the steady flow state that results from the lock exchange problem without rotation. In the first, one simply assumes that the flow at the original position of the dam becomes hydraulically critical ($G^2 = 1$). This along with the symmetry properties of the full lock exchange are sufficient to find the final steady state, namely the aforementioned hydraulic solution. Since the layer depths at the critical section will be equal, the exchange will be maximal for a contraction. The second approach is based on an energy balance for the evolving solution (Yih, 1980). The calculation is straightforward when the channel is considered uniform in y^*, and perhaps linked to reservoirs far upstream and downstream of the initial barrier. Destruction of the barrier results in the formation of the intrusions suggested in Figure 5.8.1a. An idealization of the lower intrusion is that it consists of a nose region, followed by a steady flow with uniform depth $D/2$ and uniform velocity v^*. The upper layer intrusion has the same depth but equal and opposite velocity. The energy argument attempts to calculate v^*, and thereby the layer fluxes, by equating the potential energy lost to kinetic energy gained as a result of the adjustment. The procedure is unable to account for dissipation that might occur in the nose regions, or for the detailed structure of those regions.

At the time frame shown in Figure 5.8.1a, the steady portions of the intrusions occupy a distance L. The two-layer density distribution in this region (Frame c) can be created from the initial distribution (Frame b) by interchanging regions I and II. Doing so lifts each parcel in II a distance $D/2$, thereby increasing its potential energy by amount $g\rho_1 D/2$. The volume of region II is $w^*(D/2)(L/2)$

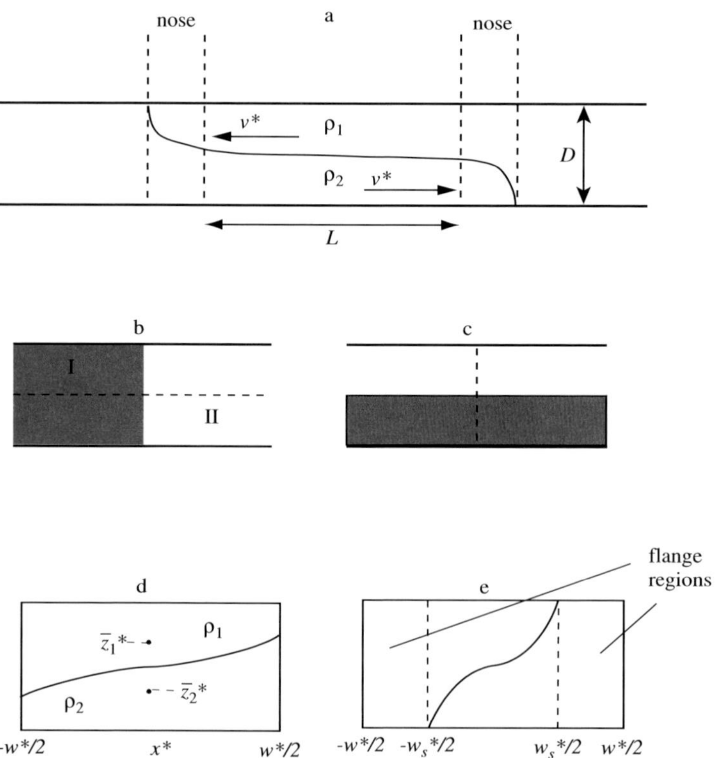

FIGURE 5.8.1. The gravitational advance of two mutual intrusions of differing density in a channel (Frame a). Side views of the initial (b) density distribution and an idealization of the adjusted density (c). Schematic cross-section for the adjusted flow in the lock exchange with rotation, with and without separation (d and e).

and thus the total increase in potential energy is $g\rho_1 D^2 L/8$. Similarly, region I sees a $D/2$ drop in elevation, and a total decrease in potential energy equal to $g\rho_2 D^2 L/8$. The net decrease in potential energy is therefore $g(\rho_2 - \rho_1)D^2 Lw^*/8$. The kinetic energy gained in the same volume is clearly $(\rho_1 v^{*2} + \rho_2 v^{*2})DLw^*/4$. Equating the potential energy lost to the kinetic energy gained leads to

$$v^* = \sqrt{\frac{gD(\rho_2 - \rho_1)}{2(\rho_1 + \rho_2)}}, \tag{5.8.1}$$

which is equivalent to $G^2 = 1$. Thus the energy balance and the hypothesis of critical flow lead to the same result.

If the fluid is rotating, a similar energy balance can be invoked, but now the cross-stream variation of speed and depth must be included. The following generalizes a calculation due to Hunkins and Whitehead (1992). It is again assumed that the channel is uniform, the adjusted flow is steady and that there is no energy loss. As we will see later, the assumption of steadiness becomes questionable when the channel width exceeds the internal radius of deformation L_I, as given by (5.1.12). The latter is simplified by $D = D_{1\infty} = D_{2\infty}$, leading to $L_I = \sqrt{g'D/2f^2}$. Since the upper and lower layer should behave as mirror images, the interface height will be antisymmetric with respect to x^*. The solutions to (5.1.11) with this property are

$$d_1^* = \frac{D}{2} - A \sinh \frac{x^*}{L_I}, \tag{5.8.2}$$

and

$$d_2^* = \frac{D}{2} + A \sinh \frac{x^*}{L_I}. \tag{5.8.3}$$

The corresponding along-channel velocities are determined by (5.1.8) and (5.1.9) as

$$v_1^* = V^* - \frac{fx^*}{2} - \frac{fL_I}{D} A \cosh \frac{x^*}{L_I} \tag{5.8.4}$$

$$v_2^* = V^* - \frac{fx^*}{2} + \frac{fL_I}{D} A \cosh \frac{x^*}{L_I}, \tag{5.8.5}$$

where V is a constant. The net (barotropic) flux $Q_b = \int_{-w^*/2}^{w^*/2} (v_1^* d_1^* + v_2^* d_2^*)dx$ can be shown to be equal to V^*Dw^* and thus $V^* = 0$ for zero net exchange.

It remains only to determine the coefficient A in terms of the channel width w^*/L_I. Following the previous argument, the loss of potential energy is equated with the gain in kinetic energy of the hypothetical steady state. The calculation is a bit more involved because both the layer depths and velocities vary across the channel. The interested reader may wish to work through Exercise 1 to see the details and to verify the result:

$$g' \int_{-w^*/2}^{w^*/2} \left(\frac{D^2}{2} - d_2^{2*} \right) dx^* = \tfrac{1}{4} \int_{-w^*/2}^{w^*/2} \left(v_2^{2*} d_2^* + v_1^{2*} d_1^* \right) dx^* \qquad (5.8.6)$$

Use of (5.8.2—5.8.5) then yields

$$\frac{w^{3*}}{24 L_I^3} + \frac{A^2}{D^2} \left[\frac{w^*}{L_I} \left(1 - \cosh \frac{w^*}{L_I} \right) + 4 \sinh \frac{w^*}{L_I} \right] = \frac{w^*}{L_I} \left(1 + \frac{2A^2}{D^2} \right) \qquad (5.8.7)$$

The volume flux of each layer is given by

$$Q_1^* = -Q_2^* = \int_{-w^*/2}^{w^*/2} d_1^* v_1^* dx^* = f A L_I \left(\frac{w^*}{2} \cosh \frac{w^*}{2L_I} - 2 L_1 \sinh \frac{w^*}{2L_I} \right). \qquad (5.8.8)$$

The layer fluxes can be calculated as a function of w^*/L_I by eliminating A between this equation and (5.8.7). The resulting relation is shown as a dashed curve in the left-hand portion of Figure 5.8.2. It is not difficult to verify that the limiting case of zero rotation (5.8.1) is approached for $w^*/L_I \to 0$. For comparison, the flux predicted by zero potential vorticity theory (Equation 5.7.22) is shown as a dotted curve.

The foregoing analysis is valid as long as the interface stays in contact with both channel walls. Separation first occurs when $d_1^*(w^*/2) = d_2^*(-w^*/2) = 0$, which corresponds to $w^* = w_s^*$, where

$$A \sinh \frac{w_s^*}{2L_I} = \frac{D}{2},$$

in view of (5.8.2). If this expression is then used to eliminate A from (5.8.7), one obtains $w_s^*/L_I = 1.459$ and $A/D = 0.628$ (correcting a numerical mistake in Hunkins and Whitehead, 1992). The corresponding value of Q_2^* is given by $.208 g' D^2/f$.

As an alternative to the energy argument, one might invoke a critical condition for the adjusted flow. The latter has been derived for the case of the flow in question by Rabe *et al.* (2007) as

$$A^2 \frac{8L_I}{w^*} \cosh^2 \frac{w^*}{2L_I} \sinh \frac{w^*}{2L_I} - 2 \cosh \frac{w^*}{2L_I} + \frac{w^*}{2L_I} \sinh \frac{w^*}{2L_I} = 0 \qquad (5.8.9)$$

The mechanics of this calculation are discussed in Exercise 3. The flux is obtained by eliminating A between (5.8.8) and (5.8.9) and is shown by the solid curve in the left-hand portion of Figure 5.8.2. The limiting separation width and flux are given by $w^*/L_I = 1.63$ and $Q_2^* = .176 g' D^2/f$. Note that, unlike the case of zero rotation, these results differ from what is predicted by the energy argument.

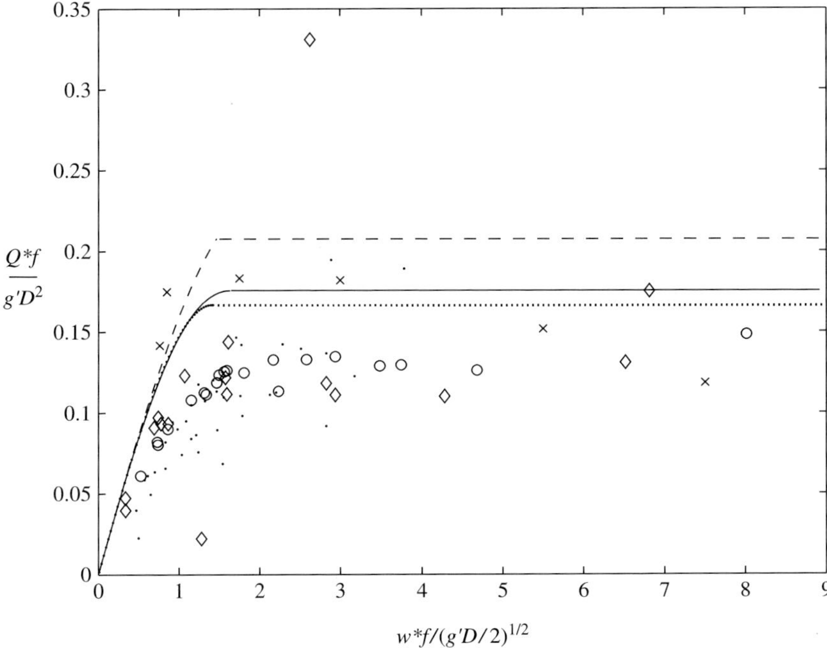

FIGURE 5.8.2. The dimensionless volume flux *vs* dimensionless channel width. The three curves give predictions based on the theories of Whitehead *et al.*, 1974 (dotted); Hunkins and Whitehead 1992 (dashed); and Rabe *et al.*, 2007 (solid). The termination of each curve has been extended as a horizontal line to indicate the width-independent flux hypothesized for the case in which the flow detaches from the sidewalls. The data points correspond to laboratory experiments by Hunkins and Whitehead 1992 (small dots); Dalziel 1988 (circles); and Rabe *et al.*, 2007 (diamonds). Note the factor $2^{1/2}$ change in the scale for w^* relative to Section 2.7.

For $w^* > w_s^*$ the flow becomes separated from both walls resulting in central, baroclinic region $(-w_s^*/2 < x^* < w_s^*/2)$ occupied by both layers and flange regions occupied by only layer (Figure 5.8.1e). In the case of zero potential vorticity, the flange regions contain horizontal shear equal to $-f$. In the present setting, where the flange region depth D is also the potential depth of the occupying layer, the shear is zero. If the velocity remains continuous throughout each layer, the uniform velocity v_n^* in a flange region must equal the velocity at the edge of the barotropic region. For example, the value of v_2^* in the right-hand flange $w_s^*/2 < x^* < w^*/2$ must equal its value at $x^* = w_s^*/2$. The corresponding flow must be included in the total flux for layer 2 and this added flux increases in proportion to the channel width w^*. An alternative idea, proposed by Whitehead et al. (1974) and also used by Hunkins and Whitehead (1992), is that the flange regions are motionless. In this case the layer velocities are discontinuous at the edges of the central region. The layer fluxes then become independent of w^* and are just those given by the above theories in the limiting case $w^* = w^*_s$. These values are summarized by

$$\frac{Q_2^* f}{g' D^2} = \begin{cases} 0.208 \ (w^*/L_I > 1.46.. \ \text{WH92}) \\ 0.176 \ (w^*/L_I > 1.63 \ldots \text{Rabe et al., 06}) \\ \frac{1}{6} \ (w^*/L_I > \sqrt{2}, \ \text{WLK74}) \end{cases} \qquad (5.8.10)$$

The w^*-independent flux values given by (5.8.10) are indicated by the dashed horizontal lines that extend the three curves in Figure 5.8.2. These lines give the hypothetical flux determined by the respective theory for $w^* > w^*_s$, assuming that the flange regions are quiescent.

Rotating lock exchange flow has been simulated in a number of laboratory experiments, including Whitehead *et al.* (1974), Dalziel (1988), and Hunkins and Whitehead (1992). In addition Rabe *et al.* (2007) carried out both laboratory and numerical simulations. Some of the experiments make use of a uniform length of channel, as assumed in the energy argument, that separate two broad reservoirs. In other cases the reservoirs are separated by a smoothly varying channel that has a miminun width at a single section. Photographs from the Hunkins and Whitehead experiment show four realizations of the experiment for different Rossby radii (Figure 5.8.3), all in the range of predicted separation. Each member of a pair of images shows the same parameter values: in the first frame the (leftward moving) fresh layer is dyed and in the second the (rightward moving) salt water is dyed. Eddies and filaments are evident on the interface between the two fluids in all cases. None of the flows exhibit complete separation from both sidewalls.

The value of $Q_1^* (= Q^*)$ can be measured indirectly by replacing the barrier after a set time, mixing the two fluids within each reservoir, and calculating the density change between the two mixed reservoirs. In the case of Rabe *et al.* (2006) direct velocity measurements based on digitally imaged drifting particles were also used to estimate the flux. The fluxes measured in all the experiments show increased scatter (as do the predictions) as the channel width is increased relative to L_I (Figure 5.8.2). The three theories are quite close, and do a good job of predicting the flux, when w^*/L_I is small. The experimental data are generally overestimated by predictions. The most problematic region over which to compare theory and observation is that where flow separation is predicted, roughly $f w^*/(g' D/2)^{1/2} > 1.5$ in Figure 5.8.2. Although double separation of the flow is predicted, it is not observed in any of the experiments. In some cases, separation of just one layer is observed, in others the picture is clouded by the presence of eddies. The predictions for $Q_2^* f / g' D^2$ are given by the horizontal lines, and these capture the weak dependence on the dimensionless channel width that is observed. Improvement in the prediction of the fluxes themselves in this regime will require a better understanding of the flow itself.

We also note that the $\frac{1}{6}$ coefficient in equation (5.8.10) has been used as an estimate for flux in applications to Spencer Gulf, South Australia, (Bye and

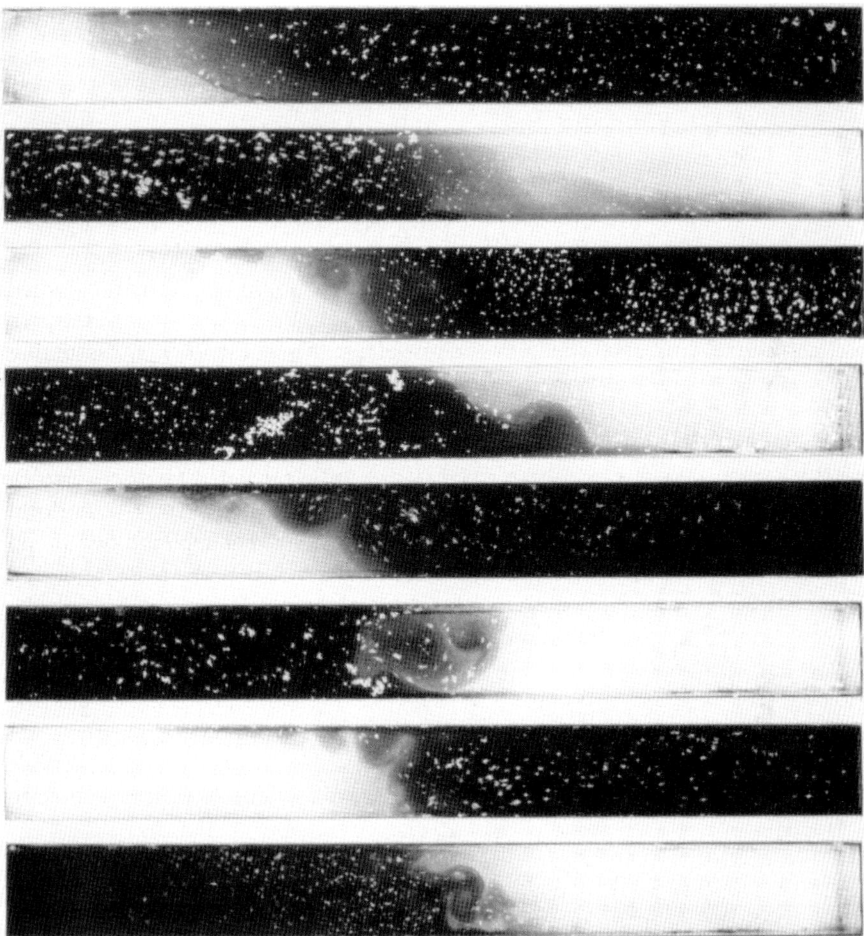

FIGURE 5.8.3. Top view of dye spreading after approximately 60 seconds. Starting from the top, the photographs show successively the layer of dyed freshwater (originating on the right) and then dyed saltwater (originating on the left). Each pair downwards has more rapid rotation, with $w +^* f/(g'D)^{1/2} = 1.2$, 2.5, 5.0 and 10.0, respectively. [Based on the experiments of Hunkins and Whitehead (1992).]

Whitehead, 1975) and Chesapeake Bay, North America (Whitehead, 1989b). Hunkins and Whitehead (1992) used the top version of (5.8.10) to estimate an exchange of freshwater for the Arctic Ocean. They found that the exchange was consistent with an observed salinity difference of about 1 psu between outflowing water and inflowing water at Fram Strait. The agreement is necessarily very crude because of factors such as wind and rafted sea-ice. Moreover, it is not known whether hydraulic control, let alone maximal exchange, occurs in any of these applications.

Exercises

(1) *Energy balance for a rotating lock exchange.* Consider the hypothetical adjusted state that results from lock exchange in a rotating channel. The steady exchange flow has a cross-section suggested in Figure 5.8.1d and the flow itself extends over a distance L, as in frame a.

(a) Show that the levels of centers of mass, $\bar{z}_1$ and $\bar{z}_2$, of the two moving layers are given by

$$\bar{z}_2 = \frac{1}{wD} \int_{-w/2}^{w/2} d_2^2 \, dx$$

$$\bar{z}_1 = \frac{1}{wD} \int_{-w/2}^{w/2} (D^2 - d_2^2) \, dx$$

(b) Argue, perhaps using a sketch, that that loss of potential energy of the flow within the length L is given by

$$\tfrac{1}{2} g(\rho_2 - \rho_1) L w^* (\bar{z}_1 - \bar{z}_2)$$

(c) Equate this loss with the gain in kinetic energy:

$$\tfrac{1}{2} L \int_{-w^*/2}^{w^*/2} (\rho_1 d_1^* v_1^{*2} + \rho_2 d_2^* v_2^{*2}) \, dx^*$$

and thereby verify, after use of the Boussinesq approximation, Equation (5.8.6).

(2) Show that using (5.8.7) in (5.8.8) produces a formula that approaches equation (5.8.1) as rotation becomes small.

(3) *Derivation of critical condition for two-layer flow with both potential depths equal to D:* Consider an attached, two-layer flow with $D = D_{1\infty} = D_{2\infty}$, as occurs in the lock exchange.

(a) By referencing the appropriate relations in Section 5.1, show that the velocity and depth profiles are given by

$$d_1{}^* = \frac{D}{2} - A \sinh \frac{x^*}{L_I} - C \cosh \frac{x^*}{L_I}, \qquad (5.8.11)$$

$$d_2{}^* = \frac{D}{2} + A \sinh \frac{x^*}{L_I} + C \cosh \frac{x^*}{L_I}, \qquad (5.8.12)$$

$$v_1{}^* = V^* - \frac{f x^*}{2} - \frac{f L_I}{D} \left[A \cosh \frac{x^*}{L_I} + C \sinh \frac{x^*}{L_I} \right], \qquad (5.8.13)$$

and

$$v_2{}^* = V^* - \frac{fx^*}{2} + \frac{fL_I}{D}\left[A\cosh\frac{x^*}{L_I} + C\sinh\frac{x^*}{L_I}\right].\qquad(5.8.14)$$

(b) Show using these expressions that the layer fluxes are given by

$$Q_1{}^* = \tfrac{1}{2}DV^*w^* + \tfrac{1}{2}AfL_Iw^*\cosh\left[\frac{w^*}{2L_I}\right]$$

$$- 2L_I(AfL_I + CV^*)\sinh\left[\frac{w^*}{2L_I}\right] + (ACfL_I^2/D)\sinh\left[\frac{w^*}{L_I}\right]$$

and

$$Q_2{}^* = \frac{1}{2}DV^*w^* - \frac{1}{2}AfL_Iw^*\cosh\left[\frac{w^*}{2L_I}\right]$$

$$+ 2L_I(AfL_I + CV^*)\sinh\left[\frac{w^*}{2L_I}\right] + (ACfL_I^2/D)\sinh\left[\frac{w^*}{L_I}\right]$$

$$(5.8.15)$$

and therefore that the net (barotropic) flux is given by

$$Q_b{}^* = Q_1{}^* + Q_2{}^* = DV^*w^* + (2CAfL_I^2/D)\sinh\left[\frac{w^*}{L_I}\right].\qquad(5.8.16)$$

b. Evaluate the internal Bernoulli function along the right wall and show that

$$\Delta B^*\Big|_{x^*=w^*/2} = \frac{fL_I}{2D}\left\{\begin{array}{l} 2DfL_I + (4CfL_I + 4AV^* - Afw^*)\cosh\left(\dfrac{w^*}{2L_I}\right) \\[2mm] + (4AfL_I + 4CV^* - Cfw^*)\sinh\left(\dfrac{w^*}{2L_I}\right)\end{array}\right\}$$

$$(5.8.17)$$

c. Note that (5.8.15–5.8.17) define three hydraulic functions in the three variables A, C, and V^*. Apply (1.5.14) to these functions to obtain the critical condition

$$f^2L_dw^*\left[(C^2+A^2)L_d\sinh\left(\frac{w^*}{L_d}\right) - D^2w^*\right]\cosh\left(\frac{w^*}{L_d}\right)$$

$$= \left[\begin{array}{c} w^*\left((A^2-C^2)f^2L_d{}^2 + D^2(f^2(2L_d^2+w^{*2}/8) - 2V^{*2})\right) \\[2mm] -4fL_d^2\left((C^2+A^2)fL_d - 2CAV^*\right)\sinh\left(\dfrac{w^*}{L_d}\right) \end{array}\right]\sinh\left(\frac{w^*}{L_d}\right)$$

$$(5.8.18)$$

(Note that a symbolic manipulation program will be helpful in doing the bookkeeping.)

    d. For conditions of zero net flux, and an asymmetrical interface (w.r.t. x^*), show that (5.8.18) reduces to the desired condition (5.8.9).

5.9. Strangulation of the Baroclinic Flow by the Barotropic Flow

Our treatment of the rotating, two-layer hydraulics is concluded with a discussion of the case in which the channel is dynamically wide, at least on the scale of the global internal Rossby radius of deformation. This limit is quite the opposite of what is assumed in the theory of zero potential vorticity flow. As we shall see, the internal or 'baroclinic' dynamics occur within right- and left-wall boundary layers that are physically separated from each other. Although the theory for this flow has not been widely applied, it is useful in illustrating a physical process that was either lacking or hidden on our previous discussions. This process is the forcing of the baroclinic boundary layers by the barotropic (depth independent) part of the flow. Unlike the former, the latter extends all the way across the channel and is altered by changes in the channel width or depth. The baroclinic flow is forced directly through interactions with the barotropic flow, leading to some novel and unexpected behavior. For one thing, a new type of hydraulic control emerges. The following treatment is based on a model developed by Pratt and Armi (1990) and having uniform potential vorticity in each layer.

    Consider a channel with varying width but constant bottom elevation ($h^* = 0$), so that $z_T{}^*$ is the total depth (Figure 5.9.1a). The potential vorticity is uniform in each of the two layers, the corresponding potential depths are $D_{1\infty}$ and $D_{2\infty}$, and the internal Rossby radius of deformation L_I is given by (5.1.12). We will only treat flows for which the layer depths are finite all across the channel, so that the interface contacts the two sidewalls and not the top or bottom. Solving (5.1.11) for the lower layer depth $d_2{}^*$, and using the rigid lid constraint $d_1{}^* = z_T{}^* - d_2{}^*$ yield the following depth profiles

$$d_1(x, y, t) = -\left[\frac{(\eta_+(y, t) + \eta_-(y, t))\cosh(x)}{2\cosh(w(y)/2)} + \frac{(\eta_+(y, t) - \eta_-(y, t))\sinh(x)}{2\sinh(w(y)/2)}\right] + \Delta$$
$$\tag{5.9.1}$$

$$d_2(x, y, t) = \left[\frac{(\eta_+(y, t) + \eta_-(y, t))\cosh(x)}{2\cosh(w(y)/2)} + \frac{(\eta_+(y, t) - \eta_-(y, t))\sinh(x)}{2\sinh(w(y)/2)}\right] + \Delta/\hat{\delta}$$
$$\tag{5.9.2}$$

where $\Delta = D/(D_{1\infty} + D_{2\infty})$, $\hat{\delta} = D_{1\infty}/D_{2\infty}$ and where the nondimensional variables

$$d_n = d_n{}^*/D_{1\infty}, \quad v_n = v_n{}^*/fL_I, \quad \text{and} \ (x, y) = (x^*, y^*)/L_I.$$

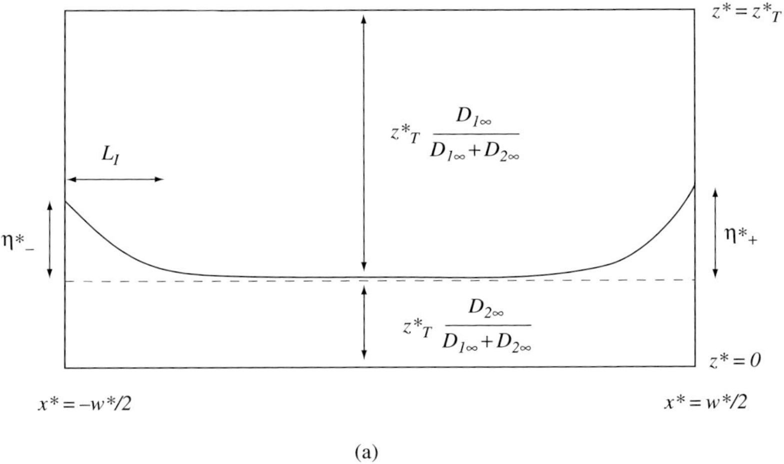

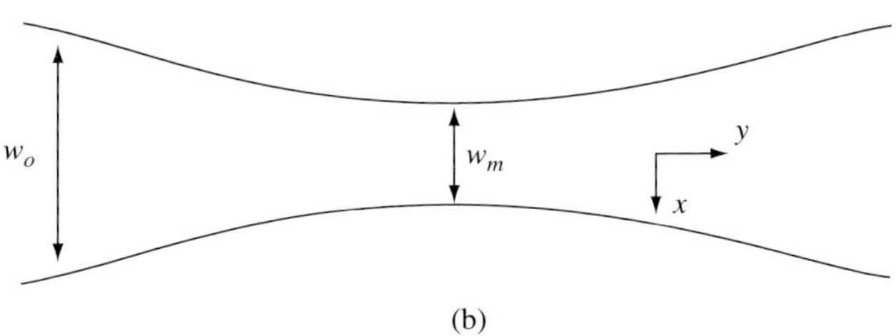

FIGURE 5.9.1. Definition sketches for two-layer flow in a wide channel with a horizontal bottom.

have been introduced. Also η_+ and η_- denote the deviations of the interface elevation at the right $(x = w/2)$ and left $(y = -w/2)$ walls, as shown in the figure. It is not difficult to show that the conditions that the interface remains attached to the sidewalls is $-\Delta/\tilde{\delta} \le \eta_\pm \le \Delta$.

The depth profiles have a boundary layer structure that is apparent when the channel width is much wider than the internal deformation radius $(w \gg 1)$. In this limit the layer thicknesses in the interior of the channel, away from the boundary layers, become

$$d_1 = \Delta, \text{ or } d_1{}^* = z_T{}^* \frac{D_{1\infty}}{D_{1\infty}+D_{2\infty}}$$

and

$$d_2 = \Delta/\hat{\delta}, \text{ or } d_2{}^* = z_T{}^* \frac{D_{2\infty}}{D_{1\infty}+D_{2\infty}}.$$

Thus the ratio of an interior layer depth to the total depth is equal to the ratio of its potential depth to the sum of the potential depths. If the potential depths are equal then the interior layer thicknesses are also equal. Unless the sum of the potential depths happens to be equal the total depth, the interior layer thicknesses are not equal to their potential depths. The interior flow can therefore have horizontal vorticity even though the interface is level.

The last remark can be verified by examining the velocity profiles for the two layers:

$$v_1(x, y, t) = -\left[\frac{(\eta_+(y, t) + \eta_-(y, t))\sinh(x)}{2\cosh(w(y)/2)} + \frac{(\eta_+(y, t) - \eta_-(y, t))\cosh(x)}{2\sinh(w(y)/2)}\right]$$
$$+(\Delta - 1)x + V(y, t) \tag{5.9.3}$$

and

$$v_2(x, y, t) = \hat{\delta}\left[\frac{(\eta_+(y, t) + \eta_-(y, t))\sinh(x)}{2\cosh(w(y)/2)} + \frac{(\eta_+(y, t) - \eta_-(y, t))\cosh(x)}{2\sinh(w(y)/2)}\right]$$
$$+(\Delta - 1)x + V(y, t), \tag{5.9.4}$$

which can be obtained from the potential vorticity equation (5.1.9) with the known thickness profiles. Each velocity is composed of a baroclinic component (term in large brackets) having opposite sign in each layer, and a barotropic, or depth-independent component $(\Delta - 1)x + V(y, t)$. The latter consists of a part that has a constant vorticity $\Delta - 1$, but no cross-sectional mean, and a part that is x-independent.[7] It is clearly the first of these that accounts for the presence of vorticity in the interior. Only when the sum of the potential depths is equal to the channel depth $(\Delta = 1)$ does this vorticity vanish. In this case the interior layer thicknesses equal their potential depths. The physical mechanism responsible for the presence of barotropic shear can be described by imagining that the channel is fed from a reservoir where the fluid is at rest and where the layer depths are therefore equal to $D_{1\infty}$ and $D_{2\infty}$ throughout. If the total depth $D_{1\infty} + D_{2\infty}$ in the reservoir is different from the channel depth $z_T{}^*$, then the water column as a whole undergoes squashing or stretching upon entering the channel. The result is that barotropic vorticity is spun up or down. Note that the lock exchange calculation of the previous section had $\Delta = 1$ and $V = 0$, so that the barotropic component of velocity was zero.

The total volume flux is given by

$$Q = \int_{-w/2}^{w/2} (v_1 d_1 + v_2 d_2)dx = \tfrac{1}{2}(1 + \hat{\delta})(\eta_+^2 - \eta_-^2) + z_T w V, \tag{5.9.5}$$

where $z_T = z_T{}^*/D_{1\infty} = \Delta(1 + \hat{\delta})/\hat{\delta}$. The two terms on the right-hand side measure the contributions from the baroclinic and barotropic parts of the flow.

---

[7] If the fluid had a free surface and thus an external deformation radius, the shear would be confined to boundary layers having that width, as in the Gill (1977) model. The surface on our channel is rigid, which is equivalent to having a infinite external deformation radius. The result is that the barotropic shear is uniform.

We now restrict attention to a channel that is very wide compared to L_I, or $w \gg 1$. The baroclinic boundary layers are then well separated and the layer velocities at the sidewalls reduce to

$$v_1(\pm\tfrac{1}{2}w(y), y, t) = \mp\eta_\pm \pm \tfrac{1}{2}(\Delta - 1)w + V \qquad (5.9.6)$$

and

$$v_2(\pm\tfrac{1}{2}w(y), y, t) = \pm\tilde{\delta}\eta_\pm \pm \tfrac{1}{2}(\Delta - 1)w + V. \qquad (5.9.7)$$

The dimensional characteristic speeds of the internal Kelvin waves that propagate along the walls at $\pm w/2$ are given by

$$c_\pm{}^* = \frac{v_1{}^*(\pm\tfrac{1}{2}w^*(y^*), x^*, t^*)D_{2\infty} + v_2{}^*(\pm\tfrac{1}{2}w^*(y^*), x^*, t^*)D_{1\infty}}{D_{1\infty} + D_{2\infty}}$$
$$\pm\left(\frac{g'D_{1\infty}D_{2\infty}}{D_{1\infty} + D_{2\infty}}\right)^{1/2} \qquad (5.9.8)$$

(see Exercise 2.) The dimensionless equivalent $c_\pm = c_\pm{}^*/(g'D_{1\infty})^{1/2}$ is given by

$$c_\pm = \frac{\pm(\hat{\delta} - 1)\eta_\pm \pm \tfrac{1}{2}(\Delta - 1)w + V \pm 1}{(1 + \hat{\delta})^{1/2}}. \qquad (5.9.9)$$

We now restrict attention to steady flow, for which the total volume transport Q is constant and the internal Bernoulli function ΔB^* is conserved along the sidewalls. In the limit $w \gg 1$, Q as given by (5.9.5) is dominated by the contribution from the barotropic velocity V, which is uniformly distributed across the channel. Thus

$$Q = Vwz_T \qquad (5.9.10)$$

in this limit. Also the dimensionless internal Bernoulli function $\Delta B^*/(g'D_{1\infty})^{1/2}$ on the left and right side walls (see 5.1.17) are given by

$$\Delta B_\pm = \tfrac{1}{2}(\hat{\delta} - 1)\eta_\pm{}^2 + [\tfrac{1}{2}(\Delta - 1)w \pm V + 1]\eta_\pm + \Delta/\hat{\delta}$$

where, again, '+' or '−' denote evaluation at $x = w/2$ or $x = -w/2$.[8] Elimination of V between these last two equations and rearrangement of the result leads to the functional relation

$$\mathcal{G}_\pm(\eta_\pm; w) = \eta_\pm{}^2 + s_\pm\eta_\pm + \mu_\pm = 0 \qquad (5.9.11)$$

---

[8] In equations with $\pm$ subscripts the top or bottom '+' or '−' signs are used together. Thus $a_\pm = b_\mp$ means that $a_+ = b_-$ and $a_- = b_+$.

where

$$\mu_{\pm} = \frac{2}{\hat{\delta}-1}\left(\frac{\Delta}{\hat{\delta}} - \Delta B_{\pm}\right)$$

and

$$s_{\pm} = \frac{2}{\tilde{\delta}-1}\left[\frac{1}{2}(\Delta-1)w \pm \frac{Q}{wz_T} + 1\right]. \tag{5.9.12}$$

Hence there are two hydraulic functions $\mathcal{G}_+$ and $\mathcal{G}_-$ governing the independent baroclinic boundary currents along the right and left walls.

Happily, the hydraulic functions are quadratic in the dependent variables (either η_+ or η_-), and therefore simpler than the higher order or transcendental dependence we have encountered in earlier problems. The geometrical forcing is contained entirely in the coefficients $s_{\pm}(y)$, which depends on $w(y)$ in two ways. The first dependence involves the term $(\Delta-1)w/2$, which is the value of the barotropic shear velocity $(\Delta-1)x$ at the right wall (or the negative of the shear velocity at the left wall). The second involves Q/wz_T, which is the value of the x-independent barotropic velocity. As w changes, the boundary layer is displaced laterally across the barotropic shear and into regions where the barotropic shear velocity is higher or lower. At the same time, the mean barotropic velocity is increased or decreased by narrowing or widening of the channel. Both effects alter the total kinetic energy of the boundary flow, forcing the interface elevation and baroclinic velocity to adjust to maintain constant total energy. Because of the dual nature of this forcing mechanism, the maximum effective constriction of a particular boundary current need not occur at the point of minimum width. We will refer to the parameter $s_{\pm}(y)$ as the *strangulation* in order to distinguish it from pure geometrical contraction. The most interesting solutions occur when both of the strangulation effects are in play, and since $w >> 1$, this requires further parameter restrictions. Examination of (5.9.12) suggests that $\frac{1}{2}(\Delta-1)w$ and Q/wz_T must balance, or

$$\Delta - 1 = O(Q/w^2). \tag{5.9.13}$$

Thus the magnitude of the barotropic shear $\Delta-1$ must be suitably small, preventing large shear velocities from forming at the sidewalls.

Conditions for critical flow can be found by taking $\partial\mathcal{G}_{\pm}/\partial\eta_{\pm} = 0$ leading to

$$\eta_{c\pm} = -\frac{1}{2}s_{c\pm} = -\frac{\left[\frac{1}{2}(\Delta-1)w_c \pm (Q/w_cz_T) + 1\right]}{\hat{\delta}-1}, \tag{5.9.14}$$

where the subscript 'c' denotes evaluation at a critical section. Comparison of the result with (5.9.9) verifies that $c_{\pm} = 0$. Note that the limit of equal potential depths ($\hat{\delta} \to 1$) must generally be avoided to avoid separation of the interface from one of the sidewalls.

The compatibility condition that must hold at a critical section

$$\frac{\eta_{c\pm}}{1-\hat{\delta}}\left[\tfrac{1}{2}(\Delta-1)\mp\frac{Q}{z_T w_c^2}\right]\left(\frac{dw}{dy}\right)_c = 0 \qquad (5.9.15)$$

is obtained in the usual manner by taking $(d\mathcal{G}_\pm/dw)(dw/dy)=0$. This relation suggests three distinct types of hydraulic controls. The first occurs at a point of extreme width $(dw/\partial y)=0$ and is similar to the 'narrows' controls discussed in connection with single layer flows. The second occurs where η_{c+} or η_{c-} vanishes, depending on which boundary layer is being considered. If it is the left-hand boundary layer then (5.9.11) implies that $\mu_-=0$ and that the solution for all y is governed by the relation $\eta_-{}^2+s_-(y)\eta_-=0$. Solutions are given by $\eta_-(y)=0$ and $\eta_-=-s_-(y)$. To interpret these solutions, note that the thermal wind relation (5.1.8) applied at the left wall can be written as

$$v_1(-w/2,\,y)-v_2(-w/2,\,y)=-(1+\hat{\delta})\eta_- \qquad (5.9.16)$$

The solution $\eta_-(y)=0$ therefore has zero vertical shear for all y and shares some elements with the similarity solution discussed in Section 5.4 (and exemplified by the straight line energy curve with contour value .50 in Figure 5.4.1a). Note, however, that the nonrotating solution has $v_1^2=v_2^2$ and is therefore shear free only for unidirectional flow. The solution $\eta_-=-s_-(y)$ has $\eta_-=0$ only at the critical section but nonzero η_- elsewhere.[9] Similar properties are possible for the boundary current on the right wall. In analogy with the nonrotating solutions, we will call a control with $\eta_+=0$ or $\eta_-=0$ a virtual control. As before, there are no restrictions on the value of w at such a control.

The third type of control occurs where the strangulation $s(y)$ reaches an extreme value away from an extreme value of w. We refer to this type of critical flow as a *remote* control and note that it requires vanishing of the bracketed term in (5.9.15) for either the '+' or '−' sign. Thus

$$w_c = \pm\left(\frac{2Q}{z_T(\Delta-1)}\right)^{1/2} \qquad (5.9.17)$$

for a remote control in a right-hand boundary layer, and

$$w_c = \pm\left(\frac{-2Q}{z_T(\Delta-1)}\right)^{1/2} \qquad (5.9.18)$$

for the left-hand boundary layer. For fixed values of Q and Δ only one of these expressions has real roots and therefore only one of the boundary currents can have such a control. In physical terms the explanation for this is rather simple. If the change in barotropic velocity due to the shear effect and mean flow effect act in opposition at one wall, they must reinforce on the other wall (where the sign of the barotropic shear velocity is reversed, but that of the mean barotropic velocity is the same).

---

[9] These definitions are sensible for a dynamically wide channel, since the boundary layers can be treated independently; finite channel widths would require some rethinking.

The possibility of three types of controls leads to a rich variety of steady solutions and we explore a few examples. The situation is simplified by the fact that different types of controls cannot arise at different locations within a particular boundary flow unless a hydraulic jump or other dissipative feature occurs. We will assume that the channel contains a simple contraction as shown in Figure 5.9.1b and that the mean barotropic flow $V = Q/wz_T$ is positive. Different types of behavior can easily be illustrated using a two-step graphical approach. First, changes in w must be related to changes in strangulation $s_\pm(w)$ using (5.9.12). A sample of curves showing this relation for various ranges in Q/z_T and Δ appears in Figure 5.9.2. Next, $s_\pm$ must be related to $\eta_\pm(s)$ using (5.9.11). Examples of this relation, which depends on $\mu_\pm$, are shown in Figure 5.9.3.

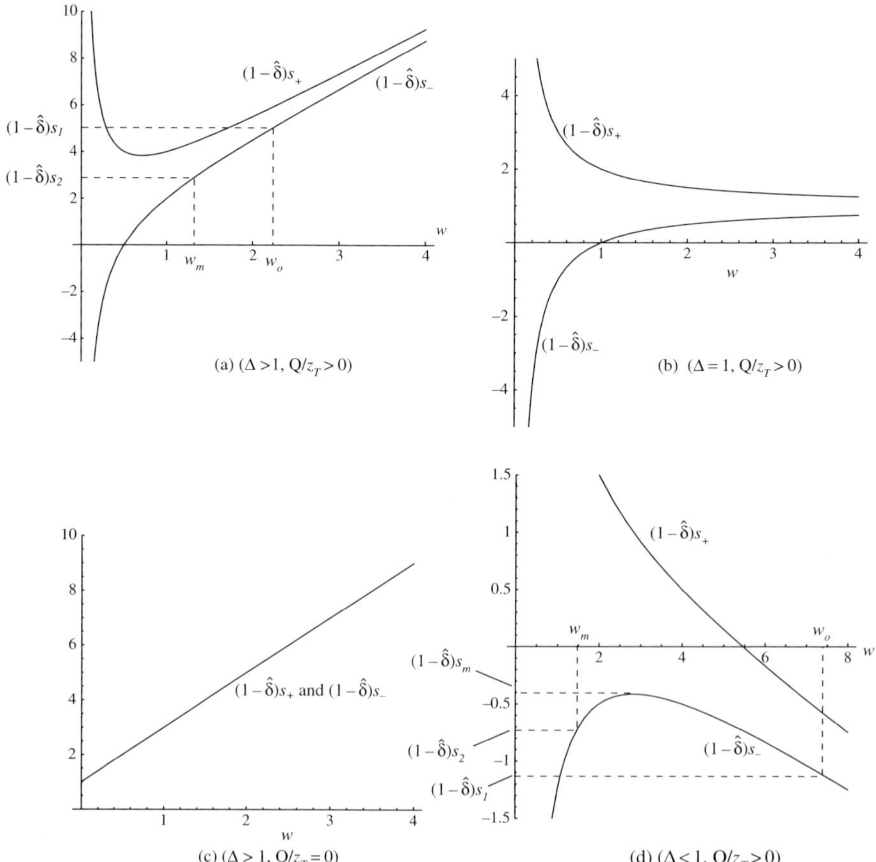

FIGURE 5.9.2. The strangulation parameter $s_\pm$ as a function of width w according to equation (5.9.12). The general geometry of the curves of s_+ and s_- takes four different forms according to the value of Δ relative to unity, and to whether Q is finite. These four forms are illustrated by (a–d). The numerical values used to obtain the plots are (a) $\Delta = 5$, $Q/z_T = 1$; (b) $\Delta = 1$, $Q/z_T = 1$; (c) $\Delta = 5$, $Q/z_T = 0$; (d) $\Delta = 0.5$, $Q/z_T = 2$.

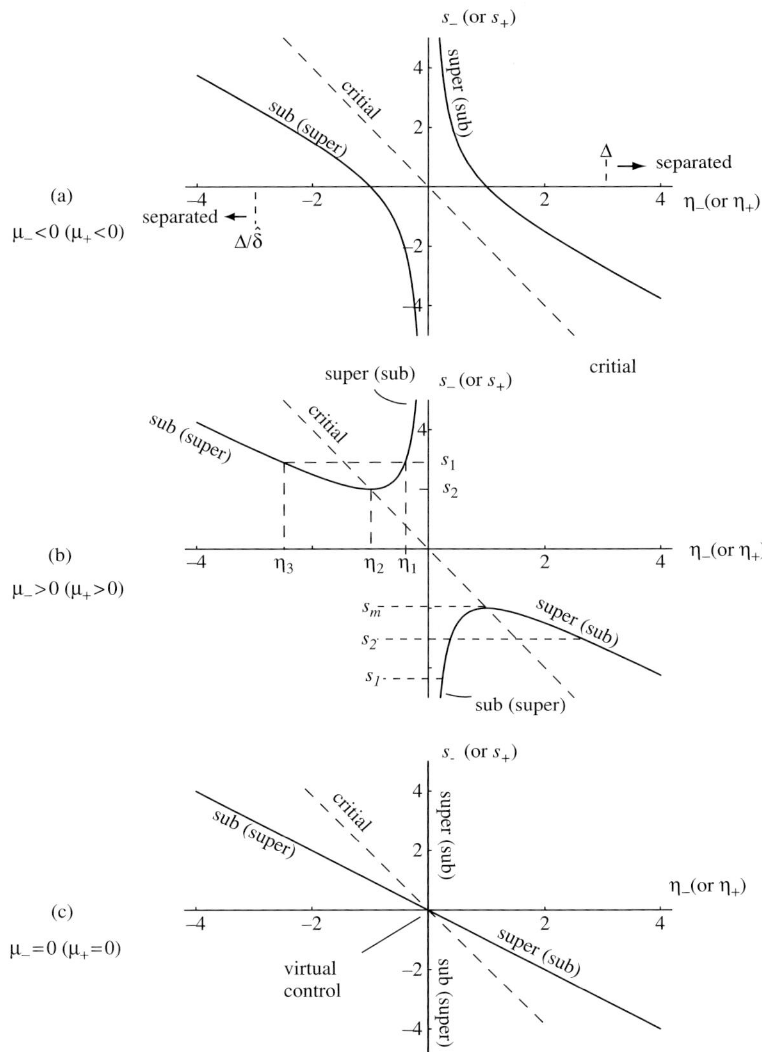

FIGURE 5.9.3. The relationship between $\eta_\pm$ and $s_\pm$ according to (5.9.11). Frames (a)–(c) give the general geometry of the solution curves for $\mu_\pm < 0$, $\mu_\pm > 0$, and $\mu_\pm = 0$. Each solid curve in each frame may represent potential solution for either boundary layer. In (c) the $s_\pm$-axis is a solution curve. The operational difference between left and right boundary layers is contained in the subcritical or supercritical designation on each curve. Here the labels have been chosen assuming that $\hat{\delta} < 1$. The leading label on each curve then corresponds to the left-hand boundary layer while the label in parenthesis corresponds to the right-hand layer. When $\hat{\delta} > 1$ this order is reversed. The dashed diagonal curve in each frame indicates the possible locations of critical flow. The numerical values of μ_- used to make frames (a), (b), and (c) are 1, -1, and 0. The quantities Δ and $-\Delta/\hat{\delta}$ marked in (a) indicate limits outside of which the flow becomes separated from one of the sidewalls. These limits are given without assigned value and apply in all three frames.

The procedure then is to use the Figure 5.9.2 curves to determine the variation of the strangulation function $s_\pm(w)$ along the channel, then use Figure 5.9.3 to find the corresponding solution. Each frame in the former shows two curves corresponding to the strangulation function for the right $(+)$ and left $(-)$ boundary layers. In the case of Frame c, the two curves coincide. A virtual control has $\eta_+ = 0$ or $\eta_- = 0$ and must, in view of (5.9.14), occur where $s_+(w)$ or $s_-(w)$ is zero. A remote control occurs where $s_+(w)$ or $s_-(w)$ has a minimum or maximum. A narrows control occurs where w itself is a minimum.

Turning to Figure 5.9.3 we see that each frame has two curves corresponding to the two solutions of the quadratic equation (5.9.11). Note that the $s_\pm$-axis in Frame c corresponds to the similarity solution. It is no longer the case that the two curves represent flow in the different boundary layers, rather each curve gives a possible solution for either boundary layer. The difference in left and right boundary layers at a particular point on a particular curve is that one solution is subcritical and the other supercritical at that point. In making these designations, we use the convention that supercritical (subcritical) flow allows wave propagation in the same (opposite) direction as the mean barotropic velocity V, here assumed to be positive. In particular, use of the definition of $s_\pm$ in (5.9.9) leads to

$$c_\pm = \pm(\hat{\delta}^{1/2} - 1)(\eta_\pm + \frac{1}{2}s_\pm),$$

and therefore

$$c_+ = (\hat{\delta}^{1/2} - 1)(\eta_+ + \frac{1}{2}s_+) \begin{cases} > 0 \text{ (supercritical)} \\ = 0 \text{ (critical)} \\ < 0 \text{ (subcritical)} \end{cases}$$

for the right-hand boundary layer, and

$$c_- = -(\hat{\delta}^{1/2} - 1)(\eta_- + \frac{1}{2}s_-) \begin{cases} > 0 \text{ (supercritical)} \\ = 0 \text{ (critical)} \\ < 0 \text{ (subcritical)} \end{cases}$$

for the left-hand boundary layer. Critical flow states must therefore occur on the dashed diagonals $\eta_\pm = -\frac{1}{2}s_\pm$ in Figure 5.9.3. The flow is supercritical or subcritical above or below this diagonal depending on the sign of $(\hat{\delta}^{1/2} - 1)$ and on which boundary layer is being examined. Figure 5.9.3 has been labeled assuming that $\hat{\delta} < 1$ and thus supercritical flow for the left boundary layer lies above. The leading label on each curve applies to the left-hand boundary layer, while the labels in parenthesis apply to the right-hand boundary layer. A virtual control requires $\eta_- = s_- = 0$ or $\eta_+ = s_+ = 0$ and these are possible only in Figure 5.9.3c. A narrows control or a remote control must occur at a minimum or maximum of the solution curve, and this is possible only in Figure 5.9.3b.

The reader who has become frustrated with the details of this narrative will benefit from exploration of a few examples, which is now undertaken. Attention is limited to the left-hand boundary layer and the channel geometry is as shown in Figure 5.9.1b with entrance and exit width w_o, and minimum width w_m.

a. A Narrows Control

Suppose that $\Delta > 1$ so that the relation between w and $s_\pm$ is given by Figure 5.9.2a. The lower curve applies to the left boundary layer. As one moves from the mouth of the channel $(w = w_o)$ to the narrowest section $(w = w_m)$, s_- decreases from values s_1 to s_2 and increases back to s_1, as indicated. To find the corresponding solution, one must apply the variation in s_- to the relevant frame and curve in Figure 5.9.3. The choice depends on the value of $\mu_\pm$. Since we are interested in the left wall boundary layer, we focus on μ_- and choose a value < 0 (Figure 5.9.3a). There are two possible solutions corresponding to the two curves. Choosing the left-hand curve gives a subcritical solution while choosing the right-hand curve gives a supercritical solution. Either curve gives a legitimate solution over the established range of s_-. In either case the flow remains supercritical or subcritical through the contraction and there is no control.

If instead we choose $\mu_- > 0$, then the solution shifts to the curves of Figure 5.9.3b and a control is possible. Beginning at $s_- = s_1$, which is > 0 for our example, one must first select a solution branch of the left hand curve. Choosing the subcritical branch places the solution at $\eta_- = \eta_3$ as indicated on the plot. Lowering s_- from this value causes η_- to increase. A sufficient increase places the solution at the minimum $(\eta_- = \eta_2)$ of the curve, where the flow is critical. If s_2 lies at this maximum, as shown in the figure, then the solution continues onto the supercritical branch in the usual manner and the flow downstream of the narrows $(\eta_- = \eta_1)$ is supercritical. The solution just described contains a narrows control. It can be shown that continuation from the subcritical branch to the supercritical branch (rather than back along the subcritical branch) is the only choice that gives a smooth solution, a matter explored in Exercise 4.

b. A Remote Control

Next suppose that $\Delta < 1$, so that the strangulation function $s_-(w)$ is given by the lower curve of Figure 5.9.2d. Suppose further that w lies in the indicated range $w_o \le w \le w_m$ such that s_- reaches its maximum value for an intermediate w, as indicated. As one moves into the channel, the value of s_- increases then decreases before the narrowest section is reached. Downstream of the narrowest section, s_- increases back to its maximum value and then decreases. Switching now to Figure 5.9.2, we again choose $\mu_- > 0$ (frame b) and note that the solution now lies along the lower, right-hand curve. We attempt to trace a solution beginning along the subcritical branch at $s_- = s_1$. Moving from the entrance into the channel causes s_- to increase and to reach its peak value s_m before the narrows is actually reached. We have arranged this point to coincide with the maximum of the solution curve. If one continues past this remote control, s_- retreats to lower values as the channel narrows but solution moves onto the supercritical branch. When the narrowest section ($w = w_m$, $s_- = s_2$) is reached, the flow is supercritical. Downstream of this point the above sequence is rewound. As the channel widens, the flow passes back through a second remote control and returns to a subcritical state. Note that the flow at this second control section is subject to the same instability discussed in connection with single layer flows; wave propagation is locally towards the critical section from upstream and downstream. We thus anticipate that a hydraulic jump or other dissipative feature may occur there.

c.   A Virtual Control

Virtual control implies $s_- = 0$ and therefore is possible only for strangulation curves which cross the w-axis in Figure 5.9.2. Suppose that $\Delta > 1$, so that the lower curve of Figure 5.9.2a applies if the range of w is such that s_- has a zero crossing. In this case, s_- will be positive at the entrance and exit of the channel, but negative at the narrowest section. Now apply this range of s_- to Figure 5.9.3c, which shows the solutions with possible virtual controls. At the channel entrance, $s_- > 0$ and the choice of subcritical there forces one to select diagonal curve. As the width narrows, the solution passes through a virtual control at the origin, and continues onto the supercritical branch until the narrowest section is reached. This sequence is retraced as the narrows is passed, with the proviso that instability may occur at the downstream virtual control. The solution therefore behaves in a similar way to that with a remote control, as just described. The similarity solution, which is represented by the s_- axis in Figure 5.9.3, is supercritical at the upstream entrance in this case.

Where the diagonal curve passes through the origin of Figure 5.9.3c, it is conceivable that the solution could jump onto the similarity solution, or vice versa. As explored in Exercise 5, such a jump would generally involve a discontinuity in $\partial\eta_-/\partial y$ or $\partial^2\eta_-/\partial y^2$ at the control and can therefore be ruled out.

These example show only some of the possible behavior that is consistent with the model. On can invent a variety of solutions by matching different Figure 5.9.2 forcing curves with different Figure 5.9.3 solutions curves. No known attempts have been made to verify any of these solutions. The main purpose has been to demonstrate the physical mechanism associated with the forcing of a boundary current by a larger scale flow and the novel features that can arise when that external flow is sheared.

Exercises

(1) Obtain the expression (5.9.5) for the total volume flux.
(2) Derive the expression for the internal Kelvin waves speeds (5.9.8) in the wide-channel limit and with attached interfaces. (You can work through the full nonlinear version of the calculation by first noting that the expressions

for velocity and layer depths derived above remain valid when the flow is time-dependent. The characteristic form of the governing equations on a certain sidewall may be obtained by evaluating the layer momentum equations there and substituting the expressions for v_n and d_n that follow from the profiles.)

(3) We noted that (5.9.14) implies separation of the critical flow from the channel sidewalls when $\hat{\delta} = 1$. An exception occurs when the numerator of the right-hand side of this equation vanishes. Explore this case and describe the flow that results.

(4) For a narrows control show that continuation from the upstream subcritical branch to the downstream subcritical branch of the lower left solution curve in Figure 5.9.3c is the only choice that avoids singularity.

(5) At a virtual control, show that the solution cannot jump from the similarity solution $\eta_- \equiv 0$ to the solution $\eta_- = -s_-(y)$, or vice versa, without incurring a discontinuity in $\partial \eta_- / \partial y$ or $\partial^2 \eta_- / \partial y^2$. (An exception occurs when virtual, remote, and narrows controls coincide at the narrowest section.)

6
Potential Vorticity Hydraulics

6.1. Introduction

To this point we have dwelt with applications in which the velocity of the current is comparable with the speed of long, internal gravity waves. This situation can arise in channels, along coastlines, in the lee of mountains, or in other special locations, but is less likely to occur in the relatively slow and broad general circulation of the ocean or atmosphere. Even the jet-like currents such as the Gulf Stream tend to be substantially subcritical with respect to long, gravest-mode, internal gravity waves. On the other hand, Rossby waves and other types of potential vorticity waves are important to the general circulation. As discussed in Section 2.1, these waves depend on lateral gradients of potential vorticity to provide a restoring mechanism. For the gradients that typically exist in geophysical applications, the waves are generally much slower than long gravity waves and can have wave speeds in the range of the current velocity.

Hydraulic behavior associated with potential vorticity waves, sometime called 'Rossby wave hydraulics', has been identified in a variety of idealized models, including those of free jets, fronts and coastal currents. The subject is reviewed by Johnson and Clarke (2001). One of the difficulties, at least at the time of this writing, is that there is very little concrete evidence of this behavior in observed flows. However, the field is relatively young and the phenomena may be present but not yet recognized. This chapter will present several examples of predicted behavior. An important departure from the hydraulics of gravity-driven systems is that the motion of the fluid is primarily horizontal (or along isopycnals) and thus classical jumps, spilling flows, and other features that require significant vertical motion are not present.

Some insight can be gained from the nonrotating flow considered in Section 2.9, where a free surface gravity wave and a discrete spectrum of potential vorticity waves were present. There, an infinite family of hydraulically controlled flows was found, one having a hydraulic transition with respect to the gravity mode and the others with respect to mixed gravity/potential vorticity modes. The hydraulic transition for the gravity mode was manifested primarily by a change in depth as the fluid crossed the sill, whereas the transition for the higher potential vorticity modes involved lateral displacements of streamlines. In order to examine the potential vorticity dynamics in more depth, it will be helpful to consider simpler systems in which the gravity wave is absent and just one or two of the potential vorticity modes are present. The gravity wave can be eliminated by making the quasigeostrophic approximation, discussed below, or by considering a homogeneous fluid bounded above by a rigid lid. The number of potential vorticity modes can be limited by considering flows with piecewise constant potential vorticity distributions.

Since gravity waves will be relatively unimportant, we need to rethink the standard hydraulic scaling in which $(gD)^{1/2}$ and $L/(gD)^{1/2}$ are chosen as scales for the longitudinal velocity and time. For larger scale flows, where Earth's rotation and the variation of rotation with latitude are of central importance, we need to select scales based on the Coriolis parameter $f = 2\Omega \sin\theta$, where Ω is Earth's angular velocity and θ is latitude. Let θ_o denote a central latitude about which the flow is centered. For the applications in mind, which include ocean and atmosphere fronts, jets and coastal currents, the variation in θ about θ_o is small compared to its full range and it is sufficient to approximate f according to

$$f(\theta) = f(\theta_o) + 2\Omega(\theta - \theta_o)\cos\theta_o$$

$$= f_o + \beta^* y^*$$

where $y^* = R(\theta - \theta_o)$, $\beta^* = 2\Omega\cos(\theta_o)/R$, and R is Earth's radius. If L represents the meridional extent of the current, then $\beta^* L/f_o \ll 1$ for this *beta plane* approximation to be valid. An obvious time scale is f_o^{-1} and we will leave the velocity scale U unspecified.

The appropriate scaling can now be deduced by reconsidering the shallow water equations (2.1.1–2.1.3) with no forcing or dissipation:

$$\frac{\partial \mathbf{u}^*}{\partial t^*} + \mathbf{u}^* \cdot \nabla\mathbf{u}^* + (f_o + \beta^* y^*)\mathbf{k} \times \mathbf{u}^* = -g\nabla\eta^* \qquad (6.1.1)$$

and

$$\frac{\partial \eta^*}{\partial t^*} + \nabla \cdot [\mathbf{u}(D + \eta^* - h^*)] = 0. \qquad (6.1.2)$$

In most applications the equations apply to a $1\frac{1}{2}$–layer system for which g should be interpreted as reduced gravity. Where the active layer lies along the bottom, η^* should be interpreted as upwards displacement of the bounding interface from its resting equilibrium position. Application to a buoyant surface layer can also be made by interpreting η^* as the downwards displacement of the lower interface and regarding h^* as zero.

Nonrotating hydraulics (Chapter 1) involves balances between the advection terms, the local time-derivative terms, and the pressure gradients terms. Semigeostrophic hydraulics includes these terms, at least in the predominant direction of the flow, and adds the Coriolis term. The scaling $U = (gD)^{1/2}$ is preserved. For the slower, broader flows subject to beta plane hydraulics, U is typically $\ll (gD)^{1/2}$ and another scaling must be sought. There are two classes of flows that one is likely to encounter depending on the size of the Rossby number $R_o = U/f_o L$. The first is characterized by $R_o = O(1)$ and includes strong jets such as the Gulf and Jet Streams and some equatorial currents. Hydraulic models of these flows are typically treated using a barotropic, rigid-lid model and this is described at the end of this section. The second class includes broad-scale flows and weaker jets in which both horizontal velocity components are in near geostrophic balance. Such flows can be treated using the quasigeostophic approximation, in which the velocities are considered weak and variations in layer thickness are assumed small. Quasigeostrophic flows have both velocity components in near geostrophic balance, whereas semigeostrophic flows have only the longitudinal velocity so balanced. Thus, if N is a scale for η^*, the quasigeostrophic approximation suggests that $N \approx f_o LU/g$. It can also be seen from (6.1.1) that the ratio of the advection terms to the Coriolis term is order R_o, which must clearly be small in the presence of near geostrophic motion. The final term to the consider in (6.1.1) is the local time derivative; its ratio to the Coriolis term is $O(T^{-1}f^{-1})$. In order that the geostrophic balance be preserved to the lowest order, the time scale T must be chosen much longer than an inertial period $(T \gg f^{-1})$. A convenient choice is $T = R_o^{-1}f^{-1}$.

For the quasigeostrophic approximation, we then use dimensionless variables $\eta = \eta^* g/f_o UL$, $(u, v) = (u^*, v^*)/U$, $(x, y) = (x^*, y^*)/L$, and $t = t^* fR_o$. Equation (6.1.1) now becomes

$$R_o\left(\frac{\partial \mathbf{u}}{\partial t} + \mathbf{u} \cdot \nabla \mathbf{u}\right) + (1 + R_o \beta y)\mathbf{k} \times \mathbf{u} = -\nabla \eta \tag{6.1.3}$$

where $\beta = \beta^* L^2/U$ and $R_o \ll 1$. The dependent variables are now expanded in powers of R_o:

$$u = u^{(0)} + R_o u^{(1)} + \cdots,$$

$$v = v^{(0)} + R_o v^{(1)} + \cdots,$$

and

$$\eta = \eta^{(0)} + R_o \eta^{(1)} + \cdots.$$

The leading order velocity is geostrophic:

$$v^{(o)} = \frac{\partial \eta^{(o)}}{\partial x} \tag{6.1.4}$$

and

$$u^{(o)} = -\frac{\partial \eta^{(o)}}{\partial y}, \tag{6.1.5}$$

showing that $\eta^{(0)}$ acts as a streamfunction.

The dimensionless version of the continuity equation (6.1.2) is

$$R_o\frac{\partial \eta}{\partial t} + \nabla \cdot \left[\mathbf{u}\left(S^{-1}\left(1 - \frac{h^*}{D}\right) + R_o \eta\right)\right] = 0, \tag{6.1.6}$$

where $S = \frac{f_o^2 L^2}{gD}$. If the typical horizontal scale of the motion is of the order of the Rossby radius $(L_d = (gD)^{1/2}/f)$, then $S = O(1)$. The lowest order approximation to (6.1.6) is then

$$\nabla \cdot \left[\mathbf{u}^{(0)} \left(1 - \frac{h^*}{D} \right) \right] = -\mathbf{u}^{(0)} \cdot \nabla \left(\frac{h^*}{D} \right) = 0.$$

Use has been made of the property $\nabla \cdot \mathbf{u}^{(0)} = 0$, which follows from (6.1.4) and (6.1.5). If h^*/D is O(1), geostrophic flow must move along contours of constant h^*. This topographic steering would imply that a current would have to move around an isolated topographic feature such as a ridge. Hydraulic effects tend to occur when the flow passes over topography and this is permissible in the current framework only when h^*/D is small. We therefore assume that $h^*/D = O(R_o)$ and so define $h = h^*/(R_o D)$. We have now constrained variations in the layer thickness to be small compared to the total thickness, an approximation that is also considered integral to quasigeostrophic theory.

At the $O(R_o)$ level of expansion, (6.1.3) and (6.1.6) are

$$\frac{\partial \mathbf{u}^{(0)}}{\partial t} + \mathbf{u}^{(0)} \cdot \nabla \mathbf{u}^{(0)} + \beta y \mathbf{k} \times \mathbf{u}^{(0)} = -\mathbf{k} \times \mathbf{u}^{(1)} - \nabla \eta^{(1)}$$

and

$$\frac{\partial \eta^{(0)}}{\partial t} + \nabla \cdot \left[\mathbf{u}^{(0)} \left(\eta^{(0)} - S^{-1} h \right) \right] = -\nabla \cdot \left[S^{-1} \mathbf{u}^{(1)} \right]$$

Taking the curl of the first equation and using the second equation to eliminate $\mathbf{u}^{(1)}$ from the result leads to the quasigeostrophic potential vorticity equation:

$$\left(\frac{\partial}{\partial t} + u^{(0)} \frac{\partial}{\partial x} + v^{(0)} \frac{\partial}{\partial y} \right) \left(\nabla^2 \eta^{(o)} - S \eta^{(o)} + h + \beta y \right) = 0. \tag{6.1.7}$$

The same result could have been obtained directly from the shallow water potential vorticity equation (2.1.8) by applying the present scaling and approximations (Exercise 1). The variable part of potential vorticity $(\zeta^* + f)/d^*$ is approximated by $\nabla^2 \eta^{(o)} - S \eta^{(o)} + h + \beta y$. The relative vorticity is $\nabla^2 \eta^{(o)}$, the stretching term resulting from departures from constant layer thickness is $-S \eta^{(o)} + h$, and the departure from constant background rotation is βy.

Now consider a plane wave of the form

$$\eta^{(0)} = \mathrm{Re} \left[N e^{i(kx + ly - \omega t)} \right],$$

propagating over a horizontal bottom and in the presence of a uniform zonal flow of velocity U_o. It is left as an exercise to show that the wave frequency is given by

$$\omega = U_o k - \frac{\beta k}{k^2 + l^2 + S}. \tag{6.1.8}$$

The resulting speed of the crests and troughs in the x-direction is

$$\frac{\omega}{k} = U_o - \frac{\beta}{k^2 + l^2 + S}.$$

A wave that is long ($k^2 \ll l^2 + S$) in the x-direction propagates in that direction at the speed $U_o - \beta/(l^2 + S)$. In addition, suppose that the wave, like the background flow, has no structure in the y-direction (i.e. $l = 0$). In order for the wave to be arrested ($\omega/k = 0$) it is necessary for that flow to be eastward ($U_o > 0$) and for $U_o = \beta/S$, or

$$U_o^*/\beta^* L_d^2 = O(1). \tag{6.1.9}$$

The dimensionless parameter can be thought of as a beta-plane Froude number, and (6.1.9) is a prerequisite for the occurrence of hydraulic effects in the quasi-geostrophic model. The specific conditions for the criticality of a particular flow with respect to a potential vorticity wave will generally be much more involved. In some applications β^* may be replaced by a potential vorticity gradient due to topography or background shear.

An alternative approach that illustrates Rossby-wave hydraulics without the complication of gravitational effects is the rigid-lid, barotropic model. No restriction is placed on the size of R_o or h^*/D, but stratified systems are excluded. The governing equation is obtained directly from (2.1.8) by regarding the depth $d^*(x^*, y^*)$ as fixed. In the absence of forcing and dissipation the result is

$$\frac{d^*}{dt^*}\left(\frac{f_o + \beta^* y^* + \dfrac{\partial v^*}{\partial x^*} - \dfrac{\partial u^*}{\partial y^*}}{d^*} \right) = 0. \tag{6.1.10}$$

Since the Rossby radius of deformation is effectively infinite, the horizontal length scale L is typically set by the topography or potential vorticity distribution. Velocity and time scales are then chosen as $\beta^* L^2$ and L/β^*.

Most of the models of Rossby-wave hydraulics involve zonal flows and it is standard to use x^* as the predominant direction of flow. We will therefore switch from the earlier convention of using y^* as the flow axis.

Exercises

(1) Beginning with the unforced shallow water potential vorticity equation (2.1.8 with $\mathbf{F}^* = 0$) apply the scaling and assumptions appropriate to the quasi-geostrophic approximation and thereby derive equation (6.1.7).
(2) Verify that the frequency of a plane wave solution to (6.1.7) is given by (6.1.8).

6.2. Potential Vorticity Front in a Channel

One of the simplest models for exploring the physics associated with potential vorticity gradients is based on a material contour separating two regions of different, uniform potential vorticity. The following development is due to Haynes *et al.* (1993) and Johnson and Clarke (1999, 2001), who considered the situation in which the potential vorticity front is confined to a channel with a mean flow. In a certain sense, the model is that obtained from the channel problem of Sec. 2.9 if the gravity mode is eliminated (using a rigid lid) and the number of potential vorticity modes is reduced to one.

Consider a zonal channel occupying $0 < y^* < w^*$ and containing a homogeneous flow capped by a rigid lid. The statement of conservation of potential vorticity is given by (6.1.10) with

$$d^* = D - h^*(x^*, y^*). \tag{6.2.1}$$

Although β^* is taken as zero, variations in ambient potential vorticity will arise due to variations in h^*. It is assumed that such variations are weak ($h^* \ll D$) and, therefore, that the horizontal velocity is approximately nondivergent:

$$\nabla \cdot (\mathbf{u}^*) = 0.$$

In this case a streamfunction ψ^* exists, with $v^* = \partial \psi^* / \partial y^*$, $u^* = -\partial \psi^* / \partial x^*$. Under these conditions the shallow water potential vorticity can be approximated as

$$\frac{f_o + \xi^*}{d^*} \cong \frac{f_o}{D} + \frac{f_o h^*}{D^2} + \frac{\xi^*}{D} = \frac{f_o}{D} + \frac{f_o h^*}{D^2} + \frac{\nabla^2 \psi^*}{D}. \tag{6.2.2}$$

Suppose that the bottom topography consists a topographic step:

$$h^* = \begin{cases} 0 & (y^* > Y_h^*(x^*)) \\ \delta h^* & (y^* < Y_h^*(x^*)) \end{cases}, \tag{6.2.3}$$

where $\delta h^*/D \ll 1$. The shallow fluid lying on the shelf to the 'right' side of the channel (facing positive x^*) implies relatively high potential vorticity there, at least in the absence of other vorticity gradients. (This configuration reverses the usual situation on a beta plane, where high potential vorticity occurs to the north.) The jump in potential vorticity across the step gives rise to a class of waves closely related to topographic Rossby waves (Section 2.1). These waves propagate forward (towards positive x^*) but can be arrested if a negative mean flow is added. For example, if the step lies at mid-channel $Y_h^* = w^*/2$ and a uniform flow $u^* = -\alpha^*$ is present, the wave frequency ω^* is given in terms of the wave number k^* by

$$\omega^* = -k^* \alpha^* + \frac{f_o \delta h^*}{2D} \tanh[\tfrac{1}{2} k^* w^*] \tag{6.2.4}$$

(see Exercise 1). Short waves ($k^* \ll 1$) are therefore advected downstream at the speed α^* of the background flow. Long waves ($k^* w^* \ll 1$) have speed $c^* = \omega^*/k^* \to -\alpha^* + f_o w^* \delta h^*/4D$ and are brought to rest when $\alpha^* = f_o w^* \delta h^*/4D$. For smaller α^* stationary, dispersive waves with downstream group velocity exist. At higher speeds, all linear disturbances are swept downstream.

We wish to consider the effects of variations in the position $Y_h{}^*$ of the step. Attention will be confined to cases where $Y_h{}^*$ experiences an isolated narrowing of the shelf, centered at $x^* = 0$ (dashed curve in Figure 1). Far upstream and downstream, it will be assumed that the step lies at channel midpoint ($Y_h{}^* \to 1/2$). In addition, the flow is assumed to be initially quiescent, so that the potential vorticity $(\Delta^2 \psi^*/D) + (f_o h^*/D^2)$ is initially zero on the deeper side of the channel ($y^* > Y_h{}^*(x^*)$) and has a value $f_o \delta h^*/D^2$ on the shallower side ($y^* < Y_h{}^*(x^*)$). At $t^* = 0^+$, a uniform velocity $u^* = -\alpha^*$ is imposed, causing some fluid to cross the step, carrying with it its initial potential vorticity. The material contour or 'front' $y^* = Y^*(x^*, t^*)$ (solid line) separating the low and high potential vorticity no longer coincides with the position of the topographic step. Conservation of potential vorticity for each fluid column then implies

$$\frac{\Delta^2 \psi^*}{D} = \frac{f_o \delta h^*}{D^2} \begin{Bmatrix} 0\ (y^* > Y^*(x^*, t^*)) \\ 1\ (y^* < Y^*(x^*, t^*)) \end{Bmatrix} - \frac{f_o \delta h^*}{D^2} \begin{Bmatrix} 0\ (y^* > Y_h{}^*(x^*)) \\ 1\ (y^* < Y_h{}^*(x^*)) \end{Bmatrix}.$$

It is important to realize at the outset that the total volume transport in the channel must remain fixed as the flow evolves. This property can be deduced from the condition of no normal flow along the channel side walls, which requires that ψ^* be uniform there. Although the sidewall values of ψ^* can generally vary with time, this is precluded by the nature of the initial-value problem posed above. The flow at $y \to \pm\infty$ remains fixed in time since disturbances generated by variations in $Y_h{}^*$ propagate upstream and downstream at finite speeds. The boundary values of ψ^* and hence the total volume flux therefore remain fixed. Other transports including that of potential vorticity may be altered.

If variables are nondimensionalized using $w^*, D/f_o \delta h^*$ and $\delta h^* f_o w^*/D$ as horizontal length, time and velocity scales (the topographic equivalent of L, L/β, and βL^2), the previous relation becomes

$$\nabla^2 \psi = \begin{Bmatrix} 0\ (y > Y(x, t)) \\ 1\ (y < Y(x, t)) \end{Bmatrix} - \begin{Bmatrix} 0\ (y > Y_h(x)) \\ 1\ (y < Y_h(x)) \end{Bmatrix}. \tag{6.2.5}$$

Initially shallow(deep) fluid that crosses the step into the deeper (shallower) region will be stretched (squashed) and its relative vorticity will be incremented by an amount $+1(-1)$. The fluid vorticity is therefore piecewise constant as shown in Figure 6.2.1b. Solutions to (6.2.5) in the various regions can be matched by requiring that ψ remain continuous across the boundaries of the regions. The potential vorticity front itself is a material boundary and its motion obeys the kinematic relation $v = \frac{\partial \psi}{\partial x} = \frac{\partial Y}{\partial t} + u(x, Y, t) \frac{\partial Y}{\partial x}$, which can also be expressed as

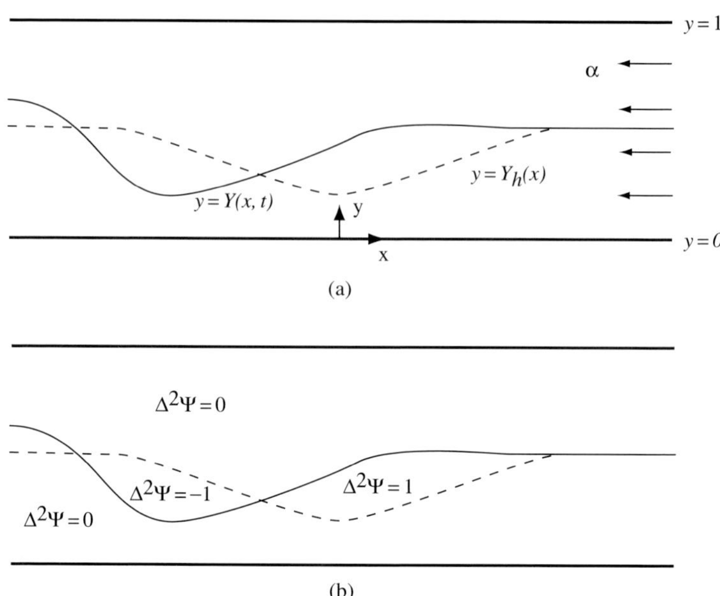

FIGURE 6.2.1. Definition sketch.

$$\frac{\partial Y}{\partial t} = \frac{d}{dx}\psi(x, Y(x, t), t). \tag{6.2.6}$$

With the present scaling, the Rossby number U/f_oL becomes $\delta h^*/D$, which is $\ll 1$ by prior assumption. The velocity is therefore geostrophically balanced and the model can be considered quasigeostrophic but with an infinite Rossby radius of deformation (i.e. $S = 0$ in 6.1.7).

a. *Long-Wave Behavior*

In the usual manner of hydraulic analysis we begin by considering the long-wave behavior of the flow induced when the along-channel variations of the step position are gradual in comparison with cross-channel variations. In particular, let

$$Y_h = \tfrac{1}{2} - \varepsilon\,\mathrm{sech}^2(\tilde{x}) \tag{6.2.7}$$

where $\tilde{x} = \mu^{1/2}x$ and $\mu \ll 1$. The stream function can be written as a sum of the contribution from the background velocity $-\alpha$ imposed at $t = 0^+$ plus a residual ϕ:

$$\psi = \alpha y + \phi(\tilde{x}, y, \tilde{t}), \tag{6.2.8}$$

where $\tilde{t} = \mu^{1/2}t$. [The slow time variable associated with gradual variations in the x-direction is suggested by (6.2.6).]

Equation (6.2.5) now becomes

$$\frac{\partial^2 \phi}{\partial y^2} + \mu \frac{\partial^2 \phi}{\partial \tilde{x}^2} = \begin{Bmatrix} 0 \ (y > Y(\tilde{x}, \tilde{t})) \\ 1 \ (y < Y(\tilde{x}, \tilde{t})) \end{Bmatrix} - \begin{Bmatrix} 0 \ (y > Y_h(\tilde{x})) \\ 1 \ (y < Y_h(\tilde{x})) \end{Bmatrix}$$

$$= H(y - Y_h(\tilde{x})) - H(y - Y(\tilde{x}, \tilde{t})) \tag{6.2.9}$$

where $H(y) = 1$ for $y > 0$ and $H(y) = 0$ for $y < 0$. The boundary conditions $\psi = (0, \alpha)$ at $y = (0,1)$ imply

$$\phi(\tilde{x}, 0, \tilde{t}) = \phi(\tilde{x}, 1, \tilde{t}) = 0 \tag{6.2.10}$$

For small μ the solution to (6.2.9) subject to (6.2.10) may be written as

$$\phi = \tfrac{1}{2}(y - Y_h)^2 H(y - Y_h) - \tfrac{1}{2}(y - Y)^2 H(y - Y) + \tfrac{1}{2}\left[(1 - Y)^2 - (1 - Y_h)^2\right] y$$

$$- \frac{\mu}{12} \frac{\partial^2}{\partial \tilde{x}^2} \{ \tfrac{1}{2}(y - Y_h)^4 H(y - Y_h) - \tfrac{1}{2}(y - Y)^4 H(y - Y) +$$

$$((1 - Y)^2 - (1 - Y_h)^2)(y^3 - y) + \tfrac{1}{2}\left[(1 - Y)^4 - (1 - Y_h)^4\right] y \} + O(\mu^2),$$

and (6.2.6) then gives

$$\frac{\partial Y}{\partial \tilde{t}} = \frac{\partial}{\partial \tilde{x}} \psi_h(Y, Y_h) + \frac{\mu}{6} \frac{\partial}{\partial \tilde{x}} \{ (Y - Y_h)^2 H(Y - Y_h) \left(\frac{\partial^2 Y_h}{\partial \tilde{x}^2}(Y - Y_h) - 3\left(\frac{\partial Y_h}{\partial \tilde{x}}\right)^2 \right)$$

$$- Y\left[\frac{\partial^2 Y_h}{\partial \tilde{x}^2}(1 - Y_h)(Y^2 - 1 + (1 - Y_h)^2) - \left(\frac{\partial Y_h}{\partial \tilde{x}}\right)^2 (Y^2 - 1 + 3(1 - Y_h)^2) \right]$$

$$- 2Y(1 - Y)\frac{\partial}{\partial \tilde{x}}[Y(1 - Y)\frac{\partial Y}{\partial \tilde{x}}] \} + O(\mu^2) \tag{6.2.11}$$

where

$$\psi_h(Y, Y_h) = \alpha Y + \tfrac{1}{2}(Y - Y_h)^2 H(Y - Y_h) + \tfrac{1}{2}\left[(1 - Y)^2 - (1 - Y_h)^2\right] Y \tag{6.2.12}$$

In the long-wave limit the $O(\mu)$ terms in (6.2.11) are neglected, leaving the hyperbolic equation

$$\frac{\partial Y}{\partial \tilde{t}} - \frac{\partial \psi_h}{\partial Y} \frac{\partial Y}{\partial \tilde{x}} = \frac{\partial \psi_h}{\partial Y_h} \frac{\partial Y_h}{\partial \tilde{x}}. \tag{6.2.13}$$

Long-wave disturbances therefore propagate at the characteristic speed

$$c = -\frac{\partial \psi_h(Y, Y_h)}{\partial Y}. \tag{6.2.14}$$

In the absence of variations in Y_h the value of Y is conserved following this speed. The presence of just one wave is another point of departure from the majority of problems we have been studying. There have typically been two waves with speeds c_+ and c_-, and the flow has been labeled supercritical or subcritical according to $c_+c_- > 0$ or $c_+c_- < 0$. Here the flow will be called supercritical if $c < 0$; that is, if wave propagation is in the same direction as the background flow. For the initial upstream state $(Y = Y_h = 1/2)$, c reduces to $1/4 - \alpha$ and therefore this flow is supercritical when $\alpha > 1/4$. A Froude number for this upstream flow can therefore be defined by

$$F_o = 4\alpha = \frac{4\alpha^* D}{f_o \delta h w^*},$$

which can be considered a form of $U/\beta L^2$ if $U = \alpha^*$, $L = w^*$, and $\beta^* = f_o \delta h / D w^*$.

If the flow is steady, the potential vorticity front is a streamline, $\psi = \psi_o$ say, and the position of the front is determined by

$$\psi_h(Y, Y_h) = \psi_o.$$

The function $\psi_h(Y, Y_h) - \psi_o$ may be treated as Gill's $\mathcal{G}$ and the critical condition $\partial \mathcal{G}/\partial Y = 0$ is clearly equivalent to $c = 0$. The relationship between Y and Y_h depends on the initial flow speed (or equivalently the transport) α, and this dependence is reflected in a the selection of contour plots of $\psi_h(Y, Y_h)$ shown in Figure 6.2.2. In the absence of jumps or other dissipative features, solutions must lie along the contours. A solution originating from an upstream condition $Y = Y_h = 1/2$ must lie along the contour that passes through the center of the plane and the corresponding contour value is $\psi_o = -\alpha/2$. The segments of contours with positive tilt correspond to subcritical $c > 0$ flow, while negative tilts correspond to supercritical flow. Maxima or minima in Y_h along a contour correspond to critical flow and it can be seen that the curves have zero, one or two such extremes. Although the presence of two extremes suggests the possibility of two control sections within the same solution, one would involve a supercritical-to-subcritical transition subject to the instability described in Section 1.4. Still, the presence of two extremes makes for a richer variety of possible steady flow configurations.

Now consider some examples in which the position of the step is given by (6.2.7). Fix the amplitude ε of the excursion of the step at 0.15 and consider a sequence of initial value problems with progressively larger transports: $\alpha = 0.05, 0.1, 0.15$, and 0.3. (The corresponding initial upstream Froude numbers are 0.2, 0.4, 0.6, and 1.2.) The families of possible steady solution curves corresponding to these transports are shown in Figures 6.2.2a, c, e, g. The potential vorticity front initially coincides with the step but will evolve in time, possibly reaching a steady state for $t \to \infty$. The expected final state for each case is indicated by a thick line overlaid on the contours. A plan view showing the corresponding front and step positions appears in the frame to the right (6.2.2b, d, f, or h). In some cases the position Y_∞ of the front far upstream is unchanged from its initial value $(Y_\infty = 0)$; that is, no upstream influence exists. In other cases this upstream state is altered. The final steady states can be grouped into four classes:

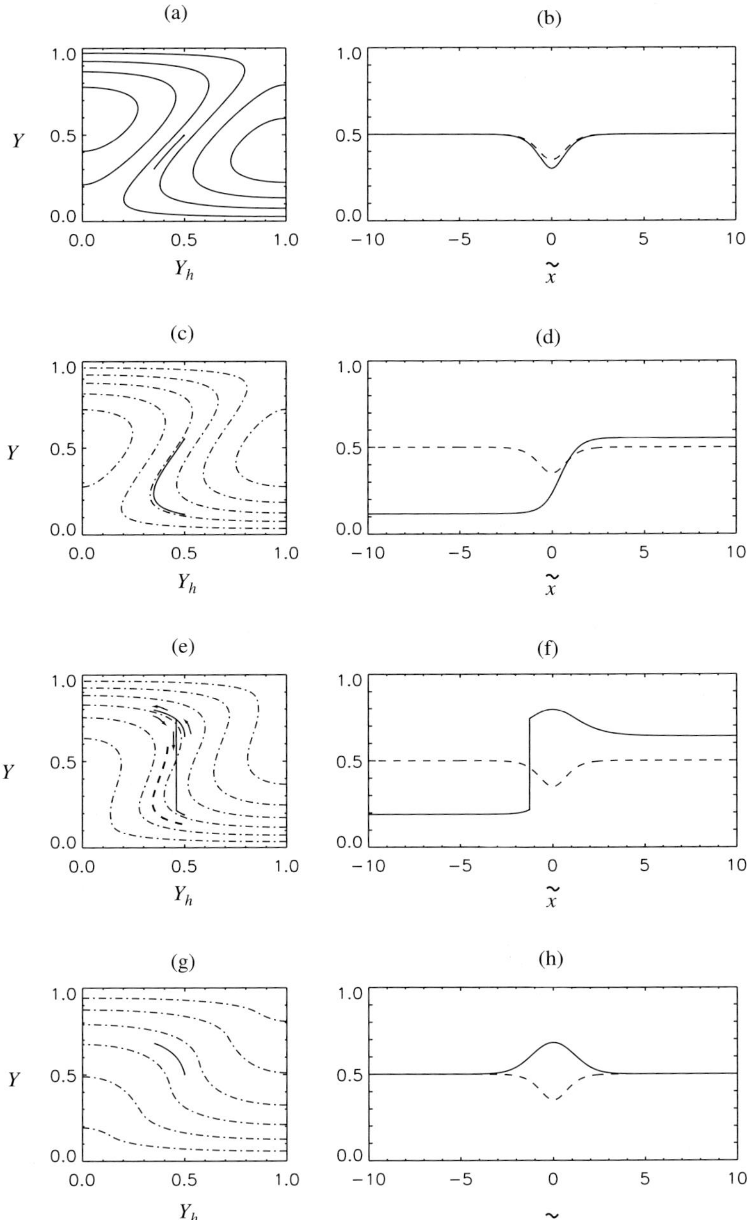

FIGURE 6.2.2. a-h Four realizations of the solution space Y vs. Y_h for different background flows: (a): $\alpha = 0.05$, (c): $\alpha = 0.1$, (e): $\alpha = 0.15$, (g): $\alpha = 0.3$. The bold overlaying curves show steady solutions for the case $\varepsilon = 0.15$ and for the initial value $Y_\infty = 0.5$. These solutions are shown in plan view in the frames immediately to the right. There the position of the step is shown as a dashed contour and the position of the front is shown as a solid contour. (From Johnson and Clarke, 1999).

Type S: Subcritical flow ($\alpha = 0.05$, *so* $F_o = 0.2$) Referring to Figure 6.2.2a, we attempt to construct a steady solution beginning upstream ($Y = Y_h = 1/2$) corresponding to the very center of the contour plot. The tilt of the contour that passes through this point is positive, so the flow is subcritical. The solution over the remainder of the channel lies along an S-shaped curve that passes through this point. Proceeding downstream from the origin, this curve is traced in the lower left direction as Y_h decreases, as suggested by the solid line in the figure. When the minimum Y_h is passed the solution is retraced back to the origin. The result is a symmetrical, subcritical solution (Frame b).

Type CC: Flow controlled at the contraction: ($\alpha = 0.1$ *so* $F_o = 0.4$) Here the larger value of α causes the excursions of the S-shaped curves to decrease. If one proceeds as in the first case, tracing along the curve that passes through the origin, the lower extreme of this curve is reached before the maximum step amplitude is encountered. Since it is impossible to continue along the same curve, we must assume that the upstream condition is no longer valid. Proceeding as in earlier examples of this type, we find the new value of Y_∞ by demanding that the flow be critical at the crest ($Y_h = 0.35$) of the step. The appropriate solution curve has an extremum at this value. The full solution is then constructed by tracing that curve along its subcritical branch upstream and along the supercritical branch downstream (vice versa would lead to an unstable flow). This trace is shown in Frame c and the corresponding solution in Frame d. The new value of Y_∞ for this solution is $> 1/2$, meaning that the upstream position lies on the deeper side of the step.

Type AC: Approach control with supercritical leap: ($\alpha = 0.15$, *so* $F_o = 0.6$) If α is further increased, the topography of the solution plane can change to the point where a contraction controlled solution is no longer viable (see Frame 6.2.2e). One begins as in the previous example by assuming the flow is critical where Y_h reaches it minimum value, in this case 0.35. The presumptive solution curve along the contour that achieves a minimum in Y_h at this value, and this is indicated by a dashed curve in Figure 6.2.2e. However, trouble arises when one attempts to trace the subcritical branch of this curve upstream. A maximum in Y_h is encountered before the upstream value ($Y_h = 0.5$) is reached. The solution cannot be continued beyond this point. Direct solution of the initial value problem for this case suggests a new type of steady solution in which the upstream flow itself is critical (and akin to 'approach control' of two-layer flow, as discussed in Chapter 5.) The corresponding solution contour has a maximum at $Y_h = 0.5$. If one proceeds downstream, a decision must be made as to which branch of the curve to follow. Selection of the subcritical branch would give rise to an unstable situation in which disturbances generated downstream would propagate upstream, only to encounter an approaching critical flow. The same accumulation of disturbance energy that marks a supercritical-to-subcritical transition would be in play. Selection of the supercritical branch avoids this difficulty and we therefore trace along this branch as the narrowest point of the shelf is passed. This trace is indicated by arrows along the solid line in Figure 6.2.2e. However, continuing downstream would require retracing the supercritical branch of the curve back to the extremum, so that the downstream flow would also be critical and the instability mentioned above would come into play. In fact, direct simulations show that the solution jumps to the other supercritical branch of the S-shaped curve, as suggested in the figure. This jump is called a supercritical leap and the corresponding solution is shown in Figure 6.2.2f. There is a range of possible Y_h values at which the leap can occur and the true value can be ascertained using jump conditions (see Johnson and Clarke, 1999).

Type SR: Supercritical flow: ($\alpha = 0.3$ *so* $F_o = 1.2$) If the transport is strong enough, the *S*-shape character of the solution curves is lost and only supercritical flow is possible. Here solutions with no upstream influence are easily constructed. A typical example is shown in Figures 6.2.2g, h.

For given α (or equivalently F_o), the predicted steady solution depends upon the value of ε as described in the previous examples. Analysis of the function $\psi_h(Y, Y_h)$ allows one to determine the boundaries between the four different flow regimes and these boundaries are indicated in Figure 6.2.3. Keep in mind that the initial upstream flow is critical ($F_o = 1$) when $\alpha = 1/4$. It is notable that flows with approach controls (AP) occupy a relatively large region of the parameter space. If the upstream position of the step is not at mid-channel, then the regime diagram is different. A key difference is the emergence of the fifth type of steady flow, the *twin supercritical leap*. These features are discussed by Johnson and Clarke (1999).

b.   *Simulations Based on Contour Dynamics*

Haines *et al.* (1993) verified the above solutions by solving the initial-value problem for the full potential vorticity equation (6.2.5) using the method

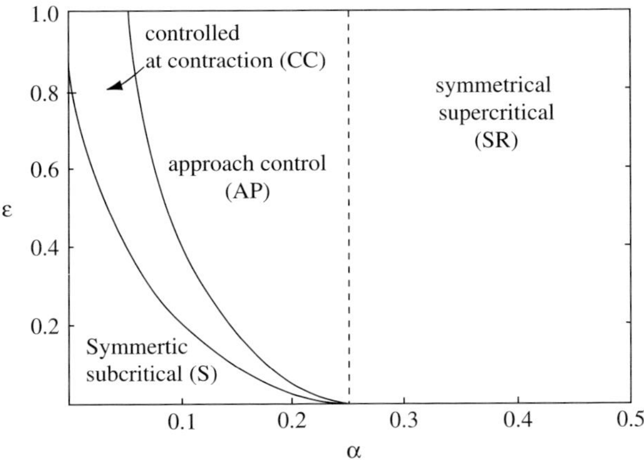

FIGURE 6.2.3. The regimes of steady, long wave solutions in terms of ε and α, all assuming the initial value $Y_\infty = 0.5$. (From Haynes, *et al.*, 1993).

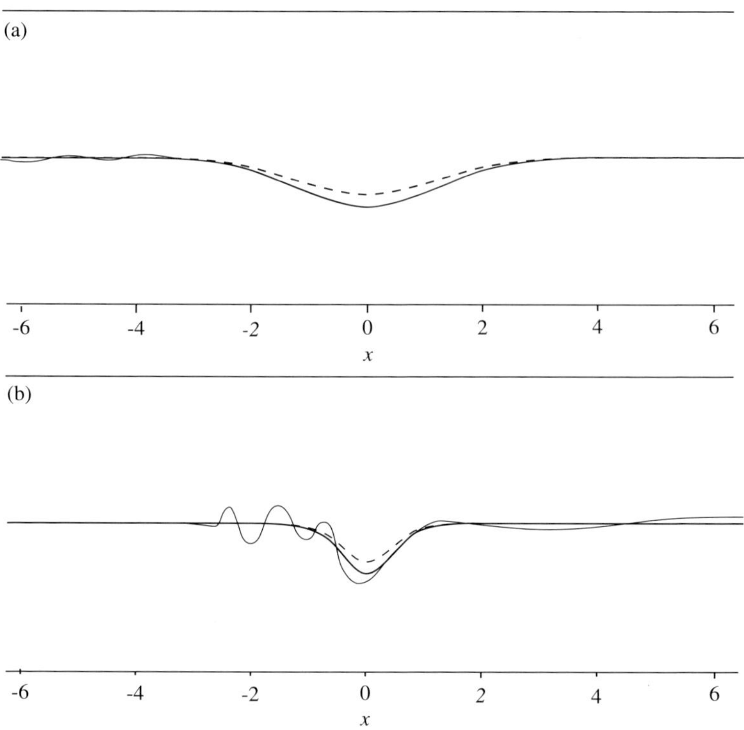

FIGURE 6.2.4. Contour dynamical solutions of the full barotropic equations for $\alpha = 1/6\pi$ and $\varepsilon = 1/8$. Frames (a) and (b) use obstacles of different lengths. The position of the topographic step is shown as a dashed line, the predicted long wave solution by a thick solid line, and the CD solution by a thin and solid line. (a): $\mu^{1/2} = \pi/(32)^{1/2}$ and $t = 102$. (b): $\mu^{1/2} = \pi/2$ and $t = 42$. Note x and t are the primitive, not the stretched, versions. (From Haynes, _et al._, 1993, but note that some parameter values differ from the published ones because of differences in scaling.)

of contour dynamics[1] (CD). The governing equation and procedure are very similar to those introduced in Section 3.2 and a sample of results is shown in Figures 6.2.4–6.2.7. In all cases the dashed line represents the position Y_h of the step.

Two examples of flow developing from initial conditions in the subcritical region of Figure 6.2.3 are shown in Figure 6.2.4. The predicted steady long-wave solution (thick line) is shown along with the result of the CD simulation (thin, solid line) as it nears a steady state. In (a) the variation in Y_h is very gradual and the final steady solution resembles that of Figure 6.2.2b. The main departure is

---

[1] The procedure is similar to that described in Section 3.3. The Green's function continues to be that defined by (3.2.13) but the potential vorticity distribution is complicated by the presence of the topographic step.

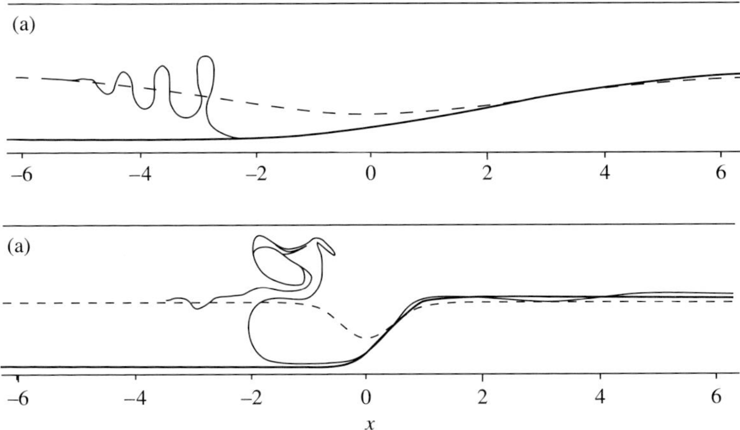

FIGURE 6.2.5. Figure 6 of HFH. Same as Figure 6.2.4 except that ε has been increased to 1/4.

the presence of small lee waves in the CD simulation. In (b) the same initial conditions and step excursion ε are used but the scale of variation of Y_h is shorter and, in fact, well outside the long-wave approximation. The scale of topographic variations in this case is closer to the scale of the lee waves, and the latter are more efficiently generated.

Solutions with a contraction control can be generated if the value of ε is increased sufficiently. The resulting upstream influence leads to an increase in Y_∞ above its value of 0.5. The characteristic wave speed in this situation increases with increasing Y and thus the disturbance is a rarefaction (Exercise 4). These expectation are born out in CD simulations (Figure 6.2.5) that also exhibit prominent short wave effects in the lee of the contraction. In Frame (a), which shows a case of gradually varying Y_h, an undular bore develops in the lee. This feature is not entirely steady and, in fact, propagates very slowly in the upstream direction. The leading wave crest is close to the point of pinching off and forming a detached eddy. In (b) the topographic feature has been shortened with the result that the undular bore has been replaced by nearly discontinuous bore, again propagating downstream slowly. A counterclockwise eddy has been ejected from the bore.

The establishment of a flow with an approach control begins in a similar manner (Figure 6.2.6a). A rarefaction wave is generated at the contraction and this disturbance moves upstream, increasing the value of Y_∞. In the contraction itself, the flow undergoes a subcritical-to-supercritical transition and a downstream bore develops. So far the evolution is similar to the previous case. However, the region of flow just upstream of the contraction begins to evolve in a new manner, as evidenced by a steepening of the front into an abrupt bend (Frame b). Although the bend first forms upstream of $x = 0$, it slowly moves downstream and settles into a fixed position (frame c). The final steady state can be compared to the long-wave solution (thicker

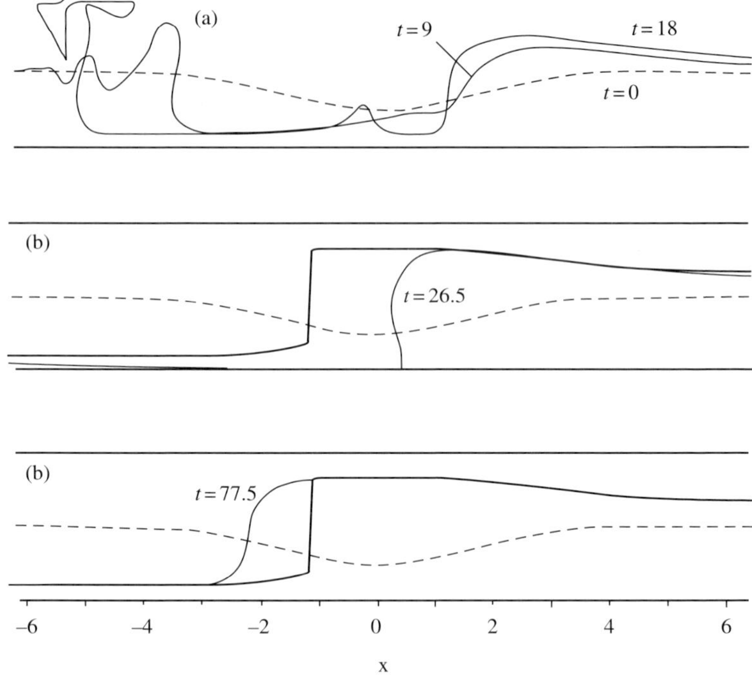

FIGURE 6.2.6. Development of an approach control with a supercritical leap. The parameters are the same as in Figure 6.2.5, but α has been increased to $1/3\pi$. (a) The front at three time intervals up to $t = 18$. The dashed curve indicates the position of the step and the initial position of the front. (b) The front (thin solid line) at $t = 26.5$ has developed an abrupt bend just upstream of the contraction. Shown by thick solid line is a version of the steady, approach-controlled, long-wave solution in which the supercritical leap occurs at the first opportunity. (c) By $t = 77.5$ the abrupt bend in the front has moved downstream of the contraction and has settled into a steady state supercritical leap. (From Haynes, *et al.*, 1993)

line in Frame c) obtained by introducing a supercritical leap at the highest value of x for which it can occur. It is notable that the supercritical leap calculated using CD is smooth and shows no signs of turbulence or energy dissipation, suggesting the energy conservation could be a basis for the formulation of the jump condition. We return to this topic later.

c. Dispersion

The CD simulations reveal the presence of lee waves, undular bores, supercritical leaps, and other features that lie outside of the realm of long-wave theory. An advantage of the present model is that these dispersive, short-scale effects can be explored with relative ease. One way to gain some insight is to include the previously neglected $O(\mu)$ corrections to the streamfunction in (6.2.11). It

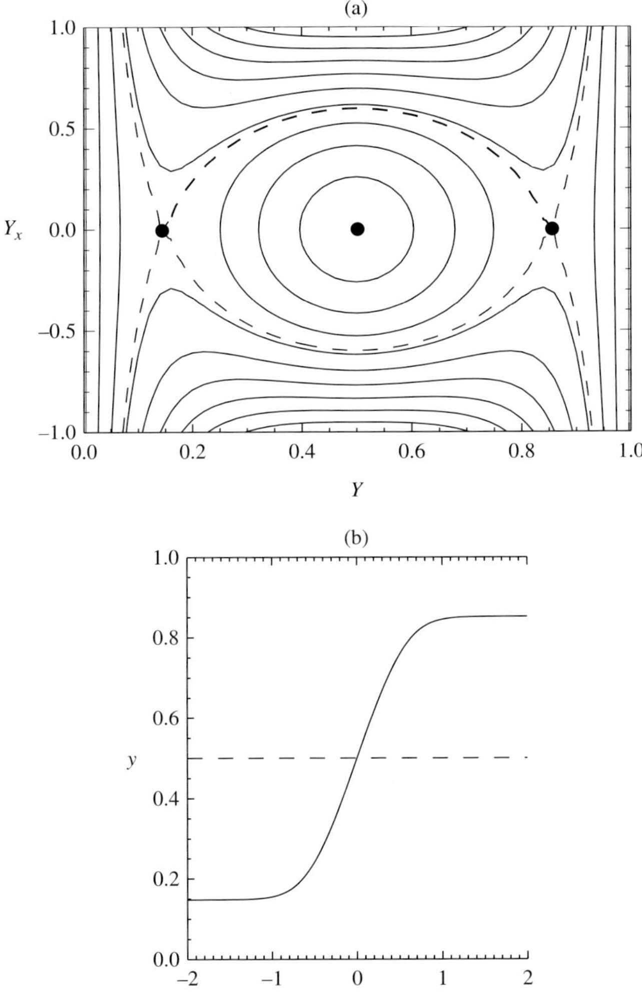

FIGURE 6.2.7. (a) Phase plane trajectories for (6.2.15) with $Y_h = 0.5$, $\alpha = 0.1$, and $\psi_0 = \alpha/2$. (b) The solution corresponding to the upper middle dashed trajectory of (a). (From Johnson and Clarke, 2001).

will be convenient to switch back to the original coordinates, with x and t replacing $\tilde{x}/\mu^{1/2}$ and $\tilde{t}/\mu^{1/2}$, but the reader should keep in mind that the resulting expressions are approximations, valid to $O(\mu)$.

One of the most important elements added by dispersion is the ability of disturbances of finite scale to remain stationary in the flow. If the position of the step is held constant, these disturbances are governed by the steady form of (6.2.11) with fixed Y_h. Integration of this relation with respect to x yields

$$\psi_h(Y, Y_h) - \frac{1}{3}Y(1-Y)\frac{d}{dx}[Y(1-Y)\frac{dY}{dx}] = \psi_o,$$

where ψ_o is a constant. If the upstream flow is independent of x, then ψ_o can be interpreted as the value of ψ on the potential vorticity front. However, in contrast to the previous analysis, the upstream flow is not constrained to be parallel. Multiplication of this relation by $\partial Y/\partial x$ and integration with respect to x leads to

$$\frac{1}{6}Y^2(1-Y)^2\left(\frac{dY}{dx}\right)^2 + V(Y, Y_h) = M, \tag{6.2.15}$$

where M is a constant and

$$V(Y, Y_h) = \psi_o Y - \frac{1}{2}\left[\alpha + Y_h - \frac{1}{2}Y_h^2 + \frac{1}{4}Y^2 - \frac{2}{3}Y\right]Y^2 - \frac{1}{6}(Y - Y_h)^3 H(Y - Y_h).$$

Solutions to (6.2.15) are represented by contours of constant M in the 'phase space' $(Y, \partial Y/\partial x)$. As an example, the contours for the case $Y_h = 0.5$, $\alpha = 0.1$, and $\psi_o = \alpha/2$ are plotted in Figure 6.2.7a. There are three fixed points $(0.5, 0)$ and $(0.5 \pm .38, 0)$ corresponding to parallel flows ($Y =$ constant). The first of these corresponds to the upstream flow assumed in the initial value problem; the front lies at the position of the step and both are at mid-channel. Since $\alpha < 1/4$, this flow is subcritical. The surrounding closed orbits represent periodic stationary waves. The small orbits in the immediate vicinity of the fixed point are essentially the linear stationary waves corresponding to $\omega^* = 0$ in (6.2.4). As one moves away from the fixed point the wave amplitude increases until a heteroclinic trajectory joining the remaining fixed points is reached. Both points represent uniform supercritical flows, a fact that can be shown directly from (6.2.14) or deduced from the property that they support no small amplitude stationary waves. The trajectories that join the two points correspond to solutions for which Y varies monotonically over $-\infty < x < \infty$ from one supercritical state to the other (Figure 6.2.7b). The solution is sometime referred to as a *kink soliton*. Since the supercritical leap shown in Figure 6.2.6c occurs over such a small interval in x, the change in Y_h from one side of the leap to the other is relatively small. For this reason, the kink soliton can be regarded as an approximation to a leap.[2]

Although the phase plane solutions are valid for a fixed step position, one can anticipate the effect of a gradually varying Y_h by allowing the actual solution trajectory to gradually move from one contour to the next in Figure 6.2.7a. (In reality, the contours change as Y_h changes, but this is more difficult to visualize.) As an example, consider a solution that is subcritical and parallel far upstream of the shelf contraction. The upstream state corresponds therefore to the central fixed point. As one moves downstream and encounters the region of variable Y_h, the solution moves away from the fixed point, crossing the closed orbits.

---

[2] Note, however, that the kink soliton of Figure 6.2.7b does not represent the supercritical leap of Figure 6.2.6c.

Downstream, where Y_h returns to its upstream value of 0.5, the solution remains on one of the closed orbits. Each orbit has a different value of M, which can be interpreted as the momentum flux of the solution (Johnson and Clarke, 2001). The change in M from the fixed point to the final closed orbit is due to wave drag (form drag) occurring in the constriction. This drag generally is at its greatest when the x-scale of the topography is comparable to the wave length of lee waves and this can result in a downstream orbit that lies quite far from the subcritical fixed point. For the solution shown in Figure 6.2.4a Y_h varies gradually ($\mu \ll 1$) and the lee waves are small, corresponding to orbits close to the fixed point. In Figure 6.2.4b, where μ is O(1), the lee wave orbits are larger.

In some cases, the form drag may be sufficient to force the solution to cross the dashed trajectory in Figure 6.2.4a, so that the downstream flow lies on an open trajectory. The latter does not represent acceptable end states since they lead to unbounded values of Y far downstream. In this case an upstream disturbance must be generated, altering the value of ψ_o and leading to a new phase plane with an expanded heteroclinic orbit. Dispersion can thereby lead to a hydraulically controlled solution even though a perfectly valid subcritical, long-wave solution exists. If the step does not lie at mid-channel, the symmetry of the phase plane is lost and the supercritical fixed points are no longer connected. Some of the consequences are explored in Exercise 6.

In order to make more explicit consideration of the effects of variations in Y_h in the presence of dispersion, one can integrate the full steady version of (6.2.11) beginning with a given upstream state. One of the difficulties in doing so is that wavy upstream states are now possible and there is no obvious reason to reject them. This matter can be handled more cleanly by solving the full initial-value problem using the full time dependent version of (6.2.11). As reported by Johnson and Clarke (2001), the typical steady state solution that develops is a dispersion-modified version of the expected long-wave solution. Among the effects of dispersion are the presence of lee waves or downstream undular bores, and the smoothing of features like supercritical leaps that are represented in the long-wave limit by discontinuities. The function $Y(x, t)$ is constrained to be single valued in x and therefore wave breaking or eddy pinching is not allowed. Solutions for values of α and ε that lie within the subcritical (S) portion of the regime diagram (Figure 6.2.3), but close to the boundary with CC, can evolve to hydraulically controlled flows of the AC type if μ is sufficiently large. The forcing of upstream influence by dispersive effects, suggested above, is thereby confirmed. Another consequence of dispersion is the failure in certain small regions of the (α, ε) parameter space for the solution to settle into steady state.

In flows with continuous variations in potential vorticity or other complexities, the hydraulic problem with dispersion becomes less tractable. In such cases, progress can be made by assuming that the topographic variations (here measured by ε) are small and gradual, and that the initial flow is close to the critical speed. The proximity to criticality means that disturbances can be resonantly excited by variations in topography. Weak nonlinearity and dispersion act at finite amplitude to limit growth and the resulting finite amplitude disturbances can be interpreted as hydraulic transitions, lee waves, and undular or monotonic jumps and bores.

Solutions bear some similarity to the cases just discussed. The precise form of the evolution equation depends on the character of the wave guide. If the flow has imposed transverse scales such as channel width or the Rossby radius of deformation, and the extent of cross-stream motion is constrained by these scales, then the dynamics of nonlinear dispersion is generally governed by an equation of the KdV type (Section 1.11). The long-wave limit in such cases can clearly be defined. A case in which this constraint is absent occurs if the wall at $y = 1$ in the Figure 6.2.1 channel is moved to infinity. Although a long-wave approximation can still be defined for the fluid lying between the wall at $y = 0$ and the potential vorticity front, it cannot be defined in the outer region. There the y scale is simply the x-scale associated with variations in Y. Such cases arise naturally in coastal applications and are governed by Benjamin-Davis-Acrivos (BDA) type equations (see Grimshaw, 1987 and Grimshaw and Yi, 1990). There is extensive literature on this subject, much of it summarized by Johnson and Clarke (2001).

d.   Related Coastal Applications

The geometry of a typical continental shelf and slope (Figure 5.2.8a) can be crudely approximated by the step topography in the above model if the wall at $y = 1$ is moved to infinity (Frame b). The topographic wave in this limit decays away from the step in the offshore direction. If an opposing uniform current is added, the waves can be arrested and hydraulic behavior similar to that discussed above can arise. A more realistic current, first proposed by Niiler and Mysak (1971), takes the form of an along-shore jet with piecewise uniform shear (Figure 6.2.8c). The jet velocity goes to zero at some offshore location and the ocean is considered quiescent further offshore. There are now two potential vorticity fronts, one associated with the topographic step and one associated with the offshelf front. As shown by Niiler and Mysak, the flow has two wave modes, the first being a 'shelf' wave that shares features with the topographic waves discussed in the above channel model. The propagation speed of this wave may be positive or negative depending on the speed and frontal configuration of the jet. As shown by Collings and Grimshaw (1980), supercritical jets tend to be narrow and subcritical jets tend to be wide, as measured by the distance between the coast and either potential vorticity front. The second mode is a 'shear' wave that depends on the existence of the offshelf potential vorticity front. It propagates in the same direction as the jet regardless of the parameter settings.

   The shelf wave is a special case of a general class of coastal trapped waves that owe their existence to the presence of the topographic potential vorticity gradients associated with continental slopes and shelves. In the absence of a background flow, the potential vorticity of the fluid increases towards the coast and Northern Hemisphere waves propagate keeping high potential vorticity (shallow water) on their right, i.e. in the same direction as a coastal Kelvin wave. Hydraulic behavior with respect to these waves can occur in the presence of an opposing flow. Hughes (1986b) investigation of the Niiler and Mysak model confirms that two alternate states can occur for a given flow rate. The first is a relatively wide,

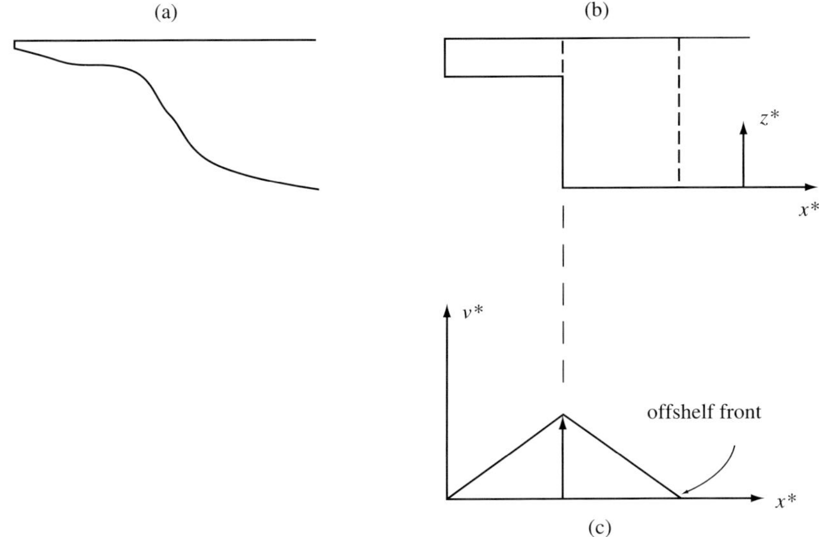

FIGURE 6.2.8. Step approximation (b) to a typical shelf topography (a). Fame (c) shows a plan view of the coastal jet considered by Niiler and Mysak (1971).

subcritical jet that allows upstream propagation of shelf waves; the second is a narrower and higher speed supercritical jet that carries shelf waves downstream. The shear wave does not appear to be directly implicated in the hydraulic behavior. However, instability can result from a resonance between the waves that can occur in the supercritical regime when their speeds become equal. It is also suggested that transitions between the two regimes can occur as a result of changes in the shelf position, as was documented in the coastal model discussed in Section 4.2.

   Similar behavior can be found if the topography and potential vorticity vary continuously (Hughes 1985a, 1986a and 1987). If the coastline is aligned with the y^*-axis (north and south) then the cross-shelf (x^*) structure of a homogeneous flow is governed by

$$\frac{\partial}{\partial x^*}\left(\frac{1}{d^*}\frac{\partial \psi^*}{\partial x^*}\right) - d^* q^*(\psi^*) = -f, \qquad (6.2.16)$$

just the barotropic potential vorticity equation for a gradually varying flow. Here ψ^* is the transport stream function ($v^* d^* = \partial \psi^*/\partial x^*$) and the depth d^* is determined by the specified bottom topography. The solution procedure for the second order equations is to choose a potential vorticity distribution $q^*(\psi^*)$ and assign a particular value of ψ^* to the coastline. A value of $\partial \psi^*/\partial x^*$ at the coast is then guessed and (6.2.16) is integrated in positive x^* until v^* vanishes. The corresponding value of x^* is then considered as the offshelf edge of the current and the fluid beyond is assumed to be at rest. A series of similar calculations with different coastal values of $\partial \psi^*/\partial x^*$ yields a family of solutions with different widths and different transports. However, it is possible to find two or more solutions with the same volume transport. Hughes has shown how pairs of solutions with the same transport may be identified as alternate states having the same transport and energy (see Exercise 7).

Although these flows are typically forced by along-shore changes in bottom topography, it is also possible to force the current by allowing the value of f to change. For a northwards (Northern Hemisphere) flow with the coast to the left, the increasing absolute value of f causes subcritical and supercritical conjugate states to approach each other, possibly merging and becoming hydraulically critical at a particular latitude. However, unlike forcing due to topographic variations, the value of f does not reach a maximum value at the critical latitude but continues to increase in the northward direction. The solutions cease to exist beyond this latitude, suggesting that the flow separates from the coastline (Hughes, 1989).

Exercises

(1) Derivation of the linear dispersion relation (6.2.4).

(a) Consider a channel with mean D and variable topography $h = h(y)$. Show that weak motion about a state of rest is governed by the topographic Rossby wave equation:

$$\frac{\partial}{\partial t}\nabla^2\psi + \frac{\partial\psi}{\partial x}\frac{dh}{dy} = 0$$

in the nondimensional units used above.

(b) For a traveling wave with form $\psi = \mathrm{Re}[\Psi(y)e^{ik(x-ct)}]$ find the equation governing the cross-channel structure function $\Psi(y)$.

(c) Now suppose that $dh/dy = 0$, except near the channel midpoint $y = 1/2$, where h changes abruptly from 1 to zero with increasing y. Using the result obtained in (b) show that

$$\left[\frac{d\Psi}{dy}\right] - \tfrac{1}{2}[h] = 0$$

where the brackets denote the change in the indicated quantity across the step.

(d) Write down the separate solutions for $\Psi(y)$ in the regions to the north and south of the step, apply the boundary conditions at the channel walls, and match the solutions, using the result in (c) and the requirement that $\Psi(y)$ remain continuous across the step, to obtain the dispersion relation

$$c = \tfrac{1}{2}k \tanh\left(\tfrac{1}{2}k\right).$$

Deduce the result (6.2.4) by adding a background velocity $-\alpha$ and making the result dimensional.

(2) Consider the linear dispersion relation (6.2.4) for the case when the step and front lie at mid-channel. According to (6.2.4) long waves are brought to rest when $\alpha^* = 1/4 f_\mathrm{o} w^* \delta h^*/D$. Try to obtain a similar estimate using the dispersion relation for waves in a channel with a constantly sloping bottom. [A good place to start is to write down a version of (2.1.30), altered to account for a mean flow and a rigid lid.]

(3) Show that equation (6.2.5) is invariant to the change in variables $\hat{y} = 1 - y$, $\hat{Y} = 1 - Y$, and $\hat{Y}_h = 1 - Y_h$. Using the fact that the initial condition $Y = Y_h$ is also invariant, conclude that if $Y(x, t)$ is a solution than $1 - Y(x, t)$ is also a solution for the case in which the position of the step is reflected about the channel axis ($y = 1/2$). By this means the solutions shown in Figures 6.2.4–6.2.7 can be used to construct the solution for the case in which the shelf width increases rather than decreases.

(4) Show that for $Y > Y_h$, the characteristic speed is given by $c = -\alpha - \frac{1}{2}[3Y^2 - 2Y + Y_h^2]$. Note the corresponding Froude number formula $F_o = -2\alpha/[3Y^2 - 2Y + Y_h^2]$. Show that an upstream disturbance that increases the value of Y over its undisturbed value 0.5 must be a rarefaction.

(5) Show that the value of the potential $V(Y, Y_h)$ is the same at the two super-critical fixed points in Figure 6.2.7a. [This result can be used as the basis for a matching condition across a supercritical leap, as discussed by Johnson and Clarke (1999).]

(6) Using a contouring routine, construct a phase plane diagram akin to Figure 6.2.7a for the case $Y_h = .55$, $\alpha = 1$, and with the upstream position of the front located at the step ($\psi_\mathrm{o} = .055$). Show that the two super-critical fixed points are no longer connected and, instead, that there is a closed (*homoclinic*) trajectory attached to one of these points. Describe the associated solution. Draw a sketch showing the shapes of the stationary waves associated with the periodic orbits as the limiting homoclinic orbit is approached. *(The resulting phase plane is given in Figure 6c of Johnson and Clarke, 2001)*

(7) Suppose that two distinct solutions to (6.2.16) are found for the same potential vorticity distribution and topography. Both have the same transport and both can be smoothly joined to a quiescent region far offshore. Show that the distribution of the Bernoulli function along streamlines is the same in each case.

6.3. Zonal Jets on a Beta Plane

The ocean and atmosphere are host to a variety of eastward jet-like flows including the Gulf and Jet Streams, the Kurishio, the Pacific equatorial counter current, and jets associated with frontal systems in the Antarctic circumpolar current. It is natural to ask whether these might be subject to hydraulic effects associated with the Rossby waves, or more general potential vorticity waves, that arise and that otherwise propagate westward. Hydraulic transitions might occur where the currents pass over ridges or other types of bottom topography or, in the case of the Antarctic circumpolar current, where it passes through the Drake passage. If hydraulic control occurs, it would imply that the effects of the topography could be felt far upstream and downstream.

The suggestion of hydraulic behavior first appears in work by Rossby (1950) who investigated an eastward jet with a half width w^* and velocity u^* independent of y^*. Rossby showed that the jet has two possible widths for the same volume and momentum fluxes. The 'conjugate' states were thought to be the two end states of a hydraulic jump of unknown form. Rossby thought that a jump might provide a mechanism for atmospheric blocking. He also showed that the two states merge when $3u_o/\beta^* w^{*^2} = 1$, which could then be interpreted as a critical condition. Armi (1989) argued that smooth, subcritical-to-supercritical transitions might also be possible by showing that two *alternate* jet states, i.e. states having the same volume flux and energy, can exist. The jet has a velocity profile that decays from a maximum velocity U_o along the centerline to zero at the edges. The conjugate states are found to be identical for $U_o/\alpha'\beta^* w^{*^2} = 1$, where α' depends on the pressure distribution across the jet. Armi was apparently able to establish a critically-controlled jet by circulating fluid through a laboratory annulus and observing that $U_o/\beta^* w^{*^2} = 1$ near the point of withdrawal and < 1 upstream.

Both of the studies mentioned above base their model on the assumption that the velocity profiles of the conjugate or alternate state are similar. The dependence of the flow on the bottom topography is not considered. As shown by Woods (1993) a complete solution with topographic forcing and self-similarity can be found provided that special upstream conditions exist and that the jet is equatorial ($f = \beta^* y^*$). His theory puts the dynamical elements of Armi's and Rossby's work in a more consistent setting, although both of these models are centered at midlatitudes. The hypothetical form of Woods' solution is suggested in Figure 6.3.1, which shows a barotropic, eastward jet, centered on and symmetrical about the equator ($y^* = 0$). The zonal velocity decays from a maximum at the center to a value of zero at the edges $y^* = \pm w^*$, where the flow is joined to a quiescent ambient region. The jet impinges on a ridge over which the depth d^* decreases gradually from the upstream value D. Self-similarity means that the fractional compression or expansion of streamlines at any x^* is the same for each streamline across the velocity profile. In other words $\psi^* = \psi^*[y^*/w^*(x^*)]$.

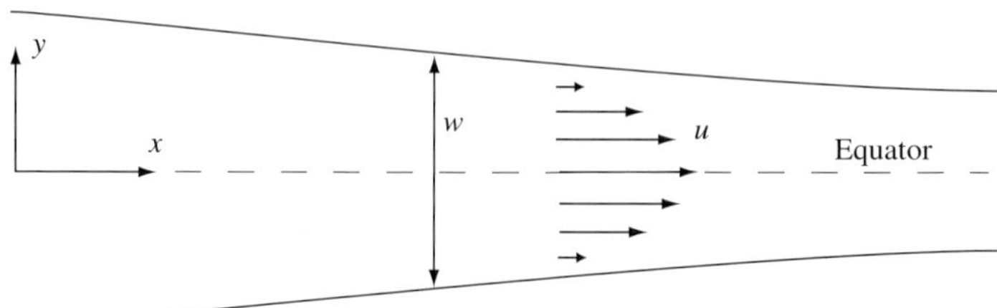

FIGURE 6.3.1. Definition sketch for the self-similar, equatorial jet of Woods (1993).

To investigate whether a dynamically consistent flow of the assumed form exists, we consider the barotropic potential vorticity equation on an equatorial beta plane. A convenient set of dimensionless variables is

$$y = y^*/w_o, \quad \psi = \psi^*/Q, \quad u = u^* D w_o/Q, \quad w = w^*/w_o, \quad \text{and } d = d^*/D,$$

where w_o is the upstream half-width and $2Q$ is the volume transport. The steady form of (6.1.10) is then

$$\frac{\beta y + d^{-1}\dfrac{\partial^2 \psi}{\partial y^2}}{d} = q(\psi). \tag{6.3.1}$$

where $\beta = \beta^* w_o^3 D/Q$ and $ud = -\partial\psi/\partial y$. The contribution to the relative vorticity from $\partial v/\partial x$ has been neglected in view of the assumed gradually varying nature of the flow.

The assumed symmetry of the flow about the x-axis means attention may be confined to the upper half plane. The boundary conditions are then

$$\psi(0) = 0, \quad \left(\frac{d\psi}{dy}\right)_{y=w} = 0, \tag{6.3.2a,b}$$

and

$$(\psi)_{y=w} = -1. \tag{6.3.3}$$

We seek a solution of the form $\psi = \psi(\zeta)$, where $\zeta = y/w(x)$. The potential vorticity can be written in terms of ζ as

$$\frac{\beta w\zeta + \dfrac{1}{dw^2}\dfrac{d^2\psi}{d\zeta^2}}{d}.$$

Since this quantity is conserved along streamlines (lines of constant ζ), its upstream $(w = d = 1)$ value can be equated with its value at any x along the same streamline:

$$\beta\zeta + \frac{d^2\psi}{d\zeta^2} = \frac{\beta w(x)\zeta + \dfrac{1}{d(x)w(x)^2}\dfrac{d^2\psi}{d\zeta^2}}{d(x)},$$

or

$$\frac{d^2\psi}{\partial\zeta^2} = \frac{\beta(d-w)dw^2}{(1-d^2w^2)}\zeta.$$

The solution satisfying the two homogeneous boundary conditions (6.3.2a,b) is

$$\psi = \frac{\beta(d-w)dw^2}{6(1-d^2w^2)}\zeta(\zeta^2 - 3). \tag{6.3.4}$$

This solution is not yet of the form sought since the coefficients w and d depend explicitly on x. However, if the final boundary condition (6.3.3) is enforced, then

$$\frac{\beta(d-w)dw^2}{6(1-d^2w^2)} = \frac{1}{2} \tag{6.3.5}$$

and therefore the desired form:

$$\psi = \frac{\zeta}{2}(\zeta^2 - 3) \tag{6.3.6}$$

is achieved.

The velocity profile corresponding to (6.3.6) is

$$u = -d^{-1}\frac{\partial\psi}{\partial y} = \frac{3}{2wd}\left(1 - \frac{y^2}{w^2}\right), \tag{6.3.7a}$$

or

$$u^* = \frac{3Q}{2w^*d^*}\left(1 - \frac{y^{*2}}{w^{*2}}\right). \tag{6.3.7b}$$

In order to track the width of the current as the jet passes over the topography, it is necessary to relate w to d. This relation is given by (6.3.5), which can be written in the form of a Gill functional:

$$\mathcal{G}(w; d; \beta) = (\beta+3)w^2d^2 - \beta w^3d - 3 = 0. \tag{6.3.8}$$

Critical states correspond to $\partial\mathcal{G}/\partial w = 0$, or

$$\frac{3\beta w_c}{2(\beta+3)d_c} = 1, \tag{6.3.9}$$

where the subscript 'c' denotes the value at the critical section.

Plots of d as a function of w for different β (Figure 6.3.2) give the standard sort of hydraulic curves with minima at $w = w_c$. All the curves connect to the assumed upstream point $d = w = 1$. The depth range in the plots has been restricted to $0 \leq d \leq 1$, corresponding to a ridge (rather than topographic trough). The left branch of each curve presumably represents supercritical states since it corresponds to relatively small widths and relatively large velocities. For $\beta > 6$, the assumed upstream flow lies on the subcritical branch of the appropriate curve and a hydraulically controlled solution can be constructed in the usual way. For $\beta < 6$, the upstream flow lies on the supercritical branch, apparently leading to an (unstable) supercritical-to-subcritical transition over the topography. However an equally valid solution for the same transport can be constructed by choosing the subcritical value of w for $d = 1$ as the upstream state. [After all, the arguments leading to the solution are equally valid if $d = w = 1$ is chosen as the downstream state.] At the value $\beta = 6$ the upstream flow is exactly critical, as can be

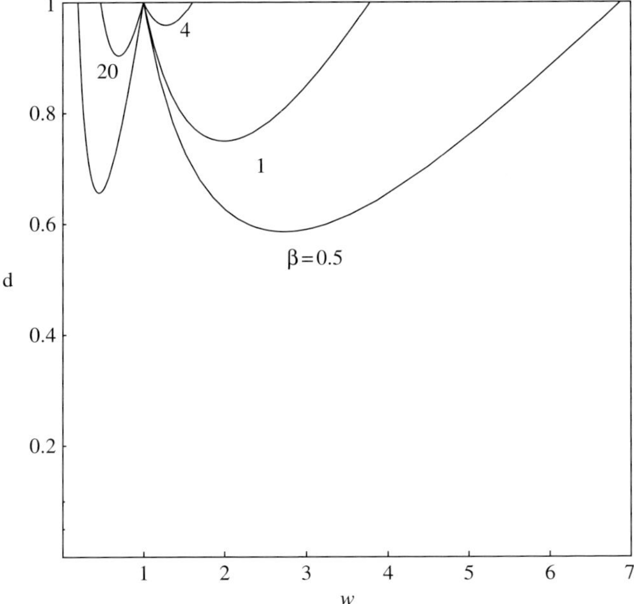

FIGURE 6.3.2. Solution curves for the Woods (1993) jet for various values of $\beta = \beta^* w_o^3 D/Q$.

verified using (6.3.9). In this case, any slight increase in topographic elevation would lead to upstream influence. In dimensional terms $\beta = 6$ corresponds to $4u^*(0)/\beta^* w_o^2 = 1$, which is also the result found by Armi (1989) for a jet with a quadratic velocity profile.

The mechanics of upstream influence in the equatorial jet is unknown. Given a hydraulically controlled flow, with a subcritical upstream state, what would happen if the minimum depth d_c over the ridge were decreased? If the resulting upstream disturbance preserves the self-similar character of the flow, then adjustments in the upstream state would be contained entirely in the value of $\beta = \beta^* w_o^3 D/Q$. Since β^* and D are fixed, the implied changes would involve the upstream width w_o or volume flux $2Q$, or both. Figure 6.3.2 implies that for $\beta < 6$, a small decrease in the depth over the ridge would lead to a decrease in β, whereas the opposite is true for $\beta > 6$. It is not clear, however, what this would mean for the width or flux. The possibility that Q could be altered distinguishes this model from the quasigeostophic model in the previous section, for which the volume flux is fixed.

Exercises

1. Calculate the form drag exerted against the equatorial jet by the ridge as a result of a subcritical to supercritical transition.

6.4.   A Mid-Latitude Jet

Unfortunately, global self-similarity is lost when the jet of the previous section is moved off of the equator ($f_o \neq 0$). Formulation of a hydraulic model of a mid-latitude jet is possible but benefits from further simplification. It will be helpful to work within the confines of the quasigeostrophic approximation, for which the layer depth is nearly constant and the Rossby number is small. In addition, it will be necessary to consider a special potential vorticity distribution.

We consider the reduced gravity, quasigeostrophic model as applied to a deep layer that flows along the bottom and is separated from the inactive upper layer by an interface. As discussed in Section 6.1, the interface displacement $\eta^{(0)}$ acts as a streamfunction ψ for the geostrophic velocity. For steady flow, the long-wave approximation to the quasigeostrophic potential vorticity (see 6.1.7) is

$$\frac{\partial^2 \psi}{\partial y^2} - S\psi + h + \beta y = q(\psi). \tag{6.4.1}$$

The topography again consists of an isolated ridge but, unlike the previous example, y-variations in the height of the ridge will be key to hydraulic transitions in the overlying jet. We therefore set $h = \mathcal{N}(x)H(y)$.

The form of the potential vorticity function $q(\psi)$ can be constrained by considering a hypothetical upstream flow consisting of a jet that is centered near $y = 0$ and that impinges on the ridge from the west. Far to the north and south of the jet, but still upstream, the flow is assumed to be westward and very broad: $u = -u_o < 0$, where u_o is a constant. Evaluating (6.4.1) in these outer regions leads to $\beta y = q(\psi) + S\psi$ or, after differentiation with respect to y, $dq/d\psi = (\beta/u_o) - S$. The corresponding potential vorticity distribution is

$$q(\psi) = [(\beta/u_o) - S]\psi + constant. \tag{6.4.2}$$

It is expected that this distribution will be carried eastward in x across the topography and therefore serve as a far field potential vorticity for the jet at all x.

The choice of $q(\psi)$ within the core of the jet can be motivated by noting that zonal currents like the Gulf Stream east of Cape Hatteras and the Jet Stream possess intrinsic potential vorticity gradients much stronger than β. These local gradients will be represented here by discontinuities in the value of q across certain streamlines. Specifically:

$$q(\psi) = [(\beta/u_o) - S]\psi + \hat{\delta}[y; L_1(x), L_2(x)] \tag{6.4.3}$$

where

$$\hat{\delta} = \begin{cases} a & y > L_1(x) \\ b & L_2(x) < y < L_1(x) \\ c & y < L_2(x) \end{cases}.$$

The flow is thus divided into three regions (I, II, and III), each containing the same potential vorticity gradient $(dq/d\psi)$ but each having a different 'background' potential vorticity a, b, or c (Figure 6.4.1). The regions are separated by potential vorticity fronts coinciding with the streamlines at $y = L_1(x)$ and $y = L_2(x)$, across which $q(\psi)$ is discontinuous but velocity and ψ are continuous.

The potential vorticity equation (6.4.1) now becomes

$$\frac{\partial^2 \psi}{\partial y^2} - \alpha^2 \psi = -\beta y - \mathcal{N}(x)H(y) + \hat{\delta}[y; L_1(x), L_2(x)], \qquad (6.4.4)$$

where $\alpha^2 = \beta/u_o$. The solution to (6.4.4) satisfying the condition of velocity continuity across the potential vorticity fronts and for which $\partial\psi/\partial y$ remains bounded as $y \to \pm\infty$ is

$$\psi = \begin{cases} (2\alpha^2)^{-1}[(b-c)e^{-\alpha\Delta L} + (a-b)]e^{-\alpha(y-L_1)} + \alpha^{-2}[\beta y + \mathcal{N}K - a](y > L_1) \\ (2\alpha^2)^{-1}[(b-a)e^{-\alpha(L_1-y)} + (b-c)e^{-\alpha(y-L_2)}] + \alpha^{-2}[\beta y + \mathcal{N}K - b](L_2 < y < L_1) \\ (2\alpha^2)^{-1}[(b-a)e^{-\alpha\Delta L} + (c-b)]e^{-\alpha(L_2-y)} + \alpha^{-2}[\beta y + \mathcal{N}K - c](y < L_2) \end{cases}$$

$$(6.4.5)$$

where $\Delta L = L_1 - L_2$ and where $K(y)$ is a solution to

$$\frac{\partial^2 K}{\partial y^2} - \alpha^2 K = -\alpha^2 H(y). \qquad (6.4.6)$$

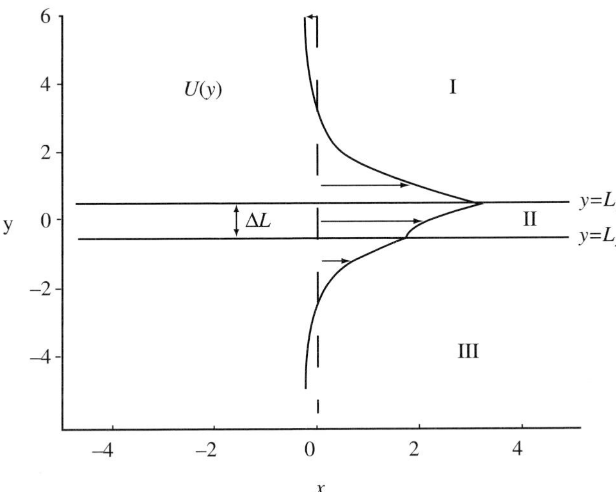

FIGURE 6.4.1. Definition sketch for a mid-latitude jet with piecewise continuous potential vorticity distribution. (Based on a figure from Pratt 1989).

The corresponding zonal velocity

$$u^{(o)} = -\frac{\partial \psi}{\partial y} = (2\alpha)^{-1}[(a-b)e^{-\alpha|y-L_1|} + (b-c)e^{-\alpha|y-L_2|}]$$

$$- u_o - \alpha^{-2}\mathcal{N}_1 \frac{dK}{dy} \tag{6.4.7}$$

consists of the sum of the ambient westward velocity u_o, a topographic contribution related to dH/dy, and a pair of eastward cusped jets centered on the potential vorticity fronts (Figure 6.4.1). The scale $\alpha^{-1} = (u_o/\beta)^{1/2}$ is the distance over which the jet velocity decays away from a front.[3] It is now clear that the meridional isolation of the jet requires westward far-field flow. (An eastward far field $(u_o < 0)$ would yield a velocity profile with sinusoidal variations in y.)

The volume transport in core region (II) of the jet is

$$Q_2 = \int_{L_1}^{L_2} u\,dy = \alpha^{-2}\{\tfrac{1}{2}(a-c)(1 - e^{-\alpha\Delta L}) - \beta\Delta L - \mathcal{N}_1[K(L_1) - K(L_2)]\}. \tag{6.4.8}$$

If the elevation of the ridge increases linearly with $y(H = H_o + sy,$ say), the solution to (6.4.6) is $K(y) = H_o + sy$ and substitution into (6.4.8) leads to a relation between $\Delta L(x)$ and N:

$$\mathcal{G}(\Delta L, \mathcal{N}) = s\mathcal{N} - \frac{\tfrac{1}{2}(a-c)(1 - e^{-\alpha\Delta L}) - \beta\Delta L - \alpha^2 Q_2}{\Delta L} = 0 \tag{6.4.9}$$

Through this relation ΔL depends only the potential vorticity difference $a - c$ across the core and not on the potential vorticity b in the core itself.

Critical flow occurs for $\partial\mathcal{G}/\partial\Delta L = 0$, or

$$\frac{2\alpha^2 Q}{(a-c)} = 1 - (1 + \alpha\Delta L)e^{-\alpha\Delta L}. \tag{6.4.10}$$

The right-hand side increases monotonically from zero to unity as ΔL increases from zero to infinity. A critical state can therefore be found if and only if

$$0 \le \frac{2\alpha^2 Q}{(a-c)} \le 1. \tag{6.4.11}$$

The numerator of this expression is proportional to $\beta Q/u_o$, the change in the planetary potential vorticity βy that would occur if the velocity had the uniform value u_o in the core. The denominator is clearly the change across the core of the potential vorticity intrinsic to the flow. For critical flow to occur, the former must be weaker than the latter. In addition, the transport Q must have the same sign as $a - c$. To an observer facing downstream, the intrinsic potential vorticity

---

[3] This scale is also the boundary layer thickness in inertial models of basin circulations (Fofonoff, 1954) and of the Gulf Stream (Charney, 1955).

must increase from right to left, meaning that the associated waves should tend to propagate upstream. We continue to assume that the jet is eastward ($Q > 0$) so that $a - c > 0$.

The topographic slope at the critical section can be found from substitution of (6.4.10) into (6.4.9). The resulting relation

$$\frac{2(\beta + s\mathcal{N}_c)}{\alpha(a - c)} = e^{-\alpha \Delta L_c} \tag{6.4.12}$$

forms the basis of a Froude number

$$F_\beta = \frac{\alpha(a - c)}{2(\beta + s\mathcal{N}_c)e^{\alpha \Delta L_c}}, \tag{6.4.13a}$$

such that $F_\beta = 1$ for critical flow. A physical interpretation of the Froude number is aided by associating with the potential vorticity difference $a - c$ an equivalent stream function difference $-\alpha^2(\psi_a - \psi_c)$, as suggested by (6.4.4). The difference $(\psi_c - \psi_a)$ is equal to the geostrophic flux Q_{ab} between two hypothetical regions with constant interface elevations a/α^2 and c/α^2. In addition, $\beta + s\mathcal{N}_c$ can be interpreted as the total ambient (planetary plus topographic) potential vorticity gradient: β_T, say. With these substitutions, the Froude number can be expressed using dimensional quantities as

$$F_\beta = \frac{Q_{ab}^{*}}{2\beta_T^{*} DL_\beta^3 e^{\Delta L^{*}/L_\beta}} \tag{6.4.13b}$$

where $L_\beta = (u_o^{*}/\beta^{*})^{1/2}$. The term $Q_{ab}^{*}/\beta_T^{*} DL_\beta^3$ is similar to the familiar scale $U/\beta^{*} L^2$, with the total ambient potential vorticity gradient substituted for β^{*}, the inertial boundary layer thickness substituted for L, and the velocity scale Q_{ab}^{*}/DL_β used for U. If the ratio $\Delta L^{*}/L_\beta$ is not much larger than unity, meaning that the two fronts are separated by a single inertial radii or less, then flow criticality requires this particular version of $U/\beta^{*} L^2$ to be O(1) as well. On of the other hand, separation of the fronts by more than a few values of L_β means that the exponential term in (6.4.13b) becomes very large and thus a large potential vorticity difference $a - c$ (and equivalent geostrophic flow) is required for critical control. The interpretation of this limit is that the two fronts lose their awareness of each other and become independent as they separate. Contact between the two can then be maintained only if the potential vorticity difference $a - b$, and thus the associated transport Q_{ac} is very large.

It follows from (6.4.9) that $\lim_{\Delta L \to \infty} s\mathcal{N} = -\beta$ whereas $\lim_{\Delta L \to 0} s\mathcal{N} = -\infty$, both limits being evident in the Figure 6.4.2 plot of $s\mathcal{N}$ vs. ΔL. The curve has a maximum value for finite ΔL provided (6.4.11) is satisfied. Flows lying to the right are presumably subcritical and those to the left supercritical. Since the upstream and downstream states correspond to $s\mathcal{N} = 0$, it is necessary that the maximum of the curve lie above the $s\mathcal{N}$ axis. Therefore a requirement for the existence of solutions, controlled or otherwise, is that $s\mathcal{N}_c$ is ≥ 0. Thus the ridge crest must slope upwards in the positive y direction.

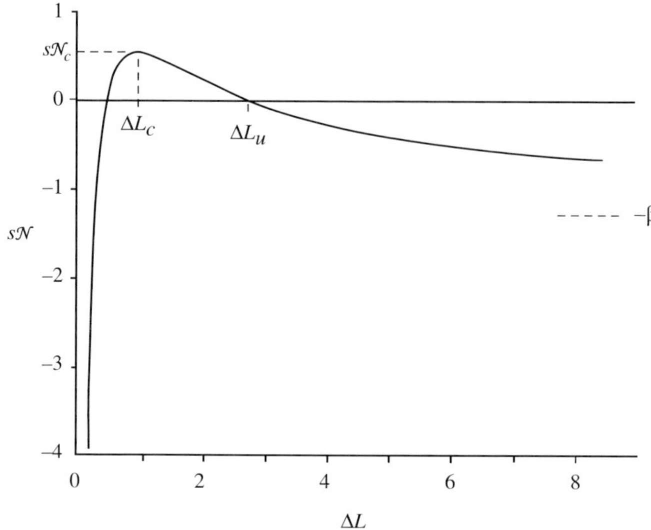

FIGURE 6.4.2. The jet width ΔL vs. the northward slope parameter $s\mathcal{N}$ for $Q = \alpha = \beta = 1$, $a - c = 8$. Based on Equation 6.4.9. (From Pratt, 1989).

For a given topographic function $s\mathcal{N}(x)$ it is possible to find the corresponding jet width $\Delta L(x)$ by tracing along the solution curve in Figure 6.4.2. It still remains to determine the individual latitudes of the potential vorticity fronts $y = L_2(x)$ and $y = L_2(x) + \Delta L(x)$. The former can be found through evaluation of (6.4.5) along the streamline $y = L_2(x)$, resulting in

$$L_2(x) = (\beta + \mathcal{N}s^{-1})[\beta L_2(-\infty) - \mathcal{N}H_o + \tfrac{1}{2}(a - b)(e^{-\alpha\Delta L} - e^{-\alpha\Delta L_{-\infty}})], \quad (6.4.14)$$

where $\Delta L_{-\infty}$ is the value of ΔL far upstream. Examples of the four standard hydraulic solutions (Figure 6.4.3) have been calculated from (6.4.9) and (6.4.14) for parameter settings close to those of Figure 6.4.2 and for $\mathcal{N}(x) = \mathcal{N}_{\max}e^{-x^2}$. The purely subcritical and supercritical solutions are distinguished by a narrowing or widening over the ridge. The subcritical-to-supercritical solution narrows as it passes the ridge, whereas the (unstable) supercritical-to-subcritical solution does the opposite.

If the ridge slopes equatorwards ($s < 0$) rather than polewards and the upstream flow is subcritical, the solution over the ridge lies to the right of ΔL_u along the Figure 6.4.2 curve. As the subcritical jet climbs the topography its width increases and the flow becomes more subcritical. If $s\mathcal{N}_{\max} < -\beta$, ΔL becomes infinite before the ridge crest is reached. Although the long-wave assumption becomes violated here, there is a suggestion that the flow may become completely blocked by the ridge through deflection to the pole and equator.

In the ocean, a possible application of the above theory is to the Antarctic Circumpolar Current (ACC) as it passes over the Kerguelen Plateau or through

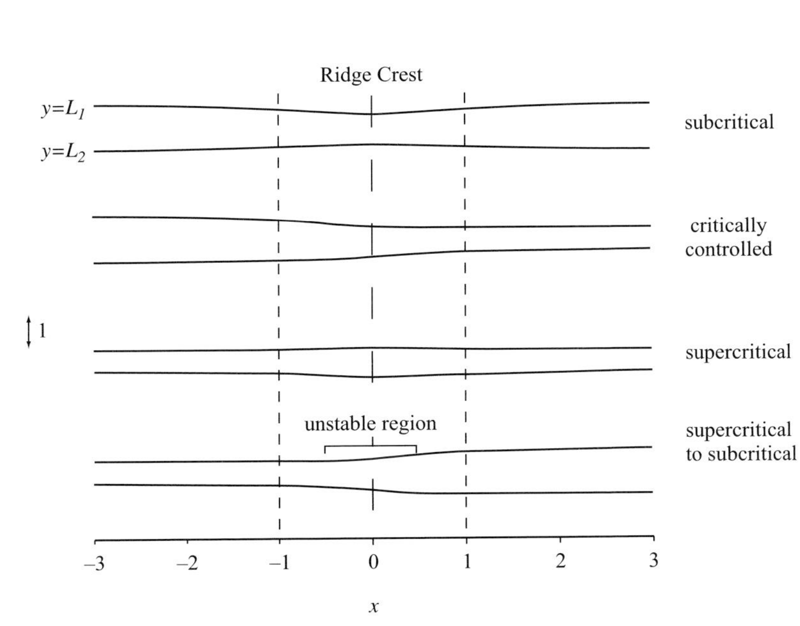

FIGURE 6.4.3. Examples of the four solution types based on the parameters of the previous figure, but with $a - c = 6.5$, $b - c = 1$, and $\mathcal{N} = \mathcal{N}_{\max} \exp(-x^2)$. $\mathcal{N}_{\max}$ has value 0.0164 (subcritical); 0.0622 (critically controlled); .0610 (supercritical); and .0622 (supercritical to subcritical). The flow is from left to right. (From Pratt 1989).

the Drake Passage. The ACC has a multi-front structure, the two most prominent features being the Subantarctic Front and the Polar Front. Pratt (1989) has made estimates of the beta Froude number (6.4.13b) at the Drake Passage and at 134W section, upstream of the Passage. The value of ΔL^* is taken as the distance between the two fronts, whereas L_β is the observed decay scale of the velocity away from the fronts and is given a range of values. As it turns out F_β ranges between 10^{-2} and 10^{-1} at the upstream section, and between 0.4 and 0.9 in the Drake Passage. This could be an indication of a hydraulic transition within the strait, but validation of this conjecture would require a more specific and sophisticated model.

The dependence of the model on the westward, far-field velocity u_o results from the fact that the theory is purely inertial. If this far-field flow is brought to zero, it can be shown (Exercise 3) that critical control is expunged. A more realistic far field in the ocean might be the wind-driven subpolar and subtropical gyres that exist to the north or south of jets such as the Gulf Stream, but this would require the addition of forcing to the model.

Comparison of the quasigeostrophic, mid-latitude jet with the barotropic, equatorial jet of the previous section yields some important similarities and differences. Critical control in each case involves a parameter $U/\beta^* L^2$ that must typically be $O(1)$. However, there are differences in how U, β^*, and L are interpreted in the two cases. In the equatorial jet U is the typical velocity in the jet, β^* is its value on the equator, and L is the jet half width. As the flow passes over the ridge, β^* remains constant while the velocity and width evolve in response to vortex squashing. In the mid-latitude jet, β^* is essentially replaced by β_T^*, the sum of the planetary and topographic potential vorticity gradients. Changes in the jet velocity and width are forced by changes in the northward topographic slope. In addition, L is the inertial boundary layer thickness and U is a velocity scale based on the difference across the jet of the intrinsic potential vorticity.

A common feature of the two models is that hydraulic behavior is reflected entirely in terms of *varicose* motions of the jet; that is, motions that effect the width. In the equatorial case, this is guaranteed by the assumption of the north-south symmetry that is integral to the similarity solution. In the mid-latitude jet, however, meandering (or *sinuous*) motions are present but are slaved to the varicose motions.

At the time of this writing, the potential hydraulic behavior of free, zonal jets has not received much attention or verification in terms of modern observations or numerically modeled flows.

Exercises

(1) Discuss the hydraulic control problem for a westward jet flowing over a northward sloping ridge in each hemisphere.

(2) Suppose that varicose motion of the jet is eliminated by taking the limit $\Delta L \to 0$. Show that the resulting meandering, cusped jets cannot undergo hydraulic transitions in the northern hemisphere.

(3) Explore the limit of a quiescent far field but taking the simultaneous limit $\beta \to 0$ and $u_o \to 0$ such that (u_o/β) remains constant. Show that no hydraulic transitions are possible.

Appendix A
List of Selected Notation

The following list includes variables that are common, that require special explanation, or that have multiple usages; is not exhaustive.

$()^*$	The asterisk always denotes the dimensional version of a variable. The star is omitted for common parameters such as g or f, or for scales such as D and L that do not have not have dimensionless versions.
$()^{\mathrm{T}}$	matrix transpose
$()_c$	a quantity evaluated at a section of critical flow
$()_R$, $()_L$	quantities measured along the right and left sides (facing downstream) of a channel
$()_s$	usually a quantity evaluated at a topographic sill
$[]$	$[AB] = A_1 B_2 - A_2 B_1$ in Appendix B
a	usually denotes the amplitude of a wave, initial condition or topographic amplitude. It can also be also the position ($x = -a$) of the left edge of a current (Sec. 2.8), the potential vorticity gradient (Sec 2.9), the basin radius (Sec. 2.13), or a background potential vorticity (Sec. 6.4)
A	cross-sectional area of flow (Sec. 2.12); Mach angle (Secs. 4.3, 4.4, and Appendices A and B); coefficient in depth profile (Sec. 5.8)
A'	angle of characteristic curve in hodograph plane (Appendix B)
$\mathcal{A}$	cross-sectional area of background flow (Sec. 3.9)
A_C	area enclosed within a contour C
b	$x = b$ generally denotes the right-hand position of the free edge, or grounding point of a current; in Sec. 6.4, b is the background potential vorticity in Region II
B	Bernoulli function
$\overline{B}$	average of the two sidewall values of B (Sec 2.2)
B_o	surface buoyancy flux (Sec 5.6)
B_f	buoyancy flux (Sec. 2.12)
ΔB	internal Bernoulli function (Sec. 5.2.)
c	generally a wave speed. Also the background potential vorticity in Region III in Sec. 6.4
$c_\pm$	characteristic wave speeds
$c^{(n)}$	the propagation velocity of a shock normal to itself (Sec. 3.5)

c_b	speed of gravity current nose (Sec. 4.4)
c_{sep}	speed of left-wall separation point in a gravity current (Sec. 4.4)
C	a contour such as a potential vorticity front (Sec. 3.2), shock (Sec. 3.5), or a contour of depth discontinuity (Sec. 6.2); a coefficient in the depth and velocity profiles (Sec. 5.8)
C_+, C_-	denote characteristic curves (Sec. 4.3, 4.4 and Appendix B and C)
C_d	dimensionless drag coefficient (Sec. 1.9)
$d,\ d_n$	layer thickness; thickness of layer n (Chapter 5)
$d_{1\infty},\ d_{2\infty}$	constant thickness of layer 1 or layer 2 in a hypothetical reservoir of infinite width (Chapter 5), used in systems with no background rotation
$\overline{d}, \hat{d}$	average of, and one-half the difference between, the wall depths in a rotating channel (Ch. 2)
$\overline{d}_1, \overline{d}_2$	mid-channel ($x = 0$) values of the layer depths (Sec. 5.7)
d_b	wall depth of a gravity current just to the rear (upstream) of the head (Sec. 4.5)
d_o	the initial depth in Long's towing experiment (Sec. 1.7); the layer thickness along a channel wall (Secs. 2.3, 5.7); the nondimensional basin depth below sill level (Sec. 2.13)
d_s	the depth (below the rigid lid) at which the plume is introduced (Sec. 5.5)
d_u	upper layer thickness in laboratory estuary basin (Sec. 5.5)
$(d\gamma_1, d\gamma_2, \ldots)$	the displacement or displacement vector associated with a small change in a steady flow due to a stationary wave (Section 1.5)
D	background current depth (Sec. 1.3) or depth scale (elsewhere)
$\mathcal{D}(x)$	depth of background current (Sec. 3.9)
$\hat{D}_f$	form drag (Section 1.6)
D_s	in a two-layer system with a rigid lid, the depth over the sill (Ch. 5)
$D_\infty, D_{n\infty}$	potential depth (Sec. 2.1) or potential depth within layer n (Chapter 5)
$D_{-\infty}$	in submerged weir flow (end of Sec. 2.4) the depth measured in a quiescent region of the downstream basin
$\hat{D}_r$	nondimensional Bernoulli function along the coast in Sec. 4.2
e_n	enstrophy (Sec. 2.9)
e_w	density of wave energy (Sec. 3.9)
E	the maximum energy or Bernoulli function (Sec. 2.10); the total energy (Sec. 3.9); the evaporation rate (Chapter 5)
$\dot{E}$	time rate of change of energy, negative for energy dissipation (Sec. 1.6)

E_b, E_m, E_w	background, mean and wave energies (Sec. 3.9)
E_n	the entrainment rate (Secs. 1.6 and 2.12)
f	Coriolis parameter, $2\Omega\sin(latitude)$, where Ω is Earth's angular rotation rate
$f_{\pm}(x - c_{\pm}t)$	linear wave solutions (Sec. 1.2)
F or $\mathbf{F}$	a scalar or vector force in the horizontal momentum equation (Sec 2.1)
F_1, F_2	layer Froude numbers (Chapter 5)
F_β	beta plane Froude number (Eq. 6.4.13)
F_d	the Froude number based on the local values of velocity and layer thickness (Sec 1.2). In Chapters 2–6, F_d can denote various generalized versions of the Froude number for rotating flow.
F_o	the Froude number of the initial flow in an initial-value problem (e.g. Secs. 1.7, 3.8, and 6.2)
F_p	Froude number for rotating flow in a parabolic channel (Sec. 2.8)
F_S	Froude number defined using Stern's criterion (Eq. 3.4.12)
F_τ	bottom drag (Sec. 2.12)
F_w	Froude number based on sidewall velocities (Sec. 3.5)
$\overline{F}_n$	pseudo Froude number for layer n of a rotating exchange flow (Sec. 5.7)
g, g'	the first can represent either the normal ($9.8 m/s^2$) or reduced gravitational acceleration depending on whether the equation in question applies to a free-surface flow or an equivalent barotropic (1.5-layer) model. The second always represents reduced gravity.
$\mathcal{G}, \mathcal{G}_n$	hydraulic function, or one of n such functions in n variables (Sec 1.5)
G	composite Froude number for two-layer flow (Chapter 5), also the Greens function (Sec. 3.2).
G_r	composite Froude number for rotating, exchange flow with zero potential vorticity (Sec 5.7)
h	topographic elevation, usually referenced to the elevation of a source basin
δh	height of topographic step or shelf (Sec. 6.2)
h_c	the minimum sill height required for upstream influence (Sec. 3.4); more generally, the bottom elevation at a critical section, usually the sill elevation
h_m	sill elevation
$H(y)$	step function (Sec. 6.2)
$\mathbf{J_n}^*$	the contribution to the vorticity flux from dissipation (Sec 2.1)
k	wave number in the cross-channel direction (Sec. 2.1) or in along-channel direction in a zonal channel (Sec. 6.2)
$K(y)$	topographic function defined by Equation (6.4.6)
$K_o(y)$	modified Bessel function of zero order

$K_{m,n}$	Helmholtz point vortex (Sec. 3.2).
l	wave number in the along-channel or along-coast direction (Sec 2.1); the width of a control volume (Sec. 3.5); the meridional wave number (Sec. 6.1)
L	generally the dimensional scale of variation in the primary direction of the current; a differential operator in Appendices B and C
L_d	the Rossby radius of deformation based on the potential depth (eq. 2.2.11)
L_I	two-layer Rossby radius (eg. 5.1.12)
ΔL	distance $L_1 - L_2$ between potential vorticity fronts in Sec. 6.4
M	total momentum (Sec. 3.9 and 6.2)
M_b, M_m, M_w	background, mean and wave momentum (Sec. 3.9)
m_w	density of wave momentum
n	In curvilinear (s, n) direction, n is normal to the current axis (Sec. 2.12), isobath (Sec. 2.13), or coastline (Sec. 4.1).
$N(x)$	cross-channel structure function for wave mode (Sec. 2.1) iN is also used as a scale for the interface displacement η (Sec. 2.13)
$\mathcal{N}(x)$	x-varying component of topographic height $h = \mathcal{N}(x)H(y)$ (Sec. 6.4)
p, p_n	fluid pressure, pressure in layer n
p_0	external pressure field (Sec. 3.8)
$p_T{}^*$	rigid lid pressure
P	precipitation rate (Sec. 5.5)
q	potential vorticity
q_2	In the Chapter 5 Froude number diagrams, q_2 is a proxy for the topographic height or channel width (see equation following 5.3.2).
$\tilde{q}$	potential vorticity anomaly (Sec. 3.2)
$q_0{}^*$	average potential vorticity (Sec. 2.9)
Δq	$|a|Q^*/2q_0{}^*$ (see equation 2.9.18)
Q, Q_n	volume transport (flux) or volume transport within layer n; in Section 2.14, $Q_0{}^*$ and $Q_1{}^*$ denote the dimensional flux for a flow with potential vorticity $q = 0$ and $q = 1$. In Section (5.2) Q denotes the net flux over two layers.
Q_b	volume flux in a gravity current (Sec. 4.5)
Q_r	ratio of top to bottom volume fluxes in a two-layer system (Sec. 5.2)
Q_R	volume flux due to river runoff or other source of fresh water (Sec. 5.5)
$\mathcal{Q}(x)$	potential vorticity of background flow (Sec. 3.9)
r	part of the Riemann function (see Eq. 2.2.24); the ratio of width to deformation radius for a parabolic channel (Sec. 2.8, 2.13); the radial coordinate (Sec. 2.13).
r_f, R_f	dimensional and nondimensional versions of a linear drag coefficient (Sec. 2.13)

r_o	radius of downwelling patch (Sec 2.13)
R	a region of the flow field
∂R	the boundary of region R
$R_\pm$	Riemann invariant functions (e.g. Secs. 1.3 and 2.2)
R_b	bulk Richardson number
R_i	Richardson number (Sec. 1.10)
R_c	reflection coefficient (Sec. 1.8)
R_o	Rossby number (Sec. 6.1)
s	in a curvilinear (s, n) coordinate system, s is tangent to the axis of the current (Sec. 2.12), isobath (Sec. 2.13), or coastline (Sec. 4.1); in Sec. 6.4, s is the topographic slope
$s_\pm$	see Eq. (5.9.12)
Δs	nondimensional salinity difference between layers (Eq. 5.5.8)
S	bottom slope (Secs. 1.9 and 2.12); ratio of horizontal length scale to Rossby radius (Secs. 6.1–6.4)
S_n	salinity in layer n
r	part of the Riemann function (see Eq. 2.2.24); the ratio of width to deformation radius for a parabolic channel (Sec. 2.8); width parameter for separated shock (Eq. 3.7.1)
t	time
$\tilde{t}$	slow time variable (Sec. 6.2)
T	$\tanh(qw/2)$ over most of the book; T is also used to denote the volume transport of a basin source (Sec. 2.13) and a time scale (Sec. 6.1)
T_e	$\tanh(qw_e/2)$
(u, v)	x- and y-velocity components in most of the book. In Section 2.13, u and v are radial and azimuthal velocities. In Section 4.1, they are the off-shore and along-shore velocities along a curvy coast line.
$u^{(n)}, u^{(s)}$	normal and tangential velocity components to a shock (Sec. 3.5)
u_o	westward velocity of far field flow (Sec. 6.4)
U_o	maximum velocity in zonal jet (Sec. 6.2)
$\bar{v}, \hat{v}$	average of, and one-half difference between, the wall velocities in a rotating channel
$\bar{v}_1, \bar{v}_2$	layer velocities at mid-channel (Sec. 5.7)
v_o	initial velocity in Long's towing experiment (Sec. 1.7); the layer velocity along a channel wall (Secs. 2.3 and 5.7)
v_e	velocity on free edge of stream (Sec. 2.3)
V	scale for v and (in Section 1.3) the velocity of the background current. V is also used to denote a control volume (Sec. 1.6) and the along-axis velocity of a plume (Sec. 2.12)
$V(x)$	the mean barotropic velocity in a two-layer system (Sec. 5.9)
$\mathcal{V}(x)$	velocity of background flow (Sec. 3.9)
(x, y, z)	cross-channel, along-channel, and upward coordinates; for coastal flows, y is aligned along the coast, or in its predominant direction
w	generally the channel or shelf width; in Section 1.1, the vertical velocity; in Sec 6.3 the jet half-width

w_b	width of a gravity current just to the rear of the head (Sec. 4.4)
w_B	variable basin width upstream of channel mouth (Sec. 5.7)
w_e	width of separated channel flow (Secs. 2.3, 5.7, 5.8); the current width (Section 4.1); the (positive downwards) entrainment velocity (Secs. 1.10, 2.13)
W	width scale
x, y	in rotating channels or coastal applications, x generally denotes the cross-channel or offshore coordinate. and y the longitudinal coordinate. In Chapter 6, where the currents are aligned east-west, this convention is reversed.
$\tilde{x}$	scaled version of x used to capture gradual variations (Sec. 6.2)
x_c	the midpoint of a separated current (Sec. 2.3)
$y_{\pm}(t)$	along-channel position along a characteristic curve
$Y(x, t)$	position of Kelvin bore (Sec. 3.6); position of potential vorticity front (Sec. 6.2)
$Y_h(x)$	position of topographic step or shelf (Sec. 6.2)
z	vertical coordinate
Δz	difference between reservoir surface or interface elevation and sill elevation
Δz_E	an upstream surface or interface, measured relative to deepest part of the sill, equivalent to E/g', where E is the maximum energy over the streamlines that pass across the sill (Sec. 2.10)
Δz_R	reservoir or surface elevation above sill, measured on right wall of reservoir (Sec. 2.6)
z_R	the source elevation above the bottom (Sec. 5.5)
z_s	elevation of source of plume (Sec. 5.5)
z_T	in a two-layer system, the elevation of the bounding rigid lid.

Greek

α	a multiplying factor (Sec. 1.3); bottom curvature (Sec. 2.8); the parameter $\lvert a \rvert^{1/2} d^*$ (Sec. 2.9); cross-axis bottom slope (Sec. 2.12); bottom slope parameter (Sec. 2.13); composite variable for the position of the edge of a coastal current (Sec. 4.2); coastline angle (Sec. 4.3); the initial velocity of the channel flow (Sec. 6.2); $(\beta/u_o)^{1/2}$ in Sec. 6.4
β	a measure of the potential vorticity gradient either due to bottom topography (Sec 2.9), or due to variations in the Coriolis parameter (Sec. 6.1.); the angle of the plume axis (Sec. 2.12) or an oblique shock (Sec. 4.3); coefficient in the expression for the speed of the nose of a gravity current (Sec. 4.5); coefficient of expansion of water due to salinity (Sec. 5.5)
β_w	coefficient in the expression for the neck width of a gravity current (Sec. 4.4)

δ	horizontal aspect ratio (e.g Sec 2.1); boundary layer thickness (Sec. 2.13)
$\hat{\delta}$	ratio of potential layer depths (Sec. 5.9)
$\delta(x)$	the Dirac delta function (Secs. 2.13 and 3.2)
δ_b	boundary layer thickness (Sec. 2.13)
δ_E	Ekman layer thickness
Δ	the change in interface elevation across the flow (Sec. 2.11); the ratio of total depth to the sum of potential layer depths (Sec. 5.9)
ε	a generic small dimensionless parameter, also the length of a control volume (Sec. 3.5)
$\phi(\tilde{x}, y, \tilde{t})$	perturbation stream function (Sec. 6.2)
γ or γ_n	a generic dependent flow variable used in a hydraulic funcion (e.g. Sec. 1.5); $\gamma = \frac{1}{2}w^*\|a\|^{1/2}D$ (Sec. 2.9)
$\Gamma_\pm$	hodograph images of characteristics (Sec. 4.4)
η	displacement of surface or interface above some reference level
$\eta_\pm$	interface displacement along right and left walls (Sec. 5.9)
η_e	interface elevation, relative to sill, at a channel entrance (Sec. 2.13)
κ	drag coefficient used in plume model (Sec. 2.12); turbulent diffusivity (Sec. 5.6)
λ	wavelength
λ_n	multiplying factor in derivation of hyperbolic forms (Appendix B)
μ	artificial 'viscosity' used in numerical code (Sec. 3.4); coordinate stretching parameter (Sec. 6.2)
$\mu_\pm$	defined by expression following Eq. (5.9.11)
θ	azimuthal coordinate (Sec. 2.13); angle of inclination of a shock (Sec. 3.5); angle of velocity vector (Sec. 4.4)
$\rho,\ \rho_n$	density, density of layer n. In Section 4.1, ρ is the radius of curvature of the coastline
$\Delta\rho$	difference in layer densities
τ	slow time variable (Sec. 3.2)
τ_B, τ_I	bottom or interfacial stress (Sec. 2.12)
ν	kinematic viscosity
ω	wave frequency; also the angle of a characteristic curve in the hodograph (Sec. 4.4)
$\sigma,\ \sigma_\pm$	parameter for characteristic curve (Appendix B); differential unit of area in integral (Sec. 3.9)
σ_θ	potential density referenced to the surface
σ_4	potential density referenced to 4000db
ξ	boundary layer coordinate (Sec. 2.13); x-coordinate of point vortex (Sec. 3.2)
ψ	transport streamfunction, but occasionally also the velocity streamfunction
ψ_i	value of ψ in quiescent interior region of the reservoir of Gill's (1977) model

ψ_o the value of the streamfunction corresponding to the potential vorticity front in a steady flow (Sec. 6.2)

ζ generally the relative vorticity; a similarity variable (Sec. 6.3)

ζ_a absolute vorticity $(f + \zeta)$ (Sec. 2.1)

Appendix B
The Method of Characteristics in Two Dimensions

a.   Mathematical Theory

Many of the flows dealt with herein involve two independent variables in a two dimensional space. Examples include wave propagation in x and t and steady shallow flow on the (x, y) plane. If we temporarily let x and y represent generic independent variables, then the governing equations take the form:

$$L_1 = A_1 \frac{\partial u}{\partial x} + B_1 \frac{\partial u}{\partial y} + C_1 \frac{\partial v}{\partial x} + D_1 \frac{\partial v}{\partial y} + E_1 = 0 \qquad (B.1a)$$

and

$$L_2 = A_2 \frac{\partial u}{\partial x} + B_2 \frac{\partial u}{\partial y} + C_2 \frac{\partial v}{\partial x} + D_2 \frac{\partial v}{\partial y} + E_2 = 0, \qquad (B.1b)$$

where u and v are generic dependent variables. For the systems we consider, the coefficients A_1, A_2, etc. may depend on x, y, u and v, but not on the derivatives of u and v. The governing equations are then *quasilinear* and may be amenable to solution using the method of characteristics, provided that further conditions hold. The development of this approach is laid out in Courant and Friedrichs (1976) and the following summary is based on their notation and exposé.

In general, we wish to take advantage of the physical property of certain systems that all information propagates in a 'forward' direction, usually meaning positive x or y, and at finite speed. Such flows are called *locally supercritical*.[1] We should then be able to construct solutions by forward integration along the paths of information travel beginning from the boundaries at which the information is generated. To make these ideas more precise, consider a path $x = x(\sigma)$ and $y = y(\sigma)$, parameterized by the variable σ. The vector $(dx/d\sigma, dy/d\sigma)$ is tangent to the path and the derivative of a function f in the same direction and

---

[1] This term should be distinguished from *hydraulically supercritical*. The latter describes a flow in which normal modes, which are felt across the width of the flow and which satisfy the boundary conditions at the edges, propagate in a single direction. Hydraulically supercritical flows need not be locally supercritical across their entire width.

with respect to σ is $df/d\sigma = \dfrac{\partial f}{\partial x}\dfrac{\partial x}{\partial \sigma} + \dfrac{\partial f}{\partial y}\dfrac{\partial y}{\partial \sigma}$. The first aim of the analysis is to manipulate (B1) to form a single equation in which the x- and y- derivatives of u and v combine to form derivatives in a particular direction. This *characteristic direction* depends on u, v, x, and y and defines a *characteristic curve* along which the derivative is taken. To this end, take $L = \lambda_1 L_1 + \lambda_2 L_2$, leading to

$$L = (\lambda_1 A_1 + \lambda_2 A_2)\frac{\partial u}{\partial x} + (\lambda_1 B_1 + \lambda_2 B_2)\frac{\partial u}{\partial y} + (\lambda_1 C_1 + \lambda_2 C_2)\frac{\partial v}{\partial x}$$

$$+ (\lambda_1 D_1 + \lambda_2 D_2)\frac{\partial v}{\partial y} + (\lambda_1 E_1 + \lambda_2 E_2) = 0. \tag{B.2}$$

In order that the directions of the derivatives of u and v along the hypothetical path be the same, we need

$$\frac{\lambda_1 A_1 + \lambda_2 A_2}{\lambda_1 B_1 + \lambda_2 B_2} = \frac{\lambda_1 C_1 + \lambda_2 C_2}{\lambda_1 D_1 + \lambda_2 D_2} = \frac{\partial x/\partial \sigma}{\partial y/\partial \sigma}, \tag{B.3}$$

which allows (B.2) to be written as

$$L = (\lambda_1 A_1 + \lambda_2 A_2)\left[\frac{\partial u}{\partial x} + \frac{\partial y/\partial \sigma}{\partial x/\partial \sigma}\frac{\partial u}{\partial y}\right]$$

$$+ (\lambda_1 C_1 + \lambda_2 C_2)\left[\frac{\partial v}{\partial x} + \frac{\partial y/\partial \sigma}{\partial x/\partial \sigma}\frac{\partial v}{\partial y}\right] + (\lambda_1 E_1 + \lambda_2 E_2) = 0,$$

or, using $\dfrac{\partial f}{\partial \sigma} = \dfrac{\partial f}{\partial x}\dfrac{\partial x}{\partial \sigma} + \dfrac{\partial f}{\partial y}\dfrac{\partial x}{\partial \sigma}$:

$$L\frac{\partial x}{\partial \sigma} = (\lambda_1 A_1 + \lambda_2 A_2)\frac{\partial u}{\partial \sigma} + (\lambda_1 C_1 + \lambda_2 C_2)\frac{\partial v}{\partial \sigma} + (\lambda_1 E_1 + \lambda_2 E_2)\frac{\partial x}{\partial \sigma} = 0. \tag{B.4a}$$

By a similar approach

$$L\frac{\partial y}{\partial \sigma} = (\lambda_1 B_1 + \lambda_2 B_2)\frac{\partial u}{\partial \sigma} + (\lambda_1 D_1 + \lambda_2 D_2)\frac{\partial v}{\partial \sigma} + (\lambda_1 E_1 + \lambda_2 E_2)\frac{\partial y}{\partial \sigma} = 0. \tag{B.4b}$$

The factors λ_1 and λ_2 are determined by rearranging (B.3) as

$$\lambda_1\left(A_1 \frac{\partial y}{\partial \sigma} - B_1 \frac{\partial x}{\partial \sigma}\right) + \lambda_2\left(A_2 \frac{\partial y}{\partial \sigma} - B_2 \frac{\partial x}{\partial \sigma}\right) = 0 \tag{B.5a}$$

and

$$\lambda_1\left(C_1 \frac{\partial y}{\partial \sigma} - D_1 \frac{\partial x}{\partial \sigma}\right) + \lambda_2\left(C_2 \frac{\partial y}{\partial \sigma} - D_2 \frac{\partial x}{\partial \sigma}\right) = 0. \tag{B.5b}$$

Setting the determinant of the coefficients of λ_1 and λ_2 to zero leads to

$$a\left(\frac{dy}{d\sigma}\right)^2 - 2b\frac{dy}{d\sigma}\frac{dx}{d\sigma} + c\left(\frac{dx}{d\sigma}\right)^2 = 0, \tag{B.6}$$

where

$$a = [AC], \quad 2b = [AD] + [BC], \quad c = [BD], \tag{B.7}$$

and $[MN] = M_1 N_2 - M_2 N_1$.

With $(dy/d\sigma)/(dx/d\sigma) = dy/dx$, the *characteristic direction* $(dx, \ dy)$ is given by

$$a \left(\frac{dy}{dx} \right)^2 - 2b \frac{dy}{dx} + c = 0.$$

This equation has two distinct, real solutions $(dy/dx)_-$ and $(dy/dx)_+$ if and only if

$$b^2 > ac. \tag{B.8}$$

If (B.8) is satisfied within a region of the (x,y) plane, the governing equations are called *hyperbolic* there. Two distinct characteristic curves C_- and C_+ can be found within this region. The curves are computed from

$$\left(\frac{dy}{dx} \right)_\pm = \frac{b \pm \sqrt{b^2 - ac}}{a}. \tag{B.9}$$

We will now use σ_- and σ_+ (formerly σ) to parameterize the two characteristic curves. Thus, a curve determined by the '+' sign in (B.9) has $\sigma_- = $ constant, and vice versa. The original intent was to obtain a form of the governing equations in which derivatives are taken in a characteristic direction, i.e. along one of the characteristic curves. Either of (B.4a) or (B.4b) provides a basis for the desired result, but λ_1 and λ_2 must first be eliminated. If one attempts to do so using (B.4a) and (B.5a), say, then it follows that

$$\begin{vmatrix} A_1 \partial y/\partial \sigma_\pm - B_1 \partial x/\partial \sigma_\pm & A_2 \partial y/\partial \sigma_\pm - B_2 \partial x/\partial \sigma_\pm \\ A_1 \partial u/\partial \sigma_\pm + C_1 \partial v/\partial \sigma_\pm + E_1 \partial x/\partial \sigma_\pm & A_2 \partial u/\partial \sigma_\pm + C_2 \partial v/\partial \sigma_\pm + E_2 \partial x/\partial \sigma_\pm \end{vmatrix} = 0.$$

or

$$T \frac{\partial u}{\partial \sigma_\pm} + \left(a \left(\frac{dy}{dx} \right)_\pm - S \right) \frac{\partial v}{\partial \sigma_\pm} + \left(K \left(\frac{dy}{dx} \right)_\pm - H \right) \frac{\partial x}{\partial \sigma_\pm} = 0, \tag{B.10}$$

where

$$T = [AB], \quad S = [BC], \quad K = [AE], \quad \text{and} \quad H = [BE].$$

A useful alternative to (B.10) can be obtained by eliminating λ_1 and λ_2 between (B.4a, b):

$$\begin{vmatrix} A_1 \partial u/\partial \sigma_\pm + C_1 \partial v/\partial \sigma_\pm + E_1 \partial x/\partial \sigma_\pm & A_2 \partial u/\partial \sigma_\pm + C_2 \partial v/\partial \sigma_\pm + E_2 \partial x/\partial \sigma_\pm \\ B_1 \partial u/\partial \sigma_\pm + D_1 \partial v/\partial \sigma_\pm + E_1 \partial y/\partial \sigma_\pm & B_2 \partial u/\partial \sigma_\pm + D_2 \partial v/\partial \sigma_\pm + E_2 \partial y/\partial \sigma_\pm \end{vmatrix} = 0. \tag{B.11}$$

b. Example 1: Steady, Irrotational, Two-Dimensional, Shallow Flow Over a Horizontal Bottom

We use (x, y), (u, v) and d to denote the nondimensional position, velocity and depth variables, as defined in Section 2.1. Dimensional versions of the following relations may be obtained by replacing d by gd, where g is the gravitational acceleration. The flow to be considered is governed by the continuity equation

$$\frac{\partial(ud)}{\partial x} + \frac{\partial(vd)}{\partial y},$$ (B.12)

and by the statmement of conservation of energy

$$\frac{u^2 + v^2}{2} + d = B.$$ (B.13)

Although the Bernoulli function B is normally a function of the streamfunction, the assumption of irrotational flow (zero vorticity) renders it a constant.

If the gradient of (B.13) is used to eliminate the gradient of d from (B.12), one obtains

$$(d - u^2)\frac{\partial u}{\partial x} - uv(\frac{\partial u}{\partial y} + \frac{\partial v}{\partial x}) + (d - v^2)\frac{\partial v}{\partial y} = 0.$$ (B.14)

Together with the condition of zero vorticity:

$$\frac{\partial u}{\partial y} - \frac{\partial v}{\partial x} = 0,$$ (B.15)

(B.14) forms a system of two quasilinear equations of the form (B1) with $A_1 = d - u^2$, $B_1 = C_1 = -uv$, $D_1 = d - v^2$, $B_2 = -C_2 = 1$, and $A_2 = D_2 = E_2 = E_1 = 0$. The two dependent variables are u and v, with d regarded as a function of u and v through (B.13). With these coefficients we have $a = (u^2 - d)$, $b = uv$, $c = (v^2 - d)$, and thus the condition for hyperbolicity (B.8) is

$$u^2 + v^2 > d.$$ (B.16)

Equation (B.6) governing the characteristic curves is now

$$(d - u^2)\left(\frac{dy}{dx}\right)^2_\pm + 2uv\left(\frac{dy}{dx}\right)_\pm + (d - v^2) = 0,$$ (B.17)

whereas (B.11) yields

$$(d - u^2)\left(\frac{du}{dv}\right)^2_\pm - 2uv\left(\frac{du}{dv}\right)_\pm + (d - v^2) = 0.$$ (B.18)

Keep in mind that du and dv represent changes in u and v measured along the curve.

A convenient expression for the orientation of the characteristic curves can be determined by rewriting (B.17), in the form

$$d = \frac{(u\,dy - v\,dx)^2}{dx_\pm^2 + dy_\pm^2} = \frac{[\mathbf{k}\cdot(u,v)\times(dx_\pm, dy_\pm)]^2}{dx_\pm^2 + dy_\pm^2} = \frac{|(u,v)|^2\,|dx_\pm, dy_\pm|^2}{dx_\pm^2 + dy_\pm^2}\sin^2 A,$$

or

$$d^{1/2} = |(u,v)|\sin(\pm A), \tag{B.19}$$

In other words, the characteristic curves at any point form an angle $A = \pm\sin^{-1}[d/(u^2+v^2)^{1/2}]$ with respect to the local flow direction (streamline). A is analogous to the Mach angle of gas dynamics, named after Ernst Mach. In shallow water theory, A is sometimes referred to as the Froude angle.

It is often helpful to consider the images of the (x, y)-plane characteristics in the (u, v)-plane, often called the *hodograph*. To this end, note that (B.17) and (B.18) together imply

$$\left(\frac{dy}{dx}\right)_\pm = -\left(\frac{du}{dv}\right)_\mp \tag{B.20}$$

[The coordination between the '+' and '−' subscripts must be established independently and can be done so using equations (B4) and (B5).] Now consider a pair of '+' and '−' characteristic curves C_+ and C_- that intersect at some point P in the (x, y)-plane (Figure 4.4.1a). If α and β are used as parameters along C_+ and C_-, then

$$\frac{\partial y}{\partial \alpha} = \left(\frac{dy}{dx}\right)_+ \frac{\partial x}{\partial \alpha} \text{ along } C_+, \text{ and } \frac{\partial y}{\partial \beta} = \left(\frac{dy}{dx}\right)_+ \frac{\partial x}{\partial \beta} \text{ along } C_-. \tag{B.21}$$

The velocity at P determines a point in the (u, v)-plane through which the images Γ_+ and Γ_- of C_+ and C_- pass. According to (B.20)

$$\frac{\partial u}{\partial \alpha} = -\left(\frac{dy}{dx}\right)_- \frac{\partial v}{\partial \alpha} \text{ along } C_+, \text{ and } \frac{\partial u}{\partial \beta} = -\left(\frac{dy}{dx}\right)_- \frac{\partial v}{\partial \beta} \text{ along } C_-. \tag{B.22}$$

It follows that $\dfrac{\partial u}{\partial \alpha}\dfrac{\partial x}{\partial \beta} + \dfrac{\partial v}{\partial \alpha}\dfrac{\partial y}{\partial \beta} = 0$ and $\dfrac{\partial u}{\partial \beta}\dfrac{\partial x}{\partial \alpha} + \dfrac{\partial v}{\partial \beta}\dfrac{\partial y}{\partial \alpha} = 0$, and thus C_+ is perpendicular to the image Γ_- of C_-, and vice versa, if the two are plotted in the same space.

The geometry of the characteristics and their images can be summarized as follows: The two characteristic curves C_+ and C_- passing through P form Froude angle $A = \pm\sin^{-1}[d/(u^2+v^2)^{1/2}]$ with respect to the streamline that passes through P (Figure 4.4.1a). If plotted in the same space the hodograph image Γ_+ of C_+ forms a right angle with C_- and vice versa, at P (Figure 4.4.1b). The relationship between A and the angle A' in the (u, v) plane between characteristics and streamlines is thus

$$A' = 90° - A. \tag{B.23}$$

It follows from (B.19) that

$$d^{1/2} = |(u, v)| \cos(A'). \tag{B.24}$$

For computational purposes, it is convenient to introduce the angle θ between the streamline and the x-axis (Figure 4.4.1a):

$$u = q \cos \theta \text{ and } v = q \sin \theta. \tag{B.25}$$

It follows that

$$\left(\frac{dy}{dx}\right)_{\pm} = \tan(\theta \pm A), \tag{B.26}$$

where we have introduced the convention that C_+ tilts to the left, and C_- to the right, as seen by an observer facing downstream. In these terms, equations (B.21) and (B.22) become

$$\cos(\theta + A)\frac{\partial y}{\partial \alpha} = \sin(\theta + A)\frac{\partial x}{\partial \alpha} \text{ along } C_+,$$

$$\cos(\theta - A)\frac{\partial y}{\partial \beta} = \sin(\theta - A)\frac{\partial x}{\partial \beta} \text{ along } C_-,$$

$$\sin(\theta - A)\frac{\partial v}{\partial \alpha} = -\cos(\theta - A)\frac{\partial u}{\partial \alpha} \text{ along } \Gamma_+,$$

$$\sin(\theta + A)\frac{\partial v}{\partial \beta} = -\cos(\theta + A)\frac{\partial u}{\partial \beta} \text{ along } \Gamma_-.$$

This set could form the basis for a numerical calculation in which characteristic curves emerge from a boundary along which u and v are known. The paths of the curves penetrating into the domain of interest are calculated by solving (B.25) simultaneously.

c.   Example 2: One-Dimensional, Time-Dependent Shallow Flow Over a Horizontal Bottom

In this example we follow the Chapter 1 notation convention that un-starred variables are dimensional. The dimensional governing equations (2.1.1) and (2.2.2), with $h = $ constant, can be expressed in the form (B1) with (t, x) in place of (x, y), d in place of v, and $A_1 = C_2 = 1$, $B_1 = D_2 = u$, $B_2 = d$, $D_1 = g$, and $C_1 = A_2 = E_1 = E_2 = 0$. Equation (B.9) then gives

$$\left(\frac{dy}{dx}\right)_{\pm} = u - (gd)^{1/2},$$

while the characteristic equations, obtained using (B.10), are

$$\frac{\partial u}{\partial \sigma_{\pm}} \pm \left(\frac{g}{d}\right)^{1/2} \frac{\partial d}{\partial \sigma_{\pm}} = 0.$$

The latter can also be written in the form $\partial R_{\pm}/\partial \sigma_{\pm} = 0$, where $R_{\pm} = u \pm (gd)^{1/2}$ is the Riemann invariant.

Appendix C
The Method of Characteristics for Steady, 2-d, Shallow Flow with Rotation

The shallow water equations for rotating, steady flow over a horizontal bottom can be written in the nondimensional forms

$$L_1 = u\frac{\partial u}{\partial x} + v\frac{\partial u}{\partial y} + \frac{\partial d}{\partial x} - v = 0, \tag{C.1a}$$

$$L_2 = u\frac{\partial v}{\partial x} + v\frac{\partial v}{\partial y} + \frac{\partial d}{\partial y} + u = 0, \tag{C.1b}$$

$$L_3 = d\frac{\partial u}{\partial x} + d\frac{\partial v}{\partial y} + u\frac{\partial d}{\partial x} + v\frac{\partial d}{\partial y} = 0. \tag{C.1c}$$

The additional dependent variable d is the depth of the flow. In contrast to the cases discussed in Appendix B, the system contains three dependent variables and this number cannot be reduced without further assumptions. A characteristic form can nevertheless be sought using the same reasoning; one would like to combine the constituent equations linearly such that differentiation of each variable takes place in a single direction. Taking the combination $L = \lambda_1 L_1 + \lambda_2 L_2 + \lambda_3 L_3$ leads to

$$L = (\lambda_1 u + \lambda_3 d)\frac{\partial u}{\partial x} + \lambda_1 v\frac{\partial u}{\partial y}$$

$$+ \lambda_2 u\frac{\partial v}{\partial x} + (\lambda_2 v + \lambda_3 d)\frac{\partial v}{\partial y}$$

$$+ (\lambda_1 + \lambda_3 u)\frac{\partial d}{\partial x} + (\lambda_2 + \lambda_3 v)\frac{\partial d}{\partial y}$$

$$- \lambda_1 v + \lambda_2 u = 0. \tag{C.2}$$

If the direction of differentiation of u, v, and d is along the curve $(x(\sigma),\ y(\sigma))$ then it is necessary that

$$\frac{(\lambda_1 u + \lambda_3 d)}{\lambda_1 v} = \frac{\lambda_2 u}{\lambda_2 v + \lambda_3 d} = \frac{\lambda_1 + \lambda_3 u}{\lambda_2 + \lambda_3 v} = \frac{\partial x/\partial \sigma}{\partial y/\partial \sigma}. \tag{C.3}$$

The characteristic directions are then determined by the solvability condition for (C.3). Writing the three equations in the form

$$\begin{pmatrix} u\dfrac{\partial y}{\partial \sigma} - v\dfrac{\partial x}{\partial \sigma} & 0 & d\dfrac{\partial y}{\partial \sigma} \\[2mm] 0 & u\dfrac{\partial y}{\partial \sigma} - v\dfrac{\partial x}{\partial \sigma} & -d\dfrac{\partial x}{\partial \sigma} \\[2mm] \dfrac{\partial y}{\partial \sigma} & -\dfrac{\partial x}{\partial \sigma} & u\dfrac{\partial y}{\partial \sigma} - v\dfrac{\partial x}{\partial \sigma} \end{pmatrix} \begin{pmatrix} \lambda_1 \\ \lambda_2 \\ \lambda_3 \end{pmatrix} = 0$$

and setting the determinant of the coefficient matrix to zero leads to

$$\left(u\frac{\partial y}{\partial \sigma} - v\frac{\partial x}{\partial \sigma} \right)\left[\left(u\frac{\partial y}{\partial \sigma} - v\frac{\partial x}{\partial \sigma} \right)^2 - d\left(\left(\frac{\partial x}{\partial \sigma} \right)^2 + \left(\frac{\partial y}{\partial \sigma} \right)^2 \right) \right] = 0. \tag{C.4}$$

One of the characteristic directions is the flow direction itself:

$$\frac{\partial y/\partial \sigma}{\partial x/\partial \sigma} = \frac{v}{u}. \tag{C.5}$$

The corresponding characteristic curves are just the streamlines of the flow and the characteristic equation expresses the conservation of the Bernoulli function along this path. [This constraint is absent in the *irrotational* case discussed in Example 1 of Appendix B because the Bernoulli function is uniform throughout the domain.] We will denote these curves using the subscript ψ.

The remaining two characteristic directions are obtained by setting the bracketed term in (C.4) to zero. Rearrangement of this expression leads to (B.19), and thus the second and third characteristic directions are the same as for the case of irrotational flow. The corresponding characteristic curves are again denoted C_+ and C_- with corresponding parameters σ_+ and σ_-. These curves cross streamlines at the Froude angle $\pm A$, where $d = |(u, v)|^2 \sin^2 A$, as shown in Figure 4.4.1a. The characteristic equations are obtained by multiplying (C.2) by $\partial x/\partial \sigma_\pm$, leading to

$$\frac{\partial x}{\partial \sigma_\pm} L_1 = (\lambda_1 u + \lambda_3 d)\frac{\partial u}{\partial \sigma_\pm} + \lambda_2 u \frac{\partial v}{\partial \sigma_\pm} + (\lambda_1 + \lambda_3 u)\frac{\partial d}{\partial \sigma_\pm} + (\lambda_2 u - \lambda_1 v)\frac{\partial x}{\partial \sigma_\pm} \tag{C.6}$$

$$= \frac{\lambda_3}{u\dfrac{\partial y}{\partial \sigma_\pm} - v\dfrac{\partial x}{\partial \sigma_\pm}} \left\{ d\left(u\frac{\partial v}{\partial \sigma_\pm} - v\frac{\partial u}{\partial \sigma_\pm} \right)\frac{\partial x}{\partial \sigma_\pm} + \left[u\left(u\frac{\partial y}{\partial \sigma_\pm} - v\frac{\partial x}{\partial \sigma_\pm} \right) - d\frac{\partial y}{\partial \sigma_\pm} \right]\frac{\partial d}{\partial \sigma_\pm} + d\left(u\frac{\partial x}{\partial \sigma_\pm} + v\frac{\partial y}{\partial \sigma_\pm} \right)\frac{\partial x}{\partial \sigma_\pm} \right\} = 0.$$

The relationships $\lambda_1 = -d\dfrac{\partial y}{\partial \sigma_\pm}\lambda_3 \Big/ \left(u\dfrac{\partial y}{\partial \sigma_\pm} - v\dfrac{\partial x}{\partial \sigma_\pm} \right)$ and $\lambda_2 = d\dfrac{\partial x}{\partial \sigma_\pm}\lambda_3 \Big/ \left(u\dfrac{\partial y}{\partial \sigma_\pm} - v\dfrac{\partial x}{\partial \sigma_\pm} \right)$, both obtained from (C.3) have been used to obtain the second step. A convenient form of the characteristic equation is obtained by writing (C.6) in terms of the variables $|u|$ and θ, where $(u,\, v) = |u|\,(\cos\theta,\, \sin\theta)$. As shown in Figure 4.4.1a, the characteristic curves C_+ and C_- are inclined at the angle $\theta \pm A$ with respect to the x-axis and

$$u\frac{\partial x}{\partial\sigma_\pm} + v\frac{\partial y}{\partial\sigma_\pm} = |\mathbf{u}|\left|\frac{d\mathbf{x}}{d\sigma_\pm}\right|\cos A, \tag{C.7a}$$

$$u\frac{\partial y}{\partial\sigma_\pm} - v\frac{\partial x}{\partial\sigma_\pm} = \pm|\mathbf{u}|\left|\frac{d\mathbf{x}}{d\sigma_\pm}\right|\sin A \tag{C.7b}$$

and

$$u\frac{\partial v}{\partial\sigma_\pm} - v\frac{\partial u}{\partial\sigma_\pm} = |\mathbf{u}|^2\frac{d\theta}{d\sigma_\pm}, \tag{C.7c}$$

where

$$\left|\frac{d\mathbf{x}}{d\sigma_\pm}\right| = \left[\left(\frac{\partial x}{\partial\sigma_\pm}\right)^2 + \left(\frac{\partial y}{\partial\sigma_\pm}\right)^2\right]^{1/2}.$$

Use of (C.7) in the second equality of (C.6) and division of the result by $d|\mathbf{u}|(\partial x/\partial\sigma_\pm)$ leads to

$$|\mathbf{u}|\frac{\partial\theta}{\partial\sigma_\pm} + \frac{1}{|\mathbf{u}|\,d}C_L\frac{\partial d}{\partial\sigma_\pm} + \left|\frac{d\mathbf{x}}{d\sigma_\pm}\right|\cos A = 0 \ \text{(on } C_\pm\text{)}, \tag{C.8}$$

where

$$C_L = \frac{\left[\pm u\,|\mathbf{u}|\left|\frac{d\mathbf{x}}{d\sigma_\pm}\right|\sin A - d\frac{\partial y}{\partial\sigma_\pm}\right]}{\frac{\partial x}{\partial\sigma_\pm}} = \frac{\left[\pm v\,|\mathbf{u}|\left|\frac{d\mathbf{x}}{d\sigma_\pm}\right|\sin A + d\frac{\partial x}{\partial\sigma_\pm}\right]}{\frac{\partial y}{\partial\sigma_\pm}}, \tag{C.9}$$

and where the second equality follows from (B.19). The term C_L can be further simplified by writing it as the linear combination $C_L = \alpha C_L + \beta C_R$ where C_R represents the final expression in (C.9), $\alpha + \beta = 1$, and α is chosen so that the terms proportional to $|\mathbf{u}|$ cancel. Equation (C.8) then becomes

$$|\mathbf{u}|\frac{\partial\theta}{\partial\sigma_\pm} \pm \frac{\cos A}{|\mathbf{u}|\sin A}\frac{\partial d}{\partial\sigma_\pm} + \left|\frac{d\mathbf{x}}{d\sigma_\pm}\right|\cos A = 0 \ \text{(on } C_\pm\text{)}. \tag{C.10}$$

If $\sigma_\pm$ is chosen to be arclength measured along the characteristic curve, then $|d\mathbf{x}/d\sigma_\pm = 1|$.

A more traditional form of (C.10), one having roots in the field of aerodynamics, uses the intrinsic long-wave speed $d^{1/2}$ as a variable. The middle term in (C.10) would then be written

$$\pm\frac{\cos A}{|\mathbf{u}|\sin A}\frac{\partial d}{\partial\sigma_\pm} = \pm 2\frac{d^{1/2}\cos A}{|\mathbf{u}|\sin A}\frac{\partial(d)^{1/2}}{\partial\sigma_\pm} = 2\cos A\frac{\partial(d)^{1/2}}{\partial\sigma_\pm}, \tag{C.11}$$

where (B.19) has been used in the last step.

In summary, the three sets of characteristic curves include the streamlines (see C.5) and the curves that cross the streamlines at the Froude angle $\pm A$ (see B.19). The characteristic equation that applies along streamlines is simply the Bernoulli equation:

$$\frac{|\mathbf{u}|^2}{2} + gd = B(\psi), \tag{C.12}$$

while the equations that apply along the Froude lines are given by (C.8) or (C.10). The dimensional versions of these three can be obtained by replacing d by gd and $\left|\frac{d\mathbf{x}}{d\sigma_\pm}\right|$ by $f\left|\frac{d\mathbf{x}}{d\sigma_\pm}\right|$, where f is the Coriolis parameter representing the background rotation.

An example of the use of these relations to compute a supercritical coastal current emerging from a river mouth is given by Garvine (1987).[1]

---

[1] A misprint in this reference lists θ in place of A in the middle term of (C.10). The calculations presented are based on the correct formulation, however.

References

Abbott, M. B. 1961. On the spreading of one fluid over another. part ii. The wave front. *La Houille Blanche* **6**, 827–846.

Alavian, V. 1986. Behavior of density currents on an incline. *J. Hydraul. Eng.* **112**, 27–42.

Armi, L. 1986. The hydraulics of two flowing layers of different densities. *J. Fluid Mech.* **163**, 27–58.

Armi, L. 1989. Hydraulic control of zonal currents on a β-plane. *J. Fluid Mech.* **201**, 357–377.

Armi, L. and D. M. Farmer 1986. Maximal two-layer exchange through a contraction with barotropic net flow. *J. Fluid Mech.* **164**, 27–51.

Armi, L. and D. M. Farmer 1987. A generalization of the concept of maximal exchange in a strait. *J. Geophys. Res.* **92**, 14679–14680.

Armi, L. and D. M. Farmer 1988. The flow of Mediterranean water through the Strait of Gibraltar. *Prog. Oceanogr.* **21**, 1–105.

Assaf, G. and A. Hecht 1974. Sea straits: A dynamic model. *Deep-Sea Res.* **21**, 947–958.

Astraldi, M., G. P. Gasparini, L. Gervasio and E. Salusti 2001. Dense water dynamics along the Strait of Sicily (Mediterranean Sea). *J. Phys. Oceanogr.* **31**, 3457–3475.

Baines, P. G. 1984. A unified description of two-layer flow over topography. *J. Fluid Mech.* **146**, 127–167.

Baines, P. G. 1987. Upstream blocking and airflow over mountains. *Ann. Rev. Fluid Mech.* **19**, 75–97.

Baines, P. G. 1995. *Topographic Effects in Stratified Flows*. Cambridge University Press, 482 pp.

Baines, P. G. and P. A. Davies 1980. Laboratory studies of topographic effects in rotating and/or stratified fluids. In *Orographic Effects in Planetary Flows,* GARP publication no. 23, WMO/ICSU, 233–299.

Baines, P. G. and B. P. Leonard 1989. The effects of rotation on flow of a single layer over a ridge. *Quart. J. Roy. Met. soc.* **115**, 293–308.

Baines, P. G. and J. A. Whitehead 2003. On multiple states in single-layer flows. *Phys. Fluids.* **15**, 298–307.

Batchelor, G. K. 1967. *An Introduction to Fluid Dynamics*. Cambridge University Press, 615 pp.

Beardsley, R. C., C. E. Dorman, L. Rosenfeld and C. D. Winant 1987. Local atmospheric forcing during the coastal ocean dynamics experiment 1. A description of the marine boundary layer and atmospheric conditions over a northern California upwelling region. *J. Geophys. Res.* **92**, 1467–1488.

Benjamin, T. B. 1968. Gravity currents and related phenomena. *J. Fluid Mech.* **80**, 641–671.

Borenäs, K. M. and P. A. Lundberg 1986. Rotating hydraulics of flow in a parabolic channel. *J. Fluid Mech.* **167**, 309–26.

Borenäs, K. M. and P. A. Lundberg 1988. On the deep-water flow through the Faroe Bank channel. *J. Geophys. Res.* **93**(C2), 1281–1292.

Borenäs, K. M. and P. A. Lundberg 2004. The Faroe Bank Channel deep-water outflow. *Deep-sea Res. (II)*, **51**, 335–350.

Borenäs, K. M. and A. Nikolopoulos 2000. Theoretical calculations based on real topography of the maximum deep-water flow through the Jungfern passage. *J. Marine Res.* **58**, 709–719.

Borenäs, K. M. and J. A. Whitehead 1998. Upstream separation in a rotating channel flow. *J. Geophys. Res.* **103**, C4, 7567–7578.

Bormans, M. and C. Garrett 1989a The effects of nonrectangular cross section, friction, and barotropic fluctuations on the exchange through the Strait of Gibraltar. *J. Phys. Oceanogr.* **19**, 1543–1557.

Bormans, M. and C. Garrett 1989b A simple criterion for gyre formation by the surface outflow from a strait, with applications to the Alboran Sea. *J. Geophys. Res.* **94**, 12,637–12,644.

Broecker, WS. 1991. The great ocean conveyor belt. *Oceanography* **4**, 79–89.

Bryden, H. L. and T. H. Kinder 1991. Steady two-layer exchange through the strait of Gibraltar. *Deep-Sea Res.* **38 (Suppl. 1)**, S445–S463.

Bryden, H. L. and H. M. Stommel 1984. Limiting processes that determine basic features of the circulation in the Mediterranean Sea. *Oceanologica Acta.* **7**, 289–296.

Burk, S. D., T. Haack and R. M. Samelson 1999. Mesoscale simulation of supercritical, subcritical and transcritical flow and coastal topography. *J. Atmos. Sci.* **58**, 2780–2795.

Bye, J. A. T. and J. A. Whitehead, Jr. 1975. A theoretical model of the flow in the mouth of Spencer Gulf, South Australia. *Estuarine and Coastal Mar. Sci.* **3**, 477–481.

Cahn, A. 1945. An investigation of the free oscillations of a simple current system. *J. Meteor.* **2**, 113–199.

Cenedese, C., J. A. Whitehead, T. A. Ascarelli and M. Ohiwa 2004. A dense current flowing down a sloping bottom in a rotating fluid. *J. Fluid Mech.* **34**, 188–203.

Charney, J. G. 1955. The Gulf stream as an inertial boundary layer. *Proc. Natl. Acad. Sci. USA* **41**, 731–740.

Charney, J. G. and M. Stern 1962. On the stability of internal baroclinic jets in a rotating atmosphere. *J. Atmos. Sci.* **19**, 159–172.

Chow, V. T. 1959. *Open Channel Hydraulics*. McGraw-Hill, New York, 680pp.

Cole, S. L. 1985. Transient waves produced by flow past a bump. *Wave Motion* **7**, 579–587.

Collings, I. L. and R. Grimshaw 1980. The effect of topography on the stability of a barotropic coastal current. *Dyn. Atmos. Ocean.* **5**, 83–106.

Courant, R. and K. O. Friedrichs 1976. *Supersonic Flow and Shock Waves.* Springer, New York, 464 pp.

Csanady, G. T. 1978. The arrested topographic wave. *J. Phys. Oceanogr.* **8**, 47–62.

Crocco, L. 1937. Eine neue Stromfunktion fur die Erforschung der Bewegung der Gas emit Rotation. *Zeits. f angew. Math. u. Mech.* **17**, 1.

Cushman-Roisin, B. 1994. *Introduction to Geophysical Fluid Dynamics*, Prentice-Hall, Eaglewood Cliffs, NJ, 320 pp.

Cushman-Roisin, B., L. Pratt and E. Ralph 1993. A general theory for equivalent barotropic thin jets. *J. Phys. Oceanogr.* **23**, 92–103.

Dale, A. C. and J. A. Barth 2001. The hydraulics of an evolving upwelling jet flowing around a cape. *J. Phys. Oceanogr.* **31**, 226–243.

Dalziel, S. B. 1988. *Two-layer Hydraulics: Maximal Exchange Flows*, PhD thesis. University of Cambridge, England.

Dalziel, S. B. 1990. Rotating two-layer sill flows. In: Pratt, L. J. (ed) *The Physical Oceanography of Sea Straits. NATO-ASI Ser.*, Kluwer Academic Publishers, Dordrecht, 587 pp.

Dalziel, S. B. 1991. Two-layer hydraulics: A functional approach. *J. Fluid Mech.* **223**, 135–163.

Dalziel, S. B. 1992. Maximal exchange in channels with nonrectangular cross sections. *J. Phys. Oceanogr.* **22**, 1188–1206.

Davies, P. A., Y. Guo and E. Rotenberg 2002. Laboratory model studies of Mediterranean outflow adjustment in the Gulf of Cadiz. *Deep-Sea Res.* II **49**, 4207–4223.

Davies, P. A., A. K. Wåhlin and Y. Guo 2006. A combined laboratory and analytical study of the flow through the Faroe Bank channel. *J. Phys. Oceanogr.* **36**, 1348–1364.

Dickson, R. R., E. M. Gmitrowicz and A. J. Watson 1990. Deep water renewal in the northern Atlantic, *Nature* **344**, 848–850.

Donato, T. O. and G. O. Marmorino 2002. The surface morphology of a coastal gravity current. *Continental Shelf Research,* **22**, 141–146.

Dorman, C. E. 1987. Possible role of gravity currents in northern California's coastal summer wind reversals. *J. Geophys. Res.* **92**, 1497–1506.

Dorman, C. E. 1985. Evidence of Kelvin waves in California's marine layer and related Eddy generation. *Mon. Wea. Rev.*, **113**, 828–839.

Dorman, C.E., D. P. Rogers, W. Nuss and W. T. Thompson 1999. Adjustment of the summer marine boundary layer around Pt. Sur, California. *Mon. Wea. Rev.* **127**, 2143–2159.

Drazin, P. G. and W. H. Reid 1981. *Hydrodynamic Instability*. Cambridge University Press, 527 pp.

Duncan, L. M., H. L. Bryden and S. A. Cunningham 2003. Friction and mixing in the Faroe Bank channel. *Ocean. Acta.* **26,** 473–486.

Dyer, K. R. 1997. *Estuaries, A Physical Introduction*, 2nd Edition. John Wiley & Sons, Chichester, 195 pp.

Eady, E. T. 1949. Long waves and cyclone waves. *Tellus* **1**, 33–52.

Edwards, K. A., P. MacCready, J. N. Moum, G. Pawlak, J. M. Klymak and A. Perlin 2004. Form drag and mixing due to tidal flow past a sharp point. *J. Phys. Oceanogr.* **34**, 1297–1312.

Edwards, K. A., A. M. Rogerson, C. D. Winant and D. P. Rogers 2001. Adjustment of the marine atmospheric boundary layer to a coastal cape. *J. Atmos. Sci.* **58**, 1511–1528.

Ellison, T. H. and J. S. Turner 1959. Turbulent entrainment in stratified flows. *J. Fluid Mech.* **6**, 423–448.

Farmer, D. M. and L. Armi 1986. Maximal two-layer exchange flow over a sill and through a combination of a sill and contraction with barotropic flow. *J. Fluid Mech.* **164**, 53–76.

Farrell, B. F. and P. J. Ioannou 1996. Generalized stability theory. Part I: Autonomous operators. *J. Atmos. Sci.* **53**, 2025–2040.

Federov, A. V. and W. K. Melville 1996. Hydraulic jumps at boundaries in rotating fluids. *J. Fluid Mech.,* **324**, 55–82.

Finnigan, T. D. and G. N. Ivey 1999. Submaximal exchange between a convectively forced basin and a large reservoir. *J. Fluid Mech.* **378**, 357–378.

Finnigan, T. D. and G. N. Ivey 2000. Convectively driven exchange flow in a stratified sill-enclosed basin. *J. Fluid Mech.* **418**, 313–338.

Fjørtorft, R. 1950. Application of integral theorems in deriving criteria of stability for laminar flows and for the baroclinic circular vortex. *Geophys. Publ.* **17**, 1–52.

Fofonoff, N. P. 1954. Steady flow in a frictionless homogeneous ocean. *J. Mar. Res.* **13**, 254–262.

Fornberg, B. and G. B. Whitham 1978. A numerical and theoretical study of certain nonlinear wave phenomena. *Phil. Trans. Roy. Soc. A* **289**, 373–404.

Fratantoni, D. M., R. J. Zantopp, W. E. Johns and J. L. Miller 1997. Updated bathymetry of the Anegada-Jungfern passage complex and implications for Atlantic inflow to the abyssal Caribbean Sea. *J. Marine Res.* **55**, 847–860.

Freeman, J. C., Jr. 1950. The wind field of the equatorial East Pacific as a Prandtl-Meyer expansion. *Bull. Amer. Meteor. Soc.,* **31**, 303–304.

Freeman, N. C. and R. S. Johnson 1970. Shallow water waves on shear flows. *J. Fluid Mech.* **42**, 401–409.

Garrett, C. 2004. Frictional processes in straits. *Deep-Sea Res. II* **51**, 393–410.

Garrett, C. and F. Gerdes 2003. Hydraulic control of homogeneous shear flows. *J. Fluid Mech.* **475**, 163–172.

Garvine, R. W. 1981. Frontal jump conditions for models of shallow, buoyant surface layer hydrodynamics. *Tellus* **33**, 301–312.

Garvine, R. W. 1987. Estuary plumes and fronts in shelf waters: A layer model. *J. Phys. Oceanogr.* **17**, 1877–1896.

Gerdes, F., C. Garrett and D. Farmer 2002. On internal hydraulics with entrainment. *J. Phys. Oceanogr.* **32**, 1106–1111.

Gill, A. E. 1976. Adjustment under gravity in a rotating channel. *J. Fluid Mech.* **77**, 603–621.

Gill, A. E. 1977. The hydraulics of rotating-channel flow. *J. Fluid Mech.* **80**, 641–671.

Gill, A. E. 1982. *Atmosphere-Ocean Dynamics*. Academic Press, San Diego, 662 pp.

Gill, A. E. and E. H. Schumann 1979. Topographically induced changes in the structure of an inertial jet: Application to the Agulhas current. *J. Phys. Oceanogr.* **9**, 975–991.

Girton, J. B., T. B. Sanford and R. H. Käse 2001. Synoptic sections of the Denmark Strait overflow. *Geophys. Res. Lett.* **28**, 1619–1622.

Girton, J. B., L. J. Pratt, D. A. Sutherland and J. F. Price 2006. Is the Faroe Bank channel overflow hydraulically controlled?. *J. Phys. Oceanogr.* **36**, 2340–2349.

Greenspan, H. P. 1968. *The Theory of Rotating Fluids*. Cambridge University Press, 327 pp.

Gregg, M. C. and E. Özsoy 2002. Flow, water mass changes, and hydraulics in the Bosporus. *J. Geopys. Res.* **107**, 10.1029/2000JC000485.

Griffiths, R. W. and E. J. Hopfinger 1983. Gravity currents moving along a lateral boundary in a rotating fluid. *J. Fluid Mech.* **134**, 357–399.

Griffiths, R. W., P. D. Killworth and M. E. Stern 1982. Ageostrophic instability of ocean currents. *J. Fluid Mech.* 117, 343–377.

Grimm, Th. and T. Maxworthy 1999. Buoyancy-driven mean flow in a long channel with a hydraulically constrained exit condition. *J. Fluid Mech.* **398**, 155–180.

Grimshaw, R. H. J. 1987. Resonant forcing of Barotropic coastally trapped waves. *J. Phys. Oceanogr.* **17**, 53–65.

Grimshaw, R. H. J. and Z. Yi 1990. Finite-amplitude long waves on coastal currents. *J. Phys. Oceanogr.* **20**, 3–18.

Grimshaw, R. H. J. and N. Smyth 1986. Resonant flow of a stratified fluid over topography. *J. Fluid Mech.* **169**, 429–464.

Hacker, J. N. and P. F. Linden 2002. Gravity currents in rotating channels. Part 1. Steady-state theory. *J. Fluid Mech.* **457**, 295–324.

Hall, M., M. McCartney and J. A. Whitehead 1997. Antarctic bottom water flux in the equatorial western Atlantic. *J. Phys. Oceanogr.* **27**, 1903–1926.

Hannah, C. G. 1992. Geostropic conrol with wind forcing: Application to the Bass Strait. *J. Phys. Oceanogr.* **22**, 1596–1599.

Hansen, G., W. R. Turnbull and S. Østerhus 2001. Decreasing overflow from the Nordic Seas through the Atlantic Ocean through the Faroe Bank channel since 1950. *Nature* **411**, 927–930.

Hayashi, Y.-Y. and W. R. Young 1987. Stable and unstable shear modes of rotating parallel flows in shallow water. *J. Fluid Mech.* **184**, 477–504.

Haynes, P. H., E. R. Johnson and R. G. Hurst 1993. A simple model of Rossby-wave hydraulic behavior. *J. Fluid Mech.* **253**, 359–384.

Helfrich, K. R., A. C. Kuo and L. J. Pratt 1999. Nonlinear Rossby adjustment in a channel. *J. Fluid Mech.* **390**, 177–222.

Helfrich, K. R. and J. C. Mullarney 2005. Gravity currents from a dam-break in a rotating channel. *J. Fluid Mech.* **536**, 253–283.

Helfrich, K. R. and L. J. Pratt 2003. Rotating hydraulics and upstream basin circulation. *J. Phys. Oceanogr.* **33**, 1651–1663.

Henderson, F. M. 1966. *Open Channel Flow.* Macmillan. New York 522 pp.

Herman, A.J., P.B. Rhines and E.R. Johnson 1989. Nonlinear Rossby adjustment in a channel: Beyond Kelvin waves. *J. Fluid Mech.* **205**, 469–502.

Hogg, A. McC. and G. O. Hughes 2006. Shear flow and viscosity in single-layer hydraulics. *J. Fluid Mech.* **548**, 431–443.

Hogg, N. G. 1983. Hydraulic control and flow separation in a multi-layered fluid with application to the Vema Channel. *J. Phys. Oceanogr.* **13**, 695–708.

Hogg, N. G., G. Siedler and W. Zenk 1999. Circulation and variability at the southern boundary of the Brazil basin. *J. Phys. Oceanogr.* **29**, 145–157.

Holland, D. M., R. R. Rosales, D. Stefanica and E.G. Tabak 2002. Internal hydraulic jumps and mixing in two-layer flows. *J. Fluid Mech.* **470**, 63–83.

Houghton, D. D. and A. Kasahara 1968. Nonlinear shallow fluid flow over an isolated ridge. *Comm. Pure Appl. Math.* **21**, 1–23.

Houghton, D. D. and E. Isaacson 1970. Mountain winds. *Stud. Numer. Anal.* **2**, 21–52.

Howard, L. N. 1961. Note on a paper of John W. Miles. *J. Fluid Mech.* **10**, 509–512.

Hughes, R. L. 1985*a*. On inertial currents over a sloping continental shelf. *Dyn. Atmos. Oceans* **9**, 49–73.

Hughes, R. L. 1985b. Multiple criticalities in coastal flows. *Dyn. Atmos. Oceans* **9**, 321–340.

Hughes, R. L. 1986a. On the role of criticality in coastal flows over irregular bottom topography. *Dyn. Atmos. Oceans* **10**, 129–147.

Hughes, R. L. 1986b. On the conjugate behaviour of weak along-shore flows. *Tellus* **38A**, 277–284.

Hughes, R. L. 1987. The role of the higher shelf modes in coastal hydraulics. *J. Mar. Res.* **45**, 33–58.

Hughes, R. L. 1989. The hydraulics of local separation in a coastal current, with application to the Kuriosho meander. *J. Phys. Oceanogr.* **19**, 1809–1820.

Hugoniot, P.-H. 1886. Sur un théorème relatif au mouvement permanent et à l'écoulement des fluids. *C. Roy. Acad. Sci. Paris*, **103**, 1178–1181.

Hunkins, K. and J. A. Whitehead 1992. Laboratory simulation of exchange through Fram Strait. *J. Geophys. Res.* **97** (C7) 11,299–11,321.

Iacono, R. 2006. Critical flow solution to Gill's model of rotating channel hydraulics. *J. Fluid Mech.* **552**, 381–392.

Jiang, Q. and R. B. Smith 2001a. Ideal shocks in two-layer flow. Part I: Under a rigid lid. *Tellus* **53A**, 129–145.

Jiang, Q. and R. B. Smith 2001b. Ideal shocks in two-layer flow. Part II: Under a passive layer. *Tellus* **53A**, 146–167.

Johnson, E. R. and S. R. Clarke 1999. Dispersive effects in Rossby-wave hydraulics. *J. Fluid Mech.* **401**, 27–54.

Johnson, E. R. and S. R. Clarke 2001. Rossby wave hydraulics. *Ann. Rev. Fluid Mech.* **33**, 207–230.

Johnson, G. C. and T. B. Sanford 1992. Secondary circulation in the Faroe Bank channel outflow. *J. Phys. Oceanogr.* **22**, 927–933.

Johnson, G. C. and Ohlsen D. R. 1994. Frictionally modified rotating hydraulic channel exchange and ocean outflows. *J. Phys. Oceanogr.* **24**, 66–78.

Jonsson, S. and H. Valdimarsson 2004. A new path for the Denmark Strait overflow water from the Iceland Sea to Denmark Strait. *Geophys. Res. Lett.* **31**, L03305, doi:10.1029/2003GL019214, 2004.

Jungclaus, J. H. and J. O. Backhans 1994. Application of a transient reduced gravity plume model to the Denmark Strait overflow. *J. Geophys. Res.* **99**, 12375–12396.

Käse, R. H. and A. Oschlies 2000. Flow through Denmark Strait. *J. Geophys. Res.* **105** (C12), 28,527–28,546.

Kelvin, L. 1879. *Proc. Roy. Soc. Edin.* **10**, 92. (Also see *Phil. Mag.* **10** (1980), 97; *Math. Phys. Papers*, vol. IV, p. 141, Cambridge University Press 1910.)

Killworth, P. D. 1977. Mixing on the Weddell Sea continental slope. *Deep-Sea Res.* **24**, 427–448.

Killworth, P. D. 1992a. On hydraulic control in a stratified fluid. *J. Fluid Mech.*, **237**, 605–626.

Killworth, P. D. 1992b. Flow properties in rotating, stratified hydraulics. *J. Phys. Oceanogr.* **22**, 997–1017.

Killworth, P. D. 1994. On reduced-gravity flow through sills. *Geophys. Astrophys. Fluid Dynamics* **75**, 91–106.

Killworth, P. D. and N. R. McDonald 1993. Maximal reduced-gravity flux in rotating hydraulics. *Geophys. Astrophys. Fluid Dynam.* **70**, 31–40.

Klemp, J. B., R. Rotunno and W. K. Skamarock 1994. On the dynamics of gravity currents in a channel. *J. Fluid Mech.* **269**, 169–198.

Klemp, J. B., R. Rotunno and W. K. Skamarock 1997. On the propagation of internal bores. *J. Fluid Mech.* **331**, 81–106.

Klinger, B. A. 1994. Inviscid separation from rounded capes. *J. Phys. Oceanogr.* **24**, 1805–1811.

Koop, C. G. and F. K. Browand 1979. Instability and turbulence in a stratified fluid with shear. *J. Fluid Mech.*, **93**, 135–159.

Kösters, F. 2004. Denmark Strait overflow: Comparing model results and hydraulic transport estimates. *J. of Geophys. Res.* **109**, (C10), doi:10.1029/2004JC002297.

Krauss, W. and R. H. Käse 1998. Eddy formation in the Denmark Strait overflow. *J. Geophys. Res.* **103** (C8), 15,525–15,538.

Kubokawa, A. and K. Hanawa 1984a. A theory of semigeostrophic gravity waves and its application to the intrusion of a density current along a coast. Part 1. Semigeostrophic gravity waves. *J. Oceanographical Society of Japan* **40**, 247–259.

Kubokawa, A. and K. Hanawa 1984b. A theory of semigeostrophic gravity waves and its application to the intrusion of a density current along a coast. Part 2. Intrusion of a density current along a coast of a rotating fluid. *J. Oceanographical Society of Japan* **40**, 260–270.

Kuo, H.-L. 1949. Dynamic instability of two-dimensional nondivergent flow in a barotropic atmosphere. *J. Atmos. Sci.*, **6**, 105–122.

Lake, I., K. M. Borenäs and P. A. Lundberg 2005. Potential-vorticity characteristics of the Faroe-Bank channel deep-water overflow. *J. Phys. Oceanogr.* **35**, 921–932.

Lamb, H. 1932. *Hydrodynamics*, 6th edn. Cambridge University Press.

Lane-Serff, G. F., D. A. Smeed and C. R. Postlethwaite 2000. Mult-layer hydraulic exchange flows. *J. Fluid Mech.* **416**, 269–296.

Lawrence, G. A. 1990. On the hydraulics of Boussinesq and non-Boussinesq two-layer flows. *J. Fluid Mech.* **215**, 457–480.

Lawrence, G. A. 1993. The hydraulics of steady two-layer flow over a fixed obstacle. *J. Fluid Mech.* **254**, 605–633.

Limeburner, R., J. A. Whitehead and C. Cenedese 2005. Variability of Antarctic Bottom Water Flow into the North Atlantic. *Deep-Sea Res. II*, **52**, 495–512.

Lipps, F. B. 1963. Stability of jets in a barotropic, divergent fluid. *J. Atmos. Sci.* **20**, 120–129.

Long, R. R. 1954. Some aspects of the flow of stratified fluids. II. Experiments with a two-fluid system. *Tellus* **6**, 97–115.

Long, R. R. 1955. Some aspects of the flow of stratified fluids. III. Continuous density gradients. *Tellus* **7**, 341–357.

Long, R. R. 1970. Blocking effects in flow over obstacles. *Tellus* **22**, 471–480.

MacCready, P. W. E. Johns, C. G. Rooth, D. M. Fratantoni, and R. A. Watlington, 1999. Overflow into the deep Caribbean: Effects of plume variability. *J. of Geophys. Research* **104**, C11, 25913–25935.

Martin, J. R. and G. F. Lane-Serff 2005. Rotating gravity currents. Part 1. Energy loss theory. *J. Fluid Mech.* **522**, 35–62.

Martin, J. R., D. A. Smeed and G. F. Lane-Serff 2005. Rotating gravity currents. Part 2. Potential vorticity theory. *J. Fluid Mech.* **522**, 63–89.

Mauritzen, C., J. Price, T. Sanford and D. Torres 2005. Circulation and mixing in the Faroese channels. *Deep-Sea Research I* **52**, 883–913.

McCartney, M. S., S. L. Bennett and M. E. Woodgate-Jones 1991. Eastward flow through the Mid-Atlantic Ridge at 118N and its influence on the abyss of the eastern basin. *J. Phys. Oceanogr.* **21**, 1089–1120.

McClimans, T. A. 1994. Entrainment/detrainment along river plumes. In: Davies, P. A. and M. J. Valente Neves. (eds) *Recent Research Advances in the Fluid Mechanics of Turbulent jets and plumes*. Kluwer Academic Publishers. Dordrecht. pp. 391–400.

Mercier, H. and K. G. Speer 1998. Transport of bottom water in the Romanche Fracture Zone and the Chain Fracture Zone. *J. Phys. Oceanog.* **28**, 779–790.

Middleton, J. F. and F. Viera 1991. The forcing of low frequency motions within the Bass Strait. *J. Phys. Oceanogr.* **21**, 695–708.

Munchow, A. and R. W. Garvine 1993. Dynamical properties of a buoyancy-driven coastal current. *J. Geophys. Res.* **98**, 20063–20077.

Nielsen, M. H., L. J. Pratt and K. R. Helfrich 2004. Mixing and entrainment in hydraulically-driven, stratified sill flows. *J. Fluid Mech.* **515**, 415–443.

Nielson, J. N. 1912. Hydrography of the Mediterranean and adjacent waters. *Report on the Danish Oceanographical Expeditions 1908–1910* **1**, 77–191.

Nikolopoulos, A., K. Borenäs, R. Hietala and P. A. Lundberg 2003. Hydraulic estimates of Denmark Strait overflow. *J. Geophys. Res.* **108(C3)**, 3095, doi:10.1029/2001JC001283.

Niiler, P. P. and L. A. Mysak 1971. Barotropic waves along an eastern continental shelf. *Geophys. Fluid Dyn.* **2**, 273–288.

Nof, D. 1983. The translation of isolated cold eddies on a sloping bottom. *Deep-Sea Res.* **30**, 171–182.

Nof, D. 1984. Shock waves in currents and outflows. *J. Phys. Oceanogr.* **14**, 1683–1702.

Nof, D. 1986. Geostrophic shock waves. *J. Phys. Oceanogr.* **16**, 886–901.

Nof, D. 1987. Penetrating outflows and the dam-break problem. *J. Marine Res.* **45**, 557–577.

Siedler G., J. Church and J. Gould, (ed.) 2001, *Ocean Circulation and Climate: Observing and Modeling the Global Ocean.* Academic Press, 715pp.

Ou, H. - S. and P. M. De Ruijter 1986. Separation of an inertial boundary current from a curved coastline. *J. Phys. Oceanogr.* **16**, 280–289.

Paldor, N. 1983. Stability and stable modes of coastal fronts. *Geophys. Astrophys. Fluid Dyn.* **27**, 217–218.

Pedlosky, J. 1968. An overlooked aspect of the wind-driven oceanic circulation. *J. Fluid Mech.* **43**, 809–821.

Pedlosky, J. 1987. *Geophysical Fluid Dynamics.* Springer-Verlag, New York, 710 pp.

Pedlosky, J. 1996. *Ocean Circulation Theory,* Springer-Verlag, New York, 453 pp.

Pedlosky, J. 2003. *Waves in the Ocean and Atmosphere: Introduction to Wave Dyamics.* Springer-Verlag, 260 pp.

Peregrin, D. H. 1966. Calculations of the development of an undular bore. *J. Fluid Mech.* **25**, 321–330.

Perlin, N., R. M. Samelson and D. B. Chelton 2004. Scatterometer and model wind and wind stress in the Oregon-northern California coastal zone. *Monthly Weather Review* **132**, 2110–2129.

Phillips, O. M. 1966. On turbulent convection currents and the circulation of the Red Sea. *Deep-Sea Res.* **13**, 1147–1160.

Poincaré, S. 1910. *Théorie des Marées. Leçons de Mécanique Celeste*, vol. 3, Paris: Gauthier-Villars.

Pratt, L. J. 1983*a*. A note on nonlinear flow over obstacles. *Geophys. Astrophys. Fluid Dyn.* **24**, 63–68.

Pratt, L. J. 1983*b*. On inertial flow over topography. Part 1. Semigeostrophic adjustment to an obstacle. *J. Fluid Mech.* **131**, 195–218.

Pratt, L. J. 1984*a*. On inertial flow over topography. Part 2. Rotating channel flow near the critical speed. *J. Fluid Mech.* **145**, 95–110.

Pratt, L. J. 1984*b*. A time-dependent aspect of hydraulic control in straits. *J. Phys. Oceanogr.* **14**, 1414–1418.

Pratt, L. J. 1984*c*. On nonlinear flow with multiple obstructions. *J. Atmos. Sci.* **41**, 1214–1225.

Pratt, L. J. 1986. Hydraulic control of sill flow with bottom friction. *J. Phys. Oceanogr.* **16**, 1970–1980.

Pratt, L. J. 1987. Rotating shocks in a separated laboratory channel flow. *J. Phys. Oceanogr.* **17**, 483–491.

Pratt, L. J. 1989. Critical control of zonal jets by bottom topography. *J. Mar. Res.* **47**, 111–130.

Pratt, L.J. 1991. Geostrophic versus critical control in straits. *J. Phys. Oceanogr.* **21**, 728–732.

Pratt, L. J. 1997. Hydraulically drained flows in rotating basins. Part I: Steady flows. *J. Phys. Oceanogr.* **27**, 2522–2535.

Pratt, L. J. and L. Armi 1987. Hydraulic control of flows with nonuniform potential vorticity. *J. Phys. Oceanogr.* **17**, 2016–2029.

Pratt, L. J. and L. Armi 1990. Two-layer rotating hydraulics: strangulation, remote and virtual controls. *Pure Appl. Geophys.* **133(4)**, 587–617.

Pratt, L. J. and M. Chechelnitsky 1997. Principles for capturing the upstream effects of deep sills in low resolution ocean models. *Dynamics of Atmospheres and Oceans* **26**, 1–25.

Pratt, L. J., H. E. Deese, S. P. Murray and W. Johns 2000. Continuous dynamical modes in straits having arbitrary cross sections, with applications to the Bab al Mandab. *J. Phys. Oceanogr.* **30**, 2515–2534.

Pratt, L. J. and K. R. Hefrich 2005. Generalized conditions for hydraulic criticality of oceanic overflows. *J. Phys. Oceanogr.* **35**, 1782–1800.

Pratt, L. J., K. R. Helfrich and E. P. Chassignet 2000. Hydraulic adjustment to an obstacle in a rotating channel. *J. Fluid Mech.* **404**, 117–149.

Pratt, L. J., W. Johns, S. P. Murray, and K. Katsumata 1999. Hydraulic interpretation of direct velocity measurements in the Bab al Mandab. *J. Phys. Oceanogr.* **29**, 2769–2784.

Pratt, L. J. and S. G. Llewellyn Smith 1997. Hydraulically drained flows in rotating basins. Part I: Method. *J. Phys. Oceanogr.* **27**, 2509–2521.

Pratt, L. J. and P. A. Lundberg 1991. Hydraulics of rotating strait and sill flow. *Ann. Rev. Fluid. Mech.* **23**, 81–106.

Pratt, L. J. and M. E. Stern 1986. Dynamics of potential vorticity fronts and Eddy detachment. *J. Phys. Oceanogr.* **16**, 1101–1120.

Price, J. F. and M. O. Baringer 1994. Outflows and deep water production by marginal seas. *Prog. Oceanogr.* **33**, 161–200.

Princevac, M., H. J. S. Fernando and D. C. Whiteman 2005. Turbulent entrainment into natural gravity-driven flows. *J. Fluid Mech.* **533**, 259–268.

Rabe, B., D. A. Smeed, S. B. Dalziel and G. F. Lane-Serff 2007. Experimental studies of rotating exchange flow. *Deep Sea Research I* **54**, 269-291.

Rayleigh, J. W. S. 1880. On the stability, or instability of certain fluid motions. *Proc. Lond. Math. Soc.* **9**, 57–70.

Rennie, S., Largier, J. L. and S. J. Lentz 1999. Observations of low-salinity coastal pulses downstream of Chesapeake Bay. *J. Geophys. Res.* **104**, 18227–18240.

Reynolds, O. 1886. On the flow of gases. *Philos. Mag.* **21(5)**, 185–198.

Ripa, P. 1983. General stability conditions for zonal flows in a one layer model on a beta-plane or the sphere. *J. Fluid Mech.* **126**, 463–487.

Röed, L. P. 1980. Curvature effects on hydraulically driven inertial boundary currents. *J. Fluid Mech.* **96**, 395–412.

Rogerson, A. M. 1999. Transcritical flows in the coastal marine atmospheric boundary layer. *J. Atmos. Sci.* **56**, 2761–2779.

Rossby, C. G. 1938. On the mutual adjustment of pressure and velocity distributions in certain simple current systems II. *J. Marine. Res.* **2**, 239–263.

Rossby, C. G. 1950. On the dynamics of certain types of blocking waves. *J. Chinese Geophys. Soc.* **2**, 1–13.

Rudnick, D. 1997. Direct velocity measurements in the Samoan passage, *J. Geophys. Res.* **102**, 3293–3302.

Samelson, R. M. 1992 Supercritical marine-layer flow along a smoothly varying coastline. *J. Atmos. Sci.* **49**, 1571–1584.

Salmon, R. 1998. *Lectures on Geophysical Fluid Dynamics*, Oxford, 400 pp.

Saunders, P. M. 1987. Flow through Discovery Gap. *J. Phys. Oceanogr.* **17**, 631–643.

Saunders, P. M. 1990. Cold outflow from the Faroe Bank channel. *J. Phys. Oceanogr.* **20**, 29–43.

Schär, L. and R. B. Smith 1993. Shallow-water flow past isolated topogrophy. Part I: Vorticity production and wake formation. *J. Atmos. Sci.* **50**, 1373–1400.

Schmitz, W. J. 1995. On the interbasin-scale thermohaline circulation. *Rev. Geophys.* **32**, 151–173.

Shen, C. Y. 1981. The rotating hydraulics of open-channel flow between two basins. *J. Fluid Mech.* **112**, 161–88.

Siddall, M., L. J. Pratt and K. R. Helfrich 2004. Testing the physical oceanographic implications of the suggested sudden Black Sea infill 8400 years ago. *Paleoceanography*, **19**, PA1024, doi:10.1029/2003PA000903.

Simson, J. E. 1997. *Gravity Currents in the Environment and the Laboratory*, 2nd edn. Cambridge University Press.

Shi, X. B., L. P. Röed and B. Hackett 2001. Variability of the Denmark Strait overflow: A numerical study. *J. Geophys. Res.* **106(C10)**, 22,277–22,294.

Slagstad, D. and T. McClimans 2005. Modeling the ecosystem dynamics of the Barents Sea including the marginal ice zone: physical and chemical oceanography. *J. Mar. Sys.* **58**, 1–18.

Smith, P. 1975. A streamtube model for bottom boundary currents in the ocean. *Deep-Sea Res.* **22**, 853–873.

Smith, R. B., A. C. Gleason, P. Gluhosky and V. Grubisic 1997. The wake of St. Vincent. *J. Atmos. Sci.* **54**, 606–623.

Smith, T. 1684. A conjecture about an under-current at the streights mouth. *Phil. Trans.*, **14**, 30–31.

Stalcup, M. C., W. G Metcalf and R. G. Johnson 1975. Deep Caribbean inflow through the Anegada-Jungfern passage. *J. Mar. Res.* **33**, 15–35.

Stern, M. E. 1974. Comment on rotating hydraulics. *Geophys. Fluid Dyn.* **6**, 127–130.

Stern, M. E. 1980. Geostrophic fronts, bores, breaking and blocking waves. *J. Fluid Mech.* **99**, 687–703.

Stern, M. E., J. A. Whitehead and B. L. Hua 1982. The intrusion of a density current along the coast of a rotating fluid. *J. Fluid Mech.* **132**, 237–265.

Stoker, J. J. 1957. *Water Waves.* New York: Interscience. 567 pp.

Stommel, H. M. 1948. The westward intensification of wind-driven currents. *Trans. Amer. Geophys. Union* **99**, 202–206.

Stommel, H. M. 1960. *The Gulf Stream.* University of California Press, 202 pp.

Stommel, H. M. and H. G. Farmer 1952. Abrupt change in width in two-layer open chanel flow. *J. Marine Res.* **11**, 205–214.

Stommel, H. M. and H. G. Farmer 1953. Control of salinity in an estuary by a transition. *J. Marine Res.* **12**, 13–20.

Thorpe, S. A. 1973. Experiments on instability and turbulence in a stratified shear flow. *J. Fluid Mech.*, **61**, 731–751.

Timmermans, M.-L. E. 1998. Hydraulic control and mixing in a semi-enclosed reservoir. *Geophys. Fluid Dyn. Summer Study Program Tech. Rep. #WHOI-98-09.*

Timmermans M.-L. E. and L. J. Pratt 2005. Two-layer exchange flow between two deep basins: Theory and application to the Strait of Gibraltar. *J. Phys. Oceanogr.* **35**, 1568–1592.

Tomasson, G. G. and W. K. Melville 1992. Geostrophic adjustment in a channel: Nonlinear and dispersive effects. *J. Fluid Mech.* **241**, 23–57.

Toulany, B. and C. J. R. Garrett 1984. Geostrophic control of fluctuating flow through straits. *J.Phys. Oceanogr.* **14**, 649–655.

Tsimplis, M. N. and H. L. Bryden 2000. Estimation of the transports through the Strait of Gibraltar. *Deep-Sea Res. Part I* **47**, 2,219–2,242.

Turner, J. S. 1973. *Buoyancy Effects In Fluids.* Cambridge University Press, 367 pp.

Vallis, G. K. 2006. *Atmospheric and Oceanic Fluid Dynamics,* Cambridge University Press, 745 pp.

Wåhlin, A. K. and G. Walin 2001. Downward migration of dense bottom currents. *Environmental Fluid Mechanics* **1**, 257–279.

Warren, B. A. 1981. Deep circulation of the world ocean. In: *Evolution of Physical Oceanography, Scientific Surveys in Honor of Henry Stommel*, B. A. Warren and C. Wunsch, editors, The MIT Press, Cambridge, Massachusetts; pp. 6–41.

Wells, M. G. and J. S. Wettlaufer 2005. Two dimensional density currents in a confined basin. *Geophys. Astro. Fluid Dyn.* **99**, 199–218.

Whitham, G. B. 1974. *Linear and Nonlinear Waves.* J. Wiley and Sons, New York, 636 pp.

Whitehead, J. A. 1989. Internal hydraulic control in rotating fluids—applications to oceans. *Geophys. Astrophys. Fluid Dyn.* **48**, 169–192.

Whitehead, J. A. 2003. Constant potential vorticity hydraulically controlled flow-complexities from passage shape. *J. Phys. Oceanogr.* **33**, 305–312.

Whitehead, J. A. 2005. The effect of potential vorticity on flow rate through a gap. *J. Geophys Res.* **110(C7)**, C0700710.1029/2004JC002720.

Whitehead, J. A., A. Leetma and R. A. Knox 1974. Rotating hydraulics of strait and sill flows. *Geophys. Fluid Dyn.* **6**, 101–125.

Whitehead, J. A. and A. R. Miller 1979. Laboratory simulation of the gyre in the Alboran Sea. *J. Geophys. Res.* **84**, 3733–3742.

Whitehead, J. A. and J. Salzig 2001. Rotating channel flow: Control and upstream currents. *Geophys. Astrophys. Fluid Dyn.* **95**, 185–226.

Whitehead, J. A., M.-L. Timmermans, W. Gregory Lawson, S. N. Bulgakov, A. M. Zatarian, J. F. A. Medina and J. Salzig 2003. Laboratory studies of thermally and/or/salinity driven flows with partial mixing 1. Stommel transitions and multiple flow states. *J. Geophys. Res.* **108(C2)**, 3036, doi:10.1029/2001JC000902.

Wilkinson, D. L. and I. R. Wood 1971. A rapidly varied flow phenomenon in a two layer flow. *J. Fluid Mech.* **47**, 241–256.

Wilkinson, D. L. and I. R. Wood 1983. The formation of an intermediate layer by horizontal convection in a two-layered shear flow. *J. Fluid Mech.*, **136**, 167–187.

Williams, R. T. and A. M. Hori 1970. Formation of hydraulic jumps in a rotating system. *J. Geophys. Res.* **75**, 2813–2821.

Winant, C. D., C. E. Dorman, C. A. Friehe and R. C. Beardsley 1988. The marine layer off Northern California: An example of supercritical channel flow. *J. Atmos. Sci.* **45**, 3588–3605.

Woelk, S. and D. Quadfasel 1996. Renewal of deep water in the Red Sea during 1982–1987. *J. Geophys. Res.*, **101(c8)**, 18155–18166.

Wood, I. R. 1970. A lock exchange flow. *J. Fluid Mech.* **42**, 671–687.

Woods, A. W. 1993. The topographic control of planetary-scale flow. *J. Fluid Mech.* **247**, 603–621.

Worthington, L. V. 1969. An attempt to measure the volume transport of Norwegian Sea overflow water through the Denmark Strait. *Deep-Sea Res.* **16 (supp)** 421–432.

Worthington, L. V. and W. R. Wright 1970. *North Atlantic Ocean Atlas*. Woods Hole Oceanographic Institution Atlas Series, Volume II.

Wright, D. G. 1987. Comments on "Geostrophic control of flustuating barotropic flow through straits." *J. Phys. Oceanogr.*, **17**, 2375–2377.

Yih, C. S. 1980. *Stratified Flows*. Academic Press, San Diego, 418pp.

Zabusky, N. J., M. H. Hughes and K. V. Roberts 1979. Contour dynamics for the Euler equations in two dimensions. *J. Comput. Phys.* **30**, 96–106.

Zenk, W., G. Seidler, B. Lenz and N. G. Hogg 1999. Antarctic bottom water Flow through the Hunter Channel. *J. Phys. Oceanogr.* **29**, 2769–2784.

Zhu, D. Z. and G. A. Lawrence 2000. Hydraulics of exchange flows. *J. Hydr. Engrg.*, ASCE **126**, 921–928.

Zhu, D. Z. and G. A. Lawrence 1998. Nonhydrostatic effects in layered shallow water flows. *J. Fluid Mech.* **355**, 1–16.

Index

ATMOSPHERIC AND OCEANOGRAPHIC SCIENCES LIBRARY

1. F.T.M. Nieuwstadt and H. van Dop (eds.): *Atmospheric Turbulence and Air Pollution Modelling*. 1982; rev. ed. 1984   ISBN 90-277-1365-6; Pb (1984) 90-277-1807-5
2. L.T. Matveev: *Cloud Dynamics*. Translated from Russian. 1984   ISBN 90-277-1737-0
3. H. Flohn and R. Fantechi (eds.): *The Climate of Europe: Past, Present and Future. Natural and Man-Induced Climate Changes: A European Perspective*. 1984
    ISBN 90-277-1745-1
4. V.E. Zuev, A.A. Zemlyanov, Yu.D.Kopytin, and A.V. Kuzikovskii: *High-Power Laser Radiation in Atmospheric Aerosols: Nonlinear Optics of Aerodispersed Media*. Translated from Russian. 1985   ISBN 90-277-1736-2
5. G. Brasseur and S. Solomon: *Aeronomy of the Middle Atmosphere: Chemistry and Physics of the Stratosphere and Mesosphere*. 1984; rev. ed. 1986
    ISBN (1986) 90-277-2343-5; Pb 90-277-2344-3
6. E.M. Feigelson (ed.): *Radiation in a Cloudy Atmosphere*. Translated from Russian. 1984
    ISBN 90-277-1803-2
7. A.S. Monin: *An Introduction to the Theory of Climate*. Translated from Russian. 1986
    ISBN 90-277-1935-7
8. S. Hastenrath: *Climate Dynamics of the Tropics*. Updated Edition from *Climate and Circulation of the Tropics*. 1985; rev. ed. 1991
    ISBN 0-7923-1213-9; Pb 0-7923-1346-1
9. M.I. Budyko: *The Evolution of the Biosphere*. Translated from Russian. 1986
    ISBN 90-277-2140-8
10. R.S. Bortkovskii: *Air-Sea Exchange of Heat and Moisture During Storms*. Translated from Russian, rev. ed. 1987   ISBN 90-277-2346-X
11. V.E. Zuev and V.S.Komarov: *Statistical Models of the Temperature and Gaseous Components of the Atmosphere*. Translated from Russian. 1987   ISBN 90-277-2466-0
12. H. Volland: *Atmospheric Tidal and Planetary Waves*. 1988   ISBN 90-277-2630-2
13. R.B. Stull: *An Introduction to Boundary Layer Meteorology*. 1988
    ISBN 90-277-2768-6; Pb 90-277-2769-4
14. M.E. Berlyand: *Prediction and Regulation of Air Pollution*. Translated from Russian, rev. ed. 1991   ISBN 0-7923-1000-4
15. F. Baer, N.L. Canfield and J.M. Mitchell (eds.): *Climate in Human Perspective: A Tribute to Helmut E. Landsberg (1906–1985)*. 1991   ISBN 0-7923-1072-1
16. Ding Yihui: *Monsoons over China*. 1994   ISBN 0-7923-1757-2
17. A. Henderson-Sellers and A.-M. Hansen: *Climate Change Atlas: Greenhouse Simulations from the Model Evaluation Consortium for Climate Assessment*. 1995
    ISBN 0-7923-3465-5
18. H. R. Pruppacher and J.D. Klett: *Microphysics of Clouds and Precipitation*, 2nd rev. ed. 1997   ISBN 0-7923-4211-9; Pb 0-7923-4409-X
19. R.L. Kagan: *Averaging of Meteorological Fields*. 1997   ISBN 0-7923-4801-X
20. G.L. Geernaert (ed.): *Air-Sea Exchange: Physics, Chemistry and Dynamics*. 1999
    ISBN 0-7923-5937-2
21. G.L. Hammer, N. Nicholls and C. Mitchell (eds.):*Applications of Seasonal Climate Forecasting in Agricultural and Natural Ecosystems*. 2000   ISBN 0-7923-6270-5
22. H.A. Dijkstra: *Nonlinear Physical Oceanography: A Dynamical Systems Approach to the Large Scale Ocean Circulation and El Niño*. 2000   ISBN 0-7923-6522-4
23. Y. Shao: *Physics and Modelling of Wind Erosion*. 2000   ISBN 0-7923-6657-3
24. Yu.Z. Miropol'sky: *Dynamics of Internal Gravity Waves in the Ocean*. Edited by O.D. Shishkina. 2001   ISBN 0-7923-6935-1
25. R. Przybylak: *Variability of Air Temperature and Atmospheric Precipitation during a Period of Instrumental Observations in the Arctic*. 2002   ISBN 1-4020-0952-6
26. R. Przybylak: *The Climate of the Arctic*. 2003   ISBN 1-4020-1134-2
27. S. Raghavan: *Radar Meteorology*. 2003   ISBN 1-4020-1604-2
28. H.A. Dijkstra: *Nonlinear Physical Oceanography: A Dynamical Systems Approach to the Large Scale Ocean Circulation and El Niño*. 2nd Revised and Enlarged Edition. 2005
    ISBN 1-4020-2272-7 Pb; 1-4020-2262-X
29. X. Lee, W. Massman and B. Law (eds.): *Handbook of Micrometeorology: A Guide for Surface Flux Measurement and Analysis*. 2004   ISBN 1-4020-2264-6

30. A. Gelencsér: *Carbonaceous Aerosol.* 2005   ISBN 1-4020-2886-5
31. A. Soloviev and L. Roger: *The Near-Surface Layer of the Ocean: Structure, Dynamics and Applications.* 2006   ISBN 1-4020-4052-0
32. G.P. Brasseur and S. Solomon: *Aeronomy of the Middle Atmosphere: Chemistry and Physics of the Stratosphere and Mesosphere.* 2005
       ISBN 1-4020-3284-6; Pb 1-4020-3285-4
33. B. Wozniak and J. Dera: *Light Absorption and Absorbants in Sea Water.* 2006
       ISBN 0-387-30753-2
34. A. Kokhanovsky: *Cloud Optics.* 2005   ISBN 1-4020-3955-7
35. T.N. Krishnamurti, H.S. Bedi, V.M. Hardiker, and L. Ramaswamy: *An Introduction to Global Spectral Modeling.*2nd Revised and Enlarged Edition. 2006
       ISBN 0-387-30254-9
36. L.J. Pratt and J.A. Whitehead: *Rotating Hydraulics: Nonlinear Topographic Effects in the Ocean and Atmosphere.* 2008   ISBN 0-387-36639-3
37. Y. Shao: *Physics and Modelling of Wind Erosion.* 2007.
38. S. Massel and T. Petelski: *Ocean Waves Breaking and Marine Aerosol Fluxes.* 2007.
       ISBN 0-387-36638-8
39. H.M. van Aken: *The Oceanic Thermohaline Circulation.* 2006.   ISBN 0-387-36637-1

Printed in Singapore